CHEMICAL REACTIVITY IN LIQUIDS

Fundamental Aspects

CHEMICAL REACTIVITY IN LIQUIDS

Fundamental Aspects

Edited by
Michel Moreau and Pierre Turq

Pierre and Marie Curie University
Paris, France

Co-editors
Jacqueline Belloni
Roger Prud'homme
Clément Troyanowsky

Springer Science+Business Media, LLC

Library of Congress Cataloging in Publication Data

Société française de chimie. Division de chimie physique. International
    Meeting (42nd: 1987: Paris, France)
    Chemical reactivity in liquids.

    Bibliography: p.
    Includes index.
    1. Chemical reaction, Rate of—Congresses. 2. Solution (Chemistry)—Con-
gresses. I. Moreau, Michel. II. Turq, Pierre. III. Title.
QD502.S62   1987   541.3'94   88-17960
ISBN 978-1-4612-8297-6   ISBN 978-1-4613-1023-5 (eBook)
DOI 10.1007/978-1-4613-1023-5

Proceedings of the Forty-Second International Meeting of the
Société Française de Chimie, Division of Chimie Physique Paris,
held September 7–11, 1987, in Paris, France

FOREWORD

Understanding chemical reactivity has been the permanent concern of chemists from time immemorial. If we were able to understand it and express it quantitatively there would practically remain no unsolved mystery, and reactions would be fully predictable, with their products and rates and even side reactions.

The beautiful developments of thermodynamics through the 19th century supplied us with the knowledge of the way a reactions progresses, and the statistical view initiated by Gibbs has progressively led to an understanding closer to the microscopic phenomena. But is was always evident to all that these advances still left our understanding of chemical reactivity far behind our empirical knowledge of the chemical reaction in its practically infinite variety.

The advances of recent years in quantum chemistry and statistical mechanics, enhanced by the present availability of powerful and fast computers, are very fast changing this picture, and bringing us really close to a microscopic understanding of chemical equilibria, reaction rates, etc.... This is the reason why our Society encouraged a few years ago the initiative of Professor Savo Bratos who, with a group of French colleagues, prepared an impressive study on "Réactivité chimique en phase liquide", a prospective report which was jointly published by the Société Française de Chimie and the French Centre National de la Recherche scientifique in March 1986. The wealth of results and proposals for future research in this report convinced us that the time was ripe for an international conference which would assess the present state of our knowledge -also of our ignorance in some fields- and expand the assessment already achieved by the French group.

We were able to bring together an impressive array of the best specialists in the field, in fact so many of them that we had some difficulty in keeping nearly sixty contributions, of which more than thirty were full lectures, within four and a half days of intense work.

It was quite a lively meeting, and the discussions enriched the contents of the lectures. Agreement was naturally far from complete on all points -in fact disagreements are among the essential points of any good scientific meeting- and this is why we left the conference perhaps not wiser, but certainly more knowledgeable on the present state of our knowledge, and on the questions that urgently require further and more elaborate research.

The meeting covered quite fully the study of very fast reactions, down to the femtosecond range, the progress in quantum chemistry, and notably the influence of solvation on activation barriers, the statistical mechanics of chemical equilibria, reaching a microscopic formulation of reaction rates in a number of cases, the rapid advances in our knowledge of systems far from equilibrium, and finally the contributions of hydrodynamics to"chemical understanding" along with a variety of contributions to chemical and process engineering. We certainly benefited from a wealth of fresh results, both experimental and theoretical.

This meeting owes very much to the initial action of Savo Bratos, and to the organising committee -Jacqueline Belloni, Michel Moreau, Roger Prud'homme and Pierre Turq- who worked actively to set up a very satisfactory conference, and contributed personally to the quality of its contents. Our thanks go also to all the authors, with a special mention of Professor Sebastien Candel: he replaced at very short notice professor Frank Marble, who was prevented from attending at the last moment, and gave a brilliant presentation of Frank Marble's research.

This meeting was held at the Ministry of Research, which we have to thank for a very generous hospitality and for its financial support. Support was also received from a number of organizations and firms: Centre National de la Recherche Scientifique -secteur Chimie-, Commissariat à l'Energie Atomique -Département de physicochimie- Direction des Recherches, Etudes et Techniques, U.S. Army European Research Office, I.B.M. France and Ministère de l'Education Nationale. This support is gratefully acknowledged.

Clément Troyanowsky
Scientific affairs Officer
Division de Chimie physique
Société Française de Chimie

CONTENTS

II - THEORY

IIa - QUANTUM STUDIES

IIb - STATISTICAL MECHANICS COMPUTER SIMULATION

TECHNIQUES FOR STUDYING FAST REACTIONS IN LIQUIDS[*]

Charles D. Jonah

Chemistry Division Argonne National Laboratory

Argonne, IL 60439 U.S.A.

Introduction

The study of chemistry is the study of chemical reactivity, — of how reactions take place. It is known that chemical transformations take place through a sequence of elementary reactions. These elementary reactions determine the outcome of the reaction sequence. In the past one determined the overall reaction, possibly determined the kinetics of the reaction, and the dependence on parameters such as the dielectric constant or the proton affinity of the solvent. From these data one then created a mechanism which explained the reactions. Now it has become possible to measure the elementary reactions that make up mechanisms. These measurements have become sufficiently routine so that the chemist, who wants to learn more about a particular chemical reaction, can find a technique to help him.

The goal of this paper is to summarize the many techniques which can be used for making fast measurements so that the reader will be able to see which techniques are suitable for his or her particular purpose. In this paper, I will consider any reaction that is faster than 1 second to be a fast reaction. Also because there is a paper in this volume which reviews techniques which have time resolutions in the picosecond and femtosecond time range, I will only make a few comments based on our recent work at Argonne in picosecond pulse radiolysis. The ultimate limits of these techniques will not be discussed in detail, because 1) they keep changing as the techniques are improved and 2) the novice user is unlikely to be well-served by knowing what is possible after 20 years of refining the technique in a given laboratory. The limitations and major difficulties of the techniques will be mentioned. These observations will be purely personal and may not be agreed to by all practioners of the art. Two books give a much more complete descriptions of all of the techniques that will be discussed here. (Hammes 1974, Bernasconi 1986)

There are three steps in studying elementary reactions and in trying to understand chemical kinetics: 1) creation of the starting products for the reaction to be studied in sufficient purity and sufficient concentration to be observed (particularly when they are unstable); 2) identification of the products of the reaction; and 3) measurement of the evolution the chemical species. Table 1 summarizes some of the chemical species that one might want to study and the techniques which are applicable. To simplify the discussion, the creation and detection techniques will be discussed separately. I will not discuss electrochemical tech-

---

[*] Work performed under the auspices of the Office of Basic Energy Sciences, Division of Chemical Science, US-DOE under contract number W-31-109-ENG-38

niques in this paper because I have so little familiarity with them that it would be more profitable to read a review such as those in the volumes Hammes 1974 and Bernasconi 1986.

Table 1

Experimental Techniques for Studying Transient Chemistry		
Species	Generation Technique	Detection Technique
Ionic Reactions	Pressure Jump Temperature Jump Flow Techniques Flash Photolysis Pulse Radiolysis	Conductivity Absorption Spectra
Excited States	Flash Photolysis Pulse Radiolysis	Emission Spectroscopy Absorption Spectroscopy Microwave Conductivity
Radicals	Flash Photolysis Pulse Radiolysis Flow Techniques	ESR Absorption Spectroscopy
Triplet States	Flash Photolysis Pulse Radiolysis	ESR Optical Absorption
Polymers	Flash Photolysis Pulse Radiolysis	Light Scattering

An alternative approach to chemical kinetics uses the broadening of spectral lines to determine lifetimes of states and thus kinetics. In NMR, exchange processes can be used to determine lifetimes in the range of $1-10^{-5}$ sec while in optical spectra, lifetimes can be determined which are on the order of picoseconds. Pulsed NMR techniques can be used to determine lifetimes which are on the order of nanoseconds. A recent review of the possibilities of NMR and references are given in the chapter on NMR and some in the chapter on ESR in the Techniques in Chemistry volumes (Swift 1974).

Techniques for the creation of the chemical system to be studied

To make kinetic measurements, it is necessary to create the chemical system to be studied in a non-equilibrium configuration and observe the relaxation to equilibrium. This can mean the determination of the rate of a simple reaction between two stable reactants or the creation of the hypothesized intermediates of a chemical reaction and watching their reaction. To clarify the discussion, it is useful to separate the techniques into two categories: 1) perturbation of pre-existing equilibria and 2) the creation of a reacting system where the reactants are not normally all in the solution.

EQUILIBRIUM PERTURBATION TECHNIQUES

These techniques start with a system in equilibrium. For the reaction $A + B \rightleftharpoons C$ the following equation holds:

$$A \quad + \quad B \quad \overset{K}{\rightleftharpoons} \quad C$$

The system is perturbed, by temperature, electric field, pressure or other and K changes. Now the system is no longer in equilibrium and a reaction takes place to return the system to equilibrium and that reaction is measured. The techniques are very useful for measuring ionic reactions, proton reactions, reactions of binary mixtures, etc. Since K is normally changed by only a small amount, the relaxation back to equilibrium can be described as a first order reaction.(Eigen and DeMaeyer 1974, Schwarz 1986). Thus data analysis is relatively simple. However relaxation techniques have an inherent disadvantage in that only reactions in which the equilibrium is not far to one side can be studied.

TEMPERATURE JUMP TECHNIQUES In this technique, a change in temperature is used to shift the equilibrium constant. The fraction by which the equilibrium constant will be changed depends on ΔH. The advantages and limitations of this technique depends on how the experiment is done. The possible approaches, as summarized by Turner (Turner 1986) are shown in the Table 1.

Table 2

Methods for Temperature Jump Studies			
Technique	Temperature range	Time scale	Limitations as to solvent
Bath	any	10 - 1 sec	any solution
Electric discharge	2-10 C	10^{-6} - 10^{-8} sec	Solution must conduct
Microwaves	1 C	10^{-6} - 10^{-8} sec	Polar solution
Laser	1-10 C	10^{-8} - 10^{-9} sec	Solution must absorb at laser wavelength

Most of the limitations are relatively obvious and are listed in the limitations as to solvent. Laser techniques require a strong absorption of laser light. Fortunately, water has a very strong absorption (although not at the most convenient wavelength) and absorption of light can take place. The longest times that can be observed using transient heating techniques — lasers, discharge and microwave heating — are limited by the the insulation of the observation system.

ELECTRIC FIELD PERTURBATION   Equilibria can be perturbed by putting a high voltage on plates across the cell. Voltages must be in the range of 10^4 - 10^5 volts. The solution must not conduct. With such high voltages, the possibility of discharge is always present. This technique is said to be good for protic equilibria. For more details see the recent review by Eyring and Hemmes 1986.

PRESSURE JUMP   A change in pressure on a chemical system will shift the equilibrium constant an amount which depends on $\Delta V°$. The time resolution depends on the technique used (Knoche 1974,1986) If a valve is used to introduce the pressure change, time resolution is $\approx .05$ seconds. Since the pressure can then remain constant, the reaction can be as long as desired. If the rupture of a membrane is used to create the pressure jump, the time resolution is ≈ 1 μsec. However the pressure jump will only hold for approximately 2 ms in that case. Because only the equilibrium is shifted and no other energy is added to the system, the system can be recycled as often as desired so that very efficient signal averaging is possible.

The shift of the equilibrium depends on $\Delta V°$. For $\Delta V° = 5$ cm^3/mol, the equilibrium constant is shifted 2% for a change in pressure of 100 bar. For reference, $\Delta V°$ for $H_2O \rightarrow H^+ + OH^-$ is -23.5 cm^3/mol. Unfortunately, the experiments are normally done under adiabatic rather than isothermal conditions so the pressure change also induces a temperature change. For the same reaction, the shift is $\approx .15$ K which will lead to a .6% change in the equilibrium if ΔH is 7 kcal/mole. This means that the shift in equilibrium constant due to temperature is only a fraction of that due to the pressure change.

There is no limitation to solvent or ionic strength with the pressure jump technique. Ionization reactions are conveniently studied using this technique because the electrostriction that occurs in solution causes $\Delta V°$ to be large. The technique can be difficult to use under certain conditions because artifacts created in the detected signals due to the pressure change. Among the processes studied are hydration of ions and inner and outer ligand reactions.

ULTRASONICS   Ultrasonics works on the same principle as the pressure jump experiment in that pressure is changed. The system to be studied is irradiated with an ultrasonic transducer which causes a repeating pressure fluctuation(Stuehr 1986). If the period of the ultrasonics is much longer than the time of the chemical reaction being studied, the chemical system will shift as the equilibrium constant shifts and there will be no absorption of energy. If however the reaction that is being studied occurs on the same time scale as the pressure fluctuates, the chemical composition will oscillate out of phase and power will be absorbed. Finally, if the reaction occurs much slower than the period of the pressure fluctuation, again no energy will be absorbed. Thus to determine the kinetics one must only measure the energy absorption as a function of frequency. This technique can work over a very large time range: 10^{-4} - 10^{-10} seconds. Unfortunately different variants and different techniques are needed for different time range which differ by only an order of magnitude or so. Another disadvantage is that kinetics are not determined directly. Only by using a large range of frequencies can the actual kinetic process be determined. The technique is however very useful for very fast processes in solution such as solvation, solvent interaction and proton transfer independent of the optical absorptions of the species involved.

CREATION OF NEW SPECIES

Often the starting materials for reaction do not exist in an equilibrium in which both have reasonable concentrations. It is then necessary to somehow create the species to be studied in the solution. If both reactants are stable, flow techniques are good; however if one or both are unstable or are excited states, techniques such as flash photolysis or pulse radiolysis are good.

FLOW TECHNIQUES   Flow techniques are the technique of choice if both reactants are stable. There are several variants of the technique - stopped flow, quenched flow, and fast flow kinetics (Chance 1974). In stopped flow, two solutions flow together in a cell. When properly designed, the solutions are mixed during the flow into the cell. A cartoon of the technique is seen in figure 1. The kinetics can then be observed in the cell. Standard methods such as conductivity measurements, absorption and emission spectroscopy can be used. Time resolution is normally about 1 msec. In fast flow kinetics, the two solutions mix and travel along a tube. The technique is shown in cartoon form in figure 2. With a constant flow rate, the position in the tube will define the time after mixing. Thus by making measurements as a function of distance, the time profile is determined. The major limitation to time resolution is the mixing of the solutions. Time resolution is in the vicinity of 10 μsec. Even faster reactions can be studied if the solutions are cooled so as to slow down the reactions. A variant of this technique is mixing the reacting solution with another solution which quenches the reaction. The products can then be measured as a function of flow distance and thus time. This is useful if the products are not easily detectable. Even faster reactions can be studied if the solutions are cooled so as to slow down the reactions.

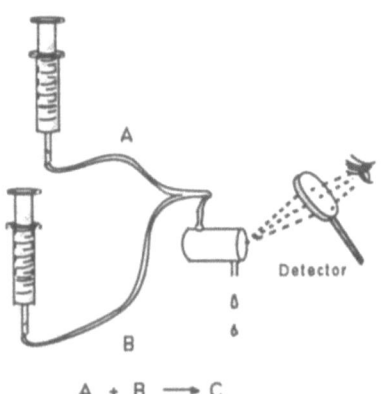

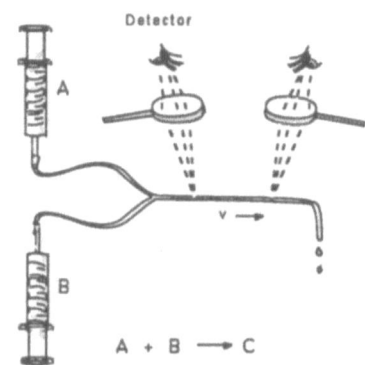

Figure 1 Diagram of a stopped flow apparatus. The eye is used as a generic detector since many detection techniques could be used.

Figure 2 Schematic diagram of the fast flow technique. The eye is again a generic detector. In practice one detector is moved along the flow path.

Flow techniques allow one to use solutions in which the concentrations are well understood and well measured. Conversely the technique will not work if the the starting materials are not stable. Time zero is often not well-determined because of the finite mixing time and disturbance in the cells during the stopping of flow. These techniques can be a valuable addition to other techniques in that a reactant mixture which has limited stability can be created and then another perturbation can be be used - flash photolysis, pulse radiolysis, temperature or pressure jump.

The next two techniques, flash photolysis and pulse radiolysis, are similar in that unstable species can be formed and studied. Many of the same limitations exist for both techniques. Beck has recently reviewed the capabilities of available laser and pulse radiolysis facilities for creating chemical species.(Beck 1986) The current available from the picosecond pulse of the Argonne Linac is about a factor of 3 higher than given in that article (25 nC).

FLASH PHOTOLYSIS   In flash photolysis, light is used to create the reactant species (West 1986). The reactant species could be either an excited state; the products from photodissociating or photoionizing a molecule; or the subsequent reaction of such a primary process. The source of light can be from a flash lamp or a pulsed laser. The flash lamp can provide a very high intensity but little wavelength selectivity and normally the time resolution is not better than a microsecond. A laser provides much better time resolution, better wavelength resolution but generally less power than a flash lamp. However the modern high-power high-energy excimer lasers can create substantial concentrations of usable species.

With modern lasers, flash photolysis has very good time resolution and important reactions can be studied with ease because significant reactants can be formed. However the formation of such states depends on finding precursors which can be photolyzed by existing lasers to give the desired states. If any other species in the solution can absorb light at that wavelength, the photochemistry can be complex. Even the product of the the photolysis could absorb light and create difficulties.

The limitations to the time resolution are the length of the excitation pulse, the time resolution of the detection equipment and the speed of the chemistry creating the reactants of interest. With lasers presently available with pulse widths less than a picosecond, the pulse length is not a major limitation. Using a pulse-probe detection technique, the only limitation of the time resolution may be the formation of the desired reactant.

PULSE RADIOLYSIS   A pulse of ionizing radiation is used to create the reactant species. The ionizing radiation interacts with the solution, creating radicals, ions and excited states (Dorfman and Sauer 1986). The creation of reactants does not depend on the absorption coefficients because the ionizing radiation deposits its energy in the solvent and either the primary species are studied or the products of the reaction with the primary species are used. In the latter case, the rate of formation of the species to be studied depends on k, the rate constant with the primary species, and c, the concentration of the reactant species. As an example, in water one can study the reactions of the the primary species, H^+, $OH\cdot$ and e^-_{aq} or a product of their reactions. For example, if one wanted to measure CO_2^- reactions, N_2O can be added to the solution to react with e^-_{aq} to form OH in less than 10 ns. The OH reacts with added formate ions to form CO_2^-.

The time resolution is limited by the same factors as is the time resolution in flash photolysis. The shortest present pulse is slightly greater than 10 ps; however using a cavity and dispersive beam optics, a pulse of less than 10 picoseconds has been measured at Argonne. An ultimate pulse length of 5 ps is expected. The major limitation of pulse radiolysis is the size and cost of the accelerator. While this can limit the acquisition of an instrument for a laboratory, it does not preclude the use of the technique since almost any laboratory which has pulse radiolysis equipment is willing to collaborate. Radiation chemistry is unfamiliar to most chemists, however it is no more difficult or abstruse than the better-known branches of chemistry.

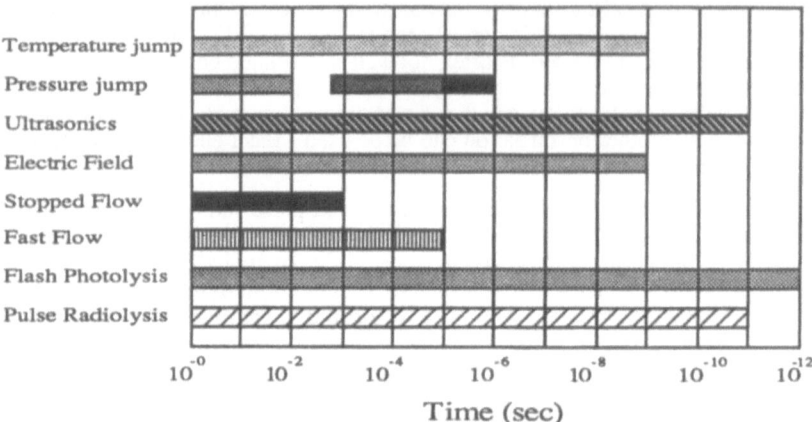

FIGURE 3 Summary of the time resolution expected for different creation techniques.

SUMMARY OF TECHNIQUES   This has been a short summary of the possible techniques for creating a nonequilibrium state from which the relaxation takes place. The equilibrium perturbation techniques are very good for ionic reactions. The kinetics are quite simple because the small change in concentration means that all reactions are first order. The techniques which create systems which are far from equilibrium can study a much wider variety of reactions, with the complication that the reaction kinetics can be quite complex. A summary of the time resolution available is given in figure 3. Many of these techniques can be used together. For example, we have used a stopped flow to create a solution of Ru^{III}(bipyridine)$_3$ in neutral solution where it decomposes quickly and then used pulse radiolysis to study the formation of the excited state of Ru^{II}(bipyridine)$_3$ from the reaction of e^-_{aq} with the Ru^{III} (Jonah et al. 1978). Another example is the use of pulse radiolysis to create an aryl radical by the dissociative electron capture of an aryl methyl halide and then laser flash photolysis to study the photochemistry of the excited state(Bromberg et al. 1985). There are many other possibilities; the options are only limited by the imagination (and money).

Detection Techniques

The choice of detection techniques depends primarily on what species are being studied and only minimally on the technique used to create the nonequilibrium situation. Each technique has its own strong and weak point - new techniques are developed to solve old problems. The matching of detection technique requires an understanding of the intermediates to be studied, their concentration, the possible interferences and the time resolution needed.

OPTICAL ABSORPTION AND EMISSION

Optical absorption is one of the most common methods for detecting and measuring kinetics. It is very general in that all intermediates absorb light and most absorb some place where measurements can be made. A generalized apparatus is shown in figure 4. A lamp is focussed through the cell which contains the system to be studied. Generally a filter or monochromator is used to select the wavelength of light to be used. If there is danger of photolysis of the system, the monochromator is before the cell; otherwise it is often directly before the photodetector to limit the stray light striking the photodetector. A change in the amount of light passing through the sample is detected with a photodetector. The output of the photodetector is then measured.

The limits of sensitivity and signal to noise are primarily determined by the lamp. One observes a change in a light level, so any unsteadiness in the lamp will be directly reflected into the signal. Lamp fluctuations set the lower limit to the size of a signal that can be observed. This is particularly true at times greater than a millisecond where the frequency of mechanical vibrations and power supply ripple on the lamp overlap the time of the measurement. Even with a perfect lamp, there is quantum noise - the effect of photon nature of light. Because light is statistical with a Poisson distribution, the noise will be proportional to √light intensity and thus the signal to noise ratio is proportional to √light intensity. This noise source is usually not serious for times greater than a few microseconds but will become the dominant noise source at much shorter times. The usual method of alleviating this problem is to pulse the lamp(Luthjens 1973, Beck 1974). With a xenon arc source, it is reasonable to increase the intensity by more than a factor of 20 (depending on wavelength). A fairly complete description of how to use different monochromators, optical designs and lamp configuration for nanosecond spectrophotometry has been given by Hunt (Hunt *et al.* 1972).

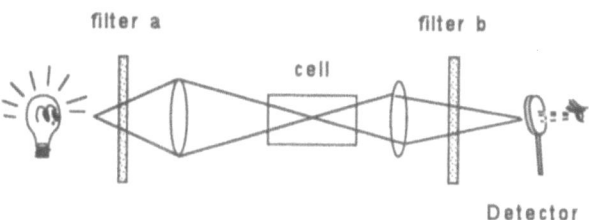

Figure 4 Generalized optical system for optical absorption. The filters select the wavelength and could be filters or monochromators. The detector could be a photomultiplier, photodiode or streak camera.

Light emitting states can be most conveniently measured by measuring the intensity of the fluorescence (or phosphorescence). Since the measurement is a zero background technique — if there are no excited states there is no emission — the sensitivity is very high. One does not have to measure the difference between light with an absorbing species and without the absorbing species but the intensity of light relative to no light. The ultimate sensitivity limit is usually other sources of light which contaminate the observation. Optical emission can be used to study reactants which don't fluoresce if the reactant can be excited to a fluorescent state.

Optical emission is very selective because usually only the excited state one is interested in will emit at the observation wavelength. The greater generality of optical absorption leads to one of its major disadvantages. Often both the reactant and product will absorb light and thus limit the sensitivity of the method and in many cases, making it impossible to determine the kinetics with any significant accuracy.

Absolute concentrations are relatively easy to determine with optical absorption, assuming that one can determine an extinction coefficient, and even without an extinction coefficient, good relative measurements can be accurately made. The amount of absorption will not depend on the optics (unless a very broad wavelength light source is used). In contrast, it is very difficult to obtain absolute measurements from emission experiments. The optics train must be calibrated, and the wavelength dependence of the monochromator and the photodetector must be determined.

Photomultipliers and photodiodes can measure changes of light at a single wavelength. This is sufficient when the absorption spectra of the compounds in question are known. Better time resolution is available in emission experiments because single photon counting techniques can be used. In the single photon counting technique the time resolution of a photomultiplier depends primarily on the stability of the transit time through the photomultiplier tube. With microchannel-plate photomultipliers used in a single photon counting experiment, time resolution can be better than 100 picoseconds. However the experiment must be repeated many hundreds of thousands of times to obtain a good signal to noise ratio. For analog measurements the resolution depends both on the transit time through the tube and on the spreading of the electron pulse in the tube. Optical absorption experiments using fast photodiodes have been done with reported time resolutions of 60 picoseconds (Beck 1976); however in my experience the detectors respond to a change in light in two stages, one very fast and one considerably slower. For a photodiode the secondary response appears to be about 1 nanosecond. A nice example of fast optical measurements is that of Closs *et al.* on electron transfer (Closs 1986).

Multiwavelength-multitime experiments using streak cameras allows the measurement of both the spectrum and the kinetics of the products of a reaction in a single experiment (Schmidt *et al.* 1976). Although very convenient for many experiments, multitime-multiwavelength detection is particularly useful for studying biological compounds where samples degrade and it is difficult to obtain enough material to run the multiple experiments necessary to determine spectra or for radioactive compounds where the more times a sample is handled the higher the probability of a radioactive spill becomes. Streak camera measurements can be used both for emission and absorption. Because one must measure the difference between the light level with and without the reaction to do absorption measurements, the limited dynamic range and the noise level of the streak camera limits the sensitivity for absorption measurements to about .004 OD. In emission, the only requirement is enough intensity to activate the camera. A time resolution of about 2 ps for emission is possible for commercially available streak cameras. New cameras with femtosecond resolution have been reported(Kinoshia *et al.* 1987). The time resolution for absorption measurements is not as good because the high amount of light needed to do the absorption measurement causes space-charge broadening in the streak tube. In our experience at Argonne, very fast streak cameras come equipped, as standard equipment, with a complete supply of artifacts. Caution and experience are needed to obtain meaningful data from them.

Another technique that has been used for measuring kinetics from emission data is the phase shift technique(Gratton *et al.* 1984, Gratton 1986). The difference in phase between an excitation source and the fluorescence intensity is measured. I have not had experience with the new instruments described in those publications; however previous experience with earlier implementations of the technique was not good. Because one is not seeing the original data but only a phase shift, interferences can be a problem. While I may be a bit dogmatic, I feel that caution must be used any time one does not directly observe the kinetics but only observes a single parameter from the fluorescence (such as a phase shift). These caveats may be out of date, but are probably worth mentioning. An experienced user may avoid the problems but the novice may create marvelous nonsense.

Pump-probe techniques use similar pulses to excite the sample to create the chemistry and to observe the reaction (see figure 5). The observation pulse is delayed so that time can be varied (see figure). The technique can be used both for absorption and emission (up conversion). The time resolution is then only limited by the width of the pulse that does the excitation. Multiple experiments must be done because each experiment will only measure one time. The approach can be used both for flash photolysis and for pulse radiolysis.

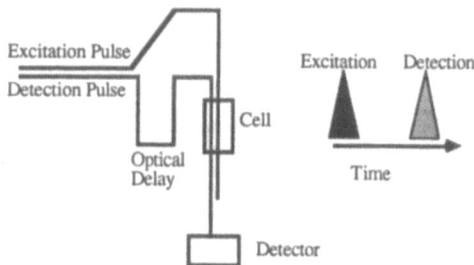

Figure 5 Diagram of pump-probe technique. Since the chemical system is only observed while the probe pulse is present, the time resolution of the detector doesn't matter. The detector could be a photodetector or an upconversion detector. The pump beam could be light or electrons.

RAMAN SPECTROSCOPY   In Raman spectroscopy the sample is irradiated with an intense light beam. This beam excites the molecules to virtual states which then can radiate during a transition to a different vibrational level of the ground state. The radiated light then is at a different frequency than the exciting light, and this frequency difference is the frequency of a vibrational transition in the molecule(see figure 6). Because Raman spectroscopy measures the vibrational bands of the molecule, the bands are narrower and more informative than the optical absorption spectra. A very nice example is the study of the oxidation of 1,2,4-benzenetriol where optical absorption showed very little change as a function of time but the resonance Raman clearly showed the decay of one molecule and the growth of a second(Qin *et al.* 1987).

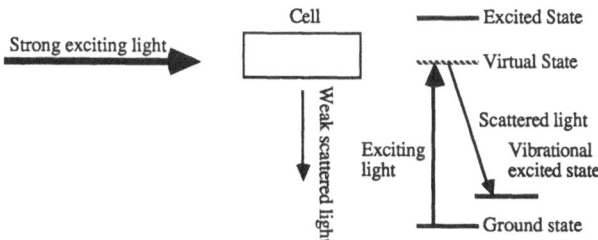

FIGURE 6 Schematic description of Raman experiment and the energy levels involved.

Raman spectroscopy is usually much less sensitive than other techniques. In resonance Raman spectroscopy, the wavelength of the excitation light source is close to an absorption band of a molecule. This increases the intensity of the Raman scattering by many orders of magnitude. A pulsed laser is often used as the scattering light since a very intense and very monochromatic light source is needed. Because a pulsed light source is used, a measurement is made at only a single time — the time of the laser flash. To develop a time profile, it is necessary to repeat the process many times. Since a very high power light source is needed and since the light must be near an absorption maximum of a molecule, photochemical reactions can be a problem.

The time resolution of the detection system is limited by the pulse width of the excitation source and is usually around a nanosecond although with very high powered picosecond lasers, much better time resolution is possible. The compounds that can be studied are those which have a strong absorption at a wavelength near a strong laser line and which don't fluoresce.

CONDUCTIVITY Conductivity techniques measure the concentration and mobility of ions in solution. They are particularly well adapted to the measurement of ions such as protons which are formed in pressure jump or temperature jump perturbation techniques. Because only the mobility of an ion is relevant, no information is available about the chemical structure of the species.

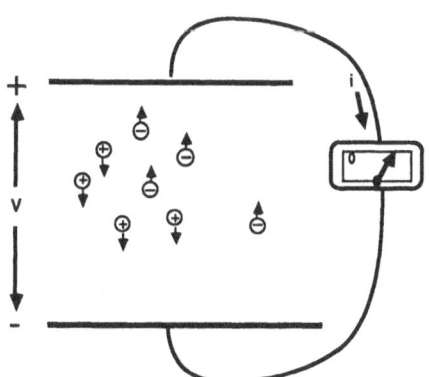

Figure 7 Conductivity experiment. Note that it is the movement of charges that lead to a signal and not the discharge at the electrodes. If the voltage is constant, D.C conductivity is measured, if it is varied at moderately low frequencies, A.C. conductivity, and if at very high frequencies, microwave conductivity.

The general principles of the conductivity experiment are shown in figure 7. There are three variants of the conductivity technique — D. C., A.C. and microwave conductivity.

D.C. conductivity measures the motion of ions in a solvent (Schmidt 1972, Maughan 1978). For reactions lasting longer than 500µsec, an alternating voltage (A.C.) should be used to avoid polarizing the solution. By perturbing equilibria, not only rate constants but also hydrolysis constants can be measured(Schmidt *et al.* 1983). The absorption of microwave power is measured in the microwave conductivity technique. Microwave conductivity can measure not only ions but also the large-dipole, excited states of molecules. The excited states of aromatic molecules and electron transfer intermediates have been measured using microwave conductivity (Warman *et al.* 1986).

The sensitivity and time response of the conductivity techniques depend on the mobility of the species being measured. For example, a very highly mobile species, the electron in iso-octane can be measured with a time resolution of less than 100 picoseconds (Beck 1979) while the mobilities of ions in water can only be measured with time resolutions much greater than a nanosecond. The ultimate limit of the microwave technique is the frequency of the microwave excitation frequency and is about 100 picoseconds normally.

ELECTRON SPIN RESONANCE   Electron spin resonance measures the effect of microwaves on a molecule with spin (usually a free radical or triplet) in a magnetic field. The detection can be either through the absorption of microwave energy (conventional ESR) or the effect of a microwave frequency on the emission of light (Fluorescence Detected Magnetic Resonance FDMR). Because the transition energy of the electron in a molecule depends on the interaction of that electron with much of the molecule, spectra have many lines and contain substantial information about the structure of the species being studied.

The time resolution of the technique depends on many factors. Direct kinetic measurements using ESR has a time resolution of less than a microsecond. Multiple pulse and FDMR techniques can have time resolution of less than 100 nanoseconds. Because the radicals are often formed in a way in which a polarization of the radicals exist, it is difficult to determine the concentration of the species, since the polarization of the radicals leads to an increase in detection sensitivity. Also, the relaxation of the polarization changes the proportionality between signal and concentration. This makes the kinetics quite complicated and difficult to interpret. A recent review by Trifunac, *et al.* 1986, gives a more complete description of the techniques an limitations (Trifunac *et al.* 1986).

SUMMARY

The large variety of techniques available make it possible to do many different kinds of experiments. There are many others which I have not discussed. For example, light scattering has been shown to be very useful in polymer studies (Beck *et al.* 1977). Figure 8 gives the time resolution for each of the techniques discussed here.

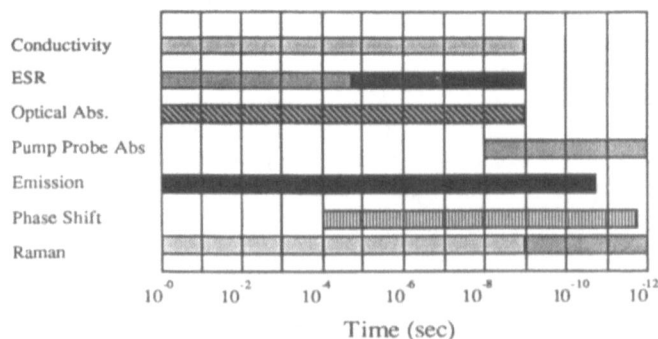

Figure 8 Summary of time resolution of the detection techniques discussed in this paper.

Methods of Acquiring the Data

The traditional technique for measuring data has been an oscilloscope (scope or CRO depending on the nationality of the experimenter). The electronic signal, proportional to a physical quantity of interest, is displayed on an oscilloscope and then photographed. This is relatively cheap since an oscilloscope is almost always necessary to correctly set up a fast kinetics experiment. However the bleary eyes that occur when the data from a photograph are digitized plus the large number of pictures which seem to be taken when the film has run out make this technique obsolete. Digitizing cameras have recently become commercially available which can digitize a trace from an oscilloscope screen. They can provide an economical technique for collecting data with high time resolution - less than a nanosecond on the presently available oscilloscopes.

Transient digitizers provide an easy method of acquiring data, saving it and transferring it to a computer. Presently available digitizers have time resolutions of 5 nanoseconds per channel for the high priced models with up to 1 part in 256 resolution. (Note that all the vertical resolution is not significant at short times because of trigger uncertainties - ask the salesman. If he or she can't explain it then worry about his or her ability.) These instruments allow methods of data handling which were not possible using oscilloscopes. For instance, part of a data trace can be taken at one rate and part at a second rate. This makes it possible to observe the concentration of an intermediate B in a reaction scheme such as:

$$A \quad \xrightarrow{k_g} \quad B \quad \xrightarrow{k_d} \quad C$$

in a single experiment. It also makes it possible to measure the growth of a species to a plateau where most of the data points are taken during the growth process and fewer data points are taken over a longer period of time to determine the plateau value accurately. Caution must be exercised when reading the specifications of digital sampling oscilloscopes because the listed specifications for time resolution and frequency response assume that a repetitive waveform is being observed.

The ultimate limit in digitizers are the ultrafast digitizers by Intertechnique of France and sold by Tektronix. These expensive devices can take data at rates up to 1000 Ghz and with a response time of less than 60 picoseconds. These instruments are not easy to use, can be easily damaged and do not offer the many options that traditional digitizers offer such as dual channel or dual time bases. However they do provide speed.

Data Treatment

The experiment is not over when the data are acquired. The final results can only be as good as the data analysis (and of course no better than the original data). The advances in computer technology have made real differences in how data are analyzed and not just made the same tasks easier.

The first stage of data analysis is the postulation of a reaction mechanism for the system that is being studied(Laidler 1965). For example, if we are observing a species B decay, is the decay due to a first order or a second order reaction? This can partially be checked by seeing if a straight line is obtained if either concentration is plotted vs. time or 1/concentration is plotted vs. time. The former would be expected to be true for a first order reaction while the second is what would be expected for a second order reaction. Further information can be gained - do the rates depend on the initial concentration of B (second order): is the rate a linear function of the concentration of a reactant? Such experiments help confirm the mechanism.

Once a mechanism is tentatively identified, actual rates can be extracted. While data can be linearized and then fit, this approach is questionable at best. The best approach is to derive the signal as a function of time and parameters such as the rate constants and initial concentration and then fit the data to the appropriate expression. This was first pointed out nearly 30 years ago (Sullivan *et al.* 1959). That is, for a pseudo first-order reaction:

$$A \quad + \quad B \quad \xrightarrow{k} \quad C$$

where B is at large excess and has concentration c, fit signal = $S_0\exp(-kct)$ where S_0 is the initial concentration rather than fitting $\ln(\text{signal}) = -kct$. The former preferable because it is usually easy to decide what the noise is on a particular datum point; it is much more complicated when the data are transformed. Common weightings are constant error for absorption measurements and $\sqrt{\text{signal}}$ for emission measurements. The data can be fit using conventional non-linear least square techniques. Complex mechanisms, such as second order approach to equilibrium can be used to explain the data.(Beitz *et al.* 1986) Interesting variants of this approach are fitting several data sets with a single set of parameters. For instance, for the mechanism:

$$A \overset{k_{growth}}{\rightleftharpoons} B + R \overset{k_{decay}}{\rightleftharpoons} C$$

where B is observed and R is an added reactant, one could fit an experimental data set where R is very small and thus emphasize the growth kinetics and a data set where R is large which emphasizes the decay portion of the kinetics.

The major disadvantage of non-linear fitting is that one must obtain reliable programs to do the fitting. A minor disadvantage is that one can fit almost any mechanism so that insufficient thought is given on exactly how to prove the reaction mechanism. Kinetics, like all computer techniques, is subject to GIGO - garbage in - garbage out.

Conclusions and Summary

Modern techniques of fast kinetics make it possible to study almost any reaction. Many of the techniques are relatively similar; some are quite complicated. The major difficulty is in properly defining the problem so that the results of the study are significant. Modern equipment and modern techniques can not create order out of chaos. Presently it is possible to purchase commercially much of the equipment necessary to make many of these fast measurements. Flash photolysis systems are commercially available. Temperature jump, pressure jump, stopped flow, fast flow and quenched flow instruments can be bought. In fact attachments to stopped flow equipment are commercially available to do many additional types of experiments.

It is now possible to do many measurements easily and quickly which would have been impossible 10 years ago. This opens up fast kinetic measurements to a whole new group of chemists; those chemists who do not want to devote a large amount of time to experimental development but who want experimental answers to chemical questions.

If this paper has just provided a view of the possibilities for understanding chemical reactions, it will have served a purpose. Hopefully it will have also given the reader an idea of what techniques might be applicable to her purpose.

Acknowledgements

Many of the references to techniques and data come from work that has been done at Argonne. This has been done because it is possible for me to know the limitations of those measurements. Most of the detection examples have come from pulse radiolysis, primarily because of my familiarity with those techniques.

I would like to thank Dani Meisel, Klaus Schmidt, Anita Chernovitz, John Miller, Alex Trifunac and Jim Sullivan for input on this paper. The errors are due to my misunderstanding their input. Some of the figures were drawn by Patricia Walsh.

References

Beck, G., J. Radioanal. Nucl. Chem. 101, 151 (1986).
Beck, G., Rev. Sci. Instrum. 47, 849 (1976).
Beck, G., Rev. Sci. Instrum. 45, 318 (1974).
Beck, G., Rev. Sci. Instrum. 50, 1147 (1979).
Beck, G., Lindenau, D. and Schnabel, W., Macromolecules 10, 135 (1977).
Beitz, J., Jonah, C. and Sullivan, J. C. Radiochimica Acta 40, 7 (1986).

Bernasconi, C. F., Ed. in "Investigations of Rates and Mechanisms of Reactions", Techniques of Chemistry, Vol. 6, Part II, 3rd Edition, Wiley Interscience: New York, 1986.

Bromberg, A., Schmidt, K. H. and Meisel, D., J. Am. Chem. Soc. 107, 83 (1985).

Chance, B. in "Investigations of Rates and Mechanisms of Reactions", Techniques of Chemistry, Hammes, G. G., Ed. Vol. 6, Part II, 3rd Edition, Wiley Interscience: New York, 1974, p. 5.

Closs, G. L., Calcaterra, L. T., Green, N. J., Penfield, K. W. and Miller, J. R. J. Phys. Chem. 90, 3673 (1986).

Dorfman, L. M. and Sauer, M. C. in "Investigations of Rates and Mechanisms of Reactions", Techniques of Chemistry, Bernasconi, C. F., Ed., Vol. 6, Part II, 3rd Edition, Wiley Interscience: New York, 1986, p. 493.

Eyring, E. M. and Hemmes, P. in "Investigations of Rates and Mechanisms of Reactions", Techniques of Chemistry, Bernasconi. C. F., Ed., Vol. 6, Part II, 3rd Edition, Wiley Interscience: New York, 1986, p. 219.

Eigen, M. and De Mayer, L. in "Investigations of Rates and Mechanisms of Reactions", Techniques of Chemistry, Hammes. G. G., Ed., Vol. 6, Part II, 3rd Edition, Wiley Interscience: New York, 1974, p. 63.

Gratton, E., Jameson, D. M., and Hall, R. D., Ann. Rev. Biophys. Bioeng. 13, 105 (1984).

Gratton, E., Spectroscopy 1 #6, 28 (1985).

Hammes, G. G., Ed. in "Investigations of Rates and Mechanisms of Reactions", Techniques of Chemistry, Vol. 6, Part II, 3rd Edition, Wiley Interscience: New York, 1974.

Hunt, J. W., Greenstock, C. L. and Bronskill, M. J. Int. J. Radiat. Phys. Chem. 4, 87 (1972).

Jonah, C. D., Matheson, M.S. and Meisel, D. J. Am. Chem. Soc. 100, 1449 (1978).

Kinoshia, K., Ito, M., Suzuki, V. Rev. Sci. Instrum. 58, 932(1987).

Knoche, W. in "Investigations of Rates and Mechanisms of Reactions", Techniques of Chemistry, Hammes, G. G., Ed. Vol. 6, Part II, 3rd Edition, Wiley Interscience: New York, 1974, p. 187.

Knoche, W. in "Investigations of Rates and Mechanisms of Reactions", Techniques of Chemistry, Bernasconi, C. F., Ed., Vol. 6, Part II, 3rd Edition, Wiley Interscience: New York, 1986, p. 191.

Laidler, K. J. "Chemical Kinetics", second edition, McGraw-Hill: New York, 1965.

Luthjens, L. H., Rev. Sci. Instrum. 44, 1661 (1973).

Maughan, R. L., Michael, B. D. and Anderson, R. F. Radiat. Phys. Chem. 11, 229 (1978).

Qin, L., Tripathi, G. N. R. and Schuler, R. H. J. Phys. Chem. 91, 1905 (1987).

Schmidt, K. II. Int. J. Radiat. Phys. Chem. 4, 439 (1972).

Schmidt, K. H., Gordon, S., and Mulac, W. A. Rev. Sci. Instrum. 47, 356 (1976).

Schmidt, K. H., Gordon, S., Thompson, M., Sullivan, J. C. and Mulac, W. A. Radiat. Phys. Chem. 21, 321 (1983).

Schwarz G. in "Investigations of Rates and Mechanisms of Reactions", Techniques of Chemistry, Bernasconi, C. F., Ed., Vol. 6, Part II, 3rd Edition, Wiley Interscience: New York, 1986, p. 27.

Stuehr, J. E. in "Investigations of Rates and Mechanisms of Reactions", Techniques of Chemistry, Bernasconi, C. F., Ed., Vol. 6, Part II, 3rd Edition, Wiley Interscience: New York, 1986, p. 247.

Sullivan, J. C., Rydberg, J., Miller, W. F. Acta Chem. Scand. 13, 2029(1959).

Swift T. J. in "Investigations of Rates and Mechanisms of Reactions", Techniques of Chemistry, Hammes, G. G., Ed., Vol. 6, Part II, 3rd Edition, Wiley Interscience: New York, 1974, p. 521.

Trifunac, A. D., Lawler, R. G., Bartels, D. M. and Thurnauer, M. C., Prog. Reaction Kinetics 14, 43 (1986).

Turner, D. H. in "Investigations of Rates and Mechanisms of Reactions", Techniques of Chemistry, Bernasconi, C. F., Ed., Vol. 6, Part II, 3rd Edition, Wiley Interscience: New York, 1986, p. 141.

Warman, J. M. in "The Study of Fast Processes and Transient Species by Electron Pulse Radiolysis", Baxendale, J. H., and Busi, F. Eds., D. Reidel: Dordrecht, Holland, 1982, p. 129.

Warman, J. M, De Haas, M. P., Oervering, H., Verhoeven, J. W., Paddon-Row, M. N., Oliver, A. M. and Hush, N. S., Chem. Phys. Lett., 128, 95 (1986).

West, M. A. in "Investigations of Rates and Mechanisms of Reactions", Techniques of Chemistry, Bernasconi, C. F., Ed., Vol. 6, Part II, 3rd Edition, Wiley Interscience: New York, 1986, p. 391.

DISCUSSION

WILSON - In addition to the long list of detection methods you listed, one would at least dream about fast neutron scattering and X-ray diffraction. Since you are at a national laboratory where one can dream expensive dreams can one look forward to such types at measurements ?

JONAH - I would hope so. The development of pulsed neutron sources at Argonne (INPS) and elsewhere make such measurements possible. These types of studies might be particularly desirable for understanding the changes in configuration that can occur in chlorophyll reaction centers. At present the discussions of such techniques are most often catalyzed by alcohol.

RIVAIL -   1°) I would like to add another example to Dr. Jonah's reply. Synchrotron radiation is naturally pulsed radiation and the development of fast multichannel detectors makes possible the study of some reacting systems by using EXAFS spectroscopy.

        2°) A question. Can you give us more details on the experiment using microwave conductivity ?

JONAH - The experiments were done by John Warman and co-workers at Delft in the Netherlands. X band (10 GHz) microwaves were measured in absorption. Thus the ultimate time resolution is 200-300 p. sec.

ADVANCES IN FEMTOSECOND OPTICAL TECHNIQUES: RECENT EXPERIMENTAL INVESTIGATIONS IN THE FIELD OF REACTION DYNAMICS

Y. Gauduel, A. Migus and A. Antonetti

Laboratoire d'Optique Appliquée, INSERM U275

Ecole Polytechnique-ENSTA, 91120 Palaiseau, France

1. Summary

The recent advances in the generation and measurement of ultrashort laser pulses has allowed to push most of the optical methods of investigation such as absorption, polarization and chemiluminescence spectroscopy down to the picosecond (10^{-12} s) and even femtosecond (10^{-15}s) time scale. These new techniques allow now to monitor the molecular dynamics of polar liquids and provide new results in the field of chemical or biochemical reaction dynamics: ultrafast free-radical reactions, intramolecular and intra-ionic dynamical processes.

The information available from the femtosecond optical techniques will be illustrated by recent results on electron trapping and solvation in aqueous media at room temperature. In pure liquid water, the existence of a precursor of hydrated state have been obtained by femtosecond absorption spectroscopy. These studies provide unique experimental basis for testing recent theoretical results obtained by newly developed techniques (Monte Carlo, molecular dynamics simulations, path integral) and to permit the obtention of a consistent picture of electron trapping and solvation in aqueous solutions.

2. Introduction

The elucidation of detailed mechanisms of ultrafast complex events occuring in molecular dynamics, charges transfers or reaction dynamics have been made possible by recent advances in spectroscopy techniques using ultrashort laser pulse generation (Fork et al. 1982; Migus et al. 1982,1985). Most of the extent optical methods of investigations such as absorption, luminescence spectroscopy are now being used in the femtosecond time domain for investigations of a great variety of new phenomena in chemistry, biology and solid state physics (Hilinski and Rentzepis, 1983; Fleming, 1986; Fleming and Siegman, 1986).

High intensity femtosecond pulses have already be demonstrated to be of prime importance in the study of ultrafast biological and chemical reactions and specially in the monitoring and analysis of initial events during photodissociation or photoionization processes and photosynthesis (Martin et al., 1983; Gauduel et al., 1984, 1985; Breton et al., 1986). In this paper, we will center on the implication of the femtosecond spectroscopy in the monitoring of the fast elementary steps of electron solvation in homogeneous aqueous solution at ambient temperature.

Numerous works using flash photolysis of solute molecules and pulse radiolysis of polar fluids have been extensively performed to generate excess electron and study electron solvation (Rentzepis, 1973; Baxendale, 1977; Kenney-Wallace and Jonah, 1982). In these experiments, electron is used as microprobe of the local dynamics of molecular reorganization induced by the presence of local dynamics of molecular reorganization induced by the presence of local electric field linked to the presence of an excess electron. The structure and dynamics of formation of the solvated electron is of considerable interest for understanding electron transfer processes in condensed matter. So, picosecond dynamics studies in alcohols have demonstrated the existence of an intermediate state of electron (infrared absorbing trapped state) which is identified as a precursor of the fully relaxed solvated state (Huppert et al. 1981). In pulse radiolysis and photolysis studies, the subsequent relaxation of this localized species have been observed to occur in a temporal range comparable to the dielectric relaxation time (T_2) and consequently is correlated to molecular rotation (Chase and Hunt, 1975; Kenney-Wallace and Jonah, 1982). These picosecond results have permitted to rule out the role of the predominent existence of pre-existing deep traps during the solvation process in alcohols and to propose a hydrodynamical model.

In the specific case of aqueous solutions at ambient temperature, since the discovery of a short living hydrated electron (Hart and Boag, 1962), very little was known about the mechanisms governing the electron-water molecules interactions during the solvation process. Up to now the great majority of electron solution studies in aqueous solutions has been realized by injecting electron using pulse radiolysis methods, providing at best a 10 ps accuracy (Chase and Hunt, 1975; Baxendale, 1977; Jonah et al., 1977; Lewis and Jonah, 1986). In this case, the appearance of the hydrated electron was always limited by the instrumental resolution. The availability of femtosecond pulses enables electron solvation to be time-resolved in aqueous solutions and it becomes possible to clearly demonstrate whether the electron solvation in liquid water proceeds as in alcohols through an intermediate presolvated species. New spectroscopic evidence concerning the dynamics of electron solvation in aqueous media is presented and integrated with recent data of molecular dynamics simulations in liquid water.

3. Generation of tunable femtosecond optical pulses and time-resolved spectroscopy

The topic of generating subpicosecond pulses has been discussed at length elsewhere. This section summarizes the basic methods used for generating ultrashort pulses in our laboratory (Figure 1).

Generation of high power femtosecond pulses is primarily based on colliding pulse mode-locked dye lasers (CPM) and further amplification. This set up and its principle have been already discussed in detail in recent publication (Migus et al., 1985) but improvements have been added. The CPM oscillator simply contains two flowing jets, one conventional jet of rhodamine 6G and one very thin jet of saturable absorber and a four prism arrangment used for compensating groups velocity inside the cavity. The pulses generated in this laser are as short as 50 fs ($5 \ 10^{-14}$ s) and the wavelength is centered around 620 nm. These pulses are amplified in a four stages pumped by a frequency doubled Q-switched Nd-Yag laser and separated by saturable absorber jets. The short pulse duration introduces new problems in the amplifier design. It should be realized that 100 fs pulses are associated with about 60 Angströms wide spectrum. When propagating in the solvents and lens media, due to the linear index dispersion, low frequencies travel faster than higher frequencies. The time delay between the different frequency components imply then a temporal broadening. However, when this broadening is due only to velocity dispersion, it can be compensated by means of an ajustable delay line composed of a grating pair (GP) or a four prism arrangment. In our case we have used the latter solution at the amplifier exit, allowing output pulses of energy above 1 mJ and typically 100 fs duration when using a pumping energy of 300 mJ at 530 nm with a pump duration of 6 ns and at a 10 Hz repetition rate.

These high intensity femtosecond pulses are fundamental for spectroscopic purpose in that they can generate white light pulses of comparable duration. This can be done by focussing the amplified 620 nm pulse of 100 fs duration into many different transparent liquids such as water, alcohols, ethylene glycol and so on. In most experiments this continuum light is used as a weak delayed probe beam while the intense pulse required for perturbing the analyzed medium consist in the initial fundamental pulse or with less energy in its second harmonic. These experiments have obvious advantages over those done with an identical frequency for the pump and the probe: small variations of the sample transmission or reflection can be simultaneously resolved all over a broad spectrum extending from 0.2 µm to a few microns. Furthermore we avoid the so-called coherent artifact which prevents usually a clear determination of the initial kinetics.

A limited choice of excitation wavelengths may however be a severe restriction if, for instance, the studied medium is transparent or on the contrary too much absorbing at these specific wavelengths. The possibility of obtaining intense femtosecond excitation, tunable all over the optical spectrum, is the result of combining the properties of continuum short duration and dyes broad spectral coverage. The principle is the following: a spectral part of the continuum is selected with an interferential filter and further amplified in a spectrally matched flowing dye cell. Fine spectral tuning is realized by slightly tilting the interferential filter.

The general scheme of time-resolved absorption experiments is also shown in figure 1. The initial pulse at 620 nm is split in two parts. One is the test beam while the other beam, the pump, can be delayed in time using a translation stage mounted on a stepping motor with an accuracy of 0.1 µm. The pump beam may be used at 620 nm, or may generate harmonics or continuum to further amplification (tunable excitation box). In the case developed here, the initial pulse is focussed into a 1-5 mm thick KDP crystal to produce up to 20 µJ at 310 nm. The test beam, a weak white light generated in CG (F is a filter to remove unwanted wavelenghts) is split in two parts; one probes the excited region of the sample, while the other, missing the sample is used as a reference. These two continuum beams are then focussed on the 400 µm entrance slit of a 0.25 m spectrometer used in the first order of a 600-groove-per-mm grating, and directed to two diodes. The two signals, corresponding to the probe and reference pulses energy at a given wavelenght, are then sent through an electronic chain (pulse shaping, sampling holder, and digitizer) to a computer; their ratio is determined and used in a multichannel analyzer program. In one sweep 100 positions of the delay line are examined, and 5 of these ratio integrated for each position. Shots for which energy varied more than 50% from the mean were rejected.

The absorption change induced by a pulse of intensity $I_P(t)$ on a test pulse of intensity $I_T(t)$ with the temporal delay between the two pulses (as determined by the variable spatial delay line) is given in our small signal case by:

$$\Delta\alpha(\tau) = \int_{-\infty}^{+\infty} A(\tau - \tau') \int_{-\infty}^{+\infty} I_P(t)\ I_T(t+\tau')\ dt\ d\tau'\ \alpha_o \quad (1)$$

In this expression where $A(t)$ is the impulse response of the molecular absorption, the influence of group velocity dispersion is not taken into account. However, a time broadening factor occurs when the UV and the test (here in the visible or the infrared) overlap along the 2 mm long sample used in the following experiments, but travel with different group velocities. We have computed this broadening to be of the order of 200 fs, so that in the equation (1) the pulses may be equivalently assumed to be of 200 fs duration. In case of instantaneous molecular response $A(t)$ is a constant, while in case of a non-null molecular response time T it becomes:

$$A_{(t)} = 1 - \exp(-t/T) \text{ for } t>0 \quad (2)$$

In femtosecond experiments a key point is the exact knowledge for each probe wavelength of both the pulse shape and the position of the zerotime delay i.e. the position of the delay line for which the pump and probe pulses coincide exactly.

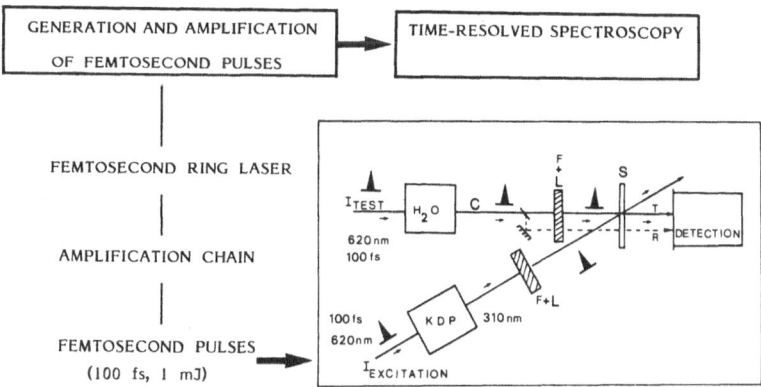

Fig. 1. General scheme for generation of femtosecond pulses, amplification and time-resolved spectroscopy (F:filter, L:lens, T:test pulse, R:reference pulse, S:sample, C:continuum).

This is of prime importance owing to the fact that the risetime responses are expected to be of the order of the pulse duration. In this case, the rise of the signal does not change significantly in shape but is rather time shifted (Gauduel et al.,1985). This determination must be repeated for each probe wavelength because of the chirp in the continuum induced group velocity dispersion in the different optics before the sample (chirp consist in the red part of the continuum arising on the sample before the blue part). The principe of the determination is the following: we replace the sample cell (2mm) with an identical cell filled with pure n-heptane solution in which we have observed a weak induced transient absorption when excited with 100fs pulse duration at 310 nm (E=4 eV). The figure 2 shows the risetime of this signal at 720 nm. Similar induced absorption curves are obtained in the infrared. This signal, which is the fastest we ever found in the pump-probe configuration of figure 1, can be due to a two photon ionization of the hydrocarbon (equations 3,4) as previously suggested by Kenney-Wallace and Jonah (1982).

$$\text{n-Heptane} + 2 \text{ h}\mu \text{ (4eV)} \longrightarrow [\text{Heptane*}] \longrightarrow \text{Heptane}^+ + \text{e}^-_{qf} \qquad (3)$$

$$\text{e}^-_{qf} + \text{n-Heptane} \longrightarrow \text{e}^-\text{Heptane} \qquad (4)$$

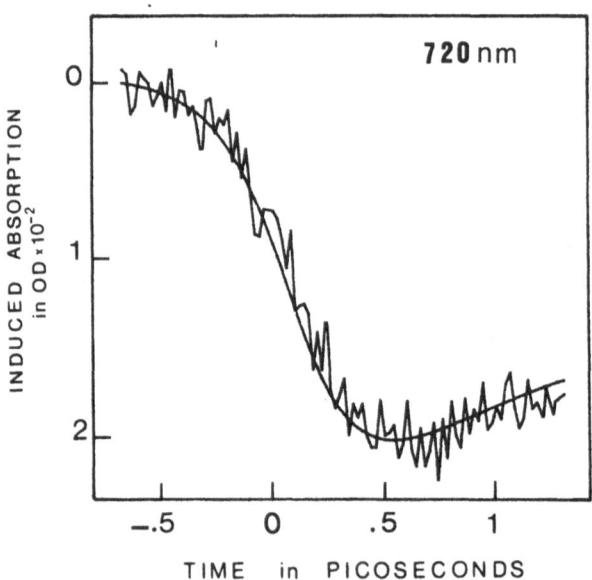

Fig.2. Femtosecond induced absorption (720 nm) at 294°K following excitation of pure n-heptane with 100 fs laser pulses (λ =310 nm E=4 eV).

From the available data of the literature, the induced absorption which is obtained at 720 nm can be well fitted with an "instantaneous" kinetics which correspond precisely to the theoretical estimate which takes into account the pulse duration and the group velocity dispersion between the UV excited and the probe along the 2mm sample path. A partial relaxation with a time constant of 800 fs can be observed with n-heptane. This ultrafast relaxation can be attributed, as it has been previously suggested at low temperature, to a very fast "spur" recombination process between electron and positive ion (Baxendale et al., 1977).

4. Reactivity of electron with aqueous solutions

The hydrated electron continues to be of considerable interest in chemistry and biology because it is the most significant species in radiation chemistry. Moreover, the energetics and the time dependence of the electron-medium interaction play an important role in the formation of this radical in condensed matter. The knowledge of the dynamics of electron localization and solvation is of prime importance for understanding electron transfer processes in condensed water. They has been continuing discussion over the relative role in the long-range and short-range electron-medium interactions in the sequence of electron thermalization, trapping and solvation in aqueous solutions. Experimental research including picosecond experiments does not permit to obtain indication on the primary species involved in the formation of hydrated electron. Recently, important informations have been obtained with the development of experimental and theoretical researches. Theoretical breakthroughs have been realized by using computer simulations and several groups have deduced some informations on the structural aspect of the hydrated electron by employing Monte-Carlo and path integral techniques (Jonah et al., 1986, Schnitker et al., 1986, 1987, Wallqvist et al., 1986). A molecular dynamic simulation of water using a model electron-molecule potential has permit to identify favorable sites for the initial electron trapping (Schnitker et al., 1986). These computer simulations lacked however experimental grounds to propose a consistent picture of electron solvation in liquid water.

One year ago, employing femtosecond spectroscopy technique we have obtained unique informations on the dynamics of both trapping and solvation of electron in aqueous solutions (Gauduel et al. 1986; Migus et al. 1987). Direct photoionization of water molecules by laser light does not seem possible since water is transparent in the UV spectral region up to 190 nm. However, recent developments on the non-linear interaction of powerful laser UV radiation ($I > 10^8$ W/cm^2) with neat and pure polar fluids have permit to investigate the mechanism of two photon photolysis and the subsequent formation of solvated electron (Nikogosyan et al., 1983). In our case of ultrashort pulse, light can be absorbed through a non linear process, namely two photo absorption. Indeed the two photon absorption coefficient of pure water at 310 nm has been estimated to be 4 10^{-13} m/W and non negligible absorption becomes possible when dealing with multigigawatt peak power pulses. In this case our 8 eV two-photon excitation is above the ionization threshold for liquid water estimated to be around 6.5 eV.

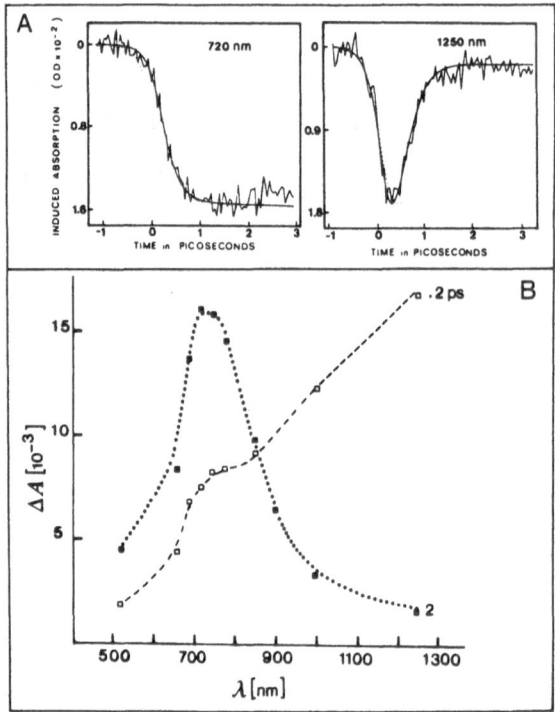

Fig. 3.   A: Risetime of absorption at 720 and 1250 nm. The smooth lines represent the computer best fits assuming an appearance time of 110 fs for the infrared specie (1250 nm) and its relaxation toward the solvated state in 240 fs. B: Transient absorption spectra of the electron at 200 fs and 2ps following photoionization of pure liquid water with 100 fs laser pulse (λ=310 nm, E=4 eV).

Each 100 femtosecond pulse produces a homogeneous solvated electron concentration of about 5 µM and futhermore permits the dynamics of electron solvation to be resolved with femtosecond accuracy, starting from initially quasi free state up to full solvation (equations 5-7).

$$H_2O + h\mu \longrightarrow \longrightarrow H_2O^* \xrightarrow{\ \tau_0\ } H_2O^+ + e^-_{qf} \qquad (5)$$

$$e^-_{qf} + n\ H_2O \xrightarrow{\ \tau_1\ } e^-_{presol} \qquad (6)$$

$$e^-_{presol} \xrightarrow{\ \tau_2\ } e^-_{sol} \qquad (7)$$

Induced absorption kinetics measured over the whole visible spectrum and in the near infrared up to 1250 nm have been used to reconstitute and analyse the transient spectra of the electron at different delays after photogeneration of solute (figure 3). A high energy tail of an infrared band extending above 1250 nm appears within the excitation and is fully developed after a delay of 0.2 ps. Two picoseconds after the pulse excitation, the infrared band has disappeared leaving a broad absorption band very similar to the structureless absorption spectrum assigned to solvated electrons. This asymetric band peaks around 1.7 eV and its high energy tails extends above 3 eV.

The absorption spectroscopy kinetics summarized in the figure 2A have been analyzed by a kinetical model which takes into account the existence of primary steps (electron trapping and electron solvation). This model is represented in the figure 4. Starting from the excited state of water molecules (N_1) we assumed the existence of two separate species during the electron solvation process: one absorbing in the infrared (N_2) appears with a time constant of T_1 and then relaxes toward the solvated electron state (N_3) following a first order kinetics with a time constant T_2. From the instantaneous response obtained in n-heptane for each test wavelength, we deduce the pulse shape and the actual position of the zero time delay i.e. the position of the dealy line for which the pump and the probe pulse overlap exactly. The exact procedure is as follow: a test pulse of temporal shape $I_T(t)$ experiences at one wavelength an absorbance (ΔA) due to a concentration of presolvated and solvated electrons created at time t through a two-photon absorption of an excitation pulse of profile $I_E(t)$. Following the kinetical model we get in case of small signal:

$$\Delta A^\lambda_P(\tau) = \epsilon^\lambda_P\ l\ \int_{-\infty}^{+\infty} N_2(\tau - \tau')\ C^\lambda(\tau')\ d\tau' \qquad (8)$$

$$\Delta A_S^\lambda(\tau) = \epsilon_S^\lambda \, 1 \int_{-\infty}^{+\infty} N3(\tau - \tau') \, C(\tau') \, d\tau' \qquad (9)$$

$N_2(t)$ and $N_3(t)$ are defined in figure 4 and

$$C^\lambda(\tau') = \int_{-\infty}^{+\infty} I_T(t+\tau') \, I^2_E(t) \, dt \, d\tau' \qquad (10)$$

In these expressions, 1 is the interaction length, A the final solvated electron concentration, ϵ_S (ϵ_p) the molar extinction coefficient of the solvated (presolvated) species at wavelengthλ.$C(\tau')$ which is a third order correlation between the probe at and the pump pulse is normalized. The total absorbance of the test pulse is therefore $\Delta A^\lambda(\tau) = \Delta A_P^\lambda(\tau) + \Delta A_S^\lambda(\tau)$, expression which takes into account all the population evolution occuring during the excitation and probe. The instantaneous kinetics in n-heptane determines strictly $C^\lambda(\tau')$ in shape and position along the time axis.

As shown in figure 3A, the kinetics at 1250 and 720 nm reflect the evolution of the infrared and solvated species respectively. The infrared band appears with a time constant of 110 +/- 30 fs and quickly relaxes following a first order kinetics with a time constant of 240 +/- 40 fs towards the fully solvated species. These data are the first evidence of at least one precursor of hydrated electron.

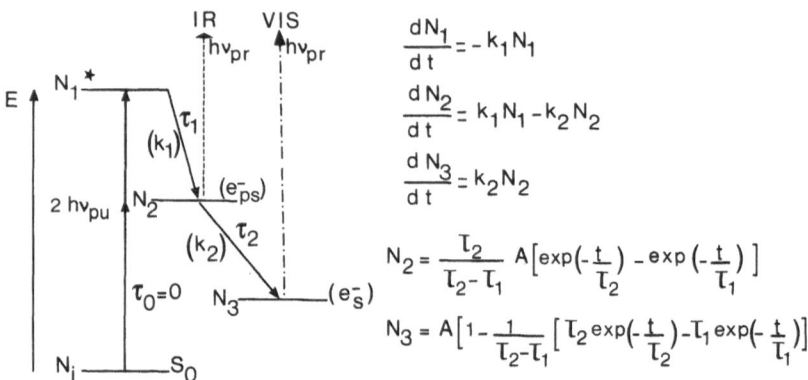

Fig. 4. Kinetical model of electron trapping and solvation.

5. Electron trapping and solvation in anionic aqueous solution

The femtosecond ultraviolet photochemistry of aqueous solution of ferrocyanide ($Fe(CN)_6^{4-}$, 0.45M) have permit to study the process of electron solvation in aqueous concentrated ionic solutions. The excited state of ferrocyanide in water can be produced by femtosecond UV pulses and may dissociate in a thermally activated process to yield a solvated electron according to the following equation:

$$Fe(CN)_6^{4-} + h\mu \quad [Fe(CN)_6)]^{4-*} \xrightarrow{\tau_0} [Fe(CN)_6)]^{3-} + e^-_{qf} \qquad (11)$$

$$e^-_{qf} + H_2O \xrightarrow{\tau_1} e^-_{presol} \xrightarrow{\tau_2} e^-_{sol} \qquad (12)$$

This photochemical process competes with a reversible photoaquation reaction leading to the formation of a pentacyanoaquo complex.

The time-resolved spectroscopy using 310 nm pulses of 100 fs duration has demonstrated that the electron solvation subsequent to a photodetachment of an electron proceeds through one intermediate state. At 1250 nm an intermediate state of electron appears with a time constant of 100 +/- 20 fs and relaxes towards a fully solvated species following a first order kinetics with a time constant of 220 +/- 30 fs. The signal observed in the red spectral region (720 nm) follows the kinetics defined by the equation of N_3 (figure 4).

The femtosecond spectroscopic data summarized in the table 1 demonstrate that the reactivity of electron trapping in concentrated ionic aqueous solutions is identical to what has been observed in pure liquid water. These experiments, which involved electron photodetachement from the ferrocyanide ion demonstrate that electron solvation does not proceed through a direct electron capture by pre-existing deep traps suggested as previously (Wiesenfeld and Ippen, 1980).

So, the mechanism of solvation of electron in concentrated ionic aqueous solutions procceeds through at least two transitions involving a transient state infrared absorbing and a subsequent fully hydrated state.

Table 1. Risetime of induced absorption in the red and infrared spectral regions (720, 1250 nm) following femtosecond excitation of aqueous media at 20°C.

Aqueous media	720 nm , 1250 nm	
	T_1 (fs)	T_2 (fs)
Pure liquid water	110 +/- 30	240 +/- 40
Ferrocyanide solution (0.48 M)	100 +/- 20	220 +/- 30

6. General discussion

The infrared data obtained in pure liquid water and anionic aqueous solutions are in agreement with an initial electron-medium interaction inducing the formation of a negatively charged cluster $(H_2O)_n^-$ i.e. a presolvated state of electron.

The very first risetime for the appearance of a trapped state is short compared to any nuclear motion, solvent dipole orientation and thermal motion of water molecules (Frank Condon principle). This implies that an efficient rate of energy loss occurs in the medium following electron detachment and that mechanisms involved in the localization process of electron do not require large molecular and dynamical reorganization. This initial trapping time ($1.1 \ 10^{-13}$s) is longer than the different theoretical estimates of electron thermalization ($2-5 \ 10^{-14}$ s) in liquid water (Frohlich and Plazman, 1953). This would suggest that in liquid water and ionic aqueous solutions, electron thermalization is able to occur before electron trapping. Such fast thermalization is quite possible if we take into account the existing numerous mode of vibration of the solvent and the high energy loss rate. This would imply that the excess energy kinetics has been transferred to the water molecules in less than 110 fs i.e. the risetime of the infrared spectrum. In this case, the local temperature rise can be estimated to be negligible since the solvated electron spectrum taken only a fraction of picosecond after creation and is not different to that obtained at 20°C.

During the 110 fs, it can be suggested that the electron either creates its own trapping site (self trapping mechanism) or searchs for pre-existing shallow traps. If we take into account the recent works of Schnitker et al., 1986, possible initial pre-existing trapping sites are arising from statistical fluctuations and molecular clustering. These authors find a distribution of such sites monotonically in energy down to -1.4 eV, which is compatible with our IR data. In this way, we have proposed that this precursor is a state (presolvated state) where the electron is still spatially extended. Recent molecular dynamics simulation of pre-existing traps give a trap size of about 4 angströms. This size is greater than the radius of the fully relaxed solvated electron (2.5 angströms). It can be suggested that the ultrafast relaxation of the infrared band corresponds to a contraction of the trap size associated with a modification of the charge distribution inside the cavity of hydated electron.

The kinetical model that we have developed to fit the experimental femtosecond IR and visible curves considers the existence of two distinct electronic and configurational states (figure 5). The correspondence that we observed between the time relaxation of the IR absorbing species (1250 nm) and the risetime of the visible absorbance (720 nm) suggest that the initial IR trapped electron is the direct precursor of the configurationaly relaxed final quantum state of solvated electron. The time constant for the appearance of the IR precursor and its relaxation are similar in pure water and an anionic aqueous solution. This implies that the existence of the trapped electron is neither influenced by the method of electron photoejection nor by the ionic strength of the polar medium.

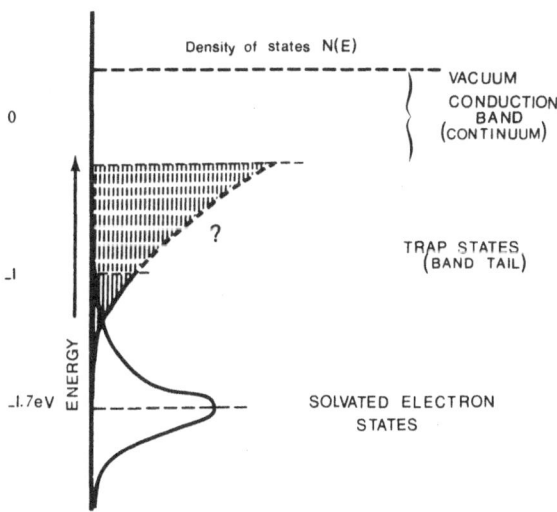

Fig.5. Representation of the energy diagram and density states for presolvated (trapped) and solvated electron in pure liquid water and anionic aqueous solution.

Moreover, the presence of negative ion [Fe(CN)$_6$)$^{4-}$] does not seem to significantly restucture the immediate water environment and does not provide more favorable spatial distribution of pre-existing deep trapping sites. From previous experiments and assuming that thermalization and solvation distance of electron with exogenous energy around 1 eV is about 40 angströms, it can be inferred that many of the quasi free electron with excess energy kinetics escape from the vicinity of the cation (H$_2$O$^+$) or the anion [Fe(CN)$_6$)$^{3-}$] and that long range Coulomb type interactions influence neither the initial presolvation nor the configurational relaxation of the medium in the electronic field of the charge.

It can be noticed that the solvation time of the trapped electron (presolvated) is surprisingly in excellent agreement with an extrapolation of measurements done in 5 M CaCl$_2$ cooled to 236 °K (Garett, 1980). In this work, where an activation energy was assumed, a 0.2 ps time was anticipated for the solvated electron time in water at ambient temperature. We may wonder whether a long-range interaction with the medium such as a dielectric relaxation can explain the ultrafast 240 fs decay of the localized species toward the solvated one in pure water and anionic aqueous solution. It is admitted that the solvation characteristic time can be predicted by the relaxation time from the fixed charge approximation (Mozumder, 1969, Calef et Wolynes, 1982). This estimate time is defined by the Debye equation: $T_L = T_D \, \varepsilon_\infty / \varepsilon_0$ where T_D is the water Debye relaxation time (10 ps), ε_0 (ε_∞) the static (high frequency) dielectric constant. This value of T_L in the range (0.2-0.4 fs) is consistent with our experimental T_2 value. However, a dielectric relaxation, when treated as a diffusive type of process (Debye relaxation) implies a continuous modification of the medium interacting with the electron. This should appear as a continuous spectral shift of the electron absorption band from the infrared towards the visible (Zusman et Helman, 1985). Such an effect is not in accordance with our data from which we observe a well defined absorption IR spectra for the localized species. However, we cannot completely rule out a dielectric relaxation of the medium. We may for instance assume a continuous shift of the energy of the electron associated with a decrease of Frank Condon factors while the solvent has not reached its equilibrium. In this case, the absorption of light by the electron would be negligible in the dynamic phase. An other hypothesis to explain fast trapped electron relaxation will be the role of local reorientation of the lattice medium through proton jumps. Up to now no pertinent experimental or theoretical data permit to support this hypothesis.

In conclusion, the photoionization of water molecules or solute $[Fe(CN_6)^{4-}]$ by light power UV femtosecond pulses permits to obtain fundamental information on the time evolution of the electron populations between the electron photodetachement and the configurationnaly relaxed final quantum state of hydrated electron. The electron solvation in pure liquid water at room temperature proceeds through, at least, two transient steps with in particular an infrared absorbing presolvated state. The time-resolved spectral evolution of the initialy non-hydrated state rule out the assumption of the predominent role of high density of pre-existing deep traps into which the electron falls directly after photodetachement from solvent molecules. Femtosecond data do not support also the existence of a pure dielectric relaxation mechanism for the final rearrangement energy linked to the solvation step. The experiments involving electron photodetachement from a ionic solute (ferrocyanide) supports the notion of a solvation time comparable to what observed in polar fluid which is not pertubed by the presence of a solute (pure water).

From the two steps model defined in these experimental work, we may speculate on the existence of a complex process in which the electrons are initially thermalized and then localized in numerous available pre-existing shallow traps and finally solvated through an energetic contraction of traps and local rearrangements of electronic cloud of water molecules.

7. References

Baxendale, J.H., 1977, Can.J. Chem., 87: 1996.

Breton, J., Martin, J.L., Migus, A., Antonetti, A. and Orszag, A., 1986, Proc. Nat. Acad. Sci. USA, 83:5121.

Calef, D.F. and Wolynes, P.G., 1983, J. Phys. Chem., 87: 3387.

Chase, W.J. and Hart, J.W., 1975, J. Phys. Chem., 79: 2835.

Fleming, G.R., 1986, Ann. Rev. Phys. Chem., 37: 81.

Fleming, G.R. and Siegman, A.E., 1986, In "Ultrafast Phenomena", Springer Verlag.

Fork, R.L., Shank, C.V. and Yen, R.T., 1982, Appl. Phys. Lett., 41: 223.

Frohlich, H. and Platzman, R.L., 1953, Phys. Rev., 92: 1152.

Garett, W.R., 1980, J. Chem. Phys., 73: 5721.

Gauduel, Y., Migus, A., Martin, J.L., Antonetti, A. 1984, IEEE QE 20: 1370.

Gauduel, Y., Migus, A., Martin, J.L., Lecarpentier, Y. and Antonetti, A., 1985, Ber. Bunsen-Ges. Phys. Chem., 89: 218.

Gauduel, Y., Martin, J.L., Migus, A., Yamada, N., Antonetti, A., 1986, In "Ultrafast Phenomena V", Springer Verlag, 308.

Hart, E.J. and Boag, J.W., 1962, J. Am. Chem. Soc., 84: 4090.

Hilinski, E.F. and Rentzepis, P.M., 1983, Nature, 302: 481.

Huppert, D. Kenney-Wallace, G.A. and Rentzepis, P.M., 1981, J. Chem. Phys., 75: 2265.

Jonah, C.D., Miller, J.R. and Matheson, M.S., 1977, J. Phys. Chem., 81: 1618.

Jonah, C.D., Romero, C. and Rahman, A., 1986, Chem. Phys. Lett., 123: 209

Kenney-Wallace, G.A. and Jonah, C.D., 1982, J. Phys. Chem., 86: 2572.

Lewis, M.A. and Jonah, C.D., 1987, J. Phys. Chem., 90: 5367.

Martin, J.L., Migus, A., Poyart, C., Lecarpentier, Y., Astier, R. and Antonetti, A., 1983, Proc. Nat. Acad. Sci., USA, 80: 173.

Migus, A., Antonetti, A., Etchepare, J., Hulin, D. and Orszag, A., 1985, J. Opt. Soc. Am. B2: 584.

Migus, A., Gauduel, Y., Martin, J.L. and Antonetti, A., 1987, Phys. Rev. Lett., 58: 1559.

Migus, A., Martin, J.L., Astier, R., Antonetti, A. and Orszag, A., 1982, In "Picosecond Phenomena III", Springer Verlag, 6.

Mozumder, A.J., 1969, Chem. Phys., 50: 3153.

Nikogosyan, D.N., Oraevsky, A.O. and Rupasov, V.I., 1983, Chem. Phys., 77: 131.

Rentzepis, P.M. and Jones, R.P., 1973, J. Chem. Phys. 59: 766.

Schnitker, J. and Rossky, P.J., 1987, J. Chem. Phys., 86: 3462.

Schnitker, J. and Rossky, P.J. and Kenney-Wallace, P.J., 1986, J. Chem. Phys., 85: 2986.

Shirom, M. and Stein, G., 1971, J. Phys. Chem., 55: 3372

Wang, Y., Crawford, M.K., Mc Auliffe, M.C. and Eisenthal, P.M., 1981, J. Chem. Phys., 75: 2265.

Wallqvist, A., Thirumalai, D. and Berne, B.J., 1986, J. Chem. Phys., 85: 1583.

Wiesenfeld, J.M. and Ippen, P., Chem. Phys. Lett., 86: 2572.

Zusman, L.D., Helman, A.B., 1985, Chem. Phys. Lett., 114: 301.

DISCUSSION

<u>BRATOS</u> - Apart from the electron solvation processes are there other reactive or prereactive processes where the use of femtosecond techniques is necessary ?

<u>GAUDUEL</u> - Femtosecond laser pulses allow to monitor molecular dynamics of polar liquids and to provide new developments in the field of chemical or biochemical reaction dynamics : initial events occurring in photosynthesis, ultrafast free radical reactions, intramolecular and intraionic dynamic processes.

<u>ROSS</u> - I question one of your early graphs which indicates the availability of microsecond techniques available by 1850 !

<u>GAUDUEL</u> - This schematic graph is approximative and does not disagree with the recent advances in femtosecond optical techniques.

Editor's note : see text of Pacault about time resolutions of 10^{-4} s in 1858, and measurement of fluorescence lifetimes of 10^{-8} s by E. Becquerel.

<u>HYNES</u> - Did I understand correctly that you infer from your experiments that the solvation time is 240 femtoseconds ? If so, it is interesting to note that this is very close to the longitudinal dielectric relaxation time of water, the result predicted by simple continuum theories.

<u>GAUDUEL</u> - Yes the measured solvation times in pure liquid water (240 fs) is in excellent agreement with the characteristic longitudinal dielectric relaxation time T_L (0.2 - 0.4 ps) predicted by continuum theory at constant charge. However, theories developed in such framework predict that this relaxation should be reflected by a continuous shift of the electron absorption spectrum, which is in contradiction with our observation of a stepwise process. Nevertheless, the correspondence between T_L and T_2 may not be just fortuitous and the observation of a quite deep initial localization suggests only modest configurational changes of the water molecules around the electrons during the liquid relaxation. This should have therefore little direct effect on the energy and absorption spectrum of the electron.

<u>NEWTON</u> - Even if one adopts the concept of T_L based on a simple Debye dispersion model, one must recognize the ambiguity in evaluating T_L, associated with the choice of a value for ϵ_∞ (infrared or optical ?). Furthermore, Wolynes has recently argued for the likely occurrence of an additional relaxation time of magnitude larger than T_L, arizing from the molecularity of the solvent in the immediate vicinity of the ion.

GAUDUEL - I agree with you, there is a continuous debate about the choice of a value for the high frequency limit (ϵ_∞) of the dielectric constant. So, if ϵ_∞ equals the square of the refractive index ($\simeq 2$) the estimate of the dielectric relaxation time T_L is about 250 fs at 20°C ($T_D \simeq 10$ ps). In the case where ϵ_∞ is defined by infrared considerations (4) T_L equals 500 fs. The two estimates of T_L are in agreement with the experimental time of electron solvation in water. However, this comparison between theoretical estimate of T_L and T_2 does not permit to definitively conclude about the role of the dielectric relaxation during relaxation of trapped electrons.

JONAH - Is the energy of the electron photodissociated from $Fe(CN)_6^{-4}$ large enough to move it from the aligned water near the ion ? If the electron does not move far enough, one would not expect such similar results, for the two measurements.

GAUDUEL - In pure liquid water and ferrocyanide solution, the kinetic energy of photoelectrons is expected to be about $0.5 - 1$ eV. The estimate of the thermalization distance being more than 40 Å, electron trapping and solvation would occur far from the H_2O^+ or $Fe(CN)_6^{3-}$ ions.

This hypothesis can explain the similar trapping and solvation time obtained in water and ionic aqueous solution.

CHEMICAL REACTIVITY IN WEAK CHARGE-TRANSFER COMPLEXES :

ANALYSIS OF INDUCED FAR INFRARED PROFILES AND RAMAN SCATTERING PROFILES

Marcel Besnard, Nathalie Del Campo, and Jean Lascombe

Laboratoire de Spectroscopie Moléculaire et Cristalline
UA 124 CNRS, 351, Cours de la Liberation
33405 Talence Cedex, France

INTRODUCTION

Vibrational studies of very fast chemical reactions, with rate constants ranging from 10^{10} to $10^{13} s^{-1}$, have already been performed since a long time. For instance, one may refer to proton transfer kinetic studies in strong acidic solutions [1,3], and, in a more general overview, to conformational dynamics studies about the internal rotation in methyl groups and the ring puckering motion in small cycles [4,5].

Actually great progress in this field may be expected due to the both improvement of experimental technics and the development of more precise theories which take into account the kinetic aspect of the reactions : this allows one to perform a deeper analysis of spectral lineshapes [6,7].

Weak donor-acceptor complexes, such as benzene/iodine (Bz/I_2), are simple chemical systems well adapted for this kind of studies [8]. Indeed, their weakness is reflected by both a small value of the equilibrium constant ($K_c \sim 0.2 \ \ell \ mole^{-1}$ at 298 K for Bz/I_2) and an enthalpy of reaction of the order of magnitude of the thermal activation energy ($\Delta H \sim - 3RT$ for Bz/I_2). This might suggest a very short life time of the complexes and a very fast reactional dynamics governed by the diffusion of molecules in the liquid ; characteristic times in the picosecond range might be invoked and, so far, reached by vibrational studies [9].

These systems exhibit the further advantage that they can be studied along a double approach :

- the first one, well known in other spectroscopic technics (like NMR), is based on the analysis of the spectral profile changes observed on a selected normal mode of vibration in both the "free" and the complexed molecules.

- the second approach is based on the lineshape analysis of normal modes
which are in principle forbidden by selection rules in the isolated molecules.
Indeed, in gaseous state under very low pressure conditions, homonuclear
diatomic molecules (such as iodine) have no infrared activity. In contrast,
in dense or condensed state, this is no longer true and infrared spectra can
be observed. For instance, for the Bz/I_2 complex, in the far infrared region
(FIR), the ν_{II} vibration becomes active giving rise to a sharp peak at about
205 cm^{-1} together with a very broad band in the very far infrared at about
100 cm^{-1} (fig.1). This infrared activity results from intermolecular interactions and therefore these induced spectra are influenced in particular by
collective translational motions of the interacting molecules. They constitute
an unique probe to get a better insight into the relevant time and lenght
scales involved in reacting fluids [10].

This paper is aimed at the presentation of a study that we have performed on the Bz/I_2 system using the two above approaches. For clarity, we shall
present firstly according to the second approach, the results obtained from
analysis of induced far infrared profiles and after that those extracted from the Raman profiles. In each of these sections, we will develop
the basic theoretical elements and present the experimental results relevant
to this discussion.

Finally, we will discuss of the merits and limits of both approaches
and it will be shown that they can provide a complementary insight into this
system in reaction.

FAR INFRARED SPECTROSCOPY

Theoretical Background

Relationship between absorption coefficient and dipole moment spectral
density. The linear absorption coefficient (expressed in Neper -cm $^{-1}$) is
related to the infrared spectral density $I(\omega)$ following the relationship[11]

$$\alpha(\omega) = \frac{4\pi^2}{3\hbar cV} \frac{1}{n(\omega)\, D(\omega)} \omega \ (1 - \exp(-\frac{\hbar\omega}{kT})) \ I\ (\omega) \tag{1}$$

where V is the volume of the sample, c the velocity of light, $n(\omega)$ the real
part of the complex refractive index and $D(\omega)$ is the "internal field" correction[12].

The spectral density is related to the time Fourier transform of the
sample total dipole moment auto-correlation function (a.c.f.)

$$I\ (\omega) = \frac{1}{2\pi} \int_{-\infty}^{+\infty} < \vec{\mu}(o) \cdot \vec{\mu}(t) > \exp\ (-i\omega t)\ dt \tag{2}$$

Two-body and Three-body contributions[10]. For a binary mixture of non
polar molecules, the sample total dipole moment may be written in the approximation of pairwise additive contributions :

$$\vec{\mu} = \sum_I \sum_B \{ \vec{\mu}_{BI} + \vec{\mu}_{IB} \} \tag{3}$$

where $\vec{\mu}_{BI}$ is the dipole moment induced in molecule I (Iodine) by its inter-action with molecule B (benzene) and $\vec{\mu}_{IB}$ is the dipole moment induced by the iodine molecule on the benzene molecule.

The indexes I and B of the sum run over the respective numbers N_I and N_B of molecules in the sample.

For a very diluted solution of iodine in benzene, i.e. if one neglects the iodine-iodine interactions, a.c.f. takes the following form :

$$\langle \vec{\mu}(o) \cdot \vec{\mu}(t) \rangle = N_I \langle \sum_{B,B'} (\vec{\mu}_{BI}(o) + \vec{\mu}_{IB}(o)) \cdot (\vec{\mu}_{B'I}(t) + \vec{\mu}_{IB'}(t)) \rangle \quad (4)$$

In fact using standard properties of classical a.c.f. it is easy to show that:
$\langle \vec{\mu}(o) \cdot \vec{\mu}(t) \rangle =$

$$N_I N_B \{\underbrace{\langle \vec{\mu}_{BI}(o) \cdot \vec{\mu}_{BI}(t) \rangle}_{(a)} + \underbrace{\langle \vec{\mu}_{IB}(o) \cdot \vec{\mu}_{IB}(t) \rangle}_{(b)} + \underbrace{2\langle \vec{\mu}_{BI}(o) \cdot \vec{\mu}_{IB}(t) \rangle}_{(c)}\}$$

$$+ N_I N_B (N_B-1) \{\langle \vec{\mu}_{BI}(o) \cdot \vec{\mu}_{B'I}(t) \rangle + \langle \vec{\mu}_{IB}(o) \cdot \vec{\mu}_{IB'}(t) \rangle + 2 \langle \vec{\mu}_{BI}(o) \cdot \vec{\mu}_{IB'}(t) \rangle\} \quad (5)$$

The expression contained between the first curlies brackets represents the two-body contribution to the total dipole moment a.c.f. ; in each of the three terms enclosed in it (a to c) the same pair of iodine and benzene molecules are involved at time zero and at time t. Term (a), which will be referred in the following as the "benzene contribution" represents the a.c.f. related to the induced dipole moment in the iodine molecule by the benzene molecule. Correspondingly, term (b) will be referred as the "iodine contribution". Finally, term (c) is a cross term which correlates the dipole moment induced in iodine by the benzene molecule at time zero whith the dipole moment indu-ced in the same benzene molecule by the iodine molecule at time t.

The three-body contribution to the sample total dipole moment a.c.f. is given by the expression contained in the second curlies brackets and it presents the same structure as the previous one. The enclosed terms involve now at time zero and at time t two different benzene molecules and the same iodine molecule.

Let's stress that a four-body contribution is not taken into account due to negligible iodine-iodine interactions.

Expression (5) can be further simplified if one assumes that there is no orientational correlation between the molecules ; it comes out that :

$$\langle \vec{\mu}(o) \cdot \vec{\mu}(t) \rangle = N_I N_B \{\langle \vec{\mu}_{BI}(o) \cdot \vec{\mu}_{BI}(t) \rangle + \langle \vec{\mu}_{IB}(o) \cdot \vec{\mu}_{IB}(t) \rangle\}$$

$$+ N_I N_B (N_B-1) \{\langle \vec{\mu}_{IB}(o) \cdot \vec{\mu}_{IB'}(t) \rangle\} \quad (6)$$

The cross terms have desappeared and only one term (which involves the iodi-ne molecule as inductor) is responsible of three-body interactions.

Extension of the theory[8]. Under the following assumptions : (i) the benzene and iodine molecules interact throught a quadrupolar-scalar polari-zability induction mechanism.

(ii) In the induced spectra associated with the ν_{II} stretching vibration, linear variations of the polarizability and the quadrupole moment of the iodine molecule with its vibrational coordinate q, $\dfrac{\partial \alpha_I}{\partial q}$ and $\dfrac{\partial \eta_I}{\partial q}$ are considered.

(iii) The translational, rotational and vibrational degrees of freedom are uncoupled, the following total dipole moment a.c.f. is obtained [8] :

$$\langle \vec{\mu}(o) \cdot \vec{\mu}(t) \rangle = A_{LF}(t) + B_{HF}(t) \, \langle q(o) \, q(t) \rangle \tag{7}$$

with

$$A_{LF}(t) = \frac{2}{15} N_I N_B \, \{ (\alpha_I \eta_B)^2 \, \langle P_2 \left[\cos \theta_B(t) \right] \rangle + (\alpha_B \eta_I)^2 \langle P_2 \left[\cos \theta_I(t) \right] \rangle \}$$

$$\times \, \langle T^{(3)} \left[r_{BI}(t) \right] \, \vdots \, T^{(3)} \left[r_{BI}(o) \right] \rangle$$

$$+ \frac{2}{15} N_I N_B \, (N_B - 1) \, (\alpha_B \eta_I)^2 \, \langle P_2 \left[\cos \theta_I(t) \right] \rangle$$

$$\times \, \langle T^{(3)} \left[r_{BI}(t) \right] \, \vdots \, T^{(3)} \left[r_{B'I}(o) \right] \rangle \tag{8}$$

$$B_{HF}(t) = \frac{2}{15} N_I N_B \{ \left(\frac{\partial \alpha_I}{\partial q} \eta_B \right)^2 \langle P_2 \left[\cos \theta_B(t) \right] \rangle + \left(\alpha_B \frac{\partial \eta_I}{\partial q} \right)^2 \langle P_2 \left[\cos \theta_I(t) \right] \rangle \}$$

$$\times \, \langle T^{(3)} \left[r_{BI}(t) \right] \, \vdots \, T^{(3)} \left[r_{BI}(o) \right] \rangle$$

$$+ \frac{2}{15} N_I N_B \, (N_B - 1) \, (\alpha_B \frac{\partial \eta_I}{\partial q})^2 \langle P_2 \left[\cos \theta_I(t) \right] \rangle \, \langle T^{(3)} \left[r_{BI}(t) \right] \vdots \, T^{(3)} \left[r_{B'I}(o) \right] \rangle \tag{9}$$

where $P_2 \left[\cos \theta(t) \right]$ is the second order Legendre polynomial, the argument of which depends upon the angle θ of reorientation of the unit vector directed along the main symmetry axis of the molecule.

$T^{(3)}(r_{BI}) = \vec{\nabla} \cdot \vec{\nabla} \cdot \vec{\nabla} \, (\frac{1}{r_{BI}})$ is a symmetric traceless tensor. $\vec{r}_{BI}$ is the intermolecular vector connecting the centres of B and I molecules.

$$\langle q(o) \, q(t) \rangle = \langle q^2(o) \rangle \, G_v(t) \, \exp(-i \omega_{II} t) \tag{10}$$

where $G_v(t)$ is the normalized vibrational correlation function associated with the ν_{II} vibration, the angular frequency of which is ω_{II}.

It is noteworthy that :

i) The correlation functions $A_{LF}(t)$ and $B_{HF}(t)$ (and consequently their corresponding spectral density obtained after Fourier transformation) contain basically the same information on the dynamical processes involved in the sample total dipole moment a.c.f. Therefore they should decrease on a comparable time scale.

ii) In both $A_{LF}(t)$ and $B_{HF}(t)$ terms, the translational two- and three-body contribution appears. This shows clearly as previously discussed in introduction that relevant informations on the chemical reaction (which involves the r_{BI} coordinate) should be obtained from the corresponding profiles.

Finally, the time Fourier transform of expression (7) leads to the following spectral density :

$$I(\omega) = I_{LF}(\omega) + I_{HF}(\omega) \otimes I_{vib}(\omega) \otimes \delta(\omega - \omega_{II}) \qquad (11)$$

where $\otimes$ denotes the convolution operation.

On one hand, the spectral density $I_{LF}(\omega)$ gives rise after transformation into the $\alpha(\omega)$ representation (through formula 1) to the broad very low frequency profile previously mentionned. On the other hand, the convolution of $I_{HF}(\omega)$ by the vibrational profile of iodine must be responsible of the sharp ν_{II} profile observed at about 205 cm^{-1}.

Experimental results

The FIR induced spectra of iodine in pure benzene and of iodine/benzene diluted in a ternary "inert " solvent (n-heptane, carbon tetrachloride) were obtained using the technics and experimental procedures described elsewhere[8]. However here we have used a high quality detector, namely a doped Ge bolometer cooled at liquid helium.

In figure 1, we present a typical FIR spectrum (3-250 cm^{-1}, resolution ~ 2 cm^{-1}) obtained for iodine diluted in pure benzene after removal of the Bz contribution. Due to the very good signal to noise ratio, we have extracted the spectral densities $\Gamma(\bar{\nu}) = \dfrac{\alpha(\bar{\nu})}{\bar{\nu}(1-\exp-hc\bar{\nu}/kT)}$ corresponding to the ν_{II} profile and to the low frequency part (fig.1). Let's emphasize however, that in this representation the absorption coefficient is divided by a factor which influences very differently the two above profiles. The low frequency one is strongly affected by a factor which is approximately proportional to $\bar{\nu}^2$ and ranges from 10^{-3} (in the dielectric region at 0.03 cm^{-1}) to 10^4 (at 100 cm^{-1}).

Therefore, the very low frequency region (from 0.03 to 20 cm^{-1}) is strongly enhanced meanwhile the broad band at about 100 cm^{-1} is reduced into a weak tail in the spectral density. As a consequence, it is hard to obtain with accuracy this very low frequency spectral density. So, we have combined different sources of measurements : the dielectric value reported in the literature[13] at 0.03 cm^{-1}, measurements in the range 3 to 20 cm^{-1} performed at the Durham University[14] and our measurements. Note a very good agreement in the last two independent measurements in the overlapping region (10 to 20 cm^{-1}).

In contrast, the shape of the ν_{II} profile is only slightly affected by the previous factor (fig.1).

The shapes of the low and high frequency spectral densities (see also figure 3) can be nicely compared, a result in perfect qualitative agreement with our theoretical predictions. This supports also our previous conclusion that, basically, they take a common origin and must contain information on the same time scale.

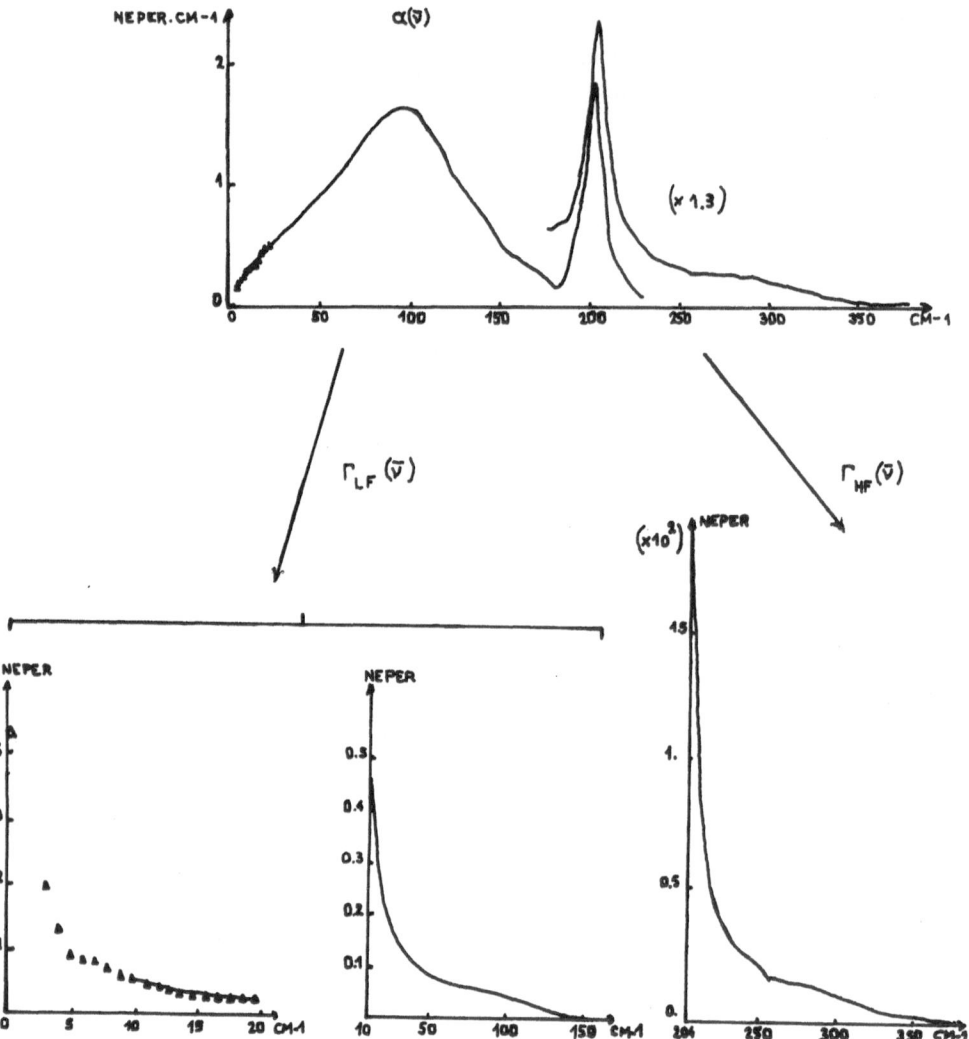

Fig.1. FIR spectra of iodine (0.2 mole ℓ^{-1}) diluted in benzene; (top of the figure). The spectral densities (see text) corresponding to the ν_{II} $\left[\Gamma_{HF}(\bar{\nu})\right]$ and the very far infrared $\left[\Gamma_{LF}(\bar{\nu})\right]$ profiles are indicated by arrows. For the sake of clarity, the "central" region and the high frequency tail of $\Gamma_{LF}(\bar{\nu})$ are displayed separately. Note that the former spectra is obtained from : Δ dielectrics 9 GHz [13] , $\blacktriangle$: Durham University measurements [14] , — : our measurements.

Discussion of the far infrared results

In the following, we will present only relevant aspects of our FIR spectra to the discussion of reaction kinetic leaving their detailed analysis for elsewhere[15].

Low frequency region. We shall discuss only the "central" region (i.e. from dielectrics up to 20 cm^{-1}) of the low frequency spectral density of iodine in benzene. Indeed, we know from our studies that the very high frequency tail of $I_{LF}(\bar{\nu})$ which is related to the short time behaviour of the total dipole moment a.c.f. depends upon other short range induction mechanisms as well[15].

In a first approximation, if one neglects the three-body contribution, $I_{LF}(\bar{\nu})$ reduces to :

$$I_{LF}(\bar{\nu}) = \frac{2}{15} N_I N_B \{(\alpha_B \eta_I)^2 \ I_{rot}^{I_2}(\bar{\nu}) + (\alpha_I \eta_B)^2 \ I_{rot}^{Bz}(\bar{\nu})\} \otimes I_{trans}^{pair}(\bar{\nu}) \qquad (12)$$

From other own measurements in Raman spectroscopy, we know that the rotational dynamics of iodine diluted in benzene obeys a rotational diffusion model; the corresponding profile is Lorentzian (L_{I_2}) with a FWHH : $\Delta\bar{\nu}_{rot}^{I_2} \simeq 1.2$ cm^{-1}.

The rotational profile of benzene diluted in iodine is not available and can be approximated by that obtained for neat liquid benzene. From literature data [16], it is well known that the tumbling motion of this molecule is diffusive and the corresponding rotational profile is Lorentzian (L_{Bz}) with a FWHH : $\Delta\bar{\nu}_{rot}^{Bz} \simeq 4$ cm^{-1}.

Finally, the ratio $R = (\alpha_B \eta_I / \alpha_I \eta_B)^2 \simeq 0.36$ according to reference[17].

We have thus compared the experimental spectral density $\Gamma_{LF}(\bar{\nu})$ with the following profile :

$$\Gamma(\bar{\nu}) = \lambda \{ L_{Bz} + R \ L_{I_2} \} \otimes L_{trans} + BKG \qquad (13)$$

where λ is an adjustable scaling factor.

L_{trans} corresponds to the two-body translational profile which is taken as a simple Lorentzian[18], the FWHH ($\Delta\bar{\nu}_{trans}^{pair}$) of which has to be adjusted.

A constant background (B K G) is added to the resulting profile to take into account the contribution of the high frequency tail.

It comes out that a good fit is obtained (fig.2) with the value $\Delta\bar{\nu}_{trans}^{pair} \simeq 2$ cm^{-1}. Several remarks can be drawn from this result :

i) The "central" region of $I_{LF}(\bar{\nu})$ can be described using a simple quadrupolar-scalar polarizability mechanism between I_2 and Bz. This conclusion is also reinforced by the fact that the intensity of $\Gamma_{LF}(\bar{\nu})$ (i.e. the zero moment) in this "central" region is close to the one calculated under this assumption and the neglect of three-body contributions.

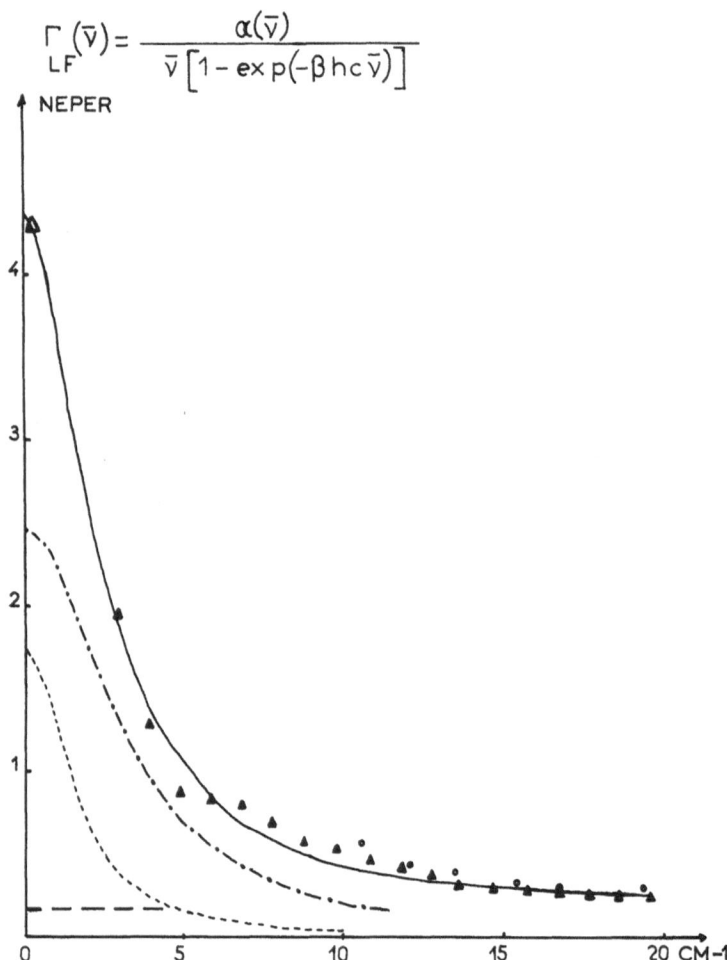

$$\Gamma_{LF}(\bar{v}) = \frac{\alpha(\bar{v})}{\bar{v}\left[1 - \exp\left(-\beta hc\bar{v}\right)\right]}$$

Fig.2. Comparison of the experimental $\Gamma_{LF}(\bar{v})$ with the calculated one (——) using formula 13 (see text). The benzene (_.__.__) and iodine contributions (-----) and the background (— — —) are also reported.

ii) The "benzene contribution" is dominant in this region (see figure 2). This observation, added with the fact that three-body effects are involved only in the iodine contribution (see formula 6 and 8), could explain why the previous analysis is valid. However, this remains an open question and dilution experiments of Bz/I_2 in a ternary solvent should provide new information. Such difficult measurements are actually in progress.

iii) Finally, the main conclusion is that the relative translational dynamics of interacting molecules influences the $I_{LF}(\bar{\nu})$ spectral density and therefore valuable informations on the chemical reaction are contained in this spectral domain. The time scale involved, which is associated with the long time behaviour of the total dipole moment a.c.f., is about 5 picosecondes and it corresponds to the time probed in dielectrics [13].

Analysis of the High Frequency Region

For a qualitative purpose we have compared on figure 3 the "central" region of the spectral density $\Gamma_{HF}(\bar{\nu})$ corresponding to the ν_{II} profiles with the corresponding one of $\Gamma_{LF}(\bar{\nu})$. The difference in shape could be interpreted as resulting from the effect of the convolution by the vibrational profile of iodine on $\Gamma_{HF}(\bar{\nu})$ (formula 8,9). As emphasized previously this indicates also qualitatively that both spectral densities provide the same information on the long time dynamics of the total dipole moment a.c.f.

From our measurements of Bz/I_2 diluted in a ternary "inert" solvent, we know that the shape and intensity of the ν_{II} profile is affected by three-bodies (see also figure 4). Therefore, for a more quantitative study, we will use the ν_{II} profile recorded for iodine (0.1 mole ℓ^{-1}) at very low benzene concentration (2.2 moles ℓ^{-1}) in n-heptane. Its intensity (zero moment) is given by :

$$\int_{-\infty}^{+\infty} \Gamma_{HF}(\omega)\, d\omega = \frac{4\pi^2}{3\hbar c V}\, B(o)\, \langle q^2(o)\rangle \tag{14}$$

with $B(o)\langle q^2(o)\rangle = 12 N_I N_B \{(\frac{\partial\alpha_I}{\partial q}\, \eta_B)^2 + (\alpha_B\, \frac{\partial\eta_I}{\partial q})^2\}\, \langle r_{BI}^{-8}\rangle\, \frac{kT}{\mu_I \omega^2_{II}}$ \hfill (15)

μ_I is the reduced mass of the iodine molecule and $\langle r_{BI}^{-8}\rangle$ can be calculated using a Lennard-Jones potential for interaction between I_2 and Bz [8].

Using the experimental value of $\int_{-\infty}^{+\infty} \Gamma_{HF}(\omega)$ together with $\frac{\partial\alpha_I}{\partial q}$ obtained from absolute intensity Raman measurements [19] it comes out that :

$$\left(\frac{\alpha_B\, \frac{\partial\eta_I}{\partial q}}{\frac{\partial\alpha_I}{\partial q}\, \eta_B}\right)^2 \gg 1$$

This shows that the spectral density associated with the ν_{II} profile is very strongly dominated by the "iodine contribution". As a result this spectral density reduces to :

$$I_{HF}(\bar{\nu}) = \frac{2}{15} N_I N_B \ (\alpha_B \frac{\partial\eta_I}{\partial q})^2 \ I_{rot}^{I_2}(\bar{\nu}) \otimes I_{trans}^{pair}(\bar{\nu}) \otimes I_{vib}^{I_2}(\bar{\nu}) \otimes \delta(\bar{\nu}-\bar{\nu}_{II}) \qquad (16)$$

We have therefore undertaken the analysis of the previous profile under these conditions. The vibrational profile obtained from our Raman measurements is Lorentzian with a FWHH $(\bar{\nu}_{vib}^{I_2}) \simeq 4.2$ cm^{-1}. The rotational profile of iodine was taken as the previous one. Implicitly, we have thus assumed that the rotational profile of Bz/I$_2$ diluted in a ternary solvent is the same as the Bz/I$_2$ one. Of course, this is a crude approximation but as shown from our Raman measurements (vide infra) a precise analysis of this aspect is not possible since a chemical exchange between "free" and "complexed" iodine leads to complicated profiles.

Finally, the translational profile is taken as a Lorentzian one with an adjustable FWHH $:\Delta\bar{\nu}_{trans}^{pair}$. The comparison of the experimental and cal-

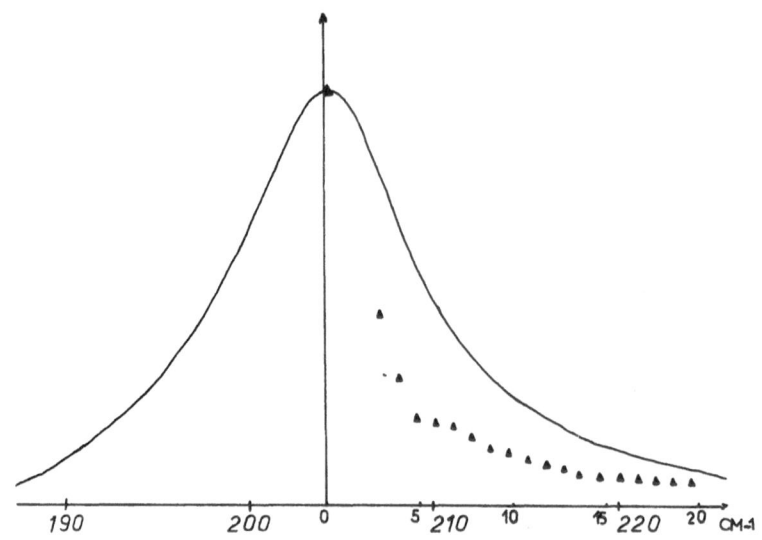

Fig.3. Comparison of $\Gamma_{HF}(\bar{\nu})$ (——) and $\Gamma_{LF}(\bar{\nu})$ for I$_2$ (0.2 mole ℓ^{-1}) in benzene. The maxima of the curves are scaled together.

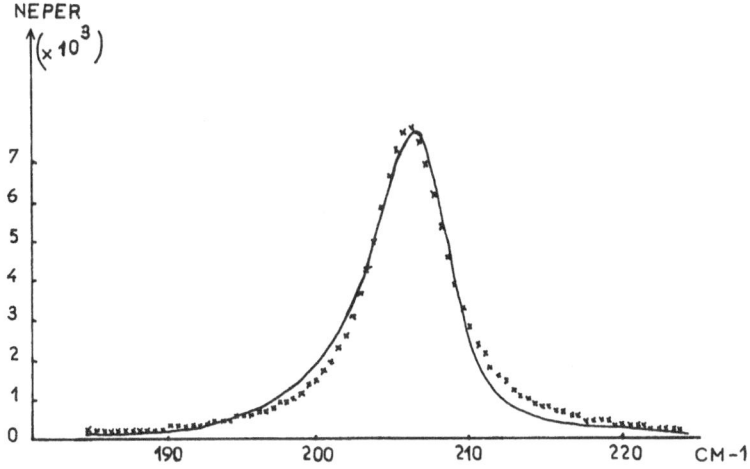

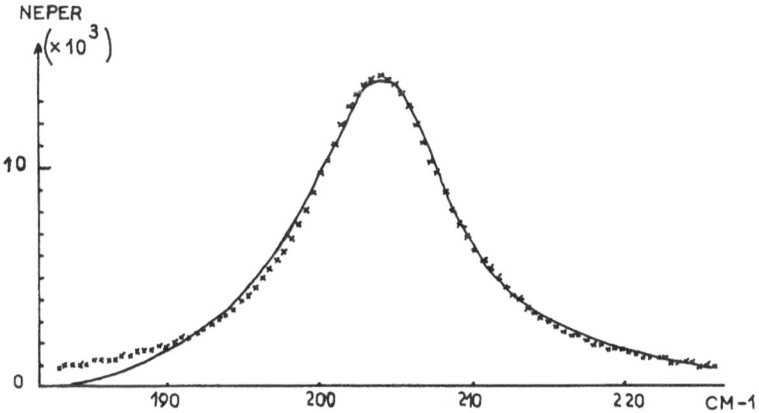

Fig.4 . Comparison of the calculated (×) spectral dendities $\Gamma_{HF}(\bar{\nu})$
(using formula 16) with the experimental ones (—).
Top of the figure I_2 (0.1 mole ℓ^{-1})/Bz (2.2 moles ℓ^{-1}) in
n-heptane.
Bottom : I_2 (0.2 mole ℓ^{-1}) in benzene.

culated profiles leads to a value $\Delta\bar{\nu}^{pair}_{trans} \simeq 1$ cm^{-1} (fig.4).

For iodine diluted in benzene, we have performed the same analysis. However in this case, the three-body contribution is no more negligible and the translational profile involved in expression (16) depends upon this contribution as well. Nevertheless, we have obtained a good fit, using a simple Lorentzian of FWHH $\Delta\bar{\nu}_{trans} \simeq 6$ cm^{-1} to take into account of this collective translational profile (fig.4).

In spite of its crudeness, this analysis reveals that :

i) The ν_{II} spectral density is influenced , as was the low frequency one., by the relative translational dynamics.

ii) In contrast to $I_{LF}(\omega)$, the previous spectral density is more strongly affected by three-body effects. Indeed they affect only the "iodine contribution" which is mostly involved in $I_{HF}(\omega)$.

iii) Finally, the main conclusion is that the low and high frequency spectral densities in their "central" region give insight into the long time dynamics of the sample total dipole moment a.c.f. and both contain information on the kinetic of the chemical reaction, i.e. on the "life time" of the complex. The time scale is about 10 picosecondes for a two-body dynamics and is shorter, as expected, about 2 picosecondes for a more collective dynamics involving three-bodies.

RAMAN SPECTROSCOPY

Theoretical background

Several recent approaches, including Bloch optical equations and Mori -Zwanzig formalism, have been proposed to derive equations for the vibrational spectra of reacting molecules [6,7]. We briefly summarize here the basic assumptions and the final results of one of these theories which is more rigorous and is concentrated on the analysis of isotropic Raman profiles of such fluids [7].

Let us picture the reaction between iodine and benzene immersed in an "inert" solvent like n-heptane as :

$$I_2 + Bz \underset{k_{21}}{\overset{k_{12}}{\rightleftharpoons}} I_2 * Bz$$

"free" "bound"

where the iodine molecule can exist in two states, namely the "free" (site 1) and the "bound" one (site 2) . Making use also of the following assumptions :

i) The iodine molecules perform random jumps among these two sites, the exchange process being a dichotomic Markovian one.

ii) The solvent fluctuations are faster than the time scale involved in the chemical reaction,

the isotropic Raman spectral density associated with
the normal coordinate of iodine takes the expression :

$$I(\omega) \propto \mathrm{Re} \left\{ \sum_{i,j=1,2} \xi_i \, \alpha_i \, \alpha_j \, \left(\left[i \, (\omega \underline{I} - \underline{\omega}) - \underline{\pi} \right] \right)^{-1}_{ij} \right\} \tag{17}$$

where α_i is the semi classical mean polarizability derivative of iodine
in state i, the concentration of which is ξ_i.
$\underline{I}$ is the unity matrix.

The elements of the frequency matrix $\underline{\omega}$ are :

$$(\underline{\omega})_{ij} = \delta_{ij} \, (\omega_j + i \, \Gamma_j)$$

where ω_j is the peak frequency and $\Gamma_j = 1/T_{2j}$ is the inverse vibrational
relaxation time corresponding to iodine in site j.

The exchange matrix $\underline{\pi}$ follows the property $\sum_{j=1}^{2} \pi_{ij} = 0$ and the two
different terms are $\pi_{12} = k_{12} \, [Bz]$ and $\pi_{21} = k_{21}$ where $[Bz]$ is the benzene
concentration in mole ℓ^{-1} and k is the rate constant of the reaction. These
two terms are related to the detailed balance condition :

$$K' = \frac{\pi_{12}}{\pi_{21}} = \frac{\xi_2}{\xi_1} = K_c \, [Bz] \ , \ \text{where} \ K_c = \frac{[I_2 * Bz]}{[I_2] \, [Bz]}$$

is the equilibrium constant of the reaction.

From the general equation (17) several limiting cases of great interest for experimental purposes can be recovered. Using the equation

$$\Delta\omega = \left[(\omega_1 - \omega_2)^2 + (T_{2,1}^{-1} - T_{2,2}^{-1})^2 \right]^{1/2} \ \text{and} \ \tau = (k_{12} \, [Bz] + k_{21})^{-1}$$

and according to whether the product $\Delta\omega \times \tau$ is very large or very small
compared to unity, a behaviour of the Raman profile familiar to NMR spec-
troscopists is expected : in the former case, namely in the slow exchange
regime , the isotopic Raman profile must display a doublet structure ; in
the latter case, the fast exchange regime, this profile may present a simple
Lorentzian (if the difference between α_1 and α_2 is small), the FWHH of
which depends only upon the vibrational relaxation of iodine in each site.

However let's emphasize that the previous theory has been proposed
for a detection of chemical kinetics rather for a precise analysis.
Indeed, it can be shown that the "two sites" vibrational optical Bloch
equations lead to an expression of $I(\omega)$ analogous to the above one (formula 17). The validity and limitations of those equations have been recently
addressed on the ground of the Zwanzig-Mori formalism by Mac Phail and
Strauss [16]. The most important conclusion is that on the vibrational time
scale (characterized by $\Delta\omega$), the requirements of a high barrier to have
quite distinct sites, of a rapid transit time and of a rapid reaction
time (the inverse reaction rate) are incompatible. As a consequence, to
observe any effect in the vibrational spectra, the barrier between the two

sites should be low but, under these circumstances, the reactive and vibrational contributions are necessarily correlated ; the reaction then influences the observed profiles but cannot be separated from the other processes.

Experimental conditions and results

Experimental conditions. The resonant Raman spectra of the ν_{II} vibration have been recorded on a Coderg T800 triple monochromator spectrometer using the 514.5 nm exciting line of a Spectra-Physics argon-ion laser, the power of which was about 300 mW.

Solutions of iodine diluted in benzene and Bz/I_2 diluted in "n-heptane" at different benzene concentrations (ranging from 1 mole ℓ^{-1} to 7 moles ℓ^{-1}) were used. The classical rotating cell technic was used for room temperature (RT) measurements. For variable temperature measurements (over the 233 to 360 K range) the sample was kept in a sealed glass tube inside a Dylor or Coderg cryostat.

In all these measurements, the I_2 concentration was about 2.10^{-3} mole ℓ^{-1}; we found that this low concentration allows to record spectra with correct signal to noise ratios and a good reproductibility.

The polarized $I_{VV}(\bar{\nu})$ and depolarized $I_{VH}(\bar{\nu})$ profiles were recorded using a classical scattering geometry in the spectral range 180-240 cm^{-1} using a spectral resolution of 1.6 cm^{-1}. These weak profiles have been digitalized with a spectral step of 0.1 cm^{-1} and accumulated (60 times for $I_{VV}(\bar{\nu})$, 120 times for $I_{VH}(\bar{\nu})$) on a Digital PDP 11/03 micro computer. Then the spectra were corrected from the base line and the isotropic profiles $I_{iso}(\bar{\nu}) = I_{VV}(\bar{\nu}) - \frac{4}{3} I_{VH}(\bar{\nu})$ were calculated.

Experimental results. We present only here some typical results postponing for a further paper their complete presentation including the study of overtones and other complexes as well[20].

On figure 5, one reported the spectra recorded for a ternary mixture $I_2/Bz/n$-heptane at low benzene concentration ($\simeq$ 1 mole ℓ^{-1}). At low temperature (233 K), this system is still in a liquid state and two bands respectively at 212.5 cm^{-1} and 209 cm^{-1} are clearly resolved. The first one is attributed to "free" iodine (i.e. iodine mainly surrounded by n-heptane molecules) as indicated by the band peak maximum which is close to that known for I_2 vapour. The second one, which is associated with iodine in interaction with its benzene surrounding, corresponds to "bound" iodine and it appears as expected, at a lower frequency[21]. Upon increasing the temperature, these two bands merge and collapse at about RT, leading finally to an unique profile almost centered at the "free" iodine frequency, the FWHH of which decreasing slowly at higher temperatures and reaching a value of about 4 cm^{-1} at 350 K.

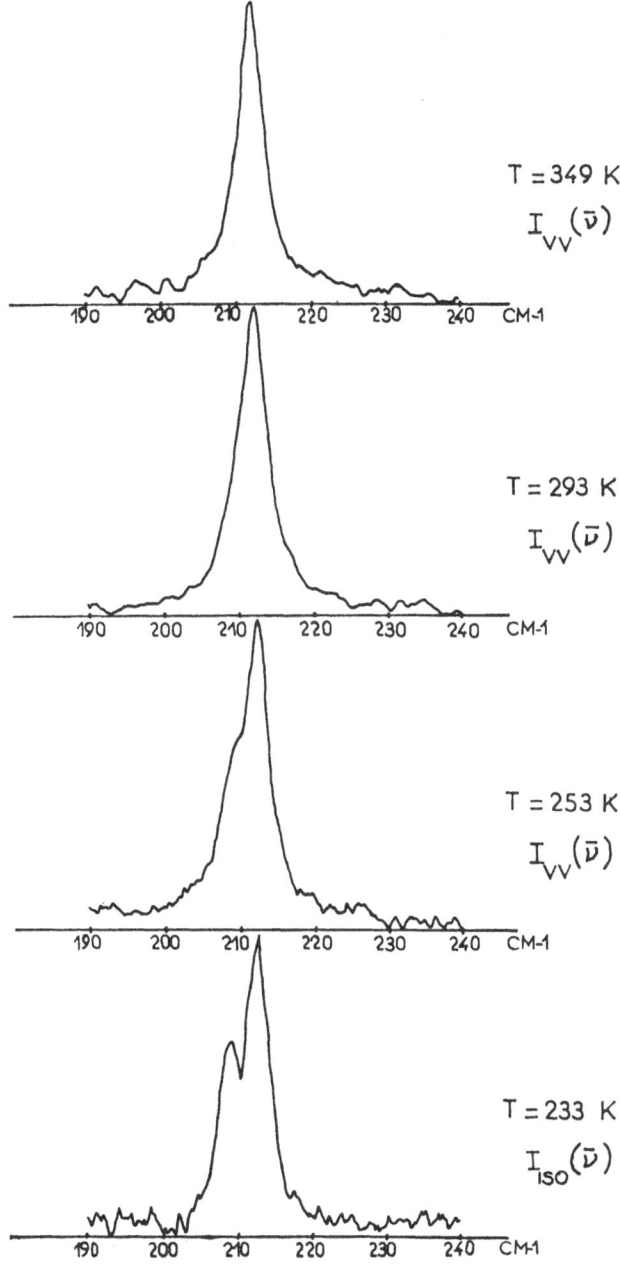

Fig.5. Resonant Raman spectra of I_2 (2.10^{-3} mole ℓ^{-1})/Bz (1.12 moles ℓ^{-1}) in n-heptane at various temperatures.

Using higher benzene concentrations, the freezing of the mixtures presents the observation of the previous doublet structure in the Raman profile. Nevertheless, at room temperature for a benzene concentration of about 3 moles ℓ^{-1}, we have observed a large structureless isotropic profile at 210.5 cm^{-1} with a FWHH of about 7 cm^{-1} and upon increasing the temperature it narrows drastically and shifts slightly to higher wavenumbers : at 368 K, this profile is situated at about 212 cm^{-1} with a FWHH of 4.2 cm^{-1}.

Finally, using a large Benzene concentration (about 7 moles ℓ^{-1}) at RT we have observed a narrow profile (FWHH $\simeq$ 4.8 cm^{-1}) centered at about 209 cm^{-1}. The band centre of this profile shifts again slightly to higher wavenumbers with increasing temperature (210 cm^{-1} at 343 K) while within experimental uncertainties, its FWHH remains constant.

Analysis and discussion of the Raman results

Analysis. The previous observations give a strong qualitative support for evidencing a chemical exchange process occurring on the time scale involved in Raman spectroscopy [6,7].

Indeed, at low temperature, the doublet structure observed on the isotropic Raman profile shows that the conditions of a slow exchange regime are fulfilled. Upon increasing the temperature, the kinetic of the complexing reaction occurs on a shorter time scale and fast exchange regime conditions are approached : this leads to the observation of a collapsing profile at RT and to a "dynamical narrowing" type of behaviour at high temperature.

We can also explain on the same ground the results at different benzene concentrations. Indeed, the exchange term π_{12} being proportional to the benzene concentration (see § "theoretical background"), varies here by about an order of magnitude. Therefore, if conditions close to the ones involved in fast exchange regime are observed at room temperature and low benzene concentration, they should be better fulfilled with increasing benzene concentration. When C_{Bz} = 7 moles ℓ^{-1}, this leads to a narrow profile which should not change very much in FWHH with increasing temperature as observed.

Let us emphasize that even if a qualitative behaviour of the isotropic Raman profiles seems in agreement with the expected theoretical one, a more quantitative analysis remains necessary[22]. Therefore, we have undertaken the analysis of our results on the basis of formula (17). However, in a first attempt we have neglected the exchange terms involved in this equation in order to detect any departure from a simple temperature and concentration dependence behaviour of the isotropic Raman profiles which could thus be explained only on the ground of the equilibrium constant. Even with such a simple model a large number of parameters is involed and some of them must be constrained. We made the following assumptions :

- The equilibrium constant and its variation with temperature was taken from UV/visible measurements[9].

- The ratio of the mean polarizability derivatives α_1 / α_2 was fixed to unity as suggested by absolute Raman intensity measurements[19].

- The "intrinsic" vibrational linewidth contribution $\Delta\bar{\nu}_{vib} = \dfrac{1}{\pi c T_2}$ of the "free" and "bound" iodine were taken respectively as $\Delta\bar{\nu}_{vib}^{free} = 2.6 \text{ cm}^{-1}$ and $\Delta\bar{\nu}_{vib}^{bound} = 4.2 \text{ cm}^{-1}$. These values come from our Raman measurements on I_2/heptane and I_2/Bz using the same resolution as in this study. We have also found that, within experimental uncertainties, the previous FWHH are not temperature dependent (233 to 368 K in heptane, 298-323 K in benzene). Thus, to calculate the profiles, we have only adjusted three parameter, namely the frequencies corresponding to iodine in its two sites and a general scaling factor. In doing so, we have unsuccessfully attempted to fit the complete set of experimental data. Systematic deviations of the experimental profiles from the calculated ones were observed. This was particularily obvious at RT and intermediate benzene concentrations for the broad observed profile which is expected to be the more appropriate one for a detection of the kinetic involved in the complexing reaction. It is worthwhile to point out that:

- The fitted frequency associated with the band centre of the "bound" iodine profile is closer to that of I_2 in pure benzene as the benzene concentration is higher in the ternary mixtures.

This frequency increases with increasing temperature. The later observation is in agreement with results reported for IBr/Bz/n-decane system[22].

- The same trend has been observed for the fitted "free" iodine frequency and has to be contrasted to the corresponding one reported for the previous system.

Finally, the shift $\Delta\bar{\nu} = \bar{\nu}^{free} - \bar{\nu}^{bound}$ is found to be almost temperature independent and ranges between 2 cm^{-1} (C_{Bz} = 7 moles ℓ^{-1}) and 3 cm^{-1} (C_{Bz} = 3 moles ℓ^{-1}).

In a second approach, we have used the full expression (17), involving the three previous fitted parameters together with an adjustable rate constant k_{12}. A complete analysis which will not be presented here reveals that the fits are substantially improved but a systematic desagreement between experimental and calculated profiles still remains. In particular, as previously, we do not succeed in fitting the experimental results obtained at RT and intermediate benzene concentrations. However, let us point out, that the previous observations corresponding to the frequencies of iodine in its two sites are still valid in this new analysis which led again to the same value of $\Delta\bar{\nu}$. Therefore, the relevant vibrational time scale given by $\tau = \frac{1}{2\pi c \Delta \bar{\nu}}$ ranges between 2 to 3 picosecondes. On this time scale, the reaction time τ, defined as the reciprocal of the rate constant, affects our Raman profiles. This is indicated by the value of the rate constant obtained from these fits and it is also illustrated on the simulation presented on figure 6 : this clearly shows that rate constants ranging from 10^{11} to $10^{12} s^{-1}$ at RT are involved in this reacting system.

Discussion. In the preceding analysis, we have shown that the isotropic Raman profiles are affected by kinetic contributions involved in the complexing reaction. However, existing theories predict only qualitatively the temperature and donor (Bz) concentration effects on these profiles, failing in reproducing their lineshapes in details. One might argue that the use of these theories is not well suited here as they have been developped for non-resonant Raman scattering. One might also invoke that the assumptions made in our analysis are not fully justified. This seems to us very unlikely. Indeed, we have performed under resonant Raman conditions a similar study on another complex involving also iodine, namely the I_2/P=dioxane/n-heptane ternary mixture. We have successfully accounted for the whole set of experimental results for this slightly stronger complex ($K_c \simeq 1\ell$ mole^{-1}) using the same assumptions as in the first analysis. In contrast to the system studied here, this indicated also that no kinetic effect is detectable within experimental uncertainties. Let stress also that, very recently we have observed at low temperature and low benzene concentration, the doublet structure presented here under pre-resonant conditions, using the 647.1 nm exciting line of a krypton ion laser. These findings show that the failure in quantitative application of these theories has to be found elsewhere, namely in their underlying assumptions, as previously discussed in section § "Theoretical Background."

The more crude hypothesis is that the transit time, i.e. the time needed for iodine to jump from one site to another, must be shorter than the vibrational time scale, which is here of the order of a few picosecondes, in order to prevent any phase change of the vibration. Transit time estimated on the basis of a diffusion controlled reaction are also of the order of a few picoseconds[6]. This kind of regime is certainly effective here in view of the agreement obtained on the order of magnitude of the rate constant estimated in this study as compared with the one calculated under the assumption for a bimolecular collision reaction using known diffusion constants of I_2 and Bz. Furthermore, let us emphasize also, that the transit time is closely connected to the relative translational time obtained in our FIR study which is also in the few picosecond range[18]. As a consequence, the reactive term cannot be simply desentangled from the other processes so that such fast reactions, cannot be simply analyzed.

T = 298 K

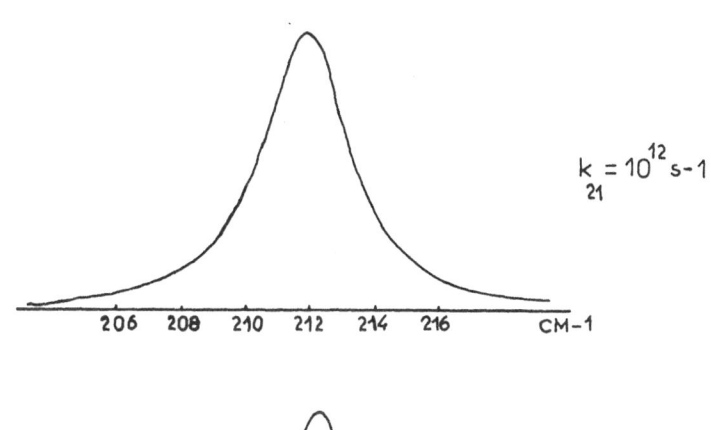

$k_{21} = 10^{12} s{-}1$

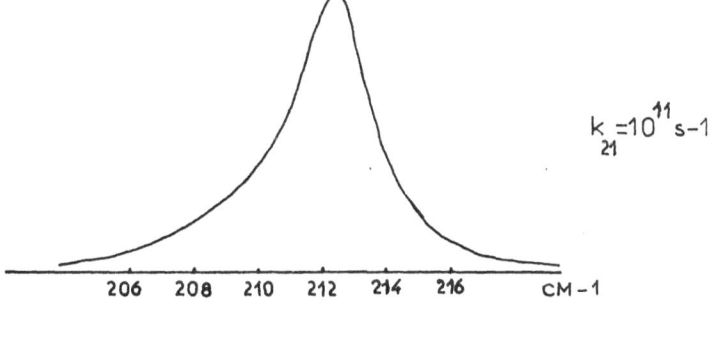

$k_{21} = 10^{11} s{-}1$

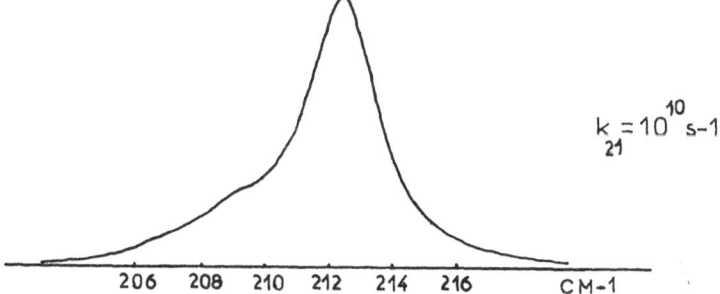

$k_{21} = 10^{10} s{-}1$

Fig.6. Influence of the rate constants on the calculated isotropic
Raman profiles (formula 17) associated with I_2 (2 10^{-3} mole ℓ^{-1})
in Bz (1.12 mole ℓ^{-1}) and n-heptane. All the other parameters
are fixed as indicated in the text.

CONCLUSION

From far infrared and Raman experimental results, a value of the "life time" of the Bz/I_2 complex ranging from two to ten picosecondes at room temperature has been estimated. However, it is noteworthy that characteristic times have been simply obtained in FIR spectroscopy from the width of the translational profiles without going into a detailed analysis of the different contributions. In contrast, in Raman spectroscopy, a characteristic time of about 2 picosecondes qualitatively accounts for the observed collapsing profiles. However, as presented before, we have unsuccessfully attempted to fit the whole set of experimental spectra with the theoretical ones under the two following assumptions : uncorrelated reactions and vibrational dynamics and a Markovian jump process between two sites. These remarks emphasize a huge difficulty encountered in such studies ; as far as very fast reactions (in the picosecond time scale) are considered, it becomes certainly very difficult to separate the reactive dynamics from the dynamics of the liquid itself. This is no longer true for slow reactions involving a few order of magnitude lower characteristic times.

REFERENCES

1. M.M.Kreevoy and C.A.Mead, J.Am.Chem.Soc. 84 : 4596 (1962).
2. M.M.Kreevoy and C.A.Mead, Discuss.Farad.Soc. 39 : 166 (1965).
3. J.C.Lassegues and J.Devaure, in : "Protons and Ions Involved in Fast Dynamic Phenomena", P.Lazlo, ed., Elsevier, Amsterdam (1978)
4. D.Cavagnat and J.Lascombe, J.Chem.Phys. 76 : 4336 (1982).
5. J.C.Lassègues, M.Besnard, D.Cavagnat and J.Lascombe, in "Proceedings of the IXth International Conference on Raman Spectroscopy" Tokyo (1984).
6. R.A.McPhail and H.L.Strauss, J.Chem.Phys. 82 : 1156 (1985).
7. S.Bratos, G.Tarjus and P.Viot, J.Chem.Phys. 85 : 803 (1986).
8. J.Lascombe and M.Besnard, Mol.Phys. 58 : 573 (1986).
9. "Spectroscopy and Structure of Molecular Complexes" J.Yarwood ed., Plenum Press, London and New-York (1973).
10. "Phenomena Induced by Intermolecular Interactions" (N.A.T.O. A.S.I. Series B : Physics Vol.127) G.Birnbaum, ed., Plenum Press, New-York (1985).
11. R.C.Gordon, J.Chem.Phys. 43 : 1307 (1965).
12. S.R.Polo and K.Wilson, J.Chem.Phys. 23 : 2376 (1955).
13. G.W.Nederbragt and J.Pelle, Mol.Phys., 1 : 97 (1959).
14. J.Yarwood and B.Catlow, J.Chem.Soc., Faraday Trans II, 83 : (1987).
15. N.Del Campo, M.Besnard and J.Lascombe (to be published).
16. J.E.Griffiths, in : "Vibrational Spectra and Structure" Vol.6 J.R.Durig Ed., Elsevier, Amsterdam (1977).
17. C.G.Gray and K.E.Gubbins, in : "Theory of molecular fluids" Vol.1 : Fundamentals,Clarendon Press, Oxford (1984).
18. B.Guillot and G.Birnbaum, in : "Phenomena induced by intermolecular interactions" (N.A.T.O. A.S.I. Series B : Physics Vol.127) G.Birnbaum, ed.,Plenum Press, New-York (1985).
19. J.Put, G.Maes, P.Huyskens and Th.Zeegers-Huyskens, Spectrochimica Acta 37A : 699 (1981).
20. M.Besnard, N.Del Campo, R.Cavagnat and J.Lascombe (to be published).
21. H.Rosen, Y.R.Shen and F.Stenman, Mol.Phys. 22 : 33 (1971).
22. R.J.Sension and H.L.Strauss, J.Chem.Phys., 86 : 6665 (1987).

DISCUSSION

KAPRAL - 1°) What experimental evidence justifies the use of the two-state model ?

2°) The analysis of the low frequency part of the spectrum assumes that the translational and rotational modes are decoupled. Is this justified ?

BESNARD - 1°) The isotropic Raman profile recorded at low temperature for Bz/I_2 diluted in a ternary solvent (n-Heptane) exhibits a doublet structure. This experimental observation suggests that the reaction might be pictured using a two state model. This is reinforced by the good qualitative agreement between our experimental results with the theories recently proposed for vibrational spectroscopy which are based upon such a simple description of the reaction.

2)° In a pragmatic way, this assumption as well as others made in our theoretical approach of the far infrared results, can be justified "a posteriori" in view of the very good qualitative agreement of the behaviour of the low and high spectral densities with the predicted one. Let's also stress that computer simulation indicates that the decoupling approximation works "amazingly" well (see for instance Steele work on that topic).

BRATOS - Is the assignment of the components of the collapsing $\nu +_2$ doublet entirely certain ? Your theory heavily relies on the answer to this question.

BESNARD - This assigment is in agreement with the previous one obtained in both Raman and infrared spectroscopy on similar charge-transfer complexes like for example IBr/Bz in n-Decane.

HYNES - What is the structure of the iodine-benzene complex ? Is the iodine perpendicular to the benzene ring ? If it is, it is not easy to see how rotation and translation are decoupled, as you have assumed.

BESNARD - 1°) The question dealing with the structure of the Bz/I_2 complex is an old one which has been discussed at length in the literature and which is still quite controversial. In his pioneering work, Mulliken proposed two energetically favorable structure for this complexe, namely the "resting" and "axial" iodine structure models and suggests that the former one was to most probable. Later on, the "axial" structure was supported by infrared spectroscopic measurements. This conclusion was then criticized by Person who argues that this technique is not adapted to check here this particular structural aspect. More recently, a quantum chemistry calculation has reinforced the spectroscopic point of view.

2) The answer to the second part of this question has to be found in my reply to Dr. Kapral intervention.

WILSON - It would seem that it would be possible to compute the spectrum from molecular dynamics with the proper simultaneous involvement of rotation, dephasing, translation and reaction. Do you agree ?

BESNARD - Yes, indeed. Molecular dynamics should provide valuable informations to discuss our experimental observations and the assumptions made in our theoretical approach. This type of study is aimed at a collaboration with Drs. Y. Guissani and B. Guillot (Laboratoire de Physique Théorique des Liquides, Université de Paris VI).

KINETIC PROCESSES IN ELECTROLYTE SOLUTIONS IN HIGH

FREQUENCY ELECTRIC FIELDS

J. Barthel, K. Bachhuber, and R. Buchner

Institut für Physikalische und Theoretische Chemie der Universität Regensburg, Federal Republic of Germany

Summary: High frequency permittivity measurements (1 to 40 GHz) are analyzed to yield information on kinetic processes in aqueous and non-aqueous electrolyte solutions. Information is obtained on the low and high frequency relaxation times of the solvent and their shifts by addition of electrolytes, rotational relaxation times of ion pairs, association constants, rate constants of ion-pair formation and decomposition, and kinetic depolarization. Examples are given for each process.

1. INTRODUCTION

The application of electromagnetic fields to material systems yields a wealth of information on kinetic processes taking place in the systems investigated. Figure 1 shows the usual scales for the definition of frequency ranges, frequency [Hz], wave length [m], and wave number [cm^{-1}] scales. Below these scales the phenomena are quoted which typically are observed when electromagnetic waves of the corresponding frequencies are applied to electrolyte solutions and solvents. Energy dissipation may occur by conductance, relaxation or absorption processes.

The phenomenological description of the interaction between electromagnetic fields and material systems is based on a set of four differential equations, the Maxwell and field equations, and three material equations

$$\vec{rot} \, \vec{H} = \vec{j} + \partial \vec{D}/\partial t; \quad \vec{rot} \, \vec{E} = -\partial \vec{B}/\partial t; \quad div \, \vec{D} = \varrho; \quad div \, \vec{B} = 0; \quad (1a\text{-}d)$$

$$\vec{D} = \varepsilon_0 \, \varepsilon \, \vec{E}; \quad \vec{B} = \mu_0 \, \mu \, \vec{H}; \quad \vec{j} = \varkappa \, \vec{E} \qquad (1e\text{-}g)$$

The polarization $\vec{P}$, which has both a macroscopic and a molecular definition, is the link between experiment and molecular interpretation of the experimental results.

$$\text{macroscopic:} \qquad \varepsilon_0 (\varepsilon - 1)\vec{E} = \vec{P} = \langle \vec{P}_\mu \rangle + \langle \vec{P}_\alpha \rangle \qquad \text{: molecular} \qquad (2)$$

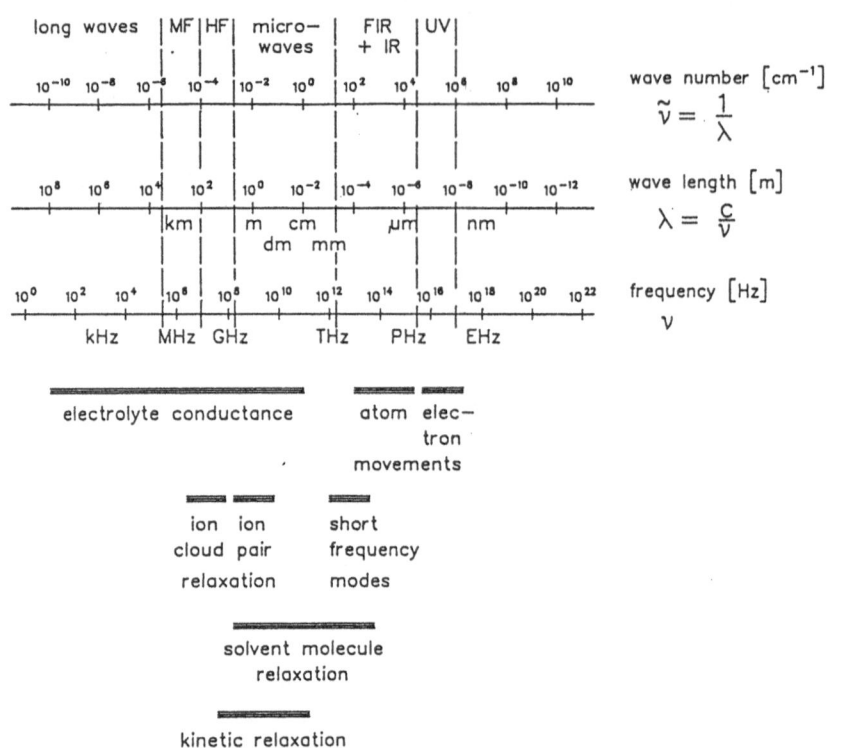

Figure 1. Frequency regions and respective processes contributing to the permittivity of electrolyte solutions and solvents.

The symbols in Eqs. (1) and (2) have their usual meaning: $\vec{E}$ and $\vec{H}$, electric and magnetic field strengths; $\vec{D}$, electric displacement; $\vec{B}$, magnetic flux density; $\vec{J}$, electric current density; ε_o and ε, vacuum permittivity and relative permittivity; μ_o and μ, vacuum permeability and relative permeability; $\varkappa$, conductivity; ϱ, charge density. The vacuum permittivity and permeability are linked by the relation: $\varepsilon_o = (\mu_o c^2)^{-1}$, c being the speed of light. Relative permeability μ may be set equal to unit for the organic solvent systems investigated in this paper.

The quantities $\vec{P}_\mu$ and $\vec{P}_\alpha$ used for the molecular interpretation of polarization are the contributions due to the alignment of the molecular permanent moments $\vec{\mu}$ and of the induced moments $\vec{\mu}_{ind}$ resulting from the polarizability α of the molecules.

It is common use to characterize dissipating systems with the help of complex quantities. Vectorial quantities $\vec{A}$ [$\vec{A}=\vec{A}(\vec{r},t):\vec{E}$, $\vec{D},\vec{J},\vec{B},\vec{H},\vec{P}$] are used in the form ($i^2 = -1$)

$$\vec{A}(\vec{r},t) = \vec{A}_o(\vec{r}) \exp(i\omega t) \tag{3}$$

where $\omega = 2\pi\nu$ is the circular frequency (ν, linear frequency) of the monochromatic wave applied. Material properties of non-isotropic (homogeneous) media are given by complex numbers

$$\varepsilon^* = \varepsilon'-i\varepsilon''; \qquad \mu^* = \mu'-i\mu''; \qquad \varkappa^* = \varkappa'+i\varkappa'' \tag{4a,b,c}$$

The Maxwell equations for electrolyte solutions then are transformed into the wave equation

$$\nabla^2 \vec{H}_o + k^{*2}\vec{H}_o = 0; \quad k^* = k_o (\varepsilon^* - \frac{i\varkappa^*}{\varepsilon_o \omega})^{1/2}; \quad k_o = (\mu_o \varepsilon_o \omega^2)^{1/2} = \frac{\omega}{c} \qquad (5a,b,c)$$

In Eqs.(5) the propagation coefficient k^* is referred to the propagation coefficient of vacuum, k_o. The complex quantity η^*

$$\eta^* = (\varepsilon^* - \frac{i\varkappa^*}{\omega\varepsilon_o}); \quad \eta' = \varepsilon' + \frac{\varkappa''}{\omega\varepsilon_o}; \quad \eta'' = \varepsilon'' - \frac{\varkappa'}{\omega\varepsilon_o} \qquad (6a,b,c)$$

is called the "generalized permittivity" of the electrolyte solution. Its real and imaginary parts, η' and η'', are measurable quantities. They show the fundamental feature that the separation of permittivity ε and conductivity $\varkappa$ is not possible in the strictest acceptation of the theory.

The dissipated energy W is given by the relation

$$\dot{W} = \frac{1}{2} \varepsilon_o \omega \eta'' \vec{E}_o \qquad (7)$$

At zero frequencies

$$\lim_{\omega \to 0} [\mathrm{rot}\ \vec{H}_o] = \varkappa \vec{E}_o; \quad \varkappa^* = \varkappa = N_A e \sum_k \varrho_k |z_k| \lambda_k \qquad (8a,b)$$

$\varkappa^*$ is reduced to the conventional conductivity $\varkappa$; N_A is Avogadro's number; ϱ_k, z_k and λ_k are the density, valency and conductance of the ions k in the solution. Ion conductances λ_k, when formulated with the help of force transfer (relaxation effect) and hydrodynamic (electrophoretic effect) interactions, are the base of the commonly used conductance equations.

Theory shows that increasing frequency entails a relaxation process due to the tendency to re-establish local electroneutrality in the disturbed system. The corresponding relaxation time for the solution of a binary electrolyte

$$\tau^{el} = [\frac{1}{2} \varkappa_D^2 (\omega_1 + \omega_2) k_B T]^{-1}; \quad \varkappa_D = \frac{e_o^2}{\varepsilon_o \varepsilon k_B T} (\varrho_1 z_1^2 + \varrho_2 z_2^2) \qquad (9a,b)$$

is known as the Debye relaxation time. In Eqs.(9) $\varkappa_D$ is the Debye parameter; ω_1 and ω_2 are the ion mobilities; k_B is the Boltzmann constant, T is the temperature; e_o is the elementary charge; ϱ_1 and ϱ_2 are the ion densities.

Extended experimental studies up to 50 MHz in our laboratory[1] have confirmed the well known result that the dependence of conductivity on frequency is very small. For lack of more precise information conductivity may be considered to be independent of frequency in the whole frequency range. This approximation is tolerable, especially at high frequencies which are the object of investigation. Eqs.(6) then are reduced to

$$\eta^* = (\varepsilon^* - \frac{i\varkappa}{\omega\varepsilon_o}); \quad \eta' = \varepsilon'; \quad \eta'' = \varepsilon'' - \frac{\varkappa}{\omega\varepsilon_o} \qquad (10a,b,c)$$

Figure 2 shows the Argand diagram (η'', η') and its reduction to a diagram $(\varepsilon'', \varepsilon')$ using conductivities $\varkappa$ at zero frequency. Such Argand diagrams $(\varepsilon'', \varepsilon')$ are the base of the following discussions.

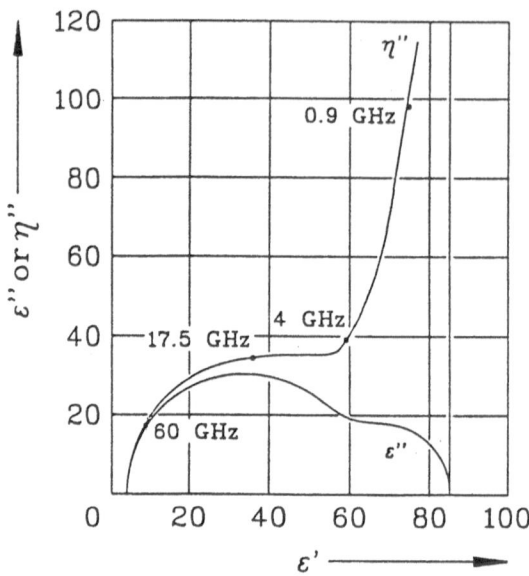

Figure 2. Diagrams η''vs.η' and ε''vs.ε' for an aqueous solution of CdSO$_4$ (1.1 mol dm^{-3}) at 25^{0}C. Reduction of (η'',η') to (ε'',ε') is carried out according to Eqs.(10b,c).

2. EXPERIMENTAL METHODS AND DATA ANALYSIS

The measured permittivity data (ε',ε'') are linked via Eq.(2) with molecular data of solvents and solutes; $\vec{P}_\mu$ is due to the orientation of the molecular dipoles of the particles k in the local direction field, $\vec{E}_k^{dir}$, whereas $\vec{P}_\alpha$ results from particle interaction with the local internal field $\vec{E}_k^{int}$ via their polarizability α_k. $\vec{P}_\alpha = \alpha\vec{E}^{int}$. Internal field and direction field differ by the reaction field $\vec{R}$, $\vec{E}^{int} = \vec{E}^{dir} + \vec{R}$. Single charges in the solution (ions) can only indirectly contribute to $\vec{P}_\mu$ by interaction with dipoles (solvent molecules or ion aggregates).

At zero frequency (static case) or slowly varying (quasistatic case) frequencies polarization is in phase with the field and is independent of time. Dipole orientation shows an equilibrium distribution at every moment, see the (static) polarization theories of Debye[2], Onsager[3], Kirkwood[4], Fröhlich[5] and Cole-Cole[6]. Increasing frequency of the applied field yields the dynamic case. The motions of the particles have characteristic times to reach a certain polarization, at room temperature generally in the GHz region, depending on their size. Polarization is no longer in equilibrium with the field. The presupposition of linear isotropic dielectrics permits the description of polarization by electric fields of arbitrary time dependence with the help of response functions. If an electric field applied to a linear dielectric is switched off at time t = 0 (time domain method), polarization $\langle P(0)\rangle$ breaks down in two steps: induced polarization $\langle P_\alpha(0)\rangle$ not depending on dipole orientation collapses without time lag; orientation polarization $\langle P_\mu(0)\rangle$ tends toward zero controlled by a step response function $F_P(t)$, e.g. an exponential decay function

$$F_P(0) = 1; \quad F_P(\infty) = 0; \quad F_P(t) = \frac{\langle \vec{P}_\mu(0)\,\vec{P}_\mu(t)\rangle}{\langle \vec{P}_\mu(0)\,\vec{P}_\mu(0)\rangle} \qquad (11a,b,c)$$

Application of a harmonically varying field, $\vec{E} = \vec{E}_0 \exp(i\omega t)$, of sufficient high frequency (500 MHz to 100 GHz, frequency domain method) results in a phase shift δ of polarization $\vec{P}$ against $\vec{E}$ and an amplitude attenuation with regard to static polarization $\langle\vec{P}^{eq}\rangle$. Attenuation is expressed through the real part $\varepsilon'(\omega)$ of the complex permittivity. The phase shift δ is given by the relation $\tan \delta = \varepsilon''/\varepsilon'$; for the energy dissipation see Eq.(7).

Time-domain and frequency-domain measurements are linked via the relationship

$$\varepsilon^*(\omega) = \varepsilon(\infty) + [\varepsilon(0) - \varepsilon(\infty)]\, L_{i\omega}[f_P(t)]; \quad f_P(t) = -\frac{d\,F_P(t-t')}{d(t-t')} \quad (12a,b)$$

where $L_{i\omega}[f_P(t)]$ is the Laplace transform of the so-called pulse-response function $f_P(t)$.

Despite the efficiency of time-domain measurements their reduced accuracy at high frequencies favours the use of frequency domain methods. In our laboratory five experimental arrangements of the travelling wave method (frequency domain method) introduced by Buchanan[7] are used, covering the range from 0.95 to 40.0 GHz. A sixth set-up extending the frequency range to about 90 GHz is under construction. For details of this equipment see Refs. 8 - 13. With this equipment precision is better than 0.4 % for $\varepsilon'(\omega)$ and 0.8 % for $\varepsilon''(\omega)$ at all frequencies and electrolyte concentrations, if purification of solvents, preparation of solutions, and measurements are carried out under an atmosphere of solvent saturated nitrogen gas such that the solutions are screened against evaporation and are never exposed to the atmosphere.

The real and imaginary parts of permittivity at frequency ν are obtained from the measurements of the medium wave lengths λ_m and the attenuation exponents α of the samples. The solution of the wave equation under boundary conditions imposed by the geometry and dimension of the measuring cells yields the relations

$$\varepsilon'(\omega) = \left(\frac{c}{\nu\lambda_m}\right)^2 - \left(\frac{c\alpha}{2\pi\nu}\right)^2 + \left(\frac{ck_c}{2\pi\nu}\right)^2; \quad \varepsilon''(\omega) = \left(\frac{\alpha}{\pi\lambda_m}\right)\left(\frac{c}{\nu}\right)^2 \quad (13a,b)$$

where c is the speed of light and k_c is the cut-off wave number of the measuring cell.

The full information ($\nu = 0.95$ to 40 GHz) obtained from such measurements is shown in the computer plots given by Figs. 3a (pure water) and 3b (pure ethanol)

Fig. 3a shows the existance of one, Fig. 3b of two relaxation processes; τ_i (i = 1,2) are the corresponding relaxation times. Data analysis in Figs. 3 is based on Eq. (12a) using the pulse response function (general case of n separable processes)

$$f_P(t) = \sum_{j=1}^{n} \frac{g_j}{\tau_j} \exp\left(-\frac{t}{\tau_j}\right); \quad \sum_{j=1}^{n} g_j = 1 \quad (14a,b)$$

yielding

$$\varepsilon^*(\omega) = \varepsilon(\infty) + (\varepsilon(0) - \varepsilon(\infty))\sum_{j=1}^{n} \frac{g_j}{1 + i\omega\tau_j}; \quad g_j = \frac{\varepsilon_j(0) - \varepsilon_j(\infty)}{\varepsilon(0) - \varepsilon(\infty)} \quad (15a,b)$$

In Eq.(15b) $\varepsilon(0)$ and $\varepsilon(\infty)$ are the permittivities of the solution at zero and infinine frequencies, i.e. the upper ($\nu \rightarrow 0$) and lower ($\nu \rightarrow \infty$)

limit of permittivity. The weight factor g_j of the j-th relaxation process is given by the ratio of its partial $(\varepsilon_j(0)-\varepsilon_j(\infty))$ and the total dispersion amplitude $(\varepsilon(0)-\varepsilon(\infty))$. The overall Argand diagram $(\varepsilon'',\varepsilon'$ projecton) is a sequence of semicircles of radius $(\varepsilon_j(0)-\varepsilon_j(\infty))/2$ having in common the point $\varepsilon_{j-1}(\infty) = \varepsilon_j(0)$ on the ε'-axis; $\varepsilon_1(0) = \varepsilon(0)$ and $\varepsilon_j(\infty) = \varepsilon(\infty)$.

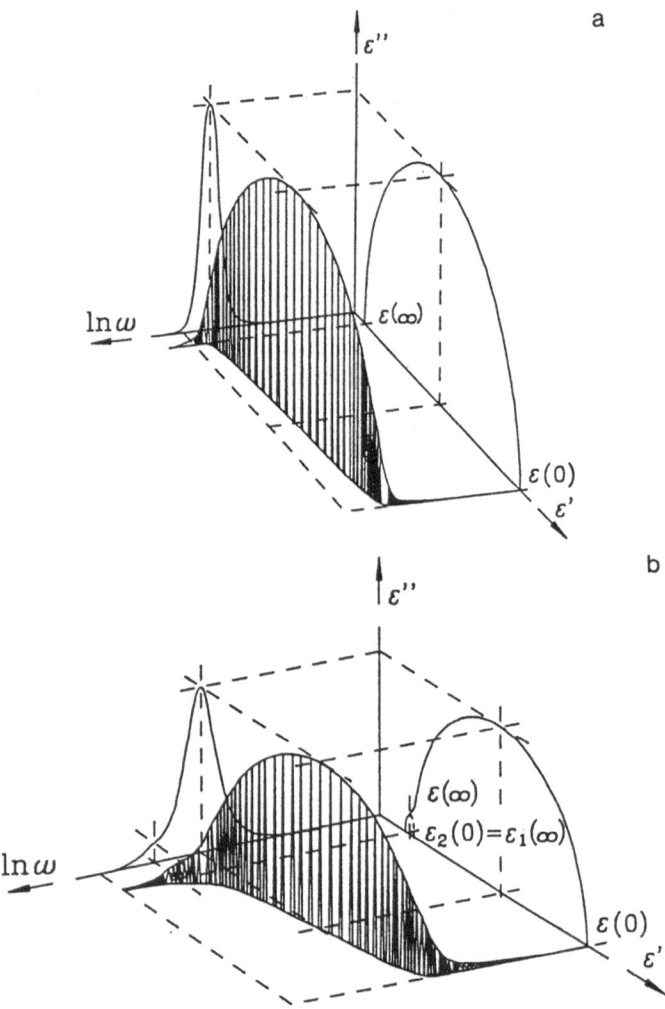

Figure 3. Spatial $(\varepsilon',\varepsilon'',\ln\omega)$ diagrams and their $(\varepsilon'',\varepsilon')$-, $(\varepsilon',\ln\omega)$-, and $(\varepsilon'',\ln\omega)$ projections in the frequency range from 1 to 40 GHz: (a) water at 25^0C, (b) ethanol at 25^0C (from Ref. 15)

Figures 2 and 5 show Argand diagrams (i.e., $\varepsilon'',\varepsilon'$ projections of space diagrams of the type given in Fig. 3), Fig. 6 shows absorption (i.e., $\varepsilon'',\ln\omega$ projections) curves.

Relaxation models with continuous relaxation time distributions are based on the pulse response function

$$f_p(t) = \int_{\tau=0}^{\infty} \frac{G(\ln\tau)}{\tau} \exp(-\frac{t}{\tau}) \, d \ln\tau \, ; \quad \int_{\tau=0}^{\infty} G(\ln\tau) d \ln\tau = 1 \qquad (16a,b)$$

yielding

$$\varepsilon^*(\omega) = \varepsilon(\infty) + (\varepsilon(0){-}\varepsilon(\infty)) \int_{\tau=0}^{\infty} \frac{G(\ln\tau)}{1+i\omega\tau} \, d \ln\tau \qquad (17)$$

Figure 4 shows an example of relaxation time distributions with the help of the Kirkwood-Fuoss distribution function

$$G(\ln\tau) = \frac{1}{2 \cosh \ln(\tau/\tau_o) + 2} \qquad (18)$$

where τ_o is the relaxation time at the maximum of the distribution.

3. KINETIC PROCESSES

A survey on the dielectric properties of electrolyte solutions has been published recently[14]. The reader will find there the fundamental features and a survey on the recent literature.

The actual contribution, in addition, is devoted to some kinetic aspects of the shift of relaxation times, dispersion amplitudes, and permittivity depression caused by addition of electrolytes.

Relaxation times obtained from permittivity measurements are macro-scopically based quantities, Eq.(11c) being the underlying correlation function. Microscopic dipole correlation functions are given by the relation

$$\Phi(t) = \frac{\langle \vec{\mu}(t) \, \vec{\mu}(0) \rangle}{\langle \vec{\mu}(0) \, \vec{\mu}(0) \rangle} \qquad (19)$$

yielding relaxation times τ_i' needed for the discussion of molecular processes. Actually there is no satisfactory general theory permitting the conversion of measured τ_i to molecular τ_i' relaxation times; In the special case of rotational diffusion the Powles-Glarum approximation

$$\tau_i' = \frac{3\varepsilon(0)}{2\varepsilon(0)+\varepsilon(\infty)} \tau_i \qquad (20)$$

may be used.

3.1. Relaxation Processes of the Solvent

The results shown in Figs.(3), i.e., one or two relaxation times in the frequecy range 1 to 100 GHz, is a general feature of the solvents commonly used for electrolyte solutions. In pure solvents the main relaxation time τ_1 is due to the re-establishment of the bulk solvent structure; values of τ_1 for some frequently used solvents at 25°C are given in Table 1.

Information on the second (high frequency) relaxation regions actually is limited to a few solvents, but measurements at high frequencies are now feasible. From recent investigations in our laboratory high frequency data are available for methanol (τ_2 = 3.3 ps [9]), ethanol (τ_2 = 5.2 ps[15]) and their electrolyte solutions.

61

Ions may shift the main relaxation time τ_1 to higher or lower frequency regions, depending on the orientation attributed to molecular motions of the solvent molecules.

Figure 4 shows the relaxation time distributions calculated under the assumption of continuous Kirkwood-Fuoss distributions[16] of pure methanol and pairs of almost equally concentrated solutions of NaI and Bu$_4$NI solutions in methanol. Addition both of the alkalimetal and the tetraalkylammonium salt broaden the relaxation time distribution. The center of gravity of the distributions is shifted to lower frequencies (higher relaxation times) by NaI, and to higher frequencies (lower relaxation times) by Bu$_4$NI, due to different kinds of cation-solvent interaction.

For solvents with simple structures, such as propylene carbonate[43], the Stokes-Einstein-Debye relation

$$\tau_i' = \frac{3V}{k_B T}\, \eta\, f + \tau_i'^0 = \tau_i^{SED} \qquad (21)$$

may be used to represent the shift of molecular relaxation times τ_i' as a function of solution viscosity η. Plots τ_1' vs. η yield straight lines from which the volume V of the relaxing solvent species can be calculated. In Eq.(21) f is a friction coefficient calculated according to Budo et al[17], and $\tau_i'^0$ is the free rotor relaxation time.

Addition both of alkalimetal and tetraalkylammonium salts decreases $\varepsilon(0)$, $\varepsilon(0;c) < \varepsilon(0;0)$, see section 3.3.

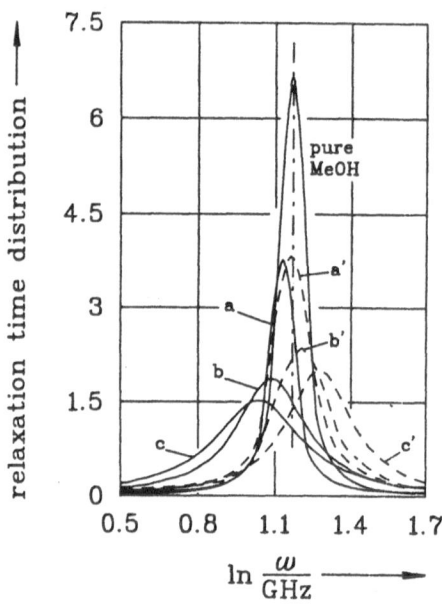

Figure 4. Kirkwood-Fouss relaxation time distributions for methanol solutions of NaI (a': 0.0637, b': 0.250, c': 0.534 mol dm^{-3}) and Bu$_4$NI (a : 0.0635, b : 0.250, c : 0.536 mol dm^{-3}). (from Ref. 46)

A rough estimation, attributing τ_2 to the movement of terminal OH-groups of alcohol molecular chains, indicates that about 18 % of the methanol molecules are -OH terminal. The study of the dispersion amplitude [$\varepsilon_2(0) - \varepsilon_2(\infty)$] shows[9] that the addition of NaCl, NaI, NaBr and NaClO$_4$ up to concentrations of 0.5 mol dm^{-3} does not change the number of terminal

OH groups, in contrast to Bu_4NCl, Bu_4NBr, Bu_4NI and Bu_4NClO_4 which yield a significant increase. This result again reflects the different types of cation-solvent interactions in methanol: Na^+ ions break the molecular chains and interact with the free oxygen electron pairs of methanol molecules to form solvate structures with more or less "irrotationally" bound methanol. On the other hand Bu_4N^+ ions break the methanol chains without irrotationally bounding solvent molecules and increase the number of terminal OH-groups. The assumption of relaxing terminal OH groups of methanol associates is in agreement with the prediction of Eq.(21).

Table 1. Dielectric parameter of several polar solvents at 25° C showing one or two separated relaxation processes

solvent	$\varepsilon(0)$ $=\varepsilon_1(0)$	τ_1 (ps)	$\varepsilon_2(0)$ $=\varepsilon_1(\infty)$	τ_2 (ps)	$\varepsilon_2(\infty)$ $=\varepsilon(\infty)$	Ref.
Water	78.02	8.30	---	---	5.39	(19)
Methanol	32.47	51.01	5.74	3.29	3.68	(9)
Ethanol	24.29	161.97	4.45	5.19	3.19	(15)
Propylene carbonate	64.71	39.37	---	---	6.23	(18)
N,N-Dimethyl-formamide	37.4	11.01	---	---	4.4	(13)
N-Methyl-formamide	182	123	---	---	5.9	(13)
Formamide	109.5	36.9	---	---	7.0	(13)
Aceto-nitrile	35.8	3.5	---	---	4.0	(20)

3.2 Kinetics of Ion-Pair Formation and Dissociation

The relaxation region in Fig.2 at the low frequency side of the main relaxation region of the solvent is unambiguously attributable to ion-pair $[Cd^{2+}(H_2O)_x SO_4^{2-}]$ relaxation modes[21,22]. Ion-pair relaxation regions were found for various aqueous solutions of 2,2 electrolytes[21-23], some 2,1 electrolytes[24,25] and 1,1 electrolytes in moderate and low permittivity solvents and solvent mixtures[26-30]. The general Argand diagram for an electrolyte solution with an ion-pair relaxation region following the main relaxation region is sketched in Fig 5.

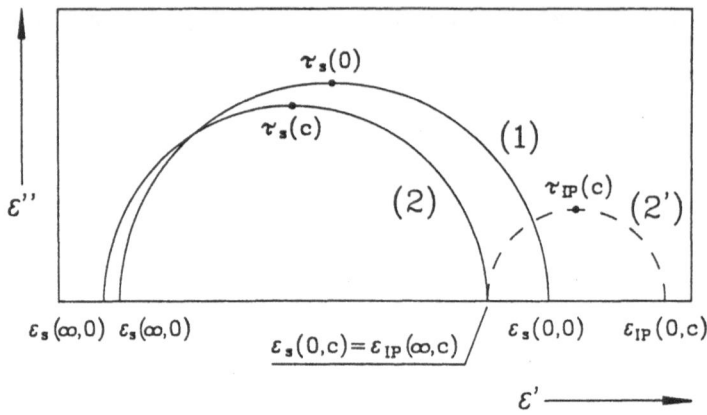

Figure 5. Schematic representation of the relaxation diagrams of an electrolyte solution and its solvent. Curve 1: solvent, Curve 2: electrolyte solution without ion-pair formation, Curve (2+2'): electrolyte solution with ion-pair formation.

Curve $N^\circ 1$ in Fig. 5 represents the ε''vs.ε'-diagram of the pure solvent, here water with a single relaxation time $\tau_1(0)=\tau_s(0)$. Curve (2+2') shows the two relaxation regions for the solvent and the ion-pair of an electrolyte solution at concentration c. The solvent relaxation time is shifted to $\tau_s(c)$; $\tau_{IP}(c)$ is the relaxation time of the ion-pair. The static permittivity of the pure solvent, $\varepsilon_s(0)$, is shifted to $\varepsilon_s(0;c)$. As usual, the permittivity depression $(\varepsilon_s(0;0) - \varepsilon_s(0;c))$ is due to ion-solvent interactions. The appearance of the ion pair yields the static permittivity of the solution $\varepsilon_{IP}(0;c)$. With the help of the Onsager equation[28] the dispersion amplitude $[\varepsilon_{IP}(0;c) - \varepsilon_s(0;c)]$ may be related to the dipole moment, μ_{IP}, and the particle density, ϱ_{IP}, of the ion-pairs

$$\varepsilon_{IP}(0,c) - \varepsilon_s(0,c) = \frac{\varepsilon_s(0;c)\, \varrho_{IP}\, \mu_{IP}^2}{2[\varepsilon_s(0;c)+1]\,[1-f_{IP}\alpha_{IP}]^2\,\varepsilon_o k_B T} \qquad (22)$$

where f_{IP} and α_{IP} are the reaction field factor and the polarizability of the ion-pair.

According to Eigen and Tamm [31,32] 2,2 electrolytes in aqueous solutions yield the sequence of equilibria

$$M^{2+}_{aq} + SO^{2-}_{4,aq} \rightleftarrows \underset{3}{[M^{2+}(H_2O)_2 SO_4^{2-}]^o} \rightleftarrows \underset{2}{[M^{2+}(H_2O)SO_4^{2-}]^o} \rightleftarrows \underset{1}{[M^{2+}SO_4^{2-}]^o}$$

requiring three ion-pair relaxation times for the species 1,2 and 3 caused by different quantities $\mu_{IP}^{(i)}$, $\alpha_{IP}^{(i)}$ and $f_{IP}^{(i)}$, (i=1,2,3). For ion pairs of species 1,2 and 3 we used as model prolate ellipsoids of long axis $a^{(i)}=a_+ + a_- + (i-1)a_w$ (i=1,2,3), a_+, a_- and a_w being the radii of cation, anion and the length of an orientated water molecule, respectively. Data analysis based on this model shows that essentially only species 2 contributes to ion-pair relaxation[33]. The absorption curves of aqueous CdSO$_4$ solutions at various concentrations are given in Fig. 6. Two maxima are observed in the frequency region from 1 to 40 GHz attributable to solvent relaxation (at high frequencies) and ion-pairs (at low frequencies).

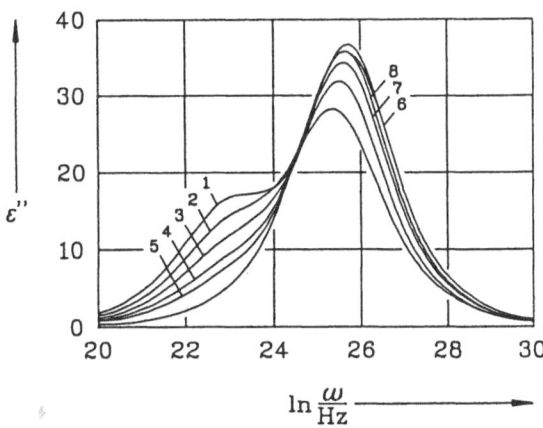

Figure 6. Absorption curves ε'' vs. $\ln\omega$ of water (8) and aqueous $CdSO_4$ solutions. 1: 1.1455 mol dm^{-3}; 2: 0.4916 mol dm^{-3}; 3: 0.2502 mol dm^{-3}; 4: 0.1171 mol dm^{-3}; 5: 0.052 mol dm^{-3}; 6: 0.02502 mol dm^{-3}; 7: 0.00915 mol dm^{-3}. (From Ref. 33)

Ion-pair concentrations calculated from the dispersion amplitudes $\varepsilon_{IP}(0,c) - \varepsilon_S(0,c)$ yield the association constants of $CdSO_4$ and $MgSO_4$ at infinite dilution, reported in Table 2.

The analysis of the observed concentration dependence of relaxation times $\tau_{IP}(c)$ of the ion-pair relaxation region, reveals two superimposed modes, a diffusional rotation mode of the ion-pair dipoles and a kinetic mode due to the re-establishment of the undisturbed equilibrium ($M^{2+} = Cd^{2+}, Mg^{2+}$)

$$M^{2+} + SO_4^{2-} \underset{k_{21}}{\overset{k_{12}}{\rightleftarrows}} M^{2+}(OH_2)SO_4 \tag{23}$$

The rotational mode yielding the relaxation time τ^{or} may be compared to the relaxation time for diffusional rotation as given by the Stokes-Einstein-Debye relation, Eq. (21), where V is the volume of the prolate ellipsoid representing the ion-pair, η and f having their usual meaning. The kinetic mode results from a normal mode analysis[35,36]

$$\tau^{kin} = k_{21} + k_{12}(c-c_{IP}) \tag{24}$$

where k_{21} and k_{12} are the rate constants corresponding to Eq.(23), c and c_{IP} are the total and the ion-pair concentration, respectively.

Data analysis of the observed relaxation times is based on the relation

$$\frac{1}{\tau_{IP}(c)} = \frac{1}{\tau^{or}} + \frac{1}{\tau^{kin}} \tag{25}$$

The results are summarized in Table 2.

Table 2. Rate constants k_{12} for ion-pair formation and k_{21} ion-pair decomposition, association constants K_A from dispersion amplitudes compared with values from the literature; experimental (τ_{IP}^{or}) and theoretical (τ_{IP}^{SED}) relaxation times for aqueous solutions of Cd- and MgSO$_4$ at 25°C

salt	k_{12} 10^{-9} dm^3(mol s)$^{-1}$	k_{21} 10^{-7} s^{-1}	K_A/dm^{3}mol^{-1} Ampl.[a]	Lit.	τ_{IP}^{or} ps	τ_{IP}^{SED} ps
CdSO$_4$	1.8±0.4	0.7±0.2	280±50	245[b](33) 239[c](34)	136±8	146
MgSO$_4$	1.80±0.01	1.03±0.01	150±40	156[b] (1) 161[c](34)	136±1.2	133

[a] calculated from the dispersion amplitude; [b] conductance measurements; [c] heat of dilution measurements

3.3 Permittivity Depression and Kinetic Depolarization

The permittivity depression of the solvent, $\varepsilon_s(0;c)-\varepsilon_s(0;0)$, shown in Fig.5 (in the case of the absence of ion pairs only curve 2 is observed), has been the object of various investigations[11-14,23,37]. Precise measurements at low electrolyte concentrations always show the permittivity depression to be non-linear and specific for the electrolyte. In the literature, permittivity depression is reproduced by

$$\varepsilon_s(0;c) - \varepsilon_s(0,0) = -\delta c + \beta c^n, \quad n = 2 \text{ or } \frac{3}{2} \tag{26}$$

The limiting slope ,$-\delta$, the so-called dielectric decrement, is the quantity considered in theoretical concepts. The underlying models roughly can be classified into equilibrium approaches based on dielectic saturation, and dynamical theories based on kinetic depolarization.

Dielectric saturation produces "irrotational bounding" of the solvent molecules surrounding the ion and hence yields solvation numbers. Pottel et al.[25,38] use the empirical Bruggeman relation for the estimation of the "effective" volume fraction of the solvent. Comparision with its "analytical" value yields solvation numbers Z_P. Lestrade et al.[30,39] use the Kirkwood Fröhlich equation [4,5], with the Kirkwood parameter assumed independent of electrolyte concentration, to calculate solvation numbers Z_L by means of the number of molecules per unit volume required to explain the limiting slope of the permittivity depression. A survey of solvation numbers Z_P and Z_L obtained from these methods and their critical discussion is given in Ref. 14.

The theories of kinetic depolarization link the decrease of the static permittivity to ionic motions. Hubbard, Colonomos and Wolynes[40] showed that the corrected original continuum theory[41,42] predicts proportionality of $\varepsilon_s(0;0)-\varepsilon_s(0;c)$ and the specific conductance $\varkappa$

$$\varepsilon_s(0;0)-\varepsilon_s(0;c) = \xi \varkappa; \quad \xi = p \frac{\varepsilon_s(0;0)-\varepsilon_s(\infty;0)}{\varepsilon_s(0;0)} \frac{\tau_s(0)}{\varepsilon_o} \tag{27a,b}$$

The constant of proportionality ξ is solely determined by the dielectic data of the solvent and the friction factor p (p=1 for sticking, p=2/3 for slipping movement of the ions). Linear relations, $\varepsilon_s(0;0)-\varepsilon_s(0;c)$ vs. $\varkappa$, are found in various solvents[9,14,23,43], the experimentally determined depolarization factor ξ_{exp}, however, generally exceeds the value predicted by Eqs.(26) and is ion specific, in contrast to the theory, see Fig.7. Superimposition of kinetic depolarization and dielectric saturation yields reasonable results for some aprotic solvents[44] but is not convincing for protic solvents[14].

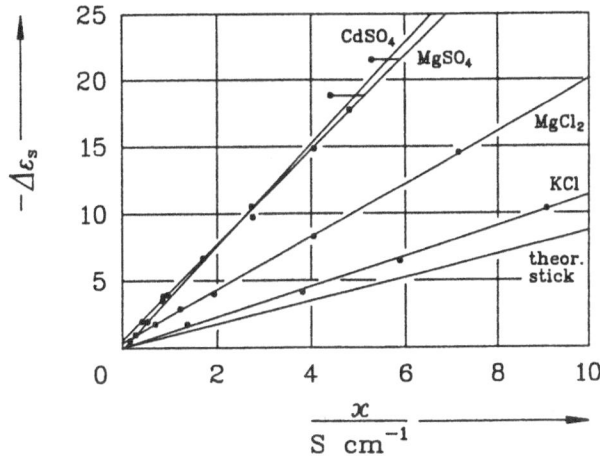

Figure 7. Kinetic depolarization of aqueous solutions
Theor. stick: Equs. 27 with p = 1. Superimposition of kinetic depolarization and dielectric saturation would require 2±2 (KCl), 12±5 ($MgCl_2$), 26±5 ($MgSO_4$) and 16±5 ($CdSO_4$) irrotationally bound water molecules.

Other theories[40,45] actually cannot be discussed since experimental results for comparision of a sufficient variety of electrolyte solutions are lacking.

REFERENCES

1. H.-J. Wittmann, "Die Dispersion der elektrischen Leitfähigkeit ver-dünnter Elektrolytlösungen im Megahertz-Bereich", Thesis, Regensburg (1985)
2. P. Debye,"Polar Molecules", Dover Publications, Inc. (1929)
3. L. Onsager, "Electric Moments of Molecules in Liquids", J. Amer. Chem. Soc. 58:1486 (1936)
4. J.G. Kirkwood, "The Dielectric Polarization of Polar Liquids", J. Chem. Phys. 7:911 (1939)
5. H. Fröhlich, "Theory of Dielectrics", 2nd ed., Oxford University Press (1958)
6. R.H. Cole, "Induced Polarization and Dielectric Constant of Polar Liquids", J. Chem. Phys. 27:33 (1957)
7. T.J. Buchanan, "Balance methods for the measurements of permittivity in the microwave region" Proc. Instn Elect. Engrs, Part III 99:61 (1952)
8. J. Barthel, "Electrolytes in non aqueous solvents", Pure Appl. Chem. 51:2093 (1979)
9. J. Barthel, K. Bachhuber, R. Buchner, and B. Kaukal, in preparation
10. J. Krüger, E. Schollmeyer, and J. Barthel, "The Influence of Higher Order Modes upon the Accarancy of Dielectric Constant Determination Using Transmission Measurement Cells for Electrolyte Solutions Having Strongly Depressed Dielectric Constants in the Gigahertz Region", Z. Naturforsch. 30a:1476(1975)

11. J. Barthel, F. Schmithals, and H. Behret, "Untersuchungen zur Dispersion der komplexen Dielektrizitätskonstante wäßriger und nichtwäßriger Elektrolytlösungen I. Auswahl der Meßmethoden und Messungen an wäßrigen Alkalichlorid- und Alkalinitratlösungen bis zur Sättigungskonzentration bei 25°C im Bereich der cm-Wellen", Z. Phys. Chem. NF 71:115 (1970)

12. J. Barthel, H. Behret, and F. Schmithals, "Dielectric Behaviour of Solutions of Alkali Chlorides and Nitrates in the Microwave Region", Ber. Bunsenges. Phys. Chem. 75:305 (1971)

13. H. Behret, F.Schmithals, and J. Barthel, " Untersuchungen zur Dispersion der komplexen Dielektrizitätskonstante wäßriger und nichtwäßriger Elektrolytlösungen II. Mikrowellenmessungen von Dielektrizitätskonstante und Relaxationszeit an Lösungen der Alkalinitrate und Chloride in polaren Lösungsmitteln", Z. Phys. Chem. NF 96:73 (1975)

14. J. Barthel and R. Buchner, "Dielectric properties of nonaqueous electrolyte solutions", Pure & Appl. Chem. 58(8):1077 (1986)

15. H. Steger, Thesis, Regensburg (in preparation)

16. J.G. Kirkwood and R.M. Fuoss, "Anomalous Dispersion and Dielectric Loss in Polar Polymers", J. Chem. Phys. 9:329 (1941)

17. A. Budo, E. Fischer and S. Migamoto, "Einfluß der Molekülform auf die dielektrische Relaxation", Phys. Z. 40:336 (1939)

18. F. Feuerlein, "Dielektrische Eigenschaften von Elektrolytlösungen in Propylencarbonat und in Propylencarbonat-Dimethoxyethan Mischungen", Thesis, Regensburg (1983)

19. J.Barthel, R. Buchner, H. Hetzenauer, K.H. Popp, H. Steger, unpublished data

20. J. Barthel and M. Kleebauer, unpublished data

21. R. Pottel, "Die komplexe Dielektrizitätskonstante wäßriger Lösungen einiger 2-2-wertiger Elektrolyte im Frequenzbereich 0.1 bis 38 GHz", Ber. Bunsenges. Phys. Chem. 69:363 (1965)

22. E.A.S. Cavell and S. Petrucci, "Dielectric Relaxation Studies of Solutions of 1:2, 2:1 and 2:2 Electrolytes in Water", J. Chem. Soc. Faraday Trans. 2 74:1019 (1978)

23. R. Buchner, "Der Einfluß von Elektrolyten und apolaren Substanzen auf die dielektrischen Eigenschaften von Methanol und Wasser im Gigahertz-Bereich", Thesis, Regensburg (1986)

24. U. Kaatze, V. Lönnecke, and R. Pottel, "Dielectric Spectroscopy of Aqueous Solutions of Zinc(II)-Chloride. Evidence of Ion Complexes", J. Phys. Chem. 91:2206 (1987)

25. U. Kaatze, "Dielectric Effects in Aqueous Solutions of 1:1, 2:1, 3:1 valent Electrolytes: Kinetic Depolarization, Saturation and Solvent Relaxation", Z. Phys. Chem.NF 135:51 (1983)

26. J. Barthel, "Ionen in nichtwäßrigen Lösungen", Dr. Dietrich Steinkopff Verlag, Darmstadt (1976), Chap. 4

27. M. Delsignore, H. Farber, and S. Petrucci, "Ionic Conductivity and Microwave Dielectric Relaxation of LiAsF$_6$ and LiClO$_4$ in Dimethyl Carbonate", J. Phys. Chem. 89:4968 (1978)

28. E.A.S. Cavell, P.C. Knight und M.A. Sheikh, "Dielectric Relaxation in Non Aqueous Solutions" J. Chem. Soc., Faraday Trans. 67:2225 (1971)

29. J. Barthel,H.J. Gores, G. Schmeer and R. Wachter, "Non-Aqueous Electrolyte Solutions in Chemistry and Modern Technology", in:"Topics in current chemistry", F.L. Boschke (ed.), Springer, Berlin (1983)

30. J.-C. Lestrade, J.-P. Badiali, H. Cachet, in: M. Davies (ed.), "Dielectric and Related Molecular Processes", Vol. 2, Chap. 3, The Chemical Society, London (1975)

31. M. Eigen and K. Tamm, "Schallabsorption in Electrolytlösungen als Folge chemischer Relaxation I. Relaxationstheorie der mehrstufigen Relaxation", Ber. Bunsenges. Phys. Chem. 66:93 (1962)

32. M. Eigen and K. Tamm, "Schallabsorption in Electrolytlösungen als Folge chemischer Relaxation II. Meßergebnisse und Relaxationsmechanismen für 2-2 -wertige Electrolyte", Ber.Bunsenges. Phys. Chem. 66:107 (1962)

33. J. Barthel, R. Buchner, and H.-J. Wittmann, "Leitfähigkeit und dielektrische Eigenschaften wäßriger CdSO₄-Lösungen",Z. Phys. Chem. NF 139:23 (1984)

34. R. Wachter and K. Riederer, "Properties of Dilute Electrolyte Solutions from Calorimetric Measurements", Pure Appl. Chem. 53:1301 (1981)

35. H. Strehlow and W. Knoche, "Fundamentals of Chemical Relaxations", Verlag Chemie, Weinheim (1977)

36. P. Turq, L. Orcil, M. Chemla, and J. Barthel, " Influence of Ion-Pair Formation in Diffusional Transport of Symmetrical Electrolytes", Ber. Bunsenges. Phys. Chem. 85:535 (1981)

37. J. Barthel, J. Krüger, and E. Schollmeyer, "Untersuchung zur Dispersion der komplexen Dielektrizitätskonstanten wäßriger und nichtwäßriger Elektrolytlösungen III.Kritische Untersuchung zur Meß- und Auswertemethode. Alkalifluoride, -bromide, -iodide und -perchlorate in wäßriger Lösung", Z. Phys. Chem. NF 104:59 (1977)

38. K. Giese, U. Kaatze und R. Pottel, "Permittivity and Dielectric and Proton Magnetic Relaxation of Aqueous Solutions of the Alkali Halides", J. Phys. Chem. 74:3718 (1970)

39. J.-P. Badiali, H. Cachet und J.-C. Lestrade, "Compartement Dielectrique de Solutions Electrolytiques dans des alcools et un Melange Eau-Dioxane" J. Chim. Phys. 25:1350 (1967)

40. J.B. Hubbard, P. Colonomos und P.G. Wolynes, "Molecular theory of solvated ion dynamics. III. The dielectric decrement", J. Chem. Phys. 71:2652 (1979)

41. J.B. Hubbard und L. Onsager, "Dielectric dispersion and dielectric friction in electrolyte solutions. I" J. Chem. Phys. 67:4850 (1977)

42. J.B. Hubbard, "Dielectric dispersion and dielectric friction in electrolyte solutions. II", J. Chem. Phys. 68:1649 (1978)

43. J. Barthel and F. Feuerlein, "Dielectric properties of Propylene Carbonate - 1,2-Dimethoxyethane Mixtures and their Electrolyte Solutions of NaClO₄ and Bu₄NClO₄", Z. Phys. Chem. NF 148:157 (1986)

44. J. Barthel and F. Feuerlein, "Dielectric Properties of Propylene Carbonate and Propylene Carbonate Solutions", J. Sol. Chem. 13:393 (1984)

45. B.U. Felderhof, "Dielectric decrement of electrolyte solutions", Mol. Phys. 51:801 (1984)

46. B. Kaukal, "Dielektrische Relaxation und Dispersion methanolischer Elektrolytlösungen im Mikrowellengebiet", Thesis, Regensburg (1982)

DISCUSSION

HYNES - Your analysis requires the calculation of the dipole moments of the various ion pairs. How, for example, was the dipole moment of the ion pair with two intervening water molecules calculated ? Also, why did you not worry about more than two H_2O molecule-containing ion pairs ?

BARTHEL - The dipole moments of the ion pairs, needed in Eq. (22), are calculated with the help of the relation (see Ref. 33)

$$\mu_{Ip} = 2ed - \frac{8\pi\epsilon_o ed^4(\alpha_+ + \alpha_-) + 8ed\,\alpha_+\alpha_-}{(4\pi\epsilon_o d^3)^2 - 4\alpha_+\alpha_-} \tag{28}$$

where α_+, α_- are the polarizabilities of cation and anion, and d is the center-to-center distance of the positive and negative charges in the ion pair, $d = a_+ + a_- + na_w$ (a_+, a_- : radius of cation and anion, respectively ; a_w : length of an orientated water molecule ; n = 0,1,2. The water molecules in the ion pairs are orientated with the free electrons of oxygen towards the cation and with the hydrogen atoms towards the anion.

Ion pairs of the type $M^{2+}(H_2O)_2A^{2-}$ (n = 2) are not detectable in the plots ϵ^*vs. ω ; ion pairs with n > 2 hence must not be considered.

KAPRAL - To what extent is the extraction of the rate coefficient data for ion pair formation dependent on the particular model used in the data analysis ; in particularly, the model used for the solvent reorientation ?

BARTHEL - The existence of superimposed orientational and kinetic modes follows from the concentration dependences of dispersion amplitude, $\epsilon_{Ip}(o,c)$ - $\epsilon_{Ip}(\infty,c)$ and ion-pair relaxation time, $\tau_{Ip}(c)$. Models of the ion pair used in connexion with Eqs. (22) and (28) to calculate particle densities make assumptions on the shape and dipole moment of the ion-pair dipoles. For the systems investigated, $M^{2+}(H_2O)_n SO_4^-/H_2O$, the contributions of ion pairs with n=0 and n=2 may be neglected (33). The assumption of prolate ellipsoides corresponding to n=1 yields association constants, $K_A = k_{12}/k_{21}$, in agreement with the association constants obtained from conventional conductance and calorimetric heat of dilution measurements, and relaxation times $\tau_{Ip}(exp)$ comparable to those calculated with the help of Eq.(21) where V is the volume per mole of the prolate ellipsoïdes. No special model is used for solvent reorientation, the solution is considered as a mixture of solvent molecules, ion-pair dipoles and free ions exposed to the high frequency electromagnetic fields (1-40 GHz, $10^{-3} < c/mol\ dm^{-3} < 1$).

RIVAIL - You mentioned the Powles Glarum equation as the only one which links the macroscopic relaxation time to the microscopic one. Is there a special reason, specific of electrolyte solutions, to prefer this equation to other ones such as KKVR[*], except simplicity ?

(*) Ref. ?

BARTHEL - Unfortunately, there is no general theory linking macroscopic and microscopic correlation functions and relaxation times. Some approximations are useful for special cases, such as the Powles Glarum equation for the rotational movements of dipole species. An estimation of the accuracy of the approximations is hardly possible. The Powles Glarum equation has the advantage of simplicity, I agree.

NEWTON - It is clear that you have accumulated a large body of data regarding the dependence of dielectric constant on ionic strength, a topic of great interest and past confusion. I take it that your high-frequency technique allows you to separate any extraneous dynamical effects (conductivity) from the desired screening part of the dielectric constant for electrolyte solutions.

BARTHEL - The traveling wave method as realized in our laboratory permits to attain the highest actually available precision for permittivity measurements on electrolyte solutions over large concentration ($10^{-4} < c/\text{mol dm}^{-3} <$ saturation) and high frequency ($\nu/\text{Hz} > 2.10^9$) ranges. The assumption of frequency independent specific conductance does not affect the investigations on the kinetic processes in high frequency fields.

FAST TECHNIQUES IN ELECTROCHEMISTRY

APPLICATION TO THE STUDY OF CHEMICAL REACTIVITY

Christian Amatore

Ecole Normale Superieure - Laboratoire de Chimie
UA 1110 CNRS "Activation Moleculaire"
24, rue Lhomond - 75231 Paris, Cedex 05 France

INTRODUCTION

Electron transfers and the reactions they initiate have received a considerable attention in the recent years both in the organic and organometallic fields [1]. Indeed it has been recently recognized that a number of so-thought elementary reactions may involve - or - may be triggered by - single electron transfers. The spreading of these new concepts coincided with the development of new experimental and conceptual tools in electrochemistry. Indeed up to the middle of the seventies most of electrochemical studies were related to the delineation of the basic and elementary chemical acts associated with electron transfers [2]. The method was then mostly limited to problems involving a single reactive path occuring within time scales larger than a few milliseconds. Yet the basic concepts thus developped have allowed, in the past decade, the extension of the electrochemical approaches to situations which interest and complexity make them particularly adequate to investigate a variety of essential problems in chemical reactivity. Concomitantly the time scale window of the method has been enlarged to the sub-nanosecond region. This together with the inherent simplicity - and low cost aspects - of the technique explain its recent diffusion and adoption by many research groups in the fields of organic or organometallic chemical reactivity.

An obvious and unique advantage of electrochemistry in the investigation of chemical reactivity derives from the intrinsic possibility of continuously tuning the electrode potential. This allows the facile variation, within several electron volts, of the driving force of an electron transfer reaction. Thus highly reactive intermediates are generated under precise and mild conditions. This is particularly true for neutral radicals or paramagnetic charged species such as cation or anion radicals, which are obtained via electron intake or uptake from stable molecular structures. Moreover, taking advantage of the fact that electron transfers are often followed by bond-breaking or bond-creation reactions, controlled cascades of chemical reactions may be devised to produce highly basic anions or acidic cations or other non paramagnetic intermediates.

A less apparent but obvious advantage is related to the kinetic information contained in the current flowing through the electrode. Indeed the current relaxation following a potential perturbation depends on the exact concentration profiles of the energetic species electrogenerated in the very close proximity (within few micrometers) of the electrode. These time dependent concentration profiles, being imposed by the transport-reaction Fick's laws, are then specific of the kinetic path(s) followed by the considered intermediate [3]. Thus the current constitutes <u>per se</u> a simple measure and characterization of the kinetics taking place in the solution. In this respect one should emphasize this unique aspect of electrochemistry, where the same device (the electrode) is used to generate highly unstable species (potential) and monitor their chemical evolution (current).

The principle of most electrochemical methods is based upon the competition between transport to and from the electrode and creation or disappearance via chemical reactions. Indeed the electrons are supplied at, or taken from, the solution/electrode interface whereas the substrate and products evolve in the solution volume adjacent to the interface. Thus the principle of the <u>direct</u> electrochemical methods consists in opposing the rate of transport, usually by means of diffusion-migration, to that of chemical reactions. Extraction of the sought information then requires the resolution of the pertinent transport-reaction, time and space dependent equations. A more serious limitation of <u>direct</u> electrochemical methods is that the interfacial nature of the electron exchange limits the time scale domain accessible, when the kinetics have to be studied in conditions matching those encountered in homogeneous chemical situations. Indeed, because of the potential difference between the electrode and the solution, an interfacial charged region develops near the electrode and extends over 10 to 50 Å [4]. In this region, the "double layer", an important electrical field exists (some tenths of a volt over few angströms) which would preclude electro-chemical kinetics to be transposed to usual homogeneous conditions. Thus meaningfull data related to intrinsic chemical reactivity need that the concentration profiles extend far from this electrically perturbed re-gion. In other words the shortest times to be investigated must be large enough for diffusion-migration being able to transport the species of interest out of the double layer region. In the practice this results in the lower time limit being of the order of several nanoseconds [5].

This time resolution is however greatly improved by the use of indirect electrochemical methods [6]. These amount to oppose the investi-gated chemical path to another, well known, reaction. The resulting competition is then monitored via its interference with the diffusion-migration process, as for the direct approach. With these more elabora-ted techniques one is able to obtain kinetic information on species with time-lifes of few picoseconds.

DIRECT ELECTROCHEMICAL METHODS
Cyclic voltammetry

Transient electrochemistry consists in applying an electrical per-turbation (choosen as the independent variable) to the electrode/solu-tion system and monitoring a response signal (the dependent variable). In most methods the independent variable is the potential, E, or the current, i, the dependent variable being then the other. Yet for particular situations any function $f(i,e)=0$ can be selected. Figure 1 gives a schematic presentation of the various electrochemical methods under the form of a family-tree [7]. Among the various techniques in figure 1, cyclic voltammetry is certainly the most adequate for the investigation of chemical reactivity problems. In cyclic voltammetry the potential perturbation is a triangular signal as sketched in figure 2. The potential is varied linearly with time from a value, E_i, where the investigated molecule is not electroactive up to a value, E_f, exceeding the standard redox potential, E^o, of the molecule. The potential is then scanned back to its initial value. The slope of the potential variations with time is commonly refered to as the scan rate, v. The current response, i(t), is usually plotted as a function of the potential rather than of the time, the pattern then obtained being designed as the cyclic voltammogram of the compound under study. Figure 2 presents such a cyclic voltammogram featuring the reduction in eqn (1), together with

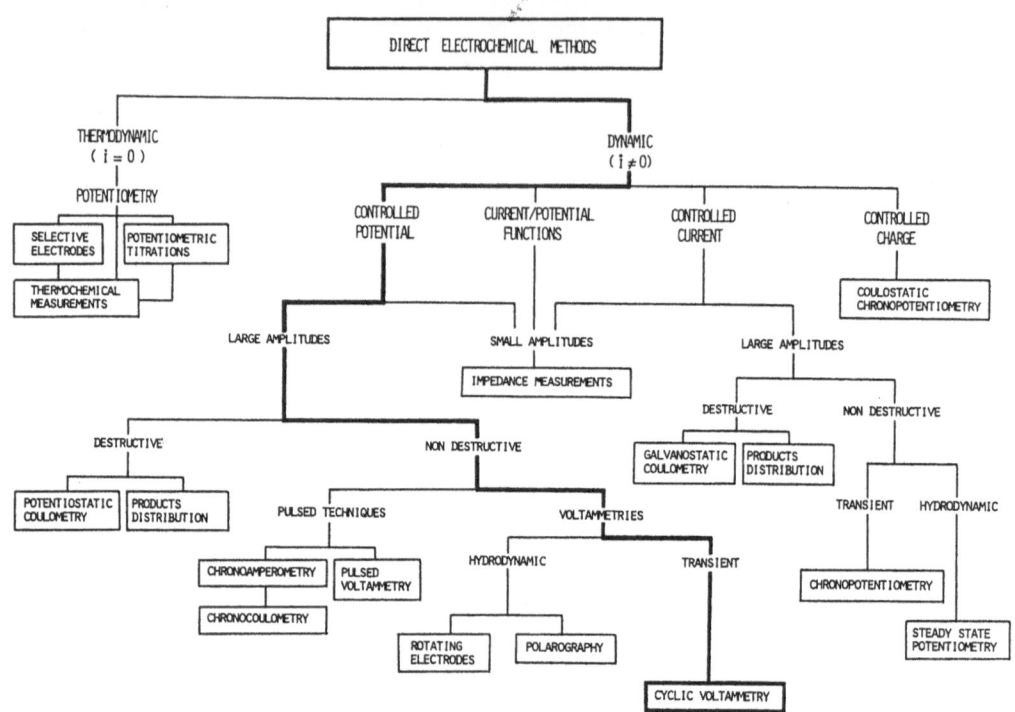

Fig.1. *Relationships between the main direct electrochemical methods. The characteristics of Cyclic Voltammetry are indicated along the thick line.*

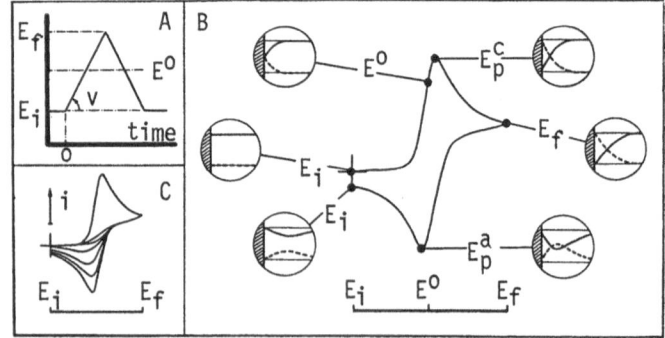

Fig.2. *Cyclic Voltammetry. (A): Potential variations vs time (see text for the definitions). (B): Cyclic Voltammogram. The concentration profiles are represented for different potential/time locations, for A (solid line), and A$^{\pm}$ (dashed line). E_p^c and E_p^a : cathodic and anodic peak potentials. (C): Effect of the scan rate on the overall shape (see text).*

$$A \quad + \quad e \quad \rightleftharpoons \quad A^{\pm} \qquad\qquad (1)$$

the A and $A^{\pm}$ concentration profiles in the vicinity of the electrode. It is thus seen that A is depleted at the electrode when the potential of the latter reaches the region corresponding to the voltammogram cathodic peak. After the peak, A concentration at the electrode tends to be negligible and the current slowly decays because of the propagation of the concentration perturbation into the solution (note in this respect that the current flowing through the electrode is proportional to the concentration profile gradient at the electrode surface). Thus the peak shape of the cyclic voltammogram arises from the time convolution of two phenomena : (i) increasing depletion of the electroactive material at the electrode surface, which tends to increase the current, the concentration gradient being steeper, and (ii) the smoothing of the concentration profile because of the propagation of the perturbation toward the bulk of the solution, which tends to decrease the concentration gradient.

The depletion of A in the vicinity of the electrode is accompanied by a building up of $A^{\pm}$ concentration. When the potential is switched back, at E_f, the redox equilibrium at the electrode surface starts to be inverted. When sufficient anodic values are attained oxidation of $A^{\pm}$ back to A occurs significantly, which then causes the current to peak and to decay because of the same antagonizing processes described above for the forward scan.

When a smaller scan rate is considered, the $A^{\pm}$ anion radical formed upon reduction may evolve as in the sequence of eqns (2) + (3). Indeed when the scan duration is small enough as compared to the half

$$A \quad + \quad e \quad \rightleftharpoons \quad A^{\pm} \qquad\qquad (2)$$

$$A^{\pm} \quad \rightarrow \quad \cdots \qquad\qquad (3)$$

life time $t_{1/2}$ of $A^{\pm}$, the overall voltammogram corresponding to the system is akin to that in figure 2B. In the converse situation, ie. when the scan duration is larger than $t_{1/2}$, $A^{\pm}$ chemically evolves before it can be oxidized back to A during the reverse scan. Thus no peak is observed upon potential reversal. When now both times are of comparable magnitudes a peak is detected during the reverse scan, its importance depending on the amount of $A^{\pm}$ which has not yet evolved (see figure 2C).

It is thus seen, from this semi-quantitative presentation that the variations of a cyclic voltammogram overall shape as a function of its time duration, i.e. with the scan rate v, are directly related to the kinetics triggered by the initial electron intake or uptake. Dimensionless analysis of the transport-reaction equations pertinent to cyclic voltammetry establishes that the characteristic time of the method is $\theta = (RT/Fv)$, e.g. at room temperature $\theta \sim (25 \ 10^{-3}/v)$ where θ is in seconds and v in volts per seconds. Thus a scan rate of 1 Vs^{-1} corresponds to 25ms, whereas a scan rate of 250 000 Vs^{-1} corresponds to 100ns. In the pratice a reactive intermediate is observable when $\theta \leqslant 10xt_{1/2}$. The fastest scan rates accessible being of the order of 250 000 Vs^{-1} (vide infra), it is seen that cyclic voltammetry has a time resolution of few nanosecond, and allows then the direct observation of extremely short lived species formed upon electron transfer to or from the electrode. Note that in this context "observation" does not rely on the spectral characteristics of the intermediate but on its redox properties. Thus the interpretation of the data does not require an a-priori knowledge of the intrinsic spectral properties, such e.g. as the extinction coefficient, of the observed intermediate.

Ultrafast cyclic voltammetry at ultramicroelectrodes

From the above presentation it can be infered that increasing the scan rate, v, should allow the observation of transient intermediates with shorter and shorter lifetimes. In the actual practice this has long remained an utopy since the use of large scan rates ($v < 100vs^{-1}$, ie. $\theta > 0.25ms$ at 20°C., for most electrochemical systems) introduces considerable distorsion of the idealistic cyclic voltammogram in figure 2B because of the involvement of capacitive currents and ohmic drop.

Ohmic drop results from the fact that the current flowing through the electrode must be carried out through the low conducting solution, to the other electrode. Thus the effective potential, $E_{eff.}$, probed by the electronated molecule is different from the true potential, E, of the electrode : $E_{eff.} = E - Ri$, where R is the solution/electrode resistance and i is the instantaneous current. Positive feedback compensation and numerical deconvolution procedures have been widely used to overcome this difficulty [8]. Yet in the practice, owing to electronic instabilities arising from phase lags, the method was limited to the low kilovolt per second scan rate domain ($\theta > 10 \mu s$).

A recent approach to overcome this limitation is based upon the use of ultramicroelectrodes, ie. of electrodes which dimensions are in the micrometer or submicrometer range. Indeed the resistance at a disk electrode is invertely proportional to its radius, r, whereas the current flowing through it is porportional to its surface area ie to r^2. Thus the ohmic drop, Ri α r, decreases with the electrode radius. The recent availability of gold, platinum or carbon fibers of micrometric radii has thus allowed the recording of totally undistorted cyclic voltammograms up to scan rates of 10 000Vs^{-1} [9].

Capacitive phenomena results from the existence of the charged interfacial region, the double layer, at the limit between the electrode and the solution. This region, schematically pictured in figure 3A, behaves as two series capacitors. One can be thought as being located between the electrode surface and the plane of closest approach of the electroactive compound, the other being related to the remaining portion of the double layer, is to be viewed as connected in parallel with the faradaic impedance, as sketched in figure 3B. The capacitor values being proportional to the electrode surface area, the faradaic/capacitive current ratio is not affected by the electrode size. Yet the charging time constant, RC, being then proportional to the electrode radius, r, decreases with the dimension of the latter. Thus although not cancelled the capacitive currents are simplified at ultramicroelectrodes because of the considerable decrease of their corresponding time constants. Thus natural deconvolution [9] between the capacitive and faradaic currents occurs at ultramicroelectrodes, without resorting to sophisticated a-posteriori procedures. This is illustrated in the following by two typical examples.

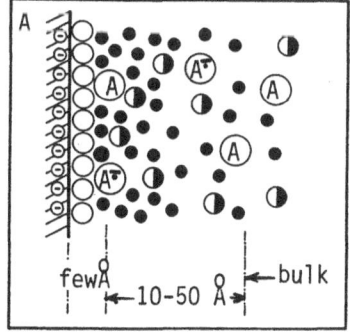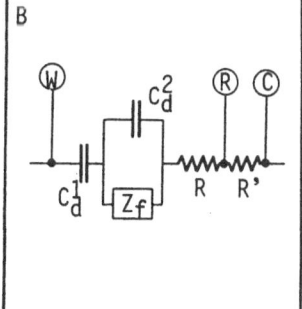

Fig.3. (A): Schematic representation of the double layer region adjacent to the electrode, while A is reduced to A$^{\pm}$. ($\bigcirc$): Solvent molecules , ($\bullet$): Supporting electrolyte cation, ($\mathbf{\Phi}$): Supporting electrolyte anion. (B): Equivalent electrical circuit; Z_f : Faradaic impedance.

Direct observation of an electrogenerated radical with bimolecular diffusion controlled decay

Although scan rates of few ten thousand volts per second (ie. $\theta \sim 1$ μs) seem to be a limit for quantitative analytical purposes (without using deconvolution procedures) for electrodes of few micrometers radii it is possible to use cyclic voltammetry in the low megavolt per second domain ($\theta \sim 25$ ns) provided that the target is only identification of transient intermediates and determination of thermodynamic figures. Besides its intrinsic attractive value, such an order of magnitude of scan rates constitutes a milestone in electrochemical kinetics. Indeed bimolecular reactions in liquids cannot proceed faster than molecules encounter. Thus bimolecular rate constants are limited by diffusion limit rate constant, k_{dif}, that is ca. 10^9 to 10^{10} $M^{-1}s^{-1}$ for most solvents [10]. Owing to the millimolar to centimolar concentrations used in electrokinetic experiments this enables to evaluate to ca 10ns the smallest lifetimes for electrogenerated intermediates evolving via bimolecular processes.

To establish the experimental achievability of this target the dimerization mechanism of 2,6-diphenyl-pyrylium radicals electrogenerated by reduction of the corresponding cation, in eqn.4, was selected. The mechanism in eqn.4 was first postulated for the zinc powder reduction of pyrylium cations in acetonitrile [11] and further confirmed by

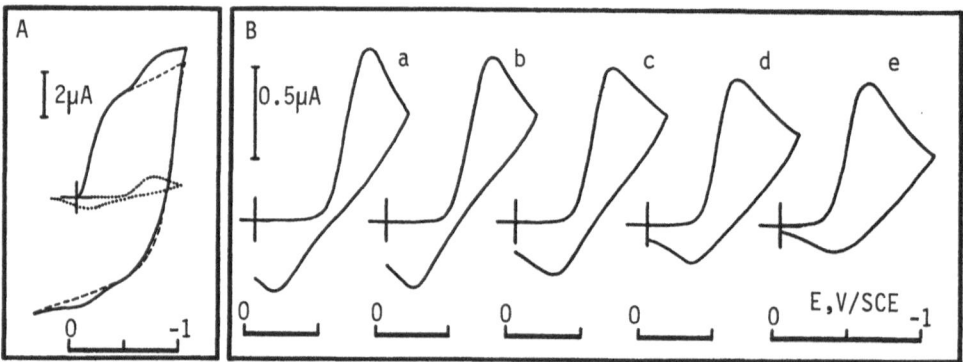

Fig.4. (A): Cyclic voltammograms of an acetonitrile (0.1M NBu₄BF₄) solution containing (solid line) or not (dashed line) a 10mM concentration of 2,6-diphenyl-pyrylium at a scan rate of 250 000 V.s⁻¹. Dotted line: substracted voltammogram. (B): Substracted voltammograms as in (A) at v= 250 (a), 200 (b), 150 (c), 100 (d), 75 (e) kV.s⁻¹. Gold electrode; diameter : 10 μm. [13]

electrochemistry. A rate constant of ca $1.2 \ 10^9 \ M^{-1}s^{-1}$ has been independently estimated for k_{dim} on the basis of flash photolysis experiments [12].

$$(4,5)$$

The voltammograms obtained at $v=250 \ 000 \ Vs^{-1}$ are shown on figure 4A for an acetonitrile solution containing (solid line) or not (dashed line) a 10mM concentration of 2.6-diphenyl-pyrylium perchlorate salt [13]. Simple arithmetic substraction of the two voltammograms [9] yields the voltammogram represented in dotted line on the same figure. An identical treatment was repeated for different scan rates ranging from 250 000 to 75 000 Vs^{-1} (i.e. θ from 100ns to 333ns) to afford, after current magnification, the voltammograms represented in figure 4B. The latter show that the chemical reversibility of the pyrylium wave – compare to the schematic figure 2C –is progressively restored when the scan rate increases, to reach ca 50% at 250 000 Vs^{-1}. This allows the determination of the standard redox potential, $E^o = -0.435$ V vsSCE of the transient radical in eqn.5, as well as that of $k=2.5 \ 10^9 \ M^{-1}s^{-1}$ for its dimerization rate constant [13].

Reduction of aromatic halides in liquid ammonia

Aromatic halides are easily reduced to give an usually short-lived anion radical, $ArX^{\pm}$, which evolves by cleaving the carbon-halogen bond to afford the σ-aryl radical, $Ar^{\cdot}$, in eqn.7. The latter is more easily

$$ArX + e \rightleftharpoons ArX^{\pm} \qquad (6)$$

$$ArX^{\pm} \xrightarrow{k} Ar^{\cdot} + X^- \qquad (7)$$

$$Ar^{\cdot} + e \rightleftharpoons Ar^- \xrightarrow{H^+} ArH \qquad (8)$$

reduced than the starting halide, which results in a second electron exchange at the electrode or in the solution via an homogeneous electron transfer from the aryl halide anion radical in eqn.9 [14]. Thus although electron transfers occur sequentially, the overall reduction wave

$$Ar^{\cdot} + ArX^{\pm} \xrightarrow[k_{diff.}]{} Ar^- + ArX \qquad (9)$$

appears as if involving a two-electron transfer [15] as soon as the rate constant of the halide expulsion in eqn.7 is sufficiently fast as compared to the characteristic time, $\theta = RT/Fv$, of the voltammogram. When the potential scan is extended to more cathodic regions, the aromatic formed in the sequence of eqns 6-9 is reversibly reduced as in eqn.10, which is further confirmation of the reaction sequence [10].

$$ArH \; \overset{e}{\rightleftharpoons} \; ArH^{\pm} \qquad\qquad (10)$$

Increasing the scan rate, v, shortens the time scale, Θ , thus allowing to reoxidize the frangible anion radical before it cleaves off the halide ion in eqn.7. Thus the organic halide reduction wave tends to reversibility and its peak heigth decreases to reach that corresponding to 1e per molecule. Concomittantly the reversible reduction wave of ArH disappears since the latter is no more formed in this time scale. Figure 5 present an experimental illustration of the overall phenomena as a function of the scan rate. With ultra microelectrodes first order rate constants up to few 10^8 s^{-1} can thus be determined. Indeed the rate constant for the cleavage of the anion radical can be derived from the variations of the apparent number of electron consummed at the ArX reduction wave [3], as shown in figure 6 for 2-chloroquinoline in liquid

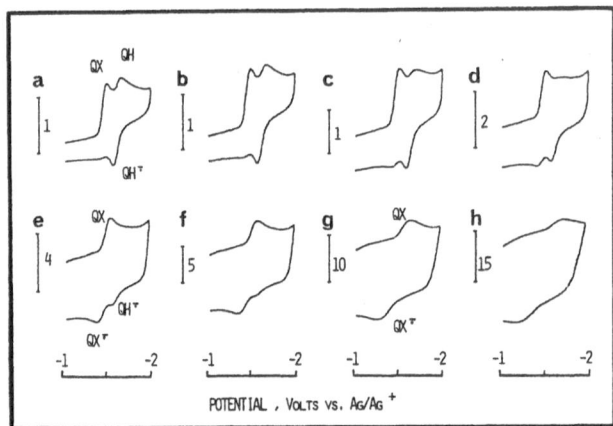

$$(11)$$

ammonia. Thus a rate constant of 1.7 10^4 s^{-1},[14] is determined for the chloride ion expulsion in eqn.11.

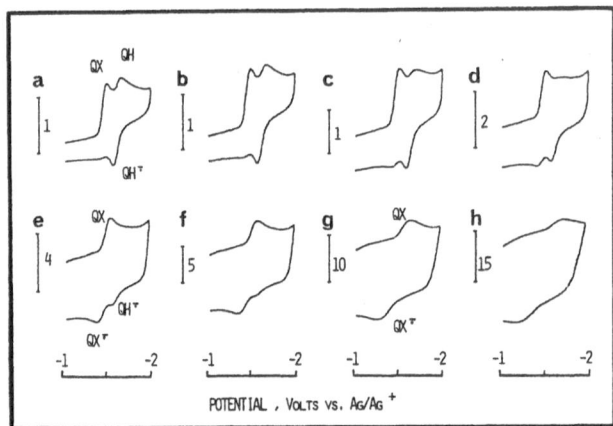

Fig.5. _Cyclic voltammetry of a 2-chloroquinoline (QX, 10mM) in liquid ammonia at -40°C (KBr 0.1M), at 2.2 (a), 3.0 (b), 5.3 (c), 10.2 (d), 28.7 (e), 51.8 (f), 100. (g) and 198. (h) kV.s^{-1}.[20] QH: quinoline. Current scales indicated in μA. Gold electrode, diameter 10 μm._

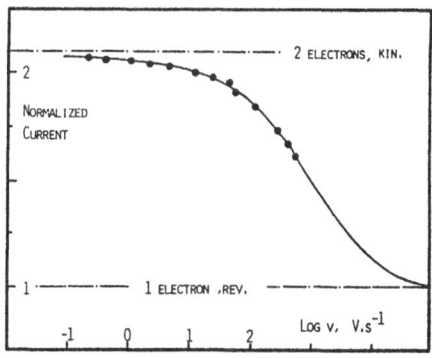

Fig.6. *Variations of the normalized current peak of the 2-chloroquinoline reduction wave as a function of the scan rate v. Solid line: theoretical variations for the mechanism in eqn. 11, and a rate constant k= 1.7 10^4 s^{-1} for the carbon-chloride bond cleavage in the 2-chloroquinoline anion radical.*

INDIRECT ELECTROCHEMICAL METHODS

As pointed out in the Introduction the electrochemical act takes place at an interfacial region. Thus when the electrogenerated species is too short-lived, it cannot survive to escape the double layer charged space region. Although significant data can be obtained in such cases, they are not directly transposable to usual homogeneous conditions. Indeed the investigated reaction is then observed in a space region in which a large electrical potential gradient operates, that is in conditions quite different as those prevailing in an homogeneous solution. Such a difference is particularly important for chemical reactivity of charged species or intermediates with strongly polarisable bonds.

Redox catalysis

A way to overcome this built-in feature consists in transporting the reaction site into the solution. This is easily done using an electron transfer mediator, as schematized on figure 7A. The latter is a reversible redox system which standard potential is located within 100 to 300 mV in front of the peak potential of the investigated molecule [6]. Thus in the absence of the latter, the mediator, M exhibits a perfect reversible redox system, featuring e.g. the electron transfer in eqn.12.

$$M \; + \; e \; \rightleftharpoons \; M^{\underline{+}} \tag{12}$$

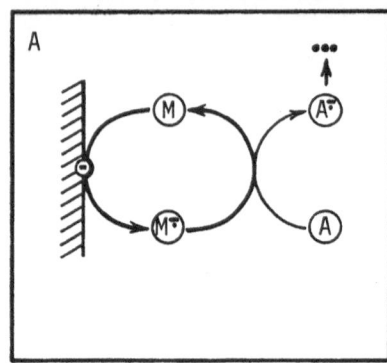
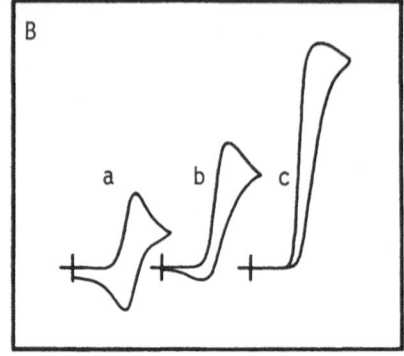

Fig.7.   Redox Catalysis of the reduction of a substrate, A, via a mediator, M. (A): Schematic illustration of the mediator recycling in the presence of A. (B): Cyclic voltammograms of the mediator reduction in the absence (a),or in the presence of increasing excesses of A (b,c).

When the substrate A is added to the solution the reversible cyclic voltammogram is distorted (figure 7B) : the forward current peak increases and the reverse peak tends to disappear, these effects being the larger, the larger the excess factor of substrate vis-à-vis the mediator. This behavior corresponds to the uphill electron transfer in eqn.13

$$M^{\pm} + A \rightleftharpoons M + A^{\pm} \tag{13}$$

$$A^{\pm} \longrightarrow \ldots \tag{14}$$

which restores the mediator concentration in the vicinity of the electrode. The thermodynamically unfavorable electron transfer in eqn.13 takes place because of the continuous removal of the unstable anion radical $A^{\pm}$ which evolves **via** the fast chemical reaction in eqn.14. Different kinetic regimes are thus observed according to the degree of competition between the backward electron transfer (usually diffusion controlled) and the chemical reaction in eqn.14, in the fast consumption of $A^{\pm}$ [6].

When the forward electron transfer is the rate limiting step, the distortion of the mediator wave depends on the rate of the uphill electron transfer in eqn.13. This allows to investigate electron transfer kinetics in the endergonic region. Conversely when the electron transfer in eqn.13 acts as a rapid pre- equilibrium, the mediator cyclic voltammogram is function of the apparent rate constant $k^{ap} = k_{14} \cdot K_{13}$, where k_{14} is the rate constant of the chemical reaction in eqn.14 and K_{13} the equilibrium constant of the electron transfer in eqn.13. Determination of the latter allows then the determination of the rate constant, k_{14} of the investigated chemical reaction.

Application of this basic procedure allows the experimental characterization of very short lived intermediates. An extended series of aromatic halides reduction (vide supra, eqns. 6-8) has thus been investigated which anion radicals have time lifes from ca 10 s to picoseconds [17]. Similarly aliphatic halides were examined as a class of reactions in which the initial electron transfer is concerted with the breaking of the carbon halogen bond. Thus the driving force of the reaction involves two contributions : (i) the usual solvent reorganization operative in outer sphere electron transfers (Marcus theory), and (ii) a bond breaking contribution equal to ca. one fourth of the bond dissociation energy of the neutral halide [16].

Pertubed redox catalysis. Bond formation between a σ-aromatic radical and a nucleophile

The bond formation between a nucleophile and a σ-aryl radical in eqn.15 can be viewed as the reverse of the carbon halogen bond

$$Ar^{\cdot} + Nu^{-} \longrightarrow ArNu^{\cdot -} \tag{15}$$

cleavage of aromatic halides anion radicals as e.g. in eqns.7 or 11. This bond formation reaction is the crucial step in electrocatalyzed aromatic nucleophilic substitutions ($S_{RN}1$), and can be viewed as a concerted electron transfer-bond forming reaction.

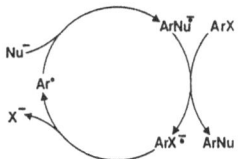

The mechanism of $S_{RN}1$ reaction was established on the basis of direct electrochemical methods [14]. Indeed the nucleophilic attack in eqn.15 competes with the σ-aryl radical reduction in eqn.8. When the former constitutes the main route for $Ar^{\cdot}$ evolution, the apparent electron number involved in the halide wave tends to zero (when $ArNu^{\cdot -}$ reoxidizes) or to one (when $ArNu^{\cdot -}$ is stable) electron. This enables then to appreciate, in a general case, the degree of competition between the two reactions, and thus to determine the rate constant of the bond formation reaction in eqn.15. Yet this direct approach does not allow the determination of coupling rate constants larger than 10^7 to 10^8 $M^{-1}s^{-1}$ [14].

Perturbation of redox catalysis of the aromatic halide by the nucleophilic addition in eqn.15 allows a facile overcoming of this limit and permits the determination of rate constants up to the diffusion limit (3 to 4.10^{10} $M^{-1}s^{-1}$ in liquid ammonia) [17]. The perturbation arises again because the nucleophilic attack in eqn.18 competes with the diffusion controlled reduction of the σ-aryl radical $Ar^{\cdot}$ by $M^{\pm}$ which is the reductant in largest concentration in the solution where $Ar^{\cdot}$ is produced.

Yet not all nucleophiles can be studied along this procedure, because of requirements on the potential location of the ArNu wave as compared to the ArX one. In the practice these nucleophiles not attainable this way are investigated on the basis of competitive experiments involving a known reference nucleophile [17]. Thus for a given radical the

$$Ar^{\cdot} \quad \overset{k_1[Nu_1^-]}{\nearrow} \quad ArNu_1^{\cdot \pm} \overset{-e}{\rightleftharpoons} ArNu_2 \quad (16)$$

$$\overset{k_2[Nu_2^-]}{\searrow} \quad ArNu_2^{\cdot \pm} \overset{-e}{\rightleftharpoons} ArNu_2 \quad (17)$$

peak heigths of the two substituted aromatics, $ArNu_1$ and $ArNu_2$, as determined upon repetitive cyclic voltammetry, are proportional to their respective apparent first order rate of formation, i.e. $k_1[Nu_1^-]$ and $k_2[Nu_2^-]$. Thus the determination of k_2 ensues if k_1 is known.

CONCLUSION

This short presentation of several aspects of direct and indirect electrochemical kinetic methods, as applied to chemical reactivity problems, was voluntary limited for lack of space and for unity to few typical examples selected in organic chemical reactivity. Yet it must be born in mind that the described procedures constitutes the frame in which a large variety of important reactions from organic and organometallic chemistry can be examined and fully characterized. This is obviously true for electron transfer induced reactions, but also for other situations not directly related to electron transfers, but of great interest for the elaboration of new structure reactivity relationships.

In this connection let us recall that the recent availability of ultramicroelectrodes has opened new frontiers to electrochemistry, besides the obvious direct access to subnanosecond time scales. Indeed owing to the considerable decrease in ohmic drop, a large variety of media (low ionic strength [18], glasses, ...) or highly resistive solvents [19] can be used although this was though impossible just few years ago.

ACKNOWLEDGMENTS

It is a considerable pleasure to acknowledge the contributions to the above described work of many coworkers whose names are found in the reference list. Yet I want to particularly acknowledge the central role of Drs. J.M. Savéant, A. Thiebault in the elaboration of the conceptual tools presented in this lecture.

REFERENCES

1. See e.g. (a): J.K. Kochi, J. Organomet. Chem., 300 (1986) 139., and (b): M. Chanon, Bull. Soc. Chim. Fr., (1985) 209, and references therein.

2. See e.g. A.J. Bard and L.R. Faulkner, in "Electrochemical Methods", Wiley, New-York, 1980.

3. C.P. Andrieux and J.M. Savéant, "Electrochemical Reactions", in "Investigation of Rates and Mechanism of Reactions", Vol.6, 4/E, Part 2, C.F. Bernasconi, Ed., Wiley, New-York, 1986 ; pp. 305-390.

4. P. Delahay, in "Double Layer and Electrode Kinetics", Wiley, New-York, 1965.

5. (a) J.M. Savéant, J. Electroanal. Chem., 112 (1980) 175 ; (b) 143 (1983) 447 ; (c) C. Amatore, unpublished results.

6. C.P. Andrieux, C. Blocman, J.M. Dumas-Bouchiat, F.M'Halla and J.M. Savéant, J. Am. Chem. Soc., 10 (1980) 3806 and refs. therein.

7. Adapted from an original presentation by P.T. Kissinger, in "Laboratory Techniques in Electroanalytical Chemistry", P.T. Kissinger and W.R. Heineman, Eds., M. Dekker, New-York, 1984 ; p.6.

8. J.M. Savéant and D. Tessier, J. Electroanal. Chem., 77 (1977) 225.

9. J.O. Howell and R.M. Wightman, Anal. Chem., 56 (1984) 524.

10. H. Kojima and A.J. Bard, J. Am. Chem. Soc., 97 (1975) 6317.

11. C. Fabre, R. Fugnitto, H. Strzelecka, C.R. Acad. Sc., ser. C, 282 (1976) 175.

12. H. Kawata, Y. Suzuki and S. Niizuma, Tet. Lett., 27 (1986) 4489.

13. C. Amatore, A. Jutand and F. Pflüger, J. Electroanal. Chem., 218 (1987) 361.

14. See e.g. C. Amatore, J. Chaussard, J. Pinson, J.M. Savéant and A. Thiébault, J. Am. Chem. Soc., 101 (1079) 6012, and refs. therein.

15. C. Amatore, M. Gareil and J.M. Savéant, J. Electroanal. Chem., 147 (1983) 1.

16. J.M. Savéant, J. Am. Chem. Soc., in press.

17. See e.g. C. Amatore, M. Gareil, M.A. Oturan, J. Pinson, J.M. Savéant and A. Thiébault, J. Am. Chem. Soc., 107 (1985) 3451.

18. C. Amatore, M.R. Deakin and R.M. Wightman, J. Electroanal. Chem., in press.

19. L. Geng, A.G. Ewing, J.C. Jernigan and R.W. Murray, Anal. Chem., 58 (1986) 852.

20. C. Amatore, C. Combellas, F. Pflüger and A. Thiébault, unpublished results (1987).

DISCUSSION

<u>LAUNAY</u> - Could you tell us more about these ultra microelectrodes and why do they allow such very high scan rates ?
What size of microelectrodes do you typically use ?

<u>AMATORE</u> - We found that electrodes with radii of few micrometers are convenient for our purposes. All the data presented were obtained with platinum or gold electrodes of 5μm radii. Yet other groups (R.M. Wightman, A.M. Bond, Fleischman and Pons, ...) use smaller electrodes for more analytical purposes. 5μm seems a good compromise for our chemical reactivity purposes, when taking into account the case of preparation, of polishing the electrodes as well as the current magnitude detected (μA at scan rates of 200.000 V.s^{-1}).
As explained in the text the resistance, R, for a disk electrode varies like $\rho/(\pi \Gamma_0 \, tg\,\theta)$ where ρ is the solvent/electrolyte resistivity, r the radius of the disk electrode, and 2θ the angle of the current tube. On the other hand the current following through the electrode is at maximum proportional to its surface area. Thus the ohmic drop is proportional to r and tends to zero with the latter.
On the other hand capacitive currents are also proportional to the electrode surface area. Thus the ratio of faradaic over capacitive current is not affected. Yet the time constant, RC, of the capacitive current is proportional to the radius, r, of the electrode since $R\alpha(1/r)$ and $C\alpha(r)^2$. Thus natural deconvolution occurs between capacitive and faradaic currents occurs when r is made smaller and smaller.

<u>MARX</u> - Qu'est ce qui limite les durées de vie étudiables ? Est-ce l'électronique, ou bien la physique de diffusion des espèces ?

<u>AMATORE</u> - Electronics involve two stages : (i) purely electrochemical apparatus and (ii) acquisition of data. The electrochemical apparatus are home built to fulfill the required time constants wanted (imposed by other aspects developed hereafter). Acquisition of data is now our limit, mainly for economic reasons owing to the current price of 200 MHz digital scopes. In the time scales, we are investigating (up to 25 ns with <u>direct</u> and few ps with <u>indirect</u> methods) our limit is really related to the interpretation of our kinetic data in terms of classical diffusion-reaction models. indeed under 25 ns in direct methods electrical fields in the double layer play a significant role on the kinetics and thus on the exact meaning of the rate constants obtained. For the indirect methods the limit is smaller and corresponds to the reality of classical rate constants when the examined molecule is too short lived to achieve motions larger than a few Å. In such situations our feeling is that our interpretation of data becomes non valid.

IN VIVO N.M.R.

J.C. Beloeil

CNRS-ICSN
91190 Gif-sur-Yvette, France

INTRODUCTION

Nuclear Magnetic Resonance Spectroscopy has seen an appreciable development since the appearance of supraconducting high field magnets and the rapid evolution of computer hardware. The applications of N.M.R. in biology developed more slowly in line with the development of new techniques such as 2D N.M.R. This delay can be explained for technological reasons : classical supraconducting magnets were very ill adapted in particular to living systems, the average diameter of the working area for the sample being restricted to 40mm. There are, however, some more "psychological" reasons : the main characteristic of a good sample being its perfect homogeneity, it is obvious that living systems correspond poorly to this criterion. Medical imagery, which takes advantage of this inhomogeneity, has since brought N.M.R. spectroscopy in its wake. Hence amongst the developments being perfected spectroscopic imagery appears principally in assaying the doses of metabolites in a living being. N.M.R. gives access, <u>in a non-invasive manner</u> and in real time, to information on the intracellular medium from cell cultures [1,2] and perfused organs or even whole organisms including man.

Simplified theory

Nuclear Magnetic Resonance (N.M.R.) spectroscopy consists in studying atomic nuclei with a magnetic moment. The existence of a quantum mechanical eigenvalue, spin, characterises the nuclei with a magnetic moment. The main nuclei studied in Biology are collected in Table 1 :

Nucleus	Spin Quantum Number	Resonance frequency in MHz at 9.4 Teslas	Natural Abundance %
^{1}H	1/2	400	99.98
^{2}H	1	61	0.0156
^{13}C	1/2	100	1.1
^{15}N	1/2	40	0.36
^{19}F	1/2	376	100
^{23}Na	3/2	106	100
^{31}P	1/2	162	100
^{39}K	3/2	19	9.1

When a nucleus of spin $S=1/2$ is placed in a static magnetic field B_o, two states of differing energy a and b appear (fig.1), corresponding to the two distinct orientations of the moment with respect to the field B_o. The populations of these two levels are regulated according to Boltzmann's law. Only the nuclei corresponding to this very slight net difference are detectible by N.M.R. e.g. we can calculate, in the case of a proton, that only 1 in 10000 nuclei will be detected for a 1,4 Tesla magnetic field (proton resonance at 60 MHz). It is thus easy to understand that in spite of considerable progress in acquisition and treatment of signals, N.M.R. remains relatively unsensitive compared with mass spectrometry, fluorescence techniques or radioactivity. The latter involve a large proportion of the molecules in the sample in effect.

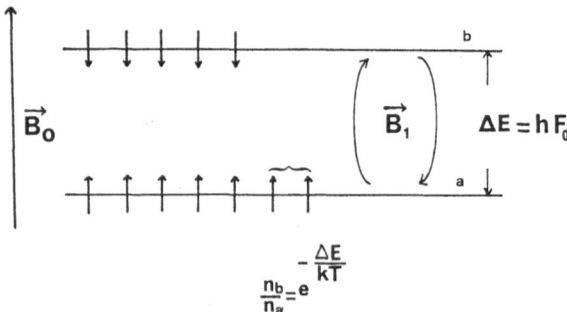

Fig. 1. Under the influence of a static magnetic field B_o, the magnetic moments take up antiparallel (b) and parallel (a) orientations with respect to B_o. The distribution of nuclei between these two levels obeys Boltzmann's law. A magnetic field B_1 of frequency F_o such that $\Delta E = h.F_o$ permits the nuclei in state a to be excited to state b.

According to the rule general to spectroscopy, a transition between two states at different energies can be induced on applying an electromagnetic field B_1 of the appropriate frequency such that :

$\Delta E = h.F_o$ (h : Planck's constant).

For N.M.R. F_o is in the radiofrequency range. With :

$\Delta E = \gamma.h.B_o$ (γ: the gyromagnetic constant, characteristic of the nucleus)

we can deduce : $F_o = \gamma.B_o$.

The resonance frequency is thus proportional to the static magnetic field and, via the gyromagnetic constant, is characteristic of the nucleus being studied. In practice, the nucleus is part of a molecule, it is hence surrounded by electrons which create a local screening magnetic field opposed to the principal state field B_o. This screening varies according to the nucleus' position whithin the molecule, such that nuclei in general experience difference effective fields :

$B_o(\text{eff.}) = (1 - \nabla).B_o$: (∇ : screen constant).

The resonance frequency of a nucleus (Larmor frequency) will hence be indicative of its nature, but also of its location within the molecule (Fig. 2) and even of the environment of the latter, indeed the presence of paramagnetic ions in the environment of the nuclei will apreciably alter their resonance frequency.

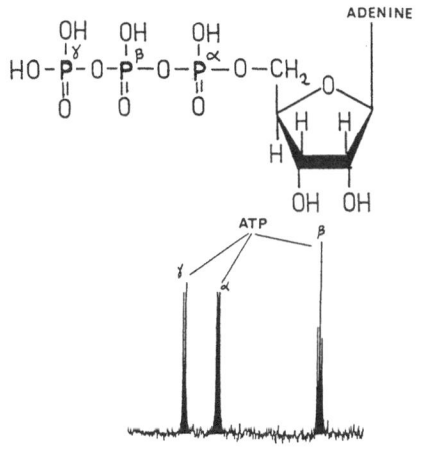

Fig. 2. ^{31}P N.M.R. spectrum of ATP, the three phosphorus atoms have different environments and resonate at different frequencies.

In order to characterise the resonance frequency, it is preferable to use a dimensionless number, i.e. the chemical shift, which is independent of B_o :

$$\delta = 10^6 \cdot (F - Fr)/Fr$$

(δ in ppm parts per million)

(Fr : the frequency of a reference signal).

In a real sample, there are groups of nuclei which resonate at different frequencies. In the vast majority of current spectrometers, a simultaneous excitation is obtained by applying a sinusoïdal field B_1 over a very short time t (fig. 3a) at a frequency equal to the middle frequency of the domain to be excited. A pulse of this type permits a frequency band 2/t wide to be excited. When the system returns to equilibrium, each nucleus emits an oscillating field at its own frequency F_o in turn. This field is detected by a simple tuned coil, the signal received comprises a damped sine wave (Fig. 3b).

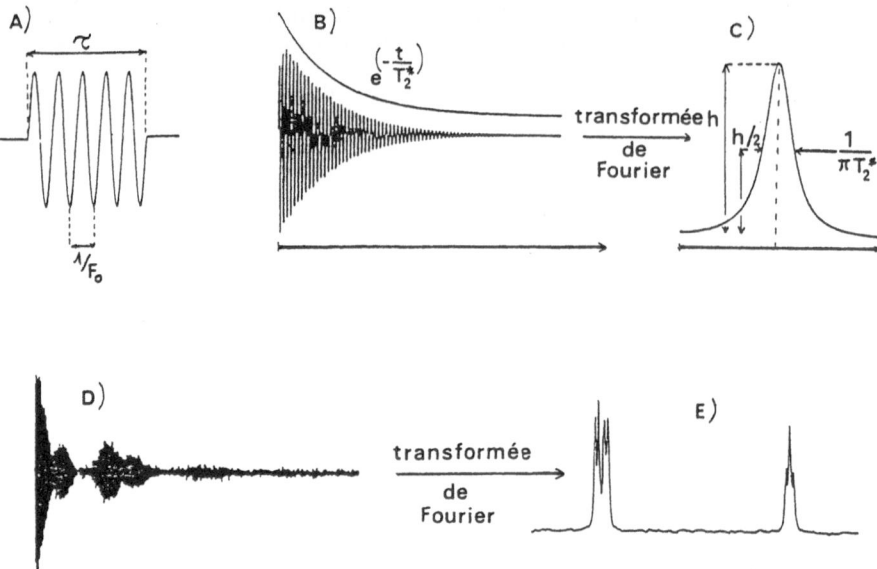

Fig. 3. a) exciting pulse (τ) of the radiofrequency field B_1 at the frequency F_o.

b) Free Induction Decay (F.I.D.) free precession signal corresponding to nuclei which resonate at this particular frequency.

c) Fourier Transformed spectrum of b).

d) FID corresponding to sets of nuclei resonating at distinct frequencies (interferogram).

e) Fourier Transform of d).

The sine waves corresponding to different groups of nuclei when superposed yield an interferogram which constitutes the F.I.D. (Fig. 3d). The best method of identifying the different frequencies present is to carry out a mathematical operation known as the Fourier Transform on the signal S(t). We thus obtain the resonance frequency spectrum of the sample (Fig. 3c and 3e).

The area of the peaks is proportional to the number of nuclei so that it is possible to dose by N.M.R. Without going into detail, we can say that the relaxation of the spin system (return towards equilibrium) being linked with the molecules' mobility, will give rise to broadened spectral lines if this mobility is decreased (and also if the homogeneity of the static field B_o decreases). These phenomena can lead to the complete disappearance of the lines. N.M.R. thus does not permit all the molecules of a cell to be detected but rather the mobile ones. Measurement of the characteristic relaxation parameters gives access to this mobility.

In the case of exchanged molecules, we must distinguish between two limiting cases, slow exchange (slow exchange rate compared with the resonance frequencies of the exchange sites) and fast exchange (faster exchange than this frequency difference). In the former case the resonance peaks of the entities exchanging will be separate, whereas in the latter we observe a unique peak at a chemical shift corresponding to the weighted average of the shifts of the exchanging sites (Fig. 4) ; this phenomenon applied to the phosphate ion detected in ^{31}P NMR, provides us with a pH probe.

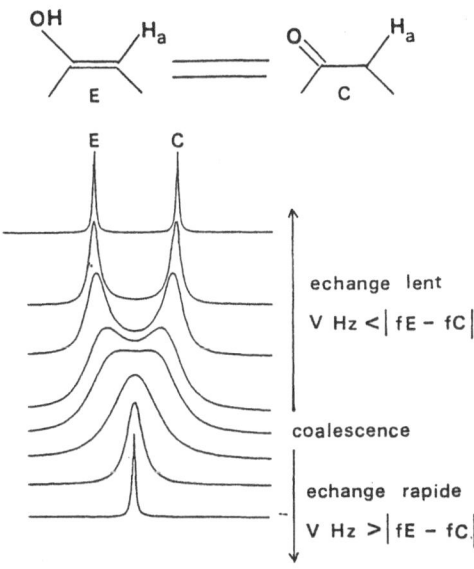

Fig. 4. Example of molecules in a structural exchange (ketone-enol). If the speed of exchange is low in comparison to the resonance frequencies of proton a in the two sites, we will observe two peaks in the spectrum. Conversely, if the exchange speed is high, we observe a single peak in the position corresponding to the weighted average of the resonance frequencies (weighted according to the abundances of C and E).

Special characteristics of "in vivo" N.M.R.

One consequence of the previous paragraph is that owing to the heterogeneity of the living medium, to the high viscosity of the intracellular medium, the spectra of living tissues will not be as well resolved as solution spectra (Fig. 6). Only small molecules (highly mobile in solution) and in high concentration (low sensitivity) will be accessible.

It will, of course be necessary to solve any problems specific to maintainaing the sample alive within the magnet.

The nuclei most frequently studied in biology are : ^{31}P, ^{13}C, ^{1}H, ^{23}Na, ^{19}F.

Phosphorus ^{31}P :

- relatively sensitive

- suitable for energy metabolism studies : ATP, creatine phosphate, inorganic phosphates, phosphoesters (frequently inseparable), ADP (under certain circumstances by difference).

- the spectra are simple

- pH measurement via the peak positions of the inorganic phosphate signal(s) (compartmentation)

Carbon ^{13}C :

- low natural abundance (1.1%)

- spectra in natural abundance possible but it is more usually used as a marker. As distinct from radioactivity measurements carbon ^{13}C N.M.R. gives access simultaneously to the location and structure of different metabolites of the marked compound, "in vivo".

The proton ^{1}H :

- has the greatest sensitivity

- necessitates the development of specialized techniques, in order to attenuate selectivity of the H_2O resonance (55M) to see the protonated metabolites (1 to 10mM)

- spectra are very complex, necessitating the development of advanced identification techniques (2D), this complexity arises from the large number of small molecules in sufficient concentration (>1mM) : lactate, N-acetylaspartate, GABA, creatine phosphate and creatine glutamate, taurine...

Sodium ^{23}Na :

- reasonably sensitive

- can measure the flux of sodium across a membrane (so long as displacement reactants are used, intracellular and extracellular sodium counterparts are separable).

Fluorine ^{19}F :

- highly sensitive

- essentially used as a marker, unlike carbon ^{13}C it does not provide access to the structure of the metabolites of the marked compound

- can be used as a pH probe.

To summarise, the information obtainable from N.M.R. studies of living systems such as : cell cultures, tissues, perfused organs, and complete organisms are :

- identification of components

- pH measurement

- detection and/or fixation of metal ions

- concentrations, kinetics, equilibria

- localisation of components and their pH (cellular compartmentalisation)

- molecular mobility

- imagery

A series of examples of applications carried out in our laboratories :

Plant cells [1] (in collaboration with the plant physiology laboratory at Gif-sur-Yvette).

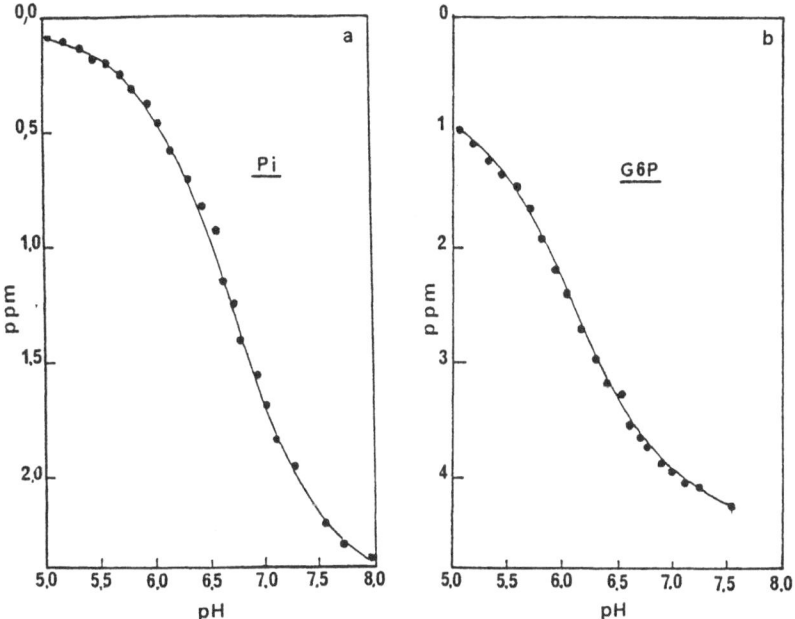

Fig. 5. a) Chemical shift of the inorganic phosphate signal versus pH (^{31}P N.M.R.)

b) Chemical shift of the signal due to the phosphate group in glucose-6-phosphate versus pH.

For this type of experiment, the extracellular medium is not renewed, whereas the oxygenation and maintenance in suspension of the cells is assured by a stream of air bubbles. For N.M.N. studies of the ^{31}P, the medium is depleted in phosphorus so that the intracellular phosphorus signal is not masked, and the cells cultivated in a manganese-free medium (a paramagnetic ion). Prior to studies on cells, calibrating curves providing the pH versus chemical shift of the ^{31}P of inorganic phosphorus and the phosphate group of glucose-6-phosphate were plotted from solutions as close as possible to the intracellular medium values for composition (Fig. 5). As demonstration we see that under anaerobic conditions, the cell consumes its glucose-6-phosphate, the cytoplasmic pH becoming acid, whilst the pH of the vacuole remains unchanged (Fig. 6).

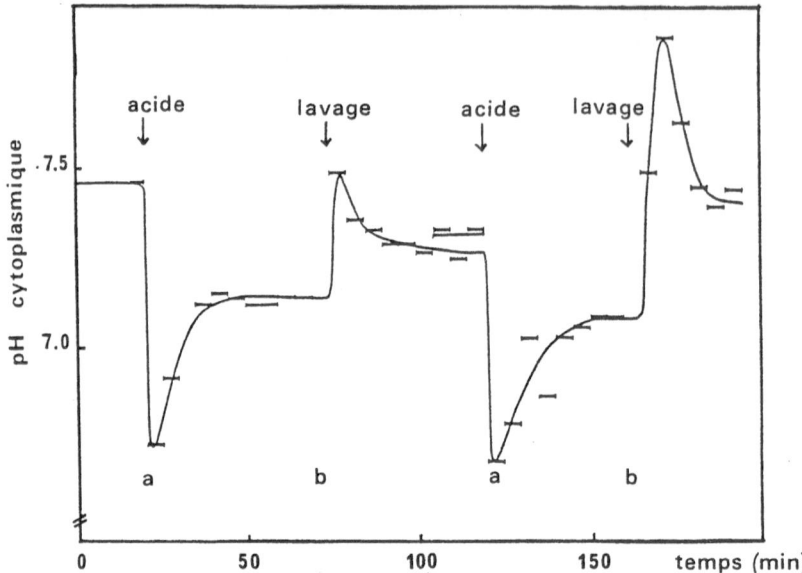

Fig. 6. a) ^{31}P N.M.R. spectrum of a normally oxygenated suspension of cells (Acer Pseudoplatanus)

b) The same suspension in the absence of oxygen. We observe that the G6P signal disapears and the cytoplasmic Pi signal alters its position corresponding to an acidification, whereas that of the vacuole does not move.

Note that, due to their different environment, the cytoplasmic and vacuole phosphates are completely separate, hence we have a means of access to cellular compartmentalisation. G6P is only present in the cytoplasm and can thus provide a confirmation of the pH measurement from the inorganic phosphate, when in sufficient concentration. A study of cytoplasmic pH regulation (in Fig. 7) shows that this pH can be selectively monitored during external pH evolutions.

In order to separate the sodium ^{23}Na of the external medium from the intracellular sodium, a chemical shift reagent must be used, i.e. a complex paramagnetic ion here DyTTHA is used, which contains Dysprosium (Fig. 8) which strongly displaces the resonance frequencies of Na ions which would be close by (^{23}Na N.M.R.). This complex cannot cross the membrane, hence we have access to the sodium flux.

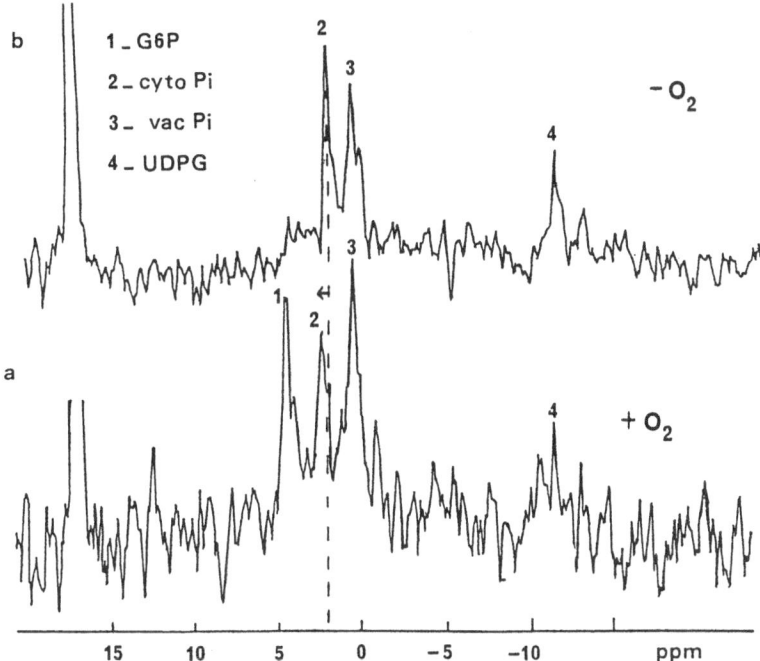

Fig. 7. Effect of introducing propionic acid on the cytoplasmic pH of Acer Pseudoplatanus.
a) propionic acid injection into the extracellular medium.
b) cells are filtered and washed in a medium devoid of acid.

Superfused brain slices of rat

(Collaboration with the nerve physiology laboratory at Gif-sur-Yvette).

Brain slices 0.5mm thick can be maintained alive by superfusion over several hours. This model allows the action of drug on neurones to be studied. There are no blood-brain barrier problems and the links between neurones are conserved, at least in part, e.g. we have studied the action of neurotransmitter amino acids (NMDA, kainic acid, glutamate...) on neurone metabolism (in Fig. 9). We have brought to light a dephosphorylation of phosphocreatine by creatine during the application of NMDA, with the parallel appearance of intracellular acidose and a hypersensitivity to hypoglycemia. The time evolution of spectra is represented in the form of a surface (Fig. 9f) permitting gains in effective sensitivity and allowing a minimum time of about 20s between spectra to be reached, allowing relatively rapid phenomena to be followed.

With the selective suppression of the H_2O signal it is possible to record proton spectra, alternating with ^{31}P spectra which permits to follow the evolution of other metabolites in parallel with phosphorylated ones (Fig. 10).

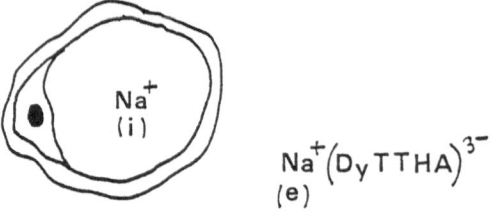

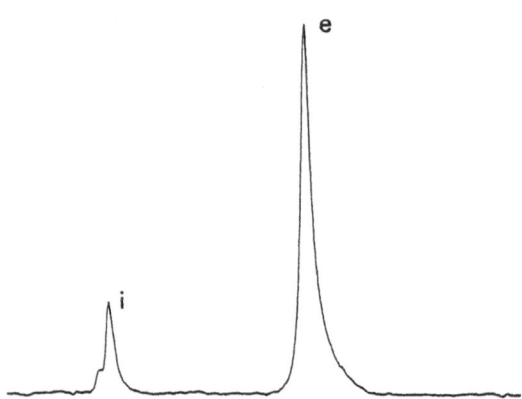

Fig. 8. N.M.R. spectrum of ^{23}Na in a suspension of isolated vacuoles, in the presence of a shift reagent in the external medium. The chemical shift of the sodium contained in the vacuoles (i) is different from that in the external medium (e) which is affected by the shift reagent (DyTTHA).

Study of the brain in the live rat

(Collaboration with the laboratory of physiology and cerebrovascular physio-pathology, Univ. Paris VII).

The study of living animals requires measurements apparatus and working technique appreciably different from those described above, their detailed description is beyond the scope of this paper. The measurement must be volume-selective, which is assured by an emission-reception coil said to be "a surface coil". This model can be compared to the previous one. The whole brain is observed in this case and we must take into account that the injected compounds may be metabolised before acting on the brain cells. All N.M.R. observation techniques ^{31}P, ^{1}H, ^{13}C, ^{23}Na..., are possible, although more difficult to carry out. In this way, we were able to study the action of hystotoxic hypoxia (KCN), of hypercapnia on the metabolism of brain cells, as well as certain aspects of ageing.

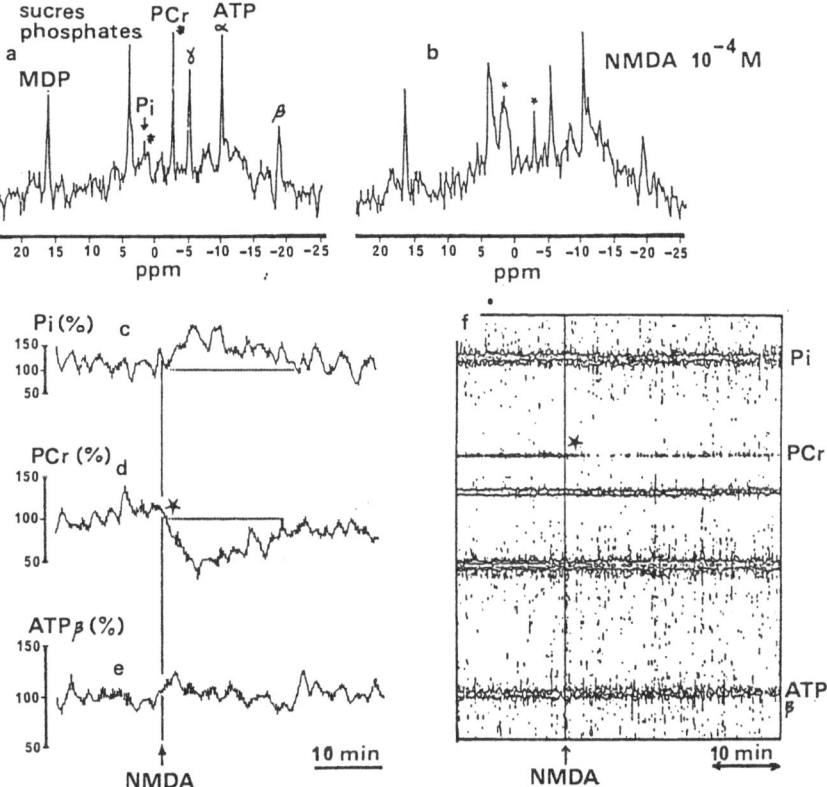

Fig. 9. a) N.M.R. ^{31}P spectrum of superfused brain slices of rat

b) effect of adding N-methyl-D-aspartic acid (NMDA) to the perfusion medium : the PCR signal diminishes, the Pi signal increases and the signals corresponding to ATP remain constant.

c) d) e) f) c, d and e correspond to horizontal sections in the f

representation of the time evolution of ^{31}P spectra. This representation (f) known as a contour plot is in fact a "plan view" of successive spectra placed one behind the other. A vertical gives the curves c, d, e of the evolution of the Pi, PCr, ATP signals of b. The representation f allows access in a single figure to complete time evolution of the experiment.

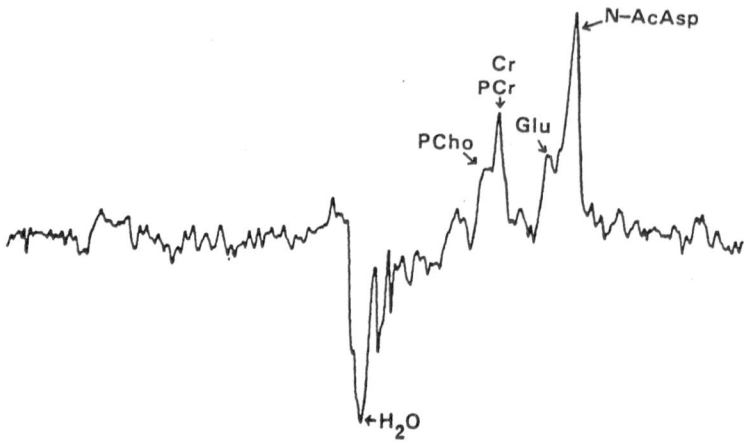

Fig. 10. Proton N.M.R. spectrum of rat brain slices (with specific suppression of the H_2O signal).

CONCLUSION

The selection of examples described above show the possibilities and limitations of N.M.R. spectroscopy applied to life studies. Its main advantage is without doubt its non-destructive and non-invasive character. In the present state of the art, its main limitation stems from its relatively low sensitivity. In the near future, we can expect a development of the spatial selectivity of N.M.R. spectroscopy, so that a study of intracellular metabolism will soon be possible for an internal part of an organism, i.e. in an animal or in man.

BASIC BIBLIOGRAPHY

Nuclear Magnetic Resonance and its Applications to Living Systems, David G. Gadian, Clarendon Press, Oxford, 1982.

REFERENCES

[1]   a. J. Guern, Y. Mathieu, M. Pean, C. Pasquier, J.C. Beloeil and J.Y. Lallemand, "Cytoplasmic pH Regulation in Acer pseudoplatanus Cells". I. A ^{31}P NMR Description of Acid-load Effects. Plant Physiology, 82, 840-845 (1986).
        b. Y. Mathieu, J. Guern, M. Pean, C. Pasquier, J.C. Beloeil and J.Y. Lallemand, "Cytoplasmic pH regulation in Acer pseudoplatanus Cells." II. Possible Mechanisms involved in pH Regulation during Acid-load. Plant Physiology, 82, 846-852 (1986).

[2]   C. Ducrocq, M. Lenfant, G.H. Werner, B. Gillet and J.C. Beloeil, "Fluoride Effects on ^{31}P NMR Spectra of Macrophages. Biochem. Biophys. Res. Com., 147, 519-525 (1987).

[3]   B. Barrère, P. Méric, M. Pérès, J. Seylaz, J.C. Beloeil, C. Pasquier and J.Y. Lallemand, "In vivo ^{31}P NMR Study of Cerebral Metabolism during Histotoxic Hypoxia in Mice." Metabolic Brain Disease (1987), to be published.

DISCUSSION

PRUD'HOMME - L'exposé montre l'intérêt de la RMN pour étudier les réacteurs chimiques que sont en particulier les êtres vivants.
Quelle résolution spatiale peut-on espérer pour les mesures locales ?

BELOEIL - La résolution spatiale, dans le cas de "l'imagerie spectrale" est limitée, entre autres, par la sensibilité de la mesure R.M.N. Dans le cas d'un appareil à très haut champ (9,4 T), disposant d'un volume de mesures relativement faible (cylindre de diamètre 8cm), pour un temps d'acquisition raisonnable (<15 mn) on peut avancer les valeurs de 10 mm³ pour le ^{31}P et 5 mm³ pour ^{1}H. Les études portant sur l'homme nécessitent des aimants de grande ouverture dont les performances sont, bien sûr, moindres ; la résolution spatiale y sera moins bonne.

AMATORE - 1) Dans la détection de molécules dans le cerveau, pouvez-vous distinguer entre les volumes extracellulaires et l'intérieur des neurones ?

2) Quelles gammes de concentration pouvez-vous atteindre, en particulier vis à vis de la détection de neurotransmetteurs dans les processus synaptiques ?

BELOEIL - 1) Il est possible de séparer les signaux RMN provenant d'une même espèce moléculaire, selon la localisation de celle-ci (intra ou extra cellulaire par exemple), à condition que l'environnement de la molécule change profondément (pH très différents, absence/présence d'ions paramagnétiques...) ; On peut provoquer cette différence par l'introduction d'ions paramagnétiques dans le milieu extracellulaire, à condition que ceux-ci ne perturbent pas le fonctionnement des cellules.

2) Très grossièrement, les concentrations minimum sont de l'ordre de 10^{-3} à 10^{-4} M en ce qui concerne ^{31}P et ^{1}H, la plus forte sensibilité de ^{1}H étant compensée par la difficulté d'observer les signaux des métabolites en présence de l'immense signal de H_2O. Par exemple, le GABA ou le glutamate sont détectables par RMN ^{1}H dans des tranches de cerveau perfusées. Le carbone ^{13}C peut être utilisé, dans des gammes de concentration similaires, mais à condition d'observer des molécules marquées préalablement, la molécule observée pouvant être un métabolite de la molécule introduite dans l'être vivant. Ce qui est important, ce n'est pas la concentration, mais le nombre de molécules présentes dans le volume de mesure ; on pourra donc observer des concentrations plus faibles, à condition d'augmenter le volume de l'échantillon si cela est possible.

NAKACHE - Vesicles are good models for biological membranes. Do you think that the NMR techniques can provide a means to measure the pH difference between the internal and external media of vesicles, provided that they countain P^{31} compound ? What is the threshold of the concentration of vesicles in that case ?

BELOEIL - Provided that vesicles contain molecules with ionisable phosphory-lated functions, measurement of intra and extra-vesicle pH is possible by ^{31}P NMR with a maximum precision of 0.05 on differential measures. In fact, the important parameter is not the concentration of ^{31}P but the total amount in the detection volume. For example, in a 9.4T wide bore magnet, when using a 20cm diameter tube, the minimum global concentration in the detection volume (5 cm³) would be about 10^{-4} M.

STUDY OF THE SYSTEM ACETONITRILE-WATER

IN THE ONE-PHASE PRETRANSITIONAL REGION

René Diguet

Université Nancy I
Chimie Théorique, U.A. 510 CNRS
F-54506 Vandoeuvre-lès-Nancy Cedex (France)

INTRODUCTION

Acetonitrile-water mixtures have often been employed as solvent media. Although their behaviour has previously been investigated by use of a variety of experimental techniques at several temperatures, relatively little is known of their general properties in the one-phase precritical and critical regions. The coexistence curve has been previously published. The system acetonitrile-water has an upper critical solution temperature at 272.1K when the mole fraction of water is 0.62 (Schneider, 1964; Armitage et al., 1968). Densities, viscosities and static dielectric constants of acetonitrile-water mixtures have only been determined up to 278.15K (Moreau et al., 1975, 1976). An extension of such measurements into the precritical and critical regions was desirable in order to gain a better understanding of the structural interactions in these binary hydroorganic solutions. Thus it was undertaken to measure the density (ρ), the viscosity (η) and the static dielectric constant (ϵ) at normal pressure up to the transition temperature, within 4 degrees of the critical temperature, for the acetonitrile-water mixtures with the composition in the mole% of water range $40 < x < 82$.

EXPERIMENTAL

Density measurements were carried out using a commercially available digital precision densimeter DMA 10 (Anton Parr) with an estimated accuracy of ± 0.03%.

The static dielectric constant has been measured with a conventional dipolemeter DM 01 (WTW) working at 2MHz with a vertical cylindrical condenser especially designed to allow a great temperature stability (better than 0.03°C) and correct determination in situ of the temperature of the solution using either a thermocoax calibrated to 0.05°C or a conventional mercury thermometer. The estimated accuracy is ± 0.5%.

The viscosity was determined with a thermostated Ubbelhode viscometer with an accuracy to 0.5%.

Some measurements of surface tensions have also been performed using a commercially available tensiometer (Prolabo). Their reproducibility is better than 0.5%.

In making the refractive index measurements, a thermostated Pulfrich refractometer (Zeiss) with a sodium lamp was used. Near the transitional region the concentration gradient in the solution produces the formation of shadows and rays on the image.

PMR spectra were obtained with a spectrometer operating at 60MHz and without spinning the sample tube. Temperature were controlled within ± 0.5°C by adjusting a flow rate of nitrogen gas stream precooled.

In addition sealed glass tubes containing weighed amounts of the two component fluids (MeCN Uvasol;bidistilled water) were slowly cooled until phase separation occurs. In this way the transition temperature has been determined by visual observation.

RESULTS AND DISCUSSION

Coexistence curve   (Fig.1)

Very similar results can be easily obtained using the apparent discontinuities observed at the phase separation, in the temperature dependence of the density or static dielectric constant. The top critical point of the coexistence curve corresponds to a mixture containing x_c=63.5 ± 0.1 mole% of water and to the temperature (272.05 ± 0.10)K from the visual observations and the density measurements. The static dielectric constant measurements however led to a slightly upper critical temperature of (272.20 ± 0.10)K. The value of the experimental critical temperature is dependent of the chosen order parameter (Yvon, 1972). The difference in density $\Delta\rho$,   volume fraction $\Delta\varphi = \Delta\rho/\rho$(water)$-\rho$ (acetonitrile), as suggested by Green (1978), and static dielectric constant $\Delta\varepsilon$ are well fitted for the reduced temperature $t=(T_t-T_c)/T_c$< 0.01 by the expression Δp= Bt^β. The critical exponent β characterises the approach to zero of the difference in order parameter (p) between the two mutual compositions for which transition occurs at the same temperature T_t< T_c. The results of the analysis are given in Table 1.

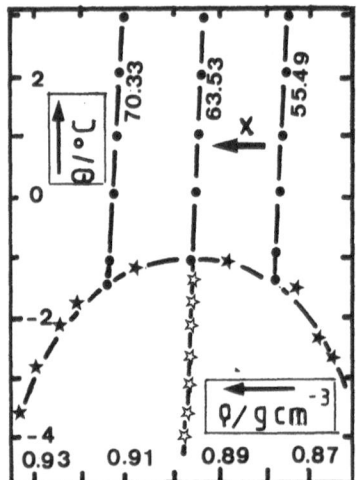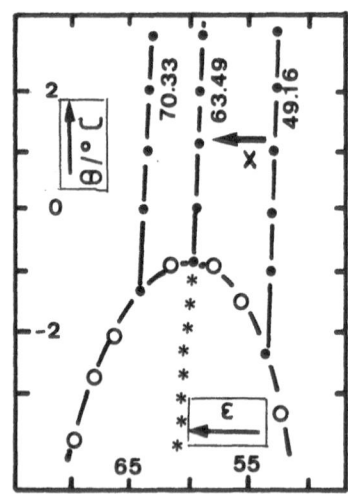

Fig.1. Coexistence curve of acetonitrile + water: densities ρ (left) and static dielectric constants ε (right) at the transition temperature ($\star$,O) and in the one-phase region ($\bullet$). Diameter of the coexistence curve ($\star$, $*$). Composition x: mole% of water.

Table 1. The fits of the difference in the order parameter (p) to the simple scaling function $\Delta p = Bt^{\beta}$

Δp	β	B	T_c/K
$\Delta\varphi$	0.360±0.030	1.95±0.20	272.05
$\Delta\rho$	0.360±0.030	0.41±0.07	272.05
$\Delta\varepsilon$	0.390±0.010	8.66±1.12	272.20

Along the diameter of the coexistence curve, the density data can be fitted to a straight line $\sum\rho/2 = At + \rho_c$ with $\rho_c = 0.8967\pm0.0001$ g/cm^3 and $A = 0.275\pm0.020$. The static dielectric constant data show an anomaly $\sum\varepsilon/2 = Ct^{1-\alpha} + \varepsilon_c$. $\varepsilon_c = 59,7$, $C = 47\pm13$ and $1-\alpha = 0.86\pm0.09$. $\sum$: sum of the conjugate compositions.

Density measurements (Fig.2)

The temperature dependence of the density far from the consolute point can be well approximated by a linear law within the accuracy of the data; a very small positive deviation is to be noticed in a pretransitional range, several degrees above the phase separation. The latter is manifested by a sudden break (increase or decrease depending on the composition of the mixture). The results are similar to those previously reported for other systems in our laboratory (Hollecker et al., 1975). Along the constant x curves, the ρ values can be well fitted to the linear function $|\rho - \rho_t| = Ct$ for $t < 0.01$ (see Table 2).

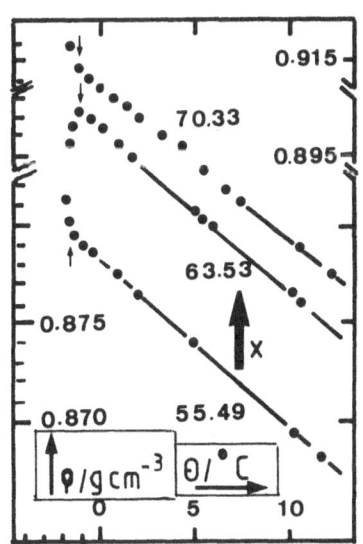 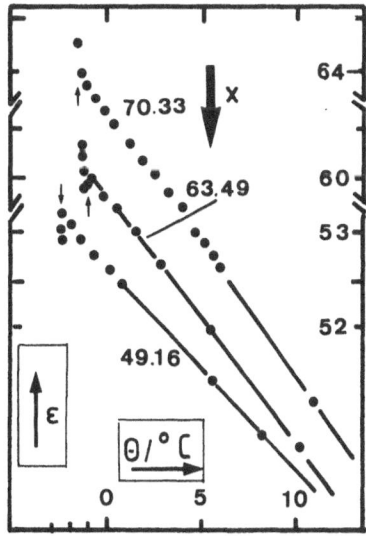

Fig. 2. Density ρ (left) and static dielectric constant ε (right) versus temperature θ for acetonitrile + water solutions. x: mole% of water. ↑: estimated transition point.

Table 2. Results of fits $|\rho - \rho_t| = Ct$

x	T_t/K	$\rho_t/g\ cm^{-3}$	$C/g\ cm^{-3}$
63.53	272.05±0.05	0.8967±0.0001	0.230±0.020
70.33	271.75±0.05	0.9150±0.0001	0.245±0.020
55.49	271.65±0.05	0.9800±0.0001	0.275±0.020

Static dielectric constant measurements (Fig.2)

Far from the transitional point, the results can also be accurately fitted with a linear law (just as for regular solutions). Again a very significant positive deviation (reaching a maximum at the critical temperature) become perceptible about 2°C above the phase separation, which is likewise manifested by a sudden decrease of ε if the water mole% x is less than the critical one x_c or a sudden increase of ε if $x > x_c$. The latter phenomena reflect macroscopic inhomogeneities induced by gravity and depend on the geometry of the cell. Similar results with other systems have been previously published (Hollecker et al, 1975). The concomitant analysis of the pretransitional variations of ρ and ε supports quite well the assumption of a strong increase of dipole-dipole interactions. The simple approach to the asymptotic dielectric behaviour, on the basis of a phenomenological droplet model, has been useful to provide a simple physical picture of the macroscopic nature of fluids in the precritical region (Goulon et al., 1979). Within the accuracy of the data, the ε results can be fitted along the x constant curves to the functional form $|\varepsilon - \varepsilon_t| = Dt^{1-\alpha}$ with $1-\alpha = 0.86\pm0.09$ (See Table 3).

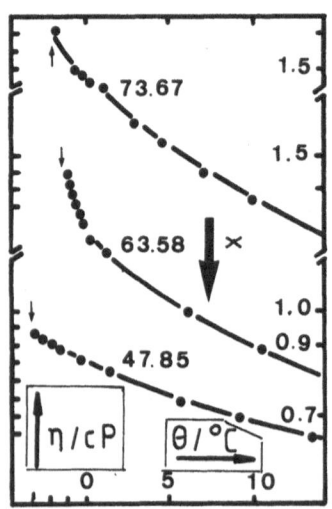
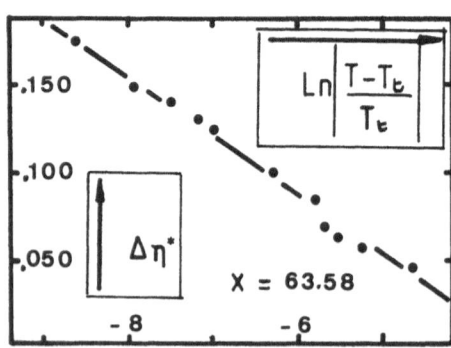

Fig. 3 a) Viscosity η versus temperature θ for acetonitrile+water mixtures. Composition x: mole% of water. ↑: estimated transition point.

3 b) The anomaly viscosity $\Delta\eta^* = \Delta\eta/\eta$id, c as a function of Ln t at the concentration x=63.58 (mole% of water) near the critical composition. t: reduced temperature.

Table 3. Results of fits to $\quad |\varepsilon - \varepsilon_t| = Dt^{1-\alpha}$

x	T_t/K	ε_t	D
63.49	272.20±0.05	59.70±0.10	44±13
70.33	271.80±0.05	64.10±0.10	44±13
49.16	270.75±0.05	53.20±0.10	34±13

Viscosity measurements (Fig. 3)

The concentration dependence of the viscosity is not linear at 273K.
The deviations from linearity are negative from x=0 to 65 and positive from
x=65 to 100 mole% of water and become pronouncedly large as the temperature
decreases. In addition, a moderate maximum occurs from x=55 to 75, which is
attributable to the circumstance that the system forms a critical solution.
For temperatures close to the critical temperature, the temperature dependence of the viscosity can be approximated by $Ln\eta = A+Bt$. Near the critical point
the viscosity shows a marked increase over the ideal viscosity η id, c to be
expected in absence of an anomaly. Within the limits of our experimental uncertainty, the anomaly of the reduced viscosity $\Delta\eta^* = \eta - \eta$ id, x/η id, c
(Sengers, 1972) may be approximated by a logarithmic divergence (Fig. 3b). On
an other hand, the viscosity activation energy exhibits a maximum in the critical concentration region, decreasing with increasing temperature.

Surface tension measurements (Fig. 4)

Within the accuracy of our data, the temperature dependence of the surface tension γ, can be well represented with a linear law to about 4°C above
the transition temperature. In the vicinity of the phase separation, a significant negative deviation occurs. However, approaching the critical point,

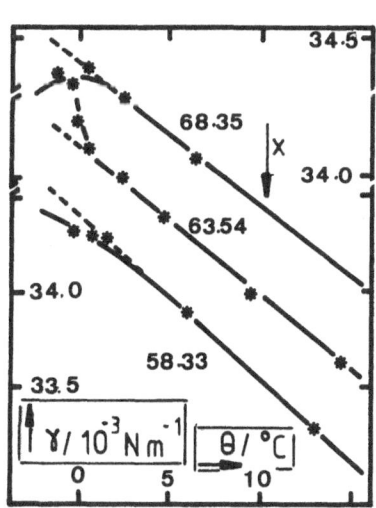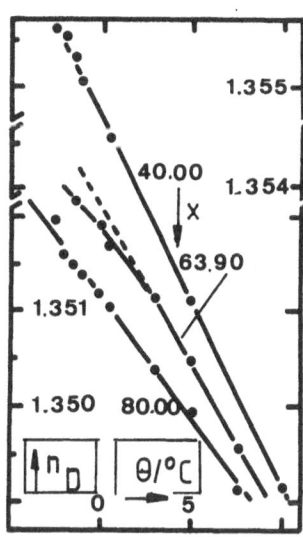

Fig. 4. Temperature dependence of the surface tension γ (left)
and of the refractive index n_D (right) for acetonitrile
and water mixtures of composition x (mole% of water).

the surface tension shows a marked increasing.

Refractive index measurements (Fig. 4)

The main observation of the study measurements of the temperature dependence of the refractive index n_D, in the homogeneous one fluid phase region has been the marked negative deviation near the critical point.

PMR Spectra (Fig. 5)

The signals due to water and acetonitrile shifted to lower fields as the temperature decreased from 20°C to -1°C. These results are in agreement with a previously published study (Morinaga et al., 1974) but our measurements did not exhibit splitting near the critical mixing temperature, although turbidity was detected with nacked eyes in the sample tube. The results concerned only a mixture with a composition in the vicinity of the critical composition.

Miscelleanous experiments

Concerning the valence absorption band C≡N of acetonitrile recorded with commercially available IR and Raman laser spectrometers, we have not observed an anomalously great width near the critical temperature, as expected (Mardaeva, 1974). The thickness of the cell containing the mixture with a composition near the critical one was 22 μ.

Except for the mixtures with x > 90 mole% of water, our estimated enthalpies of mixing at 273K were positive. These observations are in qualitative agreement with previously reported results (Morcom et al., 1969).

With regards to the above experimental data, our most probable transition temperatures T_t for the acetonitrile+water system are listed in Table 4. They correspond to the mean arithmetic value T_t (ρ) and T_t (ε), these latter

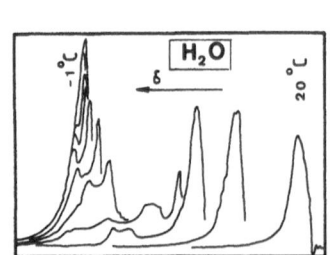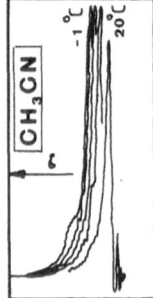

Fig. 5. Temperature dependence of the proton signals of water (left) and of acetonitrile (right) for a mixture with x=66.02 (mole% of water). δ : downfield chemical shift.

Table 4. Transition temperatures $\theta_t = T_t - 273.15$

x	θ_t	x	θ_t	x	θ_t
42.03	-4.60	58.33	-1.20	70.33	-1.40
45.72	-3.50	60.80	-1.15	72.90	-1.75
47.85	-2.75	62.08	-1.12_5	73.67	-2.00
49.16	-2.40	63.50	-1.10	74.50	-2.20
53.07	-1.60	66.02	-1.12_5	77.35	-3.55
55.49	-1.50	68.35	-1.20	78.00	-3.95

data being only different in the composition range $56 < x < 70$. The critical coordinates are $T_c = 272.10 \pm 0.20K$ and $x_c = 63.50 \pm 0.10$ mole% of water. In order to the similar functional behaviour observed for the reduced temperature t-dependence of ρ and ε in the two phase region (diameter of the coexistence curve-Fig.1) and in the one phase region (Tables 2-3), it was very interesting to observe that, within the uncertainty of the experimental data, the x-dependence curves of the physical properties considered in this work ($\rho, \varepsilon, \eta, n_D, \Delta H$, excess volumes,...) at temperatures from 298K to 273K, exhibit also a rectilinear diameter if one takes the both composition values of the conjugate points of the coexistence curve (Table 4), i. e;, to a first approximation: (63.5;55-71;50-75;45-78;40-80;35-83). This latter domain corresponds to the intermediate structural range previously reported (Moreau et al., 1976).

CONCLUSION

Coexistence curve has been investigated using density, viscosity and static dielectric constant measurements. The observations and the estimated asymptotic exponents in the precritical and critical regions are close to that recently reported (Beysens, 1982; Sengers, 1982). On an other hand, it appears that the isothermal concentration dependence of physical properties can be well interpreted in terms of the conjugate or mutual concentrations. This description agrees with the suggested existence of structural ranges to explain the interactions in the acetonitrile-water mixtures. It would be important to test the above analysis with other hydroorganic mixtures through their phase diagram and miscibility curve.

References

Armitage, D., Blandamer, M., Foster, N., Hidden, N., Morcom, K., Symons, M., and Wooten, M., 1968, Trans. Farad. Soc., 64:1193.
Beysens, D., 1982, in "Phase transitions," M. Levy, J.C. Le Gouillou and J. Zinn-Justin, ed., Plenum Press, New York.
Goulon, J., Greffe, J.L. and Oxtoby, D., 1979, J. Chem. Phys., 70:4742.
Greer, S., 1976, Physical Rewiev A, 14:1770.
Hollecker, M., Goulon, J., Thiebaut, J.M. and Rivail, J.L., 1975, Chemical Physics, 11:99.
Mardaeva, I. and Bondarev, A., 1974, Opt. Spectrosc., 36:713.
Morcom, K. and Smith, R., 1969, J. Chem. Thermodynamics, 1:503.
Moreau, C. and Douhéret, G., 1976, J. Chem. Thermodynamics, 8:403.
Moreau, C. and Douhéret, G., 1975, Thermochimica Acta, 13:385.
Morinaga, K., Miyaji, K. and Tsurumi, M., 1974, Chemistry Letters, 1381.
Schneider, G., 1964, Z. Phys. Chem., 41:327.
Sengers, J., 1982, in "Phase transitions," M. Levy, J.C. Le Gouillou and J. Zinn-Justin, ed., Plenum Press, New York.
Sengers, J., 1972, Ber. Bunsenges. Phys. Chem., 76:234.
Yvon, J., 1972, Ber. Bunsenges. Phys. Chem., 76:179.

EFFECT OF ELECTROSTATIC INTERACTIONS ON ELECTRON-TRANSFER REACTIONS

Bernard Hickel

CEA - CEN/SACLAY IRDI/DESICP/DPC/SCM UA CNRS 331
91191 Gif S/Yvette Cedex France

INTRODUCTION

Electrostatic interactions play an important role in the rate of ionic reactions, not only through the activity coefficients which depend of the ionic strength of the medium (1) but more directly the coulombic forces between the ionic reactants change the collision frequency. The first treatment of the effect of electrostatic interactions was given by Christansen and Scatchard (2) (equation I)

$$(I) \qquad k = k_o \ e^{-Q} \qquad\qquad Q = \frac{N Z_A Z_B \ e^2}{\varepsilon \ d \ R \ T}$$

where N is the Avogadro number, Z_A and Z_B the charge of the reacting species e the electronic charge, ε the dielectric constant of the solvent and d the reaction distances k_o would be the rate in a medium of infinite dielectric constant. This equation predicts that a plot of log k against $\frac{1}{\varepsilon}$ is a straight line and the slope gives d the reaction distance.

When the reaction is diffusion controlled equation I is no longer valid and is replaced by the Debye equation (3) (equation II).

$$k = k_d \ x \ \frac{e^Q}{e^Q - 1} \qquad \text{where} \qquad k_d = \frac{4 \ \pi \ N}{1 \ 000} \ (D_A + D_B) \ d$$

is the Smoluchowski equation for diffusion controlled reactions in absence of electrostatic interactions (4) and the second member of equation II is a correction factor which accounts for the effect of electrostatic interactions. D_A and D_B are the diffusion coefficients of the ions. The comparaison between the two equations for the same reaction distance is given in figure 1.

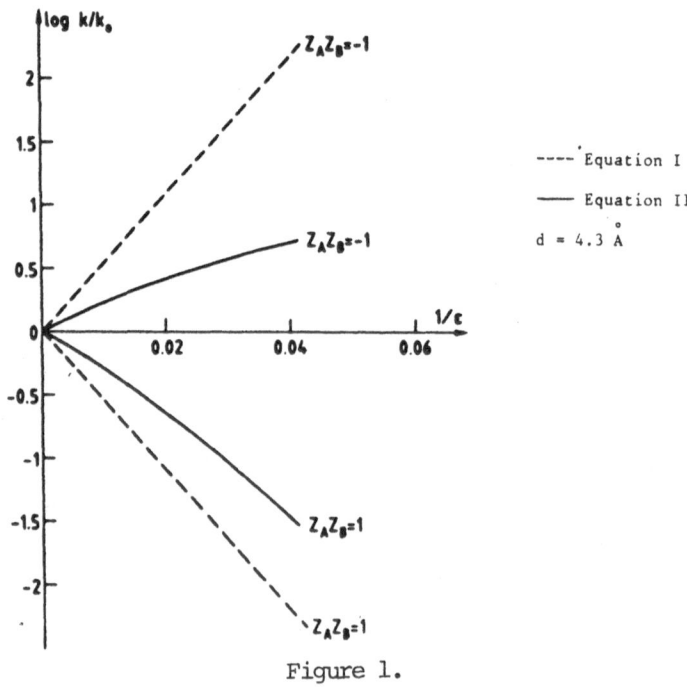

Figure 1.

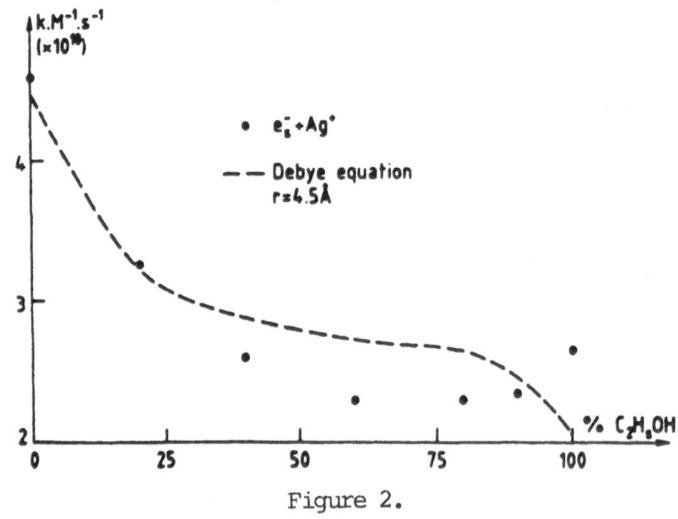

Figure 2.

We have studied the effect of electrostatic interaction on the rate constants of solvated electrons with anions and cations in water-ethanol mixtures where the dielectric constant change from 80 to 25 at room temperature. Water-ethanol mixtures have two advantages : first the diffusion coefficients of solvated electron are known (5) and the diffusion coefficients of other ions are often available from conductivity measurements. Secondly the absorption spectra of the solvated electron change very little with the composition of the mixture (6). This means that the solvation energy of the electron, which can play an important role in the rate, (7) remains approximatly constant for the whole range of composition.

EXPERIMENTAL SECTION

Solvated electrons were produced by pulse radiolysis of the solution using a modified Febetron 707 delivering single pulses of electrons in the energy range 1.6 - 1.8 MeV. The total duration of the electron pulse is less than 20 ns. The irradiation cell of high purety silica was rectangular and the optical path was 2.2 cm. Experimental details on the fast spectrophotometric detection system are given in previous publications (8) (9). The dose determined from the initial absorption of the solvated electron varied from 5 to 15 krads.

The concentration of solvated electrons was always much lower than the solute concentration to ensure that the bimolecular reaction follows pseudo first order kinetics. The rate constants were determined from the decay of the absorption of the solvated electron in solutions containing various concentrations of solutes after subtracting the decay arising from the solvent. Extrapolation to zero ionic strength was made according to the Brönsted Bjerrum equation (10).

RESULTS AND DISCUSSION

1/ Reaction with Ag^+

The reaction

$$e^-_s + Ag^+ \rightarrow Ag^0$$

is diffusion controlled in water and in ethanol (11). In water ethanol mixtures the rate remains diffusion controlled (figure 2) and the deviations from the Debye equation are smaller than 30 %. The reaction radius 4.5 Å is close to the sum of $(r_{e^-_s} + r_{Ag^+}) = 4$ Å.

The decrease in the rate constant from water to ethanol is not due to the lower thermodynamic activity of Ag^+ in alcohols than in water (12) but to the decrease of the diffusion coefficients of e^-_s and Ag^+. This effect is stronger than the increase due to the electrostatic interactions (figure 1) and the net effect is a small decrease in the rate. When $Z_A Z_B = 1$ the effect of diffusion coefficients and electrostatic interactions act in the same direction and the rate constant decreases by two orders of magnitude from water to ethanol (13) .

2/ Reaction with NH_4^+

In water the reaction

$$e^-_s + NH_4^+ \rightarrow NH_3 + H$$

is comparatively slow $k < 10^7 \ M^{-1}s^{-1}$ and becomes almost diffusion controlled in ethanol (11) .

In that part which is not diffusion controlled the rate apparently follows the equation I (figure 3 but the slope gives a reaction radius of 2 Å which is smaller than the sum of $(r_{e^-_s} + r_{NH_4^+}) = 4.2$ Å. When the concentration of ethanol increases the rate constant is faster than expected. Two experiments were performed in water-methanol mixtures and the rate constants were also faster than expected. This cannot be due to the formation of ions pairs between NH_4^+ and Cl^- which would screen the charge on NH_4^+ and hence decrease the reaction rate.

3/ Reaction with BrO_3^-

The reaction

$$e^-_s + BrO_3^- \xrightarrow{H_2O} BrO_2 + 2 \ OH^-$$

is fast but not diffusion controlled in water. When $KBrO_3$ is used deviations from equation I are observed when the dielectric constant is lower than 40. The rate constants are faster than expected. But in this case formation of ion pairs

$$BrO_3^- + K^+ \ \underset{\leftarrow}{\overset{\rightarrow}{\rightleftharpoons}} \ \left| BrO_3^- \ K^+ \right|$$

would increase the rate of reaction with the solvated electron (8) . To check this hypothesis cryptand 222 was added to the solutions. The cryptand 222 has a three-dimensionnal cavity of 1.4 Å and complexes strongly K^+ (14) . The complex which is nearly spherical has a radius of about 5 Å and does not easily form ion pairs with anions (15) .

116

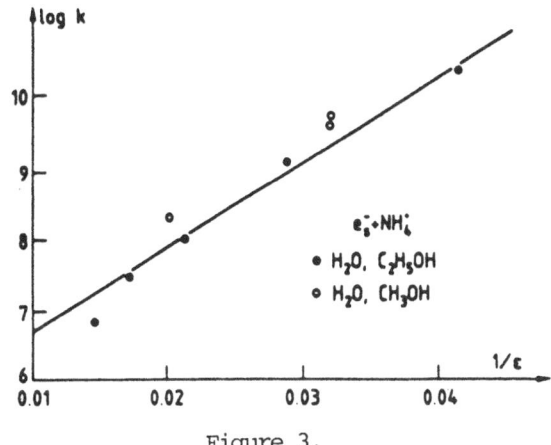

Figure 3.

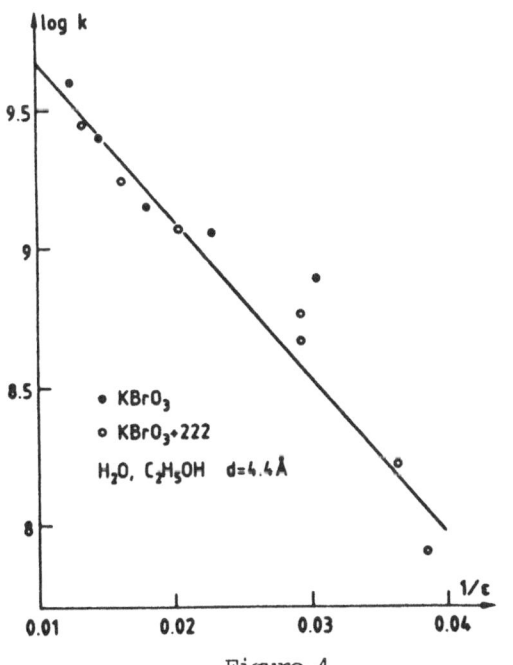

Figure 4.

The cryptand 222 does not react with e^-_s (16) . In the region of high dielectric constant addition of 222 does not change the rate constant between e^-_s and BrO_3^- (figure 4) , but for $\varepsilon < 40$ the rate is slower than in absence of 222. A plot of log k agains $\frac{1}{\varepsilon}$ gives a reasonnably straight line with a reaction distance of 4.4 $\overset{o}{A}$.

NEAR DIFFUSION CONTROLLED REACTIONS

When the rate of diffusion and the rate of the chemical reaction are of the same magnitude the rate constant k is given by the equation

(III)
$$\frac{1}{k} = \frac{1}{k_d} + \frac{1}{k_c}$$

where k_d and k_c are the rate constants respectively for the diffusion controlled reaction and for the chemical reaction (17) . For ionic reaction k_c and k_d are given by equations I and II respectively.

An example is given by the reaction

$$e^-_s + H^+ \rightarrow H$$

which is not truly diffusion controlled in water. In water-ethanol mixtures the rate is almost independant of the dielectric constant until the mixture is 99 % in weight of ethanol (13) . In this range the proton remains bonded to a water molecule and the apparent lack of electrostatic effect is due to a compensation between k_c and k_d (Table I) .

wt % C_2H_5OH	0	50	80	99
k_d in 10^{10} $M^{-1}s^{-1}$	9,5	4	3,4	3,1
k_c in 10^{10} $M^{-1}s^{-1}$	3,2	8,36	32,3	122
k equation III	2,4	2,4	3,1	3
k $_{exp.}$ (ref. 13)	2,4	2,5	2,9	2,8

$r_{e^-_s} + r_{H^+} = 4,3$ A $k_o = 6.1 \ 10^9 \ M^{-1}s^{-1}$

REFERENCES

(1) B. Perlmutter - Hayman, Progress in Reaction Kinetics, Vol. 6, G. Porter Ed. Pergamon Press 1972 p. 239

(2) G. Scatchard, Chem. Rev. 10, 229 (1932)

(3) P. Debye, Trans. Electrochem. Soc. 82, 265 (1942)

(4) R.M. Noyes, Progress in Reaction Kinetics Vol. 1, G. Porter Ed. Pergamon Press 1961 p. 129

(5) O.I. Micic and B. Cercek, J. Phys. Chem. 87, 823 (1977) (The Diffusion Coefficients of e^-_s in Methanol and Ethanol Water Mixtures have been inverted)

(6) S. Arai and M.C. Sauer, J. Chem. Phys. 44, 2297 (1966)

(7) J. Cygler and G.R. Freeman, Can. J. Chem. 62, 1265 (1984)

(8) B. Hickel, J. Phys. Chem. 82, 1005 (1978)

(9) J.L. Marignier and B. Hickel, J.Phys. Chem. 88, 5376 (1984)

(10) E.S. Amis, J.F. Hinton, Solvent Effects on Chemical Phenomena Vol. 1 Academic Press New-York 1973 p. 227

(11) H.A. Scharwz et P.S. Gill, J. Phys. Chem. 81, 22 (1977)

(12) E.J. Hart and M. Anbar, The Hydrated Electron, Wiley New-York 1970 p. 173

(13) F. Barat, L. Gilles, B. Hickel and B. Lesigne, J. Phys. Chem. 77, 1711 (1973)

(14) J.M. Lehn and J.P. Sauvage, J. Am. Chem. Soc. 97, 6700 (1975)

(15) J.M. Lehn, Acc. Chem. Res. 11, 49 (1978)

(16) P. Lardinois, Y. Rebena and B. Hickel, Nouv. J. Chim. to be published

(17) R.A. Marcus, Discuss. Faraday Soc. 29, 129 (1960)

PULSE RADIOLYSIS STUDY OF AQUEOUS SOLUTION REACTIVITY OF TRANSIENT SPECIES OF BIOLOGICAL INTEREST: REACTIONS WITH VARIOUS SUBSTRATES

A. Sekaki [*], K. Benzineb [*], G. Machtalère [*], C.Houée-Levin [*], M. Gardès-Albert [*], C. Ferradini [*] et B. Hickel [**]

[*] Laboratoire de Chimie Physique, Université René Descartes, 45, rue des Saints-Pères F-75270 Paris Cedex 06

[**] CEA IRDI/DESICP/DPC/SCM/UA 331, CEN Saclay F-91191 Gif sur Yvette Cedex

SUMMARY

We studied by pulse radiolysis three aspects of the oxydoreduction of three antitumour drugs whose mechanisms of action imply one-electron oxidative or reductive activations. The three antitumour drugs were BD40, daunurubicin, and mitomycin C.

We showed that when BD 40 is oxidized by OH free radicals the resulting BD 40 free radical reacts rapidly with superoxide anions.

When daunorubicin is intercalated in a protein, apo-riboflavin binding protein, COO-free radicals reduce both the antibiotic and a disulfide bond of the protein. However we had indications of intramolecular electron transfer between the protein and the semiquinone daunorubicin free radical.

The comparison between the reduction of mitomycin C with and without DNA allowed to identify the active transient which reacts specifically with DNA.

For each drug a kinetic scheme is proposed and the rate constants are determined.

PULSE RADIOLYSIS STUDY OF AQUEOUS SOLUTION REACTIVITY OF TRANSIENT SPECIES OF BIOLOGICAL INTEREST: REACTIONS WITH VARIOUS SUBSTRATES

A. Sekaki*, K. Benzineb*, G. Machtalere*, C. Houée-Levin*, M. Gardès-Albert*, C. Ferradini* and B. Hickel**

*Laboratoire de Chimie Physique, Université René Descartes 45, rue des Saints-Pères F-75270 Paris Cedex06

**CEA IRDI/DESICP/DPC/SCM/UA 331, CEN Saclay F-91191 Gif sur Yvette Cedex

INTRODUCTION

Un article précédent (Jonah, ce Congrès) a souligné l'intérêt de la méthode de la radiolyse pulsée pour l'étude de la réactivité des espèces à durée de vie courte. Nous avons appliqué cette méthode à l'étude cinétique de l'oxydo-réduction de trois médicaments antitumoraux dont le mécanisme d'action, in vivo, est supposé être déclenché par un échange monoélectronique. Dans ce cas, des transitoires radicalaires sont donc impliqués dans le déroulement de leurs réactions.

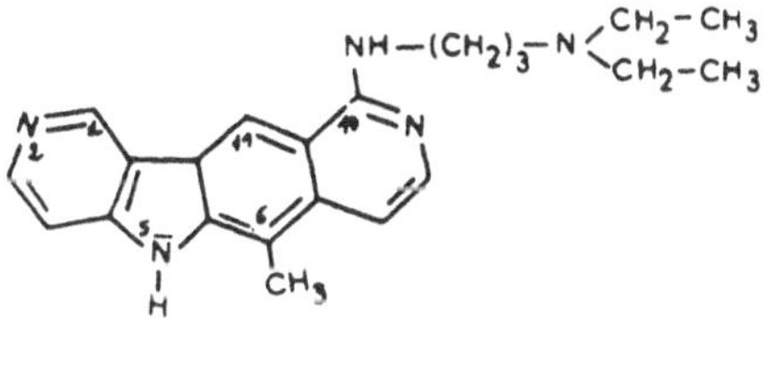

Fig. 1

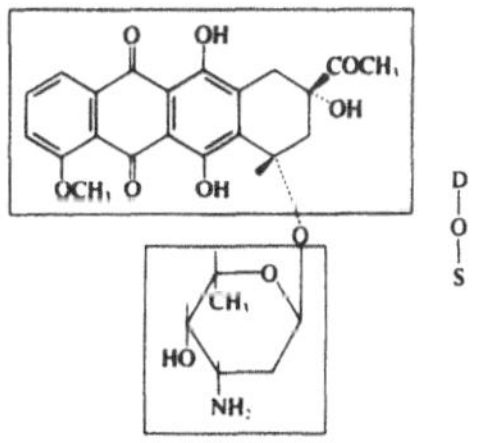

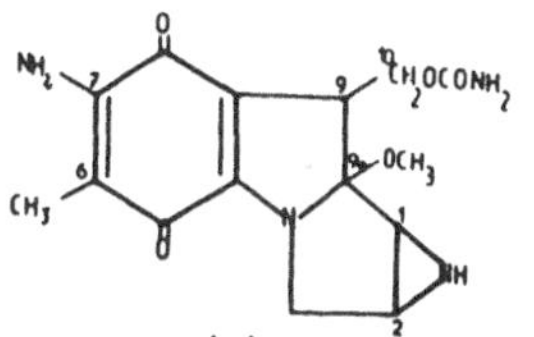

Médicaments antitumoraux étudiés : composé (1) : le BD 40; composé (2) : la daunorubicine (DOS); composé (3) : la mitomycine C (MC).

123

Dans ce travail, nous nous sommes proposés d'étudier la réactivité de chacun des transitoires radicalaires formés : a) lors de l'oxydation du BD 40 (composé (1)) par les radicaux OH·, b) lors de la réduction de la daunorubicine (composé (2)) par les radicaux COO⁻ et c) lors de la réduction de la mitomycine C (composé (3)) par les radicaux COO⁻.

Pour tester les réactions supposées se produire in vivo, nous avons étudié la réactivité de chacun des transitoires ainsi formés vis-à-vis : a) de l'oxygène et de sa forme réduite, l'anion superoxyde O_2^- , pour le BD 40, b) de l'apoprotéine de la riboflavine pour la daunorubicine, c) de l'ADN pour la mitomycine C. Pour effectuer ce travail, nous avons utilisé le Febetron 707 (DPC, CEN Saclay) déjà décrit par ailleurs[1].

Le BD 40, composé (1), est un médicament antitumoral de la famille des azaellipticines, synthétisé dans le laboratoire de E. Bisagni[2,3]. Son activité cytotoxique sur divers mutants de la levure Saccharomyces Cerevisiae ne s'observe qu'en phase de croissance et exige la présence d'oxygène[4]. C'est pourquoi il nous a paru intéressant d'étudier le rôle du couple oxydo-réducteur O_2/O_2^- vis-à-vis du transitoire R· formé lors de l'oxydation monoélectronique du BD 40 par les radicaux OH·. Ce transitoire R· a déjà été étudié en l'absence d'air au laboratoire[5].

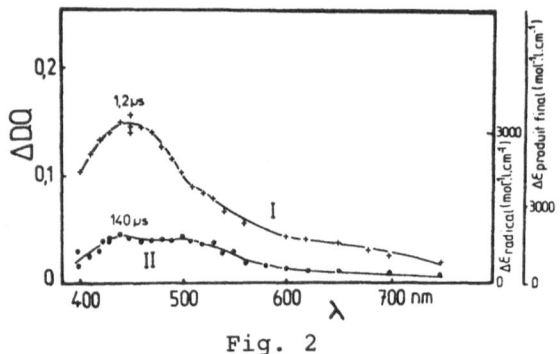

Fig. 2

Devenir du radical obtenu lors de l'oxydation du BD 40 par les radicaux OH· en présence de O_2/O_2^-. Spectres d'absorption différentiels obtenus pour deux temps après l'impulsion I(+) 1,2 µs, II(•) 140 µs.
$[\text{BD } 40]_0 = 2,7. \ 10^{-4}$mol.l⁻¹, $[O_2]_0 = 10^{-3}$mol.l⁻¹, dose = 88 Gy par impulsion, $[\text{phosphate}]_0 = 5.10^{-2}$mol.l⁻¹, $[OH·]_0 = 2,5.10^{-5}$ mol.l⁻¹, $[O_2^-]_0 = 2,9.10^{-5}$mol.l⁻¹, pH 7, l = 2,5 cm.
Echelle de droite : $\Delta \varepsilon$ pour les spectres I et II.

Les spectres différentiels représentés sur la figure 2, obtenus sous atmosphère d'oxygène, pour deux temps après l'impulsion (1,2 µs et 140 µs), correspondent respectivement, l'un au radical R· formé par oxydation monoélectronique du BD 40 par les radicaux OH·, l'autre au produit final P formé à la fin de la décroissance du radical ($[BD\ 40]_0 = 2,7.10^{-4}$mol.l^{-1}, [phosphate] = 5.10^{-2}mol.l^{-1}, pH 7, dose = 88 Gy par impulsion). Cette décroissance de R˙ est du deuxième ordre sous atmosphère d'oxygène et elle est beaucoup plus rapide que la décroissance du deuxième ordre en l'absence d'air, due à la dismutation du radical R˙. Il semblerait donc que le radical superoxyde réagisse avec R˙; ceci a été confirmé par radiolyse γ[7]. En effet, nous avons constaté qu'en présence de SOD (SOD, enzyme accélérant la dismutation des radicaux O_2^-) le rendement d'oxydation du BD 40, en milieu oxygéné, est deux fois plus faible qu'en l'absence de SOD.

Nous proposons donc le mécanisme réactionnel suivant pour expliquer la formation et la disparition du radical R˙, en présence d'oxygène et de radicaux superoxyde :

(1) $\quad$ OH˙ + BD 40 $\longrightarrow$ R˙, $\quad k_1 = (4,0 \pm 0,5)\ 10^9$mol^{-1}.l.s^{-1}

(2) $\quad$ OH˙ + OH˙ $\longrightarrow$ H_2O_2, $2\ k_2 = 10^{10}$mol^{-1}.l.s^{-1}

(3) $\quad$ $e_{aq}^- + O_2 \longrightarrow O_2^-$, $k_3 = 1,9.10^{10}$mol^{-1}.l.s^{-1}

(4) $\quad$ H˙ + $O_2 \longrightarrow HO_2^˙$, $k_4 = 1,2.10^{10}$mol^{-1}.l.s^{-1}

(5) $\quad$ OH˙ + $O_2^- \longrightarrow OH^- + O_2$, $k_5 = 10^{10}$mol^{-1}.l.s^{-1}

(6) $\quad$ R˙ + R˙ $\longrightarrow$ P + BD 40, $2\ k_6 = (2,8 \pm 0,3)10^8$ "

(7) $\quad$ R˙ + $O_2^- \longrightarrow$ P + H_2O_2, $k_7 = (5,3 \pm 0,3)\ 10^9$ "

Les valeurs des constantes de vitesse des réactions (1) (6) et (7) ont été déterminées au cours de ce travail.

LA DAUNORUBICINE INTERCALEE DANS L'APOPROTEINE DE LA RIBOFLAVINE

La daunorubicine (DOS, composé (2), offerte par Rhône-Poulenc) est un antibiotique antitumoral de la famille des anthracyclines. Ce médicament s'intercale dans les protéines et dans l'ADN où il serait activé par réduction. La réduction monoélectronique de ce médicament avait déjà été étudiée, en l'absence de protéine, dans notre laboratoire[6,7,8]. Nous nous proposons ici d'étudier cette réduction lorsque la daunorubicine est intercalée dans l'apoprotéine de la riboflavine (RSSR), préparée par V. Favaudon, pour former un complexe que nous noterons (DOS-RSSR).

L'étude de la réaction des ions radicalaires COO^- avec le complexe daunorubicine-protéine (DOS-RSSR) conduit à la conclusion que deux radicaux libres centrés respectivement sur l'antibiotique et sur la protéine sont formés (figure 3, spectre A, $[DOS-RSSR]_0 = 10^{-4}$mol.l^{-1}, [phosphate] = 6.10^{-2}mol.l^{-1}, $[HCOO^-]$ = 0,1 mol.l^{-1}, pH 7, dose = 30 Gy par impulsion, saturé de N_2O). Ces deux radicaux libres, identifiés grâce à leur spectre d'absorption, correspondent l'un à la forme semiquinonique de la daunorubicine intercalée (DOS˙-RSSR) et l'autre à un radical complexé de type RSSR˙ (DOS-RSSR˙) :

$$(8) \quad DOS\text{-}RSSR + COO^{\overline{\cdot}} \begin{array}{c} \nearrow DOS^{\overline{\cdot}}\text{-}RSSR + CO_2 \\ \searrow DOS\text{-}RSSR^{\overline{\cdot}} + CO_2 \end{array} \Bigg\} \begin{array}{l} k_8 = 2{,}4.10^8 \\ mol^{-1}.l.s^{-1} \end{array}$$

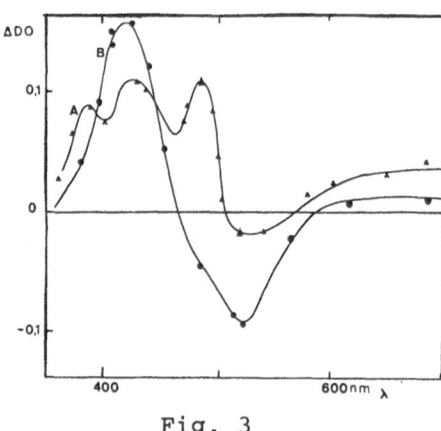

Fig. 3

Réduction par les radicaux libres $COO^{\overline{\cdot}}$ du complexe daunorubicine-apoprotéine (DOS-RSSR) en radiolyse pulsée.

Spectres d'absorption différentiels obtenus pour deux temps après l'impulsion : A 22 μ s et B 22 ms. $[DOS\text{-}RSSR]_0 = 10^{-4}$ mol.l^{-1}, $[phosphate] = 6.10^{-2}$ mol.l^{-1}, $[HCOO^-] = 0{,}1$ mol.l^{-1}, $[N_2O] = 2{,}5.10^{-2}$ mol.l^{-1}, pH 7, dose = 30 Gy par impulsion, l = 2,5 cm.

La décroissance du radical libre DOS$^{\overline{\cdot}}$-RSSR mène à la formation de l'espèce hydroquinonique correspondante (DH$_2$OS-RSSR) (figure 3, spectre B), par une réaction qui suit une loi cinétique du premier ordre. Cette réaction pourrait être un transfert d'électron intramoléculaire :

$$(9) \quad DOS^{\overline{\cdot}}\text{-}RSSR \xrightarrow{2 \ H^+} DH_2OS\text{-}RSSR^{\cdot +}, \quad k_9 = 3{,}5.10^2 s^{-1}$$

En ce qui concerne le devenir du radical DOS-RSSR$^{\overline{\cdot}}$, une étude en radiolyse γ a permis de montrer qu'il y avait tout d'abord coupure du pont disulfure RSSR et formation de fonction thiol, en compétition avec la réduction de l'antibiotique. Nous proposons donc le schéma suivant :

$$(10) \quad DOS\text{-}RSSR^{\overline{\cdot}} \xrightarrow{H^+} DOS \begin{array}{c} RS^{\cdot} \\ \diagdown \\ RSH \end{array}$$

$$(11) \quad 2 \ DOS \begin{array}{c} RS^{\cdot} \\ \diagdown \\ RSH \end{array} \longrightarrow \left[DOS \begin{array}{c} RS^{\cdot} \\ \diagdown \\ RSH \end{array} \right]_2$$

La mitomycine C (composé (3), offerte par les laboratoires Choay) est un antibiotique antitumoral utilisé dans le traitement de certains cancers gastriques. Il semble bien établi que la mitomycine C (MC) nécessite une réduction par voie chimique et/ou enzymatique pour devenir un agent alkylant dont les cibles seraient l'ADN, l'ARN et diverses protéines[9 à 12]. C'est pourquoi nous nous sommes intéressés à la réduction monoélectronique de ce médicament en présence d'ADN.

La réduction par les radicaux libres COO^- de (MC), en l'absence d'ADN, avait déjà été étudiée au laboratoire[13]. Rappelons ici les principaux résultats obtenus : MC est réduite par les radicaux CO_2^- pour donner un radical libre MC^- de type semiquinonique qui décroît ensuite par une réaction du deuxième ordre (dismutation) pour donner une forme hydroquinonique MCH_2. Ce dernier composé subit ensuite à l'échelle de la seconde, une transformation complexe pour donner finalement un mélange de deux composés finals (notés II et III dans le schéma réactionnel). Ce dernier processus se compose de trois réactions du premier ordre ((14), (16) et (17)). Les intermédiaires impliqués (notés A et B) ont été caractérisés par leur spectre d'absorption.

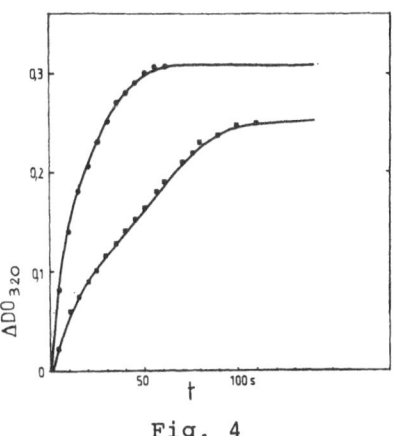

Fig. 4

Cinétiques d'évolution de l'intermédiaire A à 320 nm
- en présence d'ADN ($[ADN]_o$ = $[MC]_o$ = 10^{-4}mol.l^{-1})
- en l'absence d'ADN ($[MC]_o$ = 10^{-4}mol.l^{-1})
    $[formiate]$ = 0,1 mol.l^{-1}, pH 7, N_2O, 1 = 1 cm.

La réduction monoélectronique de la mitomycine C par les radicaux COO^- en présence d'ADN, a permis de montrer que, dans les conditions de la radiolyse pulsée, les deux premiers intermédiaires formés, à savoir MC^- et MCH_2 disparaissaient de la même manière en l'absence et en présence d'ADN. Seul le transitoire A réagit avec l'ADN (voir figure 4). Il semble que cet intermédiaire A soit la forme active du médicament.

Le schéma réactionnel de la réduction monoélectronique de la mitomycine C, en présence d'ADN, est le suivant :

127

$$(12) \quad MC + CO_2^{-} \longrightarrow MC^{-} + CO_2, \quad k_{12} = (1{,}8 \pm 0{,}1)10^9 \, mol^{-1}.l.s^{-1}$$

$$(13) \quad 2 \, MC^{-} \longrightarrow MCH_2 + MC, \quad k_{13} = (2 \pm 0{,}5) \, 10^7 \qquad "$$

$$(14) \quad MCH_2 \longrightarrow A \qquad , \quad k_{14} = 0{,}08 \; s^{-1}$$

$$(15) \quad A \xrightarrow{\; + ADN \;} A\text{-}ADN \qquad , \quad k_{15} = 500 \; mol^{-1}.l.s^{-1}$$

$$(16) \quad A \longrightarrow B \qquad , \quad k_{16} = 0{,}015 \; s^{-1}$$

$$(17) \quad B \longrightarrow II + III \; , \quad k_{17} \leqslant 0{,}008 \; s^{-1}$$

Les valeurs des constantes de vitesse des réactions (12) à (17) ont été déterminées au cours de ce travail.

CONCLUSION

Nous avons présenté trois applications de la radiolyse pulsée à l'étude du mode d'action de médicaments antitumoraux. Nous avons pu caractériser les intermédiaires formés par transfert monoélectronique et nous avons mis en évidence plusieuurs réactions qui peuvent intervenir in vivo comme la capture des radicaux libres O_2^{-} par la forme radicalaire oxydée du BD 40, la possibilité pour la daunorubicine de capter les électrons quand elle est intercalée dans une protéine. Enfin nous avons isolé la forme active de la mitomycine C responsable de son action sur l'ADN. La méthode de la radiolyse pulsée a permis l'étude cinétique quantitative et systématique de telles réactions.

REFERENCES

(1) J.L. Marignier et B. Hickel, J. Phys. Chem. 88 : 5375 (1984).
(2) E. Bisagni, C. Ducrocq, J.M. Lhoste, C. Rivalle et A. Civier, J. Chem. Soc. Perkin Trans. 1707 (1979).
(3) C. Ducrocq, F. Wendling, M. Tourbez-Perrin, C. Rivalle, P. Tambourin, F. Pochon, E. Bisagni et J.C. Chermann, J. of Med. Chem. 23 : 1212 (1980).
(4) E. Moustacchi, V. Favaudon et E. Bisagni, Cancer Res. 43 : 1300 (1983).
(5) A. Sekaki. Thèse d'Etat (juin 1987).
(6) C. Houée-Levin, M. Gardès-Albert et C. Ferradini, Febs Lett. 173 : 27-30 (1984).
(7) C. Houée-Levin, M. Gardès-Albert, C. Ferradini, M. Faraggi et M. Klapper, Febs Lett. 179 : 46-50 (1985).
(8) C. Houée-Levin, M. Gardès-Albert et C. Ferradini, J. of Free Rad. Biol. Med. 2 : 89-97 (1986).
(9) V.N. Iyer et W. Szybalsky, Science 145 : 55 (1964).
(10) M.N. Lipsett et A. Weissbach, Biochem. 4 : 206 (1965).
(11) J.W. Lown, S.K. Sim et H.H. Chen, Can. J. Biochem. 56 : 1042 (1978).
(12) M. Tomasz, R. Lipman, J.K. Snyder et K. Nakanishi, J. Am. Chem. Soc. 195 : 2059 (1983).
(13) G. Machtalère. Thèse (juillet 1987).

DETERMINATION OF THERMODYNAMIC AND SPECTRAL PARAMETERS OF A THERMOCHROMIC EQUILIBRIUM BY UV/VIS SPECTROSCOPY IN DIFFERENT SOLVENTS

L. Tan-Sien-Hee, D. Lavabre, J.C. Micheau and G. Levy

Laboratoire des IMRCP, UA au CNRS N° 470
Université Paul Sabatier, 31062 Toulouse Cedex, France

INTRODUCTION

Thermochromic reactions are of particular interest not only from the practical (1) and pedagogic (2) points of view, but also from a more fundamental standpoint (3). A number of studies drawn from a wide variety of domains have been devoted to their behavior (4-6).

The thermochromic behavior of some metallic complexes in solution stems from the existence of equilibria between different configurations of the complexes (of different colors) at different temperatures (7,8). These equilibria thus have specific thermodynamic and spectral characteristics. Practical problems are often encountered in the experimental determination of these characteristic parameters. The present study was designed to:
1) develop a new method to circumvent some of these difficulties
2) apply the method to the thermochromic equilibrium of some Ni(II) complexes (8).
3) discuss the thermodynamic and spectral data obtained.

MATERIALS AND METHODS

The following thermochromic equilibrium was investigated:

$$[\text{Ni(tmen)(acac)}]^+ + 2S \rightleftharpoons [\text{Ni(tmen)(acac)S}_2]^+ \quad (I)$$
$$\text{red} \qquad\qquad\qquad\qquad \text{green}$$

tmen: N,N,N',N'-tetramethylethylenediamine
acac: acetylacetonate
S: solvent

The ion complex $[\text{Ni(tmen)(acac)}]^+$ was prepared according to the method described by Fukada and Sone (10), and dissolved in different solvents. The solvation of the ion thus depends on the donor properties of the solvent S, and the equilibrium (I) is shifted to right or left. The equilibrium was studied at different temperatures in different solvents by UV/visible spectroscopy. The apparatus consisted of a diode-array spectrophotometer (HP 8451) equipped with a computer-controlled Peltier effect thermostated sample holder.

RESULTS

Qualitative analysis of the equilibrium

Effect of temperature. Thermochromism is observed when the green complex is heated. The equilibrium is displaced to the right, favoring the red complex. A square-planar configuration is favored at high temperatures while an octahedral arrangement (green complex) is favored at low temperatures. The reaction is endothermic to the left.

Influence of the solvent. At a given temperature, the color of the solution depends on the donor properties of the solvent S. Good donors favor the green complex ion, while the red complex is favored by the poor donors. Thermochromism is observed most readily in solvents that are intermediate donors, such as nitriles, ketones and alcohols. Mixtures of a good and a weak donor can also lead to thermochromic behavior. The binary mixtures, nitromethane/ n-butanol (70/30 mol/mol) and isoamyl alcohol/tetrachloroethane (50/50 mol/mol) have been used. Thermochromism is also observed in azeotropic mixtures such as chloroform/methanol and chloroform/ethanol. In these cases, the proportions of the two components do not require adjustment.

Quantitative analysis of the equilibrium

Spectral characteristics. Figure 1 illustrates the spectra recorded between -70°C and +45°C for a 2.10^{-2} M solution of $[Ni(tmen)(acac)]^{+}$ in acetone. Two absorption bands are observed, a relatively intense one at 490 nm corresponding to the red complex, and a much weaker one for the green complex. The existence of an isosbestic point at 593 nm demonstrates the single equilibrium R (red) $\rightleftharpoons$ G (green). It can also be seen that the intensity of the band corresponding to the red complex increases with temperature. The dotted lines represent the theoretical spectra of the pure complexes (red and green), which are independent of temperature.

Spectral and thermodynamic parameters. The following reaction scheme applies to all the thermochromic reactions studied:
   R $\rightleftharpoons$ G   with K = G/R
Given that S is both solvent and reactant, the equilibrium constant must take account of the solvent concentration. The concentrations are sufficiently low for them to be represented by their corresponding activities. In addition, each complex is characterized by its absorption spectrum ε_R^{λ} or ε_G^{λ}.
At equilibrium: $\Delta G = 0$ and $\Delta G° = \Delta H° - T\Delta S = -RT \ln(K)$
and $K = e^{(b-a/T)}$ where $a = \Delta H°/R$ and $b = \Delta S°/R$. These two parameters are assumed to be independent of temperature.
The reaction is exothermic to the right, and so $\Delta H° < 0$, $a < 0$. At a given wavelength λ, the observed optical density is given by:

   $OD = \varepsilon_R^{\lambda} R + \varepsilon_G^{\lambda} G$ for an optical path length of 1 cm.
The apparent absorption $\varepsilon^{\lambda} = OD/(R + G)$ or as a function of a and b:

$$\varepsilon^{\lambda} = [\varepsilon_R^{\lambda} + \varepsilon_G^{\lambda} \times e^{(b-a/T)}]/(1 + e^{(b-a/T)}) \qquad \underline{I}$$

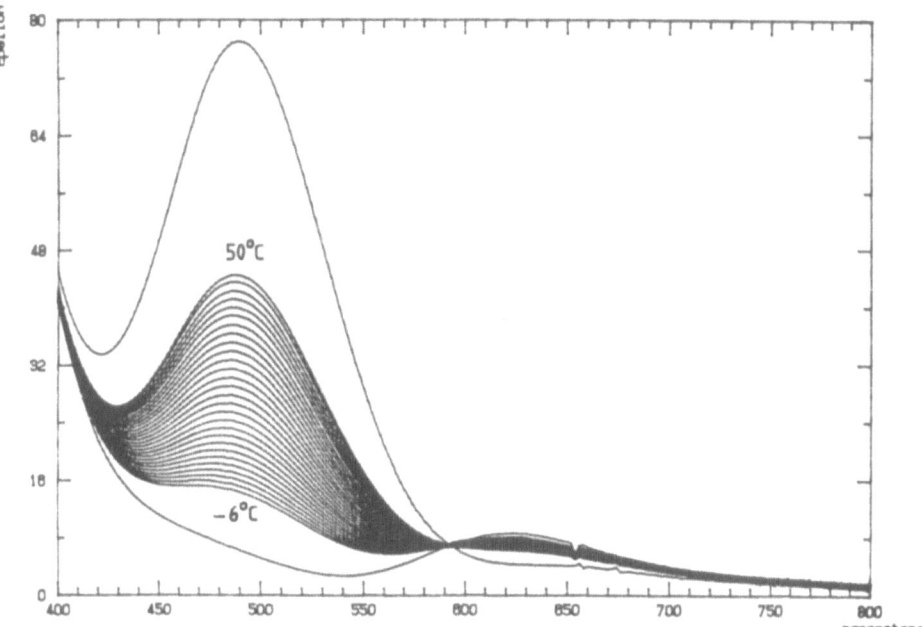

Fig.1. Thermochromism of $[Ni(tmen)(acac)]^+$ $(2.10^{-2}$ M) in acetone. Measurements were carried out every 2°C. Similar spectra were obtained with the other thermochromic solvents.

The relationship $\underline{I}$ is referred to as 'T-chrom' hereafter. It relates two thermodynamic (a, b) and two spectral parameters (ε_G^λ, ε_R^λ) to the observable absorption ε^λ (epsilon). Since these four parameters are unknown, they must be determined from data obtained from spectra recorded at different temperatures.

Mathematical analysis of the T-chrom relationship. The curve $\varepsilon = f(T)$ is sigmoidal with an inflexion corresponding to the point of maximum temperature sensitivity of the equilibrium. However, these parameters cannot be determined reliably from this curve. Rather than use a classical data fitting procedure which can theoretically determine all 4 parameters, it is preferable to separate the spectral parameters from the thermodynamic parameters and then carry out data fitting on a restricted number of parameters (only 2 in this case). Now:

$$R(T) = [d^2\varepsilon^\lambda/dT^2] \cdot [dT/d\varepsilon^\lambda] = \frac{a(1-K) - 2T(1+K)}{T^2(1+K)} \qquad \underline{II}$$

This relationship R(T) does not contain the spectral parameters and only involves the thermodynamic parameters a and b, since $K = e^{(b-a/T)}$. However, it requires storage and calculation of a large amount of data from many measurements taken at different temperatures. The spectral parameters ε_R^λ and ε_G^λ are then determined by linear regression using the T-chrom equation. This gives:

$$\varepsilon^\lambda = [\Delta\varepsilon^\lambda/(1+K)] + \varepsilon_G^\lambda \qquad \underline{III}$$

which is the equation of a straight line of ε^λ against u where u = $1/(1+K)$. The slope of this line is thus $\Delta\varepsilon^\lambda$ and the intercept on the x axis ε_G^λ. This gives the values of ε_G^λ and $\varepsilon_R^\lambda = \Delta\varepsilon^\lambda + \varepsilon_G^\lambda$.

Results (see Table 1)

Parameter a represents the enthalpy $\Delta H°/R$, and b the entropy $\Delta S°/R$. $\varepsilon_{iso}^\lambda$ is the value of ε^λ at the isosbestic point. ε_R^{max} is the value of ε_R^λ

131

if the concentration of the green complex is strictly zero. These maxima are observed around 490 nm. The equivalent ε_G^{max} are observed around 620 nm. Thus ε_R^{max} and ε_G^{max} are the standard calculated values. They are sometimes rather different from the measured values especially in the case of the strong (DMF and HMPT, green complex) or weak donors (nitromethane or nitroethane, red complex). This is discussed below.

DISCUSSION

Thermodynamic parameters. From the values a and b in table I, we traced a curve (fig.2) of b = f(a/T) for the mean temperature used in these experiments (T = 308°K).

Table 1. Thermodynamic and spectral parameters of the thermochromic equilibrium in various solvents and solvent mixtures

No.	Solvents	$a = \Delta H°/R$ (°K)	$b = \Delta S°/R$	ε_R^{max}	ε_V^{max} (1.mole^{-1}.cm^{-1})	ε_{iso}
1	Azeotrope 1	− 4168	− 11.98	53	8	5
2	Azeotrope 2	− 4256	− 12.17	46	8	5
3	Binary 1	− 4890	− 15.05	93	12	11
4	Acetone	− 3320	− 10.62	72.5	9	7.5
5	Isopropylmethyl ketone	− 4476	− 12.90	61	11.5	9
6	Propanol-1	− 4983	− 13.25	92	10	9
7	Benzyl alcohol	− 4303	− 13.56	88	8	7
8	Allyl alcohol	− 5113	− 14.60	75	7.5	6
9	Butanol-2	− 5207	− 17.28	106	11	9.5
10	Propanol-2	− 5237	− 17.50	72	10	8
11	Ethanol	− 6586	− 17.90	53	9	6
12	Hexanol-1	− 6711	− 18.92	29	8	7
13	Binary 2	− 4571	− 12.58	55	10	8
14	Nitromethane			107	0	
15	Nitroethane	red standards		100	0	
16	DMF			0	10	
17	HMPT	green standards		0	10	

Azeotrope 1 = $CHCl_3$/EtOH   84/16 mol/mol
Azeotrope 2 = $CHCl_3$/MeOH   65/35 mol/mol
Binary 1   = nitromethane/butanol-1   73/27 mol/mol
Binary 2   = nitromethane/butanol-1   63/37 mol/mol

It can be seen that the thermochromic solvents are found close to the line dividing the two zones of predominance of square-planar and octahedral complexes respectively. Figure 3 is a similar plot but for two different temperatures (T_1 = 275°K and T_2 = 318°K). It represents the displacement of the thermochromic equilibrium for the different solvents between these two temperatures. The straight lines correspond to the the different positions of the equilibrium characterized by the ratio of the concentrations G/R. There is a strong displacement of the equilibrium for the solvents situated in the lower region (eg. hexanol-1). This effect is enthalpy-driven. However, the transition from one solvent to another involves both a change in both enthalpy and entropy.

Spectroscopic parameters. The line 90/10 on curve 2 is the straight line corresponding to the sensitive color (equal intensity of red and green)

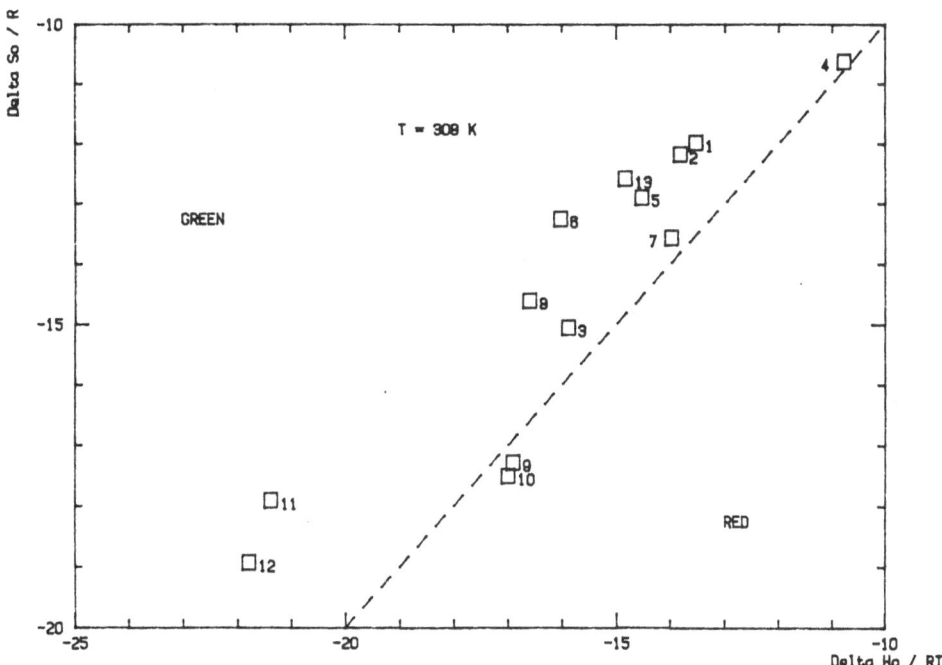

Fig.2. Distribution of different solvents with respect to $\Delta H°/R$, $\Delta S°/R$ at 308°K. The dotted line is for an equilibrium constant $K = 1$ (G/R = 50/50)

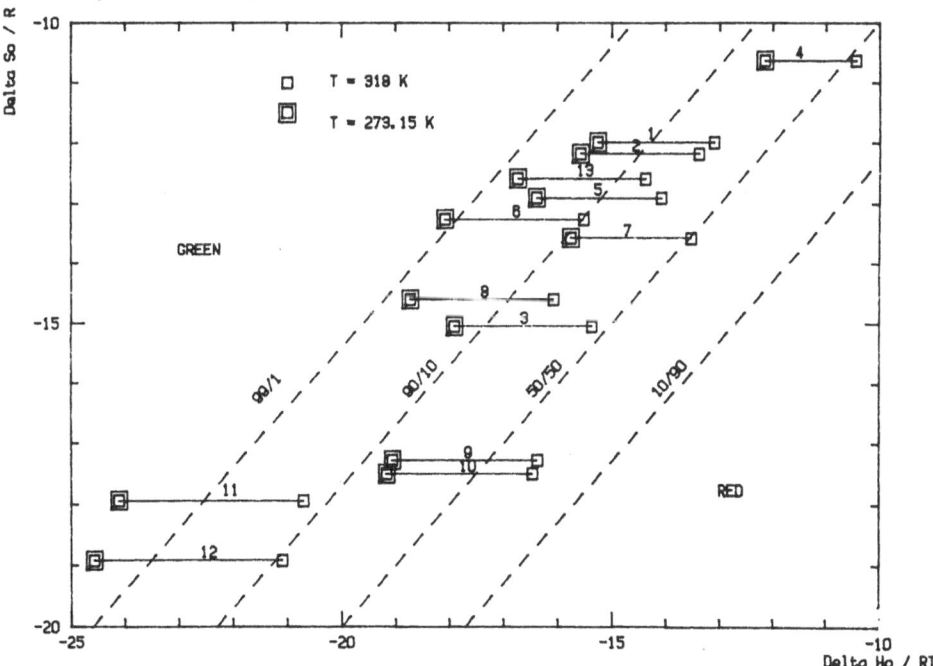

Fig.3. Displacement of the thermochromic equilibrium of $[Ni(tmen)(acac)]^+$ in various solvents between 273.15°K and 318°K. The dotted lines correspond to different molar ratios G/R.

since $\varepsilon_R^{max}.10 \cong \varepsilon_G^{max}.90$.

The high value of the ratio $\varepsilon_R^{max}/\varepsilon_G^{max}$ can be accounted for in a simplified way using ligand field theory. The following splitting diagrams show the change in degeneration of the t_{2g} and e_g energy levels as the octahedral arrangement is transformed to a square-planar complex by progressive desolvation of the molecules S on axis z. The square-planar environment is found typically in d^8 complexes of Ni(II) (11).

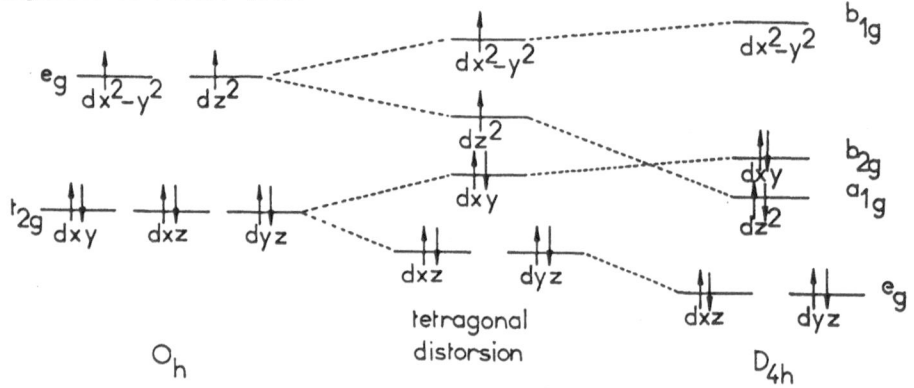

Fig.4. Splitting diagram of d orbitals (d^8 configuration of Ni(II) during transition from octahedral to square-planar environment.

It can be seen that the strictly square-planar complexes represent the limit of tetragonal distortion where there is a stabilization of levels d_{xz}, d_{yz} and d_{z^2} with respect to levels d_{xy} and $d_{x^2-y^2}$. Such complexes are found with ligands producing a strong field which favor pairing of electrons at the b_{2g} levels. The energy of level b_{1g} is too high to be occupied. It follows that electronic transitions in both the O_h and D_{4h} complexes are forbidden by Laporte's rule (g<->g). On the other hand, they are allowed for octahedral complexes presenting tetragonal distortion due to the loss of a center of inversion. This is the case for the $[Ni(tmen)(acac)]^+$ complexes in solvating media. The high value of ε_R^{max} with respect to ε_G^{max} is explained in a similar way. It may also account for the differences between the theoretical standard, and the measured values (cf. Table I). A tetragonal field is encountered in a weakly solvating medium, while in a strongly solvating medium an octahedral field is favored. It would appear as if a solvatochromic effect (depending on the donor properties of the solvent) acts in concert with the normal thermochromic behavior.

CONCLUSION

Determination of both the thermodynamic and spectral characteristics of a thermochromic equilibrium has been shown to be feasible by a computerized method of analysis of spectroscopic data. The principle feature is that unlike other methods (12) it makes no assumptions about the values of the standard spectra. Several independent trials were carried out with each solvent which produced results within 1% of each other. Another advantage is that it can be used to show up differences between the theoretical and experimental values of ε^{max}. This gives an indication of the existence of tetragonal distortion in the geometry of the complexes.

From another point of view, the determination of the thermodynamic parameters gives valuable information about the donor properties of the solvents and the behavior of binary mixtures. A particularly interesting application of this type of analysis is the ability to rank the two parameters ($\Delta H°/RT$ and $\Delta S°/R$) for intermediate solvents that are neither very strong nor very weak donors. Such scales have been described (14), but they are only applicable to the extreme cases. Intermediate solvents are not easy to classify. The thermodynamic analysis presented here produces opposite results. The extreme solvents are all lumped together as either 'green' or 'red' solvents, while the intermediate ones are readily discriminated. A particularly good illustration is that of propanol-1 and propanol-2 (cf. Fig. 2).

This methodology can also be extended to other thermochromic systems based either on nickel (13), cobalt (14) or other metals (15) and non-metals (16).

REFERENCES

1. L.G. Spears Jr. and L.G. Spears, Chemical storage of solar energy using an old color change demonstration. J. Chem. Educ. 61:3 (1984).
2. J.P. Byrne, Thermodynamic data from the thermochromic effect. J. Chem. Educ. 55:267 (1978).
3. J. Sunamoto and T. Hamada, Solvatochromism and Thermochromism of Co(II) complexes solubilized in reversible micelles. Bull. Chem. Soc. Jap. 51:3120 (1978).
4. T. Sakai and N. Ohno, Selective determination of spartein by thermochromism of associates with tetrabromophenylphthalein ethyl ester. The Analyst (London) 107:634 (1982).
5. H.D. Hardt and A. Pierre, Fluorescence thermochromism of pyridine copper iodides. Z. Anorg. Allg. Chem. 402:107 (1973).
6. V.S. Marevtsev, N.L. Zaichenko, V.D. Ermakova, S.I. Beshenko, V.A. Linskii, T. Gradyushuko and M.I. Cherkashin, Effect of electron donor and electron acceptor substituents on photo and thermochromic properties of indoline spiropyranes. Isv. Akad. Nauk. SSR, Ser. Khim. 10:2272 (1980).
7. J.H. Day, Thermochromism of inorganic compounds. Chem. Rev. 68:649 (1968).
8. D. Lavabre, J.C. Micheau and G. Levy, Comparison of thermochromic equilibrium of Co(II) and Ni(II) complexes. J. Chem. Educ. to appear (Oct. 1987).
9. Y. Fukada, A. Shimamura, M. Mukaida, E. Fujita and K. Sone, Mixed copper (II) chelates with N,N,N',N',-tetramethylethylenediamine and beta-diketones. J. Inorg. Nucl. Chem. 36:1265 (1974).
10. Y. Fukada and K. Sone, Mixed Nickel (II) chelates with N,N,N',N',-tetramethylethylenediamine and beta-diketones. J. Inorg. Nucl. Chem. 34:2315 (1972).
11. J.E. Huheey, Inorganic Chemistry. Harper and Row, p.334, Edit. 1972.
12. J.R. Long and R.S. Drago, The rigorous evaluation of spectro-photometric data to obtain an equilibrium constant. J. Chem. Educ. 59:1037 (1982).
13. Y. Ihara, A. Wada, Y. Fukada and K. Sone, Preparation and thermal square planar-octahedral transformation of nickel complexes containing 1,2-butanediamine or 3,3-dimethyl-1,2- butanediamine in solid phase. Bull. Chem. Soc. Jap. 59:1037 (1986).
14. R. Hammer and H.F. Klein, The tetrakis(triphosphine)cobalt anion. Z. Naturforsch. B 32B(2):148 (1977).
15. I. Grenthe, P. Paoletti, M. Sandstroem and S. Glikberg, Thermochromism in copper (II) complexes. Structure of the red and blue-violet forms of bis(N,N-diethylethylenediamine copper(II) perchlorate and the non-thermochromic violet bis(N,ethylenediamine copper(II) perchlorate. Inorg. Chem. 18:2687 (1979).
16. I. Rosenthal, P. Peretz and K.A. Muskat, Thermochromic and hyperchromic effects in rhodamine B solutions. J. Phys. Chem. 83:350 (1979).

THE INFLUENCE OF THE SOLVENTS ON THE ELECTRONIC SPECTRA OF SOME AROMATIC MOLECULES AND THEIR DIPOLE MOMENTS IN THE EXCITED STATE. THE SOLVENT EFFECTS ON THE ANILINE ELECTRONIC BAND AT 34,870 cm^{-1}

Vasile Macovei and Ana Sasu

The Physical Chemistry Laboratory, The Faculty of
Technological chemistry, The Politechnical Institute
Iassy, Rumania

INTRODUCTION

In a preceeding series of papers [1-8], was investigated the influence of the solvents on the spectral bands of benzene, chlorobenzene, bromobenzene, toluene, anisole, benzaldehyde, nitrobenzene, and benzonitrile with the aim to establish the behaviour differences between monosubstituted benzene molecules with the first and second order susbstituents in solutions, i.e. to contribute to the elucidation of the interactions in the solute-solvent systems determining the spectral displacements, and to evaluate the dipole moments of the solute molecules in their excited states as indices of reactivity. In the series of monosubstituted benzenes the aniline molecule has a place apart due to the presence of the amino group, and it is the object of this investigation.

RESULTS AND DISCUSSION

The physico-chemical properties of monosubstituted benzenes were discussed in terms of inductive and resonance effects [9-14], the last being very important in the interpretation of electronic spectra and of dipole moments. In vapour phase, the near ultraviolet spectra of monosubstituted benzenes are shifted towards longer wave lengths as compared to the benzene spectrum [15]. Sponer et al [16] have established such a shift of 4,055 cm^{-1} for aniline. We established the position of the electronovibrational component, v_{max}, in the vapour phase, at 39,640 cm^{-1} for benzene, and at 34,870 cm^{-1} for aniline, a shift of 4,770 cm^{-1}. The vapour spectrum of aniline, resulting from isolated molecules [17], has a well defined vibrational structure, and a pronounced rotational structure, as shown in figure 1. It represents the aniline transition [18] $^1A_1 \longrightarrow {}^1B_1$, deriving from the forbidden $^1A_{1g} \longrightarrow {}^1B_{2u}$ benzene transition, which becomes allowed in the monosubstituted benzene spectra, on the basis of C_{2v} symmetry of these in the ground and excited states [19], that is by the perturbation of the D_{6h} symmetry of benzene, a fact theoretically established by Matsen [9]. The benzenoid spectral bands shift on the benzene substitution [18-20-22]. The aniline vapour spectrum is more complex [22] than that of benzene [23] and has its origin displaced. Our results, as one sees in figure 1, confirm these assertions. In solution, the aniline spectrum is shifted again, but in a few nonpolar solvents preserves some traces of vibrational structure, in contrast with toluene [4].

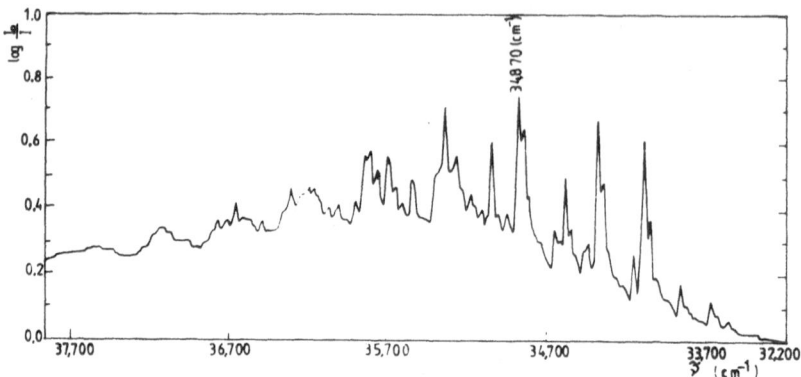

Fig. 1. The electronic spectrum of aniline in vapour phase.

In the previously studied molecules [1-4], in the $\pi \longrightarrow \pi^*$ transitions, the characteristic benzene ring vibrations were efficient, the C-C vibrations being prominent [19]. The vibrational structure of aniline spectrum in solution, in the great majority of solvents, is obliterated, as in figure 2, where are presented the effects of some of the solvents used. (The numbers in figure 2 correspond to the solvents in Table 1, and the notation is preserved throughout in the paper). This fact indicates, besides the interactions with the solvents, a pronounced interaction between the substituent and the benzene ring.

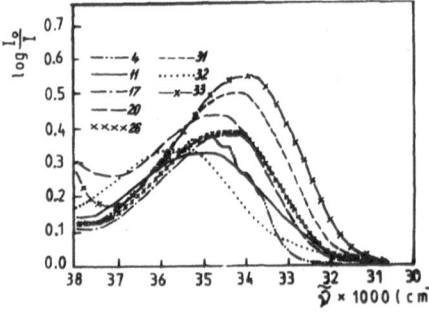

Fig. 2. The aniline spectrum in the solvents :

4. - n-Hexane ; 11.-Methanol ; 17.-Ethyl ether ; 20-Chloroform ; 26.-Acetonitrile ; 31-.-1,4-Dioxane ; 32.-Water ; 33.-N,N-Dimethyl-formamide.

The spectral shift in solution, $\Delta \nu_{exp.}$ (cm^{-1}), represents the difference between the solvation energies of the ground and excited states [24], that is the difference in the positions of the electronovibrational component $\nu_{max.}$ in the vapour spectrum of aniline and the peak of the spectral band in solution, the corresponding values being listed in Table 1. Considering as positive the shifts in solution towards smaller wave numbers, one notes a variation over a long range of values, from - 605 cm^{-1}, in water, to + 1313 cm^{-1}, in dimethylsulphoxide. In non-polar solvents such as hydrocarbons, the shifts values for aniline are smaller, as compared with those of toluene, due to the absence of a hydrogen atom in the substituent group, which determines the packing of a smaller number of solvent molecules around the solute molecule, and consequently, a reduction in the solvation energies, that is a diminished red polarization shift [24]. As the spectral shifts in hydrocarbons represent a susbstantial part as compared with the other solvents (excepting water and alcohols) we consider that the dispersion forces of the London [25] type have an important contribution to the solvation energies, and then to the red shift, because the polarizability of solute molecule in the excited state increases [26], and an instantaneous redistribution of the electric charge will take place. From the McRae's [27] theory results a formula giving the spectral shift under the solvent influence (in terms of solute polarizability and dipole moment of the solute molecule and in terms of the refractive index and dielectric constant of the solvent), which, for nonpolar solvents, reduces to :

TABLE 1. The characteristics (n, D and b.p.)[*] of solvents, the position of aniline spectral band, ν_{max}, the shifts, $\Delta\nu_{exp}$ in solutions, and the theoretical shifts values $\Delta\nu_{theor}$.

SOLVENT	n	D	b.p.	ν_{max} (cm^{-1})	$\Delta\nu_{exp}$ (cm^{-1})	$\Delta\nu_{theor}$, of.eq. (1)	$\Delta\nu_{theor}$, of.eq. (2)
0 Aniline (vapour)				34,870	0	-	-
1. Cyclopentane	1.404	1.965	49.30	34,651	219	507.48	414.00
2. n-Pentane	1.357	1.844	36.10	34,688	182	461.72	377.32
3. Cyclohexane	1.426	2.040	80.70	34,600	270	524.53	428.03
4. n-Hexane	1.375	1.890	68.80	34,738	132	478.14	390.66
5. n-Heptane	1.387	1.925	98.40	34,683	187	489.97	400.25
6. n-Octane	1.397	1.948	125.65	34,626	244	498.96	407.45
7. 2,2,4-Trimethylpentane	1.403	1.940	114.22	34,693	177	493.06	403.06
8. n-Decane	1.412	1.983	174.05	34,690	180	511.61	417.90
9. n-Dodecane	1.422	2.020	216.00	34,690	180	520.36	425.07
10. Cyclohexene	1.446	2.220	83.00	34,483	387	533.67	441.91
11. Methanol	1.329	33.790	64.60	34,931	-61	105.08	354.34
12. Ethanol	1.360	25.800	78.40	34,784	86	164.19	380.35
13. 1-Propanol	1.385	21.800	97.20	34,818	52	201.32	398.68
14. 2-Propanol	1.375	26.000	82.40	34,832	38	177.25	390.77
15. 1-Butanol	1.399	19.200	117.70	34,800	70	226.72	408.76
16. 3-Methyl-1-butanol	1.408	15.640	132.00	34,460	410	253.81	415.49
17. Ethyl ether	1.353	4.335	34.50	34,194	676	339.06	373.60
18. Isoamyl ether	1.408	2.820	173.10	34,377	493	463.40	415.47
19. Methylene chloride	1.425	9.090	40.60	34,370	500	429.93	428.87
20. Chloroform	1.445	4.810	61.10	34,460	410	429.93	441.64
21. Carbon tetrachloride	1.460	2.238	76.80	34,425	445	546.43	452.23
22. 1,2-Dichloroethane	1.446	10.130	83.60	34,350	520	338.82	442.32
23. Tetrachloroethylene	1.506	2.450	120.80	34,307	563	586.10	480.29
24. 1-Chloropentane	1.412	6.460	108.40	34,440	430	586.95	417.82
25. Bromoethane	1.425	9.300	38.40	34,524	346	558.00	426.51
26. Acetonitrile	1.345	37.500	82.00	34,305	565	120.27	367.51
27. Methyl acetate	1.362	7.080	59.60	34,205	665	288.70	380,82
28. Ethyl acetate	1.372	6.110	77.10	34,145	725	314.19	388.49
29. Butyl acetate	1.394	5.010	126.00	34,095	775	367.06	404.94
30. Amyl acetate	1.402	4.750	148.90	34,125	745	383.54	410.58
31. 1,4-Dioxane	1.422	2.235	101.10	34,058	812	508.00	425.07
32. Water[**]	1.333	80.360	100.00	34,475	-605	79.99	357.89
33. N,N-Dimethylformamide	1.426	37.000	153.00	33,713	1157	559.18	428.50
34. Tetrahydrofuran	1.407	7.600	65.50	34,095	775	581.78	414.61
35. Dimethylsulphoxide	1.448	48.900	100.00 (decomp).	33,557	1313	605.88	443.00

(*)   n is the refractive index ; D is the dielectric constant and b.p. is the boiling point of the solvent.

(**)   The refractive index and the dielectric constant of water are the reference constants.

$$\Delta\nu_{abs. (disp.)} (cm^{-1}) = (A+B) \left[(n^2-1) / (2n^2+1) \right] \tag{1}$$

where : A and B are characteristic parameters of the solute molecule, and n is the refractive index of the solvent. Hence, the spectral shift must linearly correlate with the function of refractive index $f(n^2)=(n^2-1)/(2n^2+1)$. This correlation is given in figure 3, where one sees that the representative points are approximately oriented along a straight line, but very spread to explain the solvent influence only by the dispersion forces. The theoretical shifts are calculated with the equation (1), after the evaluation of the parameter sum as : $(A+B)=2,097.792$ cm^{-1}, and the value are listed in Table 1. An attempt to correlate the $\Delta\nu_{theor.}$ values, calculated with equation (1), gathers the points along a straight line, much more than the experimental values, but being greater or smaller cannot explain the spectral shifts on the sole basis of dispersion forces. The dipole moment of the solute molecule has an important role in the spectral shift in solution. The calculated dipole moment of aniline [28], with pyramidal N is 1.44 D, and with planar N is 1.30 D, but the experimental value [29-30] is 1.54 D. In this case, to the formation of solvation energies contribute, both the dispersion and the dipolar interactions, being simultaneoustly considered in the equation of McRae [27] of the general form [31]:

$$\Delta \tilde{\nu}_{abs.} \ (cm^{-1}) \ = \ (A+B+C) \ [n^2-1)/(2n^2+1)] \ +$$
$$E \ [(D-1)/(D+2) - (n^2-1)/(n^2+2)] \ +$$
$$F \ [(D-1)/(D+2) - (n^2-1)/(n^2+2)]^2 \tag{2}$$

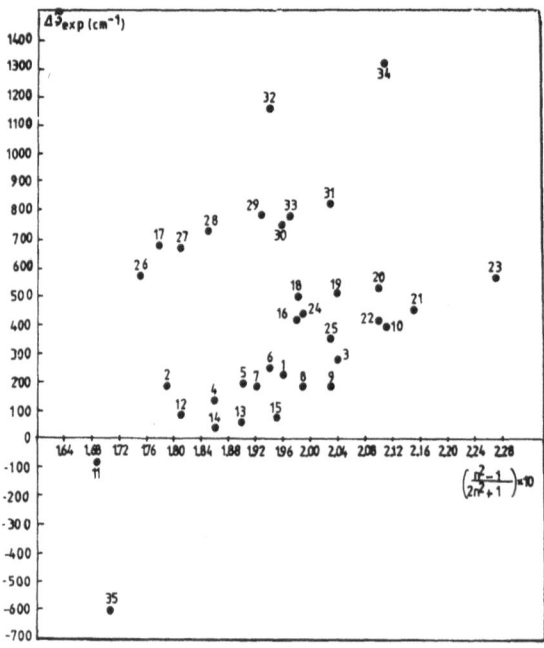

Fig. 3. The dependence of the $\Delta \nu_{exp.}$ of the aniline spectrum in solution on the function $f(n^2)$ of the solvent.

where n is the refractive index, and D is the dielectric constant of the solvent; the parameters (A+B+C), E, and F are characteristic constants of the solute molecule, and their values were evaluated* by the least - squares method [32] as :

$(A+B+C)=2,567.481 \ cm^{-1}$,
$E = - 318.704 \ cm^{-1}$, and
$F = - 202.817 \ cm^{-1}$.

The theoretical displacements, calculated by means of relation (2) are a little closer to the experimental ones (see Table 1) indicating that, besides the dispersion forces, also the dipolar forces contribute to the spectral shifts in solution.

Then the spectral shift of aniline must correlate linearly with the function [33] f(D)=(D-1)/(D-2) of the solvent. This correlation is given in figure 4, where one sees that the general orientation of the representative points is along a straight line , but with a very pronounced spread, which may be ascribed to greater complexity of the solute-solvent interactions. This correlation results if one notes that the range of variation of refractive indices of solvents is from 1.331, for methanol, to 1.460, for carbon tetrachloride, that is a difference of $\delta n = 0.127$ units, while the range of variation of dielectric constants is from 1.844 for n-pentane to 80.360, for water, that is a difference of $\delta D = 78.516$ unit (see Table 1). Hence, in relation (2) the terms containing the refractive index function $f(n^2)$ may be considered as constants, because of very small variation of refractive indices from a solvent to another. Due, to the lack of a good correlation between the $\Delta \nu_{exp.}$ and f(D) function, we try to interpret the spectral shifts of aniline under the solvents influence according to the Franck-Condon principle [17-24], knowing the

---

* The calculation program was established at the Territorial Center of Electronic Computing, Iassy.

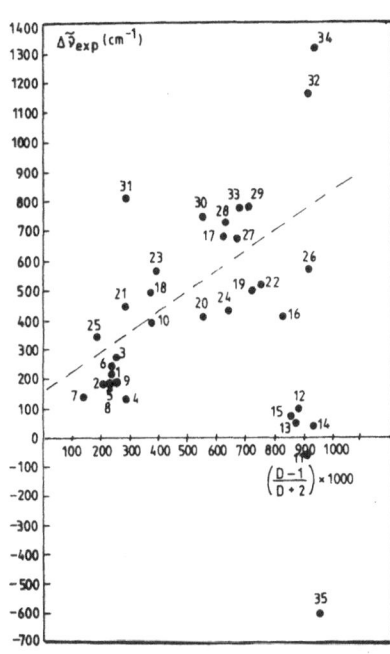

Fig. 4. The dependence of the $\Delta \nu_{exp.}$ of aniline spectrum in solution on the function f(D) of the solvent.

change of the dipole moment in going from the ground state to the excited state. One evaluates such a variation using the calculated value and the explicit form of the parameter E from the equation (2) i.e. :

$$E = (2/h\ c)\ [M_{oo}\ (M_{oo} - M_{ii})/a^3] \qquad (3)$$

where : h is the Planck constant ; c is light velocity ; M_{oo} and M_{ii} are, respectively, the dipole moments in the ground and excited states, and a is the solvent cavity radius. For the ground state of aniline we have M_{oo} = 1.54 D [29-30]. The average value of the solvent cavity radius, a, is calculated for spherical molecular model, assuming that it is equal to the mean radius of the molecule in the liquid state, or in pure crystal [34], according to the expression :

$$a^3 = 3\ M\ /\ 4\pi N_A \rho \qquad (4)$$

where : M is the molecular weight ; N_A is the Avogadro's constant, and ρ is the material density. Using the known values [35] of M and ρ (at 20°) for aniline, one calculates : $a^3(\text{Å}^3)$=36.136. With the corresponding values in equation (3), one evaluates the dipole moment of aniline in the excited states M_{ii}=2.3092 D ; the increase of the dipole moment is of 0.7192 D, but Prabhumirashi et al [36] have calculated an increase of 0.470 D for the $S_o \longrightarrow S_1$ transition ; Mataga [37] has evaluated the same increase as 2 - 2.5 D, and Suppan [38] states that this increase is firmly established at 3.8 D. Our results are more reasonable due to the use of a large variety of solvents. If the dipole moment of aniline increases in the excited state, then the solvation energy of this state augments by dipolar interactions, and the spectrum is red shifted. Hehre et al [28] showed that the π - charge distribution, with large negative charges at the ortho and para positions, agrees with those expected on the basis of contributions from usual valence structures (I), (II), (III), (IV), etc. (including polarization structure) for aniline. Also, the π_{Ph-X} overlap population indicates a considerable double bond character in the C-N bond. Our results show an increase in the dipole moment in the excited state, and we consider that the structure III is preferentially stabilized in this state. But, in the same structure, the formation of the double bond diminishes the size of solute molecule. Then, the orientation strain [24] is present in the excited state, increasing the solvation energy and shifting the spectrum to the red. The reduction of the molecular dimensions by the formation of the double bond in the C-N group acts inversely, reducing the number of solvent molecules which interact with the solute molecule, i.e. diminishing the solvation energy in the excited state and shifting the spectrum to the blue. Thus, the spectral shifts in solutions are the resultants of several individual effects which sometimes reinforce one another and sometimes cancel out [24]. But, in comparison with the aniline spectral shifts in hydro-

(I) (II) (III) (IV)

(A) (B)

carbons, one sees, in table 1, that exist very great displacements in the oxigenated solvents (excepting water and alcohols) and in acetonitrile ; also, in the halogenated solvents the shifts are approximately double as in hydrocarbons. To explain these facts we must take into account the formation of a complex between the aniline and the solvent molecule.

The aniline has two possibilities of hydrogen bonding formation [39-40] as the structures (A) and (B). In the case (A), the aniline acts as a proton donor, and in the case (B) it acts as a proton acceptor. Nagakura and Baba [41] have found abnormal great spectral shifts of aniline in dioxane and ethyl ether, typical proton acceptors. Our results agree with the previous findings [41]. In the structure (A) the valence electrons in the N-H bond are displaced with a limited quantity towards the nitrogen atom, determining a decrease of the ionization energy of the hydrogen atom [41]. On the basis of the molecular orbitals theory [41] one considers that takes place an increase of the displacement velocity of the electrons from the substituent to the benzene ring having as a result an augmentation of the spectral shift. In aniline the resultant dipole moment lies not in the benzene ring plane [41], and the displacement of the electrons diminishes the angle between the resultant dipole moment and the C-N bond. The maximum overlap of the nitrogen lone pair with the carbon π-electrons of the phenyl group occurs if all the atoms of the C-NH$_2$ group are in the same plane, a configuration in which the nitrogen uses sp^2 orbitals with angles of 120°, and the lone pair occupies a pure p orbital. Overlap between lone pair electrons and the ring π-electrons is larger the more p character has the lone pair. Hence, resonance tends to open the HNH angle toward 120° at the expense of promotion energy required to raise the lone pair from an sp^3 hybrid orbital, as in ammonia molecule [42], to a pure p orbital. Accordingly, it is expected that the HNH bond angles of aniline should lie between 109° and 120° the angles expected for sp^3 and sp^2 hybridization [17]. The angle between the resultant dipole moment and the ⟩C-N bond was evaluated by Few and Smith [43] as 53°, and by Nagakura and Baba [44] as 31°-35°. Experimentally, Brand et al [45] have found the value of 46°. These values are enough inexact to indicate that the dipole is perpendicular to or in the prolongation of the benzene ring, the conclusion being only the decrease of this angle, facilitating the displacement of the nonbonding electrons of the nitrogen towards the benzene ring. The nonplanarity of aniline molecule was proved by the studies of the ultrashort waves spectra [46], and of the infrared spectra [47-48]. Lister et al [49], in a microwave study, show that the NH$_2$ group is not planar, the HNH angle being of 113.9°, and existing a small inversion barrier. The variation of these two angles, between the resultant dipole moment and the benzene ring plane, and the HNH angle modifies the spatial configuration of aniline molecule in the excited state due, probably, to the geometry of the solvent molecules, the angles having different values in various solvents and determining the alteration of the number of the solvent molecules which interact with the solute molecule. In water and alcohols, the aniline forms hydrogen bonding of the (B) type. As Hehre et al [28] shows that in planar form the dipole moment is smaller, and if the nitrogen atom is in the sp^3 hybridization, then the formation of the hydrogen bond of this type is more favoured. The blocking of the nonbonding electrons of the nitrogen greatly lowers the ground state of the aniline, reducing, also, the polarity, and determining the interaction with a smaller number of solvent molecules, the spectrum being strongly shifted to the blue.

In the halogenated solvents the spectral shifts of aniline are comparable with those of the previous studied molecules [1-4] indicating the formation of a weak complex [50] by means of charge transfer from the aromatic ring to the halogenated solvent. According to Rao et al [51] the nature of such a complex is donor-acceptor. Briegleb [52] shows that the spectral studies [50] and the freezing points studies [53-55] are arguments in the favour of complex formation. Goates et al [56] suggests a π-bond formation between benzene ring and the 3d (4d) vacant level of the halogen atom in the solvent. The complex molecule is greater in the excited state and interacts with an increased number of solvent molecules, due to the packing strain [24], resulting in a greater solvation energy and the spectrum being red shifted.

The above discussion shows that the "universal interactions" and "specific interactions" intervene [57] in the spectral shifts of a solute molecule. But, one sees that the same molecule can form two kinds of hydrogen bonds and a donor-acceptor complex, at different positions of the molecule, justifying the denomination of "local specific interactions", in this case.

EXPERIMENTAL

The aniline p.a. (Merck) was fresh distilled, and the solvents were purified by known methods [58]. The refractive indices n, were determined with an Abbé refractometer, of the G type, for the yellow sodium light. The dielectric constants, D, were evaluated with an Oehme apparatus of 600 RL Type, and together with the n's and the boiling points, b.p. as purity criteria, are given in Table 1. The temperature was maintained constant with an Ultrathermostat U 10, at $20 \pm 0.1°$. The spectra were recorded with a S'PECORD UV-VIS spectrophotometer, Carl Zeiss, Jena, having an accuracy of ± 20 cm^{-1}, with the "nonius". The vapour spectrum of aniline was obtained with a suitable gas cell, of our own construction, with quartz windows, and of 9.762 cm lengths, at 22.3°, and a pressure of 0.382 mm, that is a concentration $c=2.070 \times 10^{-5}$ M. The concentrations of aniline solutions ranged betwen 2.500×10^{-3} and 2.500×10^{-4} M to avoid the association. The cells used were of 0.1 and 1.0 cm thickness. The position of each peak is the mean of at least three measurements.

REFERENCES

[1]   V. Macovei, Rev. Roumaine Chim., 20: 1413 (1975).
[2]   V. Macovei, Rev. Roumaine Chim., 21: 193 (1976)
[3]   V. Macovei, Anal. Sti. Univ. "Al.I.Cuza", Iasi, Sect.I,b,22: 51 (1976)
[4]   V. Macovei, Rev. Roumaine Chim., 25: 651, (1980)
[5]   V. Macovei and G. Bourceanu, unpublished results
[6]   V. Macovei, Anal. Sti. Univ. "Al.I.Cuza" Iaso, Sect.I,B,22:61 (1976)
[7]   V. Macovei, Rev. Roumaine Chim., 21: 1137 (1976)
[8]   V. Macovei, Rev. Roumaine Chim., 21: 1137 (1976)
[9]   F.A. Matsen, J. Am. Chem. Soc., 72: 5243 (1950)
[10] A.L. Sklar, J. Chem. Phys., 7: 984 (1939)
[11] K.F. Herzfeld, Chem. Rev., 41: 233 (1942)
[12] A.L. Sklar, Rev. Mod. Physik., 14: 232 (1942)
[13] H. Hückel, Z. Physik, 72: 310 (1931)
[14] W. Wheland, J. Am. Chem. Soc., 64: 900 (1931)
[15] S. Nagakura and H. Baba, J. Atm. Chem. Soc., 74: 5693 (1952)
[16] H. Sponer and E. Teller, Rev. Mod. Phys., 13: 76 (1941)
[17] H.H. Jaffé and M. Orchin, "Theory and Applications of Ultraviolet
          Spectroscopy" Wiley, New York (1964)

[18]   L. Doub and J.M. Vandenbelt, <u>J. Am. Chem. Soc.</u>, 69: 2714 (1947)

[19]   P.C. Upadhya, G. Baruah, K.N. Upadhya and D.K. Rai, <u>Isr J. Chem</u>,
            9: 7 (1971)

[20]   L. Doub and J.M. Vandenbelt, <u>J. Am. Chem. Soc.</u>, 71: 4114 (1949)

[21]   L. Doub and J.M. Vandenbelt, <u>J. Am. Chem. Soc.</u>, 77:4535 (1955)

[22]   N. Fuson, C. Garrigou-Lagrange and M.L. Josien, <u>Spectrochim. Acta.</u>,
            16: 106 (1960)

[23]   F.A. Trombetti and C. Zauli, <u>Boll. Sci. Fac. Chim., Bologna</u>, 21: 202 (1963)

[24]   N.S. Bayliss and E.G. Mc Rae, <u>J. Phys. Chem.</u>, 58, 1002 (1964)

[25]   F. London, <u>Z. Physik</u>, 63: 245 (1930)

[26]   O. Sverdlova, <u>Opt. i Spektrosk.</u>, 2: 31 (1963)

[27]   E.G. McRae, <u>J. Phys. Chem.</u>, 61:562 (1957)

[28]   W.J. Hehre, L. Radom and J.A. Pople, <u>J. Am. Chem. Soc.</u>, 94: 1496 (1972)

[29]   O.A. Osipov, V.J. Minkin, and A.D. Garnovskii, "Spravochinic po Dipol'nym
            Momentam", Izd. "Vysshaia Shkola", Moskva, 1971.

[30]   A.L. McClellan, "Tables of Experimental Dipole Moments",
            W.H. Freeman, San Francisco, California, 1963

[31]   N. Mataga and T. Kubota, "Molecular Interactions and Electronic Spectra",
            M. Dekker, Inc., New-York, 1966

[32]   D. York, <u>Canad. J. Phys.</u>, 44: 1079 (1966)

[33]   P. Suppan, <u>J. Chem. Soc.</u> (A), 1968, 3125

[34]   M. Lamotte, G.A. Gerhold and J. Joussot-Dubien,
            <u>J. Chim. Phys.</u>, 67: 2006 (1970)

[35]   "Manualul Inginerului Chimist", vol.1, Ed. Tehnica, Bucuresti, 1951

[36]   L.S. Prabhumirashi, D.K. Narayanann Kutty and A.S. Bhide,
            <u>Spectrochim, Acta</u>, 39A: 663 (1983)

[37]   N. Mataga, <u>Bull. Chem. Soc. Japan</u>, 36: 1607 (1963)

[38]   P. Suppan, <u>Spectrochim. Acta</u>, 41 A: 1353 (1985)

[39]   A.E. Lutskii, S. Gol'berkova and P.M. Bugai,
            <u>Zhur.obshchei Khim</u>, 33: 1624 (1963)

[40]   C.V. Cumper and A. Singleton, <u>J. Chem. Soc.</u>, 1968 B, 649

[41]   S. Nagakura and H. Baba, <u>J. Am. Chem. Soc.</u>, 74: 5693 (1952)

[42]   W. Gordy, W.V. Smith and R.F. Trambarulo, "Microvave Spectroscopy"
            Wiley and Sons, New York, 1953

[43]   A. Few and J.W. Smith, <u>J. Chem. Soc.</u>, 1938, 1958

[44]   S. Nagakura and H. Baba, <u>J. Chem. Soc., Japan</u>, 71: 527 (1950)

[45]   J.C.D. Brand, D.R. Williams and T.J. Cook, <u>J. Mol. Spectroscopy</u>,
            20: 359 (1966)

[46]   T.J. Bhattacharyga, <u>Indian J. Phys.</u>, 36: 33 (1962)

[47]   S.F. Mason, <u>J. Chem. Soc.</u>, 1958, 3619

[48]   P. Kreuger, <u>Canad. J. Chem.</u>, 40: 2300 (1962)

[49]   D.G. Lister and J.K. Tiler, <u>Chem. Commun.</u>, 152 (1966)

[50]   R. Andersen and J.M. Prausnitz, <u>J. Chem. Phys.</u>, 39: 1225 (1963)

[51]   N.S. Rao and K.K. Jatkar, <u>Quart. J. Ind. Sci.</u>, 51: 65 (1942)
            6:1 (1943)

[52]   G. Briegleb, "Elektronen-Donator-Acceptor Komplexe",
            Springer Verlag, Berlin, 1967

[53]   A.F. Kapustinskii and S.I. Drakin, <u>Izv. Akad. Nauk SSSR</u>,
            Otdel. Khim. Nauk, 5: 435 (1947)

[54]   W.E. Wyat, <u>Trans. Faraday Soc.</u>, 25: 48 (1929)

[55]   R.P. Rostogi and R.K. Nigam, <u>Trans. Faraday Soc.</u>, 55: 2005 (1959)

[56]   J.R. Goates, R.J. Sullivan and J.B. Ott, <u>J. Chem. Phys.</u>, 63: 589 (1954)

[57]   B.S. Neporent and N.G. Bakhshiev, <u>Opt.i. Spektrosk.</u>, 8: 77 (1960)

[58]   A. Weissberger, E.S. Proskauer, J.A. Riddich and E.E. Toops,
               "Technique of Organic Chemistry", vol.VII - "Organic
               "Organic Solvents", 2-nd ed., Interscience, New-York, 1955

DISCUSSION

<u>DELAIRE</u> - Peut-on faire des calculs quantiques sur les liaisons hydrogène dans les systèmes aniline-solvant ?

<u>MACOVEI</u> - On peut le faire mais ici on a présenté uniquement la partie expérimentale des déplacements spectraux en solution dans le but d'expliquer les interactions intermoléculaires.

FISCHER-TROPSCH SYNTHESIS IN A GAS-LIQUID-SOLID MEDIUM : SELECTIVITIES INDUCED BY PHASE EQUILIBRIA

Dominique Vanhove

Laboratoire de Catalyse Organique, U.A. 231 CNRS,
E.S.C.I.L; 43, Boulevard du 11 Novembre 1918
69622 Villeurbanne Cédex

The Fischer-Tropsch synthesis of hydrocarbons, when using common iron or cobalt industrial catalysts, usually exhibits poor selectivity, especially regarding the carbon chain length.

Recent results have established that selectivity can be improved by metal deposition on porous catalysts, especially with cobalt or ruthenium, and that a liquid-suspended catalyst can induce a very different selectivity. We therefore studied the influence of the solvent under conditions where other parameters have little effect, and also interpreted results so as to take into account the gas-solid reactions on porous catalysts.

Activity is not greatly modified in most cases by using a liquid phase as reacting medium ; moreover conversion of reactants had little influence on selectivity of the reaction. The reaction in a liquid phase has led to great differences in chain length and nature of the formed hydrocarbons. A direct relation was shown between amounts produced of methane and branched hydrocarbons and the amount of metallic cobalt in the slurry. Moreover, all catalysts have given mainly C_4-C_7 hydrocarbons when tested in the slurry form, whatever the catalyst support had been. The hydrocarbon mass distribution appeared to drop markedly for a certain chain length limit and the production of light hydrocarbons was increased at the same time. The theoretical Schulz-Flory-Anderson law was thus not respected as observed in gas-solid reactions where selectivities were clearly related to the porous diameter of the support. In this latter case, capillary condensation of hydrocarbons was manifested by dynamic adsorption-desorption experiments and the Kelvin law was used successfully to describe selectivities of numerous catalysts with a large range of porous diameters. When the same samples are tested in the same reactor for gas-solid and gas-liquid-solid reactions the chain length limit was displaced from, for instance, C_{12} to C_8 ; the influence of porosity is thus masked.

These results are interpreted as an influence of the liquid-vapour equilibrium leading to increased effective residence times of products. These residence times depend on the nature of the interface : gas-solution or gas-liquid-solid. The contact time of the products with the active metal increases very rapidly with carbon number (e.g. 1 hour for octane in the liquid phase) due to the existence of a condensed phase : solution or product in the pore structure of the catalyst. This effect, in addition to the corresponding increase in concentration of heavy hydrocarbons in the condensed phase, modifies the formal kinetic scheme of this complex reaction by the interference of secondary hydrocracking ; heavy hydrocarbons are converted to methane and linear or branched light hydrocarbons. The simulation of this kinetic network has led to selectivities in excellent accordance with the experimental results.

The influence of the liquid-vapour equilibrium could be used to adjust the selectivity of complex reactions by the proper choice of reacting medium or catalyst support.

FISCHER-TROPSCH SYNTHESIS IN A GAS-LIQUID-SOLID MEDIUM : SELECTIVITIES
INDUCED BY PHASE EQUILIBRIA

Dominique Vanhove

Laboratoire de Catalyse Organique, U.A. 231 CNRS
E.S.C.I.L; 43, Boulevard du 11 Novembre 1918
69622 Villeurbanne Cédex

INTRODUCTION

La synthèse de Fischer-Tropsch, permettant de produire essentiellement des hydrocarbures saturés à partir du mélange CO/H$_2$ sur des
catalyseurs industriels classiques à base de fer ou de cobalt, ne
donne, en général, pas de bonnes sélectivités pour des gammes choisies
d'hydrocarbures. Certes, par modification du catalyseur ou par
ajustement des variables opératoires sur les unités Sasol, on peut
modifier dans une certaine mesure les proportions respectives des
différentes coupes pétrolières, mais le contrôle réel de la longueur
des chaînes carbonées formées par la réaction reste toujours un
problème.

Une série de résultats obtenus par diverses équipes dans les dernières
années [1-4], a montré que le dépôt de métal sur un support poreux permettait
d'atteindre ce résultat. Les cas les plus démonstratifs ont été obtenus
avec le cobalt [2] et le ruthénium [1],conduisant à des coupes bien
restreintes d'hydrocarbures dans des conditions stationnaires. D'autre
part, des études conduites avec des catalyseurs au cobalt en suspension
dans un liquide donnent des sélectivités très différentes [5-7] en ne formant
que des hydrocarbures à moins de 8 atomes de carbone.

Ce comportement particulier et son intérêt potentiel ont justifié
l'étude de l'effet spécifique de la présence d'un solvant en synthèse de
Fischer-Tropsch, dans des conditions éliminant au maximum l'influence des
autres facteurs, et la recherche d'une interprétation qui tienne également
compte des résultats en réaction gaz-solide sur les catalyseurs poreux.

REACTION SUR CATALYSEURS DEPOSES SUR BILLES D'ALUMINE

Influence de la Porosité sur la Sélectivité

Par dépôt à faible teneur de cobalt carbonyle sur des alumines et une
silice-alumine de porosités diverses puis en décomposant ce précurseur à
température modérée (< 200°C), on obtient des catalyseurs formés de fines
particules de cobalt métallique et dont les propriétés catalytiques
s'avèrent différentes des catalyseurs classiques.

Alors que ces catalyseurs, à forte teneur en phase active et obtenus
par réduction de sels ou d'oxydes réduits à température élevée (500°C),

forment des hydrocarbures dans une gamme très large avec des sélectivités vérifiant une loi de type polymérisation (dite de Schulz-Flory), les catalyseurs préparés par les carbonyles donnent des gammes très restreintes d'hydrocarbures. La plus grande proportion d'hydrocarbures (>70%) formés se trouve dans une coupe de 5 à 6 masses moléculaires différentes et une relation directe a pu être mise en évidence entre la masse moyenne de cette coupe et le diamètre poreux du support du catalyseur [2,8].

Essais en Phase Liquide

Les mêmes catalyseurs ont été essayés en présence d'un liquide organique (décaline) pour la même réaction ; dans ce cas, la synthèse fournit toujours des hydrocarbures saturés mais dont la longueur de chaîne ne dépasse pas 7 C. L'analyse détaillée des produits a fait apparaître une formation accrue d'hydrocarbures branchés dans ce milieu et leur proportion dépend de manière linéaire de la teneur en cobalt du milieu liquide [5] quelles que soient la proportion de métal et la nature du support dans le catalyseur. Bien que moins nette, l'évolution de la formation de méthane semble similaire (voir Fig. 1).

Ces faits ne sont pas dus à l'intervention du support par lui-même mais bien à celle de la phase active ; ils sont le signe de réactions secondaires favorisées par la présence du solvant liquide. Cependant, la nature des supports employés (billes de plusieurs mm de diamètre) ne permet pas d'éliminer l'influence des transferts de matière entre la phase gaz et la phase liquide sur la cinétique, aussi d'autres types de supports permettant de réaliser les essais dans les deux milieux et avec le même réacteur, de manière à lever toute ambiguité, ont été recherchés.

REACTION SUR CATALYSEUR DEPOSE SUR ALUMINE AEROGEL

En utilisant une alumine aérogel Degussa C, très pure et très divisée (20 microns), on obtient, après dépôt et décomposition du cobalt carbonyle, un catalyseur très fin qui forme une suspension stable dans l'orthoterphényle. Les tests catalytiques ont pu ainsi être effectués avec la même charge de catalyseur dans un même réacteur et pour les mêmes conditions expérimentales, à la fois en réaction gaz-solide et gaz-liquide-solide. Le réacteur tubulaire porté à 200°C et contenant 20 g de catalyseur est balayé par un mélange CO/H2 de rapport molaire 1/2 à pression atmosphérique ; toutes les conditions ont été gardées constantes pour les deux types d'essais. L'analyse chromatographique de l'ensemble des consti-

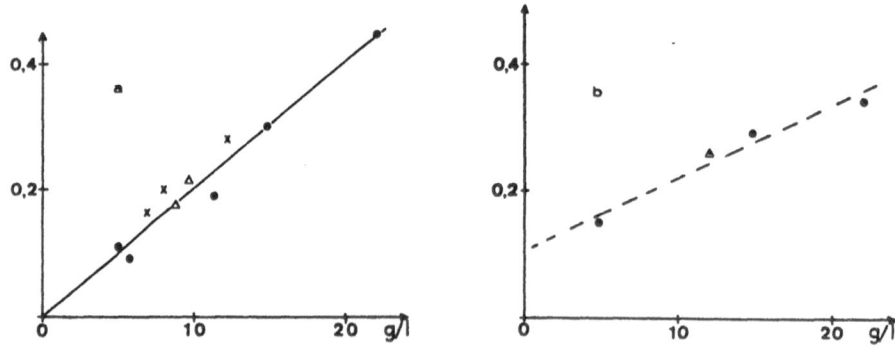

Figure 1. Effet de la teneur en cobalt dans la suspension sur le rapport isopentane/n-pentane (a) et sur la sélectivité en méthane (b) à masses, teneurs et nature du support variables ; points : alumine R.P. SCS 350, croix : alumine R.P. SCS 9, triangles : anatase

tuants de l'effluent du réacteur est effectuée en ligne au bout d'un temps de fonctionnement suffisant pour être en état stationnaire. Le catalyseur de teneur en cobalt égale à 4,6% est sous forme de particules de métal de 4 mm de dimension.

Les essais ont tout d'abord été effectués en réaction gaz-solide en lit fixe, puis 50 ml d'orthoterphényle fondu ont été introduits à l'abri de l'air sans transfert du catalyseur ; le fonctionnement du réacteur a alors été du type colonne à bulles.

Comparaison des activités

Les catalyseurs du type aérogel s'avèrent nettement plus actifs en phase gazeuse que les catalyseurs correspondants déposés sur alumine SCS Rhône-Poulenc en billes ; l'état de division important de l'aérogel doit expliquer ce résultat. Par contre, la phase liquide ramène tous les catalyseurs à des niveaux d'activité semblables (table 1). Il semble donc qu'il existe dans ce cas, soit une limitation diffusionnelle à l'arrivée des réactifs (probablement l'hydrogène) sur le catalyseur, soit une compétition avec le solvant pour l'accès ou la consommation de ce même réactif.

Comparaison des sélectivités

La comparaison des concentrations en hydrocarbures formés à même temps de contact dans un diagramme de Schulz-Flory montre, d'une part, la diminution de l'activité due à la phase liquide, d'autre part une modification importante de la sélectivité.

En phase gaz (courbe 2a), la sélectivité suit la loi linéaire de Schulz-Flory jusqu'à l'heptane; la valeur du rapport vitesse de croissance de chaîne/vitesse globale est égale à 0,8, ce qui correspond au comportement usuel de tous les catalyseurs au cobalt. Cependant, au delà de l'heptane, on observe un accroissement net de la formation des C_8-C_{11} tandis que les hydrocarbures plus lourds n'apparaissent pas en teneurs notables. De ce point de vue, ces catalyseurs se comportent donc comme les catalyseurs préparés à partir du cobalt carbonyle sur supports poreux (alumine SCS ou autres) [2,8].

En phase liquide (courbe 2b), la limite supérieure pour les hydrocarbures formés est nettement déplacée et se trouve aux C_7 avec un enrichissement en produits plus légers qui ne permet plus d'observer la

Table 1. Activités des catalyseurs ex-$Co_2(CO)_8$ supportés (mmol $CO.h^{-1}.(g\ Co)^{-1}$) ; T : 200°C, P = 100000 Pa, CO/H_2 = 1/2.

Support Teneur en Co	Alumine SCS 9 R.P. 2%	Alumine SCS 350 R.P. 5,4%	Alumine C Degussa 4,6%	Anatase 4,5%
Réaction gaz-solide	1,7		5,5	
Réaction en suspension	1,6[a]	1[a]	1,1[b]	2,4[a]

[a] Solvant : décaline
[b] Solvant : o-terphényle

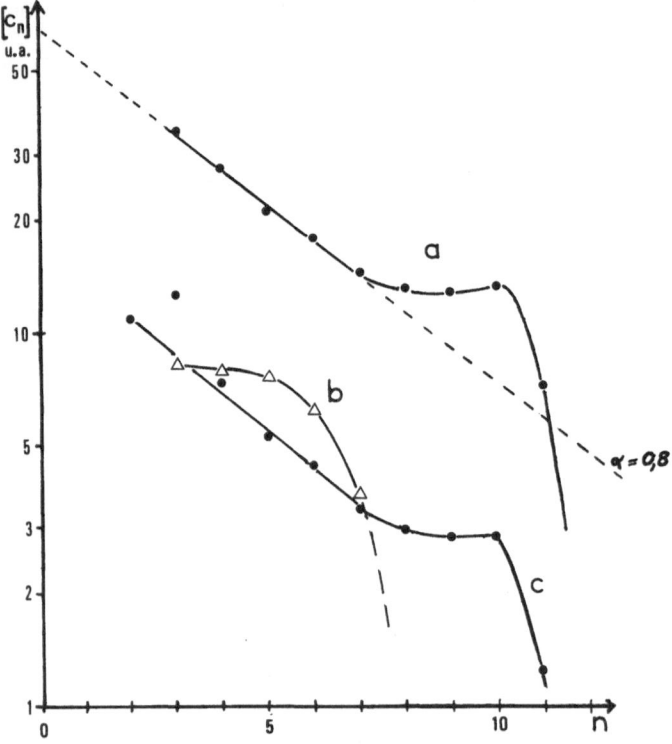

Figure 2. Représentation dans un diagramme de Schulz-Flory des concentrations molaires de chaque hydrocarbure formé en fonction du nombre de carbones de la molécule ; catalyseur déposé sur alumine C Degussa utilisé (a) en phase gaz (conversion CO = 0,419), (b) en milieu liquide (0,078), (c) en phase gaz à débit différent (0,081)

portion linéaire de la loi de Schulz-Flory. Des essais à temps de contact plus faible ont permis de vérifier que l'effet observé n'était pas dû à une différence dans la conversion des réactifs ; en effet, à même niveau de conversion, les différences de sélectivité observées entre les deux milieux réactionnels sont conservées (courbes 2b et 2c).

DISCUSSION ET CONCLUSION

Les conditions réactionnelles identiques montrent que la sélectivité en synthèse de Fischer-Tropsch peut être modifiée par d'autres facteurs que la nature et l'état du catalyseur (dimension des particules métalliques p.e.) ; ces facteurs peuvent être la dimension des pores du support, en réaction en phase gazeuse, ou la présence d'un milieu liquide.

Par des essais de mesure de coefficients de diffusion des hydrocarbures dans les supports, nous avons pu montrer précédemment [8] que l'équilibre liquide-vapeur ne s'établit plus à la surface des pores mais à l'extérieur de ceux-ci lorsque les hydrocarbures dépassent une longueur de chaîne liée à la porosité du support. Ce phénomène peut s'expliquer par la condensation capillaire des hydrocarbures. De fait, une bonne corrélation analogue à la loi de Kelvin qui règle cette condensation, est obtenue entre le logarithme de la tension de vapeur saturante du dernier hydrocarbure formé sur un catalyseur et l'inverse de son rayon poreux (voir figure 3). Les hydrocarbures condensés peuvent rester dans les pores pendant un temps beaucoup plus long que les hydrocarbures plus légers et subir des transformations secondaires à la condition que le métal actif soit effectivement à l'intérieur de ces pores. Ceci est réalisé dans des conditions de préparation douces et à faible recouvrement en métal ; dans les autres conditions de préparation et pour les hydrocarbures non condensés, l'influence de ces réactions secondaires est négligeable puisqu'il n'y a pas contact prolongé entre la phase active et les hydrocarbures formés par la synthèse.

Sur les mêmes bases, les résultats obtenus en phase liquide peuvent s'expliquer par l'influence de l'équilibre liquide-vapeur sur le temps de contact réel des hydrocarbyres avec le catalyseur. Dans ce cas, l'effet de la porosité n'existe plus puisqu'il n'y a plus d'interface gaz-liquide-solide, mais il est remplacé par l'effet du solvant qui, en raison de son volume relativement important, augmente dans de fortes proportions le temps de séjour réel des hydrocarbures : une simulation montre que, dans nos conditions de manipulation et si l'on obtenait une répartition de Schulz-Flory, le temps de séjour serait de l'ordre de 20 mn pour l'hexane,

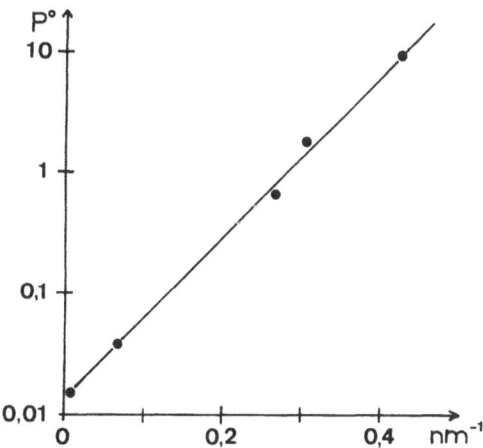

Figure 3. Tension de vapeur saturante du dernier hydrocarbure formé en fonction de l'inverse du rayon poreux du support des catalyseurs ex-$Co_2(CO)_8$ supportés (alumine SCS 9, SCS 59, SCS 250, SCS 350, silice-alumine-alumine 2,4 nm)

supérieur à l'heure pour l'octane et dépasserait 8h pour le dodécane. Dans ces conditions, des réactions secondaires même très lentes peuvent devenir prédominantes par rapport à la réaction de synthèse qui ne fait intervenir que des gaz non condensables. La même simulation montre que la fraction molaire en solution devrait augmenter d'un facteur 7 entre l'hexane et le dodécane, or, l'analyse de la phase liquide met en évidence la présence des hydrocarbures légers alors qu'aucune trace d'hydrocarbures au delà de l'octane n'est détectée.

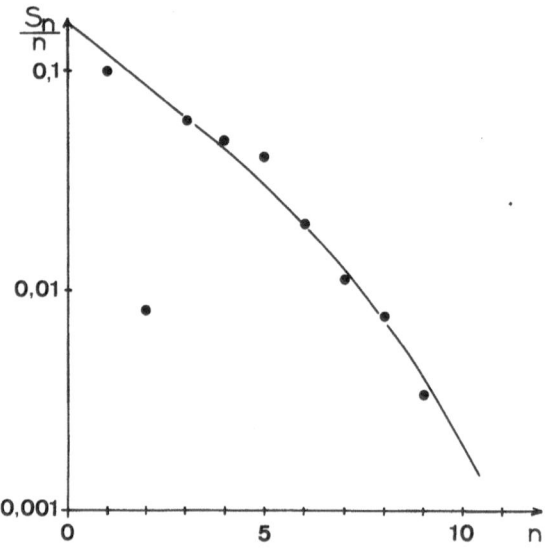

Figure 4. Sélectivités mesurées (points) et simulées dans le cas d'une réaction d'hydrocraquage ne tenant pas compte des produits formés (courbe) ; catalyseur déposé sur silice-alumine 2,4 nm

Le cobalt est un métal connu pour ses propriétés hydrogénantes et hydrogénolysantes à température relativement basse [9]. Les réactions d'hydrocraquage se manifestent en outre de l'apparition d'hydrocarbures de masse inférieure par un accroissement de la formation du méthane et des hydrocarbures ramifiés ; ce sont exactement les phénomènes observés particulièrement en réaction en phase liquide : disparition au delà d'une certaine longueur de chaîne mais formation accrue d'hydrocarbures de masses immédiatement inférieures, augmentation de la formation de méthane et de la proportion des hydrocarbures branchés, proportionnelle à la concentration de la suspension en métal et donc à la probabilité pour un hydrocarbure en solution de se transformer.

Une simulation cinétique des réactions successives de synthèse d'hydrocarbures suivant la loi de Schulz-Flory et de disparition suivant une loi d'ordre 1 par rapport à l'hydrocarbure (supposée rendre compte d'une réaction d'hydrocraquage) incorporant l'effet décrit de l'équilibre liquide-vapeur montre dans tous les cas que la pente du diagramme de Schulz-Flory passe de la valeur 0,1 à la valeur 0,32, le point de rupture de pente dépendant de la constante cinétique de la réaction d'hydrocraquage ou du diamètre poreux (dans le cas de réaction gaz-solide). Ces valeurs correspondent tout à fait aux pentes des parties linéaires du diagramme, en particulier pour les réactions en phase gazeuse dont l'analyse est obtenue avec plus de précision (voir figure 4) ; par contre on ne peut rendre compte de l'accroissement de sélectivité dû à la réaction secondaire puisque les produits de cette réaction ne sont pas incorporés dans le modèle.

La synthèse de Fischer-Tropsch sur catalyseurs au cobalt préparés de manière "douce" a fait apparaître deux moyens pour ajuster la sélectivité de la réaction sur une coupe restreinte d'hydrocarbures : le choix d'un support de porosité adaptée ou l'utilisation d'une phase liquide. Ces deux moyens utilisent le même phénomène physique : l'équilibre liquide-vapeur mais se produisant à des interfaces différentes, pour modifier profondément la sélectivité de la réaction. Le maintien en contact prolongé de certains produits primaires avec la phase active par dissolution ou condensation permet l'intervention de réactions jusqu'alors négligeables et modifie la cinétique formelle de la réaction globale. La phase liquide (solvant ou condensée) montre ainsi une intervention purement physique sur le résultat d'une réaction chimique qui laisse entrevoir des implications nombreuses au niveau des réacteurs et réactions pétrochimiques.

REFERENCES

1. R. J. Madon, J.Catalysis, 57:183 (1979).
2. D. Vanhove, L. Makambo and M. Blanchard, J.Chem.Soc, Chem.Commun. (1979) 605.
3. H. H. Nijs, P. A. Jacobs and J. B. Uytterhoeven, J.Chem.Soc, Chem.Commun, (1979) 180.
4. D. Commereuc, Y. Chauvin, F. Hughes and J. M. Basset, J.Chem.Soc, Chem.Commun. (1980) 154.
5. D. Vanhove and M. Blanchard, J.Chem.Res., (1979) S 404, M 4756,
6. M. Blanchard, D. Vanhove, F. Petit and A. Mortreux, J.Chem.Soc., Chem.Commun. (1980) 908.
7. D. Vanhove, M.Blanchard, F. Petit and A. Mortreux, Nouv.J.Chim., 5:205 (1981).
8. D. Vanhove, Zhang Zhuyong, L. Makambo and M. Blanchard, Appl.Catal., 9:327 (1984).
9. G. Leclercq, S. Pietryk, M. Peyrovi and M. Karroua, Entropie, 113-114:88 (1983).

DISCUSSION

VILLERMAUX - Peut-on relier quantitativement les observations à des mesures effectuées par la méthode chromatographique sur le catalyseur poreux afin de vérifier l'explication selon laquelle l'écart à la distribution de Schulz-Flory provient de différences de pénétration au sein du catalyseur ?

VANHOVE - La méthode chromatographique n'a été essayée que sur des supports hors réaction. L'étude du système en réaction est actuellement en cours pour la réaction gaz-liquide-solide dans le but de déterminer les temps de séjour réels et les constantes de l'équilibre.

BELLONI - Is the limitation of the chain-length in the gaseous Fischer-Troppsch also due to a competing cracking ?

VANHOVE - Yes, the retention of certain hydrocarbons by dissolution or condensation in the pores have as a consequence the contact of these hydrocarbons for a longer time with the active metallic phase and permit the hydrocracking.

THE ROLE OF ELECTRONIC STRUCTURE CALCULATION

IN MECHANISTIC ANALYSIS OF ELECTRON TRANSFER REACTIONS IN THE LIQUID PHASE

Marshall D. Newton

Chemistry Department
Brookhaven National Laboratory
Upton, New York 11973 USA

ABSTRACT

Model calculations have been employed in elucidating the mechanism of electron transfer reactions in aqueous solution. The contribution of inner shell OH bonds to activation barriers has been estimated from calculation for metal ion hydrates. Calculated electron transfer matrix elements (H_{if}) for redox processes of the type, $ML_6^{2+} + ML_6^{3+} \rightleftharpoons ML_6^{3+} + ML_6^{2+}$, M = Fe, Co, or Ru, L = H_2O or NH_3, have been analyzed in terms of various orbital concepts. The matrix elements are based on ab initio wavefunctions for model supermolecule clusters of the type, $\overline{(ML_n \cdots L_n M)}^{5+}$, with n = 1 or 3. The analysis shows that the many-electron H_{if} quantities can in fact be expressed to good approximation as effective 1-electron expressions of the type, $H_{if} \propto (\lambda')^2 N_c h_{L_\ell L_r}$, where λ' is the metal-ligand covalency parameter, $h_{L_\ell L_r}$ is a local 1-electron matrix element for ligand orbitals in contact in the transition state, and N_c is the number of such contacts. A least-squares fit of the data implies a value of ~ 5000 cm^{-1} for $h_{L_\ell L_r}$, showing that significant coupling can occur in the absence of formal bonding between reactants.

INTRODUCTION

Electronic structure plays a pivotal role in the kinetics of electron transfer reactions, as one might well expect from the name of this class of reactions.[1-10] It follows that electronic structure calculations for suitable molecular clusters can provide valuable insights into the mechanism of these reactions. If the initial (Ψ_i) and final (Ψ_f) states of an electron transfer process are characterized in terms of diabatic potential (or free) energy surfaces, then one recognizes the distinct roles of electronic structure in determining both the energy surfaces themselves (diagonal matrix elements) and the coupling between them (off-diagonal matrix elements), without which the reaction would not proceed. The diagonal matrix elements (expectation values) yield the magnitudes of activation barriers. These barriers have contributions from each structural parameter whose equilibrium value differs in the initial and final states. In the next section we shall see how cluster calculations have helped in identifying a source of activation which may be conveniently probed in terms of a kinetic H/D isotope effect.

157

Much of the current interest in the off-diagonal coupling elements (referred to below simply as transfer integrals) has been focussed on the distance dependence of their magnitude for situations of transfer over relatively large distances (tens of angstroms), where questions of electronic overlap and electronic participation by the intervening medium clearly become critical.[3d,4-6] On the other hand, even for relatively small donor-acceptor systems essentially in contact, kinetically significant electronic structural effects on transfer integrals may be expected.[3] In the present paper we pursue this notion by analyzing transfer elements for a set of prototype transition metal complex redox pairs, as modeled by ab initio electronic structure calculations for suitable super-molecule clusters.[10]

In the remainder of this paper, we shall consider the following class of reactions involving hexacoordinate complexes in aqueous solutions,

$$ML_6{}^{2+} + ML_6{}^{3+} \;\rightleftharpoons\; ML_6{}^{3+} + ML_6{}^{2+} \tag{1}$$

where M and L refer to a transition metal and a ligand, respectively. Inevitably, in modeling condensed phase reactions of this type, one defines localized donor (D) and acceptor (A) species which, typically, are immersed in a dielectric medium.[3d] In the present case (eq 1), the D and A species are the $ML_6{}^{2+/3+}$ redox partners, whose structural identity is maintained throughout the reaction (i.e., a so-called "outer-sphere" reaction, in which M-L bonds are not disrupted (see below)). Since the D and A species are taken to be in contact in the transition state, we may consider the influence of the outer aqueous medium on the transfer integrals to be relatively minor and concentrate on the role played by the electronic properties of D and A (cf. ref. 11).

Before addressing the specific systems of interest, we will find it useful to consider some general questions about electron transfer reactions. These questions will be considered again at the end, in the context of the detailed analysis reported below.

To what extent may electron transfer reactions be considered unique? Aside from specific features such as polaron coupling to a dielectric medium, one in general expects relatively "weak" coupling between initial and final states. Whereas the diabatic model invoked above is a generic one applicable to many types of chemical reactions, including those involving complex rearrangements of chemical bonds, electron transfer reactions (at least those of the outer-sphere type, to which we confine our attention) do not entail disruption of short-range chemical bonds -- hence the description of the initial state/final state interactions as "weak" (i.e., a small transfer integral).[3d] These notions may be discussed in terms of schematic energy profiles along a reaction coordinate, as depicted in Fig. 1. In Fig. 1a we see the extreme of very weak coupling, in which a transfer integral of very small magnitude yields a very "cuspy" avoided crossing of the diabatic energy surfaces. As noted above, the surfaces for Ψ_i and Ψ_f are distinguished by the fact that some structural coordinates have equilibrium values which differ in the initial and final states (hence, an activation barrier), even though bonds are not "broken" (i.e., they are only "perturbed").

The other extreme (Fig. 1b) displays a "smooth" transition state associated with strong mixing of Ψ_f and Ψ_f in the interaction region, which is seen to be relatively broad compared with the narrow reaction zone in Fig. 1a. To the extent that electron transfer reactions are indeed characterized by weak coupling (Fig. 1a), they may be considered relatively unique by comparison with the more conventional looking transition state energy profile in Fig. 1b. In the latter case the

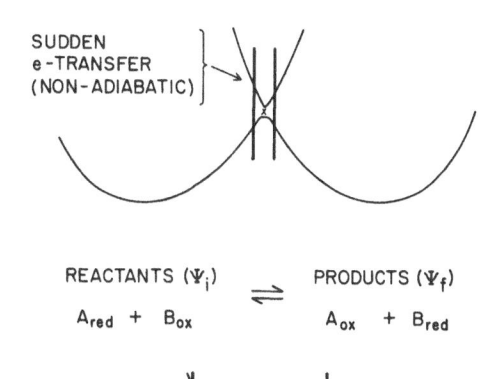

a)

SUDDEN
e-TRANSFER
(NON-ADIABATIC)

REACTANTS (Ψ_i) $\rightleftharpoons$ PRODUCTS (Ψ_f)

$A_{red} + B_{ox}$ $\qquad$ $A_{ox} + B_{red}$

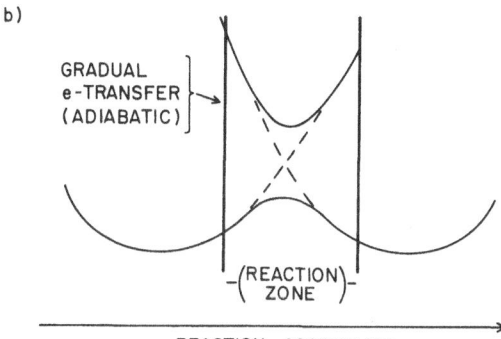

b)

GRADUAL
e-TRANSFER
(ADIABATIC)

$-$(REACTION ZONE)$-$

REACTION COORDINATE

Fig. 1. Energy profiles for reactants (Ψ_i) and products (Ψ_f) with
respect to the reaction coordinate for a generic redox process,
$A_{red} + B_{ox} \rightleftharpoons A_{ox} + B_{red}$: a) sudden (non-adiabatic);
b) gradual (adiabatic). The dashed lines refer to the diabatic
surfaces while the solid lines refer to the adiabatic surfaces.
The vertical bars define "reaction zones" where the energy is
within some interval (e.g., k_BT) of the barrier.

reaction is "gradual", whereas we designate the situation in Fig. 1a as
"sudden", corresponding to the well-known fact that many "electron
transfer" processes exhibit the same Franck-Condon control characteristic
of other fast electronic processes. We shall employ our model
calculations in attempting to place representative examples of eq 1
between the extremes defined by Fig. 1.

The connection between the energetics depicted in Fig. 1 and the rate
constant, k_{et}, for electron transfer may be illustrated in terms of the
Landau-Zener (LZ) model.[3d] If k_{et} is expressed as

$$k_{et} = \kappa_{el}\, k_{et}^{TST} \tag{2}$$

where $k_{et}^{TST} = \nu_n \exp(-E^{\dagger}/RT)$ is the transition state rate constant,
ν_n is an effective frequency for nuclear motion, and $E^{\dagger}$ is the
activation energy, then the LZ model yields,

$$\kappa_{el} = 2P_o/(1 + P_o) \tag{3}$$

$$P_o = 1 - \exp(-2\pi\gamma) \tag{4}$$

where P_0 is the probability for hopping between diabatic surfaces in the crossing region. For the case of a harmonic oscillator, we have

$$2\pi\gamma = (H_{if})^2 (\pi^3/E_\lambda k_B T)^{1/2}/h\nu_n \qquad (5)$$

where k_B is the Boltzmann constant and E_λ is the reorganization energy (equal to four times the activation energy in the case of a thermoneutral electron exchange reaction). The transfer integral H_{if} is given in general by

$$H_{if} = \frac{\int \Psi_i^* H \Psi_f - (\int \Psi_i^* \Psi_f)(\int \Psi_i^* H \Psi_i)}{1 - (\int \Psi_i^* \Psi_f)^2} \qquad (6)$$

H being the electronic Hamiltonian for the entire reactive system. In terms of eqs 2-6, the extremes represented by Fig. 1 correspond to the limits of non-adiabatic (Fig. 1a, $\kappa_{el} \ll 1$) and adiabatic (Fig. 1b, $\kappa_{el} \sim 1$) behavior. Eq 5 underscores the central role played by the transfer integral.

Aside from weakness of interaction, the possibility of being characterized as a "one-electron process" provides another criterion by which electron transfer reactions might qualify as being unique. Thus even though the transfer integral is, strictly speaking, a many-electron quantity, since it is a function of the many-electron wavefunctions (Ψ_i and Ψ_f) and Hamiltonian (see eq 6), it may be possible to express it approximately as an effective 1-electron resonance integral,

$$H_{if} \approx \beta_{DA} \qquad (7)$$

where ϕ_D and ϕ_A refer to effective 1-electron orbitals of the reactive donor/acceptor pair (eq 1). We shall explore this question both formally and computationally in the following sections, and will indeed suggest a simple, quasi 1-electron model which is in good quantitative agreement with our detailed computational results. Analysis in terms of 1-electron character is of general interest since many semi-empirical approaches for estimating transfer integrals are effective 1-electron models.[5-7]

A number of electronic structural characteristics have been employed in attempts to unravel the systematic behavior of transfer integrals. Typical questions involve the relative importance of saturated vs unsaturated ligands or bridging groups, sigma vs pi electronic manifolds, direct vs super-exchange coupling, and formally bonded vs non-bonded contact between donor and acceptor species.[4-9] One result of the present study is the demonstration that non-bonded contact between saturated species can lead to substantial transfer integrals. With regard to the important question of the effective distance for electron hopping, it is essential to understand the spatial extent of the effective donor and acceptor orbitals. In the case of metal/ligand complexes, the degree of delocalization of "metal" donor and acceptor orbitals onto the ligands which are in direct contact in the transition state is of particular importance, and as a result, we find that the nominal metal-metal separation is not necessarily a useful guide for assessing transfer integral magnitudes.

The emphasis in the present work is on the propensity for electron hopping, as controlled by the transfer integral, once the transition state has been attained. Hence, we shall not attempt to make contact with observed rate constants for the overall processes defined by eq 1. For the case of $Fe^{2+/3+}$ electron exchange, previous studies have indicated that the calculated transfer integral is consistent with the observed bimolecular rate constant.[12,13]

The transition states pertinent to eq 1 will in general involve a distribution of contact configurations for the D-A pair, as governed by the tradeoff between pair distribution function and local rate constant. The molecular clusters dealt with below include configurations which are thought to be of major importance in the overall kinetics.[5d,5e,12,13] The importance of a given configuration involves a tradeoff between effective electronic overlap and the magnitude of the pair distribution function for a configuration.[12] The face-to-face approach of reactants has been found favorable for the aqueous $Fe^{2+/3+}$ exchange, although one must take account of the restricted rotation entailed by the interpenetration of reactants in the transition state.[12,13]

HYDROGEN-BOND ASSISTED ACTIVATION DUE TO INNER SHELL OH BONDS

Cooperativity involving hydrogen bonding arises in a variety of situations in which a given species is simultaneously involved in hydrogen bonding and in some other chemical functionality, including an additional hydrogen bond.[14,15] In this section, we focus on the manner in which a water molecule in the coordination shell of a hydrated ion is affected by the joint influence of the charge of the coordinating ion and the coupling to outer solvent provided by hydrogen bonding.[16,17] This question, which has implications for both equilibrium[16] and kinetic[17] aspects of electrolyte cyemistry, has been addressed on the basis of ab initio SCF calculations for model hydrate clusters both with ($M(H_2O)_6\overline{(H_2O)_2}^{z+}$) and without ($M(H_2O)_6{}^{z+}$) outer shell solvent.

It is well known that the activation energy associated with electron transfer between coordinated metal ions may have significant contributions from differences in bond strengths in the initial and final charge or oxidation states of the ions, a consequence of the Franck-Condon control of electron transfer processes.[3d] Major attention has been focussed on the charge dependence of metal-ligand bond lengths; e.g., for first row transition metal ions, the equilibrium bond lengths for 3+ charge states are often .1 - .2 Å less than the corresponding values for 2+ charge states.[18]

Due in part to a relative dearth of accurate structural data pertaining to bonds involving hydrogen atoms, the possible role of such bonds in electron transfer mechanisms has not received much consideration to date. The model clusters described above and schematically represented in Fig. 2, have been employed to estimate the charge dependence of intra- as well as inter-molecular bonding associated with the coordinated water molecules.[16] This study reveals the striking manner in which the intrinsic charge dependence of the OH bond lengths in the coordination shell (as represented by the $M(H_2O)_6{}^{z+}$ complexes) is strongly amplified by the presence of $OH\bullet\bullet\bullet O$ hydrogen bonding between the coordination shell and the outer solvent (as represented by the $M(H_2O)_6(H_2O)_2{}^{z+}$ complexes).[16] Thus in contrast to the small variations (< 0.01 Å) in r_{OH} observed for the $m(H_2O)_6{}^{z+}$ complexes, we find an overall shift of 0.045 Å in going from z=1 to z=3 when hydrogen bonding is included. In Fig. 3 the r_{OH} values are shown to increase with z^2 in a roughly linear fashion, with the corresponding decrease in r_{MO} included for

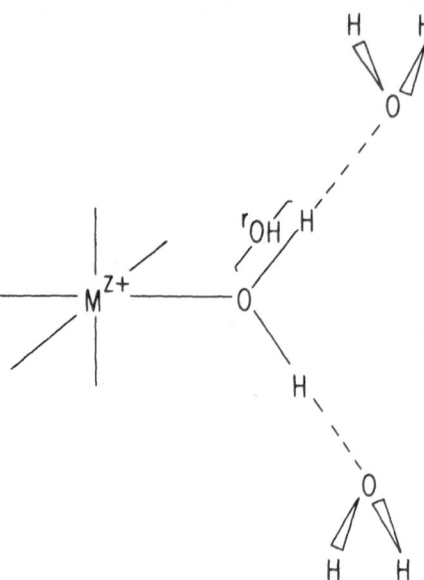

Fig. 2. Schematic representation of $M(H_2O)_6(H_2O)_2{}^{z+}$, with only one of the coordination shell waters and its hydrogen-bonded neighbors shown in detail. Details regarding the geometrical parameters are given in ref. 16.

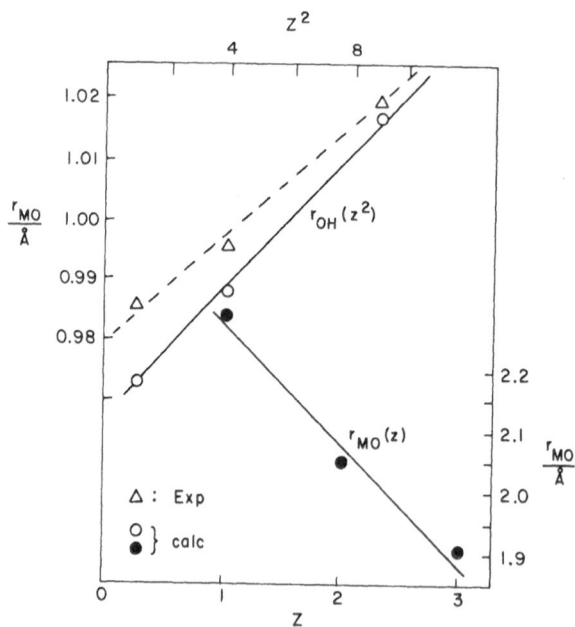

Fig. 3. Variation of equilibrium r_{OH} with z^2, and r_{MO} with z (see Fig. 1), based on $M(H_2O)_6(H_2O)_2{}^{z+}$, with M^{z+} = Na^{1+}, Mg^{2+}, and Al^{3+} [after Fig. 1 of ref. 17].

comparison. The good agreement of the calculated trend in OH distances with values obtained from experiment[19] and from molecular dynamics simulations[20] is demonstrated by Fig. 3.

The influence of the hydrogen–bond–enhanced charge–state dependence of r_{OH} on the rate constant for aqueous $Fe^{2+/3+}$ electron exchange has recently been assessed and found to account for a large fraction of the observed H/D kinetic isotope effect ($k_H/k_D \sim 2$ for aqueous $Fe^{2+/3+}$ exchange).[17]

STRUCTURAL MODELS FOR TRANSFER INTEGRAL CALCULATIONS

Transfer integrals were calculated for four redox systems, corresponding to the following M/L pairs, chosen to allow comparisons of $\sigma(e_g)$ and $\pi(t_{2g})$ electronic effects (see Table 1): Fe/H_2O, Fe^*(ligand–field excited)$/H_2O$, Ru/NH_3, and Co^*(low spin)$/NH_3$ (the asterisk denotes a ligand–field excited configuration of the reduced (2+) redox partner). The Fe and Ru systems involve $\pi(t_{2g})$ electron exchange, while the Fe^* and Co^* systems involve $\sigma(e_g)$ exchange. The isoelectronic H_2O and NH_3 ligands exemplify differing propensities for σ and π electron donation to coordinated metal ions. The ab initio calculations were carried out for two contact configurations, corresponding to apex-to-apex and face-to-face approach of the octahedral redox partners in the transition state (Fig. 4). The model clusters actually employed, $(M-L\bullet\bullet LM)^{5+}$ and $(ML_3\bullet\bullet\bullet L_3M)^{5+}$, were selected as a compromise between the full $(ML_6-L_6M)^{5+}$ encounter complexes and computational feasibility. As illustrated in Fig. 4, they include all the ligands in the important contact region between the metal ions. Fig. 4 also emphasizes the importance of $H\bullet\bullet\bullet H$ interactions in establishing the ligand–mediated contact between partners. The closest $H\bullet\bullet\bullet H$ contacts are ~ 2.2 Å (essentially van der Waals contact) and correspond to $M\bullet\bullet\bullet M$ separations of 7.0-7.3 Å (apex-to-apex) and 5.5-5.8 Å (face-to-face), depending on the particular ligand involved. The face-to-face configuration clearly reflects some interpenetration of reactants. Several types of evidence have been invoked to support the occurrence of such interpenetration.[12,13]

Table 1. Prototype Redox Systems ($ML_6^{2+/3+}$)

Ligand (L)	Donor/Acceptor Orbital Symmetry[a]	
	$t_{2g}(\pi)$	$e_g(\sigma)$
H_2O	$Fe^{2+/3+}$(h.s.)[b]	$Fe^{*2+/3+}$(h.s.)[b,c]
NH_3	$Ru^{2+/3+}$(l.s.)[b]	$Co^{*2+/3+}$(l.s.)[b,c]

[a] t_{2g} and e_g refer to octahedral point group symmetry and correspond, respectively to π and σ symmetry with respect to local ML bonds.

[b] h.s. and l.s. designate, respectively, high spin and low spin metal-ion d-electron configurations.

[c] The asterisk denotes a ligand field excited configuration for the reduced (2+) species: $t_{2g}^3 e_g^3 (Fe^{2+})$ and $t_{2g}^6 e_g(Co^{2+})$.

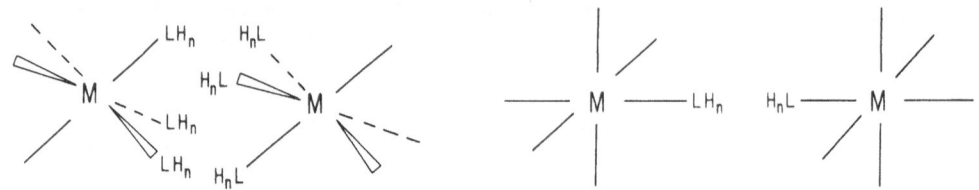

FACE-TO-FACE   APEX-TO-APEX

Fig. 4. Configuration of octahedral redox partners in a) face-to-face and
         b) apex-to-apex transition states. The atoms shown explicitly
         define the supermolecules employed in the ab initio
         calculations. In the remainder of the paper, the symbol L is
         used for the entire ligand. The H-atoms are explicitly displayed
         here to emphasize the role of these atoms in the contact between
         redox partners in the transition state.

COMPUTATIONAL MODELS FOR CALCULATED TRANSFER INTEGRALS

Our experience has shown that the diabatic states (Ψ_i and Ψ_f)
characterizing weakly interacting redox partners can be determined
variationally using spatially unrestricted self-consistent field (SCF)
calculations. The wavefunctions are charge-localized and correspond to
the desired valence bond structures appropriate to eq 1.[3] The
calculations were based on the clusters shown in Fig. 4, with extended
valence-level atom-centered basis sets for the metal (outer d and s
orbitals) and ligand orbitals.[3e,f] We feel that the basis sets employed
are sufficiently flexible for transfer integrals associated with species
in close contact. Of course for transfer over long distances, sensitivity
to long-range orbital tails becomes much more critical. The inner-shell
electrons have been included implicitly through the use of ab initio
effective core potentials.[21]

The many-electron transfer integrals, H_{if}, between the
non-orthogonal diabatic states, Ψ_i and Ψ_f, are then calculated
straightforwardly according to eq 6, using the method of corresponding
orbitals.[22] Alternatively, the calculated H_{if} integrals may be
thought of in terms of the following expression,

$$H_{if} = (E_- - E_+)/2 \qquad\qquad (8)$$

where E_- and E_+ are, respectively, the eigenvalues of the adiabatic
wavefunctions Ψ_- and Ψ_+, obtained from the configurational mixing of Ψ_i
and Ψ_f:

$$\Psi_\pm = (\Psi_i \pm \Psi_f)/(1 \pm S_{if})^{1/2} \qquad\qquad (9)$$

where $S_{if} = \int \Psi_i^* \Psi_f$. Recognizing from eq 8 that H_{if} is one-half the
energy difference of two states of similar local bonding characteristics
which are related by a 1-electron excitation leads to the conclusion that
H_{if} should be much less sensitive to electron correlation effects that
the individual energy eigenvalues.

The adiabatic functions $\Psi_\pm$ correspond to stationary electronic states (i.e., eigenfunctions) which are 50-50 mixtures of Ψ_i and Ψ_f. Clearly, the nonstationary diabatic functions (Ψ_i and Ψ_f) are a more intuitively appropriate basis for viewing a chemical reaction, even though the transfer integral can also be represented in the adiabatic basis (eq 8). Of course in cases when the coupling turns out to be large, then one no longer has a conventional "reaction", and the stationary states become the appropriate basis. In such a situation, the SCF procedure would be expected to yield $\Psi_\pm$ directly, a result which has not been observed for the systems we have investigated.

The so-called direct, double, and super-exchange contributions to H_{if} which were defined by Halpern and Orgel[1] in terms of local atomic or molecular moieties are automatically folded into our SCF-based calculational procedure. The metal-ligand mixing discussed below may be considered to provide a superexchange pathway.

ORBITAL MODELS

In this section we employ simple orbital arguments to derive expressions for H_{if} which provide a basis for evaluating the validity of eq 7.[10] The reduced reaction partner on the l.h.s. of eq 1 (ML_6^{2+}) may be thought of as containing a single electron in a donor orbital (ϕ_D), and in addition, each of the reactants has a number of occupied orbitals which we shall denote as spectator orbitals. Analogously, the reduced product species (ML_6^{2+}) on the r.h.s. of eq 1 now has a single electron in an acceptor orbital (ϕ_A), and once again, each product species has a "core" of occupied spectator orbitals. Our task is to provide criteria for defining ϕ_D and ϕ_A, and to determine the extent to which the spectator orbitals differ in the oxidized and reduced states of the complex (a many-electron relaxation effect). In previous work[3e] we have proposed a criterion based on the method of corresponding orbitals.[22] For two sets of non-orthogonal orbitals (in this case the ocupied orbitals of the initial and final states) this method defines transformations which lead to maximal correspondence between respective members of the two sets, thereby providing a suitable prescription for a donor/acceptor orbital pair (ϕ_D and ϕ_A) and a "core" of maximally overlapping spectator orbitals for the redox partners in the initial and final states. Thus we obtain Ψ_i and Ψ_f in the form,

$$\Psi_i = (\psi_i^{core}\phi_D) \tag{10}$$

and

$$\Psi_f = (\psi_f^{core}\phi_A) \tag{11}$$

where is the antisymmetrizer. The overlap integral between ψ_i^{core} and ψ_f^{core} is expected to be reasonably close to unity, with departures dependent on the degree of oxidation-state dependence of the spectator orbitals. The effective transfer orbitals ϕ_D and ϕ_A will be more or less localized on the respective redox partners. For the cases of symmetric electron exchange treated in the present work, ϕ_D and ϕ_A are related by a symmetry operation (i.e., reflection or inversion).

In the simplest possible model, we consider only a "metal-centered" orbital on each reactant, denoted M_ℓ and M_r (corresponding to the "left-" and "right-"hand partner on each side of eq 1):

$$\Psi_i = (M_\ell)^n (M'_r)^{n-1} \tag{12}$$

$$\Psi_f = (M'_\ell)^{n-1} (M_r)^n \tag{13}$$

where n = 1 or 2 and the presence or absence of the "prime" denotes the orbital M in the oxidized or reduced species, respectively. Adopting a 1-electron hamiltonian, $\underline{h}$, and ignoring overlap integrals between orbitals on different redox partners, we obtain

$$H_{if} = (S_{MM'})^2 h_{M_\ell M_r} \tag{14}$$

where

$$S_{MM'} = \int M_i M'_i \leq 1 , \qquad i \equiv \ell \text{ or } r \tag{15}$$

Thus departures from a pure 1-electron version of H_{if} (eq 7) arise as the departures of the "electronic Franck-Condon factor" $S_{MM'}$ from unity. If the relaxation of M is suppressed by constraining M = M' = M, we obtain the purely 1-electron reference expression:

$$\overline{H}_{if} = h_{M_\ell M_r} \tag{16}$$

The metal-like orbital in the reduced species, M_i, is expected to be more diffuse than M or M' (e.g., due to radial atomic orbital expansion)[23]. Accordingly, we may generally expect the unconstrained result for H_{if} in eq 14 to be <u>larger</u> in magnitude than the 1-electron reference value (eq 16), since the anticipated increase in magnitude of the 1-electron transfer integral, $h_{M_\ell M_r}$, due to enhanced spatial extension of M_ℓ and M_r should more than offset the attenuation due to the $(S_{MM'})^2$ factor.

An additional source of orbital relaxation is provided by variation of the degree of metal-ligand mixing with oxidation state. If we include a doubly-occupied orbital of predominantly ligand character, L or L', on each center,

$$\Psi_i = (M_\ell)^n (L_\ell)^2 (M'_r)^{n-1} (L'_r)^2 \tag{17}$$

and

$$\Psi_f = (M'_\ell)^{n-1} (L'_\ell)^2 (M_r)^n (L_r)^2 \tag{18}$$

and if we assume that M' and L' are related by linear transformations to M and L, the resulting expression for the transfer integral is,

$$H_{if} = (S_{LL'})^2 h_{M'_\ell M'_r} \tag{19}$$

Here the "Franck-Condon" attenuation is provided by the ligand ("core") manifold. In comparison with the delocalized reference, eq 16, this effect is once again (cf. eq 14) expected to be more than offset by the $h_{M_\ell M_r}$ factor to the extent that the M' orbitals involve more ligand mixing than the M or M orbitals, due to the stronger ligand field in the oxidized species. We recall (Fig. 4) that the primary reactant contact is made via the ligands and thus the magnitude of $h_{M_\ell M_r}$ should increase with increasing ligand field. It should also be noted that even if M' is unoccupied (n=1 in eqs 17 and 18), it is still definable by the linear relationships assumed above.[3f]

The foregoing analysis has permitted the partitioning of H_{if} into a 1-electron factor, h_{DA}, whose magnitude is controlled by the spatial extent of the metal-like donor and acceptor orbitals, and an "electronic Franck-Condon factor" arising from the spectator cores. Quantitative estimates of these factors are presented below.

LCAO ANALYSIS

Further development of the models of the last section is possible in terms of an LCAO-like analysis of the molecular orbitals. Strictly speaking, for the ML_n^{z+} complexes of interest we seek to define linear combinations of metal atomic orbitals, χ_M, and group ligand orbitals, χ_L (i.e., symmetry-adapted linear combinations of the ligand molecular orbitals which mix with the metal orbitals):[3e]

$$M = N_M(\chi_M - \lambda \chi_L) \tag{20}$$

$$L = N_L(\chi_L + \gamma \chi_M) \qquad , \tag{21}$$

where N_M and N_L are normalization constants. A common χ_M, χ_L basis is assumed for M and L and their M' and L' counterparts (defined in terms of λ' and γ'). The mixing coefficients are related through the relation,

$$\lambda = (\gamma + S_{ML})/(1 + \gamma S_{ML}) \tag{22}$$

where

$$S_{ML} = \int \chi_{L_i} \chi_{M_i} \qquad , \qquad i = \ell \text{ or } r \quad . \tag{23}$$

If the transfer integral magnitude is dominated by ligand-ligand contact as assumed above, the "LCAO" model yields

$$H_{if} \propto (\lambda'^2) N_c h_{L_\ell L_r} \tag{24}$$

where N_c is the number of close contacts (respectively 1 and 6 for the apex-to-apex and face-to-face approaches (Fig. 4);[3e,f] the 1-electron transfer integral for ligand orbitals in "contact" is given by,

$$h_{L_\ell L_r} = \int \chi_{L_\ell} h \chi_{L_r} \tag{25}$$

and the proportionality constant contains the electronic Franck-Condon factors. If the latter factors are close to unity or nearly constant, and if $h_{L_\ell L_r}$ is nearly constant, then a simple proportionality is predicted between $(\lambda')^2$ and H_{if}.

Since the ab initio calculations yield molecular orbitals (MO's) in terms of atom centered basis functions, it is straightforward to cast these MO's in the "LCAO" form of equations 20 and 21 (and their "primed" counterparts).[3e,f] These SCF MO's have been found to yield λ' values quite similar to those obtained[3e] from rigorous application of the corresponding orbital method to the ab initio wavefunctions. For the case of $Co(NH_3)_6$, in which M' (e_g symmetry) is empty, λ' has been defined by using the "primed" counterpart of eq 22 and the γ' value obtained for the occupied L' orbital of e_g symmetry.

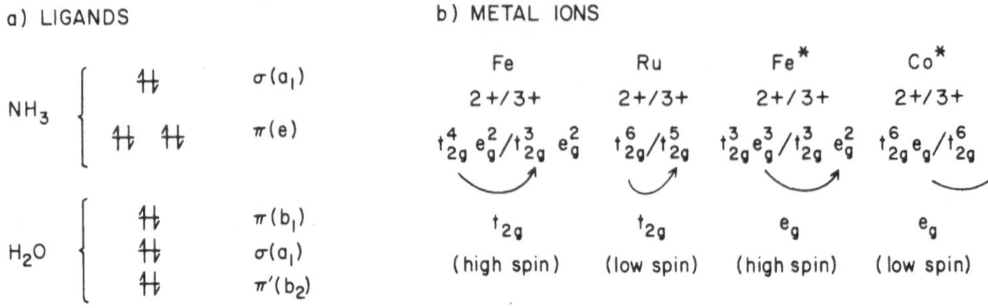

a) LIGANDS   b) METAL IONS

			Fe	Ru	Fe*	Co*
				2+/3+	2+/3+	2+/3+	2+/3+

NH₃ ligand:
- ⥮ σ(a₁)
- ⥮ ⥮ π(e)

H₂O ligand:
- ⥮ π(b₁)
- ⥮ σ(a₁)
- ⥮ π'(b₂)

Metal ions:

Fe 2+/3+	Ru 2+/3+	Fe* 2+/3+	Co* 2+/3+
$t_{2g}^4 e_g^2 / t_{2g}^3 e_g^2$	t_{2g}^6 / t_{2g}^5	$t_{2g}^3 e_g^3 / t_{2g}^3 e_g^2$	$t_{2g}^6 e_g / t_{2g}^6$
t_{2g}	t_{2g}	e_g	e_g
(high spin)	(low spin)	(high spin)	(low spin)

Fig. 5. Electronic configurations; a) ligands: schematic depiction of the highest-lying occupied sigma and pi molecular orbitals (MO's); π and π' refer, respectively, to out-of-plane (b₁) and in-plane (b₂) π-type orbitals for the C_{2v} water molecule. All the listed MO's except the H_2O π(b₁) MO involve participation of the hydrogen atom s-orbitals, corresponding to hyperconjugation in the case of π(NH₃) and π'(H_2O); the σ and π designations refer to symmetry with respect to the ligand symmetry axes: 2-fold (H_2O) or 3-fold (NH₃). The symbols in parentheses refer to irreducible representations in C_{3v}(NH₃) or C_{2v}(H_2O) symmetry. b) metal ions: the valence d-orbital configurations are listed, along with the symmetry of the transferring ("redox") electron. The e_g transfer cases involve ligand-field excited M^{2+} configurations.

RESULTS AND DISCUSSION

     We now discuss, in conjunction with the analysis of the previous sections, how the calculated transfer integrals and the associated electronic transmission factors (κ_{el}) depend on the electron donor properties of the H_2O and NH₃ ligands and on the nature of the transferring electron. In the separate $ML_6{}^{2+}$ reactants, t_{2g} and ε_g symmetries correspond, respectively, to π and σ M-L orbital interactions, and in the smaller model clusters used in the ab initio calculations ($ML_n{}^{2+}$, n = 1 or 3) the same π and σ relationships are maintained between the M and L orbitals. Electronic configurations relevant to the systems listed in Table 1 are presented in Fig. 5. At first glance (Fig. 5a), one might expect H_2O to be more effective in facilitating the transfer of π (vs σ) electrons, whereas the opposite would be anticipated for NH₃, based on the order of the highest occupied ligand orbitals for each case. On the other hand, recognizing that electronic involvement of the hydrogen orbitals should be important for good coupling in the transition state (see Fig. 4), we conclude that the lower, "in-plane" π-orbital of water (designated here as π' and corresponding to b₂ symmetry in C_{2v}) is likely to be the pertinent donor orbital.[3e,f] Accordingly, with respect to the L-M mixing relevant to the transfer integral, we would expect that both H_2O and NH₃ should be more effective as a σ (vs π) electron donor, with NH₃ being a better donor than water for both symmetry types. Note that the relevant π-electron coupling (π' for H_2O and π for NH₃) involves hyperconjugation, which can be a significant electronic effect even in formally saturated systems.

     Since the H_2O ligand has two different π-type orbitals, π and π' as discussed above, both of which are involved in L-M mixing, we extend eq (20) as follows (using the primed or oxidized version),

$$M' = N_{M'}\left(\chi_M - \lambda'_\pi\chi_L^\pi - \lambda'_\pi,\chi_L^{\pi'}\right) \tag{26}$$

In the results below, it is the λ_π' contribution which has been listed and employed in conjunction with eq 24 (detailed analysis reveals that only with this choice does one obtain a monotonic relation between H_{if} and $(\lambda')^2$).

In Fig. 5b, the four metal d-electron configurations are displayed, and the symmetry type of the transferring electron is indicated. The ligand-field excited state of Fe^{2+} has been used to provide one example of $\sigma(e_g)$ transfer, the other example being provided by the low-spin ligand-field excited state of Co^{2+}. This latter configuration has been proposed[5e] as the relevant electronic ground state species in the transition state for the hexa-ammine exchange reaction.

The calculated results for H_{if} (eq 6) and λ' (the "primed" version of λ in eq 20) are summarized in Table 2 and Fig. 6. A number of conclusions emerge. The electronic structure differences of the two isoelectronic ligands in conjunction with σ and π-type redox processes lead to a broad range of H_{if} magnitudes, spanning a range of κ_{el} from $\sim 10^{-3}$ to ~ 1.0; H_{if} magnitudes > 100 cm^{-1} generally correspond to the adiabatic limit ($\kappa = 1$; see fig 1).[3d]

Table 2. Calculated Values of H_{if} and λ'

Redox System ($ML_n^{2+/3+}$)			$H_{if}(cm^{-1})^a$	$\lambda'^{\,b}$
M	L	Donor/Acceptor Orbital Symmetry[c]		
Face-to-face[d] (R_{MM} = 5.5 Å)				
Fe	H_2O	$t_{2g}(\pi)$	16[e]	0.05
Apex-to-apex[d] (R_{MM} = 7.0-7.3 Å)[f]				
Fe	H_2O	$t_{2g}(\pi)$	25	0.06
Ru	NH_3	$t_{2g}(\pi)$	66	0.11
Fe[*g]	H_2O	$e_g(\sigma)$	312	0.25
Co[*g]	NH_3	$e_g(\sigma)$	2820	0.71

[a] Calculated from <u>ab initio</u> wavefunctions (Ψ_i, Ψ_f) according to eq 6.

[b] Calculated from the <u>ab initio</u> MO corresponding to the oxidized ("primed") version of the M orbital (cf eq 20).

[c] See footnote <u>a</u> of Table 1.

[d] See Fig 4.

[e] Magnitude of H_{if} per L···L close contact. The absolute value of H_{if} for this case (98 cm^{-1}) involves six such contacts (Fig. 4).

[f] 7.0 Å for the NH_3 cases and 7.3 Å for the H_2O cases.

[g] See footnote c of Table 1.

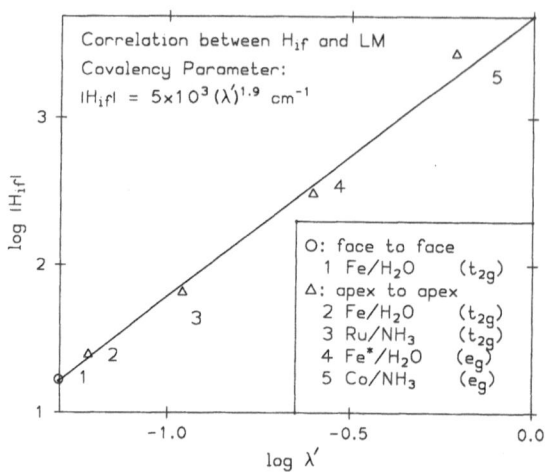

Fig. 6.   Plot of $\log_{10}|H_{if}|$ (cm^{-1}) vs $\log_{10}\lambda'$ [after ref. 10]; the solid
line and the equation refer to the linear least-squares fit
(r^2 = 0.996). The $|H_{if}|$ value for the face-to-face case has
been normalized (as in Table 2) to take account of the six
close-contact ligand pairs.

     As expected from the orbital orderings (Fig. 5a) and the special
significance of the hyperconjugative π' orbital in H_2O, the transfer
integral magnitudes are ordered according to:

$$\sigma \ > \ \pi \qquad \text{(both } H_2O \text{ and } NH_3)$$

$$NH_3 \ > \ H_2O \qquad \text{(both } \sigma \text{ and } \pi \text{ transfer)}.$$

A particularly effective combination of characteristics is provided by the
strong σ-donor properties of NH_3 in the case of σ-electron transfer, and
the transfer integral is sufficiently large in magnitude that it would be
expected to have a significant effect on the activation energy defined by
the avoided crossing region shown in Fig. 1.

     Table 2 also supports the expected monotonic relation between
$|H_{if}|^2$ and $(\lambda')^2$ (eq 24). The listed transfer integral for the
face-to-face entry is normalized through division by 6 (the value of N_c
in eq 24). Fig. 6 shows that the form of eq 24 is even justified
quantitatively since a least-squares linear fit of $\log|H_{if}|$ vs $\log\lambda$
yields an r^2-value of 0.996 and implies an exponent of λ' very close to 2,
as well as a value of ~ 5000 cm^{-1} for the effective ligand-ligand
transfer integral $h_{L_\ell L_r}$ (eq 25). The magnitude of this 1-electron
"resonance integral" indicates that substantial coupling is possible in
the absence of formal bonding, provided that the contact allows adequate
orbital overlap. Thus the degree of "weakness" of the overall interaction
is seen to depend on the mixing of the ligand and metal orbitals within
each redox partner.

     Another noteworthy feature of the results is the ability of eq 24 to
accommodate both the apex-to-apex and face-to-face data on a common basis,
even though they involve very different M-M separations (respectively

7.0–7.3 Å and 5.5 Å). This strongly supports the notion that the key factor is the amount of delocalization of the donor and acceptor orbitals onto the ligands in contact.

ACKNOWLEDGMENT

This research was carried out at Brookhaven National Laboratory under contract DE-AC02-76CH00016 with the U. S. Department of Energy and supported by its Division of Chemical Sciences, Office of Basic Energy Sciences.

REFERENCES

1. J. Halpern and L. E. Orgel, The theory of electron transfer between metal ions in bridged systems, Discuss. Faraday Soc. 29:32 (1960).
2. H. M. McConnell, Intramolecular charge transfer in aromatic free radicals, J. Chem. Phys. 35:508 (1961).
3. a) M. D. Newton, Formalisms for electron kinetics in aqueous solution and the role of ab initio techniques in their implementation, Int. J. Quantum. Chem., Quantum Chem. Symp.14:363 (1980); b) J. Logan and M. D. Newton, Ab initio study of electronic coupling in the aqueous $Fe^{2+}-Fe^{3+}$ Electron Exchange Processes, J. Chem. Phys. 78:4086 (1983); c) J. Logan, M. D. Newton, and J. O. Noell, Factors governing electronic localization in transition metal clusters and complexes, Int. J. Quantum. Chem., Quantum Chem. Symp. 18:213 (1984); d) M. D. Newton and N. Sutin, Electron transfer reactions in condensed phases, Ann. Rev. Phys. Chem. 35:437 (1984); e) M. D. Newton, Comparison of electron-transfer matrix elements for transition-metal complexes: t_{2g} vs e_g transfer and NH_3 vs. H_2O ligands, J. Phys. Chem. 90:3734 (1986); f) M. D. Newton, Ab initio models for electron tunnelling between transition metal complexes, in: "Tunneling," J. Jortner, B. Pullman, eds., Reidel, Dordrecht, Holland (1986), pp. 305.
4. a) J. R. Miller and J. V. Beitz, Long range transfer of positive charge between dopant molecules in a rigid glassy matrix, J. Chem. Phys. 74:6746 (1981); b) G. L. Closs, L. T. Calcaterra, N. J. Green, K. W. Penfield, and J. R. Miller, Distance, stereoelectronic effects, and the Marcus inverted region in intramolecular electron transfer in organic radical anions, J. Phys. Chem. 90:3673 (1986).
5. a) S. Larsson, Electron transfer in chemical and biological systems. Orbital rules for nonadiabatic transfer, J. Am. Chem. Soc. 103:4034 (1981); b) S. Larsson, π Systems as bridges for electron transfer between transition metal ions, Chem. Phys. Lett. 90:136 (1982); c) S. Larsson, Electron transfer in proteins, J. Chem. Soc., Faraday Trans. 2 79:1375 (1983); d) S. Larsson, Electron-exchange reaction in aqueous solution, J. Phys. Chem. 88:1321 (1984); e) S. Larsson, K. Stahl, and M. C. Zerner, Hexaamminecobalt electron-self-exchange reaction, Inorg. Chem. 25:3033 (1986).
6. a) J. J. Hopfield, Electron transfer between biological molecules by thermally activated tunneling, Proc. Natl. Acad. Sci. U.S.A. 71:3640 (1974); b) D. N. Beratan, J. N. Onuchic, and J. J. Hopfield, Limiting forms of the tunneling matrix element in the long distance bridge mediated electron transfer problem, J. Chem. Phys. 83:5325 (1985); c) D. N. Beratan, J. N. Onuchic, and J. J. Hopfield, Electron tunneling through covalent and noncovalent pathways in proteins, J. Chem. Phys. 86:4488 (1987).

7. a) M. J. Ondrechen, M. A. Ratner, and D. E. Ellis, The electronic structure of the Creutz-Taube ion: A Hartree-Fock-Slater study, Chem. Phys. Lett. 109:50 (1984); b) K. V. Mikkelsen and M. A. Ratner, Electron tunneling in solid-state electron-transfer reactions, Chem. Rev. 87:113 (1987).

8. a) K. Ohta, G. C. Closs, K. Morokuma, and N. J. Green, Stereoelectronic effects in intramolecular long-distance electron transfer in radical anions as predicted by ab initio MO calculations, J. Am. Chem. Soc. 108:1319 (1986); b) K. Ohta and K. Morokuma, An ab initio Mo study on electron transfer in gas-phase hydrated clusters: $O_2^-(H_2O)_n + O_2 \rightarrow O_2 + O_2^-(H_2O)_n (n = 0, 1, and 2)$, J. Phys. Chem. 91:401 (1987).

9. A. Kuki and P. Wolynes, Electron tunneling paths in proteins, Science 236:1647 (1987).

10. M. D. Newton, Electronic structure analysis of electron transfer matrix elements for transition metal redox pairs, J. Phys. Chem., submitted.

11. A. M. Kuznetsov, Faraday Discuss. Chem. Soc. 74:49 (1982); Chem. Phys. Lett. 91:34 (1982).

12. B. L. Tembe, H. L. Friedman, and M. D. Newton, The theory of the $Fe^{2+}-Fe^{3+}$ electron exchange in water, J. Chem. Phys. 76:1490 (1982).

13. H. L. Friedman and M. D. Newton, The theory of the $Fe^{2+}-Fe^{3+}$ electron exchange in water, Faraday Discuss. Chem. Soc. 74:73 (1982).

14. M. D. Newton, Theoretical aspects of the OH•••O hydrogen bond and its role in structural and kinetic phenomena, Acta Cryst. B39:104 (1983).

15. M. D. Newton, Current views of hydrogen bonding from theory and experiment -- structure, energetics, and control of chemical behavior, Trans. Am. Chem. Assoc. 22:1 (1986).

16. M. D. Newton and H. L. Friedman, A proposed neutron diffraction experiment to measure hydrogen isotope fractionation in solution, J. Chem. Phys. 83:5210 (1985).

17. H. L. Friedman and M. D. Newton, H/D isotope effect on outer sphere electron exchange, J. Electroanal. Chem. 204:21 (1986).

18. B. S. Brunschwig, C. Creutz, D. H. Macartney, T.-K. Sham, and N. Sutin, The role of inner-sphere configuration changes in electron-exchange reactions of metal complexes, Faraday Discuss. Chem. Soc. 74:113 (1982).

19. J. R. C. van der Maarel, H. R. W. M. de Boer, J. de Bleijser, D. Bedeaux, and J. C. Leyte, On the structure and dynamics of water in $AlCl_3$ solutions from H, D, ^{17}O, and ^{27}Al nuclear magnetic relaxation, J. Chem. Phys. 86:3373 (1987).

20. P. Bopp, private communication cited as ref. 8 in the paper by van der Maarel et al.[20]

21. a) C. F. Melius, B. D. Olafson, and W. A. Goddard III, Fe and Ni ab initio effective potentials for use in molecular calculations, Chem. Phys. Lett. 28:457 (1974); b) Topiol, S.; J. W. Moskowitz, and C. F. Melius, Atomic coreless Hartree-Fock pseudopotentials for atoms K through Zn, J. Chem. Phys. 68:2364 (1978).

22. H. F. King, R. E. Stanton, H. Kim, R. E. Wyatt, and R. G. Parr, Corresponding orbitals and the nonorthogonality problem in molecular quantum mechanics, J. Chem. Phys. 47:1936 (1967).

23. W. H. E. Schwarz and T. C. Chang, Multiconfiguration wave functions for highly excited states by the generalized Brillouin theorem method, Int. J. Quant. Chem. Symp. 10:91 (1976).

DISCUSSION

AMATORE - In your interesting talk you mainly focused on "spherical" inorganic molecules. How do you see the case of "flat" organic molecules such as e.g. aromatics ?

NEWTON - Your question is very pertinent and gives me an opportunity to emphasize the generality of the approach which I presented. The ab initio molecular orbital procedure which I have developed for evaluating electron transfer matrix elements has in fact been applied by Ohta, Morokuma and Closs to a study of the conformational dependence of the transfer element for some aromatic donor/acceptor systems. Furthermore, these workers estimated the degree to which the "transferring electron" was distributed among the various carbon atoms of the aromatic molecules (using ESR data), quite analogous to the procedure I employed for the inorganic systems - i.e., estimating the spatial distribution of the transferring electron by exploiting the notion of ligand-metal orbital mixing. In general, one requires a method (theoretical or experimental) for estimating the fraction of the transferring electron (in both the initial and final states of the reaction of interest) on each atomic site involved in significant orbital overlap between donor and acceptor.

LAUNAY - In a system like $Fe^{2+/3+}$ there are three t_{2g} orbitals to consider on each site. Does your model show the different conceivable pathways for electron transfer ?

NEWTON - You have raised a general question of considerable interest. Electron transfer matrix can be viewed (using, for example, simple valence bond notions) as superpositions of transfer processes occurring over various "pathways" (e.g. "direct" vs. "supercharge" pathways) which may interfere constructively or destructively with each other. Our ab initio molecular orbital approach has the "advantage" of carrying out directly this superposition, in a streamlined fashion (as dictated by the variational procedure unbiased by assumptions about relative importance of different pathways. This superposition leads to the definition of the initial (φ_i) and final (φ_f) states, and may involve some mixing of t_{2g} orbitals. However, an additional "superposition" question remains - namely, that involving the relative importance of the different initial states arising from the degeneracy or near degeneracy of the reactants in some cases (e.g., the triple spatial degeneracy expected for high-spin Fe^{2+}). The weighting in this situation requires an estimate of the relevant Boltzmann factors. A brief discussion of this point (which relates directly to your question about different t_{2g} orbitals) was given by Friedman and Newton, Disc. Faraday Soc., 74, 73 (1982).

WILSON - Could you be specific about what you think of the reasonableness of one electron approximations being used in current electron transfer reaction path integral calculations ?

NEWTON - This question can be answered at two levels. In terms of the simple orbital model which I introduced for purposes of analysis, we find that the "electronic" Franck-Condon factor, which constitutes the "many-electron" content of the transfer integral, is generally $\geqslant 0.8$: i.e., the departures from the "1-electron" model are quite modest.

A more stringent test is provided by adopting a reference level for the initial and final states (ψ_i and ψ_f), which suppresses the charge-state-dependent electronic relaxation effects - namely, symmetry - constrained SCF calculations which correspond to delocalized charge (+ 2.5, + 2.5). The transfer elements based on this representation differ by no more than a factor of two from those based on fully-relaxed variational SCF calculations. Thus once again, departures from a "1-electron" model are inferred to be modest.

ADIABATIC ELECTRON-ION RECOMBINATION

IN A POLAR SOLVENT

Michael L. Klein and Michiel Sprik

Department of Chemistry
University of Pennsylvania
Philadelphia, PA 19104-6323

INTRODUCTION

Electron transfer, such as occurs between ions in solution or between an ion and an electrode, is one of the most fundamental processes in Chemistry. Recent advances in theory[1,2] and simulation techniques[3-12] now make it possible to study different aspects of the behavior of electrons in polar fluids. In the near future, it may even prove possible to study electron transfer processes via simulation techniques. However, for the present we focus on a slightly simpler problem which will serve as a prototypic example of electron transfer: the transfer of an electron from its solvated state to an ion, in solution. We have chosen to study first the recombination reaction between a solvated electron and a lithium ion solvated in liquid ammonia. It is now a relatively routine matter to use classical simulation techniques to examine ionic solvation in polar fluids. Thus, the study of a lithium ion in liquid ammonia presents few problems or challenges other than deciding on the potential models for use in the Monte Carlo or molecular dynamics calculation.[5] In the case of the electron, the Feynman path integral formulation of quantum statistical mechanics enables us to not only treat the electron quantum mechanically but also to treat the classical ion (Li^+) and the solvent molecules (ammonia) on an equal footing.[6]

The only requirements for carrying out path integral calculations for systems of this type are the availability of generous amounts of computer time and appropriate electron-ion and electron-solvent molecule potentials.[6] In the next section, we outline how we carry out a path integral Monte Carlo study of the $Li^+ + e^-$ reaction in solution. The structural changes and the energetics of the reaction are discussed in some detail. We will see that the adiabatic pathway of the reaction involves a transition state in which the electron appears to be largely unsolvated or quasi-free. The energy barrier to recombination (or the reverse process of ionization) is rather large for our system (~ 50 k_BT). The question naturally arises as to how the electron transfer can proceed. In order to throw some light on this issue, we examine the effect of adiabatic fluctuations in the solvent on the ground state of the solvated electron. To do this, in section 3, we survey the energy states of an electron in a sample of rigid solvent configurations with the aid of a Gaussian basis set [13,14] that is optimized using the technique of simulated annealing.[10]

As mentioned in the Introduction, we have used quantum path integral Monte Carlo calculations to study the electron-ion recombination. In this approach the electron is represented by an isomorphic classical system, a ring polymer with P beads in the chain.[1,2] The Monte Carlo algorithm (Staging) which we employ to sample the electron paths of a ring polymer with P = 1024 is fully described elsewhere.[4]

We now outline the main features of our potential models. The solvent molecules interact with an effective pair potential fitted to liquid properties.[15] It consists of a Lennard-Jones (12-6) potential between each nitrogen atom plus electrostatic interactions between four charge sites on each molecule. These charges, and a neutralizing charge on the molecular symmetry axis, are chosen to yield the observed dipole moment. The Li^+-NH_3 potential consisted of a coulombic term plus a Li-N (12-6) potential whose parameters were fitted to quantum calculations on the Li^+NH_3 complex.[16]

These potential parameters were tested by carrying out a simulation of a single Li^+ ion in liquid ammonia. The solvation sheath of Li^+ was found to consist of four NH_3 molecules, which is in good agreement with deductions based on neutron scattering data and the known structure of the eutectic compound $Li(NH_3)_4$.[17]

The electron-ion and electron-ammonia interactions were treated using pseudopotentials. For Li we used a bare coulomb potential truncated at a radius (R_C) that yields essentially the correct spectroscopic term value for the 2s electron.[18] The electron-solvent molecule potential consisted of electron-charge site contributions but with R_C = 1.0 A^3 for the protons. No allowance was made for polarization or exchange contributions.[8,9] Our crude potential is justified _a posteriori_ since it yields a reasonable description of an electron in liquid ammonia.[5]

The calculations commenced with a solvated electron, a Li^+ ion and 250 NH_3 molecules in a box with V = 26.5 cm^3/mol and T = 260 K. The separation, ζ, between the center of mass of the ring polymer, representing the 2s electron charge distribution, and the solvated Li^+ ion was initially constrained to be 8 A. The recombination reaction was then followed by carrying out a sequence of path integral calculations in which ζ was constrained to successively smaller values.

At first sight, this procedure of carrying out a Monte Carlo calculation with a constrained _electron_ coordinate may seem somewhat strange. After all, normally, one defines a Born-Oppenheimer surface for, say, proton transfer by fixing the relevant _nuclear_ coordinate and averaging over the other nuclear and electronic coordinates. Unfortunately, in the electron transfer process, the relevant coordinate is far from obvious. Our procedure of constraining the center of mass of the electron polymer amounts to determining those configurations of the system which places the electron in the ground state at the chosen position. Thus, by allowing the solvent to rearrange its nuclei each time we change the value of ζ, we thereby generate the most likely adiabatic path through the configuration of the solvent.

The energetics of the recombination process and the concomitant changes in the Li^+ solvation sheath were followed on a function of ζ. The main results are collected in Fig. 1 and in Table 1. It is convenient to monitor the state of the electron using the complex time correlation function $R^2(t-t')$ where 0 < t < τ = $i\beta\hbar$.[2] The value of this function at t-t' = $\tau/2$ is expression
$$R^2 = 2 \langle (r - r_{cm})^2 \rangle.$$

Table 1. Energetics (in units of k_BT) of the Adiabatic Electron–Lithium Ion Recombination in Liquid Ammonia

ζ[a]	$-E_c$[b]	$-E_i$[c]	ΔE_s[d]	E[e]
0.0	249±5	110±26	30±10	25±15
0.5	264	110	30	15
1.0	265	140	55	5
2.0	169	185	35	40
3.0	98	225	55	90
4.0	61	290	80	85
5.0	42	330	105	90
6.0	37	355	145	110
8.0	78	385	160	60
∞	85	395	180	55
(0.5)[f]	265	115	20	0

[a]The constrained distance in Å between the electron ring polymer center of mass and the Li^+ ion.
[b]Total energy (potential plus kinetic) of the electron.
[c]Ion-solvent interaction energy.
[d]Solvent reorganization energy.
[e]Energy difference with respect to the unconstrained calculation for the polarized atom.
[f]Unconstrained path integral Monte Carlo result.

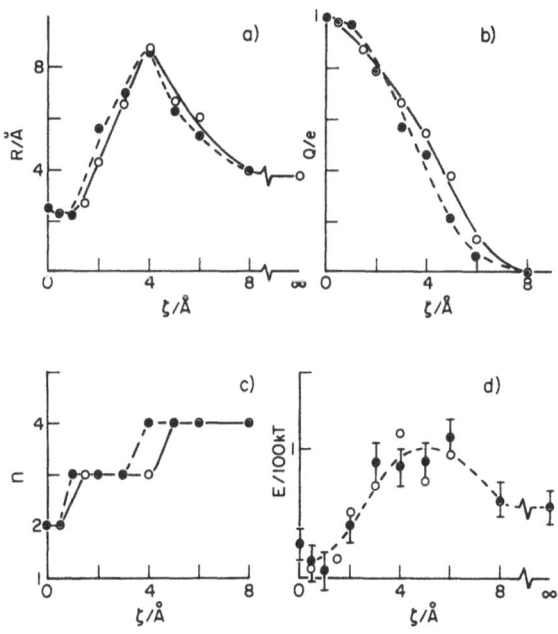

Fig. 1. Quantitites characterizing the electron–lithium ion recombination in liquid ammonia as a function of ζ. (a) R, the electron wave packet size. (b) Q, the net charge within 3 Å of the Li^+ ion. (c) n, the first solvent sheath coordination number of Li^+. (d) E, the total energy from Table 1. Open circles are for the ionization process.

There, the brackets denote an ensemble average and r is a point on the quantum path whose center of mass is given by

$$r_{cm} = \tau^{-1} \int_0^\tau r \, (t) \, dt.$$

For a free particle, R is related to the thermal wavelength $\lambda = \hbar \, (\beta/m)^{1/2}$ via $R_{free} = \sqrt{3} \, \lambda/2$ whereas with a localized state r_{cm} and R correspond to the maximum and width of the probability density. Figure 1a shows the variation of R with ζ for the electron–ion recombination. The function $R(\zeta)$ shows that recombination proceeds via a transition state, at $\zeta = 4$ Å, in which the electron is very diffuse with (R ~ 9 Å) and considerably larger than found either for the lithium atom (R ~ 3 Å) or for the solvated electron (R ~ 4Å). The total charge residing within 3 Å of the Li$^+$ ion, Q varies gradually as the electron and ion come together (Fig. 1b). At $\zeta = 4$ Å, half of the electron charge has leaked onto the Li$^+$ ion. The number of solvent molecules coordinated to Li$^+$, n changes from 4 to 3 at the transition state and finally to 2 when the electron is constrained with $\zeta = 0$ (Fig. 1c). The variation of E, the energy of the system, is shown in Fig. 1d. The electron has an energy barrier of about 50 k$_B$T to overcome before recombination can occur.

In addition to the recombination process, we also studied the reverse procedure, namely ionization. These results, which are also shown in Fig. 1 (open circles), reveal only a modest hysteresis and confirm that at the transition state (ζ ~ 4 Å), the electron is partly unsolvated or quasi free.

Three typical electron–ion configurations are shown in Fig. 2 for $\zeta = 0.5, 4.0$ and 8.0 Å which correspond, respectively, to the minimum energy state of Fig. 1d, the transition state, and to the fully solvated (ionized) state.

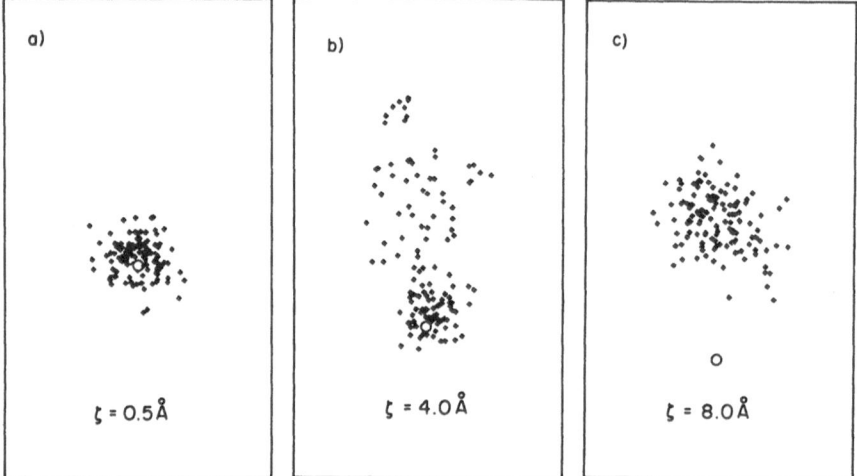

Fig. 2. Instantaneous configurations of the isomorphic ring polymer
(small squares) and the Li$^+$ ion (open circle), for
three values of the reaction coordinate ζ. The solvent ammonia
molecules are omitted for clarity.

Figure 2 reveals graphically the dramatic changes in the nature of the electron charge distribution that occur during the reaction. The instantaneous view of the polymer chain in the transition state is strongly suggestive of a superposition of localized and quasi-free states (Fig. 2b). For clarity Fig. 2 does not include the solvent ammonia molecules. An instantaneous veiw of the electron polymer and the solvent is shown in Fig. 3 for the case of the solvated electron (ζ = 8 Å). This figure and Fig. 2c emphasize the fact that although the solvated electron is characterized by R~4 Å there is also appreciable charge density at much larger distances from the center of mass of the ring polymer.

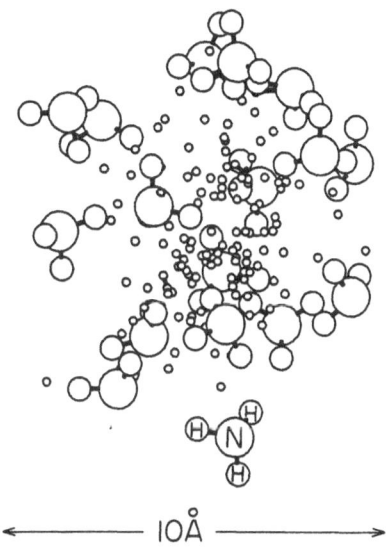

Fig. 3. Instantaneous configuration taken for the path integral calculations for the solvated electron in ammonia. The iso-morphic ring polymer is indicated by small circles; only one out of every eight particles is shown.

The nature of the electron-ion pair when ζ = 0.5 Å can be better appreciated by examination of Fig. 4. Only two solvent molecules remain in the primary solvation sheath of the Li$^+$ ion. Thus, the composite electron-ion system behaves as a polarized atom.[19]

We have already remarked that the state with ζ = 0 does not appear to be the one with minimum energy. We have confirmed this by carrying out additional calculations in which the center of mass of the electron ring polymer was unconstrained. To do this we inserted an uncharged classical particle (with the same potential parameters as Li$^+$-NH$_3$) into the ammonia solvent and equilibrated the system. The neutral particle was then replaced by a Li atom which consisted of a Li$^+$ ion and an excess electron bound to the ion by its local pseudopotential.[18] In the presence of the solvent, the Li atom was <u>unstable</u> and immediately distorted to form the polarized atom shown in Figs. 2a and 4 with ζ = 0.5 Å. We also initiated a calculation by inserting the electron in an expanded state (R~11 Å) which enveloped the Li$^+$ ion and its first solvent sheath of 4 NH$_3$ molecules. This Rydberg-like proved to be unstable, the system again evolved to the dipolar species by eventually expelling two of the primary solvent sheath NH$_3$ molecules. We conclude, therefore, that a Li atom immersed in a polar solvent spontaneously deforms to generate a dipolar excitonic state of the solvent-bound atom.[19]

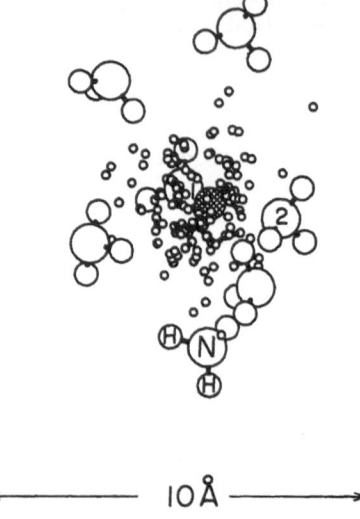

← 10 Å →

Fig. 4. Instantaneous configuration taken from the
path integral calculation for the polarized atom
(ζ = 0.5 Å). The isomorphic ring polymer
is indicated by small circles; only one out of
every eight particles is shown. The Li$^+$ ion is
shaded and the two closest solvent molecules are
indicated by Arabic numerals.

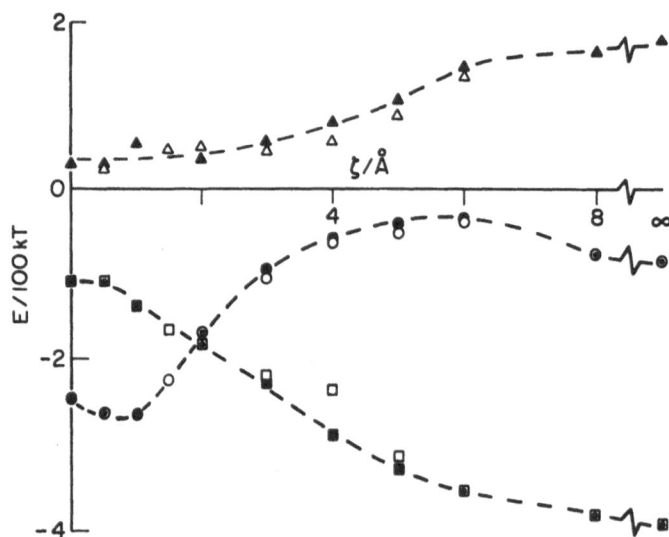

Fig. 5. Energetics of the electron-lithium ion recombination
reaction as a function of the reaction coordinate
ζ. The circles denote the electron energy (potential and
kinetic), the squares, the ion-solvent interaction energy and
the triangles, the solvent reorganization energy.

The individual contributions to the energetics of the electron–ion reaction (Table 1) are displayed graphically in Fig. 5. The total electronic energy (i.e., the sum of the electron potential and kinetic contributions) E_e. falls by about 180 k_BT during the recombination event, whereas the interaction energy of the Li^+ ion with the solvent E_i rises by about 280 kT. However, the net loss of almost 100 k_BT of energy is more than compensated by the solvent reorganization energy, $\Delta E_s = 155$ k_BT. The net result is that the polarized atom pair (recombination) is about 55 k_BT lower in energy than the separated species. Since we have not evaluated the free energy, we cannot make any remarks concerning the absolute stability of the state we find. Nevertheless, despite the lack of information on the free energy, we have been able to probe the transition state and establish that the pathway to electron–ion recombination (or ionization) involves a diffuse (quasi–free) electron. This unexpected finding may have important ramifications for understanding the metal–insulation transition[20] in metal–ammonia solutions, but such speculations take us beyond the scope of the present article.

ADIABATIC FLUCTUATIONS OF THE SOLVENT

The fact that we observed a Rydberg–like electron state (R~11 Å) evolve spontaneously to a polarized atom (R~2.4 Å) in an <u>unconstrained</u> path integral Monte Carlo calculation already suggests that adiabatic solvent fluctuations are important. The energy barrier for electron–ion recombination is about 50 k_BT (Fig. 1d). The question naturally arises as to whether or not the effect of adiabatic solvent fluctuations on the electron ground state energy is large enough to drive the electron transfer from the solvent to the ion.

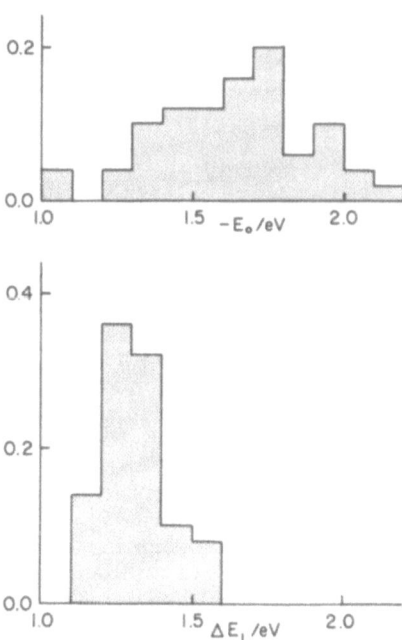

Fig. 6. Distribution of ground state energies, E_0 and excitation energy to the first excited state, ΔE_1, of the solvated electron in a set of 50 solvent configurations. The widths in the distributions arise from adiabatic solvent fluctuations.

In order to answer this question, we have calculated the lowest eigenstates of an electron in a sample of 50 different rigid solvent configurations. The latter were taken from an extensive path integral calculation of the solvated electron in liquid ammonia.[12] We employed a Gaussian basis set to represent the electron[13,14] and optimized it using the technique of simulated annealing.[10] The details of this novel approach are fully described elsewhere.[12] Figure 6 shows the effect of adiabatic solvent fluctuations on the electron ground state, E_O and the excitation energy ΔE_1. The dispersion of ground state energies is indeed very large (~25 K_BT) and likely sufficient to profoundly influence the electron-ion recombination.

The fluctuations in the value ΔE_1 are considerably smaller than those in E_O (see Fig. 6). The results for the higher excited states are very similar. This information can be used to calculate the absorption spectrum of the solvated electron. To do so, each of the eigenvalues is weighted by the

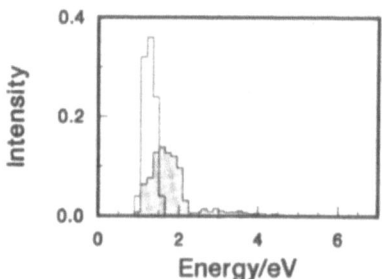

Fig. 7. Absorption spectrum for an electron-solvated in liquid ammonia based on an average over 25 solvent configurations. The shaded curve is the Franck–Condon spectrum and the other curve ΔE_1.

square of the matrix element of the transition dipole with respect to the ground state. In the adiabatic (Franck-Condon) approximation, the spectra for the various solvent configurations are simply averaged. The result is shown in Fig. 7. We find that the first three excited states contribute about 80 percent of the total intensity. The location and the width of the peak are, therefore, mainly determined by the adiabatic fluctuations in the manifold of these p-like states. Figure 7 exhibits the characteristic asymmetric shape of the experimental absorption spectrum for an electron in ammonia.[2] This width is attributed to splitting of the p-states by the solvent. Although the width of the spectrum is reproduced, the whole spectrum is, unfortunately, shifted to higher energies by about 0.8 eV. This disagreement is almost certainly due to the rather crude model used for the electron-solvent interaction.

SUMMARY

We have presented the results of a path intergral Monte Carlo study of electron-ion recombination in a polar solvent. Our method of calculation has enabled us to probe the nature of the electron's adiabatic path as it transfers from the solvent to the ion. We find that the recombination has an energy barrier of about 50 k_BT. In the transition region, the electron appears to be very diffuse. The end product of the recombination reaction is not a solvated spherical lithium atom. Instead, we find that the minimum energy state corresponds to a highly polarized atom, which is referred to as a dipolar excitonic state.[19] The separation between Li^+ and the center of gravity of the electron density is $\zeta_{min} = 0.5$ Å. Profound changes in the Li^+ solvation sheath occur during the recombination process. The reverse procedure, namely, ionization appears to follow a similar pathway with little hysteresis.

We have used a novel approach to reveal the effect of adiabatic solvent fluctuations on the electron ground state energy. We find that these fluctuations are comparable in magnitude to the energy barrier for recombination. Thus, it appears very likely that, just as in the Marcus model of electron transfer, solvent fluctuations actually drive the electron transfer from the solvent to the ion. It remains to be seen if this conclusion also holds in other systems.

REFERENCES

1. D. Chandler and P. G. Wolynes, *J. Chem. Phys.* **74**, 4078 (1981).
2. D. Chandler, Y. Singh and D. M. Richardson, *J. Chem. Phys.* **81**, 1975 (1984); A. L. Nichols III, D. Chandler, Y. Singh and D. M. Richardson, *ibid.*, **81**, 5109 (1984).
3. M. Parrinello and A. Rahman, *J. Chem. Phys.* **80**, 860 (1984); C. D. Jonah, C. Romero, and A. Rahman, *Chem. Phys. Lett.* **123**, 209 (1986).
4. M. Sprik, M. L. Klein and D. Chandler, *Phys. Rev. B* **31**, 4234 (1985); *J. Chem. Phys.* **83**, 3042 (1985).
5. M. Sprik, R. W. Impey and M. L. Klein, *J. Chem. Phys.* **83**, 5802 (1985).
6. M. Sprik, R. W. Impey and M. L. Klein, *Phys. Rev. Lett.* **56**, 2326 (1986).
7. B. J. Berne and D. Thirumulai, *Ann. Rev. Phys. Chem.* **37**, 401 (1986).
8. P. J. Rossky, J. Schnitker and R. A. Kuharski, *J. Stat. Phys.* **43**, 949 (1986); J. Schnitker and P. J. Rossky, *J. Chem. Phys.* **86**, 3462; **86**, 3471 (1987).
9. A. Wallqvist, D. Thirumulai and B. J. Berne, *J. Chem. Phys.* **86**, 6404 (1987); **85**, 1583 (1986).
10. R. Car and M. P. Parrinello, *Phys. Rev. Lett.* **55**, 2471 (1985).
11. A. Selloni, P. Carnevali, R. Car, and M. Parrinello, *Phys. Rev. Lett.* (1987).
12. M. Sprik and M. L. Klein, *J. Chem. Phys.* (1987).
13. M. J. Davis and E. J. Heller, *J. Chem. Phys.* **71**, 3383 (1979).
14. B. Hellsing and H. Metiu, *Chem. Phys. Lett.* **127**, 45 (1986).
15. R. W. Impey and M. L. Klein *Chem. Phys.* **104**, 579 (1984).
16. S. F. Smith, J. Chandrasekha, and W. L Jorgensen, *J. Phys. Chem.* **86**, 3308 (1982).
17. A. M. Stacy and M. J. Sienko, *Inorg. Chem.* **21**, 2294 (1982).
18. R. W. Shaw, *Phys. Rev.* **174**, 769 (1968).
19. D. E. Logan, *Phys. Rev. Lett.* **57** (1986).
20. J. C. Thompson, in <u>Metal-Non-Metal Transitions</u>, eds. P. P. Edwards and C. N. R. Rao (Taylor and Francis, London, 1985); N. F. Mott, <u>Metal-Insulator Transitions</u> (Taylor and Francis, London, 1974).
21. F. Y. You and G. R. Freeman, *J. Phys. Chem.* **85**, 629 (1981).

DISCUSSION

BRATOS - The model you propose is a semi-empirical model. Which of its predictions can be compared with an experiment ? What type of agreement can you reach ?

KLEIN - We do indeed use a semi-empirical model for the various interaction potentials. First, we model the ammonia intermolecular potential with an effective pair potential which ignores many body polarization. Models of this type are remarkably successful in explaining the physical properties of polar fluids. Of course, we really should include many body forces, but at this stage we ignore them. The ammonia potential is fitted to the heat of evaporation and the zero-pressure density. The electron-alkali metal (Lithium) potential is represented by the Shaw pseudo potential fitted to the ionization energy. This is the simplest and crudest model possible. We have explored the effect of using (a) Heine-Abarenkov, (b) Ashcroft, and (c) Phillips-Kleinman forms. Our results are not very sensitive to the choice of pseudo potential. (In the case of Cs metal, which I did not discuss, the sensitivity to the potential is crucial).

The most important interaction is that of the electron with the solvent molecules (ammonia). Again, this is represented as a pseudo potential. The results depend on the choice of cut-off parameter, R_c used in the Shaw pseudo potential for electron-proton (H-atom) interaction. Our choice of $R_c = 1$ Å cut off yields a reasonable binding (solvation) energy for the solvated electron in ammonia. Modest changes in the parameter do not appear to greatly affect the properties of the solvated electron, esentially because the major contribution to the solvation energy arises from the long-range Coulomb interactions.

If one changes R_c, the solvent distribution functions involving the solvated electron and the solvent also change. More realistic pseudo potentials, which include approximately the short range effects due to exchange plus the effect of polarization, suggest that the Shaw type potential is not sufficiently repulsive. Our calculated absorption spectrum for the solvated electron (Fig.7) peaks at energies that are much too high. This result also suggests that our electron solvent pseudo potential is too attractive.

LLUCH - The nonadditivity of the pair potentials can be responsible for the failure of Monte-Carlo and Molecular Dynamics simulations to give correct results in some studies of solvated ions. Inclusion of the polarisation term of the three-body correction is nowadays possible, although too costly in many cases. However, the most important source of error is often the lack of the charge transfer part of the many-body correction, whose incorporation is generally not evident. This is the case of the solvated proton.
Given that the delocalization of the electron will very much depend on the considered number of NH_3 molecules, what do you think about the validity of your electron-NH_3 pair potential in order to study the solvated electron ?

KLEIN - You raise an important question. Long-range many body forces are important. Short-range many body forces are also important. Alas, our knowlegde of the latter is very poor. We used the crudest possible model and have neglected both of these effects. Our electron-ammonia solvent potential includes the most important polarization effect namely, that caused by the solvent dipoles on the electron. We have ignored the self-consistent many body polarization of the coupled electron-solvent system. This latter effect has been discussed by Wallqvist, Thirumalai and Berne (J. Chem. Phys., 1987). I refer you to this article for details.

BELLONI - 1) Is your model predicting a solvent dependence of the optical absorption band of the solvated atom as it is found for the silver atom in various solvents by pulse radiolysis ?

2) I was interested by the estimation of the change in the ionization potential of the atom - when solvated - Could you in the same way calculate the situation for clusters of few metal atoms in interaction with a solvent ?

KLEIN - 1) For the present we have only examinal alkali metals. However, pseudo potentials exist for many metals and I see little difficulty finding potential models for other systems. Your suggestion of looking at silver in NH_3 and the comparison with H_2O could be most interesting.

2) The possibility of investigating atomic cluster in polar solvents is not yet possible. However, perhaps in the near future this will be a reality.

HYNES - Could I ask for some clarifications on the ion-electron recombination? Are you saying that there is a real barrier of 50 kT for the reaction ? In what sense do you mean that the solvent is adiabatic ? That the electron instantly adjusts to the slow solvent configuration, and that one must wait for the appropriate fluctuation in the solvent ? Also, what is the experimental rate constant for the reaction ?

KLEIN - The details of the calculation of ground-state fluctuations in the electrons-solvent system is outlined in our article to appear in J. Chem. Phys. in 1987 (M. Sprik and M.L. Klein). The term adiabatic is used in the sense that the coupled electron-solvent system is equilibrated subject to the constraint that the electron center of mass is fixed. The 50 K_BT barrier referred to is really an energy barrier and not a free energy. The coupled solvent-electron fluctuation, when the electron is in its ground state are of this order of magnitude.

JONAH - The partial molal volume of the electron is an experimental parameter which can be measured and shows a great difference between the hydrated electron and the ammoniated electron. Have you calculated or are you trying to calculate its value ?

KLEIN - Yes, indeed, we are ! My group Massimo Marchi and Michiel Sprik have written a constant-pressure Monte-Carlo program to study the electron-solvent system. To-date, only preliminary results are available ; they are sufficiently encouraging that we are actively pursuing this avenue of research. It will be necessary to compare and contrast the behaviour of hydrated and ammoniated electrons. Another quantity of interest is ΔS, the entropy change on solvation. We are also attempting to calculate ΔS for the solvated electron. This story is nearly finished and will be submitted for publication shortly.

NEWTON - Your calculated results for the local solvent structure around an electron in NH_3 or H_2O seem to be in agreement with experimental data. However, much of this latter data is based on assumed models necessary for interpretation of the magnetic resonance data (e.g., assumption of a point electron). Even through it would clearly be very difficult to calculate spin density and related hyperfine constants directly using your path integral method, you nevertheless have a representation of the wave function for the excess electron, and it seems that this data might be very valuable in testing the assumption employed in the analysis of the magnetic data.

KLEIN - I agree with your remark. The wave function we obtain from our adoptive gaussian basis set approach (M. Sprik and M.L. Klein, J. Chem. Phys., November 1987) to the solvated electron should indeed be used to probe more subtle effects than just the energy and absorption spectrum.

LIQUID STATE QUANTUM CHEMISTRY:

APPLICATION TO REACTIVITY

Jean-Louis Rivail

Laboratoire de Chimie Théorique
U.A. C.N.R.S. n° 510 - Université de Nancy I
B.P. 239 - 54506 Vandoeuvre-les-Nancy cedex (France)

INTRODUCTION

The relationship between molecular structure and chemical reactivity does not need to be emphazised, and the success of quantum chemistry in computing the electronic properties of a molecule is well established nowadays. Nevertheless it is worth recalling that quantum chemical techniques deal with small molecular systems which are either isolated or which undergo a well defined external perturbation. As soon as one extends the concept of individual molecules to a liquid, one is faced with quite a different situation, in which the modifications introduced by the liquid surroundings on the molecular structure may be far from negligible.

This problem has been tackled for many years by two main kinds of approach. In the supermolecule approach[1] standard quantum calculations are performed on a cluster containing the solute surrounded by some solvent molecules. This technique allows us to take into account the electron exchange which takes place between neighbouring molecules. Conversely it may be quite expensive, regarding computational time, even for a single configuration. Therefore it is still unrealistic to contemplate the possibility of coupling it with a statistical treatment of the cluster and the entropy variations during the solvation process are omitted. Such approach is well adapted to the situations in which strong chemical interactions occur between the solute and the solvent, such as hydrogen bonding, in which enthalpy terms are predominant.

At the other extreme several continuum models have been devised[2]. In these approaches the solvent is replaced by a continuous medium, which is characterized by macroscopic properties and which interacts physically with the molecule. From a physical point of view they are rather unrealistic since they introduce a discontinuity between the molecule treated as a microscopic system, and the surroundings. Nervertheless they may be useful in the case of small physical solute-solvent interactions. In this paper, we shall describe a model of this variety which is intended to provide us with rapid, semi-quantitative information on the average geometric and electronic structure of a molecule in solution, and, ultimately, with some thermodynamic data useful for the description of a reaction path in the liquid state.

THE MODEL

The model rests upon classical electrostatic considerations which have been developed earlier[3]. One evaluates the electrostatic potential created by the liquid surroundings into the region of space occupied by the solute molecule, in order to introduce the perturbation in the hamiltonian of this molecule. This can be achieved by imagining a cavity within a dielectric continuum, in which the molecule is placed. The electrostatic potential arises from the polarization of the continuum by the electric charge distribution of the molecule. This potential in turn keeps the molecule in a polarized state different from its equilibrium state outside the cavity so that the determination of its molecular structure in the liquid must be self-consistent.

In addition one must keep in mind that the polarization of the continuum produces a variation of its free energy which can be shown to be exactly equal to - 1/2 times the energy of the charge distribution interacting with the cavity electrostatic potential[3,4]. Therefore the free energy change of the system when the charge distribution is placed in the cavity is + 1/2 times the energy of interaction, and this quantity is easily computed by means of classical electrostatics provided that one knows :
    i - the charge distribution of the solute
   ii - the shape of the cavity
  iii - the dielectric constant of the solvent

We shall examine the computation of the equilibrium charge distribution of the solute later and focus our attention on the geometry of the cavity. Its volume may be defined unambiguously as the fraction of the total volume of the sample which is occupied by the solute e.g. the partial molecular volume of this species. The shape of the cavity is not so easily defined since it corresponds to a fictitious surface which is expected to give rise to an electric potential as close as possible to the actual potential in this region of space. On the basis of rather naive arguments one may define it as the surface which encloses, for a given volume, the largest fraction of the electric charge. This is an electronic isodensity surface (fig.1). On the other hand, if one looks for a compromise between the maximum charge included in the cavity and the minimum area, it appears that an electronic isopotential surface is a good candidate. One may notice that a nuclear isolectronic surface does not depart noticeably from the previous one and is more easily computed.

Finally if the cavity happens to be well approximated by an ellipsoid, the free energy change of the system when the solute is placed in the cavity, may be written in a simple form :

$$\Delta F = - \frac{1}{2} \sum_{\ell=0}^{\infty} \sum_{m=-\ell}^{\ell} R_\ell^m < M_\ell^m > \qquad (1)$$

where $<M_\ell^m>$ stands for a component of the ℓth multipole moment of the solute's charge distribution in the spherical tensor formalism and R_ℓ^m is the corresponding component of the reaction field which is related to this change distribution by the equation :

$$R_\ell^m = \sum_{k=0}^{\infty} \sum_{n=-k}^{k} F_{\ell k}^{mn} <M_h^k> \qquad (2)$$

The elements of the reaction field tensor are defined by the geometry of the ellipsoid and the dielectric constant of the solvent.

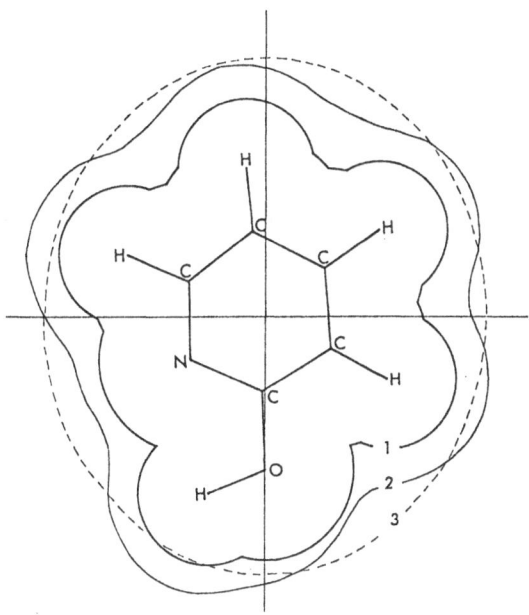

Figure 1. Cavities for 2-hydroxypyridine
1 van der Waals spheres
2 isodensity surface
3 closest ellipsoid

TESTS OF THE MODEL

Monte-Carlo calculations provide us with an alternative route
to the local properties in the bulk of a liquid which are closer to
first principles than the above model although the molecule needs to be
somewhat simplified. We performed several computations[5,6] on model
liquids in order to evaluate the electric field which a molecule undergoes from the liquid and to compare it with the values predicted by the
model. The main result of this comparison is that, due to the error
bars of the Monte-Carlo calculation, and to the uncertainties on the
dielectric constant of the medium, the model reproduces the electric
field fairly well, especially when the charge distribution reduces to a
single moment. In turn, noticeable deviations appear between the model
and the Monte-Carlo simulation when the charge distribution of the
solute is represented by more than one dominant moment (e.g. a dipole
and a quadrupole) and when the solvent is represented by point dipoles
at the centre of non polarizable molecules. This is easily understan-
dable if one bears in mind that the model replaces this medium by a
continuum. Nevertheless these discrepancies are expected to be less
important in the case of a real medium, due to the molecular polarizabilities which are nonlocal properties[7].

Finally the differences between the electrostatic potential
computed by using the various shapes of cavity considered above appear
to lie within the error bars of the Monte-Carlo computation so that an
ellipsoidal cavity appears to be an acceptable first approximation.

As explained in the introduction of this paper, the solute-solvent interaction forbids us to consider the solute independently, and its equilibrium properties must be regarded as the consequence of the equilibrium of the whole system, which is reached by minimizing its Gibbs' free energy.

Equation (1) gives us the part of the contributions of the solute-solvent interactions to the free energy of the system which changes when the charge distribution is varied within the cavity. If one neglects the variations of volume of the system induced by a modification of the solute's charge distribution (which should be due to electrostriction) this term can be assumed to be equal to the Gibbs'free energy.

On the other hand the hamiltonian of the isolated solute H_0 allows us to compute the internal energy of this molecule and if one assumes that there is no entropy variation associated with the modification of the molecular charge distribution, the equilibrium properties of the system are reached by minimizing the sum :

$$\langle H_0 \rangle - \frac{1}{2} \sum_{\ell} \sum_{m} R_{\ell}^{m} \langle M_{\ell}^{m} \rangle \qquad (3)$$

This leads us to a set of modified S.C.F. equations in which the matrix elements (μ, ν) of the Hartree-Fock matrix in the basis of the atomic orbitals used to compute the molecular wavefunction is :

$$F_{\mu\nu} = F_{\mu\nu}^{o} + \sum_{\ell=0}^{\infty} \sum_{m=-\ell}^{\ell} R_{\ell}^{m} \langle \mu | M_{\ell}^{m} | \nu \rangle \qquad (4)$$

$F_{\mu\nu}^{o}$ denoting the corresponding element of the free molecule. This set is non linear since R_{ℓ}^{m} depends on the moments $\langle M_{k}^{n} \rangle$ but the resolution of the S.C.F. equations being iterative, it can be linearized by using the wavefunctions obtained at the previous iteration to determine the R_{ℓ}^{m} factors. Then the solution of this system does not differ significantly from that of the usual S.C.F. equation of the free molecule, and does not take much longer. The minimization can be also achieved with respect to the solute's nuclear coordinates leading for this molecule to the equilibrium geometry and electron distribution as well. This computational method has been thoroughly used to analyze the solvent effects on many molecular spectroscopic properties[8] and it appears that the representation of the effect is quite satisfactory. These results will not be examined here but they suggest that the model reproduces the modification of a molecular structure in a liquid reasonably well.

The shape of the cavity being the only problem which is not entirely solved at this stage, we shall examine its influence on the electronic distribution in the molecule by comparing the results obtained in the case of a highly polarizable molecule (N-methylformamide) by using two cavities having the same volume but quite different geometries (a sphere and an ellipsoid). The results obtained with a 4,21 G ab initio computation are given in table 1. They show that the electronic distribution is not very sensitive to the shape of the cavity although solvation energies depend more strongly on this feature. Therefore, as far as the wavefunction is concerned, we can chose a reasonable shape of cavity, e.g. the ellipsoid which best approaches one of the two "physical cavities" proposed above, and defined from the free molecule. When solvation free energies are required, a more careful study is needed, which includes the use of the best possible cavity, which usually departs from a simple shape and requires the use of a numerical integration.

Table 1. N-methyl formamide, 4,31G

	Free molecule	Spherical cavity	Ellipsoidal cavity
Dimensions of the cavity (A)	...	R=2.850	A=2.642 B=2.830 C=3.096
Dipole moment (D)	4.21	4.69	4.61
Electronic O	8.660	8.696	8.691
densities on N	7.869	7.871	7.870
Electronic field gradient at O eq_{zz}	- 1.705	- 1.664	- 1.670
Solvation energy (KJ.mol^{-1})	...	18.995	15.857

It must be pointed out that the result of the calculation does not represent the whole of the solvation free energy. Since only electro-static interactions are taken into account, this quantity includes the electrostatic contribution and the induction term due to the fact that the S.C.F. computation takes the polarization of the solute into account. In turn, the dispersion term which cannot by any means be neglected is not taken into account. This term is not accessible by S.C.F. computa-tions and, therefore, it is also omitted in other results such as those obtained by the supermolecule approach. In addition one may anticipate that it would not have a substantial influence on the equilibrium electro-nic structure of the solute, although its contribution to the energetics of the process is large. In order to be able to analyze the influence of this term and, if possible, to take it into account in the model we have developed an approximate evaluation[9] based upon Linder's theory[10] and a variational calculation of multipole polarizabilities[11] which gives a good estimate of the dispersion term. In addition, this term can be included in the S.C.F.equations[12]. The resolution of these equa-tion in the case of test molecules shown that, as expected, the disper-sion term does not modify significantly the resulting electronic struc-ture.

The last contribution to the total solvation free energy is the so called cavitation term which represents the free energy variation of the solvent when the cavity is formed in the bulk of the liquid. It seems reasonable to assume that this term does not influence the electronic structure of the solute. Regarding the value of the solvation free energy, this term can be evaluated by various approximate equations.

MODIFICATION OF THE REACTIVITY OF A MOLECULE UNDER THE INFLUENCE OF A SOLVENT

An example of the use of the model in reactivity problems can be found in the papers by Bertran et al. on the nature of the frontier orbitals of a series of anions and of polar molecules. In a first study[13] they noticed that the HOMO of the formiate anion in which an interchange between the two highest occupied orbitals occurs when the molecule is in a liquid environment.

In a more recent work[14] they studied the effect of a solvent on the orbital hierarchy of larger molecules and anions. The most stri-king effect is observed in the case of the benzoate anion. The HOMO of the free molecule is a π orbital, having the symmetry a_2 and mainly located on the oxygen atoms of the carboxylate group. In a liquid of

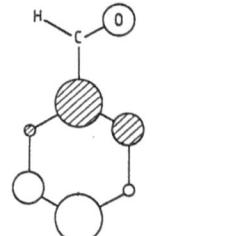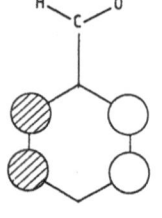

free molecule   solvated molecule

Figure 2. HOMO of benzaldehyde (after [14])

low dielectric constant ($\varepsilon < 4$) the HOMO is still a π orbital having
the symmetry b_1, and at a higher dielectric constant another π orbital
having the symmetry a_2 but mainly located on the cycle comes out.

A similar behaviour is observed with benzaldehyde as shown
on figure 2. In other cases, such as nitrobenzene, the effect of the
solvent raises the level of the HOMO increasing the nucleophilic properties of the molecule.

SOLVENT EFFECT ON A REACTION PATH

Although the model was first intended for determining the electronic structure of a solute, one may be tempted to use the energetic
data which can be computed to analyze the solvent effect on a reaction
path.

In order to be able to evaluate the model itself, we shall examine a very simple case in which there is only one reactant and one
product, having roughly the same molecular volume and the same polarizability in order to neglect the cavitation and dispersion contributions
to the variation of the solvation free energy. Tautomeric equilibria
offer us a good example of such systems.

In order to obtain a reliable value of the solute-solvent free
energy, one has to compute this quantity by using the best possible
shape of cavity which differs considerably from the ellipsoid. Therefore,
after having computed the wavefunction with the ellipsoidal cavity, the
solvation free energy is computed numerically by means of a realistic
cavity, which is limited, in this instance, by an isodensity surface.
Furthermore, in order to avoid the uncertainties on the energies of
the isolated molecules, the solvation free energies are used to predict
the modification of the equilibrium constant between the gas phase
and the liquid phase.

We report here the results obtained[15] on two tautomeric equilibria in aqueous solution :
  2 hydroxypyridine $\rightleftarrows$ 2 pyridone
  4 hydroxypyridine $\rightleftarrows$ 4 pyridone

The electronic properties of the molecules have been computed
by means of the MNDO method. The molecular volumes are assumed to be
the same for both tautomers in equilibrium. The values are 156.6 Å^3 and
156.8 Å^3 for the 2- and 4-hydroxypyridine systems respectively and the
relative electric permittivity of water is set equal to 78. Tables 2 and
3 give the equilibrium constants and the values H_s of the computed

contribution to the free energy of solvation. They also give the contribution to this quantity of the polarization energy of the solute E_{ps}. This quantity represents the difference between the electronic energies of the solvated molecule and the free molecule respectively.

These results show that the large solvent effects observed can be predicted relatively accurately, especially if one bears in mind the substantial uncertainties which affect the experimental results.

A second feature which is worth being emphazised is the importance of the polarization energy of the solute which varies strongly from one tautomer to another. This quantity can hardly be obtained without using a quantum chemical calculation.

Table 2. The 2-Hydroxypyridine system

Tautomer	ΔH_s $KJ.mole^{-1}$	E_{ps} $KJ.mole^{-1}$	Equilibrium constant (20°C)		
			gas phase exp.	aqueous solution cal.	exp.
OH	− 12.40	0.71	0.3^{16}	395	340^{17} 910^{18}
NH	− 28.21	10.92			

Table 3. The 4-hydroxypyridine system

Tautomer	ΔH_s $KJ.mole^{-1}$	E_{ps} $KJ.mole^{-1}$	Equilibrium constant (20°C)		
			gas phase exp.	aqueous solution cal.	exp.
OH	− 18.72	1.59	0.1^{16}	1667	2200^{17} 1950^{18}
NH	− 40.71	18.48			

CONCLUSION

The very simple model described in this paper offers the possibility of computing the modified electronic wavefunction of a molecule which undergoes the effect of a liquid environment. It may be used at least for exploratory studies on the rather complex phenomena which take place along a reaction path in the liquid state.

The study of tautomeric equilibria shows that the energetic data can also be used in equilibrium or ultimately kinetic studies provided that one bears in mind the approximations upon the model is based. Nevertheless, its use seems to be less restricted that one could thought. In particular, aqueous solutions seem to be well represented by this purely electrostatic approach. This fact has already been pointed out and discussed earlier[19]. It broadens somewhat the scope of applications of the model.

REFERENCES

1. A. Pullman, in Environment Effects on Molecular Structure and Properties, p.1, B. Pullman (ed.) Reidel,Dordrecht 1976.
2. R. Bonacorsi, C. Ghio and J. Tomasi, Stud.Phys.Theor.Chem. 21, 407 (1982)
   R. Constanciel, Theor.Chim.Acta, 69, 505 (1986)
3. J.L. Rivail, B. Terryn, D. Rinaldi and M.F. Ruiz-Lopez, J.Mol. Struct.THEOCHEM, 120, 387 (1985)
4. P. Claverie, in Quantum Theory of Chemical Reactions, vol.III, p. 151, R. Daudel et al. (ed.) Reidel, Dordrecht 1982.
5. B.J. Costa Cabral, D. Rinaldi and J.L. Rivail, Chem.Phys.Lett., 93, 157 (1982)
6. B. Bigot, B.J. Costa Cabral and J.L. Rivail, J.Chem.Phys., 83, 3083 (1985)
7. D.W. Oxtoby, J.Chem.Phys., 69, 1184 (1978)
8. J.L. Rivail, Stud.Phys.Theor.Chem., 21, 389 (1982)
9. B.J. Costa Cabral, D. Rinaldi and J.L. Rivail, Compt.Rend.Acad. Sci. Paris, 298 II, 675 (1984)
10. B. Linder, Adv.Chem.Phys., 12, 225 (1967)
11. J.L. Rivail and A. Cartier, Mol.Phys., 36, 1085 (1978)
12. D. Rinaldi, B.J. Costa Cabral and J.L. Rivail, Chem.Phys.Lett., 125, 495 (1986)
13. J. Bertran, A. Oliva, D. Rinaldi and J.L. Rivail, Nouv.J.Chim., 4, 209 (1980)
14. E. Sanchez Marcos, J. Maraver, M.F. Ruiz Lopez and J. Bertran, Can.J.Chem., 64, 2353 (1986)
15. B. Terryn, unpublished results
16. Estimated after the data of P.Beak,F.S.Fry Jr., J.Lee and F.Steele, J.Am.Chem.Soc., 98, 171 (1976)
17. A.Albert and J.N. Phillips, J.Chem.Soc., 1956, 1294
18. S.F. Mason, J.Chem.Soc., 1958, 674
19. E. Sanchez Marcos, B. Terryn and J.L. Rivail, J.Phys.Chem., 89, 4695 (1985)

DISCUSSION

MICHEAU - My comments deals with azeotropic binary mixtures. We have recently made some experimental measurements of the thermodynamic parameters of a thermochronic equilibrium in azeotropic liquid mixtures. Our thermochronic equilibrium (Nickel complexes : $NiR_2 + 2S \rightleftharpoons NiR_2S_2$) is sensitive to the donor number of the solvent S which is an empirical measure of the availability of the electronic doublet of S. What we have found in azeotropic mixtures (alcohol + halogenated hydrocarbons) is that near the room temperature there is a kind of natural compensation of the alcohol doublet availability by the presence of halogenated hydrocarbons molecules ; this compensation shifts the equilibrium position near 50/50 (solvated vs non solvated complex). This property is spontaneous with the azeotropes we have studied, but must to be adjusted accurately by varying the molar ratio with similar binary mixtures not giving azeotropes. So, it appears that azeotropes exhibit from this point of view some singular propertie. My question is : Do you have or do you know some results about reactivity studies in azeotropic mixtures ? Could an azeotrope be considered as a model of a particular supermolecule or cluster ?

RIVAIL - It is an interesting problem but to my knowledge, it has not been treated in the way you mention.

HYNES - Your method first assumes a solvent in equilibrium with the charge distribution to find a free energy and second uses this in a Schrodinger equation to begin the SCF process. In effect, a solvent average is performed first. A different procedure, still within a continuum picture, is the following. For a given solvent polarization, calculate the interaction with the charge distribution, use this Hamiltonian in a Schrodinger equation, and do the SCF procedure to find energies. Finally the Boltzmann average would be calculated. In this method, the solvent average comes only at the end.
Have you considered such a procedure and whether it would be equivalent of superior to the procedure you have used ?

RIVAIL - Our method consists in introducing the solute-solvent interaction in the SCF calculation so that the solvent polarization follows the optimization of the charge distribution of the solute. When one is able to use an analytical expression of the perturbation potential, this method does not take a much longer times as a usual SCF calculation and one may compute a better value of the solute-solvent free energy of interaction by using a more realistic cavity. The scheme you propose is somewhat equivalent except that if the molecular charge distribution is that of the free molecule, it is still rather different from that of the molecule in equilibrium with the liquid and one would probably need at least one more iteration to reach convergence.

THE EFFECT OF SOLVENT FLUCTUATIONS IN ELECTRON TRANSFER PROCESSES

Angels González-Lafont, José M. Lluch, Antonio Oliva and
Juan Bertrán

Departament de Química
Universitat Autònoma de Barcelona
08193 Bellaterra (Barcelona), Spain

INTRODUCTION

The homogeneous outer sphere electron transfer reactions in solution occur at a rate that is noticeably lower than the diffusion rate. This peculiar behaviour has been explained[1] through a three-step mechanism: formation of a precursor complex from the separated reactants, actual electron transfer within this complex to form a successor complex and dissociation of the latter complex into separated products. The reaction rate is usually controlled by the electron transfer step, this step being governed by the Franck-Condon principle. This principle is embodied in classical electron transfer theories using an activated-complex formalism in which the electron transfer occurs at the intersection of two potential energy surfaces, one for the reactants and the other for the products. This implies that the second step necessarily involves the reorganization of the solvent before and after the electron transfer itself is produced. So, it is obvious that solvent must play an essential role in the rate of electron transfer reactions in solution.

Up to the present, very simple solvent models[2] have been generally used in order to study electron transfer processes in solution. These simplified models imply a drastic reduction of the degrees of freedom of the system and in them the movement of the solvent is represented by harmonic oscillations. In the last years, statistical[3] methods based on numerical simulation have permitted to treat explicitly many solvent molecules in a discrete representation. These statistical methods have opened very hopeful perspectives for the study of the solvent reorganization which has to be produced before electron transfer itself takes place.

RESULTS AND DISCUSSION

In this work we have used the Monte Carlo[3] technique in order to study the electron transfer reactions $Fe^+ + Fe^+ \longrightarrow Fe^{2+} + Fe$ and $Fe^+ + Fe^{2+} \longrightarrow Fe^{2+} + Fe^+$, the first one as an example of a very fast process and the second one as a case of a self-exchange reaction. For the two processes, several simulations of water clusters around each ion species have been done by changing the number of solvent molecules and by imposing different constraints to the movement of these molecules. The

Table 1. Mean $Fe^+ - O$ distances[a] and solvation energy difference[b] associated with the $Fe^+ + Fe^+ \longrightarrow Fe^{2+} + Fe^o$ reaction at 0 K.

n[c]	R_{red}	R_{rox}	ΔE_{solv}
12	2.30	2.30	-67.2
50	2.31	2.31	57.8

(a) in $\overset{\circ}{A}$; (b) in $kJ\ mol^{-1}$; (c) number of water molecules.

solvation energy of each configuration before and after the electron transfer has been calculated in the pairwise approximation, using the MCY potential for the H_2O-H_2O interaction, and ab initio analytical potentials generated by us for the Fe^+-H_2O, $Fe^{2+}-H_2O$ and $Fe - H_2O$ interactions, although in the latter case a semiempirical potential has also been employed[6].

Study of the reaction $Fe^+ + Fe^+ \longrightarrow Fe^{2+} + Fe^o$

Statistical simulations of clusters containing 12 and 50 water molecules around the system composed by two Fe^+ ions separated by a distance of 5 $\overset{\circ}{A}$ have been carried out at 0 K and at 298 K. In order to distinguish between both Fe^+ ions, we will call "reductor" the Fe^+ ion which loses one electron and transforms into Fe^{2+}, and "oxydant" the Fe^+ ion which gains one electron and transforms into Fe^o.

At 0 K we have found that the first solvation shell for the two studied clusters is formed by six water molecules octahedrically disposed around each Fe^+ ion, the two octahedrons being staggered face-to-face. Table 1 presents for each cluster the mean distances between both Fe^+ ions and the oxygen atoms of the first solvation shell water molecules, and the

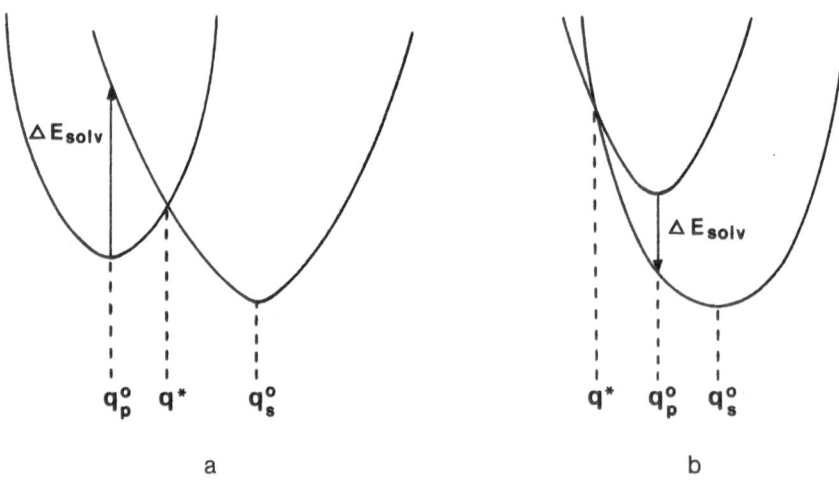

Fig. 1. Schematic representation of the normal (a) and abnormal (b) energy regions in the crossing of the potential energy surfaces of the precursor and successor complexes.

Table 2. Percentage of isoenergetic configurations and mean Fe^+- O distances[a] calculated over these configurations for the $Fe^+ + Fe^+$ $\longrightarrow Fe^{2+} + Fe^\circ$ reaction.

n[b]	%	R_{red}	R_{ox}
12	5.2	2.34	2.30
50	3.8	2.28	2.33

(a) in Å; (b) number of water molecules

solvation energy difference associated with the electron transfer process which would take place keeping unchanged the geometry structure of the precursor complex. One can observe that the first hydration shell is more expanded in the cluster containing 50 water molecules. Furthermore, the calculated values of ΔE_{solv} are very different, this magnitude being negative for the cluster containing 12 water molecules and positive for the other one. In the activated-complex formalism, this means that the clusters containing 12 and 50 water molecules provide examples of the abnormal (fig. 1b) and normal (fig. 1a) energy regions. These two situations are defined according to the relative position of the inter-section region (q*) with respect to the equilibrium configurations of the precursor (q_p°) and of the succesor (q_s°) complexes.

The calculated values of ΔE_{solv} imply that the electron transfer process between two Fe^+ ions is not possible at 0 K, since, according to the Frank-Condon principle, the electron transfer process can only be produced in the intersection region between the hypersurfaces of reactants and products. To investigate the possibility that solvent fluctuations, due to the effect of thermal agitation, lead to the existence of isoenergetic configurations, we have done statistical calcu-lations at 298 K. Among the generated configurations, we have identified as isoenergetic those for which $|\Delta E_{solv}| \leqslant 6$ KJ mol^{-1}

Table 2 presents the percentage of isoenergetic configurations along with the two mean Fe^+ - O distances between the reductor and the oxydant Fe^+ ions and the oxygen atoms of the first hydration shell water molecules, these mean distances being calculated over the isoenergetic configurations. The first thing one can observe is that the number of isoenergetic configurations in the clusters containing 12 or 50 water molecules is not at all negligible. This fact indicates that the increase in temperature strongly enhances the probability of the electron transfer process to be produced.

Let us now analyze the two Fe^+ - O distances. As it has already been mentioned, the cluster containing 12 water molecules corresponds to the inverted energy region (fig. 1b). In this case the intersection region, q*, is approached when solvent fluctuations displace the minimum energy structure of the precursor complex, q_p°, in the opposite direction to the one which would lead to the minimum energy structure of the successor complex, q_s°. In good agreement with this prediction, table 2 shows that isoenergetic configurations are reached when the first hy-dration shell of the reductor Fe^+ ion is expanded and the first hydration shell of the oxydant Fe^+ ion is contracted. On the contrary, the cluster containing 50 water molecules corresponds to the normal energy region (fig. 1a) and the same reasoning as before explains that, in isoenergetic configurations, there is now a contraction of the first hydration shell of the reductor Fe^+ ion and an expansion of the first hydration shell of the oxydant Fe^+ ion.

Finally, it is interesting to consider the energy dispersion in the

Table 3. Mean Fe - O distances[(a)] and solvation energy difference[(b)] associated with the $Fe^+ + Fe^{2+} \longrightarrow Fe^{2+} + Fe^+$ reaction at 0 K.

n[(c)]	f[(d)]	R_{red}	R_{ox}	ΔE_{solv}
	1	2.29	2.06	124.2
12	2	2.29	2.06	124.2
	12	2.29	2.06	124.6
	all	2.32	2.06	211.6
50	all	2.33	2.06	536.5

(a) in Å; (b) in $kJ\ mol^{-1}$; (c) number of water molecules (d) number of degrees
of freedom allowed to vary.

group of isoenergetic configurations. Our calculations have shown that the
60% of isoenergetic configurations for the cluster containing 12 water
molecules are found in an interval of 30 $kJ\ mol^{-1}$ in the solvation energy,
while an interval of 70 $kJ\ mol^{-1}$ is necessary in order to obtain the same
percentage for the cluster containing 50 water molecules. One can observe
that the energy dispersion is very important, this fact being a direct
consequence of the great number of degrees of freedom of the system. The
same reason explains why the energy dispersion increases as the number of
water molecules is increased. This implies that the outer hydration shells
play a non negligible role for the system to reach an isoenergetic configuration.

Study of the reaction $Fe^+ + Fe^{2+} \longrightarrow Fe^{2+} + Fe^+$

Our study of the reaction $Fe^+ + Fe^+ \longrightarrow Fe^{2+} + Fe^o$ presents the
limitation that we have only considered the solvation energy, but not the
interaction between both metal atoms and the energy associated with the
electron transfer in vacuum. The consideration of these two components would
have certainly changed the obtained results, specially in what concerns the
percentage of isoenergetic configurations. For this reason, we believe that
it is interesting to carry out an analogous study of one reaction of the
self-exchange type, as it is the reaction $Fe^+ + Fe^{2+} \longrightarrow Fe^{2+} + Fe^+$,
since in this case the energy associated with the electron transfer in
vacuum is nul and the interaction between both metal atoms is the same
before and after the electron transfer. So; in this case only the solvation
energy has to be taken into account in order to define the isoenergetic
configurations in a rigorous way.

To study the self-exchange reaction, statistical simulations of clusters
containing 12 and 50 water molecules around the system composed by Fe^+
and Fe^{2+} ions separated by a distance of 7 Å have been done at 0,298, 500
and 2000 K. In the two clusters the calculations have been carried out with
no restrictions, but in addition three different constraints to the movement
of the water molecules have also been imposed in the cluster containing 12
water molecules. In these calculations with constraints only radial movements
have been allowed. In the first case only one degree of freedom is
considered, in such a way that the six Fe^+-O distances are identical and the
same occurs with the six Fe^{2+} - O distances, the sum of the two values being
kept constant. This means that an expansion of the hydration shell around
one ion is necessarily accompanied by a contraction of the hydration shell
around the other ion. In the second case the precedent restriction on the
sum of the Fe^+ - O distances is raised, the number of degrees of freedom

ble 4. Percentage of isoenergetic configurations for the $Fe^+ + Fe^{2+} \longrightarrow$
$Fe^{2+} + Fe^+$ reaction at two different temperatures using 12 water molecules.

$f^{(a)}$	T = 500 K	T = 2000 K
1	0.06	4.60
2	0.06	4.35
12	0.02	0.16

) number of degrees of freedom allowed to vary

ing thus increased to two. Finally, the twelve iron-oxygen distances have
en permitted to vary independently.

At 0 K we have found that the water molecules of the first hydration
ell are disposed in a similar way to the one previously obtained for the
$^+$ - Fe^+ system. Table 3 presents the values of R_{red}, R_{ox} and ΔE_{solv}
tained for the two studied clusters at 0 K with and without reduction of
e number of degrees of freedom of the system. In contrast with the results
tained for the $Fe^+ + Fe^+ \longrightarrow Fe^{2+} + Fe$ reaction (see table 1), one can
serve that ΔE_{solv} is now positive for the two clusters, although the value
rresponding to the cluster with 12 water molecules is still smaller than
e other one. So, both clusters correspond this time to the normal energy
gion (fig. 1a). A second aspect which has to be remarked is that the
troduction of constraints leads to a great decrease in the value of
$_{solv}$, the Fe^+ - 0 distance being also slightly changed. This effect of the
nstraints in the solvation energy difference is due to the fact that the
position of only radial movements to water molecules avoids the free
ientation of hydrogen atoms.

As in the precedent reaction, let us now investigate the possible exis-
nce of isoenergetic configurations when the temperature is increased.
ven that the value of ΔE_{solv} is now much more positive, it is to be
pected that the percentage of isoenergetic configurations will be drasti-
lly reduced. For this reason we have weakened the criterium according to
ich one configuration is considered to be isoenergetic, this criterium
ing now $|\Delta E_{solv}| \leqslant 20$ kJ mol^{-1}. In spite of this, no one generated configura-
ons has been found to be isoenergetic at any temperature when no cons-
aints have been imposed. On the contrary, a certain amount of isoenergetic
nfigurations have been obtained when the number of degrees of freedom has
en reduced. Table 4 presents the percentage of isoenergetic configurations
500 and 2000 K, the values at 298 K being not shown since they are
tremely small. As it might be expected, the percentage of isoenergetic
nfigurations is drastically increased when the temperature is varied from
0 to 2000 K. On the other side, it tends to decrease as far as the number
dregrees of freedom allowed to vary is increased, this tendency explaining
y no one isoenergetic configuration is found when no constraints are
posed.

NCLUSIONS

We believe that this work opens hopeful perspectives for the treatment
electron transfer processes in solution. In particular, it has permitted
show that any attempt to reduce the effect of solvent fluctuations to the
rst hydration shell or the number of degrees of freedom of the system is a

very simplified model which does not permit an accurate representation of the actual movement of the solvent. However, if a great number of water molecules is used and no constraints are imposed to their movement one has to generate a big amount of configurations in order to obtain an appreciable number of solvent fluctuations which permit the electron transfer to take place.

REFERENCES

1. B.S. Brunschwig, J. Logan, M.D. Newton, and N. Sutin, A semiclassical treatment of electron-exchange reactions. Application to the hexaaquoiron (II)- Hexaaquoiron (III) system, J. Am. Chem. Soc., 102: 5798 (1980)
2. R.A. Marcus, On the theory of electron-transfer reactions. VI. Unified treatment for homogeneous and electrode reactions., J. Chem. Phys., 43: 679 (1965)
3. J.P. Valleau and S.G. Whittington, A guide to Monte Carlo for statistical mechanics: 1. Highways, in: "Statistical Mechanics", Vol. A, B.J. Berne., ed., Plenum Press, New York (1977)
4. A. González-Lafont, J.M. Lluch, A. Oliva and J. Bertrán, Analytical potentials from ab initio calculations for the Fe^+ - H_2O and $Fe°$ - H_2O systems, Int. J. Quantum Chem., in press.
5. A. González-Lafont, J.M. Lluch, A. Oliva and J. Bertrán, An intermolecular potential function for the Fe (II)- H_2O system from ab initio calculations, Int. J. Quantum Chem., 30: 663 (1986)
6. L.A. Curtis, J.W. Halley, J. Hautman and A. Rahman, Nonadditivity of ab initio pair potentials for molecular dynamics of multivalent transition metal ions in water, J. Chem. Phys., 86: 2319 (1987)

KLEIN - Do you think that 50 water molecules are really enough to reproduce solvent fluctuations in charge transfer systems ?

BERTRAN - Our purpose in this work has been to show that the models developed up to now, that imply a drastic reduction of the degrees of freedom of the solvent system, do not allow to give a complete description of the nuclear reorganisation of the solvent, previous to the electron transfer. So, we have explicitly used a number of water molecules much greater than the considered one in the standard models, in such a way that we have obtained a good representation of the fluctuations which favour the electron transfer in clusters of 12 water molecules (first shell) and 50 water molecules (at least first and second shells). In spite of the fact that the bulk really is not well introduced, we have found meaningful results. For example, in our statistical calculations, the usual separation between inner and outer solvent shells does not appear. As a matter of fact, the existence of iso-energetic configurations is generally due to a compensation between solvation contributions from both shells.

LAUNAY - You have chosen a valence of one for iron, which is not a common one. Why did you do so ?

BERTRAN - We have initially chosen the process $Fe^+ + Fe^+ \quad Fe^{2+} + Fe^0$ because it takes place at a very high rate, in such a way that Fe^+ in water is not detected experimentally. Thus it may be expected to find some isoenergetic configurations which make possible the electron transfer in this system, in spite of the great number of degrees of freedom involved when a reasonable number of water molecules are explicitly considered. In a second step, we wanted to study a self-exchange reaction. Undoubtedly, one of the most interesting examples of this type of processes would be the one constituted by the Fe^{2+}/Fe^{3+} system. However it is well known that the use of the pairwise approximation, employed in our Monte Carlo calculations, for a trivalent cation like Fe^{3+} is not valid, many body corrections being strictly necessary in this case. Moreover, the correct determination of the $Fe^{3+}-H_2O$ pair potential presents a very important intrinsic difficulty, since we have found that the ground state of the $(Fe - H_2O)^{3+}$ system, at distances very close to the equilibrium one, corresponds to the electronic configuration $Fe^{2+} - H_2O^+$. Taking into account these problems, we have finally selected the system $Fe^+ + Fe^{2+} \longrightarrow Fe^{2+} + Fe^+$. Even in this case we have not used our $Fe^{2+} - H_2O$ ab initio analytical potential but an empirical one fitted by Curtis et al., since all the $Fe^{2+} - H_2O$ ab initio potentials which have been created up to now give a coordination number of eight, probably due to the neglect of many-body corrections.

HYNES - You stressed that it was very difficult to find the equal energy configurations. One possibility to increase the probability of those configurations is to set up an intermediate charge distribution such as $Fe^{1,5+} + Fe^{1,5+}$. For equal sized ions, this would be the best charge distribution, since for it the macroscopic equilibrium for the solvent corresponds to the equal energy case.

BERTRAN - Really you have made an excellent suggestion, because the probability to find isoenergetic configurations from the macroscopic equilibrium distribution of the reactants is very low, in such a way that in some cases it is practically impossible to get them with the present calculation facilities. If we are able to obtain in our laboratory the required pair potential $Fe^{1,5+}$ - H_2O, which probably will not be an easy task, your suggestion should be the best way to reach the suitable solvent fluctuations for the electron transfer reaction. An additional problem would be to calculate the probability of these fluctuations in the equilibrium distribution of the reactants.

A SIMPLE DESCRIPTION OF THE INTERMOLECULAR PROTON TRANSFER BY TUNNELLING IN SOLUTIONS

P.H. Fries[*] and E. Belorizky[•]

[*]C.E.N.-Grenoble, DRF-G/Laboratoires de Chimie - Equipe Chimie de Coordination, 85 X, 38041 Grenoble Cedex, France
[•]Laboratoire de Spectrométrie Physique (associé au CNRS), USTMG, BP 87, 38402 Saint Martin d'Hères Cedex, France

INTRODUCTION

Intermolecular proton transfer is the subject of many theoretical and experimental studies. This process plays an important role in several chemical and biological systems in the liquid phase[1]. Here we are interested in the direct process

$$AH^+ + B \leftrightarrows A + HB^+, \tag{1}$$

between identical molecules A and B, without formation of an intermediate hydrogen bonded complex.

Besides a normal activated proton jump mechanism with a rate which obeys an Arrhenius type law

$$k_A = A \exp(- E_A/kT), \tag{2}$$

and which becomes increasingly important at higher temperatures, there is a quantum mechanical process of direct tunnelling through the potential barrier between the acceptor and donor sites.

A very interesting approach of this mechanism has been proposed by Klöffler and Brickmann[2] who assume that the proton transfer between equivalent molecules is calculated with a WKB method, while the relative A-B motion is treated as a classical diffusion. The proton transfer process was treated with the help of a one dimensional (1D) double minimum potential which is modulated by the relative random motion of the diffusing reactants.

In this paper, we present a three dimensional (3D) model in order to describe the spatial extension of the proton wave functions in a more realistic way. Furthermore, each potential well associated to the molecules A and B will be assumed to be spherical with a finite constant depth V_0. Qualitatively, the 3D nature of the wells will considerably reduce the overlap between the wave functions of the two individual wells, thus decreasing the resonant proton transfer rate. The constant depth of the wells avoids the unrealistic presence of a sharp peak at the intersection of two harmonic oscillator potentials and allows simple analytical calculations.

Otherwise, the bases of the model are identical with that of Klöffler and Brickmann and are summarized as follows :

- The two interacting molecules perform a classical random Brownian motion amidst all the other particles which are approximated as an inert continuum.

- The proton transfer is independent of the molecular rotations.

THEORY

The attractive potential of the proton with the molecule A (or B) is assumed to be an isotropic 3D well of finite constant depth :

$$V(r) = -V_0 \qquad \text{for } r < a,$$
$$V(r) = 0 \qquad \text{for } r > a. \tag{3}$$

The quantization of this well is a standard problem[3] and we assume that either there is only one bound state (s level) or that this ground s state is the only one which is appreciably populated at room temperatures. The ground state energy E_0 of a proton of mass M is the solution of

$$|\sin(\alpha_0 a)| = \left(1 + \frac{E_0}{V_0}\right)^{1/2} \tag{4a}$$

with

$$\alpha_0 = \left(\frac{2M(E_0 + V_0)}{\hbar^2}\right)^{1/2} \tag{4b}$$

and $\frac{\pi}{2} \le \alpha_0 a < \pi$. The associated wave function is

$$\Psi(r) = (4\pi)^{-1/2} \chi(r)/r \tag{5a}$$

with

$$\chi(r) = A\sin(\alpha_0 r) \qquad \text{if } r < a,$$
$$\chi(r) = A\sin\alpha_0 a \exp(-\beta_0(r-a)) \qquad \text{if } r > a, \tag{5b}$$

with

$$\beta_0 = \left(-\frac{2ME_0}{\hbar^2}\right)^{1/2} \tag{6}$$

and A, a normalization constant which can be easily expressed in terms of $\alpha_0 a$ and $\beta_0 a$.

Now, we consider the double spherical well potential represented in Fig. 1 and we denote by R the distance of the centers of the two spheres. We have $R \geq b \geq 2a$ where b is a minimal distance of approach which is characteristic of the shape and size of the two identical molecules. Let r_A and r_B be the distances of the proton of the spheres belonging to molecules A and B, and let $\Psi_A(r_A)$, $\Psi_B(r_B)$ be the wave functions of the proton in the ground s state of each potential well, V_A, V_B respectively.

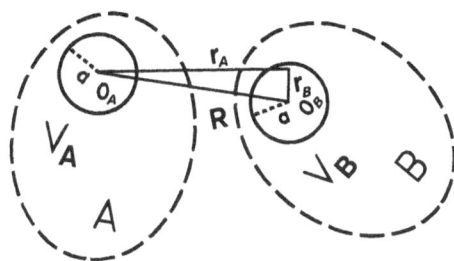

Fig. 1. Proton tunnelling between the two spherical potential wells of constant depth $-V_0$ and of radius a, which are located on equivalent positions of the two identical molecules A and B.

For the double potential well, the Hamiltonian of the system becomes

$$\mathcal{H} = - \frac{\hbar^2}{2\mu} \Delta + V_A + V_B . \tag{7}$$

In order to describe the motion of the proton in this double well, we use an LCAO method giving an energy splitting ΔE between the two lowest levels

$$\Delta E = \frac{2S\langle \Psi_A |V_B| \Psi_A \rangle - 2\langle \Psi_A |V_A| \Psi_B \rangle}{1 - S^2} \tag{8a}$$

where

$$S = \langle \Psi_A | \Psi_B \rangle . \tag{8b}$$

When the overlap integral is such as $S \ll 1$, we have to a good approximation

$$\Delta E = -2\langle \Psi_A | V_A | \Psi_B \rangle. \tag{9}$$

This energy splitting has been calculated for the potential (3) and the zero order wave functions (5). We obtain

$$\Delta E(R) = \frac{C \exp(-\beta_0 R)}{R} \tag{10a}$$

with

$$C = \frac{2V_0 \sin(\alpha_0 a) \exp(\beta_0 a)}{\left(\alpha_0^2 + \beta_0^2\right) a \left[\frac{1}{2} - \frac{\sin 2(\alpha_0 a)}{4\alpha_0 a} + \frac{\sin^2(\alpha_0 a)}{2\beta_0 a}\right]} \times$$

$$\left[\sin(\alpha_0 a)\cosh(\beta_0 a) - \frac{\alpha_0}{\beta_0}\cos(\alpha_0 a)\sinh(\beta_0 a)\right]. \tag{10b}$$

Taking into account the modulation of $\Delta E(R)$ by the random relative motion of the molecules A and B on which are located the potential wells and following the procedure of Klöffler and Brickmann[2], the average transition probability of the proton transfer from molecule A to molecule B at time t is given by

$$\overline{W_{AB}} = \frac{1}{2\hbar^2} \int_0^t \overline{\Delta E(0)\Delta E(\tau)} \, d\tau$$

$$= \frac{1}{2\hbar^2} \int_0^t d\tau \int d^3\vec{R} \int d^3\vec{R}_0 \, P_2\left(\vec{R}_0, \vec{R}, \tau\right) \Delta E(R) \Delta E(R_0), \tag{11}$$

where $P_2\left(\vec{R}_0, \vec{R}, t\right)$ is the joint probability density for the relative motion of the two spherical wells. In the long time limit,

$$P_2\left(\vec{R}_0, \vec{R}, t\right) = \frac{1}{V(4\pi D t)^{3/2}} \exp\left(-\frac{\left(\vec{R} - \vec{R}_0\right)^2}{4Dt}\right), \tag{12}$$

where D is the relative diffusion constant of the molecules and V is the sample volume. For $t \to \infty$, we obtain

$$\overline{W_{AB}} = \frac{4\pi}{\hbar^2 DV} \int_b^\infty dR \, \Delta E(R) R \int_b^R dR_0 \, \Delta E(R_0) R_0^2 \tag{13}$$

and using the expression (10) of $\Delta E(R)$,

$$\overline{W_{AB}} = \frac{2\pi}{VD} \left(\frac{\Delta E(b)}{\hbar}\right)^2 \frac{b^2}{\beta_0^2} \left(b + \frac{1}{2\beta_0}\right), \tag{14}$$

where $\Delta E(b)$ is the energy splitting $\Delta E(R)$ which corresponds to the minimal distance of approach b. For a number N of molecules (acceptors or donors, $N = N_A = N_B$) the rate constant for proton tunnelling is simply

$$k = \overline{W_{AB}} N. \tag{15}$$

DISCUSSION

The range of the energy splitting function (10) is given by $1/\beta_0$. When $\beta_0 b \gg 1$ (short range coupling) the behaviour of k is very similar to that of the 1D model. However, the energy splitting $\Delta E(b)$ is much lower than in the 1D case for the same values of the potential depth V_0 and of the potential width a. For the long range coupling case ($\beta_0 b \ll 1$), k is proportional to $(\Delta E(b))^2 b^2/\beta_0^3$ instead of $(\Delta E(b))^2/\beta_0^5$ in the 1D model.

We have considered a typical[2,4] numerical example by taking realistic values $V_0 = 0.75$ eV, $a = 0.2\ a_0 \simeq 0.1$ Å, corresponding to an energy E_0 of the ground state of each spherical potential well $E_0 = -0.080$ eV. For medium sized molecules we have chosen $D = 10^{-5}\ cm^2 s^{-1}$ and b has been fixed to an intermediate value $b = 8a = 0.85$ Å, corresponding to a barrier width $b - 2a = 0.64$ Å. With a number density of proton acceptors $\frac{N}{V} = 10^{21}\ cm^{-3}$ we obtain a tunnelling transition rate $k = 10^9\ s^{-1}$. This transfer rate is at least of the same order of magnitude as typical diffusion controlled processes at room temperature. But the main point is that k is inversely proportional to the relative diffusion constant D and behaves like η/kT, where the viscosity η is increasing exponentially with $1/T$. Thus, the proton transfer rate by direct tunnelling increases with decreasing temperature and may dominate the normal activation process which decreases with temperature. The isotope effect is considerable in this mechanism. The effect of deuteron substitution is easily evaluated and with the above values of the potential parameters, k is reduced by a factor 10^7 and becomes inobservable.

This simple 3D modle represents an improvement with respect to the usual linear models and has the advantage to be solved analytically. This theory will be improved by taking the random rotational motion of the molecules into account. As the spherical potential wells for the protons are not at the centers of the donor and acceptor molecules A and B, the distance between these wells depends on the relative position of the centers of the molecules together with their orientations.

REFERENCES

1. B. H. Robinson, in : "Proton-Transfer Reactions," E. Caldin and V. Gold, eds., Chapman and Hall, London (1975), p. 121.
2. M. Klöffler and J. Brickmann, Ber. Bunsenges. Phys. Chem. 86 : 203 (1982).
3. L. I. Schiff, Quantum Mechanics, Mc Graw Hill, London (1968), p. 83.
4. R. Nakamura and S. Hayashi, J. Mol. Struct. 145 : 331 (1986).

DISCUSSION

LAUNAY - I have a comment and a question. Your treatment of proton transfer is very analogous to the case of electron transfer when there is no temperature dependence (i.e. one does not take into account vibrations). To complete the analogy one would expect some thermally activated process, but it has not been introduced in your model. Instead of that you find that the rate constant could increase when the temperature decreases. What is the physical reason for that ?

FRIES - The rate constant k for proton tunnelling given by Eqs. (14) and (15) is inversely proportional to the relative diffusion constant D of the spherical potential wells. As the temperature decreases, so does D and thus k increases. This effect is easily interpreted by the fact that the reactants, i.e. the wells, stay longer in close contact where the proton can effectively tunnel.

TURQ - It would be relatively easy to replace your simple diffusion propagator by a slightly more raffinated $G(r,t)$ involving structural information such as $g(r)$ at $t=0$.

FRIES - The non uniform relative distribution of the reactants due to the anisotropic granularity of the solvent molecules can be taken into account by calculating the reactant-reactant solvent averaged potential of mean forces within the integral equation formalism of Blum and Torruella and by using this potential to obtain the initial value of the joint probability (12). At later times, some prescribed diffusional propagator can be employed following Blum and Oppenheim, or Harmon and Müller. More accurately, the relative dynamics of the reactants coupled by this potential of mean forces can be computed within the Smoluchowski or Langevin approaches.

HYNES - The model you describe ignores the strong potential forces in the separation coordinate for the heavy particles A and B in $AH^+ + B$. (For example, there will be a deep well in the $H_3O^+ + H_2O$ separation). These forces will dominate the motion typically, rather than the diffusion forces you include. Thus one would expect a much less pronounced dependence of the rate on diffusion than you find. Ali and Hynes, (1986) find essentially no dependence for model hydrogen atom and proton transfers when short range forces are included).

FRIES - The appearance of a strong potential between the donor and acceptor molecules A and B during the proton transfer process could be included in the relative motion of these molecules by replacing the gaussian force free joint probability (12) by the solution of a Smoluchowski diffusion equation including this potential. However, for sufficiently large particules this additional potential should become negligible. Now, if the growth of this potential is accompanied by an important decrease of the barrier height according to your theoretical results for different systems the proton will be transferred from one molecule to another without the necessity of tunnelling.

<u>JONAH</u> - How does the distance at which a proton transfers change as a function of time ? In other words if for a time t, the distance at which proton transfer occurs is d. Then for 10 t, the distance is $d+\Delta$. What is Δ ?

<u>FRIES</u> - The tunnelling rate given by Eqs. (14) and (15) varies as $\Delta E(b)^2$ which decreases very rapidly with the minimal distance of approach b of the well centers. Typical characteristic lengths of coupling, $1/\beta_o$, are of the order of 0-2 Å. Roughly speaking, one could say that the diffusion speed has to be slowed down by a factor 10 in order to compensate a variation of the minimal distance of approach of the order of $1/\beta_o$.

CONFORMATIONAL KINETICS OF ALIPHATIC CHAINS

Pier Luigi Nordio, Alberta Ferrarini and Giorgio Moro

Department of Physical Chemistry
University of Padova
35131 Padova, Italy

INTRODUCTION

The motions of a flexible aliphatic chain attached to a rigid molecular core are analyzed theoretically by means of a master equation for random walk between the sites identified with the stable conformers of the chain, as an extension in the time domain of the Rotational Isomeric State approximation.

The master equation is derived by asymptotic analysis of the multidimensional diffusion equation in the torsional angles subjected to the configurational potential, with the coupling between the variables explicitly taken into account. Molecular definitions of the kinetic rates for the conformational transitions occurring at each segment of the chain are obtained in terms of the dynamic and hydrodynamic properties of the system. The following physical quantities enter in the specification of the problem:
1) the energy difference between gauche and trans conformers;
2) the repulsive terms necessary to account for hindered sequences;
3) the elementary g $\rightarrow$ t transition rate for a single variable problem;
4) the curvatures of the potential surface at the saddle points for reactive and non-reactive modes;
5) the geometry dependence of the frictional torques opposing chain reorientations.
The correlation functions relevant for spectroscopic observables have been computed, and compared with correlation functions for the decay of trans and gauche populations of specific bonds, whose time integrals lead quite naturally to the definition of effective kinetic rates.

The present analysis can be fruitfully applied to systems up to $10 \div 12$ mobile bonds, and it is currently used to interpret NMR relaxation in tetra-alkylammonium ion salts or phospholipid model membranes.

THEORY

The starting point for the description of the conformational processes occurring in a chain formed by N methylene-methylene bonds is the multidimensional diffusion equation:

$$(\partial / \partial t) \, P \, (\mathbf{x},t) = - \, \hat{R} \, P \, (\mathbf{x},t) \qquad\qquad (1)$$

$$\hat{R} = -(\partial / \partial \mathbf{x})^{\dagger} \, D(\mathbf{x}) \, P(\mathbf{x}) \, (\partial / \partial \mathbf{x}) \, P(\mathbf{x})^{-1} \qquad (2)$$

$P(\mathbf{x})$ being the equilibrium distribution function associated with the configurational potential $V(\mathbf{x})$.

An approximate but efficient solution is achieved by assuming a time scale separation between the fast jiggering motions within the potential wells and the slow conformational jumps. Under this assumption, projection of the diffusion operator onto a set of site functions, in the same number of the potential minima, can be performed to convert the diffusion equation into a master equation for jumps between discrete sites [1]:

$$(\partial / \partial t) \, P_m(t) = - \, \sum_n W_{mn} \, P_n(t) \qquad\qquad (3)$$

The mathematical tool necessary for the projection procedure is provided by the concept of localized function $g_i(x)$. In a simple mono-dimensional problem this function resembles a square function with a value of unity at x_n, the n-th potential minima, a vanishing value outside the range $x_{n-1} < x < x_{n+1}$, and abrupt variations in correspondence of the two maxima adjacent to x_n. The functions $g_n(x) \, P(x)$ are thus peaked in correspondence of the potential minima. The diffusion operator has then non-vanishing off-diagonal elements only between functions located at adjacent minima x_m and x_n:

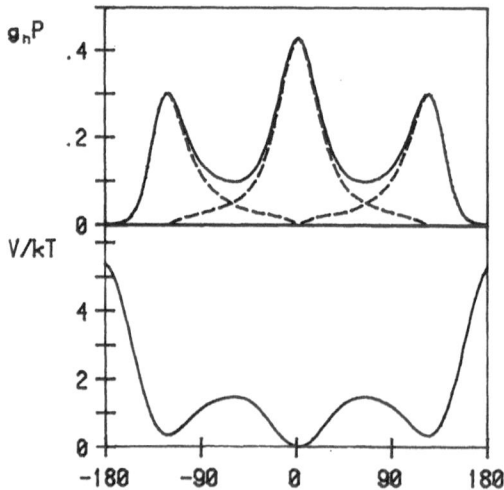

Fig.1. Distribution function for the torsional angle obtained as superposition of three localized functions corresponding to the minima of the potential given in the bottom graph.

$$R_{mn} = \langle\, g_m \,|\hat{R}|\, Pg_n \,\rangle \qquad\qquad (4)$$

the brackets denoting integration over x.

If parabolic expansions of the potential are performed about the saddle points, the well known Kramers result is recovered.

The multidimensional diffusion equation for the stochastic motions of the N torsional variables in the alkyl chain cannot in principle be factorized into single variable problems, because of the dependence upon the chain conformation of both the potential and the frictional force exerted by the surrounding viscous medium.

However, parabolic expansion of the potential surface about the saddle point x_s connecting two stable conformers m and n allows an approximate factorization of the problem, following a procedure proposed by Langer[2]. In fact, the product matrix $D_s V_s^{(2)}$ appearing in the asymptotic form of the diffusion operator, formed by the values of the N-dimensional diffusion tensor D_s for the geometry corresponding to the saddle point, and the matrix of the second derivatives $V_s^{(2)}$ calculated in the same point of the configurational space, is found to have an unique negative eigenvalue that shall be denoted by λ_1. The negative eigenvalue results from the fact that the curvature of the potential energy surface is positive in all directions except in that of maximum descent from the saddle point.

The eigenvector associated to λ_1 represents the most probable direction of crossing the saddle point amongst all the trajectories joining the adjacent minima. Along this reaction coordinate, localized functions can be defined in correspondence of the adjacent minima x_m and x_n in proximity of the saddle point x_s. The transition rate W_{mn} for the conformational jump $n \to m$ is then calculated to be:

$$W_{mn} = R_{mn} / Q_n = -(|\,\lambda_1\,| / 2\pi)\, \exp\{-(E_s - E_n) / kT\} \qquad (5)$$

with $Q_n = Z^{-1} \exp(-E_n/kT)$ the equilibrium site populations, and the following expression holding for E_s or E_n:

$$E_i = V(x_i) + (kT/2)\, \ln |\mathrm{Det}\,(V_i^{(2)}/2\pi kT)| \qquad (6)$$

This result is the generalization to the multivariate problem of the Kramers theory as found by Langer, and it confirms the validity of the site function method.

In order to implement this expression, the geometry, energetics, and hydrodynamics of the alkyl chain must be quantitatively defined.

Geometry

A molecular system formed by N methylene groups attached to a rigid core shall be considered. The chain is bound to a "ghost" methylene carbon atom labelled as C_o which is considered part of the core, and it terminates with a methyl group whose internal dynamics shall be not treated here. A Cartesian frame M_i is centered on each methylene carbon C_i, the z_i axis pointing along the $C_i - C_{i+1}$ bond. The set of Euler angles $\Omega_i = (\alpha_i, \beta_i, \gamma_i)$ that bring the axis system M_{i-1} into coincidence with M_i has the form :

$$\Omega_i = (\alpha_i, 180° - \delta_c, 180°) \tag{7}$$

with α_i the torsional angle about the $C_{i-1}-C_i$ bond, and δ_c the angle between consecutive bonds of the chain.

Any tensorial property, axially symmetric in the local frame F attached to the n-th methylene carbon atom, can be referred to a laboratory axis system L by the Wigner functions

$$D^j_{qo}(\Omega_{LF}) = \sum_p D^j_{qp}(\Omega_{LD}) D^j_{po}(\Omega_{DF}) \tag{8}$$

The diffusion frame D is chosen in the rigid core in such a way to diagonalize the diffusion tensor. The orientational angles Ω_{DF} can then be expressed by a sequence of rotations Ω_i about successive chain bonds.

Note that the conformational dynamics is entirely attributed to the torsional angles α_i, whereas Ω_{LD} fluctuates in time due to the rotational diffusion of the whole molecule. In the RIS approximation, each α_i can assume only the values 0, $\pm 120°$ in correspondence of the trans (t) and the gauche ($g_\pm$) states.

Energetics

An ensemble of indices $J = \{j_1, j_2 \ldots j_N\}$, with $j_i = 0, \pm 1$ in correspondence to t and $g_\pm$ states, is required to specify the conformation of a chain.

In terms of two scaled energy parameters

$$e_g = (E_g - E_t) / kT \tag{9}$$

$$e_p = \Delta E_p / kT \tag{10}$$

with E_g and E_t free energies of the stable conformers calculated according to Eq.(6), and ΔE_p the energy increase due to the "pentane effect"[3], the total energy of the J-th conformer is

$$E_J/kT = n_g e_g + n_p e_p \tag{11}$$

where n_g is the number of gauche states and n_p the number of sequences $g_\pm g_\mp$.

Strongly overlapping conformations such as cyclohexane-like sequences are forbidden by steric reasons. In addition, conformers with Boltzmann factors $\exp(-E_J/kT)$ less than a cutoff value q_0 can be excluded to reduce the size of the computational problem. For example, for N=7 chains the total number of conformers is 2187, reduced to 1552 after exclusion of forbidden sequences, and finally to 953 by choosing $q_0 = 10^{-3}$. Energy parameters used in the calculations were $e_g = 0.84$, $e_p = 3$.

Calculations of the transition rates require also knowledge of a parameter ρ giving the ratio of the curvatures at the saddle point of the potential surface in correspondence of reactive and non-reactive modes:

$$\rho = |V_r^{(2)}| / V_{nr}^{(2)} \tag{12}$$

High ρ values lead to strong coupling between the torsional variables and large cooperativity effects for the rotation about a specific bond. On the other hand, vanishing values of ρ imply steep slopes of the potential surface in correspondence of the non-reactive mode, resulting to effective decoupling of the segmental motions.

Hydrodynamics

The diffusion tensor $\mathbf{D}$ is written as $kT \ \underset{\sim}{\xi}^{-1}$, and the friction matrix $\underset{\sim}{\xi}$ is calculated by hydrodynamical methods by regarding the chain as composed by identical spheres centered at the carbon positions. By considering that a rotation about the $C_{k-1}-C_k$ bond causes translations of all spheres numbered from $k+1$ to $N+1$, the elements of the friction matrix are calculated to be:

$$\xi_{ki} \doteq \xi_o \ \Sigma_j^{N+1} \ |z_{k-1} \times (r_j - r_k)| \ |z_{i-1} \times (r_j - r_i)| \tag{13}$$

where z_{k-1} is a unit vector pointing in the direction of the $C_{k-1}-C_k$ bond, ξ_o is the friction opposing the translational motion of a single sphere and r_j the vector position of j-th carbon atom.

For each conformational transition $n \to m$, caused by rotation about a specific bond, the matrix $\underset{\sim}{\xi}$ is calculated given the geometry of the saddle point and the distance d of a C-C bond.

CORRELATION FUNCTIONS FOR PHYSICAL OBSERVABLES

With the ingredients presented above, the elements of a symmetrized jump matrix $\mathbf{W}^s$ are obtained as:

$$W_{mn}^s = - w|\lambda_1'| \ \exp \{ -(E_m - E_n) \ / \ 2kT \} \tag{14}$$

where $\mathbf{W}^s = \mathbf{S}^{-\frac{1}{2}} \ \mathbf{W} \ \mathbf{S}^{\frac{1}{2}}$ with $\mathbf{S}$ a diagonal matrix whose elements S_{ii} are the equilibrium fractional populations Q_i, w can be interpreted as the conformational transition frequency in the butane molecule

$$w = (|V_r^{(2)}| \ / \ 2 \pi \ d^2 \ \xi_o) \ \exp \{ -(E_s - E_g)/kT \} \tag{15}$$

and λ_1' is the negative eigenvalue scaled as:

$$\lambda_1' = \lambda_1 \ d^2 \ \xi_o \ / \ |V_r^{(2)}| \tag{16}$$

In terms of the jump matrix $\mathbf{W}$, the correlation function for the deviation of any function from its equilibrium average is calculated as:

$$\overline{\delta f(t)* \ \delta f} = \delta \mathbf{f}^\dagger \ \exp(-\mathbf{W}t) \ \mathbf{S} \ \delta \ \mathbf{f} \tag{17}$$

$\mathbf{f}$ being the vector constructed with the values of the function at each site. The corresponding spectral density is obtained as a continued fraction whose coefficients are computed by the Lanczos algorithm.

In order to analyze dielectric or magnetic relaxation data, we need to calculate correlation functions of Wigner rotation matrix components such as given in Eq.(8), relating the local interaction frame F to the laboratory system L. If the internal motions are assumed to be uncoupled with the rotational diffusion of the whole molecule, then the correlation function describing the motion of an interaction tensor component along the C_n-H is expressed by the product:

$$G_{qo}^{j \, (n)}(t) = \sum_p \, C_{qp}^{j \, (D)}(t) \, \{ \, |\overline{D_{po}^{j}}|^2 + \delta G_{po}^{j \, (int)}(t) \, \} \qquad (18)$$

with $C_{qp}^{j}(t)$ the correlation function of $D_{qp}^{j}(\Omega_{LD})$, and $G_{po}^{j \, (int)}(t)$ the correlation function for the deviations of $D_{po}^{j}(\Omega_{DF})$ from the equilibrium average, whose non-vanishing value is a measure of the degree of spatial restriction of the segmental motions. If isotropic diffusion is assumed for sake of simplicity, Eq.(18) reduces to:

$$G_n^{j}(t) = (2j+1)^{-1} \exp[-j(j+1)Dt] \, \{ \, (S_n^{j})^2 + [1-(S_n^{j})^2] g_n^{j}(t) \, \} \qquad (19)$$

with S_n^{j} a generalized order parameter for the n-th segment, and $g_n^{j}(t)$ a normalized relaxation function for tensorial quantities of rank j associated to that segment. In order to calculate the correlation function for the internal motions, the vector $\mathbf{D}_{po}^{j}$, having as elements the values of $D_{po}^{j}(\Omega_{DF})$ in correspondence of each allowed conformation, is first constructed from the proper sequence of Euler angles Ω_i.

An effective relaxation rate k_n^{j} for the tensorial properties associated to the n-th segment of the chain can then be obtained as the inverse of the time integral of the correlation function $g_n^{j}(t)$.

Alternatively, effective rate constants for the segmental motions can be calculated from the correlation functions of the configurational

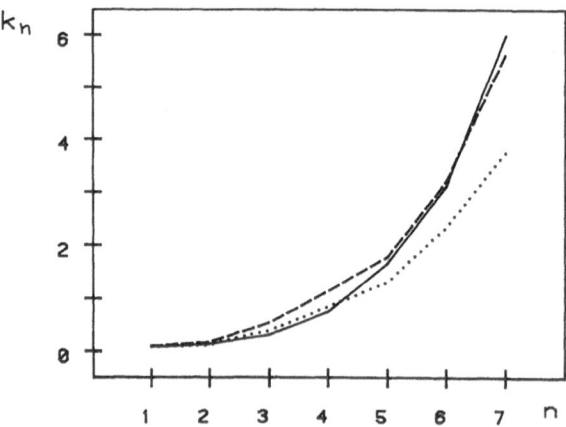

Fig.2. Effective rate constants for the motion of the different segments in a N=7 chain, calculated from time integrals of correlation functions: excess trans population (broken line); unbalanced gauche population (dotted line); second rank tensorial interaction (full line).

vectors $\mathbf{P}_n^t$ of components $|\mathbf{P}_n^t|_J = (3\,\delta_{jn0} - 1)$, and $\mathbf{P}_n^g$ of components $|\mathbf{P}_n^g|_J = j_n$, interpreted as excess trans populations and unbalanced gauche populations, respectively.

CONCLUSIONS

The theoretical method developed here provides a rigorous approach to the description of the internal dynamics of flexible aliphatic tails. The treatment is able to link the master equations used in connection with the RIS approximation to the multivariate Fokker Planck or diffusive equations, avoiding loosely defined phenomenological parameters.

Correlation functions or spectral densities for relevant properties are calculated in terms of the transition matrix between allowed conformers, the matrix elements being computed from the dynamical and hydrodynamical properties of the system under investigation.

In particular, the expressions obtained for the segmental order parameters S_n^J and the effective relaxation rates k_n^J provide a molecular justification of the model-free parameters used to analyze NMR relaxation experiments[4]. The theory also supplies detailed interpretation of the flexibility gradient determined by C-13 relaxation in tetra-octylammonium perchlorate[5] dissolved in organic solvents.

As shown in Fig.2., the kinetic rates for the decay of trans or gauche populations at different chain positions follow rather closely the behaviour of the magnetic correlation frequencies. This suggests that the kinetic rates, much easier to be calculated, can be used in many instances to describe at a qualitative level the motional features experimentally observed by different spectroscopical techniques.

This work has been supported by the Italian Ministry of Public Education, and in part by the National Research Council through its Centro Studi sugli Stati Molecolari.

REFERENCES

1. G. Moro and P.L. Nordio, Diffusion between inequivalent sites, Molec. Phys. 57:947 (1986).
2. J. S. Langer, Statistical Theory of the Decay of Metastable States, Ann. Phys. 54:258 (1969).
3. P. J. Flory, "Statistical Mechanics of Chain Molecules", Wiley, New York (1969).
4. G. Lipari and A. Szabo, Model-Free Approach to the Interpretation of N.M.R. Relaxation in Macromolecules. J. Am. Chem. Soc. 104:4546 (1982).
5. F. Coletta, G. Moro, and P.L. Nordio, Mobility of alkyl chains in quaternary ammonium salts, Molec. Phys. 61:1259 (1987).

DISCUSSION

BEAUFILS - Have you calculated correlation functions between distant nodes of the chains in order to assess the existence of the correlated motion of several segments of a chain ? What are the results ?

NORDIO - We have calculated correlation functions for nearest-neighbour and next-nearest-neighbour pairs of the chain, and they obviously do not vanish since the conformational potential in the starting diffusion equation is not factorized.
An additional source of cross correlation, the so-called "kink" or "crankshaft" motion is not considered in our theoretical model, which only takes into account transitions across single saddle points. We expect however these motions to be unimportant for relatively short chains.
On the other hand, the cooperative effects which minimize the frictional resistence to the chain rotations in the barrier crossings, enter automatically into the treatment as a consequence of the geometry dependence of the friction, and they mainly affect the calculated values of the kinetic rates.

SCEATS - In your paper you used the Smoluchowski limit of overdamped motion. Are the barrier frequencies ω_T for the trans-gauche isomerization sufficiently small that the criterion for Smoluchowski dynamics is met, namely that $\beta \gg \omega_T$ where β is the mass-weighted friction ?

NORDIO - The numerical values of the friction coefficients used to interpret the NMR relaxation data give reasonable assurance that the diffusion limit is appropriate. The same conclusion is discussed in some detail in a paper by R.M. Levy, M. Karplus and J.A. McCammon (Diffusive Langevin Dynamics of Model Alkanes, Chem. Phys. Lett., 65:4, 1979).

ERDI - I did not realize whether you tried to solve your master equation analytically or you restricted yourself for making numerical calculations only. What about the limits of the analytical treatment ?

NORDIO - Analytical expressions have been derived for the site functions, as given for example in ref. 1, Eq. (5). The calculations of dynamical properties involve manipulation of the matrix W for the transitions among allowed configurations, and they have been performed numerically on a PDP 11 mini-computer.

KAPRAL - If the diffusion equation contains monomer interactions and a diffusion tensor that incorporates hydrodynamic interactions, the calculation of the stationary density is not simple. How are the extrema in this possibility density determined in order to apply your method ?

NORDIO - According to the Rotational Isomeric State approximation, the geometry dependent interactions are simply expressed by additional terms in the total energy. On the other hand, the hydrodynamic interactions give rise to configuration dependent friction coefficients. Under these assumptions, the stationary distribution function is uniquely defined, but the construction of the site functions requires identification of a reactive path. This is done by quadratic expansion of the multivariate diffusion equation about the saddle point connecting two stable conformers, followed by a normal mode analysis.

THEORY OF ASSOCIATION AND

SUBSTITUTION REACTIONS

James T. Hynes

Department of Chemistry and Biochemistry
University of Colorado
Boulder, CO 80309-0215 USA

I. INTRODUCTION

We have interpreted the assigned title a bit liberally and selected some theoretical aspects of charge transfer reactions in solution as our topic. The type of "association" reaction we will discuss is in fact the reverse type of reaction, an S_N1 dissociation

$$RX \rightarrow R^+ + X^- \tag{1}$$

in solution. Substitutions will be discussed for an S_N2 reaction

$$Y^- + RX \rightarrow YR + X^- . \tag{2}$$

The fact that charge is transferred in these reactions means that there can be strong coupling of polar solvent molecules to the motion of the molecular reacting species, with novel and unexpected consequences for the reaction rate constant and indeed for our conceptions of the molecular level "mechanism" of the reaction.

We will first give an overview of the issues involved via a brief description of the Transition State Theory and the dynamic Grote-Hynes Theory,[1] as developed for charge transfer reactions in solution by van der Zwan and Hynes.[2-4] This will introduce the ideas of equilibrium and nonequilibrium solvation, friction and barrier recrossing. We then indicate some of the consequences and predictions for the S_N1 and S_N2 reaction types.

II. NONEQUILIBRIUM SOLVATION AND REACTION RATES

In the traditional view, polar solvent effects on reaction rates are to be understood solely in terms of free energy pictures such as those in Fig. 1. In an S_N1 dissociation, change to a more polar solvent would stabilize the more charge localized product species R^+ and X^- compared to the more charge localized reactant RX, concomitantly lowering the activation barrier and increasing the rate. In an S_N2 reaction, this same type of solvent change would be expected to favor the charge localized reactants (and products) compared to the more charge delocalized transition state. The net effect is an increase in the activation free energy barrier and a reduction in the rate in a more polar solvent.

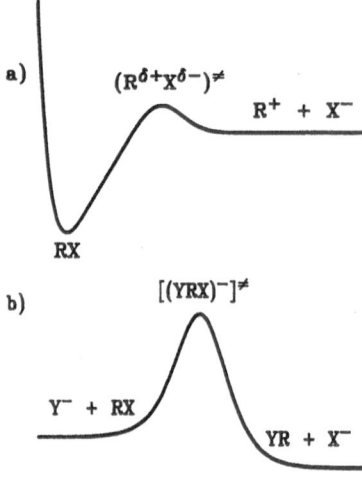

Figure 1. Schematic illustration of equilibrium solvation free energy curves for (a) an S_N1 dissociation and (b) an S_N2 substitution reaction in polar solvents.

It is implicit[2-4] in this traditional Transition State Theory (TST) view that <u>equilibrium solvation</u> applies in the reaction, i.e., that at each stage along the simple reaction coordinate (such as the $R^+ - X^-$ separation in the S_N1 case), the solvent is equilibrated to the reaction system. The resulting rate constant k^{TST} is thus independent of any solvent dynamics. But if the solvent does <u>not</u> have time to equilibrate, then instead <u>nonequilibrium solvation</u> will occur. When this happens, the resulting imbalance will lead to electrical forces acting on the reaction system, which as we will see below can be described as a "dielectric friction." Then there will be a deviation of the actual rate constant k from the TST prediction.

An equivalent contrast is the following. In the equilibrium solvation view, the reaction system is assumed to pass directly from the reactants' side to the products' side of the barrier, with no recrossing.[2-4] In the nonequilibrium solvation case, dielectic friction effects will lead to a solvent induced <u>recrossing</u> of the barrier.[2-4] This leads to a reduction in k compared to k^{TST} which is quantified by the transmission coefficient κ:

$$k = k^{TST}\kappa \quad ; \quad \kappa \leq 1 . \tag{3}$$

van der Zwan and Hynes[2-4] have developed a theoretical approach to these questions. This theory is in turn based on Grote–Hynes (GH) theory[1] for reaction rates. A brief description of the latter in terms useful for the present context is the following. The reaction system is assumed to be described by the generalized Langevin equation (GLE) for the reaction coordinate x:

$$\overset{..}{x}(t) = \omega^2_{b,eq}x(t) - \int_0^t d\tau \zeta(t-\tau)\overset{.}{x}(\tau) . \tag{4}$$

(We have ignored here the random force, whose inclusion is unnecessary for the present purposes). For an ionic dissociation, x would be the R^+-X^- separation, while for an S_N2 reaction, x would be similar to an asymmetric stretching motion of the $(YRX)^-$ system at the transition state. Here $\omega_{b,eq}^2$ is the square equilibrated barrier frequency proportional to the absolute curvature of the barrier. The first term in the GLE describes equilibrium solvation. The second term represents nonequilibrium solvation effects and involves the time dependent friction coefficient $\zeta(t)$. This is proportional to the time correlation function of the forces that the solvent exerts along the reaction coordinate in the neighborhood of the barrier top. As will be seen below, $\zeta(t)$ can be approximately determined by MD simulation or by analytic modelling. In favorable cases, some information can be obtained[5] from experimental time dependent fluorescence studies in polar solvents.

With the GLE Eq. (4), the transmission coefficient κ in Eq. (3) is given by the self-consistent GH relation[1]

$$\kappa = \left[\kappa + \int_0^\infty dt \, e^{-\omega_{b,eq}\kappa t} \zeta(t)\right]^{-1} . \tag{5}$$

When the friction is negligible, this gives $\kappa=1$, i.e., TST. If the barrier is broad and has a low frequency $\omega_{b,eq}$, then $\omega_{b,eq}$ in the Laplace transform in Eq. (5) can be neglected and

$$\kappa = [\kappa + \zeta]^{-1} , \tag{6}$$

where the friction <u>constant</u> is $\zeta = \int_0^\infty dt \, \zeta(t)$. This is the famous result of Kramers[6] and generally predicts considerable recrossing and $\kappa \ll 1$. In the general case however we need to know $\zeta(t)$ on <u>short</u> time scales associated with the actual barrier crossing step. These time scales can be quite short (vide infra) and the solvent forces are poorly represented by the zero frequency friction constant ζ. Instead the frequency dependent friction in Eq. (5) is required.

In a series of papers, van der Zwan and Hynes[2-4] used the Grote-Hynes theory Eq (5) to analyze charge transfer reactions in polar solvents. They found several important limiting regimes and analytic results for κ. In the <u>nonadiabatic</u> regime, the reaction system sees a <u>barrier</u> in its short time motion. The reaction accurs in a "frozen" solvent and κ is given by

$$\kappa = \omega_{b,na}/\omega_{b,eq} . \tag{7}$$

(This is the solution of Eq. (5) when $\omega_{b,eq}\kappa$ is large and only the initial value $\zeta(t=0)$ of the time dependent friction contributes.) In effect the reaction system sees a modified barrier frequency $\omega_{b,na}$,

$$\omega_{b,na}^2 = \omega_{b,eq}^2 - \zeta(t=0) , \tag{8}$$

which accounts for the fact that the solvent cannot clothe the reaction system with equilibrium solvation [hence the subtractive term in Eq. (8)]. In this limit — which is most likely for a slowly relaxing solvent, modest coupling to the solvent and a high $\omega_{b,eq}$ value — κ is independent of solvent dynamics (but still not equal to unity); this indicates that the barrier recrossing can be understood in terms of solvent configurations, as we will see.

In the <u>polarization caging</u> regime — most easily attained for slowly relaxing solvents, strong coupling to the solvent and low $\omega_{b,eq}$ values —

the reaction system is trapped in a well, or "polarization cage," in its initial motion off the barrier. On a longer time scale, solvent dipoles reorient, destroying the cage and allowing the reaction to proceed. Thus the rate limiting step is the solvent relaxation, and κ is predicted to decrease as this becomes progressively slower.

III. MODEL $S_N 1$ REACTION STUDIES

In the model analytic $S_N 1$ dissociation study by Zichi and Hynes,[7] the reaction system comprises two charges in a cylindrical cavity (to model the local exclusion of solvent). The cavity is surrounded by a dynamic generalized dielectric continuum model of H_2O solvent. The reaction system parameters are roughly consistent with those of methyl chloride. The charge development as the reaction $RX \rightarrow R^+ + X^-$ proceeds is a critical feature and is modelled by a function which smoothly switches the charge values from those appropriate to a polar molecule $R^{\delta+}X^{\delta-}$ to those appropriate to the ionic forms $R^+ + X^-$ in about 0.5Å. The underlying RX potential energy is taken to be a Morse curve with an added Gaussian barrier, with a "bare" barrier frequency ω_b. The equilibrium free energy curve when the solvent is equilibrated at each R-X distance can be calculated and resembles the profile displayed in Fig. 1a.

While the time dependent dielectric friction coefficient $\zeta(t)$ for the model is analytically quite complex, numerically it is close to an exponential function in time, with a decay time roughly equal to the longitudinal relaxation time τ_L for $H_2O \approx 0.25$ ps.

Since the solvent relaxation time is fixed, nonequilibrium solvation effects on κ are best studied as a function of $\omega_b \tau_L$, i.e., the bare barrier frequency is systematically varied between values corresponding to low broad barriers to those for high sharp barriers. The transmission coefficient κ is calculated from the general GH theory Eq. (5).

The analysis of this model turns out to be rather complex. Nonetheless several important features emerge. For equilibrated barrier locations that correspond to reactant-like transition states, i.e. the charges in $[R^{\delta+}X^{\delta-}]^{\neq}$ are close to those of the reactants, the coupling to the solvent is not strong. This is because the largest coupling occurs at the center of the charge switching gradient, and in the reactant-like case, this region occurs after the equilibrium barrier to dissociation. As a result, the system is in the nonadiabatic limit, as the calculations shown in Fig. 2 indicate. Shown there also is the corresponding Kramers theory prediction κ^{KR} which, since it includes the full long time frictional response, predicts too much barrier recrossing.

When the equilibrated barrier occurs at a larger separation than does the center of the charge switching gradient - a product-like transition state, something different occurs. Fig. 3 shows the results of a calculation in which a product-like solvated transition state occurs for $\omega_b \tau_L$ values above 15. Near this region the solvated barrier top occurs very close to the location of the maximum charge gradient. Here there is strong coupling, a polarization cage exists and κ drops precipitously. On further increase of ω_b, the solvated transition state shifts to large separation where there is much weaker coupling and the nonadiabatic regime is approached once again.

These theoretical results suggest that dynamic solvent effects in $S_N 1$ reactions can be a bit subtle and depend on a combination of factors. Nonetheless it is interesting to note that for a wide range of conditions, the nonadiabatic description - in which the solvent does not keep up with the reaction system - is a good approximation. Despite this lack of solvent equilibration, κ is still fairly close to unity and the

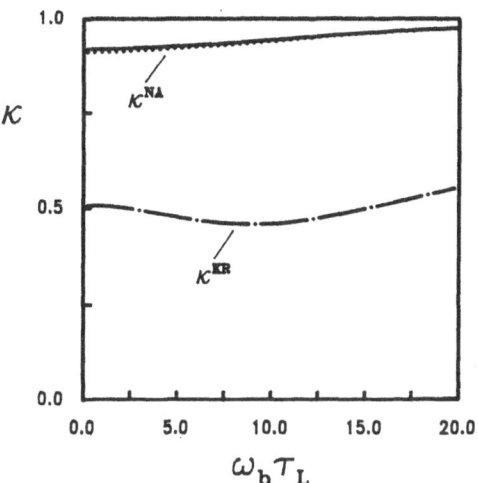

Figure 2. Calculated transmission coefficient κ
for an S_N1 dissociation with a reactant-like
transition state, in H_2O. The nonadiabatic and
Kramers predictions are included for comparison.

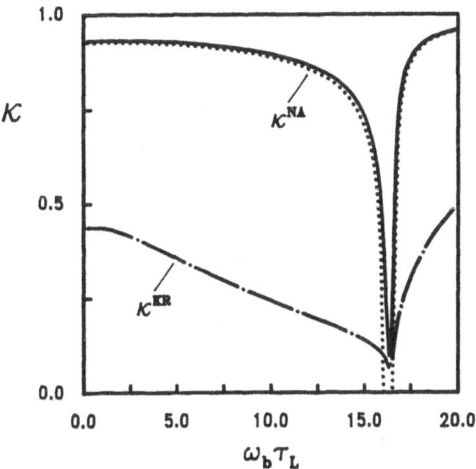

Figure 3. Same as Fig. 2, but for a product-like
transition state.

numerical consequences of the lack of equilibration are small. This is because the coupling is not very strong and most trajectories will successfully react <u>despite</u> the lack of equilibration. In the next section, we will see cases where extensive recrossing occurs in the nonadiabatic limit.

IV. MODEL $S_N 2$ REACTION STUDIES

As noted in Sec. II, the standard conception of an $S_N 2$ reaction in solution is based on the equilibrium solvation free energy picture similar to that in Fig. 1.b. As we will now see, a molecular level picture rather different from this applies.

The theoretical ideas outlined in Sec. II have been tested by comparison with the results of a Molecular Dynamics (MD) computer simulation of the model $S_N 2$ reaction $Cl^- + CH_3 Cl' \rightarrow ClCH_3 + Cl'^-$ in $H_2 O$ solvent.[8,9,10] For the details of the simulation, we refer the reader to the paper by Bergsma, Gertner, Wilson and Hynes.[8] Here we simply note that the gas phase potential energy surface of the $S_N 2$ system was modelled after the <u>ab initio</u> results of Jorgenson[11] and is characterized by a double well structure with a central barrier height of 13.9 kcal/mol with respect to the ion-molecule complex minima. There is also the very important feature that the charge switching rate in the neighborhood of the barrier top is quite large: the gradient is about $0.8e/\text{Å}$. Finally, theoretical predictions can be tested against the MD simulation results because the time dependent friction coefficient $\zeta(t)$ can be determined by a simulation of the correlation function of the force exerted by the solvent molecules on the $S_N 2$ system at the transition state.

It is found that there is considerable recrossing of the barrier top, in violation of the TST, equilibrium solvation picture. Out of 400 trajectories initially at the barrier top and aimed toward products $CH_3 Cl + Cl'^-$, only 148 are of the no-recrossing variety (cf Fig. 4a). 130 of the trajectories rapidly recross back towards reactants (Fig. 4b). 110 of the trajectories proceed directly to products but are found in fact to have originated from the products side rather than the reactants $Cl^- + CH_3 Cl'$ a short time earlier (Fig. 4c). All these recrossings lead to a transmission coefficient of $\kappa = 0.55 \pm 0.05$, a marked departure from TST.

Figure 4. Three types of transition state neighborhood trajectories state observed in the MD simulation of the $Cl^- + CH_3 Cl$ reaction in $H_2 O$. The circles indicate that at an initial time t=0, trajectories are propelled towards products from the transitions state barrier top.

It was found that the fate of a trajectory was very quickly decided, in 0.02 ps or less. On this time scale, there is negligible motion of the molecules that could change the solvent force exerted on the reaction system. Thus we are in the <u>nonadiabatic</u> regime described in general terms earlier, and the solvent is <u>effectively</u> "frozen" on the time scale during which the fate of the trajectories initiated at the barrier top is decided.

In this limit, the origin of the recrossings can be understood in terms of solvent <u>configurations</u>, rather than solvent dynamics. For trajectories of type Fig. 4b, there is nonequilibrium asymmetric solvation of the transition state (TS) such that the solvent favors formation of a full negative charge on the "left-hand" chlorine in the TS structure $[(ClCH_3Cl')^-]^{\neq}$. The solvent opposes the formation of the product and a rapid recrossing back towards reactants occurs. For trajectories of type Fig. 4c, there is asymmetric nonequilibrium solvation in the opposite sense: the solvent favors formation of a full negative charge on the "right-hand" chlorine Cl^-. This leads to the prior recrossing back towards products. For trajectories of type Fig. 4a, which are of the no-recrossing, TST type, there is no observed asymmetry in the solvation and no net solvent bias in one direction or another.

The van der Zwan-Hynes nonadiabatic limit κ prediction Eq. (7) can be tested against the MD results, and Table 1 details the results for several choices of the rate of charge switching. Clearly the agreement is quite good. Also included for comparison are the predictions of Kramers theory, Eq. (6), for κ. The friction <u>constant</u> ζ in the Kramers' Eq. (6) is in reality only established after a time on the order of picoseconds, long after the reaction fate is in fact decided. This long time friction is very high, resulting in very large amounts of recrossing in the Kramers picture and a catastrophic failure. What counts for the reaction is the <u>short time</u> friction, as evidenced by the success of the van der Zwan-Hynes κ_{NA} in Table 1. This can be stressed in another way by appying the general Grote-Hynes theory Eq. (5) to the simulation results. The short time scale of the reaction essentially "selects" the initial value $\zeta(t=0)$ as relevant for the reaction in the GH equation, an independent numerical check of the van der Zwan-Hynes analytic results Eqs. (7) and (8).

By artificially lowering the central barrier height to 4.9 kcal and altering the character of the charging characteristics,[10] the polarization caging regime can be (just barely) entered in the simulations. Table 2 shows the comparison of the MD results with the predictions of the full Grote-Hynes theory, and once again the agreement is very good.

In summary, nonequilibrium solvation effects lead to marked departures from the TST, equilibrium solvation rate prediction, and these effects are successfully accounted for theoretically. One can ask: what is <u>right</u> about the equilibrium solvation picture depicted in Fig. 1b? What is given correctly in this picture is the feature that there is a free energy cost to <u>reach</u> the transition state. But dynamics in the neighborhood of this state are quite different than is usually conceived. A further indication that the reaction system does not actually move on the depicted free energy surface is that, in the MD simulations, rapid vibrational energy transfer from the S_N2 system to the solvent is observed[8] to occur rapidly after leaving the transition state, but before the solvent molecules can equilibrate to provide the equilibrium free energy profile of Fig. 1b connecting the transition state to products. An analogous statement for vibrational activation holds in connection with the profile leading from reactants to the transition state. Understanding of the molecular level "mechanism" of the reaction requires attention to the solvent dynamics and not just equilibrium free energy, and we expect considerable further efforts in this direction in the next few years.

Acknowledgments

Acknowledgment is made to the Donors of the Petroleum Research Fund, administered by the American Chemical Society, for partial support of this research. This work was also supported by Grant CHE 84-19830 from the US National Science Foundation. We are also grateful to G. van der Zwan, D. Zichi, S. Lee, J. Bergsma, B. Gertner and K. R. Wilson for their collaboration in the work described in this article.

Table 1. Molecular Dynamics and Theoretical κ Values for the Cl^-+CH_3Cl' Reaction in H_2O

	κ_{MD}	κ_{NA}	κ_{KR}
Case A[a]	0.55±0.05	0.45	0.007
Case B[b]	0.73±0.05	0.67	0.015
Case C[c]	0.90±0.05	0.90	0.064

[a] Charge switching rate r is the standard value.[8]
[b] r is one half of the Case A value.
[c] r is one eighth of the Case A value.

Table 2. Molecular Dynamics and Theoretical κ Values for the Cl^-+CH_3Cl' Reaction in H_2O in the Polarization Caging Regime.

	κ_{MD}	κ_{GH}	κ_{KR}
Case A[a]	0.39±0.04	0.43	0.07
Case B	0.37±0.04	0.38	0.05
Case C[b]	0.32±0.04	0.30	0.01

[a] Charge switching rate r is one half that of the value for Case B, as described in Ref. 10.
[b] r is twice the value for Case B.

References

1. R. F. Grote and J. T. Hynes, J. Chem. Phys. 73:2715(1980).
2. G. van der Zwan and J. T. Hynes, J. Chem. Phys. 76:2993(1982).
3. G. van der Zwan and J. T. Hynes, J. Chem. Phys. 78:4174(1983).
4. G. van der Zwan and J. T. Hynes, J. Chem. Phys. 90:21(1984).
5. G. van der Zwan and J. T. Hynes, J. Phys. Chem. 89:4181(1985).
6. H. A. Kramers, Physica (the Hague), 7:284(1940).
7. D. A. Zichi and J. T. Hynes, J. Chem. Phys. (in press).
8. J. P. Bergsma, B. J. Gertner, K. R. Wilson and J. T. Hynes, J. Chem. Phys. 86:1356(1987).
9. B. J. Gertner, J. P. Bergsma, K. R. Wilson, S. Lee and J. T. Hynes, J. Chem. Phys. 8:1377(1987).
10. B. J. Gertner, K. R. Wilson and J. T. Hynes, (to be submitted).
11. J. Chandrasekhar, S. F. Smith and W. L. Jorgensen, J. Am. Chem. Soc. 107:154(1985).

DISCUSSION

GERSCHEL - Could you specify the definition of the force that autocorrelate in the time dependent friction ? Does it enter essentially the solvent rotational dynamics or does it include both translational and rotational contributions ?

HYNES - The force that enters the friction correlation function is the complete force arising from the solvent molecules interacting with the S_N2 reaction system held fixed at the transition state (zero asymmetric stretch). It includes all contributions from the rotational and translational (as well as intramolecular vibrational and librational) motions of the H_2O solvent molecules.

MIEHE - How do you take into account the normal viscous friction in the description of the dynamical Stokes shift ?

HYNES - I do not understand why you say that intermolecular streching motions of the solvent, expanding or compressing the first solvation shell, occur on the same time scale (20 fs) as the reactive process. We see no evidence for this. The success of the frozen solvent, non-adiabatic solvation model, which ignores all solvent motions, indicates that all solvent dynamics can be ignored for the standard Cl^- + CH_3Cl case.
However, for the hypothetical lower barrier reactions where the time scale for barrier crossing lengthens, I do believe that the intermolecular stretchings are quite important. This remains to be confirmed in the trajectory results however.

RIVAIL - Your results are obtained by using an effective potential without polarizability effects (in which the dipole moments are exaggerated). I would guess that inclusion of polarizability will reduce somewhat the magnitude of the relaxation effect since the induced moments follow immediately the modifications of the electric field. Did you examine this problem ?

HYNES - You raise an important point that concerns us. Electronic solvent polarization would presumably adjust "instantly" to the reactive motion and exert no force to induce recrossing, however we use the "bare" H_2O dipole moment in the calculations and not the solvent polarizability enhanced value. At some stage, polarizability should be included in such calculations, but we do not know how to do it.

BERTRAN - A frozen model of the solvent seems to be fully justified for the rotation movements of the solvent molecules, because its time scale is very much larger than the one of the chemical process at the top of the barrier, but what about the stretching movements of the solvent contractions or expansions of the first solvation shell, whose time scale seems to be of the same order as the one of the chemical process ?

HYNES - The relation between the dynamical Stokes shift and the friction I quoted assumes that there are only electrical effects and that there are no "viscous" friction effects that would arise for example if the absorbing solute underwent a significant conformational change subsequent to the electronic transition. The experimental results of the Fleming group quoted appear to indicate that this condition is satisfied for their solute molecule, but there will certainly be situations for which this is not true. Then a new relation needs to be derived, but we have not done it.

MOLECULAR DYNAMICS OF CHEMICAL REACTIONS IN SOLUTION

Kent R. Wilson

Department of Chemistry, B-039
University of California, San Diego
La Jolla, CA 92093

INTRODUCTION

Impressive progress has been made in the last twenty five years in understanding the detailed nature of chemical processes involving a few atoms at a time, as honored in the recent Nobel Prize to Herschbach, Lee and Polanyi. Unfortunately, most processes of interest to most chemists involve the simultaneous interaction of many atoms. Examples include all of liquid solution chemistry (and thus most of inorganic and organic chemistry) and chemistry involving large molecules (and thus most of polymer chemistry and biochemistry). A realistic assessment of many atom chemistry in general and liquid solution chemistry in particular is that we know very little as yet about the microscopic details of such processes. Given the clear importance of solution chemistry, it is surprising that so little is known about liquid state solution reactions from the fundamental viewpoint of how they take place as seen from the microscopic perspective of molecular dynamics, in other words of the atomic motions by which they happen.

Molecular dynamics calculations of solution reactions can be used i) to provide images on the microscopic molecular level of reaction dynamics in order to guide our intuitive understanding, ii) to develop and test analytic theories in order to mathematically summarize the essence of reaction dynamics, and iii) to predict reaction phenomena which can also be measured experimentally in order to test the faithfulness to nature of the dynamics calculations and thus also of analytical models and intuitive pictures.

Classical molecular dynamics is in principle an easy procedure, based on 300 year old Newtonian mechanics.[1,2] The major difficulties are that i) the potential energy surface may not be sufficiently accurately known for our needs, and ii) the number of arithmetic operations required to compute the desired properties starting with the potential surface and integrating the classical equations of motion may exceed the computer resources we can command. For liquid state chemical reactions, molecular dynamics calculations are exceedingly demanding because of the need to compute trajectories involving many atoms over and over again from differing starting points to achieve statistically sound conclusions. A typical study of the types we have carried out in recent years takes the order of a VAX 780 decade, even taking advantage of the Keck-Anderson[3,4] method of starting each trajectory in the middle, at the top of the barrier, and then integrating forward and backward in time to assemble the complete trajectory.

To represent a liquid solvent, we need to include in our calculations at minimum the order of 50-100 solvent molecules. No single molecular dynamics trajectory involving the reagents and solvent molecules can be trusted to give a reliable estimate of what happens in a real liquid reaction, as all our experimental observables are averages over many microscopic molecular events proceeding from different starting conditions. Our calculations therefore involve averages over ensembles of tens to hundreds of molecular dynamics trajectories, each trajectory begun with a set of initial atomic positions and momenta chosen from the ensemble of interest.

While in the past we carried out most of our calculations on an array processor,[5] we have also used several general purpose processors and supercomputers. The most appealing route for the near future may be a particularly simple form of parallel computation which we are now using. Since each trajectory, once initiated, runs independently for the order of a half hour or so on a fast computer before dumping a relatively small amount of data to be analyzed and averaged with the results of the other trajectories, our calculations are ideal for a loosely coupled system of parallel processors. We thus match the symmetries of the hardware to the symmetries of the problem.[6,7] We call such a computational system a *Gibbs Machine*: an ensemble of loosely coupled processors computing an ensemble of trajectories.

TYPES OF REACTIONS

We are interested ultimately in investigating a wide variety of canonical reaction types and we have already analyzed the molecular dynamics of several. First, we have studied the photodissociation and recombination of a diatomic molecule in solution. Second, we have, in conjunction with Casey Hynes, studied two simple A + BC ---> AB + C solution reaction systems. In one, the interaction with the solvent (composed of rare gas atoms) is weak. In the other, an S_N2 reaction in water, the solute-solvent as well as the solvent-solvent interactions are coulombic and strong. Third, we have studied a biochemical reaction, the diffusive passage of ions through the polypeptide channel gramicidin. We are using these reactions as test cases to explore from a microscopic molecular viewpoint some of the basic questions one might ask about solution reactions: how fast they occur, how long they last, how big they are, and how mechanically and dynamically they actually take place. We hope that such studies may help in understanding real solution reactions and perhaps catalysis and enzyme reactions as well.

Diatomic Photodissociation

We have computed the molecular dynamics of I_2 photodissociation and recombination in a variety of liquid solvents, including rare gases, hydrocarbons, and halogenated hydrocarbons, and from these dynamics we have computed the accompanying transient electronic spectra and transient x-ray diffraction which one might observe in order to follow the dynamics.[8,9,10,11] We find that dissociation, translational energy equilibration and geminate recombination are rapid events, and that solvent caging, at least in this system, is not the long term phenomenon often envisioned. We find that the I_2 vibrational energy relaxation can be a slow process, dominating the time scale for return to ground state iodine molecules, and thus able to dominate the spectral time history. Similar conclusions were also reached by Hynes and Nesbitt[12,13] and Balk, Brooks and Adelman.[14,15] Related work on Br_2 photodissociation in clusters has been carried out by Amar and Berne.[16] We were unable to meaningfully estimate the role of electronic curve crossing, which may play an important role in the recombination process, as discussed by the Kelley and Harris groups.[17,18,19]

Weak Interaction: A + BC in Rare Gas

The canonical reaction type for gas phase dynamics has been the A + BC ---> AB + C reaction.[20] We have investigated, in collaboration with Casey Hynes, two versions of this reaction in solution, the first in which the interaction between the reagents and the solvent molecules as well as the solvent-solvent interactions are weak and the second in which these same interactions are strong, involving coulombic forces. The paper in the present volume by Hynes should be consulted for more information.

The weak interaction version involves chlorine-like atoms and diatomics, reacting in various rare gas solvents at liquid densities.[21,22] We find that this prototypical atom transfer reaction can be largely understood in terms of three time periods. The first is the barrier climbing period, the few hundred femtosecond arisal time in the reactants of most of the energy and ordering necessary to reach the region of the free energy barrier top. (Vibrational energy in the BC diatomic flows in from the solvent much earlier.) Our working hypothesis to be tested is that the energy flow into the reagents takes place by a fluctuation in the distribution of translational energies of the solvent atoms, in other words the development of a translational hot spot. The middle time period is the few tens of femtoseconds spent near the barrier top during which the transition state rate is modified by whatever recrossings of the barrier occur. For the rare gas solution studies, we find recrossings are infrequent and thus that classical transition state theory in general provides a good estimate of the rate. The small deviations from the transition state rate are adequately predicted in terms of a Generalized Langevin model by the Grote-Hynes theory.[23,24] The final time period, the time reverse of the arisal period, is the decay of the energy and ordering as the reagents slide down the free energy barrier on the product side.

Strong Interaction: S_N2 in Water

Our second test A + BC ---> AB + C reaction, also studied in collaboration with Casey Hynes, is an S_N2-like reaction in water solvent, designed to represent strong interaction involving ionic and dipolar reagents and dipolar solvent.[25,26] The reaction is chosen to resemble $Cl^- + CH_3Cl ---> CH_3Cl + Cl^-$ in water solvent with the CH_3 group modelled as a united atom to preserve the simple A + BC format. We again follow the energy flow into the different modes of the reagents during the few hundred femtoseconds required to climb the barrier. At the barrier top we find that recrossings are frequent, and thus that transition state theory breaks down. We find that, in most cases studied, a simple nonadiabatic solvation model[27,28,29,30] in which the solvent is assumed to be frozen in place during the recrossing period, reproduces the rates as determined by full molecular dynamics. This is consistent with the observation that the barrier curvature is great enough that the time spent on the barrier is quite short compared with the time for water molecules in liquid water to rotate and reorient their dipole moments, which would be necessary to produce major changes in the electrical potential.

If we observe, at the top of the barrier, the electrical potential at the sites of the nascent reactant and product Cl^- ions, we find that we can reasonably well predict the outcome of the trajectories. Those trajectories which are successful in the sense of going directly from reactant to product without rate reducing recrossing are those in which the electrical potentials at the two Cl^- sites are similar. If the reactant site were electrically more felicitous, the reaction would tend to start on the reactant side, go over the barrier and return to the reactant side, producing no net reactive event, even though the top of the barrier was reached and transition state theory would have predicted a reactive event. (*Mutatis mutandis* for an attractive site on the product side.) For cases in which the frozen solvent model would predict trapping by the solvent, a more refined Grote-Hynes model for the rate is needed, as is discussed below.

Ion Transport Reaction in Gramicidin

At the other pole from the simple reactions discussed above, we have also studied a biochemical process, which may be looked upon as a chemical reaction involving the transit of an ion across a membrane.[31]   The ion motion involves water as a solvent, but the water (with the dissolved ion) is *inside* the molecule rather than outside.   The molecule is an ion channel forming dimer, each half composed of the 15mer polypeptide Gramicidin A, which is wrapped into a helix. Approximately 600 moving atoms are involved in the process, in which the helix locally distorts as the ion diffuses by.   While we would argue, as illustrated by the S_N2 reactions above, that the Kramers picture of deviations from the transition state rate is generally inappropriate for reactions which involve the making and breaking of chemical bonds because the intrinsic barriers are too sharp for slow solvent motions to follow, we have here the opposite extreme, a barrier (or series of low barriers) 25 Å long over which the motion is indeed likely to be diffusive.

We observe that the solvent is arranged as a one dimensional chain of hydrogen bonded water molecules, whose properties are considerably different than ordinary three dimensional water.   For example, this linear water is a better ion solvent than three dimensional water in terms of solvation energy as a function of distance from the ion to the water.   In addition, the 25 Å long chain of waters tends to translate as a correlated unit, thus leading us to discuss the possibility of water structures in proteins acting as the intermediary for chemical action at a distance, particularly in allosteric effects in enzymes.[32]

SOME QUESTIONS AND A FEW ANSWERS

It is surprising that so little is known about liquid state solution reactions from the fundamental viewpoint of how they take place as seen from the microscopic perspective of molecular dynamics, in other words of the atomic motions by which they happen. The investigation of the molecular dynamics of solution reactions has only begun.   This is in contrast to thirty years of progress in the gas phase,[33] characterized by a widening of vision beyond the traditional focus on reaction rates, to include the whole picture of the motions, the trajectories, by which reactions occur, and how these dynamics are produced by the potential energy surface and masses of the reacting system.

This broader picture of reactions has yet to be developed for the liquid state, where the relatively much smaller body of molecular dynamics research, ours included, has so far focused almost exclusively on rates.   We now are trying to go beyond the rate determining behavior at the barrier top or transition state region to look at the question of how we *reach* the barrier top.   The process of climbing to the top dominates the reaction time, and I hope to convince the reader that the territory to be passed through on the *way* to the barrier is not without its own bestiary of interesting creatures to be discovered, described and understood, all to be added to those already discovered lurking near the activated complex.

How Long is a Solution Reaction?

As we explore the routes from the product valley to the transition state pass, we find that there are at least three different processes going on with three different time scales.   The longest time scale corresponds to the transfer of energy from the solvent to the reaction coordinate, the intermediate time scale to development of proper phasing of the reactants and the shortest time to the period necessary for the rate to decay, as a result of recrossings, from the transition state value to the real value.

First, we can consider the process of energy flow from solvent to reaction coordinate. We find for both the rare gas and water solvent cases, that the time for the majority of the energy needed to climb the barrier to flow from the solvent into the reaction coordinate, is a few hundred femtoseconds. This time is so short because the coupling of the reaction coordinate motion to the solvent is sufficiently strong that if the energy *had* flowed into the reactants, for example, a picosecond before the barrier was to be reached, it would have flowed back out again before it could have been used. (Vibrational energy is the exception, as it can remain for very long times before equilibrating with a rare gas solvent.) This means, for a typical barrier height of a few tens of kcal/mole, that if we could synchronize a mole of reagents to react simultaneously, the power flow from solvent to reagents would be over 10^5 times all the electrical power being generated on this planet! Clearly some interesting molecular dynamics must be involved in those special trajectories which are able to effect this solvent to reagent energy flow necessary to climb the reaction barrier. We would like to discover just what causes this energy flow in different types of reactions in different types of solvents. We have already made progress in this direction in rare gas solvent reactions where we watch the time evolution of the ensemble of trajectories which will later reach the barrier top. We suspect that the ensemble begins with the solvent in an *apparent* equilibrium distribution with respect to solvent translational energy, followed by a fluctuation, i.e. the development of a translational hot spot involving one or two solvent atoms, followed by the transfer of the solvent hot spot energy to the reagents and then into reaction coordinate potential energy as the barrier is climbed. The energy flow processes in our model S_N2 reactions will perhaps involve the orientations of the water molecules and the effects of the electrical fields they produce. Thus we expect to see the development of highly nonequilibrium phenomena in systems governed by the differential equations of classical mechanics, within a set of samples which initially *appear* as if they could have been drawn from an equilibrium distribution.

Second, if we consider that the energy flow resembles a t_1 process in molecular line shape theory, there must also be t_2-like processes involving phasing and dephasing. Energy flow into the reactant molecules (energy barrier climbing) is thus a necessary but not sufficient condition for reaction as the true barrier is a free energy barrier and ordering must be achieved as well as energy flow. For example, there must be not only proper translational energy flow into the reactants, but also the translational motions of the two reactants must be so that their velocities are in the proper directions pointed toward one another, and not away. Similarly, rotations need to become properly phased for reaction to occur. These phasing processes can be studied in terms of correlation and cross correlation functions, first among the various translational, rotational and vibrational modes of the reactants themselves as they climb the barrier, and second between the modes of the reactants and those of the nearby solvent molecules, to watch the flow of "order" between solvent and reactants.

Third, there is a time scale for the dying out of the recrossings which lead to the reduction of the rate below that predicted by transition state theory. This flux-flux correlation time scale we find in the reactions we have so far studied to be a few tens of femtoseconds.[34,35] One can hope to discover, again by molecular dynamics simulation, the microscopic cause and nature of these recrossings. It is already clear that they involve the inability of the solvent to adiabatically follow the reagent motions. Thus there is a nonequilibrium distribution of states of the surrounding solvent molecules as the trajectories move away from the barrier top.[36]

How Big is a Solution Reaction?

We can ask the question, "how big a drop of liquid do we need so that the chemical reaction of interest occurring within it would behave substantially as it would in the bulk liquid?"[37] We can also ask, "how far away from the reacting molecules do we have to be in the bulk liquid reaction so that the what is happening with the distant solvent molecules is irrelevant to the reaction." The answer to the second question depends on the time scale, i.e. on how long the perturbation of the solvent molecules has to propagate to the reaction site. Both of these questions in principle can be answered by molecular dynamics calculations.

The second question is one which we have already been studying. In the rare gas A + BC ---> A + BC reaction, we have tested two types of perturbations, momentum only,[38] or both position and momentum.[39] We take a sample set of trajectories which reach the top of the barrier and then look back at those trajectories as they existed at a time τ earlier. Then we freeze in place the reagents and all the solvent molecules inside a given radius and for the solvent molecules outside that radius either choose new momenta from an equilibrium distribution or choose new coordinates and momenta by allowing the molecules to equilibrate for a time period. Finally we restart the trajectories in the forward direction, restoring the original velocities to the reagents and to the solvent molecules inside the chosen radius and look to see if the trajectories still lead to reaction, in spite of the perturbation. In this way we map out the distance versus time relationship of the perturbation induced by the solvent on the reaction. We find that if we try to define a propagation velocity for disturbance, that it is quite different for the rare gas and water cases, as one might expect from the short versus long range nature of the forces involved. The disturbance by coordinate and momentum perturbation is stronger than that induced by momentum perturbation alone, again as one might expect from the fact that momentum perturbation requires time to induce the perturbation in position needed to produce a perturbation in force on the reacting atoms which then can divert their trajectories from reaching the barrier top.

How Fast is a Solution Reaction?

It is now generally appreciated that the transition state rate for a solution reaction is only an upper limit to the actual rate, due to the possibility of recrossings of the transition barrier induced by jostling by the solvent or by other modes of the reagents themselves.[40,41] We have investigated this, as briefly described above, in recent collaborations with Casey Hynes on tests of the Grote-Hynes theory for dynamic deviations from transition state rates.[42,43,44] We find that the Grote-Hynes approach works remarkably well when compared to full molecular dynamics rates, indicating that the Generalized Langevin Equation, nonequilibrium solvation picture upon which it is based is indeed applicable.

In addition, we find that the power spectrum of the frictional forces which perturb the motion at the barrier top, thus causing the rate decreasing recrossings, can be related to the pure solvent infrared and Raman spectra, therefore helping us to understand the physical nature of the perturbing forces. We compute this friction spectrum from the molecular dynamics derived time history of the forces between solvent molecules and reagent atoms, projected along the reaction coordinate. The similarity between the features seen in the experimental pure solvent spectra and those seen in the friction spectrum raises the interesting question of whether we could learn to predict the dynamic deviations from transition state rates by using pure solvent spectra combined with physically reasonable estimates of the coupling strength to the reaction coordinate of the various types for motions observed in the pure solvent spectra.[45]

How do Solution Reactions Happen?

In addition to analytical models, we would like simple *pictures* for solution reactions. If we are to calibrate our intuition, we need to develop appropriate dynamical images, internal movies to understand how solution reactions actually take place. Many of the important questions of physical organic chemistry can now in principle be answered by molecular dynamics simulations. Some of the pictures we would like to develop include i) How does the nature of the solvent determine the branching ratio among different thermodynamically allowed products? ii) How is stereochemistry determined in solution?

What are the Design Rules for Catalysts?

Given a chemical reaction whose rate we would like to control, and given the ability to build a catalyst from any atoms bonded in any manner that is possible, what would be our ideal catalyst? In other words, what are the rules for designing catalysts? This is even more of a mystery than the microscopic goings-on by which solution reactions take place.

What we have already observed in solution reaction dynamics tells us that the simple picture of catalytic action which we teach to our students is incomplete. The lowering of the free energy barrier between reactants and products is certainly the major catalytic key to increasing reaction rates. But because the transition state rate is only an upper bound to the rate, due to recrossings of the barrier induced by forces external to the reaction coordinate, we would also like to smooth the passage over the barrier to maximize the rate. So, one would like to develop a general microscopic catalytic strategy for both lowering reaction barrier free energies and lower the recrossing rate. Perhaps a deeper understanding of solution reaction dynamics will aid us in this more complex undertaking.

How Do Enzymes Work?

The microscopic understanding of how enzymes work, the molecular dynamics of their action and specificity, will certainly remain a challenge for many years. The literature is in fact already full of suppositions,[46,47] not all of which can be true, because they are often mutually contradictory. One wonders whether progress in understanding the many atom nature of solution reactions will also help us in understanding the many atom nature of enzyme-controlled reactions.

One can ask the question as to whether enzymes are ahead of chemists. Enzymes evolved into effective and selective catalysts through hundreds of millions of years of random steps in codon space, keeping what worked as measured against a fitness criterion. They certainly became efficient lowerers of free energy barriers. Did they also random walk into ways of efficiently smoothing the passage over barriers, reducing recrossings compared to the equivalent uncatalysed reactions?[48] This is a question which can potentially be explored with molecular dynamics techniques.

CONCLUSION

In the present paper we have reviewed briefly the beginning steps we have taken to try to explore some of the aspects of solution reaction dynamics for a few of the canonical types of solution reactions. The questions which initially concern us are very general ones such as the size, duration, speed, and mechanism for the reactions. From the little which is yet known, it should be clear that the study of the molecular dynamics, the detailed microscopic picture of how solution reactions take place has barely begun. This is surprising, given the central role of liquid solutions and solution reactions in chemistry. The fundamental studies which have been carried out are overwhelmingly concentrated on the theory of reaction rates in solution, ignoring other aspects of reaction dynamics. Very few experiments have yet been done which can cleanly distinguish among alternative molecular dynamics models, or test existing theories. One would hope that this situation will change drastically in the next few years, given the expected great increases in computer speed which can be harnessed to molecular dynamics calculations and thus applied to test analytical models and intuitive pictures, as well as expected progress in more definitive experiments.

REFERENCES

1 M. P. Allen and D. J. Tildesley, *Computer Simulation of Liquids* (Clarendon Press, Oxford, 1987).
2 J. A. McCammon and S. C. Harvey, *Dynamics of Proteins and Nucleic Acids* (Cambridge University Press, New York, 1987).
3 J. C. Keck, Discuss. Faraday Soc. **33**, 173 (1962).
4 J. B. Anderson, J. Chem. Phys. **58**, 4684 (1973).
5 P. H. Berens and K. R. Wilson, J. Comp. Chem. **4**, 313 (1983).
6 K. R. Wilson, in *Computer Networking and Chemistry*, edited by P. Lykos (American Chemical Society, Washington, D.C., 1975), p. 17.
7 K. R. Wilson, in *Minicomputers and Large Scale Computations*, edited by P. Lykos (American Chemical Society, Washington, D.C., 1977).
8 P. Bado, P. H. Berens, J. P. Bergsma, M. H. Coladonato, C. G. Dupuy, P. M. Edelsten, J. D. Kahn, K. R. Wilson, and D. R. Fredkin, Laser Chem. **3**, 231 (1983).
9 P. H. Berens, J. P. Bergsma, and K. R. Wilson, in *Time-Resolved Vibrational Spectroscopy*, edited by G. Atkinson (Academic Press, New York, 1983) p. 59.
10 P. Bado, C. G. Dupuy, J. P. Bergsma, and K. R. Wilson, in *Ultrafast Phenomena IV*, edited by D. H. Auston and K. B. Eisenthal (Springer-Verlag, Berlin, 1984) p. 296.
11 J. P. Bergsma, M. H. Coladonato, P. M. Edelsten, J. D. Kahn, and K. R. Wilson, J. Chem. Phys. **84**, 6154 (1986).
12 D. J. Nesbitt and J. T. Hynes, J. Chem. Phys. **76**, 6002 (1982).
13 D. J. Nesbitt and J. T. Hynes, J. Chem. Phys. **77**, 2130 (1982).
14 M. W. Balk, C. L. Brooks III, and S. A. Adelman, J. Chem. Phys. **79**, 804 (1983).
15 C. L. Brooks III and S. A. Adelman, J. Chem. Phys. **80**, 5598 (1984).
16 F. G. Amar and B. J. Berne, J. Phys. Chem. **88**, 6720 (1984).
17 D. F. Kelley, N. A. Abul-Haj, and D. Jang, J. Chem. Phys. **80**, 4105 (1984).
18 A. L. Harris, M. Berg, and C. B. Harris, J. Chem. Phys. **84**, 788 (1986).
19 M. E. Paige, D. J. Russel, and C. B. Harris, J. Chem Phys. **85**, 3699 (1986).
20 R. D. Levine and R. B. Bernstein, *Molecular Reaction Dynamics and Chemical Reactivity* (Oxford University Press, New York, 1987).
21 J. P. Bergsma, P. M. Edelsten, B.. J. Gertner, K. R. Huber, J. T. Hynes, J. R. Reimers, K. R. Wilson and S. M. Wu, Chem. Phys. Lett. **123**, 394 (1986).
22 J. P. Bergsma, J. T. Hynes, J. R. Reimers, and K. R. Wilson, J. Chem. Phys. **85**, 5625 (1986).
23 R. F. Grote and J. T. Hynes, J. Chem. Phys. **73**, 2715 (1980).
24 R. F. Grote and J. T. Hynes, J. Chem. Phys. **75**, 2191 (1981).
25 J. P. Bergsma, B. J. Gertner, K. R. Wilson, and J. T. Hynes, J. Chem. Phys. **86**, 1356 (1987).
26 B. J. Gertner, J. P. Bergsma, K. R. Wilson, S. Lee, and J. T. Hynes, J. Chem. Phys. **86**, 1377 (1987).
27 G. van der Zwan and J. T. Hynes, J. Chem. Phys. **76**, 2993 (1982).
28 G. van der Zwan and J. T. Hynes, J. Chem. Phys. **78**, 4174 (1983).
29 G. van der Zwan and J. T. Hynes, J. Chem. Phys. **90**, 21 (1984).
30 R. O. Rosenberg, B. J. Berne and D. Chandler, Chem. Phys. Lett. **75**, 162 (1980).
31 D. H. J. Mackay, P. H. Berens, K. R. Wilson, and A. T. Hagler, J. Biophys. Soc., **46**, 229 (1984).
32 D. H. J. Mackay and K. R. Wilson, J. Biomolecular Structure & Dynamics, **4**, 491, 1986.

33 R. D. Levine and R. B. Bernstein, *Molecular Reaction Dynamics and Chemical Reactivity* (Oxford University Press, New York, 1987).

34 J. P. Bergsma, J. T. Hynes, J. R. Reimers, and K. R. Wilson, J. Chem. Phys. **85**, 5625 (1986).

35 J. P. Bergsma, B. J. Gertner, K. R. Wilson, and J. T. Hynes, J. Chem. Phys. **86**, 1356 (1987).

36 *ibid.*

37 F. G. Amar and B. J. Berne, J. Phys. Chem. **88**, 6720 (1984).

38 J. P. Bergsma, J. T. Hynes, J. R. Reimers, and K. R. Wilson, J. Chem. Phys. **85**, 5625 (1986).

39 I. Benjamin, B. J. Gertner, and K. R. Wilson, (to be published).

40 J. T. Hynes, in *The Theory of Chemical Reaction Dynamics*, edited by M. Baer (Chemical Rubber, Boca Raton, FL, 1985).

41 D. Chandler, J. Chem. Phys. **68**, 2959 (1978).

42 J. P. Bergsma, J. T. Hynes, J. R. Reimers, and K. R. Wilson, J. Chem. Phys. **85**, 5625 (1986).

43 J. P. Bergsma, B. J. Gertner, K. R. Wilson, and J. T. Hynes, J. Chem. Phys. **86**, 1356 (1987).

44 B. J. Gertner, J. T. Hynes, and K. R. Wilson, (to be submitted).

45 *ibid.*

46 G. R. Welch, *The Fluctuating Enzyme* (John Wiley & Sons, New York, 1986).

47 A. Fersht, *Enzyme Structure and Mechanism* (W. H. Freeman and Company, New York, 1985).

48 J. P. Bergsma, B. J. Gertner, K. R. Wilson, and J. T. Hynes, J. Chem. Phys. **86**, 1356 (1987).

THE STATISTICAL THEORY OF THE IONIC EQUILIBRIUM OF WATER. BASIC PRINCIPLES

AND PRACTICAL REALIZATION

S.Bratos,   Y. Guissani, and B. Guillot

Laboratoire de Physique Théorique des Liquides
Université Pierre et Marie Curie
75252 Paris CEDEX 05

SUMMARY

A statistical theory is presented to study the ionic equilibrium of water. The paper is essentially technical : basic principles of the theory and its practical realisation are discussed in detail. The main results are sketched and the role of various polarisation mechanisms is emphasized.

1) INTRODUCTION

A considerable interest has been manifested over a long period in studying the ionic equilibrium of water, $2H_2O \rightleftharpoons H_3O^+ + HO^-$, in various fields of chemistry. In fact, its equilibrium constant $K(p,T)$ determines the acidity scale and the presence of the two ions gives rise to numerous physico-chemical processes in water solutions. In spite of its importance, no up to date statistical theory of this phenomenon seems to exist, a situation due essentially to intrinsic difficulties of the problem. Strong H-bonds are present between H_3O^+, or HO^-, and water whereas weak H-bonds link two H_2O molecules to each other. Intermolecular potentials, strongly non additive, are particularly complex. Moreover, the energy required to create an ion pair is largely compensated by that due to the solvation of these ions in water : the theoretical expression for the pH thus involves a difference of two large terms of opposite signs and comparable magnitude. A considerable effort is required to overcome these difficulties.

The existing work covers two directions of investigation. (i) Ab-initio calculations have been carried out, mostly at the SCF-MO level, for a set of hydrated oxonium and hydroxyl ions. These entities were found highly flexible and their energy strongly non-additive[1-3]. Moreover, molecular dynamics and Monte Carlo methods have been applied to the study of liquid water. Both, equilibrium and dynamical properties have been examined in much detail and a number of relatively simple effective pair potentials have been elaborated[4-7]. (ii) The experimental work refers, essentially, to the investigation of $H_3O^+ (H_2O)_n$ and $HO^- (H_2O)_n$ in the gas phase, n = 1-4[8-11]. It provides basic data on the energy of hydration of oxonium and hydroxyl ions in gases. The information gathered in this way represents the basic ingredient of any theoretical work in this field.

Recently, the present authors proposed a theory of ionic equilibrium of water[12] in which the equilibrium constant $K(p,T)$ was calculated by combining molecular dynamics and test-particle method[13-24]. The purpose of the present paper is to develop in more detail a number of technical points, only briefly sketched therein. One concludes that this problem, although near to the feasibility limit, is soluble by employing modern techniques of statistical mechanics.

2) BASIC PRINCIPLES

a) Description of the model. The system submitted to investigation contains n_1 water molecules, n_2 oxonium and n_3 hydroxyl ions. The following model was employed to describe its thermodynamic behavior. (i) Molecular translations and rotations obey the laws of classical mechanics whereas their vibrations follow the laws of quantum mechanics. (ii) H_2O, H_3O^+ and HO^- are assimilated to rigid rotators. Their geometry is a mean geometry of these species in liquid water, different from that of free species. (iii) Vibrational frequencies of H_2O, H_3O^+ and HO^- are mean frequencies in liquid water. Heavily affected by hydrogen bonding they differ considerably from the corresponding gas phase frequencies. (iv) Only the ground vibrational state of H_2O, H_3O^+ and HO^- is thermally populated. For justification of these assumptions see Ref. (12).

b) Theory. The conceptually simplest way to study chemical equilibria in liquids consists in realizing a molecular dynamics or Monte Carlo simulation of a mixture of reacting molecules. If the intermolecular potentials are known, the computer simulation generates the equilibrium situation automatically. Unfortunately, this simple method is inapplicable in the present case : the fraction of dissociated molecules being of the order of 10^{-7}, the simulation would require $n_1 > 10^7$ water molecules !

The route chosen in this paper is thus indirect and consists to express $K(p,T)$ in terms of chemical potentials of reaction partners taken in appropriate reference states. Let A_1, A_2, A_3 designate H_2O, H_3O^+, HO^-, ν_1, ν_2, ν_3 positive integers, μ_1, μ_2, μ_3 chemical potentials, μ_1^o, μ_2^o, μ_3^o chemical potentials in the reference states and a_1, a_2, a_3 activities. The following equations can then be written :

$$2H_2O \rightleftharpoons H_3O^+ + HO^- \quad ; \quad \nu_1 A_1 \rightleftharpoons \nu_2 A_2 + \nu_3 A_3 \quad ; \quad \nu_1\mu_1 - \nu_2\mu_2 - \nu_3\mu_3 = 0 \quad ;$$

$$\mu_i = \mu_i^o + K_B T \ln a_i \quad ; \quad \Delta G(pT) = \nu_2\mu_2^o + \nu_3\mu_3^o - \nu_1\mu_1^o = - K_B T \ln K(p,T) \qquad (1-5)$$

where i=1,2,3. The definition of the activities a_1, a_2, a_3 is as follows. (i) The activity a_1 of H_2O is defined by taking the pure water as a reference state and by postulating that in this state $a_1 = 1$. (ii) The activities a_2, a_3 of H_3O^+ and HO^-, respectively, are defined by taking an infinitely diluted water solution of these ions as a reference state and by postulating that in this state $\lim_{c_2 \to 0} (a_2/c_2) = 1$, $\lim_{c_3 \to 0} (a_3/c_3) = 1$.

Based on these premises, the calculation can be sketched in the following way. (i) The canonical partition function $Q(V,T,n)$, $n = (n_1, n_2, n_3)$, is determined first. Let E_{01}, E_{02}, E_{03} be the energies of the ground electronic-vibrational states of H_2O, H_3O^+ and HO^- in absence of any molecular interaction and m_1, m_2, m_3 the masses of these molecules. One finds :

$$Q(V,T,n) = \left[\frac{1}{h^6 \sum_i n_i \prod_i n_i!} \iint dp\, dq\ e^{-\beta[T(p)+V(q)]} \right] e^{-\beta \sum_i n_i E_{0i}}$$

$$= \left[\prod_i \left[\frac{eV}{n_i} \left(\frac{2\pi m_i k_B T}{h^2} \right)^{3/2} Q_{Ri}(T)\ \right]^{n_i} \right] \left[\frac{1}{(\Omega V)^{\sum_i n_i}} \int dq\ e^{-\beta V(q)} \right] \qquad (6)$$

where $Q_{Ri}(T)$ represents the free rotation partition function of the species i and Ω is equal to 4π for HO^+ and to $8\pi^2$ for H_3O^+, H_2O. (ii) The Helmoltz free energy $F(V,T,n)$ and chemical potentials $\mu_i(V,T,n)$, $i = 1,2,3$, are deduced from Eqn (6). There results :

$$\mu_i(V,T,n) = -k_B T \ln \left[\frac{V}{n_i} \left(\frac{2\pi m_i k_B T}{h^2} \right)^{3/2} Q_{Ri}(T)\ e^{-\beta E_{0i}} \right]$$

$$- k_B T \frac{\partial}{\partial n_i} \left[\ln \frac{1}{(\Omega V)^{\sum_i n_i}} \int dq\ e^{-\beta V(q)} \right] \qquad (7)$$

(iii) The quantities μ_i^0 entering, according to Eqn (5), in $K(p,T)$ are calculated next. It should be pointed out that the activities a_1 and a_2, a_3, respectively, are defined differently. In fact, designating the limit $\lim\limits_{n_1 \to N} \lim\limits_{n_2 \to 0} \lim\limits_{n_3 \to 0}$ shortly by lim and putting $N = n_1 + n_2 + n_3$ one finds :

$$\mu_1^0(p,T,N) = \lim \left[\mu_1(V(p,T,n),T,n) \right] ; \qquad (8a)$$

$$\mu_{2,3}^0(p,T,N) = \lim \left[\mu_{2,3}(V(p,T,n),T,n) - k_B T \ln \frac{n_{2,3}}{N} \right] \qquad (8b)$$

μ_i^0's are generally expressed in terms of variables p,T rather than V,T. V must thus be considered as a function of these variables ; one finds $V = Nv(p,T, n/N)$ where v is the volume per molecule. In spite of the fact that the definitions (8a,b) are different, the resulting μ_i^0's are all of the same form :

$$\mu_i^0(p,T,N) = -k_B T \ln \left[\frac{k_B T}{p} \left(\frac{2\pi m_i k_B T}{h^2} \right)^{3/2} Q_{Ri}(T)\ e^{-\beta E_{0i}} \frac{pV}{Nk_B T} \right]$$

$$- k_B T \lim \left[\frac{\partial}{\partial n_i} \left(\ln \frac{1}{(\Omega V)^{\sum_i n_i}} \int dq\ e^{-\beta V(q)} \right) \right] = \mu_i^{id} + \mu_i^{ex} \qquad (9)$$

(iv) The above expression for μ_i^{ex} can conveniently be rewritten by replacing the derivative $\partial/\partial n_i\, f(n_i)$ by the difference $f(n_i+1)-f(n_i)$[14]. Then, introducing the energy $U_i(q',q)$ of interaction between a molecule of species i placed at q' and a system of molecules of composition n in configuration q, one finds readily :

$$\mu_i^{ex} = -k_BT \ln \frac{\iint dq'dq \; e^{-\beta U_i(q'q)} e^{-\beta U(q)}}{\Omega V \int dq \; e^{-\beta U(q)}} = -k_BT \ln \langle e^{-\beta U_i} \rangle \qquad (10)$$

(v) The final formula for $K(p,T)$ is obtainable by combining Eqs (9,10) and standard formulas for the free rotation partition functions $Q_{Ri}(T)^{(25)}$. Then, designating by I_{11}, I_{12}, I_{13} the principal moments of inertia of H_2O, by I_{21}, I_{22}, I_{23} the corresponding quantities for H_3O^+, by I_3 the moment of inertia of HO^- and by σ_1, σ_2 the symmetry numbers for H_2O, H_3O^+, one finds

$$K(p,T) = \exp\left[-\beta \sum_i \nu_i \mu_i^0\right] = \prod_{i=1}^{3}\left[Q_{Ti}Q_{Ri}\right]^{\nu_i} e^{-\beta \Delta E_0}\left[\frac{pV}{Nk_BT}\right]^{\sum \nu_i} \exp\left[-\beta \sum_i \mu_i^{ex}\right] \qquad (11)$$

$$Q_{Ti}(p,T) = \frac{k_BT}{p}\left[\frac{2\pi m_i k_BT}{h^2}\right]^{3/2} \qquad (12)$$

$$Q_{Ri}(T) = \frac{(2\pi)^3 (2k_BT)^{3/2} (\pi I_{i1}I_{i2}I_{i3})^{1/2}}{\sigma_i h^3} \qquad (i=1,2) \qquad (13)$$

$$Q_{R3}(T) = \frac{(2\pi)^2 (2k_BT)I_3}{h^2} \; ; \quad \Delta E_0 = \nu_2 E_{20} + \nu_3 E_{30} - \nu_1 E_{10} \qquad (14,15)$$

c) Discussion. Two important conclusions can be reached from the above analysis. The first is that the calculation of an equilibrium constant is reducible to that of chemical potentials ; an indirect method of calculation of $K(p,T)$ is thus realizable in principle. However, its practicability in dense phases relies heavily on the existence of powerful supercomputers necessary to compute μ^{ex}. The second conclusion is that, if the activities are defined as they are, the study of chemical equilibria in solutions containing a mixture of reacting molecules reduces to that of a pure solvent. No study of coupled motions of dissolved species is required.

3) PRACTICAL REALIZATION

a) General considerations. Two major difficulties are inherent to the present calculation. The first is relative to the determination of intermolecular potentials which are exceptionnally complex here. This problem was discussed in Ref. (12) and will not be rediscussed here. The resulting expressions for U_i are combinations of expressions describing the Lennard-Jones interaction between the oxygen nuclei, the coulombic interaction between the point charges placed on atoms and the interactions due to the polarization of water molecules by the ions and of strong H-bonds around the ions by water molecules :

244

$$U_{w/w} = \sum_i 4\epsilon \left[\left(\frac{\sigma}{R_{0i}}\right)^{12} - \left(\frac{\sigma}{R_{0i}}\right)^{6} \right] + \sum_{0'} \sum_i \frac{q_{0'}q_i}{R_{0'i}} \qquad (16)$$

$$U^{\pm}_{I/w} = \sum_i 4\epsilon \left[\left(\frac{\sigma}{R_{0i}}\right)^{12} - \left(\frac{\sigma}{R_{0i}}\right)^{6} \right] - \sum_{0'} \sum_i \frac{q_{0'}q_i}{R_{0'i}} - \sum_i \frac{1}{2}\alpha \vec{F}_i^2 - \frac{S}{2}\vec{F}'\cdot\vec{\alpha}'\cdot\vec{F}' \qquad (17)$$

$$S = \left[1 + e^{\gamma(u-u_0)} \right]^{-1} \qquad (18)$$

In the above expressions the index 0 refers to the extra particle and the index i to the water molecules ; $\vec{F}_i$ and $\vec{F}'$ are electric fields at the center of gravity of the i 'th water molecule and of the ion, respectively ; and α is the mean polarizability of H_2O whereas $\vec{\alpha}'$ is the effective polarizability tensor of H_3O^+, or HO^-, built into a perfectly aligned hydrate $H_3O^+ (H_2O)_n$, or $HO^- (H_2O)_n$. Finally, S is a function switching in the solute polarizability in configurations where strong H-bonds exist and switching it out elsewhere. The quantity u entering into its definition comprises the Lennard-Jones, Coulombic and water polarization energies of interaction between the ion and its n nearest neighbours. In the liquid, n was put equal to 5. See Tables I and II.

Table I

	$r_{OH}(Å)$	θ_{HOH}	$\nu_1(cm^{-1})$	$\nu_2(cm^{-1})$	$\nu_3(cm^{-1})$	$\nu_4(cm^{-1})$
H_2O	1.	$109°28'$	1650	3300	3500 .	
H_3O^+	1.1	$113°$	2300	1200	2300	1750
HO^-	1.2		2300			

The second major difficulty is to calculate μ^{ex}'s. The standard procedure consists in repeatedly inserting the extra particle, or the test particule, into the computer generated configurations of water. Unfortunately, in dense phases, random inserting places the test particule most frequently in a position already occupied by a water molecule. Moreover, H-bonding being strongly orientational, a good angular insertion is required. Every effort must thus be done to minimize the computational effort in treating the events in which the test particle is rejected. The discussion of this point is the main objective of this section.

Table II

	O-O		H₂O		H₃O⁺		HO⁻		$a(\text{Å}^3)$	$a_\parallel(\text{Å}^3)$	$a_\perp(\text{Å}^3)$	u_0 (Kcal/mole)	γ (mole/Kcal)
	$\frac{\varepsilon}{k}$(K)	σ(Å)	q_0(e)	q_H(e)	q_0(e)	q_H(e)	q_0(e)	q_H(e)					
H₂O/H₂O	78.22	3.165	-0.82	0.41									
H₃O⁺/H₂O	234.00	2.983	-0.82	0.41	0.445	0.185			1.444	3.	17.	-17.	1.5
HO⁻/H₂O	364.66	2.958	-0.82	0.41			-1.1	0.1	1.444	5.	12.5	-17.	1.5

b) Test particle method. The starting point of calculation is the generation of water molecule configurations. This was done by applying usual methods of molecular dynamics. The system submitted to the study consisted of $N = 216$ water molecules contained in a box of volume equal to L^3. Periodic boundary conditions were imposed and an Ewald summation was performed. The thermodynamic states considered are : $T = 293$ K, $\varrho = .999$ g/cm³ ; $T = 473$ K, $\varrho = .865$ g/cm³ ; $T = 593$ K, $\varrho = .667$ g/cm³. Three independent production runs of 4000 steps were carried out in the former case, and two production runs of the same length in the two latter. For other details, see Ref. (12).

Once water molecule configurations were determined, three series of calculations of increasing accuracy were performed. In the first of them the water molecule configurations were sampled uniformly every ten dynamical steps. Both, the center of gravity and angular coordinates of the test particule were chosen randomly. The center of gravity distances between the latter and water molecules were then calculated ; the remainder of calculation was as follows. (i) If in an event the distance between the test particule and a water molecule was smaller than a cut-off distance r_c, its contribution to μ_i^{ex}/k_BT was omitted. (ii) If $r > r_c$, the interaction energy was calculated for 5 nearest neighbors of the test particule. The value $U_i^{(5)}$ determined in this way was compared with a cut-off energy $U_{ic}^{(5)}$: if $U_i^{(5)} > U_{ic}^{(5)}$ the contribution of the event to μ_i^{ex}/k_BT was omitted. If not, the procedure was continued, first with 8 nearest neighbors of the test particule, and next for all molecules contained in a sphere S centered on the test particule and having a radius equal to $L/2$; the corresponding energies are $U_i^{(8)}$, $U_{ic}^{(8)}$, U_i^S, U_{ic}^S. (iii) If $U_i^S < U_{ic}^S$, the contribution of the event was retained and was calculated completely. Ewald summation was performed for charge-charge interactions and long-range corrections were evaluated for Lennard-Jones and polarization interactions. All water molecules in the box of volume L^3 were considered as contributing to the electric field on the ion. Proceeding in this way permits to minimize the effort required to treat the events dominated by strongly repulsive interactions. They do not contribute any significative contribution to μ_i^{ex}/k_BT, even if the cumulative effect due to their number is considered. The total number of events $\mathbb{P}_1$ studied in this series of calculations was equal to $P_1 p_1$ where P_1 is the number of dynamical steps which were retained and p_1 the number of insertions by step. $\mathbb{P}_1$ was equal to $4.8 \, 10^7$ for the low temperature thermodynamic point and $3.2 \, 10^7$ for the two others.

In the second series of calculations only P_2 water molecule

246

configurations were retained, where $P_2 \ll P_1$. These configuration were chosen to contribute noticeably to μ_i^{ex}/k_BT, not less than a unity to its k 'th significative figure. They were sampled uniformly $p_2 = 10p_1$ times. Proceeding in this way amounts to consider $\mathbb{P}_2 = (P_1/P_2) \, P_2p_2 = 10 \, \mathbb{P}_1$ events by treating explicitly only P_2p_2 of them. In practice, $\mathbb{P}_2$ was equal to 4.8×10^8, or to 3.2×10^8.

In the third, and the last, series of calculations P_3 water molecule configurations were only retained, where $P_3 \sim P_2 \ll P_1$. They were chosen as to contribute noticeably to μ_i^{ex}/k_BT, at least a unity to its l 'th significative figure. The calculations showed that a satisfactory insertion of a test particule was only possible if one, or several, cavities are present in a given configuration. Histograms were thus constructed in order to explore their number and size (Fig. 1). In this figure, the energy of the test particule is indicated along with its x, y, z coordinates, each point corresponding to one event. It is easily seen that one low-and one high-energy cavity exist in the case of Fig. 1. It should be noticed that, in an overwhelming majority of events, the energy of the test particule is too high to be presented in this figure ; the number of points therein is thus comparatively small. Once the cavities were located, and their dimensions

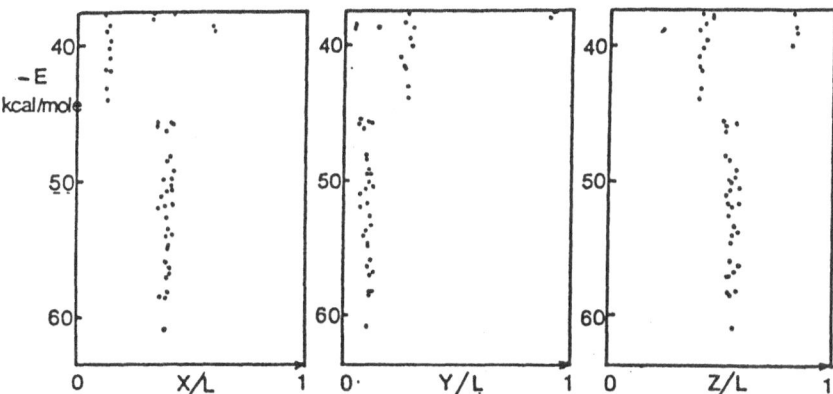

Fig.1. Localization of cavities accepting on H_3O^+ ion. Each point corresponds to one event. A low energy cavity is centered on x/L = 0.39, y/L = 0.12, z/L = 0.55 and a high energy cavity is located at x/L = 0.13, y/L = 0.29, z/L = 0.40.

determined, the water molecule configurations were sampled again, non-uniformly, p_3 times ; the test particle was inserted only into the cavities. The corresponding histograms represent an enlarged view of a small portion of histograms similar to those of Fig.1 but contain a large number of points (Fig.2). Proceeding in this way amounts to consider $\mathbb{P}_3 = (P_1/P_3) \, (V/\sum_1 v_i) \, P_3p_3$ events by treating only P_3p_3 of them ; v_i is the volume of the cavity i and $V = L^3$. In practice, $\mathbb{P}_3$ was of the order of 10^{10} in this third series of calculations.

This technique was applied in the following conditions. The excess chemical potential of water was calculated by employing the second method whereas the third method was necessary to determine those of two ions. This latter procedure permitted satisfactory angular insertion of H_3O^+ and of HO^- into the liquid water structure. The numerical precision of the results is difficult to evaluate with certitude. The usual procedure consists to construct the so called f/g plots by combining the test and the real particle methods[19]. Unfortunately, this method is difficult to apply to diluted

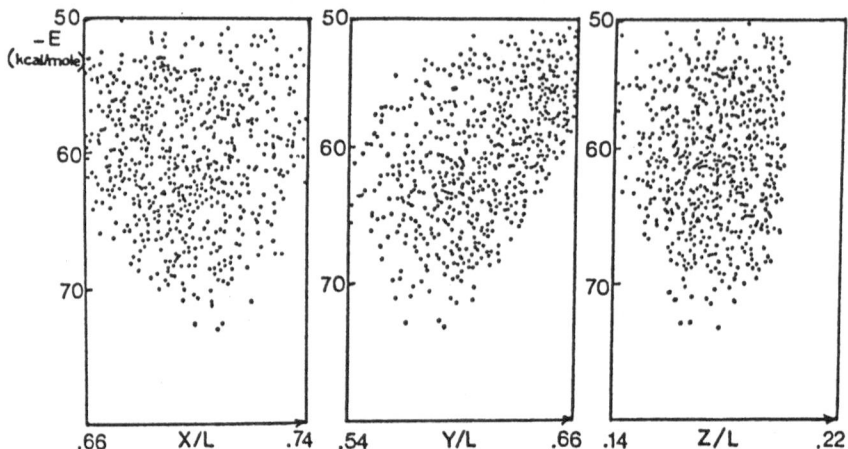

Fig.2. Exploration of a low energy cavity accepting an H_3O^+ ion
and located at x/L = 0.7, y/L = 0.6, z/L = 0.18. Only the
volume $0.66 \leq x/L \leq 0.74$, $0.54 \leq y/L \leq 0.66$, $0.14 \leq z/L \leq 0.22$
was sampled. Each point corresponds to one event.

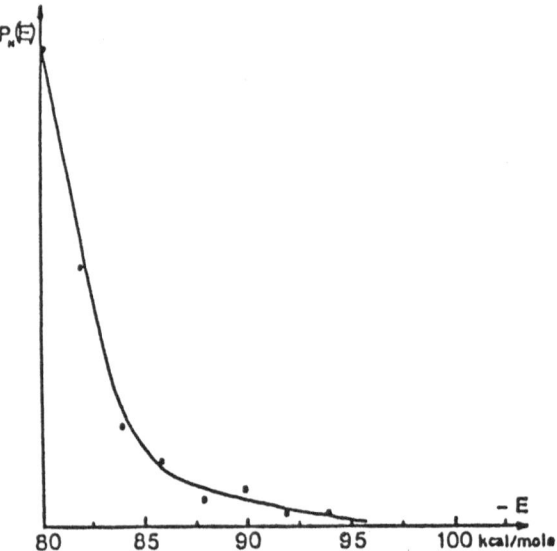

Fig.3. The general aspect of the low energy portion of $P_N(E)$.
The points indicate the number of events contained in an energy
interval equal to 2 Kcal/mole.

solutions. Another way to guess the reliability of the method is to consider the function P(E), the probability density that the test particle is inserted into the liquid with energy E. Only the low energy end of this curve needs to be considered : the high energy events do not contribute significantly to $\mu_i^{ex}/k_B T$. It was checked that if the runs are sufficiently long and independent of each other, P(E) does not change significantly when going from one of them to another (Fig. 3). As a consequence, considering these arguments as well as the over-all numerical coherence of the above results, and of those furnished by some other independent runs non-described here, the relative error is believed to be of the order of 1 % . This is less than that due to other uncertainties of the model and particularly to the uncertainty in determining the zero point vibrational energy of reaction partners.

Table III

T (^{0}K)	ϱ (g/cm^3)	pH (theor)	pH (exp)
293	0.998	6.5	6.97
473	0.865	3.6	5.45
593	0.667	3.8	5.87

4) RESULTS AND DISCUSSION

This Section contains a short description of main results of the present theory ; for more details, see Ref. 12. The following points merit attention. (i) The theory predicts a value of 6.5 for the pH of water at $T = 293$ K and $\varrho = 0.999$ g/cm^3. It would be possible, by choosing effective vibrational frequencies appropriately, to find a closer agreement with experience. However, this would not reflect the real limits of the theory. The magic number pH = 7 is only approached, but not exactly reproduced. (ii) When temperature and density are varied, the pH varies as illustrated in Table III. All essential trends are correctly reproduced, even the fact that the pH increases when going from the thermodynamic point $T = 473$ K, $\varrho = 0.865$ g/cm^3 to the point $T = 593$ K, $\varrho = 0.667$ g/cm^3. It can be shown that this effect is due mainly to the variation of the density between these two points. (iii) The presence of the two ions dominates the behavior of the excess factor of K(p,T). The water molecules contribute only 5 % to the sum $\mu_2^{ex} + \mu_3^{ex} - 2\mu_1^{ex}$. (iv) The polarization effects are enormous (Table IV). Both, the polarization of water molecules by ions as well as that of strong hydrogen bonds around the ions by water molecules must be considered. In their absence, the pH would have the value of 56.7 instead of 6.5 ! This result shows the usefulness of statistical studies of basic chemical reactions : the crucial role of polarization in the ionic dissociation of water never was correctly appreciated.

Table IV

Molecular Interaction Forces Considered	pH
None	82.4
L.J., Coulomb	56.7
L.J., Coulomb, water polarization	22.1
L.J., Coulomb, water and strong H-bond polarizations	6.5

LITERATURE

(1)   M. De Paz, S. Ehrenson, L. Friedman, J. Chem. Phys. $\underline{52}$ (1970), 3362.
(2)   M.D. Newton, S. Ehrenson, JACS $\underline{93}$ (1971), 4971.
(3)   M.D. Newton, J. Chem. Phys. $\underline{67}$ (1977), 5535.
(4)   A. Rahman, F.H. Stillinger, J. Chem. Phys. $\underline{55}$ (1971), 3336.
(5)   F.H. Stillinger, A. Rahman, J. Chem. Phys. $\underline{60}$ (1974), 1545.
(6)   D. Matsuoka, E. Clementi, M. Yoshimine, J. Chem. Phys. $\underline{64}$ (1976), 1351.
(7)   H.J.C. Berendsen, J.P.M. Postma, W.F.Van Gunsteren, J. Hermans, in "Intermolecular Forces", B. Pullman Editor, Reidel, Dordrecht 1981, p. 331.
(8)   P. Kebarle, S.K. Searles, A.Zolla, J. Scarborough, M. Arshadi, JACS $\underline{89}$ (1967), 6393.
(9)   M. De Paz, J.J. Leventhal, L. Friedman, J. Chem. Phys. $\underline{51}$ (1969), 3748.
(10) M. De Paz, A.G. Guidoni, L. Friedman, J. Chem. Phys. $\underline{52}$ (1970), 687.
(11) E.F. Ferguson, F.C. Fehsenfeld, D.L. Albritton, Gas Phase Ion Chemistry I, Academic Press, New York 1979.
(12) Y. Guissani, B. Guillot, S.Bratos, J. Chem. Phys., in press.
(13) E. Byckling, Physica $\underline{27}$ (1961), 1030.
(14) B. Widom, J. Chem. Phys. $\underline{39}$ (1963), 2808.
(15) S. Romano, K. Singer, Mol. Phys. $\underline{37}$ (1978), 1765.
(16) J.G. Powles, Mol. Phys. $\underline{41}$ (1980), 715.
(17) K.S. Shing, K.E. Gubbins, Mol. Phys. $\underline{43}$ (1981), 717.
(18) K.S. Shing, K.E. Gubbins, Mol. Phys. $\underline{46}$ (1982), 1109.
(19) J.G. Powles, W.A.B. Evans, N. Quirke, Mol. Phys. $\underline{46}$ (1982), 1347.
(20) J.G. Powles, Chem. Phys. Lett. $\underline{86}$ (1982), 335.
(21) K.S.Shing, K.E. Gubbins, Mol. Phys. $\underline{49}$ (1983), 1121.
(22) Y. Guissani, B. Guillot, F. Sokolic, Chem. Phys. $\underline{96}$ (1985), 271.
(23) D. Fincham, N. Quirke, D.J. Tildesley, J. Chem. Phys. $\underline{84}$ (1986), 4535.
(24) D.Frenkel, in "Molecular Dynamic Simulation of Statistical Mechanical Systems", G. Cicotti and W. Hoover Editors, North Holand, Amsterdam 1984, p. 151.
(25) E.D. Landau, E.M. Lifshitz, Statistical Physics, Pergamon Press London 1969.

RIVAIL — Your paper emphasizes the crucial role of polarizability of the water molecules by the ions. Since it is possible to consider the polarization of every water molecule by all other water molecules (which are more numerous than ions), don't you think that this effect may also play a role in the pH value ?

BRATOS — The polarization of water molecules by each other certainly is a non negligible effect. Nevertheless, our model is the crudest possible model of the pH of water and only the most important polarization effects are included : the polarization of water molecules by ions and that of strong H-bonds around the ions by water molecules. A more refined model should also include the effect you mention.

KAPRAL — The magnitude of the effects due to polarizability may depend on the specific details of the potential model. If the assumption of rigid molecules is relaxed, do you expect polarizability to play as large a role ?

BRATOS — The polarizability in any event plays a major role in this process. If the theory is built starting from protons and oxygen nuclei, or starting from our rigid molecules, the polarization effects would be described in different terms. Nevertheless, they must necessarily be considered whatever the description one may choose !

BELLONI — It is known that the pK of the ionization equilibrium of liquid amonia, where the hydrogen bonds are much weaker than in water, is around 33 at -33 C and 27 at room temperature. These data support strongly your conclusions on the importance of the last term E_{zundel}. Could you also do similar calculations in the case of liquid NH_3 ?

BRATOS — These calculations certainly are feasible and should be performed as soon as possible. The theory will probably be similar to that presented to-day.

HYNES — You emphasized that the pH depends strongly on the Zundel hydrogen bond polarizability. Since that might depend on proton (presumably quantum) motion in the bond, there might be a big (solvent) isotope effect. Can you comment on this and also indicate what is known experimentally for the H_2O/D_2O pH ratio ?

BRATOS — The isotope effect of the pH is included, in principle, into the "ideal" factor $K^{id}(p,T)$ of $K(p,T)$. It should thus be relatively easy to calculate it from the present theory.

TURQ – My question is related to that of Belloni and Hynes : what are the solvents for which you have enough spectroscopic informations to do the same kind of calculations ?

BRATOS – Spectroscopic informations in question are mainly those which can be obtained from IR spectra of strong H-bonds. These systems most often are solids. However, some of them can be prepared as solutions in usual spectroscopic solvents, like CCl_4, $CHCl_3$, etc. They have been studied in much details by authors such as D. Hadzi, H. Ratajczak, etc.

NEWTON – The H/D isotope effect on pK_a H_2O would be interesting to pursue. I believe the experimental data are available. As a related fact, I note that a pronounced H/D fractionalism occurs between L_2O and L_3O^+ (L=H or D) in aqueous solution. Related to these quantum effects, it would seem appropriate to include, as well as the intra-molecular vibrational modes which you listed, also the librational modes, since these latter, intermolecular modes can also have rather high frequencies (≤ 300 cm^{-1}). The OH bond lengths you have listed ranging from 1.0 → 1.2 A, involve some assumptions about the corresponding O...O separations. Our calculations for hydrated H_3O^+ and OH^-, and also Narten and Triolo's neutron scattering results for aqueous HCl, suggest that O...O separations are ≥ 2.55 indicate that OH distances would be expected to be ≤ 1.05 A.

BRATOS – The quantum mechanical effects on intermolecular vibrations are not present in our theory. They very probably are not particularly large, but are certainly not vanishing at room tempreratures. A more refined theory would be required to describe them.

WILSON – Could you please comment on the attempt by David and Stillinger to compute the pH of water, particularly in the light that they did try to include polarizability to all orders. You pointed out the large effect of polarization on pH values.

BRATOS – The polarizability used by David and Stillinger is relative to the bare O^{2-} units, whereas those employed here concern the H_2O molecules and the strong H-bonds around the ions. It is thus difficult to compare them.

MODELING MOLECULAR TRANSFORMATIONS IN SOLUTION

William L. Jorgensen, J. Kathleen Buckner, and
Jiali Gao

Department of Chemistry
Purdue University
West Lafayette, Indiana 47907 USA

INTRODUCTION

Transformations of molecular systems in solution are the central focus of chemical and biochemical sciences. The connection to processes of life itself is clear and emphasizes the importance of obtaining an atomic level understanding of structure and reactivity in solution. It would obviously be exciting to be able to computationally model complex molecular systems accurately enough to provide valid insights that yield enhanced predictive capabilities for structure and reactivity. In fact, progress is being made along these lines via molecular dynamics and statistical mechanics simulations for organic and biomolecular systems in solution. Summaries of contributions by different groups are available in two recent books.[1,2] In the present paper, the focus will be on recent results from our group on organic transformations, specifically, solvent effects on two important classes of processes, conformational equilibria and ion-pair complexation. Our work on free energy surfaces for organic reactions in solution that involve changes in covalent bonding has been reviewed recently elsewhere.[3]

METHODOLOGY

A principal objective of the theoretical studies is to obtain free energy surfaces for the transformations in solution.[3] The full, multi-dimensional problem for even moderate sized systems is beyond most available computer resources. Therefore, the work so far has concentrated on systems where a one dimensional reaction coordinate can be defined. This may be adequate for a simple dissociation process or an internal rotation about one dihedral angle, while for more general transformations a mapping of the geometric changes to one dimension may be possible. For example, in our studies of S_N2 and addition reactions,[4,5] the minimum energy reaction path (MERP), $E°(r_c)$, for the gas phase was determined as a function of a single geometrical variable, such as a bond length, which is taken as the reaction coordinate r_c. Of course, there are other geometrical changes that occur as a function of r_c and that are expressed as smooth functions, $P_i(r_c)$, for each variable i. The effect of solvation on the MERP with its multi-dimensional geometrical changes can then be computed as a function of r_c, though the possibility of a change in mechanism (MERP) in solution must be considered.[6] The interconversion of two conformational isomers might also involve changes in multiple dihedral angles, e.g. Φ and ψ for a peptide, that, however, can be mapped to a one dimensional, possibly arbitrary reaction path for computation of the relative solvation energies for the conformers.[7]

Once the reaction or transformation path has been defined, two approaches have been taken to evaluating the free energy profile in solution, $W(r_c)$. The simpler one conceptually is to use statistical perturbation theory to "step along" the reaction coordinate and compute the free energy change at each step. The procedure follows from eq 1[8] which represents the free energy difference between points i and j along

$$G_j - G_i = -k_B T \, \ln \, <\exp[-\beta(E_j - E_i)]>_i \qquad (1)$$

r_c as a function involving the energy difference for the two points. In the equation, $\beta = (k_B T)^{-1}$ and the average is based on running the simulation in the isothermal-isobaric ensemble for the system at point i. Thus, j is treated as a perturbation from i, and, as usual, the perturbation cannot be too large or convergence of eq 1 will be too slow. Typical differences in i and j might be 0.1-0.2 Å for a distance and 15° for a dihedral angle; however, twice these values, i.e. 0.2-0.4 Å and 30°, can be covered in one simulation by considering the perturbations from point i to i-1 and to i+1.

The alternative procedure involves a series of simulations with "importance sampling"[3-5,9] that each cover a range of r_c and yield $W(r_c)$ continuously. Biasing functions are used to constrain the system to a range around a point r_c^i. The reaction coordinate is then allowed to vary during the simulation just like any other variable. The occurrence of different values of r_c is recorded as a distribution function, $g_i(r_c)$, which is simply related to the relative free energy or "potential of mean force" (pmf) by eq 2 once correction is made for the

$$W(r_c) = -k_B T \, \ln \, g_i(r_c) \qquad (2)$$

influence of the biasing. Repetition for overlapping ranges of r_c or "windows" then allows splicing of the individual $g_i(r_c)$'s to yield $W(r_c)$ for the full range of r_c. The difficulties with this approach include the choice of biasing functions and concern that the sampling in each window is complete.

The remaining critical component in the simulations is the potential functions that describe the intra- and intermolecular energetics for the system. The intermolecular part is usually represented in a Coulomb plus Lennard-Jones form with the interactions occurring between sites located on the nuclei. Simple potential functions are now available that give excellent thermodynamic and structural results for many pure liquids including water,[10] hydrocarbons,[11] alcohols,[12] ethers,[13] and sulfur compounds.[14] Force fields for proteins and peptide-water interactions are also well-evolved.[15-17] The biggest problem is developing potential functions for specific reacting systems in which the potential function parameters, e.g. atomic charges, clearly vary along the reaction coordinate. A viable approach is to fit to results of ab initio molecular orbital calculations.[4,5] Other difficulties are associated with the treatment of polarization particulary for highly charged atomic sites and of long-range electrostatic interactions. The former issue is a three-body effect that has so far received little attention in large-scale molecular simulations, while the latter, related problem is particularly serious for ionic solids, molten salts, and polyelectrolytes such as nucleic acids.[18]

Given the reaction path and the potential functions, $W(r_c)$ can then be determined via eq 1 or 2 using either Monte Carlo or molecular dynamics simulations. Our work on the representative organic systems summarized below has featured Monte Carlo statistical mechanics for the reacting system plus 200-300 solvent molecules in a cubic or rectangular box with periodic boundary conditions. The isothermal-isobaric (NPT) ensemble has been employed at 25°C and 1 atm. The sampling for each step or window in the free energy calculations has typically consisted of an equilibration phase for 10^6 configurations followed by averaging for 2×10^6 configurations and would require ca. 10 days on a computer like a VAX 11/780. Additional details may be found elsewhere.[3-6]

APPLICATIONS

Conformational Free Energy Surfaces

Fluid simulations have been used to investigate the effect of condensed phase environments on conformational energy surfaces and equilibria for both pure liquids and dilute solutions.[19] In the case of rotation about one bond, the dihedral angle is the reaction coordinate for this simple isomerization. The studies so far have mostly assumed the rotation to be rigid, though concurrent variations in bond lengths and bond angles can easily be accommodated. In the latter case, a simulation for the rotation of an isolated molecule in the gas phase would yield the reference dihedral angle distribution, $S°(\Phi)$, which could be compared to the result in solution, $S(\Phi)$. The two distributions are related to the potentials of mean force for the gas phase, $W°(\Phi)$, and solution, $W(\Phi)$, by eqs 3 and 4 which are analogous to

$$W°(\Phi) = -k_B T \ln S°(\Phi) + \text{const.} \qquad (3)$$

$$W(\Phi) = -k_B \ln S(\Phi) + \text{const.} \qquad (4)$$

eq 2. $W°(\Phi)$ and $W(\Phi)$ both contain terms for sampling over the intramolecular vibrations, while $W(\Phi)$ also embodies the effects of any frequency shifts for the vibrations in solution as well as the change in free energy of solvation as a function of Φ, $\Delta G^{sol}(\Phi)$. In the simplified case where the intramolecular vibrations are ignored, $W°(\Phi)$ is just the gas-phase torsional potential, $V(\Phi)$, and $W(\Phi) = V(\Phi) + \Delta G^{sol}(\Phi)$. The solvent effect on $S(\Phi)$ is then clearly seen through eq 5 where c is a normalization constant.

$$S(\Phi) = c \, S°(\Phi) \exp(-\beta \, \Delta G^{sol}(\Phi)) \qquad (5)$$

Calculation of $S(\Phi)$ for many pure organic liquids such as alkanes, alcohols, ethers, thiols, sulfides, disulfides, and haloalkanes is easily achieved by directly sampling over the dihedral angle(s) or by umbrella sampling using reduced torsional barriers.[11-14,19] However, dilute solutions, which are modeled as a single solute plus several hundred solvent molecules, present potential convergence problems since statistics are only obtained on the one solute instead of the hundreds of identical molecules in the simulation of a pure liquid, i.e. in a Monte Carlo calculation most of the time would be spent moving the solvent molecules.

Several approaches can be used to address the problem. For one dihedral angle, umbrella sampling over a torsional surface with greatly reduced barriers is viable coupled with long simulations.[19,20] In Monte Carlo calculations, convergence of $S(\Phi)$ can also be enhanced by preferential sampling whereby moves of the solute and the nearby solvent molecules are attempted more frequently than for the more distant solvent molecules.[19,22] For example, this approach was used in several simulations of butane in water that were aimed at studying the hydrophobic effect on the trans-gauche equilibrium.[19,21,23] Though the potential functions have varied, the results have consistently revealed a 10-20% increase in the gauche population at 25°C upon transfer from the gas phase to water. This is consistent with other theoretical results[24,25] and fundamental notions about hydrophobic effects on biomolecular structure.[26] Furthermore, the effect has been shown not to just be due to the fluid environment, since no conformational shift is found for pure liquid butane.[11,19]

Statistical perturbation theory has also recently been applied to this problem.[27] The major advantage to this procedure is that the reaction coordinate, Φ, is progressively stepped along so there is no concern about the thoroughness of sampling over Φ. In addition, the convergence of the calculations was readily checked by running the perturbations from $\Phi = 0°$ to 180° and then back from $\Phi = 180°$ to 0° in 15° increments. The hysteresis amounted to only 1-2% in the conformer populations and a 12% increase in the gauche population was obtained at 25°C using the TIP4P model for water[10] and the OPLS parameters for butane.[11] The results for $S°(\Phi)$ and $S(\Phi)$ are compared in Figure 1; the trans population decreases from 68% in the gas phase using the MM2 $V(\Phi)$[21] to 56% in water.

In another recent application of the perturbation method, the interconversion of the trans and cis conformers of N-methylacetamide was

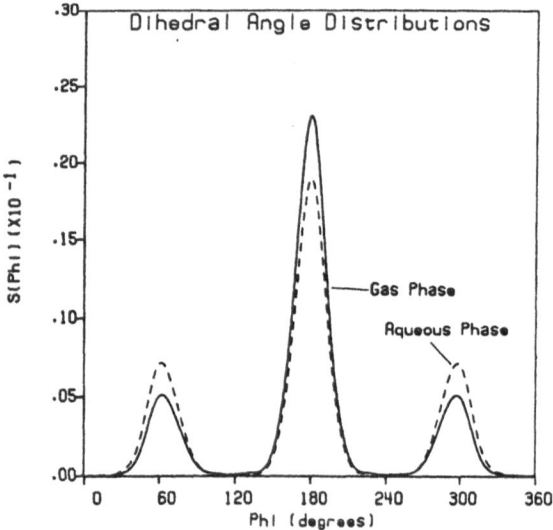

Figure 1. Population distributions for the central CC bond of butane in the gas phase and in aqueous solution at 25°C. Units for the ordinate are mole fraction per degree.

modeled in TIP4P water at 25°C.[28] The cis form was found to have a less favorable free energy of hydration by 1.6 ± 0.2 kcal/mol which adds to the gas-phase energy difference of 2-3 kcal/mol. The resultant substantial preference for the trans form in water is consistent with the extreme rarity of cis peptide bonds in proteins not involving Pro residues.

For more substantial structural differences between conformers involving multiple dihedral angles, importance sampling or the perturbation procedure are probably the best approaches using a one-dimensional reaction path with concerted variations in the angles.[7] Obtaining even a complete two-dimensional map for the solvent effect, e.g. $\Delta G^{sol}(\Phi, \psi)$, is computationally too demanding. Adequate simulations for the one-dimensional case require $(1-2) \times 10^7$ Monte Carlo configurations, though the earlier studies with umbrella sampling used less than half this number. The full N-dimensional problem would require ca. 10^{N-1} times the effort for similar grid spacings.

Complexation

Another fundamental class of molecular transformations features the complexation (association or dissociation) of two species. The potential of mean force needs to be determined as a function of the inter-species separation in this case. For two spherical atoms or ions a one-dimensional reaction coordinate is appropriate and $W(r_c)$ will reveal important information, e.g. on the existence of contact and solvent-separated ion pairs and on the intervening barriers. The treatment of non-spherical substrates can either involve some geometrical restrictions on the approach path or include appropriate orientational averaging with r_c as a characteristic distance such as the centers-of-mass separation. Since it will generally be desirable to cover substantial ranges of r_c, importance sampling or statistical perturbation theory can be used to obtain $W(r_c)$.

Few studies have been carried out along these lines in view of the computational demands of the multiple simulations. The initial investigations were for two Lennard-Jones particles or two benzene molecules in water and featured importance sampling.[9b,29] The results showed the contact free energy minima anticipated from the hydrophobic effect as well as intriguing solvent-separated minima. Importance sampling was also used more recently to compute a pmf for Na^+Cl^- in water, again revealing contact and solvent-separated minima.[30] Then, in the last two years, our group has applied the perturbation method to obtain potentials of mean force for three prototypical organic ion pairs, $(CH_3)_3C^+Cl^-$, $(CH_3)_4N^+Cl^-$, and $CH_3NH_3^+CH_3COO^-$, in water.[31,32]

The tertiary butyl chloride system was examined to help elucidate the energetics in the ion-pair region for a solvolysis reaction. The results have been described in detail elsewhere,[31] though the computed pmf is reproduced in Figure 2. The system consisted of 250 TIP4P water molecules plus the ion pair in a rectangular periodic cell. The chloride ion was constrained to lie on the three-fold axis for the planar t-butyl cation, which is the favored orientation in the absence of solvent, and the reaction coordinate was taken as the central carbon-chlorine distance. Perturbations of ±0.125 or ±0.25 Å along r_c were found to be acceptable with a total of 15 Monte Carlo simulations used to cover r_c from 2.5 to 8.0 Å. As shown in Figure 2, minima were found at 2.9 and 5.75 Å corresponding to contact and solvent-separated ion pairs. This is consistent with Winstein's ion-pair scheme, though the simulations revealed no statistically significant minima beyond 5.75 Å which suggests there is just a continuum of structures between the solvent-separated ion pair and "free ion" stages.[31] The delicacy of theresults should be noted since a remarkable cancellation of the ion-ion and ion-solvent interactions occurs to leave the residual pmf. Though the pmf varies by only 5 kcal/mol between 3 and 8 Å, the two components vary by ca. 60 kcal/mol in opposite directions. For comparison, the "primitive model" prediction is also shown in Figure 2. It is obtained from the gas-phase ion-ion interaction by dividing the Coulombic term by the experimental dielectric constant, and is also used to anchor the Monte Carlo results at 8 Å. The importance of the explicit structure of the water molecules on the pmf is apparent from comparing the two curves.

The corresponding results for the tetramethylammonium chloride ion pair are shown in Figure 3 as obtained from analogous Monte Carlo simulations. The barrier between the contact and solvent-separated minima is now very small and the deviations from the primitive model are modest. The $(CH_3)_4N^+$-water attractions are particularly weak, < 10 kcal/mol,[32,33] and there is little orientational dependence for the interactions in comparison to $(CH_3)_3C^+$. The barrier between the minima may be greater for the carbenium ion due to the orientational preference of having two water molecules or a water molecule and the counterion on either side of the central carbon, which has the largest fractional positive charge (0.4e) and smallest Lennard-Jones σ.[31] Another interesting feature in Figure 3 is the location of the contact minimum at an N-Cℓ separation of 5-5.25 Å. It has been pushed out from the primitive model result by ca. 0.5 Å due most likely to the screening of the chloride ion by the large ammonium ion at short separations. An illustration of the contact ion pair is provided in Figure 4. Though a water molecule can not fit directly between the ions at this distance, keeping the ions farther apart helps retain the number of hydrogen bonds to the chloride ion near the 7-8 that are found for the isolated anion in water.

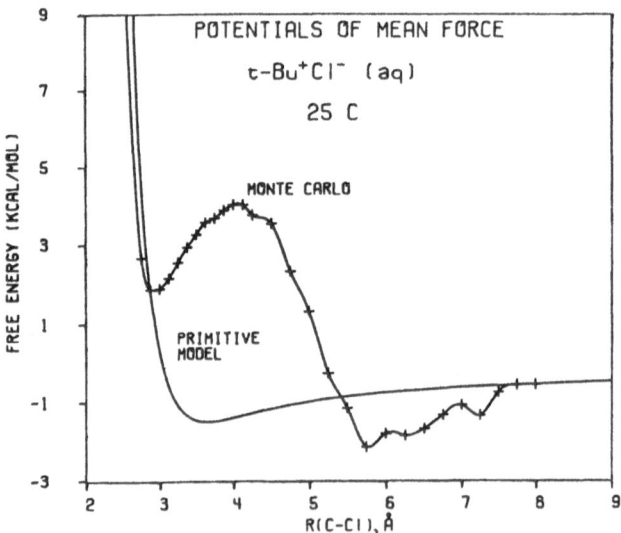

Figure 2.   Calculated potentials of mean force for separating the $(CH_3)_3C^+C\ell^-$ ion pair in water at 25°C.

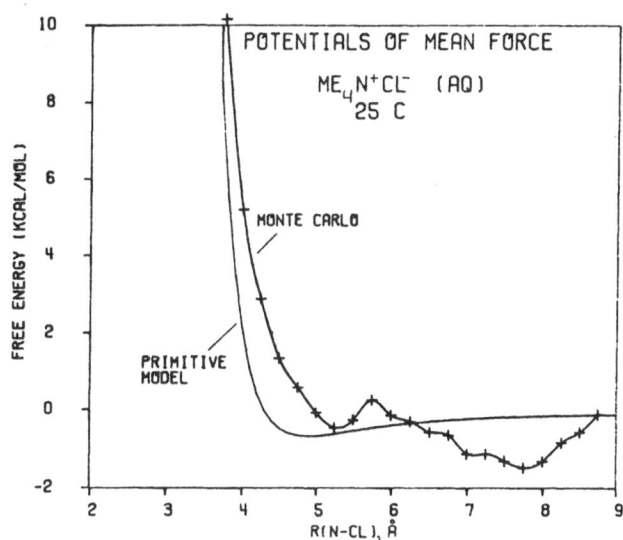

Figure 3. Calculated potentials of mean force for separating the $(CH_3)_4N^+Cl^-$ ion pair in water at 25°C.

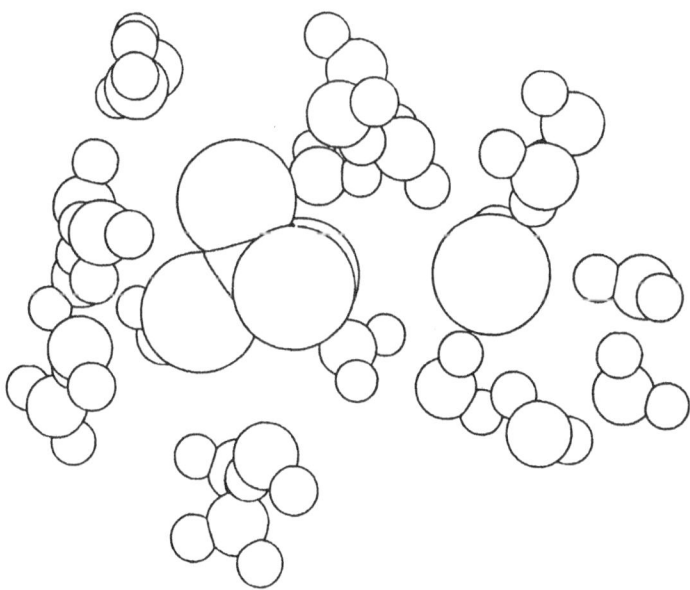

Figure 4. The $(CH_3)_4N^+Cl^-$ contact ion pair at an N-Cl separation of 5 Å. Only water molecules with oxygens within 4.5 Å of any atom in the ion pair are shown. Taken from an arbitrary configuration in the Monte Carlo simulation.

The methylammonium acetate system was studied as a simple model for salt bridges in proteins. The Monte Carlo calculations were carried out with the ion pair oriented as $CH_3NH_3^+\cdots{}^-O_2CCH_3$ and with the C-N$\cdots$C-C unit colinear, though rotation of the two ions about this axis was allowed. This approach is the lowest energy one in the absence of solvent and is consistent with the structures observed for salt bridges. The optimal interactions with a water molecule are stronger for both $CH_3NH_3^+$ (18 kcal/mol) and $CH_3CO_2^-$ (16-19) than for the other ions, $(CH_3)_3N^+$ (9), $(CH_3)_3C^+$ (16), and $C\ell^-$ (13).[31,33] The increased ion-water attractions are apparently enough to fully offset the increased ion-ion attraction, so the computed pmf continually rises from an N$\cdots$C separation of 8 Å down to 3 Å, as shown in Figure 5. Such behavior is not unreasonable for strong electrolytes, though the lack of polarization in the potential functions is always a concern. It is also not inconsistent with the thermodynamics of forming salt bridges which are estimated to only stabilize a protein by 1-2 kcal/mol.[34] The more favorable interaction in proteins follows since the protein backbone constrains the side chains and terminal ionic groups, and the salt bridges are normally formed on the protein surface where the effective dielectric constant is substantially less than for bulk water.

Other work on complexation should be noted, especially the recent computations of relative binding affinities for substrates to enzymes in water.[35,36] Statistical perturbation theory was used to compute the

$$
\begin{array}{ccccc}
E & + & S_1 & \xrightarrow{\;\Delta G_1\;} & ES_1 \\
 & & \Delta G_3 \downarrow & & \downarrow \Delta G_4 \\
E & + & S_2 & \xrightarrow{\;\Delta G_2\;} & ES_2
\end{array}
$$

free energy for mutating S_1 to S_2 and ES_1 to ES_2 which yields the relative binding affinity from the thermodynamic cycle since $\Delta G_1 - \Delta G_2 = \Delta G_3 - \Delta G_4$.

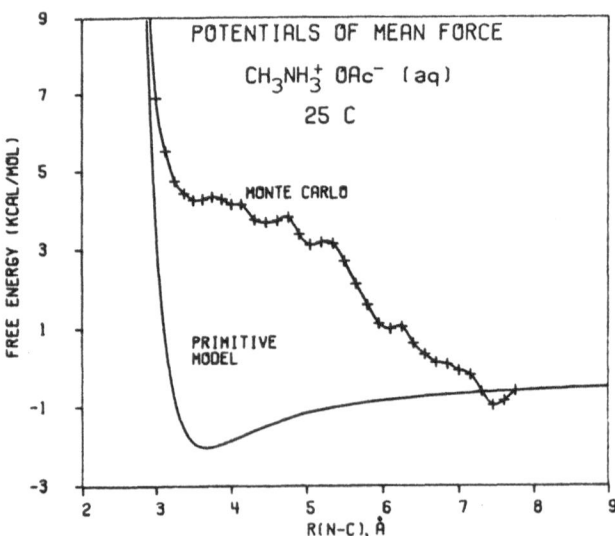

Figure 5. Calculated potentials of mean force for separating the $CH_3NH_3^+CH_3CO_2^-$ ion pair in water at 25°C.

We note here that an efficient method may be devised for computing absolute binding energies for such complexes or for anchoring the ion-pair pmfs discussed above. Consider the thermodynamic cycle below for the general complexation of A and B. ΔG_A, ΔG_B and ΔG_{AB} are the free

$$
\begin{array}{ccc}
A \ + \ B & \xrightarrow{\ \Delta G_{gas}\ } & AB \\
\Delta G_A \Big\downarrow \quad \Delta G_B \Big\downarrow & \xrightarrow{\ \Delta G_{sol}\ } & \Big\downarrow \Delta G_{AB} \\
A \ + \ B & \longrightarrow & AB
\end{array}
$$

energies of solvation of the respective species from the gas phase and ΔG_{sol} is the desired free energy of complexation in solution. ΔG_{gas} must be known from a quantum or molecular mechanics calculation. Then the free energies of solvation could be obtained by mutating A, B, and AB to nothing in solution, i.e. making them disappear.[37] Thus, eq 6 would apply. However, $0 \to AB$ can be achieved in two steps as $0 \to A$ and

$$\Delta G_{sol} = \Delta G_{gas} + \Delta G(0 \to AB) - \Delta G(0 \to A) - \Delta G(0 \to B) \qquad (6)$$

and $A \to AB$, so $\Delta G(0 \to AB) = \Delta G(0 \to A) + \Delta G(A \to AB)$. This substitution yields eq 7 which shows that only two simulations would be needed in

$$\Delta G_{sol} = \Delta G_{gas} + \Delta G(A \to AB) - \Delta G(0 \to B) \qquad (7)$$

both of which B vanishes. Clearly, choosing B to be the smaller component would be computationally desirable. The overall efficiency of the procedure in comparison to eq 6 is also evident.

CONCLUSION

Substantial progress is being made in theoretically modelling organic and biomolecular transformations in solution. The methods reviewed here in conjunction with Monte Carlo and molecular dynamics simulations are versatile and will be applied to an increasing range of problems. The high precision that is now available in computing free energy changes represents a major advance that allows meaningful quantitative predictions as well as direct comparisons between experiment and theory. The thermodynamic results are also complemented by the extreme details on structure and dynamics that are available from the simulations. This is clearly an exciting period during which the atomic-level understanding of the structure and energetics of molecular systems in solution will be greatly enhanced.

Acknowledgement

Gratitude is expressed to the National Science Foundation and National Institutes of Health for support of this research.

REFERENCES

1. "Computer Simulation of Chemical and Biomolecular Systems," D. L. Beveridge and W. L. Jorgensen, eds., Ann. N. Y. Acad. Sci., Vol. 482 (1986).
2. J. A. McCammon and S. C. Harvey, "Dynamics of Proteins and Nucleic Acids," Cambridge University Press, Cambridge (1987).
3. W. L. Jorgensen, Adv. Chem. Phys. (1987), in press.

4. (a) J. Chandrasekhar, S. F. Smith, and W. L. Jorgensen, <u>J. Am. Chem. Soc</u>. 106:3049 (1984), 107:154 (1985). (b) J. Chandrasekhar and W. L. Jorgensen, <u>ibid</u>. 107:2974 (1985).

5. J. D. Madura and W. L. Jorgensen, <u>J. Am. Chem. Soc</u>. 108:2517 (1986).

6. W. L. Jorgensen and J. K. Buckner, <u>J. Phys. Chem</u>. 90:4651 (1986).

7. M. Mezei, P. K. Mehrotra, and D. L. Beveridge, <u>J. Am. Chem. Soc</u>. 107:2239 (1985).

8. R. W. Zwanzig, <u>J. Chem. Phys</u>. 22:1420 (1954).

9. (a) G. N. Patey and J. P. Valleau, <u>J. Chem. Phys</u>. 63:2334 (1975). (b) C. S. Pangali, M. Rao, and B. J. Berne, <u>ibid</u>. 71:2975 (1979).

10. W. L. Jorgensen, J. Chandrasekhar, J. D. Madura, R. W. Impey, and M. L. Klein, <u>J. Chem. Phys</u>. 79:926 (1983). W. L. Jorgensen and J. D. Madura, <u>Molec. Phys</u>. 56:1381 (1985).

11. W. L. Jorgensen, J. D. Madura, and C. J. Swenson, <u>J. Am. Chem. Soc</u>. 106:6638 (1984).

12. W. L. Jorgensen, <u>J. Phys. Chem</u>. 90:1276 (1986).

13. W. L. Jorgensen and M. Ibrahim, <u>J. Am. Chem. Soc</u>. 103:3976 (1981). J. M. Briggs and W. L. Jorgensen, to be published.

14. W. L. Jorgensen, <u>J. Phys. Chem</u>. 90:6379 (1986).

15. For a review, see: M. Levitt, <u>Annu. Rev. Biophys. Bioeng</u>. 11:251 (1982).

16. S. J. Weiner, P. A. Kollman, D. A. Case, U. C. Singh, C. Ghio, G. Alagona, S. Profeta, and P. Weiner, <u>J. Am. Chem. Soc</u>. 106:765 (1984).

17. W. L. Jorgensen and C. J. Swenson, <u>J. Am. Chem. Soc</u>. 107:569, 1489 (1985). W. L. Jorgensen and J. Tirado-Rives, to be published.

18. D. J. Adams, <u>Chem. Phys. Lett</u> 62:329 (1979).

19. For a review, see: W. L. Jorgensen, <u>J. Phys. Chem</u>. 87:5304 (1983).

20. D. W. Rebertus, B. J. Berne, and D. Chandler, <u>J. Chem. Phys</u>. 70:3395 (1979).

21. W. L. Jorgensen, <u>J. Chem. Phys</u>. 77:5757 (1982).

22. J. C. Owicki, <u>ACS Symp. Ser</u>. 86:159 (1978).

23. W. L. Jorgensen, J. Gao, and C. Ravimohan, <u>J. Phys. Chem</u>. 89:3470 (1985).

24. L. R. Pratt and D. Chandler, <u>J. Chem. Phys</u>. 67:3683 (1977).

25. R. O. Rosenberg, R. Mikkilineni, and B. J. Berne, <u>J. Am. Chem. Soc</u>. 104:7647 (1982).

26. A. Ben-Naim, "Hydrophobic Interactions," Wiley, New York (1980).

27. W. L. Jorgensen and J. K. Buckner, <u>J. Phys. Chem</u>., in press.

28. W. L. Jorgensen and J. Gao, <u>J. Am. Chem. Soc</u>., submitted for publication.

29. G. Ravishanker, M. Mezei, and D. L. Beveridge, <u>Faraday Symp. Chem. Soc</u>. 17:79 (1982). G. Ravishanker and D. L. Beveridge, <u>J. Am. Chem. Soc</u>. 107:2565 (1985).

30. M. Berkowitz, O. A. Karim, J. A. McCammon, and P. J. Rossky, <u>Chem. Phys. Lett</u>. 105:577 (1984).

31. W. L. Jorgensen, J. K. Buckner, S. E. Huston, and P. J. Rossky, <u>J. Am. Chem. Soc</u>. 109:1891 (1987).

32. W. L. Jorgensen, J. K. Buckner, J. Gao, and J. Tirado-Rives, to be published.

33. W. L. Jorgensen and J. Gao, <u>J. Phys. Chem</u>. 90:2174 (1986).

34. M. F. Perutz, <u>Nature</u> 228:726 (1970).

35. C. F. Wong and J. A. McCammon, <u>J. Am. Chem. Soc</u> 108:3830 (1986).

36. P. A. Bash, U. C. Singh, R. Langridge, and P. A. Kollman, <u>Science</u> 236:564 (1987).

37. T. P. Straatsma, H.J.C. Berendsen, and J.P.M. Postma, <u>J. Chem. Phys</u>. 85:6720 (1986).

DISCUSSION

BRATOS - How far have you to go in carrying out a-priori calculations in order to get useful pair potentials ?

JORGENSEN - It depends on the system. First-row anions require the largest basis sets such as 6-31+G(d) which importantly includes a set of diffuse S and P orbitals on each nonhydrogen atom. Positive ions are not so demanding ; the diffuse functions are not needed in this case, though polarization functions are still desirable, e.g. 6-31 G(d). For both the cation-molecule and anion molecule complexes, correlation corrections are not substantial due to the dominance of the electrostatic interactions.

HYNES - I have two questions, first, can you comment on the simulation of the t-BuCl reaction in the transition state, rather than the ion pair, region ? Second, can you comment on the real importance in an S_N2 reaction of a weak "pre-complex", when the reaction rate is dominated by the large central barrier.

JORGENSEN - 1) Simulation of the step from t-BuCl to the contact ion pair is difficult because it corresponds to an electronic curve crossing in the gas phase. The ground state in the gas phase dissociates to t-Bu˙ + Cl˙, while a highly excited state correlates with the ion pair. The latter surface is, of course, preferentially stabilized by hydration. To model the initial step would consequently require results for both gas-phase surfaces and potential functions for water interacting with the substrate at all points along those surfaces.

2) The kinetics are still bimolecular due to the dominance of the second step. However, the notion of non-concerted S_N2 reactions in solution is novel and $k_{obs} = Kk_z$ where K is the equilibrium constant for the complexation step which emphasizes the importance K and k_z in interpreting the kinetics.

KLEIN - You utilize the primitive model and experimental bulk water dielectric constant to normalize your potential of mean force calculations. Would it not be more natural to use the value of the dielectric constant appropriate to your water model (TIP4P) ? How could this affect your results ?

JORGENSEN - There has been only reported calculation of the dielectric constant of TIP4P water with a result of about 55. Since the experimental value (79) and this result are both relatively large, the base electrostatics are still well-damped, i.e. the well-depth for a representative contact ion pair would only change from ca. 1.5 to 2.0 Kcal/mol.

MOLECULAR DYNAMICS SIMULATIONS AS A TOOL FOR THE STUDY OF
CHEMICAL PHYSICS AND CHEMICAL KINETICS : THE KINETICS OF
ADSORPTION AT A LIQUID-LIQUID INTERFACE

Michel Mareschal

Université Libre de Bruxelles
Faculté des Sciences, C.P. 231
Boulevard du Triomphe
B-1050 Brussels, Belgium

1. INTRODUCTION

The theme of this conference concerns chemical reactivity both at equilibrium and in non equilibrium situations. Although, at present, molecular dynamics studies of chemical kinetics only deals with equilibrium systems, during the recent years there has been a development of simulations in non equilibrium fluids due to the large increase of computing capacity offered by supercomputers /1,2/.

As an example, I will briefly report on a simulation of an hydrodynamical instability that has been done by non equilibrium molecular dynamics /3/. The system considered consisted of an assembly of more than 5000 hard disks enclosed in a rectangle. The vertical sides are specularly reflecting boundaries, whereas the horizontal sides, approximately 3 times longer, were thermal reservoirs /4/. An external force is acting on the fluid particles directed downwards, and the bottom reservoir's temperature was set at a higher value than the top reservoir's one. Density, temperature difference and external force's magnitude were chosen so as to have a constant density in the system and a Rayleigh number larger than the critical number obtained from a hydrodynamical stability analysis for the same system. The model fluid was then integrated in time on a computer, using standard algorithms of hard-core systems. A typical velocity field of the system is shown on figure 1 : each arrow of the graph is an average over a cell of the velocities of the particles that belong to the cell (typically 5 particles per cell). Besides, one performs also a time average, and the figure shown corresponds to an average over more than ten millions of particles collisions; this represents a time interval of 1000 picoseconds.

It is quite remarquable that a relatively small system exhibits a behaviour which can be understood in terms of macroscopic hydrodynamics. However, it should be stressed that, compared to "usual" equilibrium molecular dynamics, the system described here is an order of magnitude larger and has

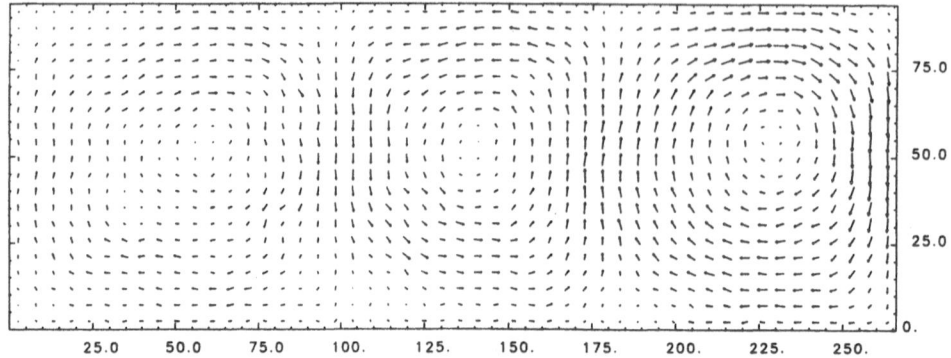

Fig. 1. Velocity field for a thermally driven convective
instability by molecular dynamics (reference 1).

been integrated over a time which is also an order of magnitude longer. We may conjecture that, not very far from now,
the time will come when such simulations will become a lot
easier (and cheaper!) to perform and that non equilibrium
steady states, whith spatial pattern formation or time oscillatory behaviour, will also be studied by atomistic simulations /1/.

We shall present here an equilibrium simulation of the
transport of a solute across a liquid-liquid interface, which
permits to measure the rate constant. This work has been
done with the same rationale than other recent molecular
dynamics studies of chemical kinetics /5,6/. The idea is to
obtain by simulation, at the same time, a computation of the
mean potential as a function of the reaction coordinate and a
direct measure of the rate constant. The mean potential can
then be used as an input for a theoretical expression of the
rate constant, using transition state /7/, Kramers /8/ or
Grote-Hynes /9/ theories for instance. The comparaison can
then be done in order to give a correct description of the
kinetics process. A distinct feature of molecular dynamics,
with respect to an experimental testing of theoretical
results, is that the numerical simulations have both aspects,
theoretical and experimental. Indeed, the computation of
mean potentials, as functions of the microscopic models used,
is simple to obtain here whereas an analytical derivation
would be a heavy task. On the other hand, the computation of
the kinetics constant is more comparable to an experimental
output.

The physico-chemical process that has been simulated is
that of the approach of an amphiphilic molecule towards an
interface where it is adsorbed. The microscopic model is
simple and, although not very realistic, it contains, we
believe, the essential ingredients that make the problem
highly non trivial. It is quite obvious from the model
chosen that a quantitative comparaison with experimental
results is out of range of the present study. However

complex and, sometimes, unexpected behavior already appears; the understanding of such simplified models should be very helpfull for the elaboration and analysis of more realistic models.

We shall first state the problem in the next section. Then we shall describe the model used and, in particular, the necessary details of the microscopic structure of the liquid liquid interface. The last part of this report will be devoted to the analysis of the adsorption's trajectories as given by the numerical simulation and to a discussion of the results obtained. Finally we end by giving some perspectives in the study of this problem.

2. Kinetics of the transport across an interface

The problem that we want to study is that of the kinetics of passage for a solute S from the bulk of a liquid phase A to the bulk of a liquid phase B, separated from phase A by a stable interface, I (see figure 2). The passage will involve different steps : first the solute, S, will have to approach the boundary between the two phases. That approach will be diffusive and, once the diffusion coefficient of S in A (or B) is known, that fixes the kinetics of approach. Then, the next step will be that of crossing of the interface itself, and depending on the mean potential that the solute will encounter, it will have to overcome a barrier or to be absorbed in a well and to escape from that well to the bulk phase B (see figure 3); both situations arise in the transport of solute across a liquid-liquid interface.

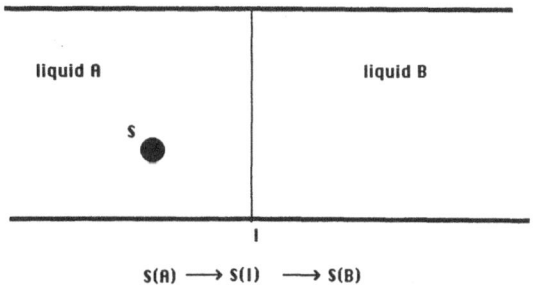

$$S(A) \longrightarrow S(I) \longrightarrow S(B)$$

Fig. 2. Decomposition of the passage of S from A to B in intermediate steps.

In both cases, the crossing of the interface itself becomes rate-determinant. For instance, in the case of a well, neglecting the direct passages from bulk A to bulk B with no adsorption - and that will be reasonable if the well depth is a few k_BT - the time to go from A to B will be equal to the time to go from A to I and then from I to B :

$$k^{-1} = k_{AI}^{-1} + k_{IB}^{-1} \tag{1}$$

where k, k_{AI}, k_{IB} are the rate constants for the interface crossing, the adsorption and escape respectively. The definition of k reads

$$k = \langle j_R^+ (0) \, \mathbb{P} \rangle \tag{2}$$

where $j_R^+ (0)$ is the flux of solute particles crossing the boundary denoted R in figure 3b towards the well, and $\mathbb{P}$ is zero or one depending on whether or not the solute trajectory has ended at the boundary P /5/.

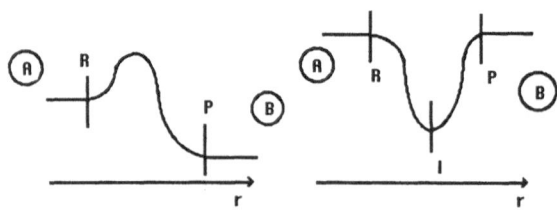

Fig. 3.   Mean potential profile in the case of a barrier (a) and of a well (b).

This problem is typical of chemical kinetics problems : the time scale of interest for the "chemical reaction" to occur is much larger than that of solved relaxation. This separation of time scales is a difficulty that has to be faced in order to perform simulations of the processes : for instance, in the case of a barrier crossing, one follows in time the trajectories of systems that are started from the barrier's top and study their subsequent evolution /6/. In the case of a well, the fall of the solute towards the interface can easily be followed in time by starting the system at R and integrate it forwards in time : that permits to measure k_{AI}. In order to measure k_{IB}, however, it is not necessary to wait for the unfrequent escape of the solute from I to P, but rather study the fall of the solute from P to I, measuring thus k_{BI}, and use then detailed balance in order to obtain,

$$k_{IB} = k_{BI} \exp -\Delta G / k_B T \tag{3}$$

where, of course, one needs to compute the free energy difference, ΔG, while moving the coordinate r from I to P.

In this discussion it has been tacitely assumed that the reaction coordinate could be simply defined as the distance of the solute from the interface. It should be stressed however that this definition may be, in particular cases, much more complex and may require more specific elaboration.

3. The interface model

The model chosen for the liquid solvents is an immiscible mixture of two atomic liquids whose intra- and interspecies interaction potential energies are different. The A and B liquids are modelled by point particles interacting through a Lennard-Jones potential :

$$V_{AA}(r) = V_{BB}(r) = (4\varepsilon) \cdot \left[\left(\frac{\sigma}{r}\right)^{12} - \left(\frac{\sigma}{r}\right)^{6} \right] \qquad (4)$$

whereas the interspecies interaction is modified by the introduction of two parameters α, β :

$$V_{AB}(r) = (4\varepsilon) \cdot \beta \left[\left(\frac{\sigma}{r}\right)^{12} - \alpha \cdot \left(\frac{\sigma}{r}\right)^{6} \right] \qquad (5)$$

A description of the physico chemical properties of the interface as a function of these α and β parameters can be found in reference 10. In the simulations done for the adsorption of the solute, β was set to 1 and α to 0,3. In that case, the A-B interaction has a very small attraction well and at temperatures and densities of the liquid phase, just above the triple point, the two species separate.

The system consisted of 1728 atoms enclosed in a parallelipiped of size (L_x, L_y, L_z), with $L_x=L_y=L$ and $L_z=2 L$. The z axis was chosen to be perpendicular to the plane of the interface, whereas x and y plane was parallel to it. We used periodic boundary conditions, so that, in fact, the system simulated contained 2 interfaces, one at z=0 and at z=±L. The shortest hydrodynamical time scale for the system is the time for a sound wave to cross the system, that is the time needed for mechanical equilibrium. This is approximately 2000 time steps of integration where, as usual for such models, the time step is one hundredth of a picosecond. The largest time scale, on the other hand, is the time needed for diffusion over a distance L, $(D_{AB}/L^2)^{-1}$, and this is approximately 500.000 time steps. This last time is out of range for our computer studies as in one hour of CPU time on a FUJITSU VP-200, on which the calculations were done, the program could integrate 40.000 time steps. However the parameters chosen for our simulation were such that the miscibility of A into B (or B into A) was smaller than the one corresponding to one atom A into liquid B.

The interface itself has been characterized by three different profiles.

- the density profile, shown on figure 4, has been obtained by a time average over more than 20.000 time steps. The dentity is constant in the bulk and it varies over a distance of thickness equal to 3σ, around z = 0. Note that there is a decrease of the total density at z = 0, decrease which is quite large, more than 30%. This decrease very much depends on the values of the (α, β) parameters in the A-B potential, equation (5). For other (α, β) values, one had an increase of the total density, or even constant density across z = 0.

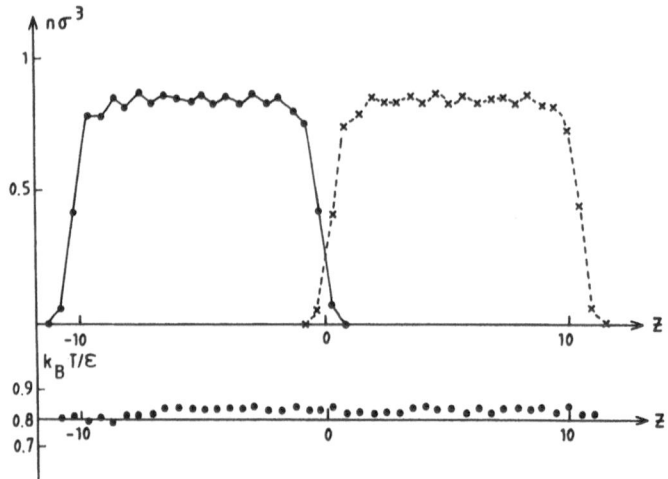

Fig. 4.   Density and temperature profiles for the steady state of the interface simulation (T=99°K, $n\sigma^3$ = 0,84 in the bulk; time average : 20.000 time steps).

- the pressure profile is in fact anisotropic /12,13/. Indeed mechanical equilibrium only requires the normal component of the pressure P_{zz}, to be constant and, in particular, independant of z.   The transverse component of the pressure, on the other hand, decreases when approaching the interface, and the interfacial tension is given by /13/

$$\gamma = \int_{-\infty}^{+\infty} dz \left[P_N(z) - P_T(z) \right] \qquad (6)$$

whith $P_T = 1/2(P_{xx} + P_{yy})$.   We have measured the pressure profile, using the definition for the local pressure tensor components given by Irving and Kirkwood /14/.   The measured value of $P_T(z)$ fluctuates much more than that of $P_N(z)$. However, at T = 99°K, one obtains, using equation (6), an interfacial tension equal to 33 (±2) ergs/cm^2 which is quite comparable to experimental values of water-benzene at 20°C for instance.   The thickness of the interface is much broader than the one obtained from the density profile, of the order of 5σ.   An interesting consequence of this variation of the pressure, is the anisotropy in the diffusion constants.   We have measured $D_\perp$ and $D_{//}$, respectively the self diffusion constant in the direction perpendicular and parallel to the interface.   Whereas $D_\perp$ remains more or less constant as function of z, $D_{//}$ increases very strongly, doubling its value for z = 0.   This can be qualitatively understood; the potential part of the transverse pressure tensor drops to nearly zero at z = 0, so that diffusion in that direction is increased to that of a "perfect gas" (the "Boltzmann" diffusion coefficient is inversely proportionnal to the pressure).
- a geometric construction of the interface has been done, and this has been used in order to characterize the fluctuations of the interface.   Indeed, as figure 5 shows, the instantaneous location of the interface is very different

from the values given by the time averaged density and pressure profiles. The surface construction, is done by computing the heights z(i,j), of grid points (i,j) in the x,y plane that are between phase A and B : the fluctuations in time

$$\langle z(i,j)^2 \rangle - \langle z(i,j) \rangle^2 \; \approx \; \sigma^2 \qquad (7)$$

at a fixed (i,j) value uniform us that the instantaneous location of the interface can vary from $-\sigma$ to $+\sigma$ during a simulation. The typical time scale of this variation is 10 picoseconds as can be measured from the decay of the correlation function, $\langle \delta z(i,j;t) . \delta z(i,j;o) \rangle$. The time behavior of this function is typically a rapid exponential decay over less than a picosecond, followed by a slow decay over a longer time scale; after 10 picoseconds the signal is still 7-8% of its initial value.

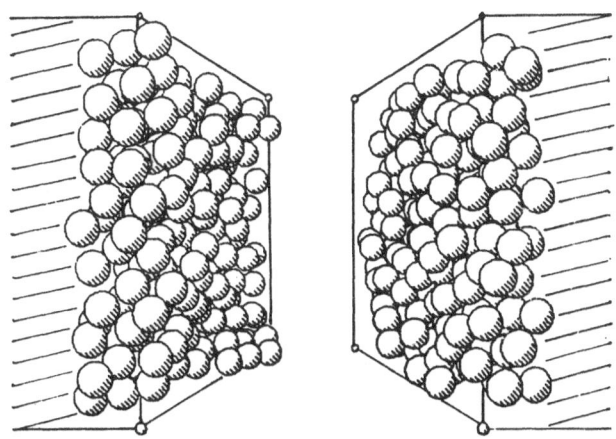

Fig. 5. Atoms A and B have been pushed apart and are shown here as spheres of radius σ .

4. The solute model

We have studied the case of the adsorption of an amphiphilic molecule, that is a molecule that has a preferable orientation at the interface (a surfactant or a combination of surfactants) /15/. The dependance of the interfacial tension on the concentration of amphiphilic has been the subject of extensive experimental studies, due to its practical importance. However, theoretical studies are more rare, since the systems of interest are too complicated for a full theoretical treatment.

Our model is very simple and, so, allows for an extensive study : it consists of a diatomic molecule, made of an atom A linked rigidly to an atom B which is at a fixed distance, σ . The number of degrees of freedom of that molecule is 5; however due to the symetry of the problem, only 2 are relevant for us, namely the z component of the solute's center of mass position and its orientation with respect to the z axis, θ (see figure 6).

We have generated by molecular dynamics a serie of trajectories of the fall of the solute towards the interface.
The procedure is the following : starting from a configuration of a equilibrated A-B mixture, we select a pair in
liquid A (B) of atoms whose relative distance is σ to within 5%, and whose center of mass is between -2σ and
$-2,5\sigma$. We then force the relative distance to be σ and
change the nature of one atom of the pair from A to B, in 100
time steps. We then follow the motion of the solute in time.
We can use different equilibrium configurations in order to
generate as many trajectories as we want. Typical trajectories are shown on figure 7.

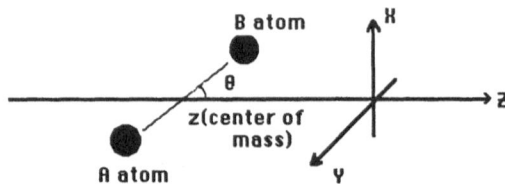

Fig. 6. Definition for z and θ variables for the amphi-
         philic solute.

One can see that, after some variable time during which
the solute is diffusing, there may be a sudden acceleration
towards the interface. The time to fall is not very large, a
few picoseconds, compared to the duration of the trajectory.
The dispersion of the times before which there is a fall, is
very large. However there seems to be a strong correlation
between the orientation of the amphiphilic molecule and the
acceleration. Each fall starts when the θ orientational angle has decreased to a value lower than $\pi/2$, that is an orientation where the B atom of the solute is nearer of the B phase than the A atom. We have plotted (on figure 8) the average velocity, $\bar{v}(z)$, as a function of the distance from the
interface, obtained by an average over all trajectories.
    A very rough theoretical model for this system can be
developped, starting from the following Langevin equation :

$$m_s \frac{d v_s}{dt} = -\gamma v_s - \frac{\partial V(z)}{\partial z} + F(t) \qquad (8)$$

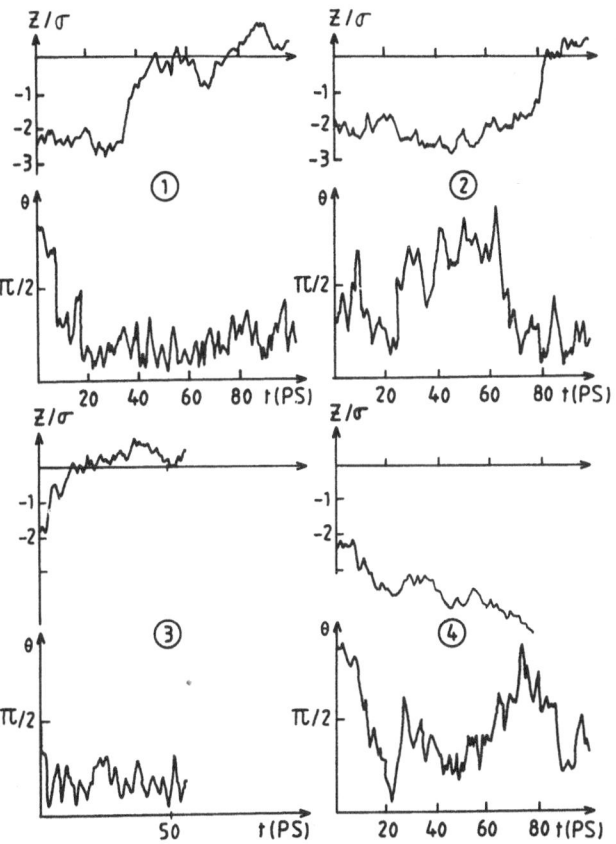

Fig. 7. z and θ as function of time for different trajectories.

where m_s is the mass of the solute molecule, γ its friction
coefficient, v_s its z component velocity. $V(z)$ is the mean
potential and $F(t)$ the random force. Applying this equation
to the values of the velocity plotted on figure 8, one reali-
ses that the system is overdamped and that equation (8) redu-
ces to a balance between the mean force, $-\partial V/\partial z$ and the friction acceleration γv_s. One may then use this in order to get
an rough estimate of the mean potential : this is shown on
figure 8b where we have plotted the potential obtained by
integrating equation (8) with respect to the z variable from
the bulk (z = -2,5σ) to the interface (z = 0). The friction
coefficient γ being obtained independently by a measure of
the mean square displacement of the solute in the bulk as a
function of time. The well depth obtained in this way is
then 8,5 $k_B T$.

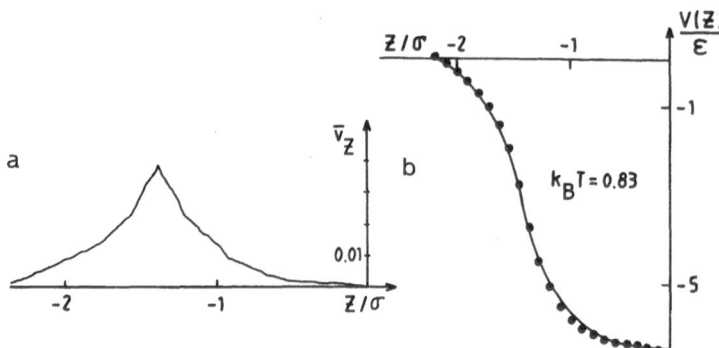

Fig. 8.   (a) profile of v obtained by an average over
             57 trajectories.
          (b) corresponding mean potential; in the energy
             units used, $k_B T$ = 0,85.

The computation of the transmission coefficient, κ ,
defined as the ratio

$$K = k/k^{TST} \qquad (9)$$

where k^{TST} is the transition state theory rate constant,
gives a result equal to 0,15 if we fix the reactant boundary
to be z = -2,5 σ (whatever θ). These results were obtained
finally by an average over 57 trajectories, out of which 18
went back to the bulk of phase A, and the remaining 39 fell
in the interface.

It is worth noting that the transmission coefficient was computed from the trajectories data in the following way; the slow diffusive motion of the solute is used to generate equilibrium configurations of solvent at a given distance. Fixing then the R boundary to be a plane at $z = -2,5 \sigma$ (whatever the θ angle), we then select from all the configurations that we have those which correspond to the crossing of the R boundary with a positive velocity ($\dot{z} > 0$), and we compute the transmission coefficient by determining which are the configurations which lead to a recrossing in the future.

Of course there is an ambiguity in fixing the R boundary at $z = -2,5 \sigma$. This value was chosen because the mean potential, obtained from the velocity profile, becomes flat at that distance from the interface (see figure 8). We could have fixed however the boundary deeper in the bulk and obtained a lower value for the transmission coefficient, due to greater diffusive behavior. This cannot be avoided and the absence of the exact location of the R boundary introduces some uncertainty in the result.

Another aspect which comes out of the analysis of the trajectories is that there is a strong dependance on the angle for the reaction to proceed. The z dependent mean potential is an average over all θ values sampled from the trajectories. When we look at the solute's behavior as a function the angle, we have seen that the solute would first fix itself around a favorable orientation (small θ angle) before it would fall down in the well. A direct measure of the mean potential as a function of z and θ seems to be necessary, and further work in this direction is currently in progress.

5. Conclusions

The equilibrium molecular dynamic simulations done have permitted to obtain usefull informations on physico-chemical properties at a liquid liquid interface. In particular, the anisotropy of the diffusion near the interface and its relation with the anisotropy of the pressure tensor is an original and interesting result.

The kinetics of the approach of the amphiphilic molecule towards the interface has been analyzed on the basis of a rough theoretical description, with a one dimensional reaction coordinate. This has permitted to obtain an estimate of the mean field in this case, where the motion is overdamped.

A deeper analysis is however necessary as the initial angle dependance in the set of trajectories will probably force us to consider a two dimensional reaction coordinate. Work in this direction is presently in progress.

This work may be considered as a necessary first step in the study of transport across and adsorption to liquid liquid interfaces. We believe that modifications of the original model may be done in order to attack different problems such as the dependance of interfacial tension on amphiphilic concentration or the kinetics of more complex molecules.

6. Acknowledgements

This work has been started during a workshop organized by the C.E.C.A.M. in Orsay and it has benefited from the financial support of C.E.C.A.M. I acknowledge the hospitality of the "Laboratoire de Physique des Matériaux", CNRS, Meudon. Many of the results reported here have been obtained in collaboration with Madeleine Meyer : I thank her for many enlightening discussions on the physical aspects of the problem and for her patience in clearing all the computational details. I am gratefull to all the participants to the CECAM workshops, and in particular G. Ciccotti and J.T. Hynes for many usefull discussions, suggestions and comments.

REFERENCES

1. D.J. Evans and W.G. Hoover, Flows far from equilibrium, Ann. Rev. Fluid Mech. $\underline{18}$, 243-264 (1986).
2. J.A. Blink, W.G. Hoover, Phys. Rev. $\underline{A32}$, 1027 (1985).
3. M. Mareschal, E. Kestemont, J. Stat. Phys. (1987).
4. A. Tenenbaum, G. Ciccotti and R. Gallico, Phys. Rev. $\underline{A25}$, 2778 (1982).
6. J.T. Hynes, The Theory of Chemical Reactions, M. Baer ed. (CRC Press, Boca Raton, Florida 1985).
a. J.T. Hynes, J. Stat. Phys. $\underline{42}$, 149 (1986).
b. similar problems are discussed in the contributions by J.T. Hynes and K.R. Wilson to these proceedings.
7. G. Ciccotti, in these proceedings.
    S. Glasstone, K. Laidler and H. Eyring, The theory of rate Processes (Mac Graw-Hill, New York, 1941)
8. H.A. Kramers, Physica $\underline{7}$, 284 (1940).
9. R.F. Grote, J.T. Hynes, J. Chem. Phys. $\underline{74}$, 4465 (1981).
10. M. Meyer, M. Mareschal (to be published).
11. Schoen M., Hoheisel C., Molec. Phys. $\underline{53}$, 1367 (1984).
12. R. Aveyard, B. Vincent, Progress in Surface Science, Vol. 8, 59, (Pergamon Press, 1977).
13. J.S. Rowlinson, B. Widom, Molecular theory of Capillarity, (Clarendom Press, Oxford, 1982).
14. J.H. Irving and J.G. Kirkwood, J. Chem. Phys. $\underline{18}$, 817 (1950).
15. M.M. Telo Da Gama, K.E. Gubbins, Molec. Phys. $\underline{59}$, 227 (1986).

KAPRAL - What is the nature of the microscopic boundary condition used in the nonequilibrium molecular dynamics simulation of the Bénard problem in order to obtain the temperature gradient ? How far into the fluid does the temperature boundary layer extend ?

MARESCHAL - In the Bénard problem, the thermal boundaries are simulated along the ways developped in non-equilibrium molecular dynamics, using stochastic boundary conditions (see G. Ciccotti). The boundary layer does not extend over more than a mean free path in the system and can hardly be seen in our measurements.

ERDI - A molecular dynamic simulation experiment leads to one realization of a stochastic process. Did you try to construct experimental probability distributions in order to get deeper insight about the qualitative behaviour of the system.

MARESCHAL - For the computation of the kinetic constant and of the transmission coefficients, one has to analyze a set of trajectories all generated from equilibrium configurations arising naturally from the dynamics of the system. In our particular case, 50 trajectories have been used and we believe that this number is quite sufficient for qualitative understanding.

MICHEAU - My questions are related to the near to equilibrium part of your talk.
The first one is : you have two immiscible phases A and B. Do you take the separation of the two phases as an hypothesis maintained all along your calculation or is it contained in the intrinsic properties of A and B species ?
My second one is : are there simulations which would have been already done to modelize a phase separation process starting with an A+B homogeneous mixture as an initial condition.

MARESCHAL - The immiscibility of the two fluids is a consequence of the microscopic model chosen, and not an additional input.
Models, similar to ours, have been described in the literature and, in particular, a molecular dynamics simulation of the demixtion of an initially homogeneous mixture of, say, A and B species, has been reported.

MOLECULAR DYNAMICS SIMULATION OF A LIQUID-LIQUID INTERFACE

M.Hayoun[1], M.Meyer[2], M.Mareschal[3], G.Ciccotti[4], and P.Turq[5]

[1]CEN Saclay, D.TECH, 91191 Gif/Yvette Cedex, France
[2]LPM-CNRS, 1, place A.Briand, 92195 Meudon, Cedex, France
[3]Université Libre de Bruxelles, Faculté des Sciences, CP231, B1050 Bruxelles, Belgique
[4]Universita "La Sapienza", Dip. di Fisica, ple Aldo Moro 2, 00185 Roma, Italia
[5]Université Pierre et Marie Curie, Lab. Electrochimie, 8, rue Cuvier, 75005, Paris, France

I.INTRODUCTION

The knowledge of the structural and dynamical properties of the Liquid-Liquid Interface (LLI) may be of great help to have a better understanding of some chemical processes such as solute transfer or interfacial adsorption. The LLI have been less systematically studied than the liquid-vapor interfaces from the experimental point of view[1]. The situation is different for the theoretical part, since the similarity of the two types of interfaces result in a common description for both of them[2]. Some recent theoretical studies, dealing specifically with the structure of the LLI, provided interesting results on interfacial thickness and tension[3-6]. Microscopic theoretical models, based on statistical theories of inhomogeneous fluids, require assumptions which are difficult to control. Moreover they cannot give a detailed description of the interface at the atomic scale. This is the reason why Molecular Dynamics (MD) and Monte Carlo (MC) simulations have been developped to model interfaces. Most of these computations deal with liquid-vapor systems[7,8], while only one liquid-liquid system has been investigated using MC technique to study a Benzene-Water interface[9].

The main objective of this work is to show that it is possible to model a LLI, using Lennard-Jones (LJ) potential and MD simulation technique. The idea is to simulate not a realistic system but a "simple" model suitable for the study of the generic properties of an interface between non miscible liquids. To do that, we have chosen the MD simulation technique and periodic boundary conditions, for a system of particles interacting via a LJ potential already used for unstable mixtures[10,11]. The results show that the LLI thus obtained is stable over the simulation time scale, as indicated by the density profiles. It is also interesting to note that the interfacial tension yielded by this model is in the range of the experimental values. The model and some computational details are described in section II. The results are reported in the following part and discussed in terms of stability and spatial extension of the LLI. The paper ends with some concluding remarks.

MD simulations have been performed in NVE ensemble on a system containing 1728 particles. The equations of motion have been integrated using Verlet algorithm[12] with a time step of 10^{-14} s. The three dimensions L_X, L_Y and L_Z, of the computational cell, are respectively equal to 37, 37 and 69 Å. The origin of the reference frame is at the centre of the simulated system and the three X, Y, Z axes are parallel to the edges of the box, as indicated in Fig. 1. The initial configuration of the LLI is obtained as follow : two subsystems, composed of 864 particles, are equilibrated separately, they contain Liquid 1 (L1) and Liquid 2 (L2). They are joined to form the computational cell (Fig. 1) with the XY plane parallel to the initial position of the interface. The normal periodic boundary conditions are applied to the simulated system. If a stable interface is to be obtained the only effect of the X and Y periodicity is to keep its average position in the initial orientation. In the Z direction the periodic boundary conditions generate a second interface in the system (Fig. 1). They have been preferred to walls because, in these conditions, it is reasonable to expect a less extended perturbation. To infer this previous point, we assume that the extension of the interfacial region is smaller in liquid-liquid system than in a liquid-gas, where we know that the size of this region is comparable to the range of perturbation introduced in a liquid by a wall[13]. The knowledge of this range is also useful to choose the size of the box. In the z direction, the length of the cell is large enough to take into account these perturbation distances.

Since our purpose is the modelling of a LLI, the potential functions, describing the interactions between the particles of the two liquids, must follow some requirements. The main property we want to obtain is the non miscibility of the two liquids. We have chosen a potential function suitable for MD simulation of liquids, namely a pairwise Lennard-Jones one :

$$u(r) = 4 \, \varepsilon_{\alpha\beta} \, [(\frac{\sigma}{r})^{12} - (\frac{\sigma}{r})^6] \qquad (1)$$

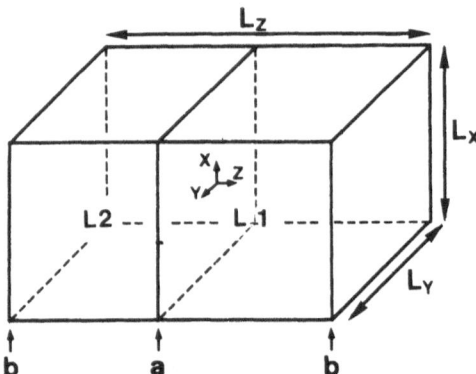

Fig. 1 - Shematic drawing of the MD simulation cell. Two adjacent boxes containing respectively L1 and L2 presented in their initial configuration. The two interfaces a and b are indicated by the arrows, b has been generated by periodic boundary conditions. $L_X = L_Y = 37$ Å, $L_Z = 69$ Å. The origin of the reference frame is at the centre of the computational cell.

Table I. Lennard-Jones parameters for the system
used to simulate the LLI.

ε_{11}/k_B	ε_{22}/k_B	ε_{12}/k_B (K)	σ (Å)
100	200	75	3.591

k_B is the Boltzmann constant.

The cutoff radius equal to 2.5σ. $\alpha=1,2$ and $\beta=1,2$ refer respectively to the atoms of L1 and L2. In order to achieve the non miscibility of L1 and L2 the Berthelot rule is not followed and ε_{12} differs from $(\varepsilon_{11} \varepsilon_{22})^{1/2}$. The potential parameters $\varepsilon_{\alpha\beta}$ and σ, selected for this simulation of a LLI, are taken from the study of Shoen and Hoheisel on unstable liquid mixtures[10]. Their values are listed in Table I, the atomic masses are the same for the two liquids $m=1.02747 \ 10^{-22}$g.

The preparation of the system is made in two steps. In the first one, two boxes with X and Y dimensions equal to L_X and L_Y are equilibrated separately. The density of one of the liquids is varied in order to get the same temperature and the same pressure in the two systems. A description of these systems is given in Table II, together with the density ρ, the temperature T and the pressure P. In the second step, they are placed side by side with the initial position of the interface in the XY plane. This procedure requires some care, since at the beginning of the equilibration of the whole system, the density of the interfacial regions is far from equilibrum. The distance between two particles of L1 and L2 may be very short, therefore, to avoid an important local heating, we arbitrarily reduce the potential interaction of such atomic pairs over a few time steps. The equilibration is then performed over 10,000 time steps, without any modification of the potential. The total length of the runs is 50,000 time steps. The first 30,000 are devoted to the interface stability study and the average values are calculated over the last 20,000. Table III contains some details relative to the model and some results of the LLI simulation.

III. RESULTS AND DISCUSSION

The spatial extension and the stability of the interface have been characterized through the calculation of the density and pressure profiles, which have been measured as a function of Z. This investigation has been completed by the computation of the interfacial tension γ. The density profiles $\rho_1(Z)$ for L1 and $\rho_2(Z)$ for L2 have been calculated in

Table II. MD results for liquid 1 (L1) and liquid 2 (L2)
equilibrated separately.

	$L_X=L_Y$ (Å)	L_Z (Å)	N	ρ (10^{23} cm^{-3})	T (K)	P (Kbar)
L1	37.0	37.0	864	0.170	133.5	0.70
L2	37.0	32.0	864	0.197	134.2	0.71

The values P and T are averaged over 4,000 time steps.

Table III. MD results for the liquid-liquid interface (LLI).

$L_X = L_Y$ (Å)	L_Z (Å)	N	ρ_1 (10^{23} cm^{-3})	ρ_2 (10^{23} cm^{-3})	T(K)	P(Kbar)	P$^+$(Kbar)
37.0	69.0	1728	0.175	0.194	135.2	0.72	1.08

$^+$ without long range correction. ρ_1 and ρ_2 reported here correspond to plateau values measured on the density profiles plotted in Fig. 2.

equally thick (ΔZ) slabs parallel to the XY plane. The two components $P_N(Z)$ and $P_T(Z)$ (normal and transverse) of the pressure tensor, for a planar interface, have been computed with the following relations[14], originally due to Irving and Kirkwood :

$$P_N(z) = (\rho_1(z) + \rho_2(z)) \, k_B T - \frac{1}{L_x L_y} < \sum_{i<j} \frac{z_{ij}^2}{r_{ij}} \frac{1}{|z_{ij}|} \frac{du(r_{ij})}{dr_{ij}} \theta\left(\frac{z - z_i}{z_{ij}}\right) \theta\left(\frac{z_j - z}{z_{ij}}\right) >_t \quad (2)$$

$$P_T(z) = (\rho_1(z) + \rho_2(z)) \, k_B T - \frac{1}{2L_x L_y} < \sum_{i<j} \frac{[x_{ij}^2 + y_{ij}^2]}{r_{ij}} \frac{1}{|z_{ij}|} \frac{du(r_{ij})}{dr_{ij}} \theta\left(\frac{z - z_i}{z_{ij}}\right) \theta\left(\frac{z_j - z}{z_{ij}}\right) >_t \quad (3)$$

The interfacial tension γ is given by the relation (4)[15] derived from the Kirkwood-Buff formula[16].

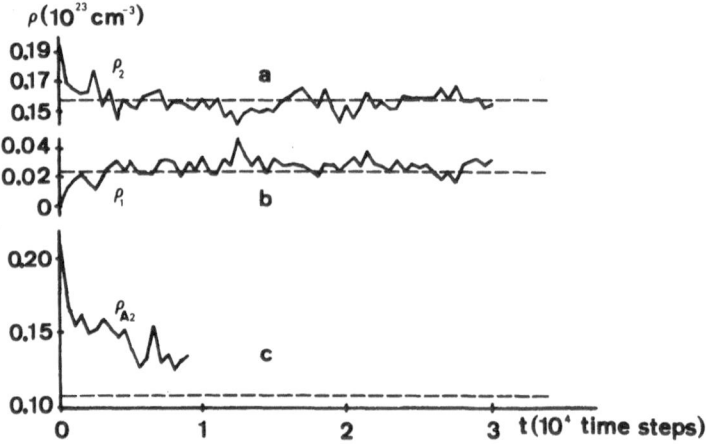

Fig. 2 - Density profiles of L1 ($\rho_1(Z)$) and L2 ($\rho_2(Z)$) together with their sum, averaged over 20,000 time steps. At the second interface (see Fig. 1) the profiles are extrapolated for Z>34.5 Å (dashed lines), using the image provided by the periodic boundary condition. The scale corresponding to one particle per slab is pointed out by the arrow.

282

$$\gamma = \frac{1}{2} \left(\frac{1}{2L_x L_y} < \sum_{i<j} \left(1 - \frac{3z_{ij}^2}{r_{ij}^2} \right) r_{ij} \frac{du(r_{ij})}{dr_{ij}} >_t \right) \qquad (4)$$

The existence of two interfaces in the simulated system requires the extra division by 2 introduced in relation (4).

Before calculating the average density profiles $\rho_1(Z)$ and $\rho_2(Z)$ reported in Fig. 2, we have checked the time dependance of the densities in a given slab. Two different slice thicknesses have been chosen to compute the densities. The average values have been obtained with $\Delta Z=1.73$ Å, while the time evolution study has been performed on a thicker slab ($\Delta Z=3.45$ A). The profiles have been computed over the usual averaging time (20,000 time steps, see §II) and the evolution of the instantaneous densities has been followed, since the beginning of the equilibration run, over 30,000 time steps (see Fig. 3). After 5,000 time steps, the local densities $\rho_1(t)$ and $\rho_1(t)$, plotted in Fig. 3a and 3b, tend to oscillate around equilibrium values. In these conditions, we may use averaged values to obtain density profiles representative of our system, they are given in Fig. 2. The values of $\rho_1(Z)$ and $\rho_2(Z)$ remain constant on some 20 Å each.

The plateau values indicate clearly the existence of bulk phases between the interfaces. This is confirmed by the comparison with the densities of the pure liquids (see Tables II and III). Each of these profiles exhibits a sharp variation in a limited range of Z. A thickness of the interface can be defined as the size of the region on which one profile does not remain constant. The analysis of the two profiles yields two comparable values. The estimated thickness is equal to 10±2 Å, whereas the atomic

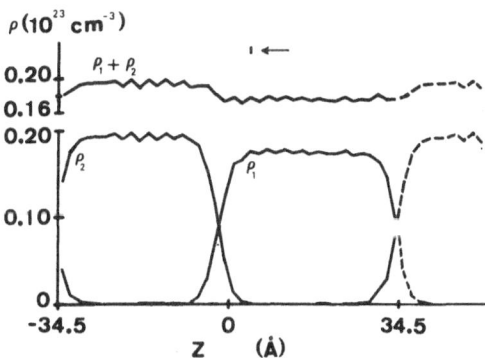

Fig. 3 - Instantaneous local densities as a function of time. The curves a and b refer to the L1-L2 interface, while c corresponds to A1-A2 system. They have been calculated in the same slab (Z±ΔZ/2) with a thickness (ΔZ) equal to 3.4 Å. The dashed lines indicate the average values for a and b, and the expected average for c.

283

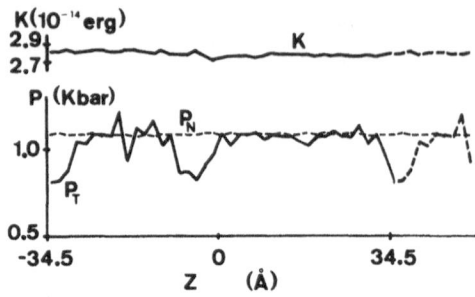

Fig. 4 - Normal and transverse pressure profiles $P_N(Z)$, $P_T(Z)$ and kinetic energy profile per particle $K(Z)$, averaged over 20,000 time steps. These profiles are extrapolated for $Z > 34.5$ Å as in Fig. 2.

diameter is equal to 3.6 Å. The total density profile $(\rho_1(Z) + \rho_2(Z))$ is plotted in Fig. 2, the variation is smooth and there is no gap of density in the interfacial regions. An other information, obtained from the analysis of the density profiles, is the existence of a slight miscibility between the two liquids. This is not obvious in Fig. 2, it is only noticeable when looking at the numerical values. When $\rho_1(Z)$ reaches its plateau value, ρ_2 is on average equal to $3 \cdot 10^{19}$ cm^{-3}, while in the lower part of the profile ρ_1 is equal to $5 \cdot 10^{19}$ cm^{-3}. These values correspond to solubilities of 0.3 % of L1 in L2 and 0.1 % of L2 in L1. Of course, we cannot ascertain that these values are equilibrium ones, since the determination of the exact solubilities would require too long runs. Nevertheless, we have been able to control that the miscibility was constant, within the accuracy of our calculation, on the time scale of the simulation. In these conditions, we are entitled to consider that the slight miscibility of the liquids is not an argument against the system stability. As indicated before, there is no drift in the time evolution of the local densities. This is the first indication showing that the average position of the interface is stable. To confirm that we have studied an other system, using the model described in §II. This one is totally miscible since it is composed of 1728 identical particles of Argon, artificially labelled A1 and A2. Then we have compared the time evolution of the density profiles of the two systems A1-A2 and L1-L2. An example of the variation of ρ_{A2}, a local density in A2, is given in Fig. 3c. The comparison of the curves a and c shows that ρ_2 reaches a stationary state after 5,000 time steps, while ρ_{A2} is still drifting after 9,000 time steps. Thus the simulation time is long enough to study the stability of the interface on a relevant time scale. Indeed, when the system is totally miscible the instability of the interface is clearly noticeable from the density profiles.

The normal and transverse components of the pressure tensor $P_N(Z)$ and $P_T(Z)$, calculated with relations (2) and (3), are plotted as a function of Z in Fig. 4. As expected, $P_N(Z)$ remains constant, while $P_T(Z)$ exhibits a noticeable variation in the interfacial regions. To analyse the interfacial structure of the P_T profile, we have studied the kinetic energy per particle K(Z), which is constant and plotted in Fig. 4. Thus the kinetic part of the pressure profiles varies as the total density $\rho_1(Z)+\rho_2(Z)$ (see Fig. 2). It means that the gap observed in $P_T(Z)$ is due to the Virial part (see relation (3)). The difference between the two pressure profiles is related to the interfacial tension γ. This quantity can be calculated directly from the pressure profiles, using its mechanical definition[17] given by the following relation :

$$\gamma = \int_{-\frac{L_z}{2}}^{\frac{L_z}{2}} [P_N(z) - P_T(z)]\,dz \qquad (5)$$

The interfacial tension has been computed with relation (4) and by integrating the pressure profile difference (relation (5)). The two results are similar and give a value of γ equal to 20 ± 5 erg/cm^2, which is of the order of magnitude of experimental values[1].

IV.CONCLUSION

Using MD simulation technique and LJ potential function, we have shown that it is possible to model an interface between two immiscible liquids. The introduction of the periodic boundary conditions generates a system which is stable over the time scale of the simulation. The interface, which has a thickness of about 3 atomic diameters, has been characterized by the density and pressure profiles. The interfacial tension has also been computed and is comparable to the values generally obtained in experimental measurements. Moreover this model gives information on the slight miscibility between the two liquids.

The model and the use of a "simple" potential makes it possible to think of larger systems. In these conditions, the range of wavelength would be large enough to obtain detailed description of the interface structure. Longer runs are also possible, from the point of view of computational time, and this is particularly interesting for a study of dynamical properties involved in chemical processes.

ACKNOWLEDGMENTS

Stimulating discussions with V.Pontikis are gratefully acknowledged. Thanks are also due to J. Mathie for her technical help. This work was supported by the Commission des Communautés Européennes through the EEC contrat No ST2J-0094-1-F. This study has been initiated during a CECAM workshop held in Orsay in 1985. Computational facilities have been provided by the Centre Inter Régional de Calcul Electronique in Orsay.

REFERENCES

1. R.Aveyard, B.Vincent, Prog. in Surf. Sci. 8:59 (1977).
2. J.S.Rowlinson, B.Widom, "Molecular Theory of Capillarity", Clarendon Press, Oxford (1982).
3. P.Tarazona, M.M.Telo da Gama, R.Evans, Mol.Phys., 49:283 (1983).
4. P.Tarazona, M.M.Telo da Gama, R.Evans, Mol.Phys., 49:301 (1983).
5. M.Robert, Phys.Rev.A, 30:2785 (1984).

6.   M.Robert, Phys.Rev.Lett., 54:444 (1985).

7.   Ref. 2 p. 173 to 189 give a review of the papers dealing with the liquid-vapor interface.

8.   S.M.Thompson, K.E.Gubbins, J.P.R.B.Walton, R.A.R.Chantry and J.S.Rowlinson, J.Chem.Phys., 81:530 (1984).

9.   P.Linse, J.Chem.Phys., 86:4177 (1987).

10. M.Shoen, C.Hoheisel, Mol.Phys., 53:1367 (1984).

11. M.Shoen, C.Hoheisel, Mol.Phys., 57:65 (1986).

12. H.J.C.Berendsen, W.F.Gunsteren, Practical Algorithms for Dynamic Simulations, in : "Molecular Dynamics Simulation of Statistical-Mechanical Systems", ed., G.Ciccotti, W.G.Hoover, North-Holland Physics Publishing, Amsterdam (1986).

13. Ref. 2 p. 178.

14. M.Rao, B.J.Berne, Mol.Phys., 37:455 (1979).

15. M.Rao, D.Levesque, J.Chem.Phys., 65:3233 (1976).

16. J.G.Kirkwood, F.P.Buff, J.Chem.Phys. 17:338 (1949).

17. Ref. 2 p. 44.

BROWNIAN DYNAMICS OF CHEMICAL REACTIONS

Pierre Turq, Jose Luis Fernandez-Abascal and
Giovanni Ciccotti

Laboratoire d'Electrochimie UA 430
Universite Pierre et Marie Curie
8 Rue Cuvier 75005 Paris, France

SUMMARY

The method of brownian dynamics appears as the dynamical equivalent of molecular dynamics, when one goes from the Born-Oppenheimer to the Mac-Millan Mayer level.

The application of brownian dynamics to chemical reactions is possible when the solvent molecules are not directly involved in the chemical process, either as reactants or as products.

Several applications of brownian dynamics were made for electrolyte solutions and for biochemical systems with the aim to computing rate constants or characterizing chemical species.

I) INTRODUCTION

Before applying brownian dynamics to the computer simulation of chemical reactions, we have to define precisely brownian dynamics, and especially its relationship to molecular dynamics simulation.

The first question to ask is then the following: what is brownian dynamics with respect to molecular dynamics? The answer is very simple: brownian dynamics is simply a dynamics for the solute in a solution involving both solvent and solute particles, since molecular dynamics describes explicitly both solvent and solute.

In order to enhance the characteristics of brownian dynamics (BD), we will first recall briefly the main features of molecular dynamics (MD), which are more often presented in the literature.

I) Molecular Dynamics

Both solvent and solute are treated at the Born-Oppenheimer level, which means separation of nuclear and electronic motions.

The system is then characterized by an Hamiltonian of the solution:

287

$$H^{BO} = \sum_\alpha \frac{p_\alpha^2}{2m_\alpha} + \sum_{\alpha,\beta} V_{\alpha,\beta} + \sum_i \frac{p_i^2}{2m_i} + \sum_{i,j} V_{i,j} + \sum_{i,\alpha} V_{i\alpha}$$

solute   solvent   solute-solvent

In this representation, the equations of motion for the solute, which are discretized for numerical solution on the computer, are simply the Newton equations:

$$\dot{p}_\alpha(t) = F_\alpha \quad \left[\begin{array}{l} \text{where } p_\alpha \text{ is the momentum of the solute} \\ \\ \\ \text{since } F_\alpha \text{ is the force } \textit{including solute-solvent term} \end{array} \right.$$

The characteristic feature of this level of description is to allow for a complete hamiltonian description of the forces:

$$\frac{\partial H^{BO}}{\partial q_\alpha} = -\dot{p}_\alpha$$

On the opposite, we have for the brownian dynamics the Mac-Millan Mayer level of description:

II) Brownian Dynamics: Mac-Millan Mayer level of description

The brownian dynamics is a solvent averaged dynamics, which allows for a dynamical representation of the static Mac-Millan Mayer hamiltonian, which has been proved to be a useful tool for the structural and excess thermodynamical properties of solutions[1].

$$\langle H^{MM} \rangle_{solvent} = \sum_\alpha \frac{p_\alpha^2}{2m_\alpha} + \sum_{\alpha,\beta} \langle V_{\alpha\beta} \rangle_{solvent}$$

solute Hamiltonian

The Mac-Millan Mayer hamiltonian is compatible with all modern statistical mechanical methods for equilibrium and static properties, such as integral equations (HNC, PY, ...), Monte Carlo computer simulations, especially for electrolyte solutions[2].

It should be noticed that for the evaluation of dynamical properties, Mac-Millan Mayer hamiltonian is not sufficient, since it does not include the solute-solvent forces which are missing in a solvent averaged model.

At the Mac-Millan Mayer level we have to introduce *solute-solvent interactions* in a *non hamiltonian description*.

This can be achieved by introducing any kind of stochastic or brownian hypothesis to mimic the short time and distance behaviour of solute solvent collisions. We are guided in this task by the existence of to limiting models for indefinitely diluted solutions:

- *Static and thermodynamical properties: ideal solution*

It is well known that for any kind of intermolecular potential the ideal solution constitutes an asymptotic model for the behaviour of extremely diluted solutions.

- *Dynamical properties: Brownian motion (heavy solute)*

Solute particles of large mass and size with respect to the solvent molecules undergo a brownian motion, which can be described by a Langevin equation at infinite dilution.

Those two limits for the static and dynamic behaviour of dilute solutions, combined with the solute-solute interactions coming from the Mac-Millan Mayer hamiltonian, are the guides for the construction of the brownian dynamics of solutions. There are naturally several (non hamiltonian) ways to realize a Mac-Millan Mayer dynamics. Basically any brownian machinery will describe the motion of the solute whereas the solute-solute interactions will be calculated from the Mac-Millan Mayer hamiltonian.

	INFINITE DILUTION	FINITE CONCENTRATION
STATICS	IDEAL SOLUTION	MAC-MILLAN MAYER
DYNAMICS	BROWNIAN MOTION	BROWNIAN DYNAMICS

II) LANGEVIN DYNAMICS OF ELECTROLYTES

an example of brownian dynamics

The simplest brownian dynamics is the so-called Langevin dynamics, described by a simple Langevin equation:

$$\overset{\cdot}{P}_i = - \zeta_i \frac{P_i}{m_i} + \sum_{j \neq i} F_{ij} + R_i$$

$$\qquad\qquad \textit{friction} \qquad\qquad\qquad \textit{random}$$

Obviously there are several algorithms to integrate numerically this equation, according to the time steps used in the consideration of the random forces.

The friction and random forces are balanced by fluctuation dissipation relationships, which characterize the thermalization process of the the solute by the solvent.

This model has been applied to electrolytes since about 10 years and has been proved to be a successful tool in the derivation of dynamical and kinetic properties of electrolyte solutions with Mac-Millan Mayer solute-solute interactions and continuous brownian solvent[3,4].

The interaction forces between the solute particles are essentially:

- repulsion forces in $\dfrac{1}{r^n}$ with n = 9 or 12

- Coulomb forces (solvent averaged)

Brownian dynamics has been applied with this kind of potentials to 1-1 and 2-2 electrolytes and to some models of polyelectrolytes.

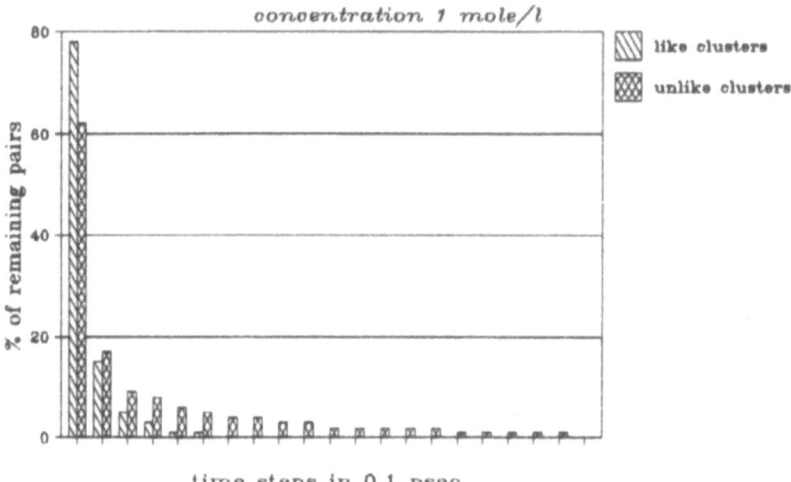

Fig. 1a. Electrolyte 2-2 life-time of pairs

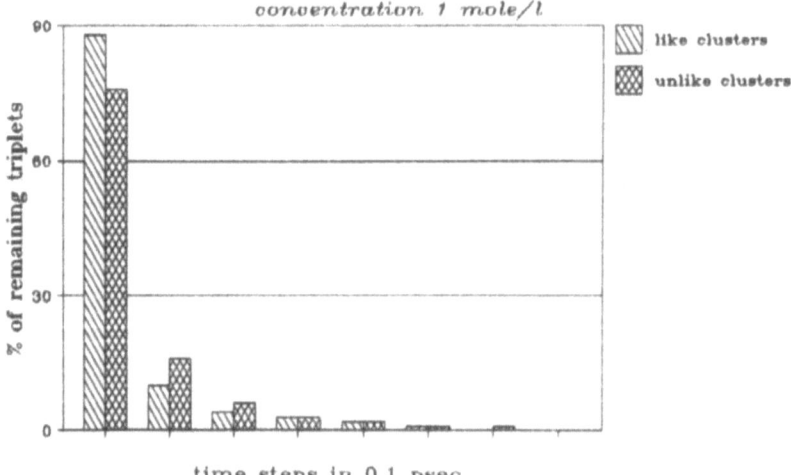

Fig. 1b. Triplets life-time

This technique allows for the study of pairs, triplets and other ionic types of clusters in electrolyte and counterion condensation in polyelectrolyte solutions.

The basic concept to achieve this task is that of cluster density: we consider that two particles form a cluster if they are separated by a distance smaller than a given distance Δ characterizing the effective range for the characterization of the cluster under consideration. The choice of a particular value of Δ can seem arbitrary, but a scanning as a function of the various values of Δ will be a powerful mean to detect the size of the different types of clusters. A peak in cluster density will be observed for a value of Δ close to the characteristic distance of the observed distance. More precisely the scanning with respect to Δ acts as a lens magnifying all the details of the structure.

The time evolution of clusters is basically described by the histograms for the life times of ionic clusters (*Fig.1a and 1b*).

It is possible to analyze the life time of the different ionic clusters[4] and to derive[5] "à la Kubo formulae" for the rate constants of the exchange reactions of the free and associated clusters. The corresponding expression involves the time correlation function $C_{AA}(t)$ for the cluster densities or its Laplace transform, as well as the equilibrium average values of those densities:

$$A^- + B^+ \underset{k_2}{\overset{k_1}{\rightleftarrows}} (AB)^0$$

$$k_1^{-1} = (n_{AB}^o)^{-1} \det(n) \, \tilde{C}_{AA}^{-1}(0) \int_o^\infty C_{AA}(t)\,dt$$

and

$$k_2^{-1} = (n_A^o \, n_B^o)^{-1} \det(n) \, \tilde{C}_{AA}^{-1}(0) \int_o^\infty C_{AA}(t) \, dt$$

with $\det(n) = n_A^o \, n_B^o + n_A^o \, n_{AB}^o + n_{AB}^o \, n_B^o$

n_A^o, n_B^o, n_{AB}^o being the averaged cluster densities at equilibrium for the cluster species A^-, B^+, and $(AB)^0$.

It should be noticed that the above formula allows for a simple evaluation of the kinetic coefficients since all quantities involved in this expression are easily obtained in computer simulation. The basic point in this determination is the fixed volume and occupancy of the computer simulation box. The total number of particles of each species is a constant value because of periodic boundary conditions. Here we are speaking of species in the chemical sense, that is for example counting all A under the two different cluster forms A^- and $(AB)^0$. The total number of A is then a constant since the computer simulation box does not present any variation in particle density: when a particle goes out, it is immediately replaced by another one coming from the opposite side of the computer simulation box.

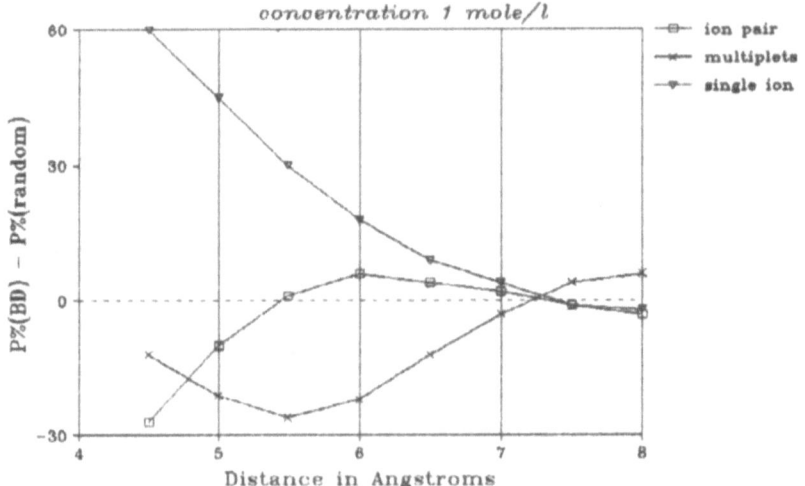

Fig. 2a. Electrolyte 1-1 cluster percentage

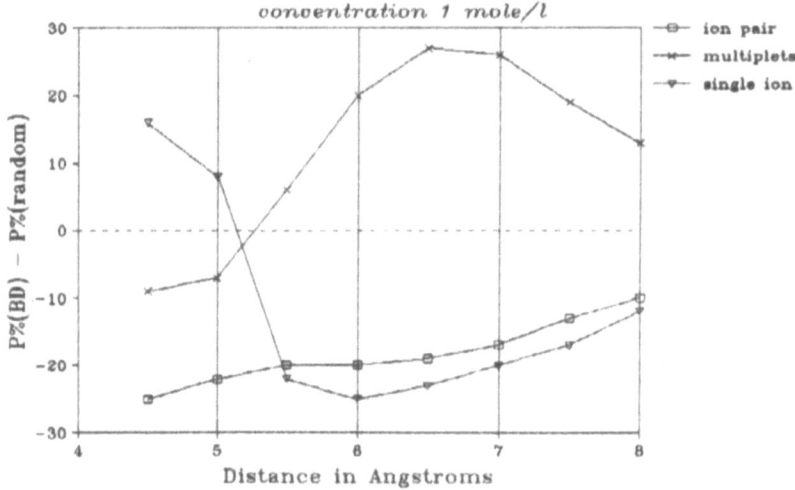

Fig. 2b. Electrolyte 2-2 cluster percentage

Among all possible dynamical and kinetic applications of the cluster densities analysis, it should be noticed that cluster approach can also be fruitful for the study of static and equilibrium properties of electrolyte solutions.

As an example, the brownian dynamics results of the simulation of associated and non associated electrolytes in water at 1M show that different cluster densities can be evaluated from this simulation.

More important is the fact that new types of clustering can be postulated from these simulations.

In order to understand these new features, we will recall briefly the basic principles of brownian dynamics cluster analysis for electrolytes.

It is obviously possible to compute single particle densities as well as pair, triplet or higher order cluster densities. In fact these cluster densities are significant only for the structure of the solution, if one substracts the corresponding values for a brownian dynamic simulation at random i.e. a simulation in wich one removed the interactions between the particles. We present on *Fig. 2a* the values for a 1-1 electrolyte in water at 1 M, for an aqueous electrolyte with a $\dfrac{1}{r^9}$ repulsive potential. Corresponding results are presented on *Fig. 2b* for a 2-2 electrolyte solution with identical concentration and short range potential.

The results presented here are the percentage of ionic clusters (single ion, pairs of ions of opposite charges, multiplets,...). In order to enhance the significant structural contributions, we substract the values at random for the same system (i.e. without interactions). We scan the percentage of clusters, according to the values of Δ. The main results are that any structural evidence of clustering disappears for a 1-1 electrolyte after 7 Å. On the opposite, 2-2 electrolytes exhibit only few simple pair association but a strong multiplet clustering tendency. In fact, a careful analysis of this clustering phenomenon shows that the corresponding clusters are fluctuating chains of pairs. These new results are not in contradiction with the well known thermodynamical and transport properties of 2-2 electrolytes, which give evidence of pair-like association, without precising the possibility of subsequent chaining of the pairs. Structural neutron-scattering studies are in progress for analogous systems, in order to confirm or destroy this new pictural scope on the structure of associated electrolytes.

III) OTHER BROWNIAN DYNAMICS OF CHEMICAL REACTIONS

There are, as indicated before, several implementations of the brownian concepts of computer simulation, we will recall here briefly the main features of some of these methods as applied to chemical reactions.

a) Intermediate dynamics

Berkowitz *et al.*[6] introduced an **intermediate dynamics**, which combines a Molecular Dynamics treatment of the relevant solvent and solute molecules in a **cage**, since the external molecules are treated as a continuum. The practical limitation of this method is that it is not possible to get reasonable solvent properties, for any practical size of the cage. The results are then cage-size dependent an cannot be connected directly to Molecular Dynamics simulation.

b) MTGLE

This method (molecular time scale generalized Langevin equation), is another kind of brownian dynamics, introduced by Adelman[7] It is also a cage-like model, in which one integrates a reduced set of equation of motion for the solute particles and the successive solvent shells.

This model has proved to be successful in the study of chemical reactions. As an example, it mimics properly the photolysis of iodine in xenon. Some of the properties of interfacial reactions, such as the recombination of H-H can also be reproduced. Even if the set of generalized Langevin equations used here is in limited number, it requires a numerical solution.

c) Smoluchowski dynamics

Langevin equation is not the only possibility to realize a brownian simulation. Smoluchowski dynamics is also possible, even if the time scale associated with this equation[8] implies that short range forces (in particular repulsion forces) have to be neglected or at least treated differently. This variety of brownian dynamics holds then for long range interactions, and the time scale associated with this kind of simulation allows for the study of reactions much slower than those approchable by Langevin simulation. We just briefly mention here the Smoluchowski equation which was extensively used by Mac-Cammon and coworkers[9,10] in order to study chemical reactions for complex systems, such as enzymatic reactions:

$$\frac{\partial \rho}{\partial t} = \vec{\nabla}.\left[- \vec{J}(r) \right] \quad \text{with} \quad \vec{J}(r) = D(r) \left[\vec{\nabla} - \vec{F}(r)/k_B T \right] \rho(r)$$

In this expression $D(r)$ is the diffusion tensor involving hydrodynamic interactions (Oseen's tensor). $\vec{F}(r) = - \vec{\nabla} U(r)$, where $U(r)$ is the potential of mean force between the solute particles.

The results of those simulations are that hydrodynamic interactions reduce the rate constants of about 20% . On the opposite coulomb forces (attractive) increase dramatically the rate constants (by a multiplicative factor of 7), whereas Debye screening reduces the rate constants of 30% .

This method can also be applied to non spherical solutes, as in the case of the reaction of superoxyde dismutase with O_2^- . The experimentally observed result is that the apparent rate constant increases with increasing ionic strength, and this effect can be taken into account by the computer simulation by considering the charge asymmetry.

Smoluchowski dynamics is now a common tool in theoretical biochemistry, and has been applied also to colloidal chemistry to study aggregation processes.

References

(1) J.C. RASAIAH and H.L. FRIEDMAN, *J. Chem. Phys.*, **48**, 2742 (1968).

(2) J.C. RASAIAH, D.N. CARD and J.P. VALLEAU, *J. Chem. Phys.*, **56**, 246 (1972).

(3) P. TURQ, F. LANTELME and H.L. FRIEDMAN, *J. Chem. Phys.*, **66**, 7, 3039 (1977).

(4) P. TURQ, F. LANTELME and D. LEVESQUE, *Mol. Phys.*, **37**, 1, 223 (1979).

(5) G. CICCOTTI, P. TURQ and F. LANTELME, *Chem. Phys.*, **88**, 333 (1984).

(6a) M. BERKOWITZ and J.A. Mc CAMMON, *Chem. Phys. Letters*, **90**, 215 (1982).

(6b) A.C. BLECH and M. BERKOWITZ, *Chem. Phys. Letters*, **113**, 278 (1985).

(7a) C.L. BROOKS III, M. BERKOWITZ and S.A. ADELMAN, *J. Chem. Phys.*, **73**, 9, 4353 (1980).

(7b) C.L. BROOKS III and S.A. ADELMAN, *J. Chem. Phys.*, **80**, 11, 5598 (1983).

(8) D.L. ERMAK, *J. Chem. Phys.*, **62**, 4189, 4197 (1975).

(9) S.H. NORTHRUP, S.A. ALLISON and J.A. Mc CAMMON, *J. Chem. Phys.*, **80**, 4, 1517 (1984).

(10) S.A. ALLISON, S.H. NORTHRUP and J.A. Mc CAMMON, *J. Chem. Phys.*, **83**, 6, 2894 (1985).

DISCUSSION

GERSCHEL - Do you have an idea of the life-time of the clusters you mentioned (in your brownian dynamics simulation) ?

TURQ - The life-time depends on the details of the short range part of the interactions. For a potentials with a repulsive part in $1/r^n$ and Coulomb attraction the life-time is of about 10 ps.
The introduction of a solvation barrier increases dramatically this life-time for ionic pairs up to 1 ms and more.

KAPRAL - The results for the equilibrium fractions of clusters in the electrolyte were obtained from long-time averages over the Brownian dynamics simulation. These results could be checked by direct Monte-Carlo simulations on the primitive model electrolyte. Has this been done ?
Is the concentration of electrolyte in your simulations sufficiently large so that the concentration dependence of the association rate constants can be detected and studied ?

TURQ - The new phenomena observed by B.D. are not observable by integral equations but should already be observed in Monte-Carlo simulations with the same potential. We will try to achieve this program in a next future. The concentration dependence of the observed phenomena is relatively small in B.D. for the relevant range of concentration (0.5 M to 3 M), since the averaged distance between the ions in this domain varies only by a factor smaller than 2.

BELLONI - You mentioned the examples of the superoxyde anion dismutation catalyzed by the superoxyde dismutase and of the aggregation of colloids which are both multisteps processes. I wonder how, without sufficient knowledge of the slowest reaction governing the mechanism, the ionic strength effects found can be explained by your predictions ?

TURQ - I dont have made myself the work on superoxyde dismutase and I am sometimes afraid by what some people are doing with Brownian dynamics.
I would summarize my position by saying that Brownian dynamics is like rugby : a sport of brutes which should be played only by gentlemen.

NONEQUILIBRIUM RATE PROCESSES

Ivan L'Heureux and Raymond Kapral

Chemical Physics Theory Group
Department of Chemistry
University of Toronto
Toronto, Ontario M5S 1A1, Canada

1. INTRODUCTION

One of the goals of a theory of condensed-phase chemical reactions is the calculation of the rate coefficient. The apparatus of linear response theory can be brought to bear on this problem for reactions taking place close to chemical equilibrium, and formal correlation function expressions for the rate coefficient can be derived. Of course, the dynamics of a liquid-state reacting mixture are not simple and these expressions are difficult to evaluate; however, molecular dynamics simulations for simple systems are now possible and provide insight into the details of the reactive event and how it couples to solvent motions, as well as numerical estimates of the rate coefficient. Theoretical treatments necessarily model the full many-particle dynamics by stochastic equations of motion, and it is in the development and utilization of such models that most progress has been made in the theory of condensed-phase reaction rates.

Reactions taking place far from equilibrium show a number of unusual features, even at the macroscopic level. Consequently, the aim of much of the research in this area is rather different, and has concentrated on characterizing and unraveling the kinetics giving rise to behavior like oscillations, chaos and bistability. So the microscopic details of the reactive event, which are presumably like those in close-to-equilibrium systems, are not studied, but rather the complex structure of the macroscopic temporal and spatial evolution is the focus of attention. However, there is a class of phenomena which is analogous in some respects to the rate problem in systems close to equilibrium: these are external noise-induced transitions between bistable states. The purpose of this article is to describe these phenomena and point out how some of the theoretical tools developed for equilibrium systems can be exploited to investigate these new rate processes.

Section 2 contains a brief sketch of the standard methods used to describe close-to-equilibrium reacting systems, with emphasis on those that are useful in the study of far-from-equilibrium systems. In Sec. 3 a one-dimensional bistable system driven by an external noise source is considered. Correlation function techniques similar to those in Sec.

2 are used to calculate the rate of the noise-induced transitions between the two steady states. Sec. 4 contains a discussion of the noise-induced rate phenomena that occur when an oscillatory state coexists with a steady state, and Sec. 5 summarizes the results and describes extensions to more-complex situations.

2. SKETCH OF EQUILIBRIUM REACTION RATE THEORY

Phenomenology

The study of close-to-equilibrium reaction rates is of prime importance in chemistry. To review the essential features of the derivation of the phenomenological rate law and the correlation approach, we consider the following isomerization process:

$$A \underset{k_r}{\overset{k_f}{\rightleftharpoons}} B. \tag{2.1}$$

Let $\bar{N}_\alpha(t)$ be the nonequilibrium average of species α population number $N_\alpha(t)$ and N_α^{eq} the corresponding equilibrium value. The phenomenological rate law for $\bar{N}_A(t)$ reads:

$$\frac{d}{dt} \bar{N}_A(t) = -k_f \bar{N}_A(t) + k_r \bar{N}_B(t). \tag{2.2}$$

Diffusion has been ignored, so that (2.2) is valid for a closed, homogenous system, where $\bar{N}_A(t) + \bar{N}_B(t) = $ constant. The detailed balance condition then gives $k_r = k_f N_A^{eq}/N_B^{eq}$. Let the number correlation function be $C_{\alpha\beta}(t) = \langle \delta N_\alpha(t) \delta N_\beta \rangle$ where the angular brackets refer to an equilibrium average and the deviations from equilibrium are

$$\delta \bar{N}_\alpha(t) = \bar{N}_\alpha(t) - N_\alpha^{eq}. \tag{2.3}$$

A variable without time argument is understood to be taken at $t = 0$. From the regression of fluctuations hypothesis [1], the long-time decay of the correlation function obeys the phenomenological law (2.2):

$$\frac{d}{dt} C_{AA}(t) = -k_f C_{AA}(t) + k_r C_{BA}(t). \tag{2.4}$$

Approximate expressions for the rate coefficient

The goals of microscopic theories of chemical rate processes are to define the range of validity of the macroscopic law, to provide microscopic expressions for the rate coefficients and to extend the kinetic description to smaller distance and time scales. Traditional approaches model the isomerization reaction by motion of a particle of mass m across a potential barrier in a thermal bath of temperature T. Transition state theory (TST) [2] assumes an equilibrium distribution at the barrier top; the rate is given by the equilibrium one-way flux across the barrier. The rate thus obtained has the form:

$$k_f^{TST} = (\omega_o/2\pi) \exp(-E_b/kT) \tag{2.5}$$

where $(\omega_o/2\pi)$ is an effective frequency and E_b is the barrier height. Kramers [3] assumed that the particle is moving in a one-dimensional double minimum potential $V(x)$ and is subjected to a friction force $-\varsigma v$. Solving for the nonequilibrium steady state distribution from the corresponding Fokker-Planck equation in phase space (x, v), he obtained:

$$k_f^K = \frac{\varsigma\omega_o}{4\pi m\omega_b}[(1 + \frac{4\omega_b^2 m^2}{\varsigma^2})^{1/2} - 1]e^{-E_b/kT} \tag{2.6}$$

where ω_o and ω_b are frequencies defined by adopting an harmonic approximation to the potential at the bottom of the well and at the top of the barrier, respectively. For intermediate friction $\varsigma/m << 2\omega_b$, (2.6) reduces to (2.5). In the high friction limit $\varsigma/m >> 2\omega_b$, (2.6) reduces to:

$$k_f^S = k_f^{TST}(\omega_b m/\varsigma). \tag{2.7}$$

(2.7) is identical with the result obtained from the Smoluchowski equation:

$$\frac{\partial}{\partial t}p(x,t) = \frac{\partial}{\partial x}[\frac{kT}{\varsigma}(\frac{\partial}{\partial x} + \frac{1}{kT}\frac{\partial V}{\partial x})]p(x,t) \tag{2.8}$$

which is a Fokker-Planck description for the dynamics contracted onto the position variable [3].

A more accurate description of dynamics can be achieved by taking into account the fact that friction coefficient is in general time dependent. In this circumstance the motion of the particle may be described by the generalized Langevin equation:

$$m\frac{dv(t)}{dt} = -\int_0^t \varsigma(t - t')v(t')dt' + F(t) + f(t), \tag{2.9}$$

where $F(t)$ is the deterministic force on the particle and $f(t)$ is the random force. Grote and Hynes [4] derived a very useful expression for the rate coefficient using (2.9) as the starting point. Assuming a parabolic barrier with associated frequency ω_b, and recognizing that the relevant time scale for the evaluation of the time-dependent friction is given by λ_r^{-1}, where λ_r is the reactive frequency, they find,

$$k_f = \frac{\lambda_r}{\omega_b}k^{TST}; \lambda_r = \omega_b^2[\lambda_r + \hat{\varsigma}(\lambda_r)/m]^{-1}, \tag{2.10}$$

with $\hat{\varsigma}(\omega)$ the Laplace transform of $\varsigma(t)$.

Chemical species

Any microscopic description of chemical rate laws needs to characterize the chemical species undergoing transitions. A formal way of doing so classically [5] is by the introduction of characteristic functions χ_i^α, which depend on the coordinates used to specify the species α associated with the molecule i. They have the properties:

$$\chi_i^\alpha \chi_i^\beta = \delta_{\alpha\beta}; \sum_\alpha \chi_i^\alpha = 1. \tag{2.11}$$

The number of molecules of species α is then:

$$N_\alpha = \sum_{i=1}^N \chi_i^\alpha, \tag{2.12}$$

where N is the total number of molecules. For the isomerization reaction (2.1), which is modeled by the escape of a molecule across a potential barrier, the characteristic function is conveniently defined as:

$$\chi_i^A = \theta(-x_i + r_o) \text{ and } \chi_i^B = \theta(x_i - r_o) \tag{2.13}$$

where $\theta(x)$ is the Heaviside step function and r_o is the position of the barrier maximum in reaction coordinate space. Other examples of characteristic functions can also be constructed for similar or more complicated types of reactions [6].

Microscopic dynamical equation

The microscopic equation of motion for the number of molecules of species α is given by:

$$\frac{d}{dt}N_\alpha(t) = i\mathcal{L}N_\alpha(t) \tag{2.14}$$

where $\mathcal{L}$ is the Liouvillian of the system. Using projection operator techniques [7], a formally exact Langevin equation for $N_\alpha(t)$ can be derived. The decay of the correlation function is then given by:

$$\frac{d}{dt}C_{\alpha\beta}(t) = -\int_0^t dt' \phi_{\alpha\gamma}(t')C_{\gamma\beta}(t-t') \tag{2.15}$$

where:

$$\phi_{\alpha\beta}(t) = \langle (e^{i(1-P)\mathcal{L}t}\delta\dot{N}_\alpha)\dot{N}_\gamma \rangle K_{\gamma\beta}^{-1} \tag{2.16}$$

is the memory kernel and $PA = \langle A\delta N_\alpha \rangle K_{\alpha\beta}^{-1}\delta N_\beta$ is the projector onto the slow variables δN_α. Here, the summation convention has been adapted and $K_{\alpha\beta} = \langle \delta N_\alpha \delta N_\beta \rangle$. For long times, the decay time scale of the correlation function is much larger than the relaxation time scale of the memory kernel. One can show [5] that (2.15) reduces to the phenomenological law:

$$\frac{d}{dt}C_{\alpha\beta}(t) = -k_{\alpha\gamma}C_{\gamma\beta}(t) \tag{2.17}$$

with:

$$k_{\alpha\beta} = \int_0^\infty dt \langle (e^{i\mathcal{L}t}R_\alpha)R_\gamma \rangle K_{\gamma\beta}^{-1}, \tag{2.18}$$

where R_α is the reactive flux:

$$R_\alpha = \delta\dot{N}_\alpha = \sum_{i=1}^N \dot{\chi}_i^\alpha. \tag{2.19}$$

For the isomerization process (2.1), (2.16) reduces approximately at very short times to [6]:

$$k_f \equiv k_{AA} = \langle \sum_i v_i \delta(x_i - r_o)\theta(v_i) \rangle \frac{N}{N_A^{eq}} \frac{N}{N_B^{eq}} \tag{2.20}$$

which is the standard TST result.

Eq. (2.15) may easily be generalized [5] when the system is open and diffusive motion of the molecules is allowed. In this case, the relevant slow variables are number densities.

The above description is formal and although the rate calculation is well-posed it is difficult to carry out for problems of chemical interest. Hence, much of the focus of recent theoretical studies of reaction rates in the condensed phases has centered on evaluation of the reaction rate for model systems [8]. Perhaps the best way to obtain information on the structure of the rate coefficient expressions is through direct molecular dynamics simulations on model systems. Since both of these approaches have been reviewed by others in this volume, we now turn our attention to far-from-equilibrium systems where some of the ideas summarized above can be used, but new phenomena appear which require special techniques.

3. NON-EQUILIBRIUM RATE PROCESS IN A ONE-DIMENSIONAL BISTABLE SYSTEM

Bistable systems

When nonlinear, dissipative systems are driven far from equilibrium, they often exhibit multistability, whereby there exist two or many coexisting steady states and/or oscillating states. Through internal or external noise, such systems may be driven to undergo transitions between these states. Moreover, the noise itself may create new states and transitions between them [9]. A wide variety of systems chosen from the fields of chemistry [10], optics [11] and biology [12] display such a behavior.

We will be mainly concerned with external noise, where the stochastic character of the system is due to random time variations in the coupling of the system to its environment. The stochastic properties of the external noise are known in this case.

The vector of dynamical variables $\mathbf{x}(t)$ which characterizes the bistable system obeys a stochastic evolution equation of the form

$$d\mathbf{x}(t)/dt = \mathbf{F}[\mathbf{x}(t); \mathbf{a}(t)] \tag{3.1}$$

where $\mathbf{F}$ is a non-linear function of the random variables $\mathbf{x}$ and $\mathbf{a}$ denotes the set of random parameters.

In the rest of this section, we will specialize to the case of one-dimensional bistable systems. The transitions then occur between steady states. By applying the ideas of section 2 we can obtain the corresponding nonequilibrium rate coefficient. The following simple model will illustrate the essential features of the calculation. Let the evolution equation with additive noise be

$$\dot{x} = f(x) + \xi(t) \tag{3.2}$$

where the deterministic contribution is conveniently chosen as deriving from a symmetrical quartic potential:

$$f(x) = -\frac{dV}{dx}; \tag{3.3a}$$

$$V(x) = \frac{ax^4}{4} - \frac{bx^2}{2}, a > 0, b > 0 \tag{3.3b}$$

and $\xi(t)$ is gaussian white noise process:

$$< \xi(t) > = 0$$

$$< \xi(t)\xi(t') >= 2D\delta(t - t') \tag{3.4}$$

D being the parameter characterizing the strength of the noise.

Species determination problem

The deterministic version of (3.2) leads to three steady states, two of which (at $x_{1,2} = \pm\sqrt{b/a}$) are stable and the third at ($x_3 = 0$) unstable. The corresponding deterministic basins of attraction are therefore $x > 0$ and $x < 0$, respectively; x_3 corresponds to the basin boundary.

Under noisy dynamics, during part of its trajectory , the system may enter a regime where bistability ceases to exist, and noise-induced transitions occur. This mechanism is the one-dimensional analog of the tangent mechanism described in [13].

It is convenient to define an intermediate region I of length 2δ centered about the unstable state. Such divisions are often used in the literature of chemical kinetics and form the basis for the stable states picture [14]. We then define the characteristic functions for the stable states in the following way:

$$\text{state A } (-\infty < x < -\delta) : \chi^A = \theta(-x - \delta)$$

$$\text{state B } (\delta < x < \infty) : \chi^B = \theta(x - \delta). \tag{3.5}$$

Dynamical equations

The stochastic differential equation (3.2), together with the noise properties (3.4) may be transformed to the Fokker-Planck equation

$$\dot{p}(x,t) = \mathcal{F}p(x,t)$$

$$\mathcal{F} \equiv \frac{\partial}{\partial x}[-f(x) + D\frac{\partial}{\partial x}]. \tag{3.5}$$

The formal solution of (3.5) is:

$$p(x,t) = e^{\mathcal{F}t}p(x,0). \tag{3.6}$$

The steady state distribution is given by:

$$p_s(x) = Z \exp[-\frac{1}{D}V(x)] \tag{3.7}$$

where Z is the normalization constant; a steady state average will be denoted by $<>$. The population number in state A, say, may be written in two equivalent ways:

$$\bar{N}_A(t) = \int_{-\infty}^{\infty} dx\theta(-x - \delta)p(x,t) = \int_{-\infty}^{\infty} dx\theta(-x - \delta)e^{\mathcal{F}t}p(x,0) \tag{3.8a}$$

$$\bar{N}_A(t) = \int_{-\infty}^{-\infty} dxp(x,0)e^{\mathcal{L}t}\theta(-x - \delta) \tag{3.8b}$$

where $\mathcal{L}$ is the operator adjoint to $\mathcal{F}$ (not to be confused with $\mathcal{L}$ in (2.14)):

$$\mathcal{L} = f(x)\partial_x + D\partial_{xx} = Dp_s^{-1}(x)\partial_x[p_s(x)\partial_x]. \tag{3.9}$$

The steady-state A-population number is denoted by η:

$$\eta_A = < \theta(-x-\delta) > \equiv \eta. \tag{3.10a}$$

We have by symmetry:

$$\eta_B = < \theta(x-\delta) > = \eta. \tag{3.10b}$$

In order to make the analogy with the correlation formulation of section 2 more explicit, we will adopt the Heisenberg picture (3.8b), which we rewrite in the form:

$$\bar{N}_A(t) = \int_{-\infty}^{\infty} dx p(x,0) N_A(x,t);$$

$$N_A(x,t) \equiv e^{\mathcal{L}t} \theta(-x-\delta) \equiv e^{\mathcal{L}t} N_A(x). \tag{3.11}$$

The equation of motion for the dynamical variable $N_A(x,t)$ is then

$$\dot{N}_A(x,t) = \mathcal{L} N_A(x,t). \tag{3.12}$$

Although the Smoluchowski equation (2.8) is formally identical to (3.5), the physics of the two cases is very different. The diffusion coefficient in (2.8) is given by $D = kT/\varsigma$ and follows a fluctuation dissipation relation. On the other hand the diffusion coefficient D in (3.5) is a fixed constant imposed by the external noise. In (2.14), the molecular evolution equation has a Liouvillian structure. While the dynamics in (3.12) is not Liouvillian.

Rate coefficient for noise − induced transitions

We define the projector P onto the dynamical variables $N_\alpha(x)$, with $\alpha = \{A, B\}$, in the following way:

$$PA(x,t) = \sum_\alpha \left(\int_{-\infty}^{\infty} dx p_s(x) A(x,t) N_\alpha(x) \right) \eta^{-1} N_\alpha(x) \tag{3.13}$$

Using standard techniques [7], one obtains a generalized Langevin equation for $N_\alpha(x,t)$:

$$\frac{dN_\alpha(x,t)}{dt} = \sum_\beta \Omega_{\alpha\beta} N_\beta(x,t) + \sum_\beta \int_0^t d\tau \Gamma_{\alpha\beta}(t-\tau) N_\beta(x,\tau) + F_\alpha(x,t) \tag{3.14}$$

where

$$\begin{aligned}
\Omega_{\alpha\beta} &= < N_\beta(x) \mathcal{L} N_\alpha(x) > \eta^{-1}, \\
\Gamma_{\alpha\beta}(t) &= < N_\beta(x) \mathcal{L} F_\alpha(x,t) > \eta^{-1} \\
F_\alpha(x,t) &= e^{(1-P)\mathcal{L}t}(1-P)\mathcal{L}N_\alpha(x).
\end{aligned} \tag{3.15}$$

We assume the memory kernel $\Gamma_{\alpha\beta}(t)$ to have a fast decay time scale τ_{tr} compared with the times of interest and we extend the upper limit of integration to ∞. Eq. (3.14) reduces to an equation for the decay of the correlation $C_{\alpha\beta}(t) = < N_\alpha(x,t) N_\beta(x) >$ which has the form (2.17). Moreover, assuming the initial distribution $p(x,0)$ to be proportional to $p_s(x)$ in each stable state α, we get $\bar{F}_\alpha(x,t) = 0$. The nonequilibrium average of (3.14), with $\alpha = A$, say, therefore leads for large times to the phenomenological law (2.2) with:

$$\begin{aligned}
k_f &= -\Omega_{AA} - \int_0^{\infty} dt \Gamma_{AA}(t) \\
k_r &= \Omega_{AB} + \int_0^{\infty} dt \Gamma_{AB}(t).
\end{aligned} \tag{3.16}$$

303

We proceed with the evaluation of each term in (3.16).

Instantaneous flux

Using (3.15) and (3.9) we obtain for the instantaneous flux Ω_{AA}:

$$\Omega_{AA} = \int_{-\infty}^{-\delta} dx D\eta^{-1} \frac{d}{dx}[p_s(x)\frac{d}{dx}\theta(-x-\delta)] = j(-\delta) \tag{3.17a}$$

where

$$j(x) = D\eta^{-1}p_s(x)\frac{d}{dx}\theta(-x-\delta). \tag{3.17b}$$

It is convenient [14] to consider x as a discrete set of points separated by a distance ϵ. The limit $\epsilon \to 0$ is to be taken at the end of the calculation. We define:

$$\theta(-x-\delta) = \begin{cases} 1, & x \le -\delta^+; \\ 0, & \text{otherwise.} \end{cases} \tag{3.18}$$

where $\delta^{\pm} = \delta \pm \epsilon/2$. Integrating (3.17b) from $-\delta^+$ to $-\delta$ leads to:

$$\Omega_{AA} = -D\eta^{-1}p_s(-\delta)\frac{2}{\epsilon}. \tag{3.19}$$

It is seen that the instantaneous flux diverges as $\epsilon \to 0$. This result is expected since a one-way flux across a surface from points in the immediate neighborhood of the surface is infinite.

Since the jump in $\theta(x - \delta)$ is at the point $x = \delta$, it is easy to see that:

$$\Omega_{AB} = 0. \tag{3.20}$$

Integrated memory term

Using (3.15) and (3.9), the integrated memory term is:

$$\int_0^{\infty} dt\Gamma_{AA}(t) = \int_0^{\infty} dt h(-\delta, t) \tag{3.21a}$$

where

$$h(x,t) = Dp_s(x)\frac{\partial}{\partial x}g(x,t) \tag{3.21b}$$

and

$$g(x,t) = e^{(1-P)\mathcal{L}t}(1-P)\mathcal{L}\theta(-x-\delta)\eta^{-1}. \tag{3.21c}$$

Since $\dot{g}(x,t) = (1-P)\mathcal{L}g(x,t) = (1-P)p_s^{-1}(x)\frac{\partial}{\partial x}h(x,t)$, a time integration gives the relation:

$$\int_0^{\infty} dt(1-P)p_s^{-1}(x)\frac{\partial}{\partial x}h(x,t) = -g(x,0) = (1-P)\mathcal{L}\theta(-x-\delta)\eta^{-1}. \tag{3.22}$$

We now introduce the stable state assumption [4,14]. The projector onto the fast variables $Q \equiv 1 - P$ has two contributions: a contribution describing the deviations from local equilibrium in each stable state well and a contribution from the intermediate region I. We will assume that the characteristic time for equilibration of each stable state τ_{int} is much smaller that the characteristic time τ_{rel} for the transition $A \to B$ to occur. Under these conditions we have approximately:

$$P = 1 \text{ for } x < -\delta \text{ or } x > \delta.$$

Therefore
$$1 - P = \hat{\theta}(x) \equiv 1 - \theta(-x - \delta) - \theta(x - \delta). \tag{3.23}$$

Multiplying (3.22) by $p_s(x)$ and integrating from $-\delta$ to $x \leq \delta$, we obtain, with (3.21):

$$\int_0^\infty dt \Gamma_{AA}(t) = \int_0^\infty dt D p_s(x) \frac{\partial}{\partial x} e^{\hat{\theta} \mathcal{L} t} \hat{\theta} \mathcal{L} \theta(-x - \delta) \eta^{-1} + \int_{-\delta}^x dx' p_s(x') \hat{\theta}(x') \mathcal{L} \theta(-x' - \delta) \eta^{-1}. \tag{3.24}$$

Finally, multiplying (3.24) by $p_s^{-1}(x)$, integrating over x from $-\delta$ to δ, and using (3.9), we obtain:

$$\int_0^\infty dt \Gamma_{AA}(t) = [\int_{-\delta}^\delta p_s^{-1}(x) dx]^{-1} \int_{-\delta}^\delta p_s^{-1}(x) dx [D p_s(x) \frac{d}{dx} \theta(-x - \delta) \eta^{-1}]_{-\delta}^x$$

$$= -[\int_{-\delta}^\delta p_s^{-1}(x) dx]^{-1} [\int_{-\delta}^\delta p_s^{-1}(x) dx \Omega_{AA} \theta(x + \delta^-)]. \tag{3.25}$$

The last line comes from the definition of the instantaneous flux (3.17b). Expanding δ^- to first order in ϵ, we finally obtain:

$$\int_0^\infty dt \Gamma_{AA}(t) = -\Omega_{AA} + \frac{\Omega_{AA} p_s^{-1}(\delta)}{\int_{-\delta}^\delta p_s^{-1}(x) dx} \frac{\epsilon}{2}. \tag{3.26}$$

It is seen that the infinite instantaneous flux vanishes from the total rate coefficient (3.16). Using the evaluation of Ω_{AA} in (3.19), the total rate coefficient is:

$$k_f = D \eta^{-1} [\int_{-\delta}^\delta p_s^{-1}(x) dx]^{-1}. \tag{3.27}$$

By using similar manipulations, it is easy to see that:

$$k_r = k_f. \tag{3.28}$$

Rate coefficient for highly peaked potential barrier

For a highly peaked potential barrier, $b^2/4aD >> 1$. Choosing $\delta \leq \mathcal{O}(\frac{1}{2}\sqrt{\frac{b}{a}})$ and $\delta^2 b/D >> 1$, the two integrals in (3.27) can be approximated by Gaussian integrals and are easy to evaluate. We obtain

$$k_f = \frac{b}{\sqrt{2\pi}} e^{-b^2/4aD} \tag{3.29}$$

which corresponds to Kramer's solution of the Smoluchowski equation (2.7). The approximation used to derive (3.27) can be checked self-consistently. With τ_{int} scaling like the free diffusion time across the stable well ($\tau_{int} \sim D^{-1} b/a$) and $\tau_{rel} \sim k_f^{-1}$, the stable state assumption is valid for high potential barriers: $\tau_{int}/\tau_{rel} << 1$. Similarly, with τ_{tr} scaling like the diffusion time across the region I ($\tau_{tr} \sim D^{-1} \delta^2$), we have $\tau_{tr}/\tau_{rel} << 1$, and the time-scale separation hypothesis is valid.

This calculation shows how simple integrations and reasonable assumptions lead from a Fokker-Planck equation to a nonequilibrium rate coefficient for the noise-induced transitions between two steady states. The calculation is straightforward and may easily be generalized to more complicated situations. For instance, an asymmetric potential barrier can be constructed by adding a constant term c on the right-hand side of (3.3a). With, $-2b^{3/2}/3\sqrt{3a} < c < 2b^{3/2}/3\sqrt{3a}$, the system exhibits bistability and noise-induced transitions of the type just described are possible. Another generalization is to multiplicative noise in (3.2). But the most interesting generalization comes from the consideration of colored noise, whereby the noise correlation has a non-vanishing relaxation time. By extending the space of dynamical variables to include the noisy variable $\xi(t)$, it is possible to construct a Fokker-Planck-like equation for the dynamics contracted onto the x variable. This is shown explicitly in [9, 15, 16] for the Ornstein-Uhlenbeck process or in [9, 17] for the Poisson dichotomous noise. In the latter case with $f(x)$ as in (3.3), calculations similar to the ones sketched here [18] show that:

$$k_f = k_r = \eta^{-1}\left\{ \int_{-\delta}^{\delta} dx[D^{eff}(x)p_s(x)]^{-1}[1+f^2(x)/\gamma D^{eff}(x)]-2f(-\delta)/\gamma D^{eff}(-\delta)p_s(-\delta)\right\}^{-1}$$

(3.30)

where

$$p_s(x) = \frac{Z}{D^{eff}(x)} \exp \int^x \frac{f(x')dx'}{D^{eff}(x')}, x\epsilon\cup;$$
$$= 0, x \notin \cup.$$

The renormalized diffusion coefficient is $D^{eff}(x) = D(1 - f^2(x)/\triangle^2)$ with $D = \triangle^2/\gamma$ and the support $\cup$ is defined so that $D^{eff}(x) > 0$ for $x\epsilon\cup$. Here $\triangle$ is the amplitude of the symmetric dichotomous noise variable and γ^{-1} is its correlation time. For a highly peaked potential barrier, the steepest descent approximation gives [17, 18]:

$$k_f = \frac{b}{\sqrt{2\pi}}e^{-\triangle\phi/D}[1 + b/\gamma]^{-1}[1 + 2b/\gamma]^{-1}$$

(3.31)

where

$$\triangle\phi = \int_0^{\sqrt{b/a}} \frac{f(x')dx'}{1 - f^2(x')/\triangle^2}$$

is the potential barrier height in units of D.

Finally, the correlation approach sketched here may be used for the calculation of corrections to the rate coefficient due to finite memory effects.

4. RATE PROCESSES INVOLVING COMPLEX SYSTEM STATES

Fixed points are not the only possible states of dissipative, far-from-equilibrium systems; more complex macroscopic attractors, like limit cycles or even strange attractors, are commonly observed [19]. Bistabilities between the different attractor types may occur and give rise to interesting transition rate processes when these systems are subjected to external noise. We examine some of the new features that enter the calculation of the transition rate by examining some specific examples of systems displaying bistability between a fixed point and a limit cycle, but the discussion can be generalized to other situations.

Consider an irreversible, exothermic chemical reaction, $A \rightarrow B$, taking place in a continuously stirred tank reactor (CSTR) thermostated by a heat bath. Assuming that the rate coefficient of the reaction has an Arrhenius form, the macroscopic kinetic equations for the (dimensionless) composition, c, and temperature, T, can be written as,

$$
\begin{aligned}
\frac{dc}{dt} &= -\epsilon c + D(1-c)e^{\gamma T/(1+T)}, \\
\frac{dT}{dt} &= -(1+\epsilon)T + BD(1-c)e^{\gamma T/(1+T)} + \eta,
\end{aligned}
\tag{4.1}
$$

which we term the GK model [20]. The system state is determined by the control parameters B, D, γ, ϵ, and η. The relevant parameters for our purpose are ϵ and η, which are related to the flow rate and bath temperature, respectively. These deterministic equations show bistability between a limit cycle and fixed point for a range of control parameter values. Figure 1 is a plot of these coexisting attractors for two sets of parameter values: $1 = (\epsilon_1 = 0.375, \eta_1 = -0.034)$ and $2 = (\epsilon_2 = 0.35, \eta_2 = -0.0312)$, with $B = 0.22$, $D = 0.25$ and $\gamma = 100.0$ in both cases.

The two-dimensional (c, T) phase space for each of these systems can be partitioned into basins of attraction of the distinct attractors. Naturally, the basin structure is more complicated than that of the one-dimensional model studied in Sec. 3, but it is still quite simple. In the one-dimensional model the two stable fixed points were separated by an unstable fixed point, which formed a boundary separating the basins of attraction of the fixed points. In the GK model for the specified parameter setting the situation differs in that one of the stable fixed points has undergone a Hopf bifurcation to yield a stable limit cycle. The basin boundary in the two-dimensional space now consists of the unstable fixed point along with its stable manifolds; these are shown in Fig. 1 and labelled B_1 and B_2.

The nature of the noise-induced transition process depends on the type of external noise applied to the system. Here, we study a Poisson-dichotomous noise process where the system parameters switch between 1 and 2 at time intervals τ taken from an exponential distribution with correlation time τ_n,

$$
p(\tau) = \tau_n^{-1} \exp(-\tau/\tau_n).
\tag{4.2}
$$

This colored noise process can induce irreversible transitions from the limit cycle region to fixed point region. The definition of these regions needs to be made precise, and constitutes the species specification problem for this system.

The mechanism of the one-way transition process is easily deduced from an examination of Fig. 1 [13]. Suppose the system is initially in the limit cycle state $L2$. Note that the basin boundary B_1 intersects $L2$; thus, if a noise event occurs while the phase point is on $L2$ and to the left of B_1, it will find itself in the basin of $F1$ and tend to evolve to $F1$ in the absence of other noise events. Alternatively, if it is to the right of B_1 when the noise acts, it will evolve to $L1$. The transition occurs by a shift of the basin boundary under the action of the parametric noise process (moving boundary mechanism) [13]. A similar argument shows that once a phase point is near $F1$ or $F2$ it can never escape since both of these fixed points lie to the left of B_1 and B_2. Hence, under the stochastic dynamics the fixed point region will appear fuzzy as a result of noise-induced hops between $F1$ and $F2$; likewise there will be noise-induced hops between $L1$ and $L2$ leading to a fuzzy limit cycle.

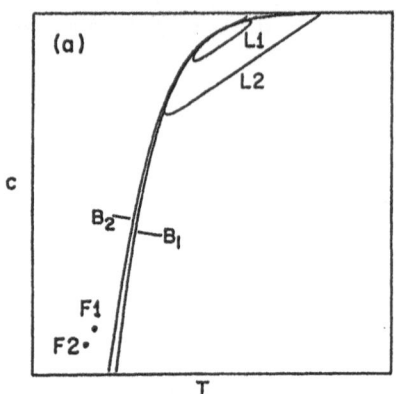 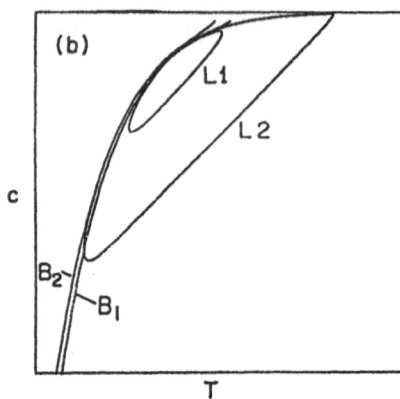

Fig. 1. (a) Plot of limit cycles $L1$ and $L2$, fixed points $F1$ and $F2$ and basin boundaries B_1 and B_2 for parameter values 1 and 2 given in the text. (b) Enlargement of the limit cycle region of Fig. 1a.

Only the portion of this fuzzy L region to the left of B_2 is vulnerable in the sense that it gives rise to transitions from L to F.

The basin boundary region has some of the characteristics of potential barrier. Phase points near B_2 and to its left, regardless of the parameter values, 1 or 2, will continue to move to the left toward F; while those to the right of B_1, again regardless of parameter values, will move to the right toward L. Within the very narrow region between B_1 and B_2 a noise event can change the fate of a potentially "reactive" trajectory. On the basis of these considerations the phase space can be partitioned into F and L regions by selecting a dividing surface to lie between B_1 and B_2. Characteristic functions defined in the two regions then provide a precise specification of the species. It is also possible to define stable F and L states by defining characteristic functions in the immediate vicinities of the stable attractors. The two fixed points lie well to the left of B_1 and B_2 so small region of phase space encompassing both $F1$ and $F2$ serves to define the stable F state. A stable L state can be defined by excluding the vulnerable region.

Given this qualitative description of the transition process, the rate of the "reaction" $L \to F$ can be computed. A complete calculation along the lines presented in Sec. 3 gives rise to a number of problems associated with the solution of the multi-dimensional Fokker-Planck equation for this system. While this is a promising route for future developments in this area, we instead show that many features of the rate can be accounted for by a simple stochastic model.

The stochastic dynamics takes its simplest form when the correlation time of the noise events is long compared to the periods of the limit cycles $L1$ and $L2$. In this circumstance, the phase points are largely confined to the limit cycles, with infrequent hops between them. The short-time decay depends on the initial preparation of the system: if the system is initially in $L2$, then those phase points in the vulnerable region will decay directly to F provided a second noise event does not act before the phase points cross B_2. If the system is initially in $L1$, then $L2$ must be populated before escape can occur. The long-time decay, which is independent of the system preparation, yields the rate coefficient of the process $L \to F$. The results of simulations shown in Fig. 2 verify the existence of a simple macroscopic rate law for the decay of the fraction of phase points in L:

$$\frac{dn_L(t)}{dt} = -k_f\, n_L(t). \tag{4.3}$$

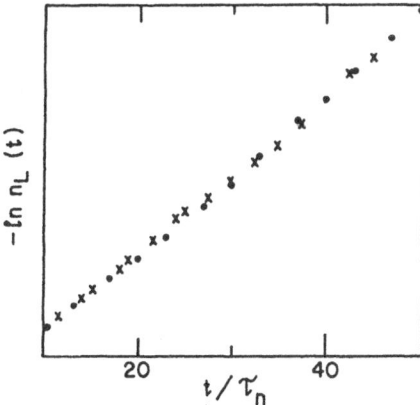

Fig. 2. Plot of $-\ell n\, n_L(t)$ versus t/τ_n for two values of τ_n : $\bullet, \tau_n = 10$; X, $\tau_n = 20$. Note the scaling with τ_n in accord with (4.5).

Furthermore, a simple estimate for the rate coefficient adequately describes the measured values for k_f. The measure of the vulnerable region can be written approximately as,

$$\beta = t_v/(\tau_1 + \tau_2), \tag{4.4}$$

where t_v is the time it takes a phase point on $L2$ to transverse the vulnerable region, and τ_1 and τ_2 are the periods of $L1$ and $L2$, respectively. The rate is then just the measure of the vulnerable region, β, times the noise frequency:

$$k_f = \beta/\tau_n. \tag{4.5}$$

While this simple description captures the gross features of the rate process, it is possible to write more elaborate stochastic models that account for the internal dynamics in L. Let $p_i(\phi, t)$ be the probability density that the system is in Li at phase ϕ at time t. A kinetic equation that should apply provided the noise correlation time is not too short is

$$\frac{\partial p_i(\phi,t)}{dt} = -v_i(\phi)\frac{\partial}{\partial \phi} p_i(\phi,t) - \tau_n^{-1} \sum_{j=1}^{2} \int d\phi' W\left(i\phi/j\phi'\right) p_j(\phi',t) - \tau_n^{-1}\theta_v(\phi)\delta_{i2} p_i(\phi,t). \tag{4.6}$$

The first term on the r.h.s. corresponds to the oscillatory motion on cycle Li with velocity $v_i(\phi)$, the second term accounts for the jumps between $L1$ and $L2$ (non-reactive events), while the last term gives rise to reactive events, with $\theta_v(\phi)$ defined to be non-zero in the vulnerable phase region on $L2$. Rather than pursue this development, we next consider another model which shows some different features.

A simple stochastic two-dimensional model whose underlying deterministic dynamics supports a coexisting limit cycle and fixed point, with a different geometry from that discussed above, is [12]

$$\frac{dx}{dt} = c(1-r)(r-r_o)x - 2\pi[1+\epsilon(1-r)]y + d\sum_{i}\delta(t-\tau_i),$$

$$\frac{dy}{dt} = c(1-r)(r-r_o)y + 2\pi[1+\epsilon(1-r)]x. \tag{4.7}$$

Here $r^2 = x^2 + y^2$. The term proportional to d is the external noise source on the deterministic dynamics; it consists of a series of impulsive events at times r_i drawn from the exponential distribution (4.2) (shot noise), that displace x by d units. The deterministic system has a stable circular limit cycle with unit radius, and a stable fixed point at the origin. These are separated by an unstable circular limit cycle with radius r_o, which acts as the basin boundary separating the two stable attractors. Fixed point, F, and limit cycle, L, species can be simply defined for this system:

$$\chi_{\ } (x,y) = \theta(r_{\ } - r), \chi_{\ } (x,y) = \theta(r - r_{\ }). \tag{4.8}$$

Stable states can be defined by considering a small-diameter disc about the origin for F, and an annulus about $r = 1$ for L.

The noise is not parametric in this case, and the effect of a noise event is to displace the phase point rather than the basin boundary. One feature that this model has in common with the GK model is the existence of a vulnerable phase region. If $d > 0$, then the noise displaces the phase point to the right along x. A set of phases on L whose measure is

$$\mu_v = \frac{1}{\pi} \arcsin\{(2d)^{-1}[4r_o^2 d^2 - (d^2 + r_o^2 - 1)^2]^{1/2}\}, \tag{4.9}$$

will lead to direct transitions into the basin of F. Once in F every noise event will lead to a transition to L, provided $d > r_o$. Thus, two-way transitions between F and L, i.e. the reversible reaction $F \rightleftharpoons L$, occur in this model.

A very crude model for the "nonequilibrium" constant of this reaction can be constructed if $d > r_o$ so that direct transitions dominate the rate process, and the noise correlation time is not too short. If only direct transitions are considered, the rate $k_{L \to F}$ for $L \to F$ may be written as $k_{L \to F} = \mu_v \tau_n^{-1}$, since, as in the discussion of the GK model, it is proportional to the noise frequency and the measure of the vulnerable region. For this to be valid the noise correlation time must be long compared to the period so that phase points are largely confined to the deterministic attractors. Once at the fixed point any noise event will cause the system to hop to L so, $k_{F \to L} = \tau_n^{-1}$. Hence, the population ratio in the nonequilibrium state, K_{neq}, is given by,

$$K_{neq} = \frac{k_{L \to F}}{k_{F \to L}} = \mu_v. \tag{4.10}$$

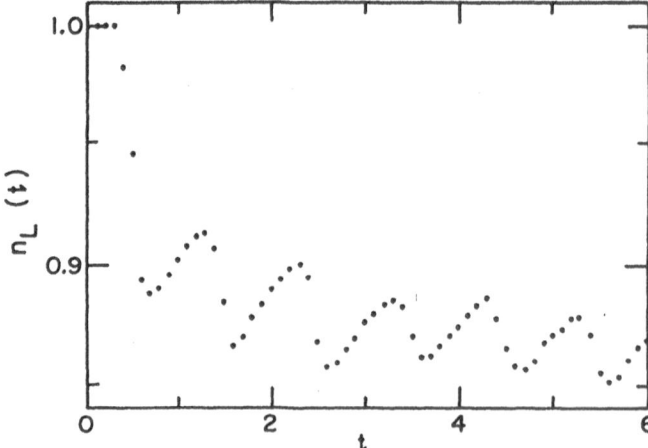

Fig. 3. Plot of $n_L(t)$ versus t for an ensemble of phase points initially at F.

In order to test this model, K_{neq} was computed from simulations of 10 stochastic trajectories which were followed for 5000 periods with $r_o = 1/2$. For $d = 0.6$ the results are $K_{neq}(\tau_n = 1.0) = 0.145 \pm 0.008$, $K_{neq}(\tau_n = 2.0) = 0.137 \pm 0.01$ and $K_{neq}(\tau_n = 5.0) = 0.129 \pm 0.019$. The measure of the vulnerable region for $d = 0.6$ is $\mu_v = 0.124$; thus, when $\tau_n >> 1$ the results of the simulation approach the predicted value.

This crude model does not take into account the possibility of correlated noise events. Even if $d < r_o$ transitions are still possible if a series of noise events act before relaxation to the stable attractors is complete. This effect is clearly manifest in K_{neq} : if only direct transitions are taken into account K_{neq} is predicted to be zero if $d < r_o$, but simulations show $K_{neq}(\tau_n = 5.0) = 0.09 \pm 0.01$ for $d = 0.4$. The simple model must be refined to describe such correlated effects.

The nature of the relaxation to the nonequilibrium asymptotic state can also be used to probe the transition process. As an example, Fig. 3 shows the decay of the fraction of phase points in L as function of time. The results were obtained from an average over 2000 stochastic trajectories with $d = 0.6$ and $\tau_n = 2.0$. The system was initially prepared in the fixed point F. The first noise event will immediately cause a transition to the L region at phase $\phi = 0$. Note that the population in L does not fall until the phase points have had time to move into the vulnerable region. Thereafter, the decay shows oscillations arising from the above mechanism: phase points which reach F are pumped to $\phi = 0$ on L and the density builds up until the vulnerable phase region is reached.

5. DISCUSSION

Rate processes occurring both close to equilibrium and far from equilibrium pose challenges for theory. For close-to-equilibrium rates, the microscopic details of the reactive event in a dense many-body system are crucial to an understanding of the rate, and the task of theory is to formulate tractable models for realistic situations. For far-from-equilibrium, noise-induced rate processes, complex macroscopic system states may be involved in the transition process, and the underlying deterministic system may itself lead to chaotic dynamics.

We showed above how some of the techniques which have proved useful in the study of close-to-equilibrium reactions may also be applied to noise-induced transitions between steady states. Here the attractive feature is the ability to control the statistics of the external noise source which induces the transition. Variations in the character and properties of the noise source permit the dynamics of the transition process to be probed.

The appearance of oscillating system states presents new features in the rate calculation. The existence of phases in the cycle which are especially vulnerable to noise events is a distinctive feature of such systems, which can have important consequences (see especially the biological literature on this topic [12]). When one of the attractors is chaotic, then the interplay between deterministic chaos and external noise must be considered in calculating the transition rate [21].

Acknowledgments

The research was supported in part by grants from the Natural Sciences and Engineering Research Council of Canada and the Petroleum Research Fund administered by the American Chemical Society.

REFERENCES

1. S.R. DeGroot and P. Mazur, **Non-Equilibrium Thermodynamics** (North-Holland, Amsterdam, 1962).
2. S. Glasstone, K.J. Laidler, and H. Eyring, **The Theory of Rate Processes** (McGraw-Hill, New York, 1941).
3. H.A. Kramers, Physica **7**, 284 (1940).
4. R.F. Grote and J.T. Hynes, J.Chem.Phys. **73**, 2715 (1980).
5. R. Kapral, Adv.Chem.Phys. **48**, 71 (1981).
6. T. Yamamoto, J.Chem.Phys. **33**, 281 (1960); D. Chandler, J.Chem.Phys. **68**, 2959 (1978). J.L. Skinner and P.G. Wolynes, J.Chem.Phys., **69**, 2143 (1978); J.C. Keck, Adv.Chem.Phys. **13**, 85 (1967); F.H. Stillinger, in **Theoretical Chemistry Advances and Perspectives**, vol. 3, eds. H. Eyring and D. Henderson (Academic Press, New York, 1978).
7. H. Mori, Prog.Theor.Phys. (Kyoto) **33**, 423 (1965); R. Zwanzig, J.Chem.Phys. **33**, 1338 (1960); Lect.Theor.Phys. **3**, 106 (1961).
8. J.T. Hynes, in **The Theory of Chemical Reaction Dynamics**, vol. 4, ed. M. Baer (CRC Press, Boca Raton, FL, 1985).
9. W. Horsthemke and R. Lefever, **Noise-Induced Transitions. Theory and Applications in Physics, Chemistry and Biology** (Springer-Verlag, Berlin, 1984).
10. J.C. Roux, Physica D**7**, 57 (1983); P. de Kepper and J. Boissonade in **Oscillations and Travelling Waves in Chemical Systems**, eds. R. Field and M. Burger (Wiley, New York, 1984); M. Alamgir and I. Epstein, J.Am.Chem.Soc. **105**, 2500 (1983); O. Decroley and A. Goldbeter, Proc.Natl.Acad.Sci.USA **79**, 6917 (1982); E. Celarier and R. Kapral, J.Chem.Phys. **86**, 3357 (1987); **86**, 3366 (1987).
11. C.M. Bowden, M. Clifton and H.R. Robb, **Optical Bistability** (Plenum Press, New York, 1981); K. Ideda, H. Diado and O. Akimoto, Phys.Rev.Lett. **45**, 709 (1980).
12. A.T. Winfree, **The Geometry of Biological Time** (Springer-Verlag, New York, 1980).
13. R. Kapral and E. Celarier in **Noise in Nonlinear Dynamical Systems**, eds. F. Moss and P.V.E. McClintock, (Cambridge University Press, Cambridge, 1988).
14. S.H. Northrup and J.T. Hynes, Chem.Phys.Lett. **54**, 248 (1978); J.Chem.Phys. **69**, 5261 (1978); J.Chem.Phys. **69**, 5246 (1978); J.Chem.Phys. **73**, 2700 (1980).
15. K. Lindenberg, B. West and J. Masoliver, in **Noise in Nonlinear Dynamical Systems**, eds. F. Moss and P.V.E. McClintock (Cambridge University Press, Cambridge, 1988).
16. J.M. Sancho, M. San Miguel, S.L. Katz and J.D. Gunton, Phys.Rev.A **26**, 1589 (1982); P. Grigolini and F. Marchesoni, Adv.Chem.Phys. **62**, 29 (1985); P. Grigolini, Phys.Lett.A **119**, 157 (1986); J. Masoliver, B.J. West and K. Lindenberg, Phys.Rev.A **35**, 3086 (1987).
17. P. Hänggi and P. Riseborough, Phys.Rev.A **27**, 3379 (1983); C. Van den Broeck and P. Hänggi, Phys.Rev. A **30**, 2730 (1984); M.A. Rodriguez and L. Pesquera, Phys. Rev. A **34**, 4532 (1986).
18. I. L'Heureux and R. Kapral, to be published.
19. Y. Pomeau, P. Bergé and C. Vidal, **Order Within Chaos** (Wiley, New York, 1986).
20. M. Golubitsky and B.L. Keyfitz, SIAM J.Math.Anal. **11**, 316 (1980).
21. R. Kapral, M. Schell and S. Fraser, J.Phys.Chem. **86**, 2205 (1982).

DISCUSSION

ERDI - The coexistence of limit cycles has been demonstrated in a Rössler model. What about the possibility for finding coexistent strange attractors in systems of polynomial ODE ?

KAPRAL - There are a number of examples of such phenomena in models for chemical and biological systems, and one can consider external noise processes that induce transitions between a chaotic state and another attractor. One example that we (R. Kapral, M. Schell and S. Frazer, J. Phys. Chem. 86, 2205, 1982) have considered is a forced non linear oscillator in the presence of external noise. The deterministic dynamics exhibits bistability between a chaotic state and a periodic state. The rates of transition between the chaotic state and the periodic state were studied as a function of the amplitude of the external noise and, in some circumstances, interesting "non-Arrhenius" behavior was observed.

BORGIS - Did you apply your ideas to other kinds of noise than a Poisson dichotomous noise and, in particular, are there any qualitative differences between a bounded and unbounded noise process ?

KAPRAL - Yes, we have examined a number of other types of noise process. A general feature of rate processes induced by external noise is the diverse character of the response : it depends on the statistics and nature of the noise process. For the Rössler model some of this work is described in E. Celarier and R. Kapral, J. Chem. Phys. 86, 3357, (1987), 86, 3366 (1987). In this paper three types of noise process are examined : Poisson-dichotomous, Poisson-uniform and Gaussian-white. Naturally there is a fundamental difference between bounded and unbounded noise processes. While sharp criteria for the onset transition cannot be established for unbounded noise, qualitatively, the dynamics is not greatly different from that for bounded noise for normal observation times.

MICHEAU - Could you make some comments about the comparison between near to and far from equilibrium phase transitions. Especially, about the universality of some critical exponents ? For example in gas-liquid phase transition and thermodynamics to flow branch transitions in a bistable CSTR ?

KAPRAL - There are a number of close analogies between phase transition phenomena close to equilibrium and far from equilibrium. There are especially close for the Schlögl model I discussed and the analogy was exploited in Schlögl's original paper on this model system. However, while the steady-state structure of the Schlögl model is like that of a Van der Waals gas, it is probably worth noticing that the phase separation process in the chemical system is governed by non-conserved order-parameter dynamics, while in the gas-liquid system the order parameter is conserved.

HYNES - You emphasized using ideas and techniques from near equilibrium kinetics to understand far from equilibrium phenomena. Have you considered now going back in the other direction ? For example, can you visualize a molecular system (probably more complex than say a simple isomerization) for which the double well plus Poisson noise far from equilibrium problem you described might apply ?

KAPRAL - Part of the motivation for the present study was, of course, the desire to realize rate processes described by simple stochastic dynamics, a feature which is not evident in the microscopic dynamics underlying the rate coefficient. However, it is possible to construct some possible examples. Consider a Poisson-dichotomous noise process. One might imagine a bistable oscillator coupled to a solvent degree of freedom that itself randomly hops between two internal states. The random hop of this bath degree of freedom when coupled to the oscillator will act as a dichotomous noise process.

INTRAMOLECULAR ELECTRON TRANSFER IN THE LIQUID PHASE

Jean Pierre Launay

Laboratoire de Chimie des Métaux de Transition
Université Pierre et Marie Curie
4 Place Jussieu 75252 Paris Cedex 05

INTRODUCTION

Intramolecular electron transfer is currently the subject of much experimental and theoretical studies, due to its wide occurence in diverse fields of chemistry, physics and biology[1]. From a fundamental point of view, it is one of the simplest conceivable chemical reactions. Its study in discrete molecular species is a very active research topic and could lead to a better understanding of the process of charge migration in condensed matter. This studies are performed on chemical systems containing two redox sites linked by some kind of bridge. In addition to their interest for fundamental research, such systems can be considered as precursors of "molecular" electronic devices[2].

Intramolecular electron transfer can be experimentally studied in two general classes of compounds:

(i): In system of the donor-acceptor type which can be prepared at $t = o$ in a far-from equilibrium excited state, e.g.:

$$D - A \xrightarrow{h\nu} D^+ A^-$$

Then the return to equilibrium is followed by a physical method (luminescence or absorption) in the time domain.

(ii): In systems like mixed valence compounds which exhibit two degenerate ground states:

$$M^{n+} - L - M^{(n+1)+} \rightleftharpoons M^{(n+1)+} - L - M^{n+}$$

The electron transfer occurs thermally (spontaneously), but in addition an optical process is possible. It gives rise to the so-called "intervalence band" which generally occurs in the near infra red[3]. These systems are thus commonly studied through the properties of the intervalence band which is recorded in a static experiment. The advantage of this indirect method is the quasi-absence of interference from the intermolecular process. Note that direct information on the thermal process is also possible, but less easily, by methods such as EPR[4].

The role of solvent in intramolecular electron transfer has been recognized for a long time. Although the solvent and more generally the external medium is not usually directly on the electron transfer path, it plays a role through the variations of its polarization in response to charge migration. Thus the electron transfer rate can be affected by the properties of the outside medium, as well as by the structure of the molecule itself.

315

In this paper we shall begin by a short historical overview of the different theories of electron transfer. This overview will be of course limited, in order to emphasize the physical principles involved in electron transfer. For additional details, exhaustive reviews of the different theoretical treatments can be found in refs[1]. Moreover, we shall restrict ourselves to theories of quantum mechanical nature. Thus, stochastic models (cf for instance the Kramers model) are not discussed here, but they are treated elsewhere in this book. We shall then focus on recent developments in intramolecular electron transfer and its solvent influence.

OVERVIEW OF THEORIES

The Hush-Marcus Model

It is well established that electron localization yields to a polarization of the intramolecular and extramolecular media. Thus, in coordination complexes, a change in oxidation state produces a change in metal ligand bond lengths. Consider for instance a binuclear mixed valence compound where an electron can exchange between two chemically equivalent sites. Since electronic transfer can occur only when the two metal sites have identical environments, a rearrangement of metal ligand bond distances must occur prior to electron transfer. It is then possible to compute in a semi-classical way the contribution E_{in} of the internal medium to the activation energy, which is given by[5] :

$$E_{in} = \frac{n \ \ f_2 \ f_3 \ (d_2^o - d_3^o)^2}{2 \ (f_2 + f_3)}$$

(1)

where n is the number of ligands, f_2 and f_3 are the force constants and d_2^o, d_3^o the equilibrium bond lengths in the two oxidation states (2+ and 3+ for instance).

A similar reasonment can be made for the case of the outside medium (solvent). It is treated as a dielectric continuum experiencing a non equilibrium polarization. This is necessary because, owing to the time scale of electron transfer processes, only the orientational part of the solvent polarization plays a role, while the electronic part is inactive. The final result E_{out} for the contribution of solvent to the activation energy is[6] :

$$E_{out} = \frac{1}{4\pi \ \varepsilon_0} \ (\Delta e)^2 \ (\frac{1}{\varepsilon_{op}} - \frac{1}{\varepsilon_S}) \ (\frac{1}{2a_A} + \frac{1}{2a_B} - \frac{1}{R})$$

(2)

where Δe is the amount of charge which is transferred in the activated state (typically 0.5 e), a_A and a_B are the radii of the two sites, R is the intersite distance, ε_S is the static dielectric constant, and ε_{op} is the dielectric constant at optical frequencies taken as the square of the index of refraction n.

Expressions (1) and (2) are the basis for the Hush-Marcus model. They allow the construction of potential energy curves of parabolic shapes, when the energy is plotted as a function of a composite reaction coordinate. These curves in turn are the basis for an elementary description of the thermal and optical processes in mixed valences complexes. In principle, it is possible to compute a rate constant from this model, using the total reorganization energy as an activation energy and introducing an electronic transmission factor calculated by the Landau-Zener formula. However this procedure is now supplanted by the quantum models.

316

Quantum Models

The classical model based on potential energy curves does not explain correctly the existence of nuclear tunneling at low temperature nor the difficulties in observing the inverted region (i.e. the region where the rate of reaction should <u>decrease</u> when the exothermicity increases) . It is thus necessary to devise quantum models taking into account the existence of vibronic levels. The electron transfer is then a non radiative process between manifolds of vibronic levels and the rate can be calculated from the Fermi Golden Rule:

$$P_{aj} = \frac{2\pi}{\hbar} < \Psi^o_{bn} \mid \hat{V} \mid \Psi^o_{aj} >^2 \rho_{(E^o_{bn} = E^o_{aj})}$$

(3)

where P_{aj} is the transfer probability per unit time from the vibronic level denoted by electronic and nuclear indexes a and j, to a set of closely spaced final states. In this expression, ψ^o_{aj} and ψ^o_{bn} are the initial and final vibronic wavefunctions, $\hat{V}$ is the electronic coupling operator and ρ is the value of the density of final states for $E^o_{bn} = E^o_{aj}$. The vibronic wave functions are then written as a product of electronic ϕ_b and nuclear χ^o_{bn} wave functions. Taking into account that $\hat{V}$ acts only on the electronic wavefunction, one obtains:

$$P_{aj} = \frac{2\pi}{\hbar} V^2_{ab} <\chi^o_{bn} \mid \chi^o_{aj} >^2 \rho_{(E^o_{bn} = E^o_{aj})}$$

(4)

where V_{ab} is now the electronic coupling matrix element. The final rate is obtained after averaging according to the Boltzmann distribution on the vibronic levels of the initial electronic configuration. The first use of the Fermi Golden Rule was reported by Levich and Dogonadze[7], using the same physical principles as in the Hush-Marcus model. However, a more general treatment has been performed later by Jortner[8], in which intramolecular and solvent relaxations are each represented by a single effective frequency and a single vibronic coupling parameter. Of course the situation is different for the internal (metal-ligand for instance) vibrational modes, because the energy levels are widely separated (ca 500 cm^{-1}) while for the outer (solvent) modes, the energy levels are closely spaced (ca 1 cm^{-1}). The final expression is rather complicated. One can distinguish three classes of behaviour: (i) at very low temperatures only the lowest vibrational levels are populated and the reaction can only occur by nuclear tunneling, (ii) in the intermediate range, the intramolecular modes behave quantum mechanically but the solvent behaves classically because kT is much greater than the corresponding energy separation , (iii) finally at high temperatures, the behaviour converges with the one of the classical model. Thus the apparent activation energy is not constant and decreases with temperature. The only simple prediction is that the rate is proportional to the square of the electronic coupling matrix element.

There are two critiques which can be made to a Quantum Model based on the Fermi Golden Rule. First it is based on the assumption of a Boltzmann equilibrium for the initial state, which may not be true for ultrafast reactions. Second, it is in fact a static model to which the dynamic effect has been added as a perturbation through the use of the Fermi Golden Rule. Thus it is not valid for strong electronic coupling[9]. Also as a consequence of its static origin, it is implicitly based on the Born-Oppenheimer approximation, which in fact breaks down in this particular case because the electronic wave functions are sensitive functions of the nuclear coordinates[9]. Incidentally, this breakdown of the Born-Oppenheimer approximation has important consequences for the dynamic behaviour; in particular it is no longer possible to define from the Hamiltonian a force acting on the system, and thus molecular dynamics calculation are not possible[10]. The Vibronic Coupling Model, which is exposed below, is an answer to this second critique.

The Vibronic Coupling (PKS) Model

The vibronic coupling model of Piepho, Krausz and Schatz (PKS) has been introduced recently in order to treat correctly electron transfer by using the complete Hamiltonian. The PKS model emphasizes in particuliar the breakdown of the Born Oppenheimer approximation.

Originally, the PKS model has been devised to describe the intervalence band profiles[11]. The complete Hamiltonian of the system is written down, without omitting nuclear kinetic energy terms. It is then used with a vibronic basis and the resulting matrix is diagonalized to provide eigenenergies and eigenvectors. Consequently the concept of potential energy curves vanishes and the system is entirely described by vibronic levels (of course potential energy curves can be calculated for pedagogical purposes, to visualize the amount of ground state trapping for instance, but this is not a necessity). From the fitting of the experimental spectrum, one determines two parameters: ε representing the electronic coupling between the two redox sites, and λ which is the vibronic coupling parameter. The latter is in fact an effective average parameter describing both the intramolecular and solvent polarizations.

From the PKS model, two attempts have been made to predict rate constants. The first one (Wong and Schatz)[9] uses the transmission method suggested by Weiner. A transmission coefficient K_n between two levels E_n and E_{n+1} is calculated as :

$$K_n = \pi^2 \left(\frac{E_{n+1} - E_n}{h\nu_0} \right)^2$$

(5)

and the jump frequency between these levels is obtained by :

$$P_n = \nu_0 K_n$$

(6)

ν_0 being the frequency factor, equal to the vibration frequency. Finally the jump frequency is averaged by the Boltzmann factors as above. Rates obtained by this procedure converge towards the values obtained by the Fermi Golden Rule when the electronic coupling is weak.

The second method (Babonneau and Livage)[12] states that the rate constant can be written for each vibronic level i as a product of three terms:

$$\nu_i = \nu_0^i \; K_n^i \; K_e^i$$

(7)

where ν_0^i is a vibrational frequency factor, K_n^i a nuclear factor corresponding to the probability for the donor and acceptor sites to be in the right configuration and K_e^i is an electronic factor representing the probability for the electron transfer to occur when the right nuclear configuration has been reached. K_n^i is then equated to the integrated nuclear probability distribution value for the symmetrical configuration, while K_e^i is taken as the square of a factor γ_i related to the overlap density. Finally the result is averaged by the usual Boltzmann distribution. The results obtained by this last procedure are comparable (but not identical) with those obtained by Wong and Schatz. Both methods predict that for weak electronic coupling, the rate increases with temperature, except at low temperatures where nuclear tunneling occurs. As the electronic coupling increases, the rate increases and becomes less temperature dependent.

To summarize, it is thus possible from the analysis of a single intervalence spectrum (i.e. from the <u>optical</u> electron transfer process) to extract the PKS parameters ε and λ, and then to compute the rate of <u>thermal</u> electron transfer.

Ultrafast Reactions

In all the previous models, the Boltzmann distribution is assumed to hold for the populations of the vibronic levels in the initial state. Thus it is assumed that the rate of electron transfer is slow with respect to the rate of equilibration between the different vibrational levels. This assumption is no longer true for ultrafast reactions which are currently studied by picosecond spectroscopy. In this case, the concept of a thermally averaged rate becomes meaningless. This situation has been recently discussed by Bixon and Jortner[13] in the frame of a quantum mechanical model and it is clearly an open field of research.

Complex Systems: Effective Hamiltonian Theory

For large complex systems of general topology S_1—L—S_2 where S_1 and S_2 are redox active sites and L is a bridging ligand involving a large number of orbitals, the dynamic behaviour can be efficiently described through the use of effective Hamiltonian theory[14]. The problem is there to replace the actual system by a two sites model in which the electronic coupling is represented by an effective coupling parameter (ECP). Thus one has to describe the system in the L_1 subspace generated by the localized kets $|s_1>$ and $|s_2>$, instead of the complete model space L generated by all the possible electronic states. This is accomplished by the following three-step procedure:

i) A transformation matrix U is defined using the Bloch construction of effective Hamiltonians. This transformation is such that in the $U^{-1} H U$ representation, the energy matrix is block-diagonalized on L_1 and L_2 (L_2 being the orthogonal subspace of L_1). In addition a criterium of minimum distance between the actual wavefunction and the projection is used to completely determine the U transformation.

ii) The resulting $U^{-1} H U$ matrix is restricted to the 2×2 matrix on L_1, in order to define the new Hamiltonian of the problem.

iii) Finally, the effective coupling parameter (ECP) is calculated as:

$$ECP = 1/2 \, (<s_1 \mid P \, U^{-1} H \, U \, P \mid s_2> \, + \, <s_2 \mid P \, U^{-1} H \, U \, P \mid s_1>) \qquad (8)$$

where P is the projector on L_1, and the definition of ECP takes into account the possibility that U could be non unitary.

The effective Hamiltonian theory, in its present stage, does not take into account the effect of vibrations and medium reorganization, i.e. it is a purely electronic description. However, it is a very useful tool to describe complex systems in which the evolution between localized states can spread on a large number of electronic degrees of freedom. In addition it yields a satisfactory description of the <u>dynamical behaviour</u> by giving the envelope of the time evolution even for long times (while treatments based on Fermi Golden Rule are only valid for short times). We presently use this model to predict the possibilities of long distance electron transfer and of switching at the molecular level.

EXPERIMENTAL ASPECTS

Early Experiments.

One of the first study of intramolecular electron transfer in the liquid phase was performed by Harriman and Maki in 1963 on organic radicals containing two identical NO_2

groups, i. e. systems of general formula[15] :

$$NO_2 - \langle O \rangle - X - \langle O \rangle - \overset{\bullet}{NO_2}$$

where the monoradical was generated electrochemically from the parent neutral dinitro compound. Electron exchange between the two equivalent NO_2 sites could be monitored by EPR, and computer simulation of the spectra yielded the electron transfer rate. It was then observed that for $X = O$ and $X = S$ the rate was sensitive to the solvent, being roughly three times higher in dimethylsulfoxide than in acetonitrile. Few analogous studies have been reported since, but recently the method has been applied to dianthryl compounds in methyltetrahydrofuran[16].

Intervalence Electron Transfer

The use of the so-called intervalence band of mixed valence compounds to characterize electron transfer has received considerable attention in the recent years. After the pioneering work of H.Taube and T.Meyer[17], it became clear that it was a convenient method to study the solvent influence on intramolecular electron transfer. The basic equation is always eq. (2), obtained in the dielectric continuum model, in which one makes $\Delta e = 1$ e. Systematic studies with different bridging ligands showed that the influence of distance is correctly described. For the solvent itself, the band energy has been found to vary linearly as a function of $(1/\varepsilon_{op} - 1/\varepsilon_s)$ for dinuclear species of the type

$$[(bipy)_2 \, Cl \, Ru - L - Ru \, Cl \, (bipy)_2]^{3+}$$

However the linearity is lost in strongly hydrogen-bonding solvents (H_2O or alcohols) or with redox groups such as $(NH_3)_5 \, Ru^{2+/3+}$ or $(CN)_5 \, Fe^{II/III}$, for which specific interactions with the solvent are suspected.

Recently we have undertaken the study of mixed valence compounds of general formula [18]:

$$[(NH_3)_5 \, Ru^{III} - N \langle O \rangle - (CH = CH)_n - \langle O \rangle N - Ru^{II} (NH_3)_5]^{5+}$$

with n = 0 to 4. In these compounds, the two Ru atoms are linked by a conjugated pathway and the distance between Ru centers can reach 20 Å . When n increases, a consequence of equation (2) is that the energy of the intervalence band increases and thus it begins to be buried beneath the tail of the intense metal-to-ligand charge transfer band which is observed near 600 nm in these complexes. Fortunately these bands do not depend on the same solvent parameters. For the metal-to-ligand charge transfer, the relevant parameter is the donor number[19]. This is a consequence of hydrogen bonding between H atoms of the coordinated NH_3 molecules and the strongly donating solvent molecules acting as Lewis bases. Thus it can be experimentally shown that the higher is the donor number, the lower is the energy of the charge transfer band. Using these observations we have selected a weakly donor solvent exhibiting at the same time a low $(1/\varepsilon_{op} - 1/\varepsilon_s)$ parameter, i.e. nitrobenzene. This allowed a marked improvement in the resolution of the intervalence band with respect to a solvent such as D_2O . We are now in a situation where the variation of the electronic coupling with the number of double bonds in the bridging ligand can be tested[18].

Recent results on mixed valence compounds have shown that the solvent influence can lead to strange and spectacular effects. Thus preferential solvation is shown by the behaviour of a symmetrical system in mixtures of dimethylsulfoxide (DMSO) and acetonitrile (AN)[20]. The band energy increases in mixtures showing that the Ru^{III} end is solvated by DMSO, while the Ru^{II} end is solvated by AN. As a consequence, for an unsymmetrical system in which the two redox isomers have almost the same energy, a change in solvent can change the ground state electronic structure, yielding dramatic modifications in the absorption spectrum[21].

Fast Electron Transfer

The advent of pulsed methods has allowed the development of dynamic studies of electron transfer in the nanosecond, the picosecond and now the subpicosecond range. In principle, by using an unsymmetrical binuclear complex, it should be possible to study dynamic electron transfer after triggering by an optical excitation. This would allow the connexion with the study of spontaneous electron transfer. However, this kind of experiment has proved to be particularly difficult with binuclear coordination compounds. The first report has been made by Creutz, Netzel, Sutin et al[22] on the unsymmetrical system:

$$[(NH_3)_5 \, Ru^{II} \, pyz \, Ru^{III} \, (edta) \,]^+ \qquad (pyz = pyrazine)$$

They found that direct excitation on the intervalence band did not give the expected transient. This was attributed to the fact that the largest part of the activation barrier for back electron transfer came from the solvent. As the optical excitation is very fast, the solvent had no time to establish the barrier. It was thus necessary to use an indirect excitation using a metal-to ligand charge transfer band. Finally the rate of electron transfer between the two sites was found near 10^{10} sec^{-1}. More recently Meyer et al have described an alternate method, based on the creation of the mixed valence state in the photolysis step[23].

However most of the current work is performed on organic Donor-Acceptor systems. In some cases they allow the study of long distance electron transfer across saturated bridges,with large solvent effects arising from the relaxation around the "giant dipole" created after excitation[24]. Another fascinating example is provided by the intriguing case of twisted internal charge transfer states (TICT). They appear in several donor acceptor compounds, the prototype of which is dimethylaminobenzonitrile (DMABN)[25]. In the ground state, the molecule is planar, owing to the conjugation between the donor and acceptor parts. Upon photochemical excitation, a planar excited state with partial charge transfer is first formed. It then gives a twisted state with full charge separation. These two excited states are easily monitored by luminescence measurements, the planar state emitting near 350 nm and the twisted state near 480 nm. The luminescence near 480 nm is strongly solvent dependent, showing the polar character of the corresponding excited state. In fact, it is the relaxation of the solvent around this excited state which provides the driving force for the twisting motion. Time resolved luminescence studies performed by J.A. Miéhé and F. Heisel allow indeed the dynamic study of this solvent relaxation by providing instantaneous luminescence spectra at different times after excitation[26].

An interesting observation is that the decay of the excited planar form is clearly non exponential, while the decay of the twisted state is exponential, at least for "long" times, i.e. after the solvent has relaxed. Thus the deactivation of the twisted state is a "classical" reaction, because it occurs from a species which has had time enough to equilibrate with its surrounding. The twisted state corresponds to a local minimum in a potential-energy-versus nuclear-and-solvent-coordinates diagram. By contrast, the deactivation of the planar excited state is an example of ultrafast reaction for which the Boltzmann equilibrium is not reached (Note that the conversion to the twisted form involves almost no activation energy).

Additional information is now available since we have prepared coordination complexes containing the DMABN molecule as a ligand[27], i.e. $[(NH_3)_5 \, Ru \, (DMABN)]^{2+/3+}$ and $[(bipy)_2Cl \, Ru \, (DMABN)]^+$. They allow the study of the influence of complexation on the TICT process, and these studies are undertaken in order to investigate the possibility of switching at the molecular level. It is found that the luminescence is weakened, particularly the one corresponding to the twisted state. In addition the lifetimes are different, with the lifetime of the planar state being longer than in the free ligand. The higher lifetime of the planar (initial) state is apparently linked to a weaker yield in the twisted state[28]. Thus, it can be said in a qualitative way that the complexation of the

DMABN molecule hinders the formation of the TICT state. This is probably a consequence of the changes in the solvation shell induced by the introduction of the $(NH_3)_5 Ru^{2+}$ group: The driving force for the twisting motion seems largely due to the relaxation of the solvent shell around the strongly polar TICT state, and this process is probably perturbed when one end of the molecule already bears a bulky and charged group.

CONCLUSION

The complexity of an exact treatment of the solvent effects on intramolecular electron transfer has precluded such an analysis until now. Thus one has to use, for some time still further, macroscopic models such as the dielectric continuum model. Such models have indeed good predictive properties but they fail to describe specific solvent effects such as the donor ability or the H bonding ability. Even the more sophisticated quantum model, in its present form, is an oversimplification since the solvent motion is described by a single vibrational mode. The quantum model has some success because the vibronic levels corresponding to solvent modes are so closely spaced that in fact they can be approximated by a continuum. There is no doubt however that the progress in computing ability will allow in the future the simulation of the exact behaviour of the solvent in these reactions.

ACKNOWLEDGEMENTS

The author is endebted to F. Babonneau, J. A. Miehé, S. Bratos, and C. Joachim for helpful discussions.

REFERENCES

1. See for instance R.D. Cannon, "Electron transfer reactions", Butterworths, London, 1980; B. Chance, D. De Vault, H. Frauenfelder, R.A. Marcus, J.R. Schrieffer, N.Sutin, "Tunneling in Biological Systems", Acad. Press, New York, 1979; Progress Inorg. Chem.: An appreciation of H. Taube 30 (1983); T. Guarr and G. Mc Lendon, Coord. Chem. Rev. 68, 1 (1985). M. D. Newton and N. Sutin, Ann. Rev. Phys. Chem. 35, 437 (1984)
2. J.P. Launay, S. Woitellier, M. Sowinska, M. Tourrel and C. Joachim. Proc. 3rd Int. Symp. on "Molecular" Electronic Devices, Ed. by F.L. Carter and H. Wohltjen, North-Holland, Amsterdam, in press.
3. C. Creutz. Progr. Inorg. Chem. 30, 1 (1983).
4. C. Sanchez, J. Livage, J.P. Launay and M. Fournier. J. Am. Chem. Soc. 105, 6817 (1983).
5. N.Sutin. Ann. Rev. Nucl. Sci., 12, 285 (1962) ; N. Sutin, Progr. Inorg. Chem. 30, 441 (1983)
6. R.A. Marcus. Disc. Farad. Soc. 29, 21 (1960) ; N.S. Hush. Trans. Farad. Soc. 57, 557 (1961) ; see also R.D. Cannon, op. cit. p.201.
7. V. Levich and R.R. Dogonadze. Dokl. Akad Nauk SSSR 133, 158 (1960).
8. N.R. Kestner, J. Logan, J. Jortner. J. Phys. Chem. 78, 2148 (1974); J. Jortner. J. Chem. Phys. 64, 4860 (1976).
9. K.Y. Wong and P.N. Schatz. Progr. Inorg. Chem. 28, 369 (1981).
10. S. Bratos, Private communication
11. S.B. Piepho, E.R. Krausz and P.N. Schatz. J. Am. Chem. Soc. 100, 2996 (1978).
12. F. Babonneau and J. Livage. Nouv. J. Chimie, 10, 191 (1986).
13. M. Bixon and J. Jortner. Farad. Disc. Chem. Soc. 74, 17 (1982)
14. C. Joachim and J.P. Launay. Chem. Phys. 109, 93 (1986); C. Joachim and J.P. Launay. Proc. 3rd Int. Symp. on "Molecular" Electronic Devices, Ed. by F.L. Carter and H. Wohltjen, North-Holland, Amsterdam, in press.
15. J.E. Harriman and A.H. Maki. J. Chem. Phys. 39, 778 (1963).
16. W. Huber and K. Müllen. Acc. Chem. Res. 19, 300 (1986).

17. G.M. Tom, C. Creutz and H. Taube. J. Am. Chem. Soc. 96, 7827 (1974);
    M.J. Powers and T.J. Meyer. J. Am. Chem. Soc. 102, 1289 (1980);
    B.P. Sullivan, J.C. Curtis, E.M. Kober and T.J. Meyer. Nouv. J. Chimie
    4, 643 (1980).
18. C.W. Spangler, S. Woitellier and J.P. Launay, work in progress.
19. J.C. Curtis, B.P. Sullivan and T.J. Meyer. Inorg. Chem. 22, 224 (1983).
20. J.T. Hupp and J. Weydert. Inorg. Chem. 26, 2657 (1987); K.S. Ennix,
    P.T. Mc Mahon, R. de la Rosa and J.C. Curtis. Inorg. Chem. 26, 2660
    (1987).
21. J.T. Hupp, G.A.Neyhart and T.J. Meyer. J. Am. Chem. Soc. 108, 5350 (1986).
22. C. Creutz, P. Kroger, T. Matsubara, T. L. Netzel, and N. Sutin J. Am. Chem.
    Soc. 101, 5442 (1979)
23. K.S. Schanze and T.J. Meyer Inorg. Chem. 24, 2121 (1985) ; K. S. Schanze,
    G. A. Neyhart, and T. J. Meyer J. Phys. Chem. 90, 2182 (1986)
24. G. F. Mes, B. de Jong, H. J. van Ramesdonk, J. W. Verhoeven, J. M. Warman,
    M. P. de Haas, and L. E. W. Horsman van den Dool J. Am. Chem. Soc. 106,
    6524 (1984)
25. Z. R. Grabowski, K. Rotkiewicz, A. Semiarczuk, D. J. Cowley, and W.
    Baumann, Nouv. J. Chimie 3, 443 (1979). W. Rettig, Angew. Chem. Int.
    Ed. Engl. 25, 971 (1986)
26. F. Heisel and J. A. Miehé Chem. Phys. Lett. 128, 323 (1986)
27. M. Sowinska, J.P.Launay, J.Mugnier, J.Pouget, and B. Valeur
    J. Photochem., 37, 69 (1987)
28. M. Sowinska, F. Heisel, J. A. Miehé and J. P. Launay To be published

DISCUSSION

BARTHEL - What is the significance to use $1/n^2$ in the Marcus formula ? Is the assumptions equivalent to adopt n^2 as the permittivity of the solvation shells ?

LAUNAY - Not exactly. The use of $1/n^2$ in the Marcus formula comes from the theory of non equilibrium polarization of the solvent considered as a dielectric continuum. One may think that the contribution of the solvent to the activation energy is broken in two terms : a term due to orientational polarization and a term due to electronic polarization. Only the first term is kept, because it is slow and the corresponding rearrangement must occur before electron transfer (cf the case of the first coordination sphere). The second term is deleted because it is fast and thus can occur during electron transfer. Since at optical frequencies, only the electronic polarization can respond, this term $1/n^2 = 1/\epsilon_{op}$ allows the separation between the fast and slow components of polarization.

NEWTON - The fact that Prof. Barthel employed 5.5 instead of $n^2=1.8$ for underscores the fact that there is an intermediate regime for liquid water (i.e., between optical and the very low frequency regime there is an infrared region, giving $\epsilon_{IR} \sim 5$, as well as many other contributions). The Marcus model assumes that only the optical modes can follow the transferring electron : hence, ϵ_∞ is set equal to n^2.

The balance between electronic coupling and solvation is very delicate and determines which "mixed valence" systems are localized (on some timescale) and which are in fact delocalized. Do you know what is the solvent effect on the degree of localization for some of the systems which have been studied experimentally, including the so-called Creutz-Taube molecule ?

The vibronic PKS has enjoyed much success, but do you agree that with its limitation to antisymmetric vibrational modes, it becomes increasingly less valid as the strength of the electronic coupling increases. For the delocalized limits one expects most of the lineshape to be associated with the symmetric vibrational modes.

LAUNAY - About the solvent effect on the degree of delocalization, one could effectively imagine a situation in which changing the solvent would change the vibronic coupling parameter (λ in PKS treatment) enough to drive the system from a rather localized to a rather delocalized state. However I do not know examples of such an effect.

About the PKS model, it is true that its limitation to antisymmetric modes is a drawback and this has already been the subject of much debate. This could be a problem for delocalized systems, and also for bridged systems (which involve in fact three sites). However I think this model has really improved our understanding of the electron. Several groups are working on extensions of this PKS model, in particular Borshch in Kishinev.

ERDI - I'm wondering about how to overcome the difficulties due to the occurrence of ultrafast reactions when the Boltzmann equilibrium can not be reached. Do you have any suggestion how to treat the problem at the level of nonequilibrium statistical mechanics ?

LAUNAY - No. To my knowledge, the problem has been evoked only recently (Bixon and Jortner, Farad. Disc. Chem. Soc. 1982, 74, 17). There is clearly much to do and stochastic models (which I have not considered at all but which are the subject of several further lectures) will certainly be useful in this respect.

HYNES - A comment. The results you mention on specific solvent effects in mixtures suggest that these systems would be interesting candidates to experimentally study solvent dynamical effects. All continuum theories for these effects say in one way or another that it is only the solvent dielectric relaxation time that matters. The specific solvation effects suggest to me that this picture would break down seriously for such systems, and reveal new and interesting dynamic solvent effects.

LAUNAY - You are perfectly right. To show the influence of distance or medium on intramolecular electron transfer (T.J. Meyer et al) it is necessarry to carefully choose the system. Hydrogen bonding solvents in particular introduce a scattering of experimental data.

RATE CONSTANTS OF REACTIONS IN THE LIQUID STATE:

PRESENT STATE OF KRAMERS AND SMOLUCHOWSKI-DEBYE MODELS

Jacques A. Delaire

Physico-chimie des Rayonnements
UA 75 of CNRS
Universite de Paris-Sud
91405 – Orsay, Cédex, France

INTRODUCTION

It is mere common sense to say that chemical reactions in solution
are more complicated to describe than in the gas phase. In dilute gases,
the rate constant can be expressed in terms of an integral. This integral
takes into account the collision rate which can be calculated from the
kinetic theory of gases, and the probability that reaction occurs upon
collision, which can sometimes be approached by quantum mechanical
calculations. In solution, the molecules are constantly jostling one
another, and the notion of an isolated binary collision no longer makes
sense. However, it is still possible to express reaction rates through
phenomenological models in which the solvent intervenes through macroscopic parameters such as viscosity, static dielectric constant, etc...
We want to review these models here and to compare them with the results
of recent experiments.

A chemical reaction can often be considered as the crossing of a
potential barrier. As already pointed out by Northrup and Hynes (1978),
there are roughly three cases for barrier crossing dynamics. In the first
case, equilibrium distributions in position and velocity are approximately maintained by collisions, and transition state theory (TST) gives
rate constants with a good approximation. In the other two cases,
reactants and high energy states are in non equilibrium : the first one
arises when high energy states react very rapidly this is for example the
case of some low pressure unimolecular reactions, and the second one, on
the contrary, occurs when reaction is slowed down by collisions of the
reactants with the solvent. There are two classical examples belonging to
this regime : intramolecular reactions like isomerizations, and diffusion
controlled bimolecular reactions like dissociation-recombination in solution. We will successively deal with these two examples.

UNIMOLECULAR REACTIONS

Let us consider here reaction A → B where a molecule A can give B
after crossing a potential barrier in the bath of the surrounding
molecules of solvent.

Kramers model

This model idealizes the reaction as motion on a one-dimensional harmonic potential surface, and describes solvent friction by means of a single (constant) parameter. thus it is clear that the chemical identity of the reactive solute and of the solvent is not taken into account. However, the principal merit of such a model is that it provides analytical expressions for rate constants of unimolecular reactions in solutions. These expressions provide important information on the influence of the solvent.

Kramers (1940) considered a particle A moving in a potential U(x) of the type shown in Fig. 1

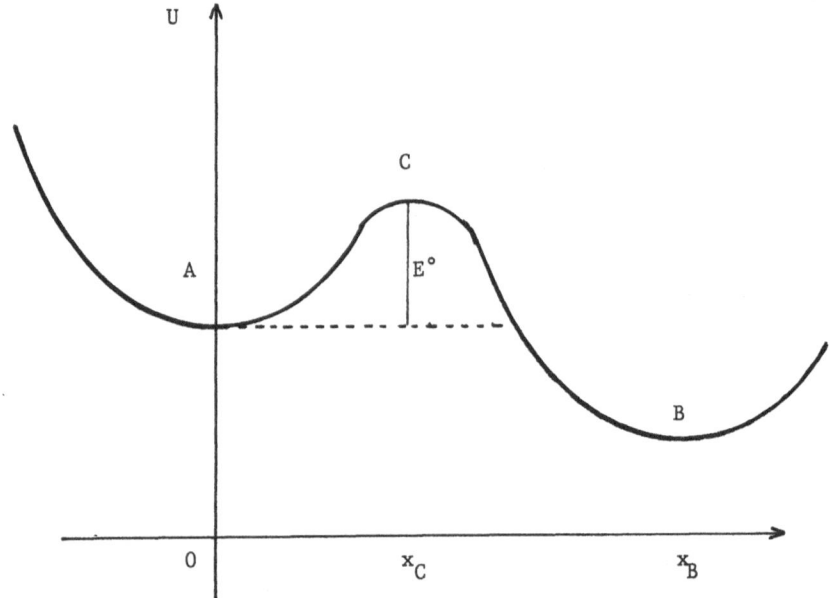

Fig.1. Potential for barrier crossing problem. The two potential wells are those of initial and final species A and B. E° is the activation energy of the reaction A → B.

Initially, all the particles are caught in the potential well A, and, as a consequence of Brownian motion, they escape over the potential barrier (E° is the activation energy). In the particular case where the height of the potential barrier is large compared to the thermal energy ($E° \gg k_BT$), a quasi-stationary flux of particles A at the barrier top C can be assumed. On the same time, a Maxwell-Boltzmann distribution of particles is assumed around A.

Another important aspect of the Kramers treatment is the fact that the differential equation describing the probability density w (x,t) for a particle A to be at position x at time t is derived from the basic Langevin equation :

$$\frac{du}{dt} = - \zeta u + f(t) \qquad (1)$$

where u denotes the velocity of the particle. According to this equation, the influence of the solvent on the motion of the particle is split into two parts : first, a systematic part $- \zeta u$ — representing a dynamical friction (ζ is called the friction coefficient) and second, a fluctuating part $f(t)$ which is characteristic of the Brownian motion.

Because of this fluctuating term, this differential equation is called stochastic and cannot be solved like an ordinary differential equation. The method consists in deriving from the Langevin equation another equation describing the probability W (r,u,t) for finding the particle at position r, with velocity u at time t. This equation, which is named the Fokker-Planck equation in position and velocity space, or the Chandrasekhar equation, reduces to the well-known Smoluchowski equation, provided we are interested in time intervals very large compared with the "time of relaxation" ζ^{-1}. In other words, this equation is mainly valid in the case of high friction. In the one-dimensional case, the Smoluchowski equation can be written in the following form (Chandrasekhar, 1943):

$$\frac{\partial w(x,t)}{\partial t} = \frac{\partial}{\partial x} \left(\frac{k_B T}{m\zeta} \frac{\partial w}{\partial x} - \frac{K}{\zeta} w \right) \qquad (2)$$

where w (x,t) is the probability for finding the particle at position x and at time t (whatever its velocity), k_B is the Boltzmann constant, T the absolute temperature, m the reduced mass of the particle, and K the force per unit mass caused by an external field.

We supposed a quasi-stationary flux Ø of particles, so it follows that

$$\emptyset = - \frac{k_B T}{m\zeta} \frac{\partial w}{\partial x} + \frac{K}{\zeta} w = \text{constant.} \qquad (3)$$

If K can be derived from a potential U so that

$$K = - \nabla U \qquad (4)$$

equation (3) can be rewritten in the form :

$$\emptyset = - \frac{k_B T}{m\zeta} \exp\left[- \frac{mU}{k_B T} \right] \nabla \left(w \exp\left[\frac{mU}{k_B T} \right] \right) \qquad (5)$$

Integrating equation (5) between any two points A and B, we obtain

$$\emptyset = \frac{\dfrac{k_B T}{m} w \exp\left[\dfrac{m U}{k_B T} \right] \Big|_B^A}{\displaystyle\int_A^B \zeta \exp\left[\frac{mU}{k_B T} dx \right]} \qquad (6)$$

This important equation, first derived by Kramers (1940), allows an easy derivation of the rate constant defined as the ratio ϕ/n_A, where n_A is the number of particles A. A harmonic potential is taken for U near A ($U_A = 1/2 \, \omega_A x^2$) and in the quasi stationary state, W is taken as zero near B. The potential near C is supposed to have a continuous curvature :

$$U \simeq \frac{E^\circ}{m} - \frac{1}{2} \, \omega_C^2 \, (x - x_C)^2 \qquad (x \simeq x_C) \qquad (7)$$

On these different assumptions, we get the following expression for the Smoluchowski limit of Kramers'equation k_{SL} :

$$k_{SL} = \frac{\omega_A \, \omega_C}{2 \pi \zeta} \, \exp(-E^\circ/k_B T) \qquad (8)$$

Let us recall that this equation is only valid in the case of large friction (or large viscosity).

In the case of mean friction, equation (2) is no longer valid, and one has to use the so-called Chandrasekhar equation (Kramers 1940, Chandrasekhar 1943). From a method of derivation similar to the one used above, a more general equation is obtained for the rate constant, named k_{KRAM} below :

$$k_{KRAM} = \frac{\omega_A}{2\pi\omega_C} \, (\left[\frac{\zeta^2}{4} + \omega_C^2 \right]^{1/2} - \left[\frac{\zeta}{2} \right]) \, \exp(- \frac{E^\circ}{k_B T}) \qquad (9)$$

For $\zeta \gg 2 \, \omega_C$, this formula reduces to equation (8) above, whereas for small friction ($\zeta \ll 2\omega_C$) it reduces to the rate constant obtained in the transition state method (k_{TST}) (Eyring, 1935; Evans et al., 1935):

$$k_{TST} \simeq \frac{\omega_A}{2\pi} \, \exp(- \frac{E^\circ}{k_B T}) \qquad (10)$$

However, at the limit of no friction (zero viscosity), the experimental rate constant is expected to tend to zero, as there are no more collisions to give the minimum energy to A particles for barrier crossing. In the case where there are few collisions during an oscillation period, the reaction rate constant will be proportional to the friction coefficient ζ governing energy transfer between particle A and the solvent. Kramers (1940), and Grote et al.(1980) derived the following expression for the rate constant, called the energy-controlled rate constant k_E.

$$k_E = \frac{\zeta \, E^\circ}{k_B T} \, \exp (- \frac{E^\circ}{k_B T}) \qquad (11)$$

Thus it appears that a turn-over can be expected for the rate constants measured at low viscosity (Kramers turn-over). Furthermore, due to repeated crossing and recrossing of the barrier under the influence of the solvent, the rate constant is always lower than the transition – state value. Finally, at high viscosities, in the Smoluchowski limit, the preexponential factor is proportional to ζ^{-1}.

Fig.2 summarizes the general friction (or viscosity) dependence expected for the rate constant of activated barrier crossing reactions.

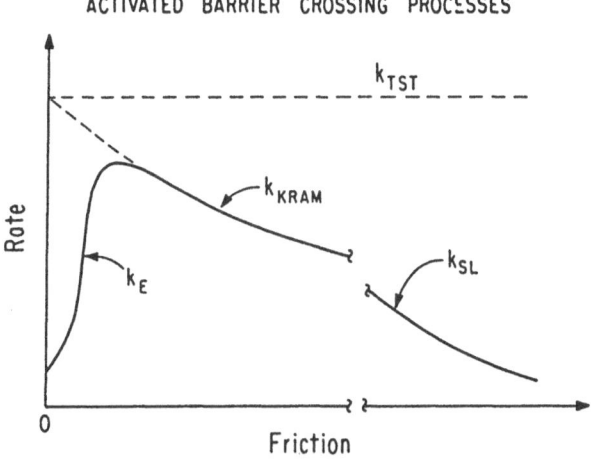

ACTIVATED BARRIER CROSSING PROCESSES

Fig.2. Viscosity dependence of the rate of activated barrier crossing. k_E is the energy controlled rate, k_{KRAM} is the rate obtained from Kramers' expression, and k_{SL} is the rate in the Smoluchowski-limit region. (from G.R. Flemming, 1986).

Experimental tests of the Kramers model

Photochemical isomerization provides a practical testing ground for the above theory. Time resolved spectroscopy has been used to study the isomerization, i.e. a large amplitude structural change, in different molecules which are listed in Table I. In these molecules, the rate of isomerization was measured as a function of temperature, pressure and solvent viscosity.

Table I. Formulae of molecules whose photochemical isomerization has been studied by time resolved spectroscopy.

Molecule	Developed formula	Reference
trans–stilbene		Rothenberger et al., 1983, Courtney et al, 1985
trans– 1,1' – biinda nylidene ("stiff stilbene")		Rothenberger et al., 1983
diphenylbutadiene (DPB)		Velsko et al., 1982 Troe et al., 1985
binaphthyl		Shank et al., 1977 Millar et al., 1985
DODCI		Velsko et al., 1983

Before giving some details on experimental results, let us show how the above one–dimensional treatment applies to isomerization rates : for example, 1,1'– binaphthyl presents a conformational change in the excited singlet state, this change involving a shift in the dihedral angle Θ between the two naphthalene groups, as shown in Figure 3.

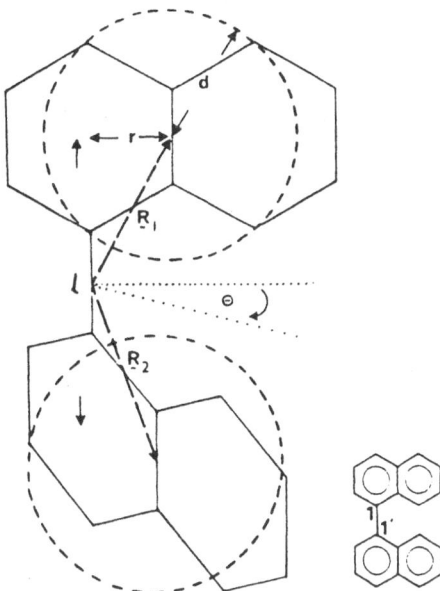

Fig. 3. Hydrodynamic model of 1,1'- binaphthyl as two spheres of radius d constrained to rotate about the 1,1' bond at a radius of gyration r and separated by a distance l along this axis. (from Mc Caskill et al., 1979)

So, the dynamics of this system can be described by the variable Θ. The Langevin equation has a form similar to equation (1) (McCaskill et al., 1979; Wilhelmi, 1982) :

$$\frac{1}{2} I \ddot{\Theta} + \frac{1}{2} \zeta_r \dot{\Theta} + \frac{dU}{d\Theta} = f(t) \tag{12}$$

where I is the moment of inertia of one naphthyl group around the rotating axis, ζ_r is the angular friction coefficient and $U(\Theta)$ is the intramolecular potential. In the case of a parabolic potential, $U = 1/4 I \omega_A^2 \Theta^2$, and the analytical expressions of the rate constants given by equations (8) and (9) are still valid, provided ζ is replaced by ζ_r/I.

In order to study the influence of viscosity, it is necessary to relate by a hydrodynamic model the friction coefficient to the viscosity η of the solvent. Generally, Stokes'formula with slip boundary conditions is chosen :

$$\zeta_r = r^2 \zeta = 4 \pi \eta d r^2 \tag{13}$$

where d is the hydrodynamic radius of the sphere and r^2 the radius of gyration (see Fig.3)

If the isomerization rate k_{exp} has the form

$$k_{exp} = F (\eta) \exp(-E^\circ/k_B T) \tag{14}$$

where $F(\eta)$ is a universal function of viscosity; then a plot of ln k_{exp} vs. 1/T at constant viscosity will give the value of the activation energy E°. Provided similar solvents are used, this procedure proved to work (Fleming, 1986). For example, the barrier height E° for trans-stilbene in alkane solution (3.5 kcal/mole ; Courtney et al., 1985) is very similar to values found by Syage et al. (1982) for isolated stilbene molecules (3.4 kcal/mole).

As concerns the fitting of the preexponential factor to equations (8) and (9), the data fall into two groups (Fleming, 1986). DPB and stilbene in alcohols, stiff stilbene in alkanes, 1,1'- binaphthyl in alcohols fit to Kramers ' expressions very well. All the other systems in Table 1, viz. trans-stilbene in alkanes or DODCI in alcohols, fit Kramers equations poorly. Figure 4 shows the deviation in the case of trans-stilbene.

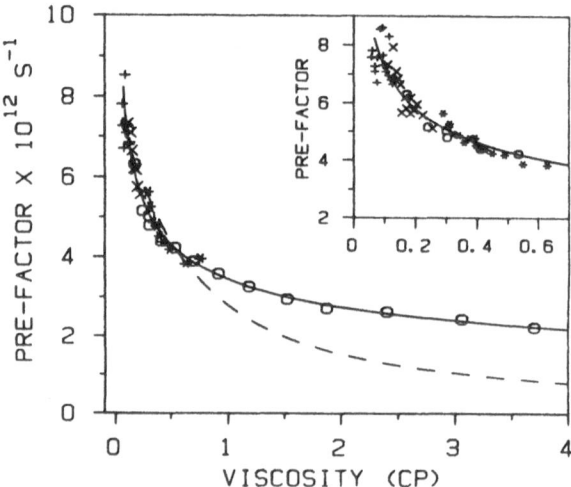

Figure 4. Plot of the preexponential factor of the isomerization rate vs. viscosity for stilbene in normal alkanes. (o,*, x, +) experimental points. The dashed line is a fit to Kramers'expression (equation (9)) (from Courtney et al.,1985)

Different possible explanations have been invoked to explain the bad fitting between Kramers theory and experimental results.

First, Kramers' model is a unidimensional one and, in most of the molecules studied, several rotational modes may ineract during the isomerization (for example rotation of phenyl groups around the single adjacent carbon-carbon bond in trans-stilbene). As a matter of fact, Kramers'model works well for 1,1'- binaphthyl, which seems to be the best molecular model (Millar et al., 1985).

Second, diffusion has been assumed to be a markovian process in Kramers'treatment,i.e. the friction coefficient has been supposed to be time (or frequency) independent. We will now consider the effect of a frequency- dependent friction on rate constants.

Frequency-dependent friction model

Grote and Hynes (1980) have reinvestigated the Kramers model in order to include non-Markovian response. Indeed, for sharp barriers, the frequency dependence of the medium response becomes important. In other words, the correlation time of the random forces exerted by the solvent can be comparable with the time during which the reactant crosses the barrier.

Grote and Hynes then used a generalized Langevin equation :

$$\frac{1}{2} I \ddot{\Theta} + \frac{1}{2} \int_0^t \zeta_r(t) \dot{\Theta} (t - \tau)d\tau + \frac{dU}{d\Theta} = f(t) \qquad (15)$$

As in Eq (1), f(t) is a Gaussian random force, but now, the angular friction coefficient ζ_r is proportional to the solvent-averaged time correlation function of f(t) :

$$\zeta_r(t) = (k_B T)^{-1} < f(0) f(t) > \qquad (16)$$

The rate constant for barrier crossing is then calculated from the fluxes out of and into the reactant and product stable states. The rate constant k_{GH} obtained with this model is the following :

$$k_{GH} = \frac{\omega_A}{2\pi\omega_C} \lambda_r \exp(- \frac{E^\circ}{k_B T}) \qquad (17)$$

where λ_r is a reactive frequency found from the iterative solution to the equation :

$$\lambda_r = \omega_C^2 / \left[\lambda_r + \hat{\zeta}_r (\lambda_r)/I \right] \qquad (18)$$

In this equation, the time dependent friction coefficient appears through $\hat{\zeta}_r(\lambda_r)$ which is the value of the Laplace transform $\hat{\zeta}_r(s)$ of the time-dependent angular friction coefficient at $s = \lambda_r$ (Rothenberger et al, 1983).

In order to apply this model, the frequency dependence of the friction coefficient must be known. A hydrodynamic frequency dependent model is generally chosen (Rothenberger et al, 1983 ; Millar et al., 1985 ; Bagchi et al., 1983). As far as this model is correct , it appears that equation (17) can fit the experimental data with more accuracy than Kramers'equations. However, Rothenberger et al. (1983) noticed that, although their data on trans-stilbene can be made to fit the Grote—Hynes model, an unphysically low frequency (8 cm^{-1}) for the barrier curvature is required for the best fit. Thus although the fitting of the experimental results in the case of sharp barriers is better with this frequency-dependent model, we need more information in order to know whether it is the dominant effect.

Concluding remarks

Recent experiments of time-resolved spectroscopy have provided a good test on the phenomenological models concerning the influence of solvent on unidimensional barrier crossing. However, we have seen that (frequency dependent) viscosity effects cannot explain all the variation of the rate constant. Recently, polarity and hydrogen bonding effects on isomerizations involving large dipole moment changes (p- dimethylamino-benzonitrile DMABN) and those involving an apolar intermediate (trans-stilbene) have also been shown to play a role (Hicks et al. 1987). An example of the effect of solvent polarity on the energy profile of the isomerization of DMABN is shown in Fig. 5 :

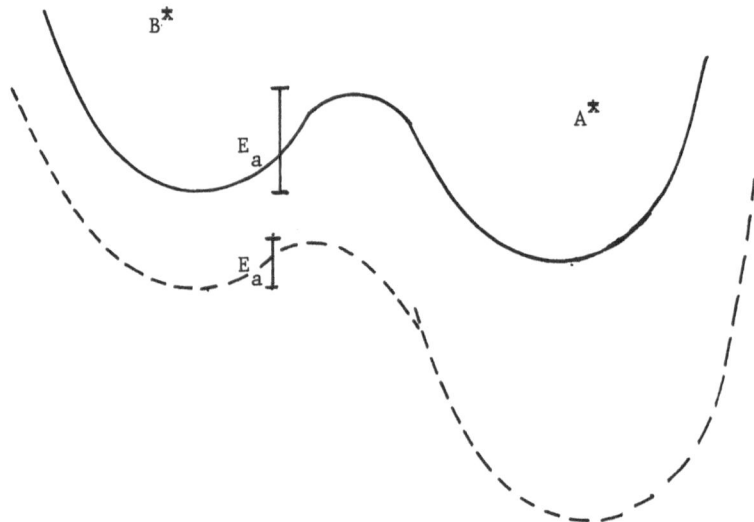

Fig.5 : Dependence of barrier E_a on solvent polarity. An increase in solvent polarity (dashed curve), achieved by lowering the temperature and/or by going to a more polar solvent, preferentially stabilizes the more polar A^* state and thereby lowers the barrier E_a (from Hicks et al., 1987)

BIMOLECULAR REACTIONS

Most theories of bimolecular reactions in solution have been based on Smoluchowski's treatment of the coagulation of colloid particles (Smoluchowski, 1917). We will briefly recall the development of this model, with its recent improvements, and then examine how recent experiments with a fast time resolution have permitted to test these theories.

Debye – Smoluchowski equation (Noyes, 1961)

We consider the reaction A + B → Products. Let w (r,t) be the probability density of species B around fixed A Species. Because of reactions, a concentration gradient of B is set up around A. The Smoluchowski equation wich describes the evolution of w in configuration space is simply the second Fick law, but, as we have shown above, this equation can also be derived from the Langevin equation, its validity being restricted to time larger than $m\zeta^{-1}$. In spherical coordinates, this equation writes :

$$\frac{\partial w\ (r,t)}{\partial t} = \nabla . D \left[\nabla - \frac{F'}{k_B T} \right] w(r,t) \qquad (19)$$

Here F' is the external force (F'=-∇W(r)), where W is the interaction energy, and D (D = D_A + D_B) is the relative diffusion coefficient, supposed to be time- and concentration - independent.

In the original Smoluchowski model, A and B pairs react upon passage through an inner cut-off spherical shell of radius R, generally taken as the sum of Van der Waals radii, as shown in Fig.6.

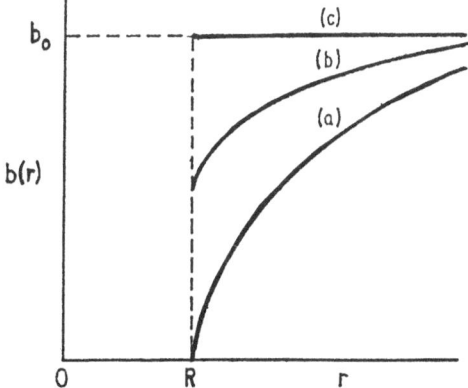

Fig.6. Concentration profiles for chemical reactions w(r) vs. r for (a) diffusion-controlled reaction, Smoluchowski condition, (b) radiation boundary condition, (c) activation controlled reaction ≡ random distribution of B around A. R is the encounter distance (from Pilling ; 1975)

Initial and boundary conditions are necessary for solving this equation. Generally, it is assumed that the bulk density distribution is random, so that w (r → ∞, t) = 1 for all t. Two kinds of boundary conditions have been taken at r = R : i) in the so-called Smoluchowski boundary condition, it is assumed that once B species are inside the sphere of radius R, they react immediately with A, and thus

$$w(r,t) = 0 \qquad r \geqslant R , \quad t \geqslant 0 \qquad (20)$$

ii) if reaction is slow at encounter, Collins and Kimball (I949) incorporated the effects of unsuccessful encounters and used a partially reflecting boundary condition, often called "radiation boundary condition" :

$$\emptyset (R,t) = k_a w (R,t) \qquad (21)$$

which states that the flux $\emptyset$ of B particles crossing the barrier is proportional to the density of these particles, the proportionality coefficient being the "activation-controlled" rate constant k_a, i.e. the rate which would be measured if A and B species were constantly maintained in close contact. The flux $\emptyset$ is given by

$$\emptyset (r,t) = 4\pi r^2 D \left[\nabla - \frac{F'}{k_B T} \right] w(r,t) \qquad (22)$$

In the steady state, it is possible to integrate equation (19) with boundary condition (21) in the case of a general potential

$$w(r) = \exp \left(\frac{w(\infty) - w(r)}{k_B T} \right) \left[\frac{4\pi D \exp(\frac{w(R)}{k_B T}) + k_a I(r)}{4\pi D \exp(\frac{w(R)}{k_B T}) + k_a I(\infty)} \right] \qquad (23)$$

$$\text{with} \quad I(r) = \int_R^r \frac{\exp(\frac{w(r)}{k_B T})}{r^2} dr \qquad (24)$$

The bimolecular rate constant k_2 is related to the stationary flux of B particles moving through the reactive barrier :

$$k_2 = \emptyset(R) = \frac{4\pi D k_a \exp(-\frac{w(\infty)}{k_B T})}{4\pi D \exp(-\frac{w(r)}{k_B T}) + k_a I(\infty)} \qquad (25)$$

In the particular case where $W(r) = 0$ (no interaction potential), equation (25) reduces to :

$$\frac{1}{k_2} = \frac{1}{k_a} + \frac{1}{4\pi DR} \qquad (26)$$

This equation was first derived by Collins and Kimball (1940) and reduces to the well-known Smoluchowski equation if $k_a \to \infty$, i.e. if the reaction is diffusion-controlled :

$$k_2 = 4 \pi D R \qquad (27)$$

In the case of a coulombic potential ($W(r) = Z_A Z_B e^2/4\pi\epsilon_0\epsilon r$ where Z_A and Z_B are the integral ionic charges, e the electronic charge, and ϵ_0 and ϵ the vacuum permittivity and the relative permittivity of the solution), the rate constant expresses :

$$\frac{1}{k_2} = \frac{1}{k_a \exp(R_c/R)} + \frac{1 - \exp(- R_c/R)}{4\pi D R_c} \qquad (28)$$

where R_c is the Onsager length, defined as $R_c = Z_A Z_B e^2/4\pi\epsilon_0\epsilon k_B T$.
This equation reduces to the Debye equation (Rice et al., 1979)

$$k_2 = 4\pi D R_c/[1 - \exp(- R_c/R)] \qquad (29)$$

in the diffusion-controlled limit. Let us note that equations (28) and (26) have the same two terms in the right member : the first is the reverse of the activation-controlled rate constant, and the second is the reverse of the diffusion-controlled rate constant.

Kinetic salt effects on bimolecular rate constants have also been examined (Logan et al., 1966), but will not be detailed here.

The transient behavior of the Debye Smoluchowski equation has also been studied extensively during recent years, in connection with experiments at short times using pulse techniques (Noyes, 1961 ; Pilling, 1975 ; Rice et al., 1979 ; Hynes et al., 1980). In the case where $W(r) = o$ (no potential), equation (19) has an exact solution which gives the following expression for $k_2(t)$ (Noyes,1961)

$$k_2(t) = \frac{4\pi D R}{1 + 4\pi D R/k_a} \left[1 + \frac{R}{(\pi D t)^{1/2} (1 + 4\pi D R/k_a)} \right] \qquad (30)$$

This expression gives equation (26) in the limit $t \to \infty$. Thus at short times ($\approx 1 - 5$ ps at room temperature in solvents with viscosities $\approx 10^{-3}$N m^{-2}s), $k_2(t)$ exceeds the steady-state value by a factor of 2 or 3. The mathematical problem is more complex in the case of charged species. However, Hummel (1974) solved equation (19) in the case of an almost diffusion-controlled reaction. Using the radiation-boundary condition (equation (21), Rice et al., (1979) or Hong and Noolandi (1978) gave an approximate expression of $k_2(t)$, where, as in equation (30), a $t^{-1/2}$ law appears at intermediate times. Numerical solutions have also been carried out (Rice et al.,1979, Delaire et al., 1981).

The above-described Debye-Smoluchowski model is subject to severe limitations (Wilemski and Fixmam, 1973). First, the choice of the coordinate system is not self-evidently valid. Second, the mutual diffusion coefficient is assumed to be constant, even for short separation distances between A and B, where some variations are expected. Third, the diffusion equation is only valid for low concentrations. A last limitation is due to the method of describing the reaction process in which it is assumed that the distribution of unreacted molecules B around A is unaffected by competition between different sinks. Felderhof and Deutch (1976) have investigated the effect of concentration : the rate coefficient is predicted to increase with concentration of sinks. Furthermore, it has been shown that in time—dependent situations it is necessary to take account of retardation effects in the building up of correlations between competing sinks. The retardation leads to a time—dependence of the rate constant (Felderhof, 1977).

In a recent paper, Sipp and Voltz (1983) have shown that, for describing the time dependence of a quenching reaction between an excited donor and an acceptor, it is necessary to consider many particle distribution functions which satisfy a set of coupled equations. This general method lifts several of the limitations quoted above, such as choice of coordinates and concentration effects.

Experimental investigation on time – dependent rate constants

In experiments where $[B]_0 \gg [A]_0$ ($[B]_0$ and $[A]_0$ being the initial concentrations of B and A respectively), $[B]_0$ can be considered as a constant and, in the limit $k_a \to \infty$ (Smoluchowski boundary condition), equation (30) leads to the following equation.

$$t^{-1/2} \ln ([A]_0/[A](t)) = 4\pi DR [B]_0 (t^{1/2} + 2R \sqrt{\pi D}) \qquad (31)$$

Hence D and R may be extracted separately from plots of $t^{-1/2} \ln [A](t)$ vs. $t^{1/2}$. This procedure has been applied to two kinds of reactions, all studied by fast transient spectroscopy. The first kind of reaction is solvated electron scavenging. For example, Buxton et al.(1975) have applied equation (31) to the reaction of solvated electrons generated by pulse radiolysis with chromate ions, and were thus able to show that the reaction $e^-_{aq} + CrO_4^{2-}$ is purely diffusion – controlled, and to determine D and R. The second kind of reaction concerns fluorescence quenching :

$$^1S + Q \to {}^0S + Q \qquad (32)$$

Experiments on nanosecond and picosecond time scales (Nemzek and Ware, 1975 ; Sipp and Voltz, 1983) have clearly demonstrated the occurence of a transient behavior. Namely, deviations from monoexponential decays are observed. However, the model of fluorescence quenching is complicated by several phenomena, such as donor—donor energy transfer at high donor concentrations (Millar et al., 1981) or complex formation between chromophore and quencher (Nemzek and Ware, 1975).

Recombination of iodine atoms

Because it looks a priori simple, recombination of iodine atoms produced by photodissociation of a molecule has been studied for a long time (Noyes, 1954). However, it appeared that the experimental study of the recombination process was a complicated problem. First, after the dissociation of a iodine molecule, the atoms have a highly non—equilibrium spatial distribution, and some of them can recombine before they escape

340

from the initial solvent cage ("cage effect"). Solvent separated atoms may recombine in a "secondary" process, although they arise from the same I_2 molecule. Both types of recombination are called "geminate", and expected to occur on a picosecond time scale (Hynes, et al., 1980). Second, the iodine molecule possesses several electronic levels which can be populated after recombination (Ali and Miller, 1983). Third, vibrational relaxation in the lowest excited state of I_2 may slow down the repopulation of the ground state of the molecule (Berg et al., 1985). Figure 7 shows the increase in transmission of the ground state and the recovery of the absorption following dissociation by a picosecond pulse at 500 nanometers.

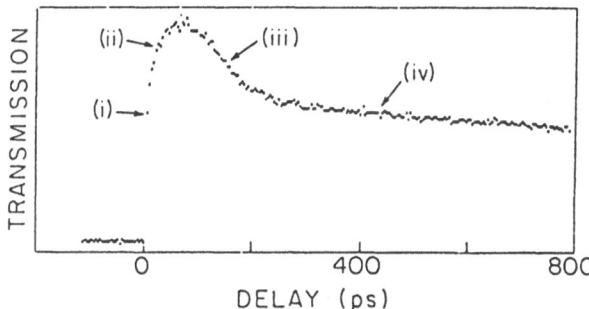

Fig. 7 : Transmission at 500 nm following photodissociation of I_2 in CCl_4 (from Berg et al., 1985)

The experimental result is similar to the one obtained ten years before by Chuang et al. (1974). The fast increase is due to predissociation and vibrational relaxation of the initially populated B state. The fast decay represents the time scale which could have been attributed to geminate recombination, but is now thought to be a consequence of vibrational relaxation in the ground state. The slow decay at "long" times is assigned to relaxation out of upper electronic (A,A') states (Berg et al., 1985 ; Fleming, 1986, p.197). By considering the rate of decay of the B state to the levels (A,A') below the dissociation limit, Berg et al. concluded that the recombination process occurs in ≤ 15 ps. Hynes et al. (1980) simulated the recombination process by either using the Smoluchowski approach or doing a stochastic trajectory simulation using the Langevin equation. The differences from both approaches are discussed. Finally, a geminate recombination in t ≤ 10 ps. has been found by these authors, in good agreement with experimental observations.

Molecular dynamics calculations have been carried out (Bado et al. 1982) : from trajectory simulations of two iodine atoms surrounded by fifty xenon atoms; these authors were able to take account of solvent caging, atomic recombination and vibrational relaxation to the solvent. Transient electronic absorption spectra of iodine were calculated during the first 800 ps following laser excitation. From these spectra, transient kinetics at any wavelength can be obtained.

CONCLUSION

We have reviewed models of unimolecular and bimolecular reactions in solution. We have shown that, if the Smoluchowski equation is used to describe the spatio-temporal evolution of the probability density, rather simple analytical expressions can be derived where the influence of the solvent is taken into account through macroscopic parameters like viscosity or dielectric constant. However, we have shown the limits of the Smoluchowski equation for reactions occurring at short times and/or distances. Time- or frequency - dependent rate constants have been derived, but the theory suffers from the difficulty of correctly expressing the response of the solvent as a function of frequency.

New theoretical developments based on modern statistic theories of liquids are now in progress. At the same time, much effort must be made experimentally to find reactive models to which the theory can be applied, and to refine the new advanced technologies of pulse radiolysis and photolysis.

ACKNOWLEDGEMENTS

The author would like to thank P. Cordier who initiated him to the Smoluchowski equation, and J. Belloni for many fruitful discussions.

REFERENCES

Ali, D. P., and Miller, W. H., 1983, Effect of electronic transition dynamics on iodine atom recombination in liquids, J.Chem.Phys., 78:6640.

Bado, P. Berens, P.H., Bergsma J.P., Wilson, S.B., Wilson, K.R., and Heller, E.J., 1982, Picosecond dynamics of I_2 photodissociation, Picosecond Phenomena, 3 : 260.

Bagchi, B., and Oxtoby, D. W., 1983. The effect of frequency dependent friction on isomerization dynamics in solution, J.Chem.Phys., 78 : 2735.

Berg, M., Harris, A. L., and Harris, C. B., 1985, Rapid solvent induced recombination and slow energy relaxation in a simple chemical reaction, Phys. Rev. Lett., 54:951.

Buxton, G. V., Cattel, F. C. R., and Dainton F. S., 1975, Application of time-dependent rate constant theory to reactions of solvated electrons, J. C. S. Faraday I, 71:115.

Chandrasekhar, S., 1943, Stochastic problems in physics and astronomy, Review of Modern Physics, 15:1.

Chuang, T.J., Hoffman, G.W., and Eisenthal, K.B., Picosecond studies of the cage effect and collision induced predissociation of iodine in liquids, Chem. Phys. Lett., 25:201.

Collins, F. C., and Kimball, G. E., 1949, Diffusion-controlled rate constants, J. Colloid Sci., 4:425.

Courtney, S. H., and Fleming, G. R., 1985, Photoisomerization of stilbene in low viscosity solvents, J. Chem. Phys., 83:215.

Delaire, J.A., Croc, E. and Cordier, P., 1981, Numerical solution of Smoluchowski equation, J.Phys.Chem., 85 : 1549.

Evans, M. C., and Polanyi, M., 1935, Trans. Far. Soc., 31:875.

Eyring, H., 1935, The activated complex in chemical reactions, J. Chem. Phys., 3:107.

Felderhof, B. U., and Deutch, J. M., 1976, Concentration dependence of the rate of diffusion-controlled reactions, J. Chem. Phys., 64:4551.

Felderhof, B. U., 1977, Frequency-dependent rate coefficient in diffusion-controlled reactions, J. Chem. Phys., 66:4385;

Fleming, G. R., 1986, Chemincal applications of ultrafast spectroscopy, The international series of monographs on Chemistry, 13:124, Oxford University Press, New York.

Grote, R. F, and Hynes, J. T., 1980, The stable states picture of chemical reactions, J. Chem. Phys., 73:2715.

Hummel A., 1974, Ionization in non polar molecular liquids by high energy electrons, Adv.Radiat.Chem. 4:1, J. Wiley, New York.

Hicks, J. M., Vandersall, H. T., Sitzmann, E. V. and Eisenthal, K. B., 1987, Polarity dependent barriers and the photoisomerization dynamics of molecules in solution, J. Phys. Chem., to be published.

Hong, K. M., and Noolandi, J. 1978, Solution if the Smoluchow ski equation with a Coulomb potential, J. Chem. Phys., 68:5163.

Hynes, J. T., Kapral, R., and Torrie, G. M., 1980, Stochastic trajectory simulation of iodine recombination in liquids, J. Chem. Phys. 72:177.

Kramers, H. A., 1940, Brownian motion in a field of force and the diffusion model of chemical reactions, Physica, 7:284.

Logan, S. R., 1966, Theory of kinetic salt effects in diffusion-controlled reactions, Trans. Far. Soc., 62:3416.

McCaskill, J. S., and Gilbert, R. G., 1979, Fokker-Planck interpretation of picosecond intramolecular dynamics in solution, Chem. Phys., 44:389.

Millar, D. P., Robbins, R. J., and Zewail, A. H., 1981, Picosecond dynamics of electronic energy transfer in condensed phases, J. Chem. Phys., 75:3649.

Millar, D. P., and Eisenthal, K. B., 1985, Picosecond dynamics of barrier crossing in solution, J. Chem. Phys., 83:5076.

Nemzek, T. L., and Ware, W. R., 1975, Kinetics of diffusion controlled reactions, J. Chem. Phys., 62:477.

Northrup, S. H., and Hynes, J. T., 1978, Reactive dynamics for diffusive barrier crossing, J. Chem. Phys., 69:5246.

Noyes, R. M., 1954, A. treatment of chemical kinetics with special applicability to diffusion controlled reactions, J. Chem. Phys., 22:1349.

Noyes, R. M., 1961, Effects of diffusion rates on chemical Kinetics, Prog. Reaction Kinetics, 1:129.

Pilling, M. J., 1975, Short time behavior of diffusion-controlled reactions, in "Lasers in Physical Chemistry and Biophysics" J. Joussot- Dubien, ed., Elsevier, Amsterdam.

Rice, S. A., Butler, P. R., Pilling, M. J., and Baird, J. K., 1979, A solution of the Debye-Smoluchowski equation for the rate of reaction of ions in dilute solutions, J. Chem. Phys., 70:4001.

Rothenberger, G., Negus, D. K., and Hochstrasser, R. M., 1983, Solvent influence on photoisomerization dynamics, J. Chem. Phys., 79:5360.

Shank, C. V., Ippen, E. P., Teschke, O., and Eisenthal, K. B., 1978, Picosecond dynamics of conformational changes in 1,1'- binaphthyl, J. Chem. Phys., 67:5547.

Sipp, B., and Voltz, R., 1983, Reactive dynamics in energy transfer problems, J. Chem. Phys., 79:434.

Smoluchowski, M. V., 1917, Z. Phys. Chem., 92:129.

Syage, J. A., Lambert, W. R., Felker, P. M., Zewail, A. H., and Hochstrasser, R. M., 1982, Picosecond excitation and trans-cis isomerization of stilbene in a supersonic jet, Chem. Phys. Lett., 88:266.

Troe, J., Amirav, A. and Jortner, J., 1985, Energy resolved and thermalized photoisomerization rates of DPB, Chem. Phys. Lett., 115:245.

Velsko, S. P., and Fleming, G. R., 1982, Photochemical isomerization in solution. Photophysics of DPB, J. Chem. Phys., 76:3553.

Velsko, S.P., Waldeck, D.H., and Fleming, G.R., 1983, Breakdown of Kramers theory description of photochemical isomerization, J.Chem.Phys., 78:249.

Wilemski, G., and Fixman, M., 1973, General theory of diffusion - controlled reactions, J. Chem. Phys. 58:4009.

Wilhemi, B., 1982, Influence of solvent viscosity on excited state lifetime and fluorescence quantum yield of dye molecules, Chem.Phys., 66:351.

DISCUSSION

SCEATS - In your work you have considered the Kramers model for 1-dimensional motion in the high and low friction limits and the Smoluchowski model for 3-dimensional motion. In my paper at this conference, I will show that the 3-d motion can be reduced to a one-dimensional problem by use of the one dimensional potential $V(R)-2kTlnR$, and that application of Kramers theory to the dynamics on this potential yields the Smoluchowski-Debye result in the high friction limit. Hence unimolecular and bimolecular reactions can be compared using the same model.

DELAIRE - As far as the same basis equation, namely the Smoluchowski equation, or the more general Fokker-Planck equation, describes both kinds of reactions, the reduction of the 3-dimensional problem to a 1-dimensional one seems indeed possible, and your development is thus very interesting.

FRIES - In your treatment of the bimolecular reactions, you have used a macroscopic constant, and isotropic diffusion tensor in your Smoluchowski equation. Don't you believe that for those separations of the reactants which are smaller than one solvent diameter, the diffusion tensor becomes space dependent, especially for charged species which are coupled by very strong electrostatic forces ? In this case, would it be possible to employ space dependent diffusion tensors obtained by molecular dynamics studies ?

HYNES - In response to Fries' question, Northrup and Hynes (J. Chem. Phys., 1979) examined the effect of mean solvent potential and distance dependent diffusion for diffusion - influenced neutral reactions. The rate constant effects were very small - order of 30% only.

KAPRAL - My comment concerns the effect on the rate coefficient of including spatial dependence in the diffusion or friction coefficient. We investigated this problem some time ago (M. Schell, R. Kapral and R.I. Cukier, Chem. Phys. Lett.) for a model bimolecular reaction using a space-dependent friction coefficient obtained from a kinetic theory calculation (R.I. Cukier, J.R. Mehaffey and R. Kapral, J. Chem. Phys.). The effects are rather small, amounting to no more than about forty per cent corrections to the rate coefficient. The small size of the effect arises from averaging over a range of spatial configurations.

JONAH - I would like to say that the term "time dependent rate constant" is a dangerous term. The time dependence arises from a change in local concentration and is not due to a change in the "true" rate constant.

DELAIRE - I agree with you. The above-defined "activation-controlled rate constant", which is the expression of the "true" reactivity of species, may depend on distance, but is time-independent. However, the experimental rate constant has generally a contribution due to diffusion (see equation (26)), and, due to the fact that a time interval is needed to establish steady-state profiles for the local concentrations of reactants, the experimental rate constant is time dependent.

AMATORE - This is just to add a comment on your answer to Jonah's question. In electrochemistry, at spherical electrode we obtain the same kind of problem in which reaction is convoluted with diffusion. Yet in such a case we are used to deconvolute the diffusional contribution to obtain a real rate constant independent of time or distance. However this is done for diffusional distances in the micrometric range. Would that be transposable for Angström diffusion distances as considered in Smoluchowski's model ?

BELLONI - I agree that the time dependence concerns the local concentration. However, the reactivity may also change due to the rearrangement of the solvent in the vicinity of the reactants (Robinson, Moore,...).

ENCOUNTER AND REACTION DYNAMICS ON THE POTENTIAL OF MEAN FORCE

Mark G. Sceats

Department of Physical Chemistry
University of Sydney
NSW 2006 Australia

I. INTRODUCTION

The purpose of this paper is to present a formulation of the rate constants for bimolecular encounter and reaction which applies over the entire range of bath densities, from low pressure gas to compressed liquid, and is applicable for an arbitrary potential along the separation coordinate R which is assumed to closely represent the reaction coordinate. This work is motivated by the increasing interest in forging connections between gas and liquid phase kinetics, both for bimolecular and unimolecular reactions[1]. Primarily, this is a problem of bridging the low pressure regime of stabilization of a complex by collisions with the bath gas molecules and the diffusion controlled reaction of the pair at high bath density. In the case of bimolecular reactions of large aerosol particles this connection, in terms of the transition from free-molecule to continuum (diffusive) behaviour has been the subject of extensive study for many years [2]. The large particle size of aerosols relative to that of the bath molecules confers two simplifying characteristics: Firstly, the transition to the diffusion regime occurs at modest bath densities characteristic of the gas phase. For small molecules interacting by modestly short ranged forces (ie. non-Coulombic) the transition occurs at densities approaching that of the liquid phase. Secondly, the large number of internal degrees of freedom of large particles provides a large density of states near the dissociation threshold that kinetic energy exchange into these from the R coordinate gives a long-lived collision complex. Stabilization of the aerosol complexes by collision with the bath molecules is readily achieved, except at interstellar bath densities, such that the sticking probability upon encounter approaches unity. For atomic recombination no internal degrees of freedom are available for redistribution and successful recombination requires immediate stabilization by bath collisions, which leads to a linear dependence of the recombination rate and thus the "sticking probability" on the bath density. For unimolecular reactions the role of the internal degrees of freedom can be treated statistically, as is done in RRKM theory [3].

In a stochastic formulation of bimolecular reactions the reaction is perceived to take place on a multidimensional potential surface under the influence of stochastic forces which mimic the collisions with the bath gas. It is usual to treat the separation coordinate R as the reaction coordinate, but this is strictly true only at large separations and fails if internal rearrangements dominate the position of the transition state in the full multidimensional potential surface. In section 11 the treatment of this potential is discussed. In section 111 the nature of the stochastic forces are considered, and this is followed by an evaluation of the recombination rates in section 1V and briefly illustrated by examples in section V.

11. EFFECTIVE POTENTIALS

It is often easier to consider the problem of bimolecular reaction rate theory from the perspective of dissociation of the particles, and then to evaluate the recombination rate from the derived dissociation rate using the equilibrium constant. The principal difference between unimolecular and bimolecular reactions is in the treatment of angular momentum J. In unimolecular reactions the transition state is regarded as fixed by the internal coordinates simply because centrifugal effects are small. In bimolecular reactions this is not the case, as is demonstrated by the behaviour of the effective centrifugal potential

$$V_e(R,\xi,J) = V(R,\xi) - J^2/2\mu R^2 \tag{1}$$

where μ is the reduced mass and ξ represents the set of internal coordinates. The transition states are defined by the saddle point of $V_e(R,\xi,J)$ for each J and it is only when $V(R,\xi)$ exhibits a sharp saddle point itself, that the centrifugal effects of the last term in (1) can be ignored. R is taken to be the reaction coordinate. The treatment becomes complicated because of the multitude of transition states located at different positions $R_T(J)$. Collisions not only change the energy but also the angular momentum, and at low bath density collisions are described by the collisional-transition probability $P(E,J,E',J')$. Thus dissociation and recombination depends not only on the particle fluxes across the transition state at each J, but also on the fluxes between populations with different J values. If complications due to the latter are neglected, then this approach leads to a microcanonical variational rate theory [4] as a generalization of RRKM theory for bimolecular reactions. The high pressure limit yields the microcanonical variational transition state rate. In this high pressure limit the internal degrees of freedom have become equilibrated, and a potential $V(R)$ is defined by the thermal average

$$e^{-V(R)/kT} = \int d\xi \, e^{-V(R,\xi)/kT} \tag{2}$$

where a multidimensional average is implied. Of course, (2) can equally well be expressed in terms of R-dependent partition functions. When (2) is combined with the centrifugal potential in (1), a transition state rate can be evaluated from the characteristics of the maxima in the one-dimensional potential $V(R) - J^2/2\mu R^2$, and a thermal average of this rate over J yields the required rate constants for dissociation and recombination. When this approach is used to study open-channel reactions at low bath pressure, as in many ion-molecule reactions, excellent agreement is obtained [5] because, in fact, J is a conserved quantity and the transition state rate is a good approximation to the encounter rate and, with unit reaction probability, the reaction rate. In general it is not reasonable to assume equilibration of internal degrees of freedom by collision at high bath density yet retain J as a conserved quantity. Indeed, rotational relaxation is exceedingly efficient and it is more reasonable to assume that the first coordinate equilibrated by collisions is the orientational coordinate. Thus the Botzmann average of (1) gives an effective potential $\tilde{V}(R,\xi)$

$$e^{-\tilde{V}(R,\xi)} = \int_0^\infty J \, dJ \, e^{-V_e(R,\xi,J)/kT} \tag{3}$$

which becomes

$$\tilde{V}(R,\xi) = V(R,\xi) - 2 \, kT\ell nR \qquad (4)$$

In general, when any set of independent coordinates say ξ_T are eliminated by assumption of thermalization, the dynamics on the remaining coordinates ξ_R occurs on the effective potential

$$\tilde{V}(\xi_R) = V(\xi_R) - kT\ell n \sqrt{|g_{ij}|} \qquad (5)$$

where $|g_{ij}|$ is the determinant of that portion of the metric tensor pertaining to the thermalized coordinates [6]. The result of (4) is obtained from (5) when orientational and centre of mass coordinates of a pair of particles are eliminated. An important point is that dynamical information, appropriate to kinetics, can be obtained from trajectories on $\tilde{V}(R,\xi)$, and is easily manipulated because $\tilde{V}(R,\xi)$ exhibits only one saddle point. When all coordinates other than R are pushed into thermal equilibrium by collisions, then the effective potential becomes

$$\tilde{V}(R) = V(R) - 2kT\ell nR \qquad (6)$$

where V(R) is seen to be the potential of mean force often discussed in the theory of liquids in terms of the equilibrium pair distribution function $g(R) \equiv e^{-V(R)/kT}$. The effective potential $\tilde{V}(R)$ accounts for the increase of phase space at large R, and gives the correct radial distribution

$$e^{-\tilde{V}(R)/kT} \equiv R^2 g(R)$$

while treating R as a one-dimensional coordinate. The formal reduction of the two particle Fokker-Planck equation to this one dimensional represent-ation [6] then allows the application of Kramers' one-dimensional barrier crossing theory [7] to evaluate a rate constant for dissociation and recombin-ation [8] discussed in section 1V. The barrier exists in $\tilde{V}(R)$ even for attractive potentials with a well-depth greater than kT as a result of the repulsive character of the logarithmic term in (6). The potential V(R) is also the free energy, and it is not surprising that the transition state rate based on this approach is equivalent to that of free-energy variational rate theory in general and canonical variational rate theory in absence of internal degrees of freedom. However the use of $\tilde{V}(R)$ gives a very simple picture of the potential on which the dynamics occurs, and this has been exploited in a wide range of applications [9], some of which are considered in this article.

In summary, at low bath densities the rates of dissociation and recombin-ation are determined by the energy dynamics for particles moving in the set of potentials $V_e(R,\xi,J)$. The rates are determined not only by the particle fluxes through the saddle points of these potentials, but also by $J \rightarrow J'$ scattering. In the other limit at high bath densities the rates of dissocia-tion and recombination are determined by spatial diffusion on the effective potential $\tilde{V}(R)$. Nevertheless, in the case of diatomic dissociation and re-combination where $\xi = 0$, it has been shown that the use of $\tilde{V}(R)$ provides an excellent approximation at low bath densities, at least in the weak impulsive collision limit that was tested [10] against simulation [11]. The reasons for this are not clear, but could be associated with the efficient $J \rightarrow J'$ thermalization. RRKM theory of unimolecular reactions [3] does not consider the thermalization of internal coordinates as the bath density increases, but rather considers the gradual establishment of a thermal distribution among all the available states above threshold. It assumes, therefore, that intra-molecular vibrational energy redistribution (IVR) is sufficiently fast relative to the collisional induced energy relaxation from internal modes. This aspect has been discussed in the weak and strong collision limits[12,13], and for systems with many internal degrees of freedom it is certain that this reduction in dimensionality occurs at bath densities where the reaction rates have attained their high pressure limits. That is, there is no observable consequence of mode thermalization in thermal experiments. For bimolecular reactions involving a loose collision complex exchange of orbital angular momentum with rotational angular momenta of the fragments allows a restricted feeding of the reaction coordinate by the pair orbital angular momentum between collisions. But this is not a prerequisite for use of the effective potential $\tilde{V}(R)$ in the weak coupling limit [10] where the discrete nature of collisions is not required to be described by virtue of the timescale arguments used to map the problem into one of energy diffusion, and the relevant timescales are those of collision induced $J \rightarrow J'$ thermalization rather than IVR. The rotational energy is a reservoir that is available on collision.

III. THE STOCHASTIC FORCES

Consider the extreme limit in which all of the atoms of the two particles are separated by distances which are large relative to the size of the bath molecules and the range of the molecular potential. The motion of each atom can be described by a generalized Langevin equation in which the influence of the bath molecules is described by a random force characterized by a frictional memory kernel $\beta(t)$. At low bath densities, $\beta(t)$ depends solely on the nature of the atom-bath potential and if the internal and rotational degrees of freedom of the bath molecule are neglected then this potential can be modeled by a strongly repulsive potential with a weakly attractive limb. A Morse potential is most convenient to work with in which case the Fourier transform of the friction $\tilde{\beta}_B(\omega)$ of atom B in a bath of molecules M, in mass weighted units of s^{-1}, is [9]

$$\tilde{\beta}_B(\omega) = \frac{2}{3} \left(\frac{2\mu_{BM}}{M_B}\right) \pi R_{BM}^2 \, g_{BM} \, B_{BM}(\omega) \qquad (7)$$

where g_{BM} is modelled by $\exp\left(-V_{BM}(R_{BM})\right)/kT$ in which R_{BM} is the transition state of the atom-bath effective potential defined similarly to (4) and $B_{BM}(\omega)$ accounts for the finite timescale of the atom-bath molecule encounter and is defined such that $B_{BM}(0) = 1$. At frequencies higher than the bandwidth of the collision duration on the repulsive limb this form yields the Landau-Teller result for the adiabatic factor [14] used to describe V-T transfer in atom-diatom collisions. At zero frequency, the friction is simply that of Enskog theory and is related to the atomic diffusion coefficient $D_B = kT/M_B\beta_B(0)$. As the bath density increases, the collective motions of the bath molecules give hydrodynamic contributions to the friction of frequencies smaller than the inverse of the structural relaxation time of the solvent [15]. The zero-frequency friction is often modelled by the Stokes-Einstein relationship at high density and an interpolation formula has been suggested to model the zero-frequency friction over the entire range of bath density [16]. However the magnitude of the deviations from Enskog theory for atomic diffusion are relatively small.

Consider next the more reasonable case in which the atoms of the particles form a bonded molecule. The collisional friction acting on the atoms of the molecule is reduced by screening in proportion to the "surface area" exposed to the bath molecules by the atom embedded in the particle[17]. This is typically about 0.5 or less. The atom frictions can be transformed into mode frictions $\beta(\xi, E(\xi))$ using the transformation to appropriate mode coordinates, where $E(\xi)$ is the energy in mode ξ. Hydrodynamic contributions are not additive and the friction associated with the R coordinate for most systems is modelled from the zero-frequency frictions of each particle. In most cases the barrier frequency ω_T of the transition state of $\tilde{V}(R)$ defined in the next section, is sufficiently small relative to the atom-atom collision bandwidths or the inverse of the structural relaxation time of the solvent, that frequency dependent friction along the R coordinate can be neglected. Excpetions will occur when intramolecular rearrangement determines the transitions state in which case R ceases to be an appropriage representation of the reaction coordinate. The non-parabolic nature of $\tilde{V}(R)$ at large R requires that the friction which describes motion across the barrier covers a wide spectral range from about ω_T near R_T to zero as $R \to \infty$. This problem has been addressed [9] but not solved. However, as mentioned above ω_T is often below the region of dispersion of $\hat{\beta}_R(\omega)$ and the zero frequency frictions can be used. This is not true for internal modes. The magnitude of β_R can be deduced from the diffusion coefficients D_A and D_B of the separated particles. With

$$\beta_X = \frac{kT}{M_X D_X} \quad \text{for } X = A, B, \text{ it is found that at short times}$$

where kinematic correlations are important

$$\beta_R = S_{AB}(R) \; [(\frac{M_A}{M_A + M_B}) \; \beta_B + (\frac{M_B}{M_A + M_B}) \; \beta_A] \tag{8}$$

where $S_{AB}(R)$ is a screening factor which is about ½ for particles near contact at low bath density [18], but which has a long range hydrodynamic contribution when the bath density approaches that of a liquid [19]. The result of (8) should apply in the low friction regime of energy diffusion. At long times, and at higher bath densities, the spatial diffusion is characterized by a friction

$$\beta_R = S_{AB}(R) \; [(\frac{M_A}{M_A + M_B}) \; \frac{1}{\beta_B} + (\frac{M_B}{M_A + M_B}) \; \frac{1}{\beta_A}]^{-1} \tag{9}$$

which is equivalent to a relative diffusion coefficient $D_A + D_B$ at large R where $S_{AB} \to 1$. In a rigorous treatment of the elimination of centre of mass and orientational degrees of freedom of a pair of coupled Brownian particles, a time dependent friction kernel should be found which encompasses the limits of (8) and (9) when $S_{AB} = 1$.

In summary, the motion of the system of variables ξ, R on the effective potential $\tilde{V}(R,\xi)$ are described by coupled generalized Langevin equations with mode frictions $\beta(\xi, E(\xi))$ associated with collisional excitation of the atoms involved in that mode, and a friction β_R which describes the relative motion of the two particles and is closely associated with the particle diffusion coefficients. Processes such as IVR are determined by off-diagonal potential and kinetic energy coupling terms incorporated within the coupled Langevin equations.

IV. THE RECOMBINATION AND DISSOCIATION RATES

The problem of evaluating the reaction rate over the entire range of bath density is, in a stochastic formulation, one of dealing with the regimes of friction relative to the barrier frequency ω_T and the frequencies $\omega(\xi, E(\xi))$ of the internal degrees of freedom. At low friction, corresponding to low bath densities, the problem is one of energy diffusion but at high friction becomes one of spatial diffusion. Only recently have attempts been successful in treating both processes together [20] in unimolecular reactions but this is restricted to one dimensional problems. For bimolecular reactions the approach has been successfully applied to diatomic recombination [21] over the entire range of bath densities. It is usual to treat energy and spatial diffusion separately, and then to blend the resultant rates to give the rate over the entire bath density range.

For unimolecular reactions a suitable form is

$$\frac{1}{k} = \frac{1}{k_{ED}} + \frac{1}{k_{TST}} + \frac{1}{k_{SD}} \tag{10}$$

where k_{ED} is the low-friction rate which is linear in bath density, k_{TST} is the transition state rate which is independent of bath density and k_{SD} is the spatial diffusion, or Smoluchowski, rate which decreases with increasing bath density. The disadvantage of this form is that it does not properly describe the fall-off regime in which the rate rolls over from k_{ED} at low bath density to k_{TST} at high bath density. This is better handled by the full RRKM theory [3], or by a broadening factor approach [22], although applications have been limited by poor knowledge of the "collisional efficiency"[3]. As discussed in the introduction, the RRKM theory does not account for the finite timescale of IVR and the consequent reduction of the dimensionality of the effective potential at high bath densities. While this process most likely occurs in the broad plateau where k is pinned to k_{TST}, it is nonetheless required to occur in order that the spatial diffusion at even higher bath density be treated on the potential of mean force.

Similar principles apply to bimolecular reactions but are complicated by the treatment of angular momentum, as discussed below.

(a)   Transition State Rates

Consider the case of a central potential V(R). Microcanonical treatments give the recombination rate k(E) and the thermal rate $k_{\mu-TST}$

$$k(E) = \frac{3\pi\mu}{(2\mu E)^{3/2}} R^2 [E-V(R)]^2 \tag{11}$$

$$k_{\mu-TST} = \frac{4}{3\sqrt{\pi}} \left(\frac{1}{k_T}\right)^{5/2} \int_0^\infty E^{3/2} k(E) e^{-E/kT} dE$$

where R is chosen for a given E by minimizing k(E). For an attractive potential of the form $-AR^{-n}$, the rate is

$$k_{\mu-TST} = \pi R_T^2 \left(\frac{8kT}{\pi\mu}\right)^{\frac{1}{2}} \left\{ 2 \left(\frac{n}{2(n-1)}\right)^2 \frac{n-1}{n} \Gamma(3-2/n) \right\} \tag{12}$$

where R_T is the transition state of the J-averaged potential in (6). The factor in brackets is denoted below by X(n). Canonical theory is equivalent to one-dimensional barrier crossing on the J-averaged potential and the general result is

$$k_{C-TST} = \pi R_T^2 \left(\frac{8kT}{\pi\mu}\right)^{\frac{1}{2}} e^{-V(R_T)/kT} = \pi R_T^2 \left(\frac{8kT}{\pi\mu}\right)^{\frac{1}{2}} e^{2/n} \tag{13}$$

The difference between (12) and (13) is that the assumption of J-thermalization opens up the density of states such that $k_{C-TST} > k_{\mu-TST}$. The difference between (12) and (13) becomes small for large n as shown in Table I.

Table 1. Comparison of Microcanonical and Canonical
Transition State Rates for attractive potentials
$- AR^{-n}$.

n.	X(n)	$\exp(-V(R_T)/kT)$	Density of States Expansion Factor
1	2.00	7.39	3.70
2	2.00	2.72	1.36
3	1.62	1.95	1.20
4	1.45	1.65	1.14
5	1.35	1.49	1.11
6	1.28	1.40	1.09
∞	1.00	1.00	1.00

The point to note is that the high pressure limit of the transition
state rate is k_{C-TST} because the presence of bath molecules will destroy
the conservation of the pair angular momentum.

(b) Spatial Diffusion

At sufficiently high bath densities the internal and orientational
degrees of freedom are thermalized and the motion takes place on the
effective one-dimensional potential $\tilde{V}(R)$ in (6). Sceats [8] has shown
that passage over this transition state at R_T in $\tilde{V}(R)$ can be treated using
Kramers theory to give the encounter rate K_E

$$k_E = k_{C-TST} \left[(1 + (\frac{\beta_R}{2\omega_T})^2)^{\frac{1}{2}} - \frac{\beta_R}{2\mu_T} \right] \qquad (14)$$

where β_R is the relative motion friction given by (9) and ω_T is the barrier
frequency given by the configuration integral

$$\omega_T = (\frac{2\pi kT}{\mu})^{\frac{1}{2}} / \int_{R_p}^{\infty} \exp[-(\tilde{V}(R)-\tilde{V}(R_T))/kT]dR \qquad (15)$$

where R_p is the minimum of $\tilde{V}(R)$.

This has the low friction (low bath density) limit of k_{C-TST}, and at high
friction gives, upon rearrangement, the exact result for diffusion in a
field of force [23]

$$k_{SD} = 4\pi D_{AB} R_T / \int_{R_p}^{\infty} \frac{R_p}{R^2} \exp[-(V(R)/kT) \, dR \qquad (16)$$

The factor in the denominator is the stability factor discussed extensively
in colloid science [24]. The important aspect of the result of (14) is that
it gives the encounter rate over the entire range of bath densities for
arbitrary potentials $V(R)$, and gives the correct limits at low and high
bath densities. With regard to the recombination reaction rate of (10) it
it noted that (14) is well approximated by $[k_{C-TST}^{-1} + k_{SD}^{-1}]^{-1}$. The encounter
rate is equal to the recombination rate if the "sticking probability" is
unity. This is an excellent approximation for large molecules and aerosols
because the relative kinetic energy is rapidly dissipated among the internal
degrees of freedom. This occurs when $k_{ED} \gg k_{TST}$ in (10) in the bath density
regime of interest.

This formulation of the encounter rate has been recently applied to a wide variety of problems [9]. Examples will be given in section V. The crossover from free-molecule to the diffusion regime occurs when $\beta_R/\omega_T > 1$, which can be rearranged to give the condition

$$(\frac{8kT}{\pi\mu})^{\frac{1}{2}} \frac{R_p W(o)\Omega*}{4D_T} > 1 \tag{17}$$

where $W(o)$ is the stability factor in the denomiator of (16) and $\Omega_{IN} = R_T^2/R_p^2 \exp(-V(R_T)/kT)$ is a dimensionless collision integral for encounter.

(c) Energy Diffisuon

At very low bath densities the dissociation rate of a pair of particles can be cast in terms of the rate of accumulation of energy up to a dissociation energy. This is well-studied for unimolecular reactions in which the transition state is fixed, but for bimolecular reactions the complexity associated with angular momentum leads to complications. Nevertheless, an assumption will be made that a threshold energy Q can be defined such that when E > Q dissociation may take place at a rate $k_{RRKM}(E)$, which depends only on the relative density of states of the reactant and the transition state [3]. The steady state dissociation rate is then given by

$$k_{DISS} = \int_Q^\infty k_{RRKM}(E) \, P_{SS}(E) dE \tag{18}$$

where $P_{SS}(E)$ is the steady state energy distribution function. At any pressure k_{DISS} depends on the relative magnitudes of $k_{RRKM}(E)$ and the rates of collisional activation and deactivation. Only at the lowest pressures does the rate become independent of $k_{RRKM}(E)$ because $P_{SS}(E) = 0$ for E > Q. In this case the dissociation rate is equivalent to that obtained from the mean time for first passage over the barrier [13]. If D(E) is the energy diffusion coefficient and $\Omega(E)$ the density of states, then the recombination rate at low pressures is found from (10) and the equilibrium contant as

$$k_{ED} = \frac{4\pi R_p^2}{\omega_p} (\frac{2\pi kT}{\mu})^{\frac{1}{2}} \left\{ \int_0^Q dE \, \frac{e^{\frac{E-D}{kT}}}{\Omega(E)D(E)} \int_0^E dE' \Omega(E') e^{-E'/kT} \right\}^{-1} \tag{19}$$

where ω_p is the dissociation attempt frequency defined below, D is the dissociation energy of V(R), $\Omega(E)$ is the density of states of the two particle system and D(E) is the energy diffusion coefficient. Now ω_p is given by the configuration integral

$$\omega_p = (\frac{2\pi kT}{\mu})^{\frac{1}{2}} \Big/ \int_0^{R_T} \exp[-(\tilde{V}(R) - \tilde{V}(R_p))/kT] dR \tag{20}$$

and

$$D(E) = \frac{kT}{\Omega(E)} \int_0^E \Omega(E') deE' [\beta_R + \sum_i \beta_i (E'_i)] \tag{21}$$

where β_i is derived from (7) by the normal mode transformations, and E_i is the average energy in mode i. The adiabatic factors B_{BM} in (7) account for the difficulty of excitation of high frequency modes. The sum of terms in (21) depend not only on these factors, but also on the surface area of the molecule through the shielding factors discussed in Section II. Given D(E), $\Omega(E)$, $k_{RRKM}(E)$ and the energies D and Q, then the recombination rate can be calculated at any bath pressure. The low pressure solution is (19) which is linear in bath density and the high pressure solution is the pressure independent transition state rate obtained when $P_{SS}(E)$ is given by the Boltzmann distribution. It is expected that this pressure dependence will approximately follow $1/k = 1/k_{ED} + 1/k_{TST}$ suggested by (10) [22].

For molecules with many degrees of freedom, the average energy per mode is small and the harmonic approximation can be made for β_i. If these modes are treated as harmonic oscillators and the reaction coordinate approximated by a truncated harmonic oscillator, then (19) becomes

$$ k_{ED} = \frac{4\pi R_p^2}{\omega_\rho} \left(\frac{2\pi kT}{\mu}\right)^{\frac{1}{2}} \frac{1}{N!} \left(\frac{Q}{kT}\right)^N e^{(D-Q)/kT} [\beta_R + \sum_i \beta_i] \qquad (22) $$

The choice of the cut-off energy Q depends on the treatment of angular momentum. Sceats [10] has shown that in the case of diatomic dissociation on a Morse potential in 3-d, Q is well approximated by $\tilde{V}(R_T) - \tilde{V}(R_\rho)$. That is, the energy accumulation can be considered as occurring on the effective potential V(R). The more important point is that K_{ED} becomes very large for large molecules as a result of the N dependence of k_{ED}. If the diatomic results can be generalized, then (22) becomes

$$ k_{ED} \tilde{} k_{C-TST} \left\{ \frac{2\pi}{\omega_p} \left(\frac{Q'}{kT}\right)^N / N! \right\} [\beta_R + \sum_i \beta_i] \qquad (23) $$

where Q' is bounded by D and Q. The difficulty of adequately testing such a formulation against experiment is that the results depend on the density of states which is not adequately modelled above. The primary assumption in this formulation of k_{ED} is that IVR is rapid relative to the collision induced energy relaxation for each mode and that the molecular phase space trajectories between collisions are chaotic. As the β_i all increase with increase in bath density, the modes will be driven into thermal equilibrium at high enough bath density and cannot effectively feed the reaction coordinate. N can be reinterpreted as the effective number of modes which are not in thermal equilibrium at a given bath density [13], in which case (23) has the property that it tends to the result expected for one-dimensional motion on the reaction coordinate at high bath density.

Further work is required to test and improve (23). That the effective potential $\tilde{V}(R)$ gave reasonable results for the dissociation of a 3-d Morse oscillator is encouraging [8]. Also, the formulation of D(E) from the adiabatic factors B_{BM} has been tested against one-dimensional diatom-atom simulations and found to be satisfactory over a wide range of energies and collider masses [9]. When $\beta_R = 0$, D(E) applies to energy diffusion in any molecular system and can be tested against energy relaxation experiments from the relationship of D(E) to the average energy transfer $\langle\Delta E\rangle$ and the energy distribution function P(E,E') per collision [25]. There is little evidence that the weak collision assumption used to define an energy diffusion coefficient is in significant error. Lastly, the results of this section are classical, and can only be applied to real molecular systems with appropriate modifications [3].

V. APPLICATIONS

(a) DIATOMIC RECOMBINATION AND DISSOCIATION

The treatment of diatomic recombination and dissociation using the theory of section IV has been previously applied, at least for impulsive collisions [21]. In the low bath density regime, an adiabatic correction factor C has been used [9] based on the form for B_{BM} for an exponentially repulsive potential [26]. The important result was that the iodine recombination rate could be accounted for over the entire range of bath densities using as input the diatomic potentials and the atomic diffusion coefficients. The rate increased linearly with bath density according to energy diffusion, reached a narrow maximum below the transition state value and then decreased as spatial diffusion set in. The tight competition between energy and spatial diffusion near the transition state regime was well accounted for by application of barrier crossing theory [20] which combined both mechanisms.

In stochastic treatments the coordinates of the bath atoms are projected out, thereby circumventing the description of the recombination processes in terms of cluster equilibria and kinetics [27].

(b) ION-ION RECOMBINATION

The long range nature of the Coulomb potential leads to a situation in which coupling to internal degrees of freedom of the ions is unimportant. Use of the effective potential $\tilde{V}(R)$ to evaluate the encounter rate gives excellent agreement with experiment above 1 atm, and accurately predicts the pressure at which turn over to the energy diffusion regime occurs. The low pressure energy diffusion rate is found from (19) using the Coulomb density of states and the impulsive β_R appropriate to ion-molecule encounters. When Q is defined by the transition state of the effective potential, the correct dependence of k_{ED} on ion charge is found but the low pressure rate is significantly larger than that of the Flannery-Bates model [28]. This is most likely due to a breakdown of rapid thermalization of the ion pair orbital angular momentum by the bath atoms assumed when using $\tilde{V}(R)$. Thermalization increases the density of states available for capture and, as indicated in Table I, the increase is largest for the Coulomb potential.

(c) AEROSOL AGGLOMERATION

Application of the encounter theory of this work to aerosols is readily made [9]. In this case the large number of internal degrees of freedom lead to a large energy diffusion rate, hence a sticking probability approaching unity. Thus the recombination state is the encounter rate and the theory of section IV B can be used directly. For uncharged particles, the dominant long-range force is the dispersion interaction which, for two large particles of radii R_A and R_B,, is approximated by the Hameka potential

$$V(R) = -\frac{A}{12kT}\left[\frac{4R_A R_B}{(R_A R_B)^2}\right]\left\{\left[\left(\frac{R}{R_A+R_B}\right)^2-1\right]^{-1} + \left[\frac{R}{R_A+R_B}\right]^{-2} + 2\ell n\left(1-\left(\frac{R_A+R_B}{R}\right)^2\right)\right\}$$

where A is the Hameka constant. The effective potential $\tilde{V}(R)$ exhibits a transition state at a separation R_T which increases with A. The important point is that $R_T - (R_A+R_B)$ is much larger than the range of short-ranged forces so that particle deformation for solid aerosols is not involved in the determination of R_T. The encounter theory applies over the entire range of bath density and at all bath densities the encounter rate exceeds the hard-sphere rate by an amount which depends on the value of A/kT. The interesting point is that the difference is largest in the free molecule regime of low bath density. For example, for A/kT = 15 the encounter rate is 225% larger than the hard sphere rate, but this reduces to 25% in the continuum regime at high bath density. The transition region between these limits is described by the condition in (17) and for large particles this occurs in the gas phase. The factor $D_T/(\frac{8kT}{\pi\mu})^{\frac{1}{2}}$ is the "mean free path" and when this becomes less than the effective particle size $R_p W(o)\Omega*/4$, then the continuum limit is approached.

An interesting point concerning the encounter theory is that the results remain well defined even in the "sticky" hard sphere limit, in which

$V(R) = 0$ for $R > R_A + R_B$. In this case $W(\infty) = \Omega^* = 1$ and the rate (14) smoothly interpolates between the Boltzmann result at low bath density and the Smoluchouwski form at high bath density, even though the transition state is not well-defined, ie. $R_T \to R_A + R_B$. The results of this theory [9] when applied to hard spheres are in good agreement with other theory for the case of equal size hard-spheres [29], and with semiempirical theory [2] in all other cases. The advantage of this theory is that it is also applicable. for arbitrary potentials and that the expression of (14) is very simple.

(d) MOLECULAR RECOMBINATION

For moderately large molecules, say with more than 6 atoms, there are enough internal degrees of freedom to ensure that the energy diffusion rate k_{ED} is large enough at typical gas pressures that the fall-off regime in which k approaches k_{TST} occurs in a different regime to that in which the onset of diffusion control sets in. ie. a broad plateau is observed in the density dependence of the recombination rate. The problem in evaluating the encounter rate is that the internal degrees of freedom are coupled to the reaction coordinate R.

The simplest example of this is the case at ion-polar molecule reactions, in which the orientation of the molecular dipole μ is the internal coordinate. If q is the ion charge and α the molecular polarizability, then the potential of mean force is obtained from (2) as

$$V(R) = - \frac{\alpha q^2}{2R^4} - kT \ln \left[\sinh\left(\frac{q\mu}{kTR^2}\right) / \frac{q\mu}{kTR^2} \right]$$

When this is used in $\tilde{V}(R)$, the resulting transition state rate is equivalent to that of free-energy variational transition state theory [5]. Application of the encounter theory shows [30] that the onset of diffusion control occurs at bath densities of several hundred atmospheres. However at these pressures dielectric screening and "solvation" effects cannot be neglected and the theory would have to be modified.

A more interesting example of molecular recombination is that of radical recombination. The case of methyl radical recombination is of historical importance [31] and the onset of diffusion control has been observed [32]. The free energy surface $\tilde{V}(R)$ has been constructed [33] in order to estimate the transition state rate, and more recently this surface was used to estimate the recombination rate in the high pressure regime using the encounter theory of this work in order to investigate the onset of diffusion control [30]. The results were in excellent agreement with experiment and the role of the restricted methyl rotations were shown to be important. The free energy surface also accounts for quantum effects by the use of partition fuctions using

$$\tilde{V}(R) = V_o(R) - kT \ln Q (R) - 2 kT\ln R$$

where $V_o(R)$ is the ground state potential (eg a Morse potential) and $Q(R) = Q_e(R) Q_{RV}(R) Q_{EZ}(R)$ is the internal partition function. $Q_e(R)$ is the electronic partition function relative to $V_o(R)$ which accounts for electronic degeneracy and electronic level splittings, $Q_{RV}(R)$ is the ro-vibrational partition function excluding rotation of the pair (dealt with by the last term) and $Q_{EZ}(R)$ is the zero-point energy contribution. The difficulty of evaluating $Q(R)$ for all R was overcome [33] by using a simple interpolation between the values for reactants ($R \to \infty$) and products ($R = R_p$) of the form

$$\ln[Q(R)] = [\ln Q(R_p) - \ln Q (\infty)] e^{\gamma(R_p - R)} + \ln Q (\infty)$$

For a wide variety of systems, it was found that a value of $\gamma = 0.75 \overset{o-1}{A}$ gave satisfactory results for the transition state rates. When $V(R)=V_0(R) - kT\ln Q(R)$ is used to evaluate the barrier frequency in (15), the free energy surface gives predictions of the reaction dynamics in the bath density regime in which diffusion control may be important. Of course, solvent packing and stabilization effects may modify $V(R)$.

V. CONCLUDING REMARKS

The use of the effective potential $V(R) - 2kT\ln R$ to describe encounter and reaction dynamics has been developed, and applications to a variety of physical processes have been discussed. Further work especially in the low bath density regime, is required to clarify the use of the J-averaging procedure and to extend its use to polyatomic systems. The theory is simple to use and in many cases where reasonable models of the potential $V(R\xi)$ can be made, a direct comparison of reaction rates can be made in both gas and liquid phases.

ACKNOWLEDGEMENT

This work was supported by the Australian Research Grants scheme.

REFERENCES

1. J. Schroeder and J. Troe, Ann. Rev. Phys. Chem. 38, 163 (1987)
2. N. A. Fuchs, "The Mechanics of Aerosols" (Pergamon, Oxford, 1964)
3. P. J. Robinson and K.A. Holbrook, "Unimolecular Reactions" (Wiley, London, 1972)
4. J. C. Keck, Advances in Chemical Physics, 13, 85 (1967)
5. W. J. Chesnavich and M.T. Bowers, Prog. React. Kinetics 11, 137 (1982)
6. P. M. Rodger and M. G. Sceats, J. Chem. Phys. 83, 3358 (1985)
7. H. A. Kramers, Physica 7, 284 (1940)
8. M. G. Sceats, J. Chem. Phys. 84, 5206 (1986)
9. M. G. Sceats, Advances in Chemical Physics 70B (1988), in press
10. M. G. Sceats, Chem. Phys. Lett. 128, 55 (1986)
11. M. Borkovec and B. J. Berne, J.Chem. Phys. 84, 4327 (1986)
12. J. T. Hynes, J. Stat. Phys. 42, 149 (1986)
13. M. Borkovec and B. J. Berne, J. Chem. Phys. 82, 794 (1985)
14. D. Rapp and T. Kassal, Chem. Rev. 69, 61 (1969)
15. R. F. Grote, G. Van der Zwan and J. T. Hynes, J. Phys. Chem. 88, 4676 (1984)
16. H. Hippler, V. Schubert, and J. Troe, Ber. Bunsenges Phys. Chem. 89, 760 (1985)
17. P. M. Rodger, R. G. Gilbert and M. G. Sceats, J. Chem. Phys. (in press)
18. R. I. Cukier, R. Kapral and J.R. Mehaffey, J. Chem. Phys. 73, 5254 (1980)
19. J. Rotne and S. Prager, J. Chem. Phys. 50, 4831 (1969)
20. B. Carmelli and A. B. Nitzan, J. Chem. Phys. 80, 3596 (1984)
21. M. G. Sceats, J. M. Dawes and D. P. Millar, Chem. Phys. Lett. 114, 63 (1985)
22. J. Troe, Ber. Bunsenges Phys. Chem. 87, 161 (1983)
23. N. A. Fuchs, Zh. Experimo i Teor. Fiz 4, 7 (1934); P. Debye, Trans. Electrochem. Soc. 82, 265 (1942)
24. E. J. Verwey and J. Th. G. Overbeek, "Theory of Stability of Lyophobic Colloids" (Elsevier, Amsterdam, 1948)
25. J. Troe in "Physical Chemistry, An Advanced Treatise" 7B - ed. W. Jost, (Academic Press, NY, 1975) 835
26. J. Keck and G. Carrier, J. Chem. Phys. 43, 2284 (1965)
27. J. Troe, Ann. Rev. Phys. Chem. 29, 223 (1978)

28. D. R. Bates and M. R. Flannery, Proc. Roy. Soc. (Lond.) A302, 367 (1968)

29. D. C. Sahni, J. Colloid and Interface Science, 91, 418 (1983); 96, 560 (1983)

30. M. G. Sceats (to be published)

31. M. Quack and J. Troe, "Gas Kinetics and Energy Transfer, Vol. 2", (The Chemical Society, London, 1977) p. 175

32. H. Hippler, K. Luther, A. R. Ravishankara and J. Troe, Z. Phys. Chem. NF 142, 1 (1984)

33. M. Quack and J. Troe, Ber. Bunsenges Phys. Chem. 81, 329 (1977)

HYNES - In your rate versus buffer gas density plot for iodine, you assumed energy diffusion. However it is believed that for most buffer gases, a radical complex mechanisms holds (and not an energy transfer mechanism). Can you comment ?

SCEATS - In a stochastic approach the bath coordinates are eliminated altogether. The radical complex mechanism in which IM species are produced is just the leading term in a cluster expansion of the rate. We have shown that the rollover of the rate at high bath pressure can be explained in terms of the density dependence of the friction, and that the complexities of cluster equilibria and kinetics are not required. Therefore, in a generalized Langevin approach the M coordinates are eliminated and the role of complex formation will appear in either the potential of mean force or the friction kernel. In so far as our model at low density did not require explicit consideration of this, my conclusion would be that it is not important to differentiate between the mechanisms. In other work on iodine vibrational relaxation the role of attractive forces is essential in explaining the temperature dependence over the range of 1K, in supersonic expansions, to 1800 K in shock tube measurements and in dissociation of the I_2M Van der Waals complex.

In summary, the elimination of the bath coordinates in a stochastic model does not allow a detailed comparison of the energy transfer and radical complex mechanisms because both are, in principle, incorporated into the stochastics model by the parameters of the stochastic model.

As Professor Kapral has pointed out, the radical complex mechanism will only be important at low temperatures. Specifically when $kT \ll \epsilon$, the well depth of the I-M complex. For I-Ar, ϵ is about 150-200 K so that recombination kinetics at 300 K should not be important. Perhaps this explains why we did not need to incorporate these effects to explain Troe's experimental data.

BESNARD - What is the value of the vibrational correlation time of iodine in the solvents presented in your slide ?

SCEATS - For the studies on energy transfer in colinear I-I+M encounters, the Morse potential used for the I_2 was fitted to the experimental dissociation energy and harmonic frequency. In our calculations, we modelled only the average energy transfer which is intimately related to the vibrational relaxation rate, and hence the T_1 relaxation time for a given level v. Except near the dissociation threshold, this time is much longer then the dephasing time T_2 of the oscillator simply because the oscillator is almost certainly "dephased" in each colinear collision, but not vibrationally relaxed. Crudely speaking, the value of the adiatic factor B(w) discussed above is the ratio of the T_1 time to the T_2 time.

BINARY REACTIONS OCCURING IN THE PROCESS OF RANDOM WALKS

A.I. Burshtein

Institute of Chemical Kinetics and Combustion
Novosibirsk 630090, USSR

Abstract

A new approach of the bimolecular reaction theories is presented, which is based on the averaging of chemical anisotropy by translational and rotational Brownian motion of the particles.The effective steric factor change in reactions with only one anisotropic reagent was found. It is shown, that it can fall down to the values experimentally observed, only if the hopping mechanism of molecules approach and reorientation is realized. But if the motion is diffusive, then both particles should be chemically anisotropic to explain the experiment.

Introduction

Classical pattern of binary reactions in liquids is the following: reagents, randomly walking in solution, occasionaly find each other and having entered in a direct contact, react. If the process of walks may be considered as continual encounter diffusion, then the frequency of reagents encounter is determined by "diffusive" rate constant $k_D = 4\pi RD$, where R is a contact sphere radius (at the closest approach distance), D is the (self) diffusion coefficient /1/. If, moreover, the reaction rate under contact with the sphere in its any point is the same and equals k_r, then

$$\frac{1}{k} = \frac{1}{k_D} + \frac{1}{k_r} \qquad (1.1)$$

where k is a rate constant of bimolecular reaction, which is considered as "diffusion-controlled" when $k \simeq k_D \ll k_r$, or "kinetic-controlled" if $k \simeq k_r \ll k_D$ /2-4/. This subdivision of bimolecular reactions into two classes is generally accepted now. If the reaction is kinetic controlled, then its constant k_r is considered to be smaller than k_D and independent of D, e.g. of viscosity η, too; if $k = k_D = 4\pi RD$, then the reaction is diffusion controlled and its rate increases proportionally to $D=kT/6\pi\eta R$. However, among fast reactions (radicals and some others) there are those having constants one or two orders of magnitude less than k_D, but, nevertheless, proportional to D (fig.1) /5/. These reactions were termed as pseudodiffusion and demanded to be explained by the revision of classification based on formula (1/1) /6/.

Both basic suppositions of the theory have been revised: diffusive approach of the particles and chemical isotropy of contact reactions. It is evident, that the reactions, even if they may be considered as occuring at contact, really occur in a spherical layer having radius R and depth $\Delta \ll R$. The probability of their realization inside this layer (W) being constant between its interior and exterior boundaries may essentially depend on angular variables. Therefore particles approach may be considered as diffusive only if a migration step λ_0 is small enough in comparison to geometrical sizes of the reaction zone, and the reaction ought to admit chemically isotropic only if W is the same everywhere. Then actually everything depends only on $D=\lambda_0^2/\tau_0$ and $K_\mu = W4\pi R^2\Delta$ where $1/\tau_0$ is an average frequency of random walks. If the situation is different, then one should take into account that encounter may occur not by sucession of small steps, but by the only jump and in this case encouter frequency constant must be equal to

$$K_j = \frac{4\pi R^2 \Delta}{\tau_0} = mK_D \qquad m = \frac{R\Delta}{\lambda_0^2} \qquad (1.2)$$

and be called hopping, as it is assumed in the theory of transfer reactions in solid solutions /7/. The encounters of chemically isotropic particles are hopping, when the number of recontacts during the only encounter, equal to m is small : $m \ll 1$. Due to that the first explanation of the pseudodiffusional reactions was that the radical reactions in liquids were hopping, but not diffusive, and so their efficiency decreases in m times. However, if one does not suppose an improbable thing, then $\lambda_0 \leq R$ and this is an essential lower limit of m possible changes range :

$$\frac{\Delta}{R} \ll m \leq \frac{R}{\Delta} \qquad (1.3)$$

One can see that $\Delta/R \simeq 10^{-1}$ for the particles of moderate size, but not less. Meantime, there are some pseudodiffusion reactions with $k/k_D \simeq 10^{-2}$. Therefore, besides the hopping kinematic it is necessary to introduce chemical anisotropy.

The simplest way to do it is to suppose that only f-part of contact surface is chemically active and looks like a round spot (or a number of spots) on the sphere, where the reaction can occur. Rotation of the reagents during the contact, as well as in the pauses between recontacts increases greatly their chances to achieve proper

mutual coordination, when the contact point appears within the spot.
So the efficient steric factor

$$f_{eff} = k/k_D \qquad (1.4)$$

turns out to be not so small as the geometrical one. Moreover, such
its definition, accepted in previous works, includes both decreasing
factors of diffusion-controlled reactions efficiency : steric as well
as kinematic.

For dividing them henceforth, it is convenient to introduce the
encounter rate constant k_0, which coincides with k_D (when $m \gg 1$) and
with k_j, (when $m \ll 1$). It is calculated quite rigorously by methods
of the encounter theory for distance reactions /8/, as well as for
contact ones /9/, the result being brilliantly approximated by a
simple formula, suggested for the latter case in /10/ (fig.2, dashed
line) :

$$\frac{1}{K_0} = \frac{1}{K_D} + \frac{1}{K_J} = \frac{m+1}{m} \frac{1}{K_D} \qquad (1.5)$$

The reaction constant coincides with k_0 only in case if both
the reagents are chemically isotropic, and $k_\mu \simeq \infty$. Otherwise,

$$k = w \, k_0 \qquad (1.6)$$

where w is the encounter efficiency which may be less than 1 not only
because of the kinetic reaction control, but also owing to steric
difficulties in conditions of kinematic (migration) control.

Supposing that $K_\mu \simeq W = \infty$ in active spots on the contact
sphere, they may be regarded as "black" ones, kinetic-controlled
reactions being neglected. But even then the encounter efficiency
which will be indicated as w_0 in this case, varies within the
following limits :

$$1 \geq W_0 \geq \frac{(m+1)f}{1 + \frac{16}{3\pi} m\sqrt{f}} = \begin{cases} f & \text{in hopping reactions} \qquad (1.7a) \\ \frac{3\pi}{16}\sqrt{f} & \text{in difusion reactions} \qquad (1.7b) \end{cases}$$

if only one reagent is anisotropic and its single active spot
occupies f-part of the contact sphere area, with $f \ll 1$. The values of
w_0 possible values decrease during the transition from hopping
reactions to diffusion ones. Its lower limit w_{min}, having been found
in /11/, corresponds to hypothetic situation when the reagents do not
rotate during the whole encounter as if the rotation rate $1/\tau_\theta = 0$.

Nevertheless, minimum encounter efficiency in diffusion limit appears
equal to $\sqrt{f}$, which is much more than f. This result, indicated
firstly by numerical analysis of the problem /12/, was explained and
confirmed by exact analytical calculation /13/, which clarified
numerical magnitude of w_0 lower limit in diffusion reactions :
$(2/\pi)\sqrt{f}$.

Since the number of recontacts in these reactions is great,
encountered partners may repeat their attempts to react many times,
untill they separate. As the motion between successive contacts is
random, their mutual orientation may occur different at any time,
despite the absence of rotation. Opposite, for a hopping encounter

the initial contact remain actually the only one, and the probability that exactly this contact would be successful, is just equal f.

Switching of the rotation increases the encounter efficiency with any way of its realization, leading the anisotropy to complete averaging up to $w_0 = 1$, when $w\tau_\theta \Rightarrow 0$. In reality molecules rotate with moderate rate and to estimate how it contributes to the reaction it is necessary to calculate $w_0(1/\tau_\theta)$

The problem in such formulation has not the uniquesolution, as the rotation may also occur either by many turns of molecule axis through minor angle (rotational diffusion) or by rare turns through angle exceeding 2π. In the latter case, all the orientations after jump proved to be equally probable, and the rotation mechanism is hopping. This alternative exists, if one uses any method of calculation of w , but there are two most well-developed and mutually supplementing methods which will be described in the two following paragraphs before summing up.

Modified model of diffusion-controlled reactions (MMDCR)

Under this unsuccessful name there is known the simplest and chronologically the first model of chemically anisotropic reaction, formed by translational and rotational motion of partners. It was called so in the original paper /10/, although applicability conditions formulated by the authors supposed limited number of recontacts, so limited that they might be considered as statistically independent. They exclude a transition to the diffusion limit when the time intervals between successive contacts shorten so that the following contact takes place quite near the previous one. Later the MMDCR appeared to be an exact asymptotic method of the encounter theory which holds true when

$$m \ll 1/\sqrt{f} \qquad (2.1)$$

It is a mathematical condition of independence of successive contacts. This condition essentially limits the generality only when the reagents are anisotropic (f<1). Fig.3 shows that MMDCR set W_{min} too higher, when $m > 1/f$. And at the same time its applicability is wider than that of hopping reaction theory, holding true with $m \ll 1$. The main result of MMDCR may be presented as

$$\frac{1}{K} = \frac{1}{K_D} + \frac{1}{eK_J} \qquad (2.2)$$

In the absence of anisotropy (e=1), it is reduced to eq.(1.5) and in the presence of that it makes possible to calculate the contact efficiency

$$e = 1 - \int_0^\infty \bar{\eta}(t) \exp(-t/\tau_0) \, dt/\tau_0 \qquad \text{where} \qquad \bar{\eta} = \int \eta \, dt \qquad (2.3)$$

It is expressed through the average density of non-reacted systems which are already in contact (at a distance R), but react with a rate $W(\theta)$, when the rotational motion represented by operator $\hat{L}_{rot}$ leads them to the region where W differs from 0:

$$\dot{\eta} = - W(\theta)\ \eta + \hat{L}_{rot}\ \eta \qquad (2.4)$$

So the task is reduced to the calculation of the simpler reaction, proceeding in a cage and may be simplified even more if to consider $W = \infty$ in the limits of a spot with an angular size θ_0. Thereby the kinetic reactions are excluded from consideration and the eq.(2.4) is reduced to the equation of random walks in angular space, which is

$$\dot{\eta} = \frac{1}{\tau_\theta \sin\theta}\ \frac{\partial}{\partial\theta}\ (\sin\theta\ \frac{\partial}{\partial\theta})\ \eta \qquad \Delta\Theta \ll 2\pi \qquad (2.5)$$

with the diffusion rotation ;

$$\dot{\eta} = - \frac{1}{\tau_\theta}\ (\eta - \bar{\eta}/4\pi\ ) \qquad (2.5a)$$

with the hopping rotation.

These equations ought to be solved with the Smoluchowski boundary condition $\eta(\theta_0) = 0$, where $\theta_0 = $ Arccos $(1-2f)$. The obtained solution subsituted in eq.(2.3) give the value $e_0(1/\tau_\theta)$

- the contact efficiency when the spot is "black". In the case of hopping re-orientation this problem has been solved in /10/, and for the rotational diffusion it has been solved in /9/, f being arbitrary (not necessarily small) in both cases. The comparison of these results (fig.4) shows that, in the same conditions, the diffusive re-orientation averages the chemical anisotropy faster and more efficiently than the hopping one.

Using eqs.(1.5), (2.2) in eq.(1.6) the connection between the encounter efficiency and the contact efficiency in MMDCR is found quite easily :

$$w = e\ \frac{1 + m}{1 + me} \qquad (2.6)$$

Naturally, w_0 and e_0 are connected just the same. The validity of these relations is proved by the encounter theory methods /9/ in limits determined by the inequality (2.1). In the absence of rotation, when e=f, a corrective term in a denominator may be neglected. Then the efficiency of the encounters, during which the anisotropy averages only by re-contacts, being independent, is confirmed to be in direct proportion to the complete number of contacts m+1. But if the rotation enlarges the contact efficiency so that it reaches its upper limit (e = 1), then the re-contacts can add nothing to the result of the first one, and w=1.

Kinematic approximation

The rate constant calculation may be essentially simplified in comparison to standard recipe used in the encounter theory, basing on the fact that Δ/R is a small parameter according to the definition of the contact reactions. This way, having approved in chemically isotropic reactions in /14/, was spread on the reactions of anisotropic particles A and B /13,15/. It turned to be very efficient approximate way to calculate the rate constant, presented as

$$\frac{1}{K} = \frac{1}{W_0 K_0} + \frac{1}{K_\mu} \qquad (3.1)$$

whereas before, $K_\mu = Wv$, but $v = f_A f_B 4\pi R^2 \Delta$. The approximation is that the first term in eq.(3.1) is determined by the formula

$$W_0 K_0 = U/\tau \qquad (3.2)$$

which, besides the reaction volume v, contains the residence time averaged over all starting points, situated in these volume limits

$$\tau = \int_0^\infty \int_0^\infty dt \; P(q,q',t) \; dq \; dq' \; / \; 64\pi \; v^4 \qquad (3.3)$$

Here $P(q, q', t)$ is a two-dimensional possibility of finding a system in the point q at time t after it started from the point q'. The arguments of this function (q and q') include, besides the distance between the particles r , the collection of angular variables Ω_1 and Ω_2 , determining the particles A and B orientation relatively to the direction r/r. Integration over q and q' in eq.(3.3) extends only to active ("black") volume of the contact layer, being decreased in comparison to its complete size in $f_A f_B$ times, where

$$f_A = \frac{1}{8\pi^2} \int d\Omega_1, \quad f_B = \frac{1}{8\pi^2} \int d\Omega_2$$

For the calculation of τ by the formula (3.3) only kinematic information about the particles relative motion (hopping or diffusion) is necessary, and it countains in $P(q, q',t)$. It is just because of that, this rate constant calculation method, utilising only geometrical and kinematic information, acquires the definition as "kinematic approximation", being proposed in /17/, if used in binary variant of the theory.

The rate constant of remote reactions for any kinematic mechanism /18/ has been presented the same way. Although this way of the rate constant calculation has been considered in all papers, addressed to isotropic reactions, as the estimation which is all right in order of magnitude and reproducing all functional dependences. Only in papers /13-15/ the kinematic approximation became a calculation method, securing a satisfactory accuracy in the contact reactions even if they are stereospecific. This conclusion has been done basing on the comparison of some exact results, derived in the encounter theory /13/ or by the numerical methods /12/ with their analogies, which have been found in the kinematic approximation. The essence of the approximation is to neglect the dispersion of residence times in the reaction zone, which are a bit different starting from zone center or from its perifery. The smaller the steric factors, the worse it is performed, but even for the smallest spots the error estimated in /13/ in the absence of rotation, does not exceed 8%. If the kinematic approximation is used in just the same limits as MMDCR, we have /9/

$$W_0 = \frac{e_0(1 + m)}{1 + me_0} \qquad (3.4)$$

where

$$e_0 = f_A f_B \tau_0 / \tau_j \qquad (3.5)$$

and $\tau_j = \tau - f_A f_B R\Delta/D$ is the average residence tim in active zone in hopping reaction. That confirms the result (2.6), but with $e = e_0 W/(W + 1/\tau_j)$. This definition of e, as well as e_0 in eq.(3.5), is approximate in accordance with the accuracy of kinematic approximation. The idea of this accuracy is given by the fig.5 in /9/, in which the averaging of chemical anistropy of one reagent by rotational diffusion, accounted by MMDCR for several f, is compared with the small spot contact efficiency, being found the kinematic approximation in /19/. If this accuracy is considered to be acceptable, then, using the kinematic approximation with $m \ll 1/\sqrt{f}$, a general formula (2.7) may be carried to

$$\frac{1}{K} = \frac{1}{K_D} + \frac{1}{e_0 K_J} + \frac{1}{K_\mu} \qquad (3.6)$$

using eqs. (2.10) and (2.11). Although it is an approximate formula in comparison with the eq.(2.2), but it is easily interpreted according to successive stages of a chemical reaction. The first stage is the diffusion approach of A and B with the rate determined by the constant k_D. The second stage is the hopping contact approaching (a penetration in the first coordination sphere), which with the probability $e_0(f, 1/\tau_\theta)$ brings the reactants to a right coordination during the contact. And at last the reaction itself, which occurs in this coordination with the rate W and which determines the kinetic control of the rate constant K_R.

When $m \gg 1/\sqrt{f}$, the contacts fail to be independant and such a simple interpretation loses its power together with the MMDCR. However the kinematic approximation is still true and using it the efficiency of the diffusion encounters with on anisotropic reagent has been calculated in /20,21/ for hopping rotation and in /20/ for the rational diffusion. As it is shown in the fig.6, rotational diffusion averages the chemical anisotropy faster and more efficiently then large angles reorientation.

Indisputable advantage of the kinematic approximation in comparison with the MMDCR is its applicability to any m. It may be used to connect the results, relating to the hopping and diffusion approach. It has been done in /9/ for hopping reorientation of the anisotropic reagents. It is clear from the foregoing that this result establishes the lower limit of w, when the rate of the reorientation $1/\tau_\theta$ is finite. The upper limit is determined by the rotational diffusion, which unfortunately can not be described so exactly with any m (fig.7). But using MMDCR it is possible to advance so far that interpolation becomes possible between the solution being got in its limits and the diffusion result, being found in the kinematic approximation. This interpolation result as well as the solution, concerning the hopping reorientation, are shown in fig.8, adopted from /9/.

Conclusion

It is obvious from formulae (1.4)-(1.6), that the value f_{eff} characterizes pseudodiffusion rate constant decreasing as compared with the diffusive one. It falls with the encounter efficiency decreasing as well as with the number of re-contacts decreasing

$$f_{eff} = \frac{W_0 m}{m + 1} \qquad (4.1)$$

The longer the migration step in angular and usual space, the smaller is f_{eff}. Its miminum value realizing at $m \ll 1$ and complete rotation stop, may hardly be less than 10^{-4} if one consideres that f and $m \geq \Delta/R$ are restricted from below by the value of the order 10^{-2}. Orientational relaxation essentially enlarges this limit, as it is shown at fig.9, concerning the case of one anisotropic reagent. If the reactions with the anisotropic reagent are characterized $f_{eff} \simeq 10^{-2}$, it obviously means that they approach the contact in hopping manner and $f_{eff} = e_0 m$, when e_0, $m \ll 1$.

So low efficient factors may be explained in terms of diffusion approach mechanism only if both the reagents are anisotropic. The theory of these reactions developed in /20/ shows that

$$f_{eff} = W_0 \geq f_A \sqrt{f_B} \qquad (4.2)$$

As usual the minimum value is reached in the absence of rotation. It is a little more for $f_A = f_B = f = 10^{-2}$ than in eq.(4.1).

The plausibility of such explanation of the pseudodiffusion reactions increases if to turn to the large and particularly biological molecules. Their Brownian motion in angular and coordinate space is realized obviously by little steps, and the steric factors may become essentially less than discussed above.

Finally, it is appropriate to mention a restriction of above stated approach, which is connected to the fact that the contact layer depth Δ has been ignored in kinematic calculations (the spots have been considered as flat). This restriction has been overcome recently in the range of diffusive description of particles motion in many-dimensional space, being compounded of translational and rotational coordinates for partners /22/. Boundary condition for such approach permits to take into account the chemical and physical interaction of the active centers on finite distance between them and to vary the spot shape. All above mentioned functional dependences $f_{eff}(f)$, obtained in the diffusion approach, has been already reproduced with this method. But it is able to go further, if its advantages are used completely.

REFERENCES

1- M.V. Smolukhovsky Phys. Z. 17 (1916) p.557

2- H.Eyring, S.H.Lin, S.M.Lin "Basic Chemical Kinetics" ; NY 1980 Ch.IX.

3- A.A.Ovchinnikov, S.F. Timashev, A.A.Belyi. Kinetics of Diffusion- Controlled Chemical Processes. Moscow, "Khimia", 1986.

4- S.A. Rice "Diffusion-Limited Reactions" in Comprehensive Chem.Kinetics v.25, Elsevier 1985 p.404.

5- A.I. Burshtein, I.V.Khudyakov, P.P. Levin. Izv. AN SSSR, ser. Khim. N°2 (1980).

6- A.I. Burshtein, I.V. Khudyakov, B.I. Yakobson. Progress in Reaction Kinetics. Pergamon Press Oxford, N.Y., v.13, N°4 (1984).

7- A.I. Burshtein. Uspekhi Fiz.Nauk, v.143, (1984) p.553.

8- A.I. Burshtein, A.B. Doktorov, A.A. Kipriyanov, V.A. Morozov, S.G.Fedorenko. Zh. Eksp. Teor. Fiz., v.88 (1985), p.878.

9- A.I.Burshtein, A.B.Doktorov, V.A. Morozov, Chem. Phys. v.104, p.1 (1986).

10- A.I.Burshtein, B.I. Yakobson. Int. J. Chem. Kinetics V.12, p.261, (1980).

11- A.B.Doktorov, B.I.Yakobson, Chem.Phys. v.60 (1981) p.498.

12- Samson R., J.M.Deuth, J.Chem.Phys. V.68 (1978) p.285.

13- A.B.Doktorov, N.N. Lukzen Chem.Phys.Lett. V.79, (1981) p.498.

14- A.B.Doktorov, A.A.Kipriyanov Khim.Fizika, v.5, (1982) p.599 v.6 (1982) p.794.

15- V.M.Berdnikov, A.B.Doktorov Teor.Eksp.Khim. v. 17 (1981) p.318.

16- Doi M.Chem.Phys.v.11, (1975) p.115.

17- G.Wilemsky, M.Fixman J.Chem.Phys. 58 (1973) 4009.

18- A.B.Doktorov, A.A.Kipriyanov, A.I.Burshtein Zh .Eksp. Teor. Fiz. v.74 (1978) p.1184.

19- A.B.Doktorov Khim.Fizika v.4 (1985) p.800.

20- S.I.Temkin, B.I.Yakobson Khim.Fizika 3 (1984) p.1658, J.Phys.Chem. 88 (1984) p.2679.

21- A.B.Doktorov, N.N.Lukzen. Khim. Fizika 4 (1985) p.616.

22- A.I.Shushin Chem.Phys.Lett. v.130 p.452 (1986).

EFFECT OF AN EXTERNAL ELECTRIC FIELD ON THE RATE OF DIFFUSION-CONTROLLED REACTIONS

M. Tachiya

National Chemical Laboratory for Industry[*]
Yatabe, Ibaraki, 305, Japan
and
Hahn-Meitner-Institut Berlin
1000 Berlin 39, Fed. Rep. Germany

I. INTRODUCTION

In a series of papers[1,2] we have studied how the dynamics of diffusion-controlled reactions is influenced by the dynamic interaction between reactants and the solvent. The strength of the interaction is represented, for example, by the mean free path of reactant for scattering by the solvent. We have calculated[2] the rate constant as a function of the mean free path of reactant.

In the course of the above study we have found a new possible mechanism for the effect of an external electric field on the rate of reactions. In this paper we describe the new mechanism and analyze the electric field effect through this mechanism.

II. GEOMETRY OF REACTANT TRAJECTORY AND THE RATE OF REACTION

Consider a reaction between reactants A and B. Assume that reactants A are randomly distributed and fixed in the space and that reactants B move with a velocity v and are occasionally scattered by the solvent. Assume also that the reation occurs whenever a reactant B approaches a reactant A to a distance a. How does the scattering of reactant B by the solvent affect the rate of reaction? We have already shown[2] that the rate constant decreases with decreasing mean free path of reactant for scattering. Here we try to explain this result intuitively in terms of reactant trajectory.

When a reactant B is seldom scattered by the solvent, its trajectory is almost straight (case (a) in Fig. 1). From the above assumption on the reaction we can consider that concerning reaction with reactants A a reactant B sweeps up to a distance a from its trajectory. The rate of reaction is proportional to the volume swept per unit time, which is given by $\pi a^2 v$.

---

[*] permanent address

On the other hand, when a reactant B is frequently scattered by the solvent, its trajectory is entangled (case (b)) and it sweeps the same place repeatedly. Consequently the effective volume swept per unit time becomes smaller than $\pi a^2 v$. This is the reason the rate of reaction or the rate constant decreases with decreasing mean free path of reactant. Case (a) corresponds to low-pressure gas phase reactions, while case (b) corresponds to (narrowly defined) diffusion-controlled reactions.

III. EFFECT OF AN EXTERNAL ELECTRIC FIELD

Assume that the trajectory of a reactant B is engangled (case (b) in Fig. 1). Assume also that the reactant B carries a charge. Reactants A need not carry any charge. If an external electric field E is applied on the system, the entangled trajectory of reactant B will become more or less straight along the electric field (case (c)). Consequently the effective volume swept per unit time should increase, and thus the rate of reaction should increase.

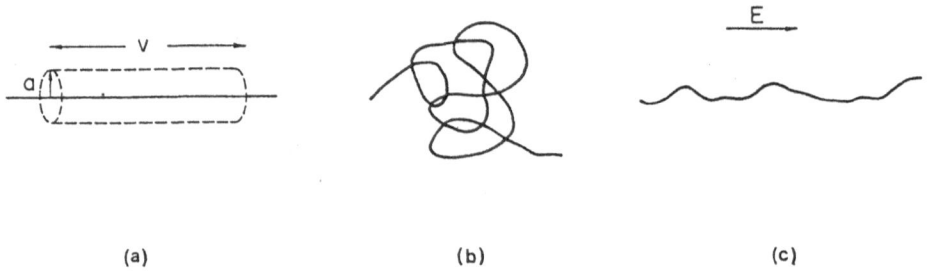

(a)   (b)   (c)

Fig. 1   Trajectory of reactant B (a) when it is seldom scattered by the solvent; (b) when it is frequently scattered; (c) when an external electric field is applied in the case of (b).

Let us consider this problem a little more quantitatively. A fully quantitative treatment will be given in the next Section. If the electric field is sufficiently high, the displacement of reactant B due to diffusion is negligible compared with that due to drift along the field. In this case the trajectory of reactant B is, roughly speaking, straight along the field direction and its speed is considered to be equal to the drift velocity μE where μ is the mobility of reactant B. Therefore, the rate constant should be given, on the analogy of case (a), by

$$k = \pi a^2 \mu E \tag{3.1}$$

That is, the rate constant should increase in proportion to the electric field strength at sufficiently high fields.

IV. QUANTITATIVE TREATMENT OF THE ELECTRIC FIELD EFFECT

The method of calculating the rate constant of diffusion-controlled reactions is well established[3]. A reactant A is put at the origin and the concentration distribution of reactant B around the reactant A is considered. Assume that a reactant B carries a charge q. We take z axis along the direction of the electric field if q is negative, and along its reverse direction if q is positive. The steady-state concentration distribution of reactant B satisfies the following diffusion equation

$$D\nabla \left[\nabla C(r) + \beta C(r)\nabla V \right] = 0 \qquad (4.1)$$

where $C(r)$ is the concentration of reactant B at point r, D its diffusion coefficient, V the potential acting on it, $\beta = 1/k_B T$, k_B the Boltzmann constant, and T the absolute temperature. Note that in the above choice of z axis V is expressed as $V = |q| Ez$ irrespective of the sign of q. On the assumption that reactants A and B react on approaching each other to distance a, we obtain

$$C(r=a) = 0 \qquad (4.2)$$

On the other hand, at infinity the concentration has to approach the bulk value C_0

$$C(r=\infty) = C_0 \qquad (4.3)$$

The solution of eqs. (4.1) – (4.3) can be obtained in the following way. Making the transformation

$$C(\underset{\sim}{r}) = e^{-\beta V/2}\psi(\underset{\sim}{r}) \qquad (4.4)$$

we find that ψ satisfies

$$\nabla^2\psi - F^2\psi = 0 \qquad (4.5)$$

where

$$F = \beta|q|E/2 \qquad (4.6)$$

In the polar coordinate system eq. (4.5) is written as

$$\frac{\partial^2\psi}{\partial r^2} + \frac{2}{r}\frac{\partial\psi}{\partial r} + \frac{1}{r^2}\frac{\partial}{\partial\eta}[(1-\eta^2)\frac{\partial\psi}{\partial\eta}] - F^2\psi = 0 \qquad (4.7)$$

where $\eta = \cos\theta$, θ being the polar angle. Equation (4.7) is separable and the solution may be written as

$$\psi(\underset{\sim}{r}) \equiv \psi(r, \eta) = \sum_{\ell=0}^{\infty} R_\ell(r) P_\ell(\eta) \qquad (4.8)$$

where $P_\ell(\eta)$ is the Legendre polynomial. The radial function satisfies

$$\frac{d^2R_\ell}{dr^2} + \frac{2}{r}\frac{dR_\ell}{dr} - [F^2 + \frac{\ell(\ell+1)}{r^2}]R_\ell = 0 \qquad (4.9)$$

Two linearly independent solutions of eq. (4.9) are given[4] by the modified spherical Bessel functions $\sqrt{\pi/2Fr}\ I_{\ell+\frac{1}{2}}(Fr)$ and $\sqrt{\pi/2Fr}\ K_{\ell+\frac{1}{2}}(Fr)$. Therefore, $C(r)$ is expressed as

$$C(\underset{\sim}{r}) \equiv C(r, \eta) = \sqrt{\frac{\pi}{2Fr}}\ e^{-Fr\eta} \sum_{\ell=0}^{\infty} [A_\ell I_{\ell+1/2}(Fr) + B_\ell K_{\ell+1/2}(Fr)]\ P_\ell(\eta) \qquad (4.10)$$

where A_ℓ and B_ℓ are arbitrary constants to be determined from the boundary conditions.

Noting the relation[4] that in the limit $r \to \infty$

$$K_{\ell+1/2}(z) \sim \sqrt{\frac{\pi}{2z}}\ e^{-z} \qquad (4.11)$$

we have from eq. (4.3)

$$\sqrt{\frac{\pi}{2Fr}}\ e^{-Fr\eta} \sum_{\ell=0}^{\infty} A_\ell I_{\ell+1/2}(Fr) P_\ell(\eta) = C_0 \qquad (4.12)$$

By comparing eq. (4.12) with the relation[4]

$$e^{z\cos\theta} = \sum_{\ell=0}^{\infty} (2\ell+1) \sqrt{\frac{\pi}{2z}} I_{\ell+1/2}(z)\ P_\ell(\cos\theta) \qquad (4.13)$$

A_ℓ is determined as

$$A_\ell = (2\ell+1) C_0 \qquad (4.14)$$

With the aid of eq. (4.12) $C(r, \eta)$ is expressed as

$$C(r, \eta) = C_0 + \sqrt{\frac{\pi}{2Fr}}\ e^{-Fr\eta} \sum_{\ell=0}^{\infty} B_\ell K_{\ell+1/2}(Fr) P_\ell(\eta) \qquad (4.15)$$

Substitution of eq. (4.15) in (4.2) yields

$$\sqrt{\frac{\pi}{2Fa}}\ e^{-Fa\eta} \sum_{\ell=0}^{\infty} B_\ell K_{\ell+1/2}(Fa) P_\ell(\eta) = -C_0 \qquad (4.16)$$

Again by comparing eq. (4.16) with eq. (4.13), B_ℓ is determined as

$$B_\ell = -(2\ell+1) C_0\ I_{\ell+1/2}(Fa)/K_{\ell+1/2}(Fa) \qquad (4.17)$$

Accordingly $C(r, \eta)$ is expressed as

$$C(r, \eta) = C_0 [1 - \sqrt{\frac{\pi}{2Fr}}\ e^{-Fr\eta} \sum_{\ell=0}^{\infty} (2\ell+1) \frac{I_{\ell+1/2}(Fa)}{K_{\ell+1/2}(Fa)} K_{\ell+1/2}(Fr) P_\ell(\eta)] \qquad (4.18)$$

The rate constant is given by the number per unit time of reactants B flowing into the reaction sphere, divided by the bulk concentration C_0.

374

$$k = \int_o^\pi \int_o^{2\pi} D[\frac{\partial C}{\partial r}]_{r=a}\, a^2\, \sin\theta d\theta d\phi / C_o$$

$$= 2\pi Da^2 \int_{-1}^1 [\frac{\partial C}{\partial r}]_{r=a}\, d\eta / C_o \tag{4.19}$$

From eq. (4.18) we have

$$[\frac{\partial C}{\partial r}]_{r=a} = C_o[(\frac{1}{2a} + F\eta)\sqrt{\frac{1}{2Fa}}\, e^{-Fa\eta} \sum_{\ell=0}^\infty (2\ell+1)\, I_{\ell+1/2}(Fa) P_\ell(\eta)$$

$$-\sqrt{\frac{\pi F}{2a}}\, e^{-Fa\eta} \sum_{\ell=0}^\infty (2\ell+1) \frac{I_{\ell+1/2}(Fa)}{K_{\ell+1/2}(Fa)}\, K'_{\ell+1/2}(Fa) P_\ell(\eta)] \tag{4.20}$$

where the prime denotes the derivative. Noting eq. (4.13), we have from eq. (4.20)

$$[\frac{\partial C}{\partial r}]_{r=a} = C_o[\frac{1}{2a} + F\eta - \sqrt{\frac{\pi F}{2a}}\, e^{-Fa\eta} \sum_{\ell=0}^\infty (2\ell+1) \frac{I_{\ell+1/2}(Fa)}{K_{\ell+1/2}(Fa)}\, K'_{\ell+1/2}(Fa) P_\ell(\eta)] \tag{4.21}$$

Substitution of eq. (4.21) in (4.19) yields

$$k = 2\pi Da^2 [\frac{1}{a} - \sqrt{\frac{\pi F}{2a}} \sum_{\ell=0}^\infty (2\ell+1) \frac{I_{\ell+1/2}(Fa)}{K_{\ell+1/2}(Fa)}\, K'_{\ell+1/2}(Fa) \int_{-1}^1 e^{-Fa\eta} P_\ell(\eta) d\eta] \tag{4.22}$$

By utilizing the relation[4]

$$e^{-z\cos\theta} = \sum_{\ell=0}^\infty (-1)^\ell (2\ell+1) \sqrt{\frac{\pi}{2z}}\, I_{\ell+1/2}(z)\, P_\ell(\cos\theta) \tag{4.23}$$

the integral in eq. (4.22) is calculated as

$$\int_{-1}^1 e^{-Fa\eta} P_\ell(\eta) d\eta = (-1)^\ell \sqrt{\frac{2\pi}{Fa}}\, I_{\ell+1/2}(Fa) \tag{4.24}$$

Accordingly k is expressed as

$$k = 2\pi Da\{1 - \pi \sum_{\ell=0}^\infty (-1)^\ell (2\ell+1) \frac{[I_{\ell+1/2}(Fa)]^2}{K_{\ell+1/2}(Fa)}\, K'_{\ell+1/2}(Fa)\} \tag{4.25}$$

With the aid of the relation[4]

$$K'_\nu(z) = \frac{\nu}{z} K_\nu(z) - K_{\nu-1}(z) \tag{4.26}$$

eq. (4.25) may be rewritten as

$$k = 2\pi Da\{1 + \pi \sum_{\ell=0}^\infty (-1)^\ell (2\ell+1)[I_{\ell+1/2}(Fa)]^2 [\frac{\ell+1/2}{Fa} + \frac{K_{\ell-1/2}(Fa)}{K_{\ell+1/2}(Fa)}]\} \tag{4.27}$$

The limiting expressions for the rate constant at high and low electric fields may be interesting. We have already shown in Sect. III that at high electric fields the rate constant should be given by eq. (3.1) which is, with the aid of the Einstein relation $\mu = |q| \, D/k_BT$, rewritten as

$$k/4\pi Da = \tfrac{1}{2} \, Fa \qquad\qquad (4.28)$$

On the other hand, we can show that for low electric fields eq. (4.21) is approximated by

$$k/4\pi Da = 1 + Fa \qquad\qquad (4.29)$$

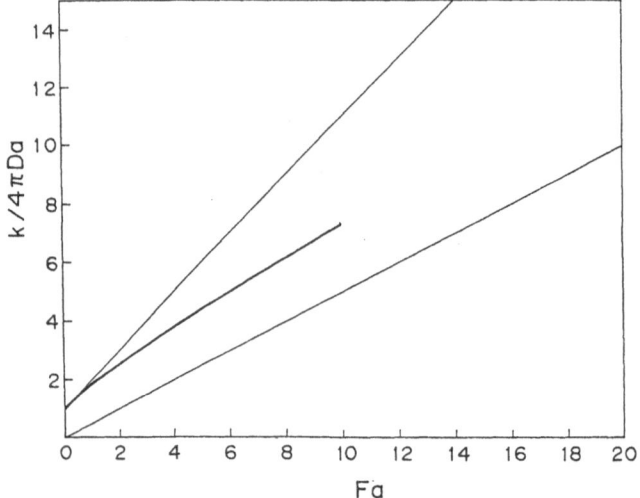

Fig. 2   Effect of the external electric field E on the rate constant k. The ordinate represents $k/4\pi Da$ where $4\pi Da$ is the rate constant in the absence of an electric field. The abscissa represents Fa where $F = |q|E/2k_BT$. The bold line represents the exact result calculated by use of eq. (4.27). The upper thin line represents eq. (4.29), while the lower thin line represents eq. (4.28).

In Fig. 2 the rate constant is plotted against Fa. The bold line represents the exact result based on eq. (4.27). The upper thin line represents eq. (4.29), while the lower thin line represents eq. (4.28). As seen from eq. (4.6) and Fig. 2, the electric field effect on the rate constant is enhanced as the reaction radius a increases. It is also enhanced by lowering the temperature T.

V. DISCUSSION

There may be several mechanisms through which the rate of reactions are affected by the external electric field. Two possible mechanisms may be immediately conceivable. One is related with the change in the potential energy surface. If ionic or polar species are involved in the reaction, the potential energy surface for the reaction may be changed by the application of the electric field. Then the rate of reaction is changed through the change in the activation energy. The other mechanism is realted with the heating of reactants. If some reactants involved are ionic, they may be heated by the electric field under some conditions. Then the rate of reaction may be changed through the change in the energy distribution of reactants.

The mechanism proposed in the present paper is distinct from the above two mechanisms. In the proposed mechanism neither the change in the potential energy surface nor the heating of reactants is important at all. In fact, in the reaction considered in Sect. II the potential energy barrier against reaction does not exist originally, and thus does not change with the electric field. Furthermore, in the model presented in Sect. IV the mobility of reactants B is assumed to be constant. This is equivalent to the assumption that the heating of reactants B does not occur; if the heating occurs, the mobility changes with the electric field. In the proposed mechanism the electric field increases the rate of reaction through the increase in the rate of encounter between reactants. Thus the effect of the electric field through this mechanism is similar to that of a stirrer, which also increases the rate of reaction by increasing the encounter rate. In this connection we point out that the influence of convective flow on the rate of diffusion-controlled reactions has been studied by several workers[5-7].

A few experiments are found in the literature which may be related with our theory. Bakale and Schmidt[8] have studied the effect of an external electric field on electron attachment to SF_6 in liquid ethane and propane. They have found that the rate constant increases when an electric field is applied. They have attributed this increase to the increase in the diffusion coefficient of electrons in the presence of a high electric field. However, this increase may be alternatively interpreted in terms of the proposed mechanism. Let us estimate the expected magnitude of the increase in the rate constant through the proposed mechanism. We assume a = 14.5 A as they did. Consider a case of T = 133K and E = 200 kV/cm, for example. Then we have Fa = 1.27. It is seen from Fig. 2 that in this case the rate constant becomes nearly twice as large as that in the absence of an electric field. This increase is comparable with the observed increase. In our theory the variation in the mobility of electrons with the electric field is not taken into consideration, whereas in the system they studied the mobility actually increases with the electric field. Therefore, a fully quantitative comparison between our theory and the experimental result may not be justified. However, we believe that the proposed mechanism is also involved in the observed effect.

Shinsaka, Nakamura, and Hatano[9] have experimentally studied the effect of an external electric field on bulk recombination between electrons and positive ions in high-density methane gases. They have found that the rate constant increases when an electric field is applied. On the other hand, Morgan[10] has studied the same problem theoretically and obtained the result that the rate constant decreases by the application of an electric field. The reason for this decrease in the rate constant is not clarified in his paper. At any rate his theory conflicts with the experimental observation.

We believe that the observed increase in the rate constant is related with the proposed mechanism, altough we cannot make a quantitative comparison between our theory and the experimental result since in our theory the Coulomb interaction between reactants is not taken into account. The extension of the present treatment to the case where a Coulomb force acts between reactants is in progress in our laboratory.

REFERENCES

1. M. Tachiya, Chem. Phys. Lett. 127:55 (1986)
2. M. Tachiya, J. Chem. Phys. 84:6178 (1986)
3. See, for example, D.F. Calef and J.M. Deutsch, Ann. Rev. Phys. Chem. 34: 493 (1983)
4. M. Abramowitz and I.A. Stegun, "Handbook of Mathematical Functions", Dover, New York (1965)
5. V.G. Levich, "Physicochemical Hydrodynamics", Prentice-Hall, N.J. (1962)
6. R. Samson and J.M. Deutsch, J. Chem. Phys. 74:2904 (1981)
7. M.A. Delichatsios and R.F. Probstein, J. Colloid Interface Sci. 51:394 (1975)
8. G. Bakale and W.F. Schmidt, Z. Naturforsch. 36a:802 (1981)
9. K. Shinsaka, Y. Nakamura, and Y. Hatano, Abstracts of the 27th Symp. on Radiat. Chem., Tokyo, p. 33 (1984)
10. W.L. Morgan, J. Chem. Phys. 84:2298 (1986)

DISCUSSION

PRUD'HOMME - You speak about the trajectory of each molecule of reactant B. What is the influence of these trajectories on the mean properties ?

TACHIYA - Of course the trajectory of each molecule differs from one another. However, the fractal geometry of trajectories, which characterizes the mean property of trajectories depends only on the mean free path. For instance, the fractal dimension of trajectories is one, if the mean free path is infinitely large. On the other hand, if the mean free path is negligibly small compared with the reaction radius a, the fractal dimension of trajectories is two on the scale of a. What I am saying is that the rate of reaction is related with the fractal geometry of reactant trajectories. For more detail I refer you to ref. 13.

(13) M. Tachiya, J. Chem. Phys. 87. October 15 issue.

JONAH - Could the increase in rate at higher voltages be due to a change in the energy of the electron ?

TACHIYA - Bakele and Schmidt originally attributed the increase in the rate to the increase in the diffusion coefficient of electrons. They estimated the latter from the mobility data by using the Einstein relation. However, in their experimental condition, the Einstein relation does not hold. When the increase in the diffusion coefficient is correctly estimated, it does not account for the increase in the rate.

TIME DEPENDENT RATE IN INTRAMOLECULAR REACTIONS WITH CONFORMATIONAL CHANGES

F. Heisel and J.A. Miehé

Groupe de Photophysique Moléculaire and
Groupe d'Optique Appliquée
Centre de Recherches Nucléaires   23 rue du Loess
67200 Strasbourg-France

Chemical reactions in liquid are complex many body events and the problem is how to construct a theoretical model which takes into account the interactions and correlated motions of solute and solvent molecules. This is possible only by using some simplifications owing to the physical situation. Since the chemical reactions occur in spatially localized regions, the system can be divided into two parts : one comprising the solute and few solvent molecules initiating the reaction, the second composed of the solvent molecules acting as a heat bath and influencing the dynamics of the primary system. Due to the interaction between the two systems there is an irreversible energy flow to the heat bath and the dynamics of the reacting molecules is considered as stochastic.

INTRAMOLECULAR MOTION

A simplified way to quantitatively analyze the process of the intramolecular motion in the excited state is to approximate the stochastic behavior by a diffusive Markov process. If

$$P_1(t,x)\Delta x = Pr[x \leqslant X(t) < x + \Delta x] \tag{1}$$

represents the probability law of the motion of the system characterized by the random variable $X(t)$ (for example the reaction coordinate) the following well-known conditions hold for the aforementioned approximation

$$p(t_{n-1},x_{n-1};t_n,x_n) \equiv Pr\left[X(t_n) < x_n/X(t_{n-1}) = x_{n-1}\right] \tag{2}$$

$$= Pr\left[X(t_n) < x_n/X(t_{n-1}) = x_{n-1},\ldots,X(t_1) = x_1\right]$$

with $t_1 < \ldots < t_n$ (Markovian condition)

$$\lim_{\Delta t \to 0} \frac{1}{\Delta t} \int_{|y-x| > \delta} p(t,x;t+\Delta t,y)dy = 0 \tag{3}$$

$$\lim_{\Delta t \to 0} \frac{1}{\Delta t} \int_{|y-x| \leqslant \delta} (y-x)p(t,x;t+\Delta t,y)dy = M(t,x)$$

$$\tag{4}$$

$$\lim_{\Delta t \to 0} \frac{1}{\Delta t} \int_{|y-x| \leqslant \delta} (y-x)^2 p(t,x;t+\Delta t,y)dy = S^2(t,x)$$

(diffusion conditions).

The condition (Eq.3) means that if the system jumps from a state x to some other state y finitely different from x during the time interval $(t,t+\Delta t)$, the transition probability tends to zero faster than Δt, as Δt goes to zero. In other words, the system cannot undergo an appreciable change during a small time interval.

The transition probability $p(s,x;t,y)$ satisfies the Fokker-Planck equation (also known as Kolmogorov's forward equation) [1]

$$\frac{\partial}{\partial t} p(s,x;t,y) = -\frac{\partial}{\partial y} M(t,y)p(s,x;t,y) + \frac{1}{2} \frac{\partial^2}{\partial y^2} S^2(t,y)p(s,x;t,y) \tag{5}$$

with $s < t$.

By integrating this expression with respect to x the probability law $P_2(t,y) = Pr\left[X(t) < y\right]$ describing the time evolution of the system is given by

$$\frac{\partial}{\partial t} P_2(t,y) = -\frac{\partial}{\partial y} M(t,y) P_2(t,y) + \frac{1}{2} \frac{\partial^2}{\partial y^2} S^2(t,y)P_2(t,y) \tag{6}$$

It can easily be shown that the probability density to find at time t the system in the state y verifies

$$\frac{\partial}{\partial t} P_1(t,y) = -\frac{\partial}{\partial y} M(t,y)P_1(t,y) + \frac{1}{2} \frac{\partial^2}{\partial y^2} S^2(t,y)P_1(t,y) \qquad (7)$$

At this stage it is of importance to examine the relationship between the Fokker-Planck and Langevin equations.

First consider the Langevin equation for a particle in a force field :

$$\frac{d}{dt} X(t) = V(t)$$

$$m \frac{d}{dt} V(t) = -\zeta V(t) + F[X(t)] + \mathcal{F}(t) \qquad (8)$$

where m is the mass of the Brownian particle, $-\zeta dX(t)/dt$ is the frictional force exerted by the medium, $\mathcal{F}(t)$ is the fluctuating force related to the collisions of the molecules of the fluid inducing the displacement of the Brownian particle, and F(X) is an external driving force independent of the random force $\mathcal{F}(t)$.

The essential assumptions concerning the random force are the following :

1. $\mathcal{F}(t)$ is a normal process with zero mean.

2. $\mathcal{F}(t)$ is a white noise, namely, its autocorrelation function is of the forme of

$$E[\mathcal{F}(t)\mathcal{F}(t + \tau)] = 2k_B T\zeta\delta(\tau) \qquad (9)$$

where k_B is the Boltzmann constant, T is the absolute temperature ; this equation is a version of the fluctuation-dissipation theorem. This expression shows how the Brownian motion is related to the thermal fluctuations occuring spontaneously in the fluid.

3. $\mathcal{F}(t)$ and $V(t) = \dot{X}(t)$ are statistically independent, and the fluctuations of $\mathcal{F}(t)$ are very much faster than those of V(t).

Let $\mathcal{W}(t)$ denote the Wiener process and let $d\mathcal{W}(t) = \mathcal{W}(t+dt) - \mathcal{W}(t)$. Note first, according to the treatment of the stochastic integration in the sense of Itô and Stratonowitch [1,2], that the white noise $\mathcal{F}(t)$ is the formal derivative of the Wiener process :

$$\mathcal{F}(t) = \frac{d\mathcal{W}(t)}{dt} \qquad (10)$$

$d\mathcal{W}(t)$ also verifies

$$E[d\mathcal{W}(t)] = 0$$
$$E[d\mathcal{W}(t)^2] = 2k_B T \zeta \, dt \qquad (11)$$

Setting $\mathcal{W}(t) = (2k_B T \zeta)^{1/2} W(t)$, the Langevin relation —describing the motion of the particle in the phase space— can be written as a two-variable stochastic differential equation system of the form of

$$dX(t) = V(t)dt$$
$$dV(t) = \frac{1}{m} \{ -\zeta V(t) + F[X(t)] \} \, dt + \frac{(2k_B T \zeta)^{1/2}}{m} \, dW(t) \qquad (12)$$

It is easy to compute the expectations (Eqs.4) and to show that they are finite valued.

Consider a n–dimensional diffusion process $X(t) = \{ X_1(t), \ldots, X_n(t) \}$ defined by a multivariate stochastic differential equation of the form of [3]

$$dX(t) = M[X(t),t]dt + S[X(t),t] \, dW \qquad (13)$$

where $W(t) = \{ W_1(t), W_2(t), \ldots, W_n(t) \}$ is an n–variable Wiener process satisfying the stochastic independence conditions

$$E[dW_i(t)dW_j(t)] = \delta_{ij}dt \qquad (14)$$

Note that Eq.13 stands for the following set of equations $(i = 1, 2, \ldots, n)$:

$$dX_i(t) = M_i[X(t),t]dt + \sum_j S_{ij}[X(t),t]dW_j \qquad (15)$$

The n–variable version of Kolmogorov's forward equation (or Fokker–Planck equation) can be written as

$$\frac{\partial}{\partial t} p = - \sum_i \frac{\partial}{\partial x_i} M_i(\mathbf{x},t)p + \frac{1}{2} \sum_{ij} \frac{\partial^2}{\partial x_i \partial x_j} [S(\mathbf{x},t)S^T(\mathbf{x},t)]_{ij} p \quad (16)$$

where $p \equiv p(t_o,\mathbf{x}_o;t,\mathbf{x})$ is the conditional probability density, $S(\mathbf{x},t)$ $[S^T(\mathbf{x},t)]$ is the matrix [the transpose] whose ij element is S_{ij} $[S^T_{ij} = S_{ij}]$.

(i) High–Friction Limit : Smoluchowski Equation. This approximation is of interest in the diffusion theory of chemical reactions and was used by Kramers [4,5] in the calculations of the reaction rate.

Let us divide the second expression of Eq.8 by ζ and let m/ζ tend to zero much faster than any of the terms on the right–hand side. It follows that Eqs.12 reduce to

$$dX(t) = + \frac{1}{\zeta} F[X(t)]dt + \sqrt{2D} \; dW(t) \quad (17)$$

Therefore, the motion of the particle is determined by a one–variable stochastic differential equation. The physical contents of $(m/\zeta)dV(t)/dt \to 0$ can be related to the fact that the velocity relaxes on a time scale much shorter than the time scale characterizing variations in position. Comparison of Eqs.15 and 17 leads to $M[X(t),t)] = F[X(t)]/\zeta$ and $S[X(t)] = (2D)^{1/2}$; then Eq.16 yields the Smoluchowski equation

$$\frac{\partial}{\partial t} p = D \frac{\partial^2}{\partial x^2} p + \frac{1}{\zeta} \frac{\partial}{\partial x} p \frac{\partial U}{\partial x} \quad (18)$$

with the field force $F(x) = - dU(x)/dx$.

For a harmonic oscillator immersed in a Brownian medium, the displacement -in the high–friction approximation- is a position Ornstein–Uhlenbeck process. With $U(x) = m \omega^2 x^2/2$, the solution of Eq.18 is [5]

$$p(t_o,x_o;t,x) = p(x_o;t-t_o,x) = (2 \pi \sigma^2)^{-1/2} \exp[-(x-y_o)^2/2\sigma^2] \quad (19)$$

where

$$
\begin{aligned}
y_o &= x_o[\exp -(t-t_o)/\tau^\omega_c] = E[X(t)] \\
\sigma^2 &= D\tau^\omega_c \{1 - \exp[-2(t-t_o)/\tau^\omega_c]\} = Var[X(t)]
\end{aligned}
\quad (20)
$$

The characteristic time τ_c^{ω} is defined by

$$\tau_c^{\omega} = \zeta / m\omega^2 \tag{21}$$

(ii) Arbitrary friction regime : Kramers equation. In the presence of inertial effects, the one-dimensional motion is determined by a bivariate Fokker-Planck equation.

Comparing Eq.12 with Eq.13, one concludes that

$$
\begin{aligned}
M_1 &= v & S_{11} &= 0 \\
M_2 &= [- \zeta v + F(x)]/m & S_{22} &= (2k_B T\zeta)^{1/2}/m
\end{aligned}
\tag{22}
$$

Hence Eq.16 for the probability law becomes

$$\frac{\partial}{\partial t} p + v \frac{\partial}{\partial x} p - \frac{1}{m} \frac{\partial U}{\partial x} \frac{\partial}{\partial v} p = \frac{\zeta}{m} \frac{\partial}{\partial v} vp + \frac{k_B T\zeta}{m^2} \frac{\partial^2}{\partial v^2} p \tag{23}$$

expression due to Kramers [4,5], where $p \equiv p(t_o, x_o, v_o; t, x, v)$. In classical mechanics the left-hand side represents the total derivative of p with respect to t and equals zero (Liouville's theorem). Therefore, the contribution of the heat bath on the motion of the particle is described by the right-hand side of Eq.23.

UNIMOLECULAR REACTION IN THE HIGH FRICTION LIMIT

This analysis concerns for instance the formation of the twisted intramolecular charge transfer state, the photoisomerization processes, etc... in solution for which the intramolecular motion is related with the solvent motion. The unimolecular reaction -the passage from the reactant well to the product well in the Kramers' treatment- is modeled by a sink term depending on the reaction coordinate. In the high-viscosity case the motion is governed by the modified Smoluchowski equation [6]

$$\frac{\partial}{\partial t} \overline{P} - D \frac{\partial^2}{\partial x^2} \overline{P} - \frac{m\omega^2}{\zeta} \frac{\partial}{\partial x} x\overline{P} = - s(x)\overline{P} - k_r \overline{P} + \delta(x+|x_o|)\delta(t) \tag{24}$$

where $\overline{P} \equiv \overline{P}(0, x_o; t, x)$ represents the probability density of finding a particle at x at time t, starting at t = 0 from the position x_o. The left-hand-side expression corresponds to the motion of a one-dimensional Brownian particle subject to a force from a harmonic potential (see Eq.18). The right-hand side includes the sink functions $s(x)$, the spontaneous decay of rate k_r, and, finally, the initial condition.

If the sink corresponds to a pinhole, it is of the form

$$s(x) = K_{nr} \, \delta(x-x_1) \tag{25}$$

where K_{nr} is related to the microscopic rate constant that determines the effectiveness of the reaction occuring at the position x_1. If a reaction takes place whenever the particle reaches x_1 ($K_{nr} \to \infty$), P satisfies the so-called Smoluchowski condition

$$\bar{P}(0,x_0;t,x_1) = 0 \tag{26}$$

For a finite value $K_{nr} \neq 0$, P verifies the following boundary condition:

$$D \frac{\partial}{\partial x} \bar{P}\Big|_{x=x_1} = K_{nr} \bar{P}\Big|_{x=x_1} \tag{27}$$

If $K_{nr} = 0$, Eq.24 describes the motion of a Brownian particle undergoing spontaneous decay and from Eqs. 19 and 24 it is easy to compute that

$$\bar{P}(0,x_0;t,x) = \bar{p}(0,x_0;t,x)\exp -k_r t \tag{28}$$

For this model, reaction schemes in the presence ($x_0 < 0$, $x_1 > 0$) or in the absence ($x_0 < 0$, $x_0 < x_1 \leqslant 0$) of a potential barrier can be distinguished (Figs.1a and 1b).

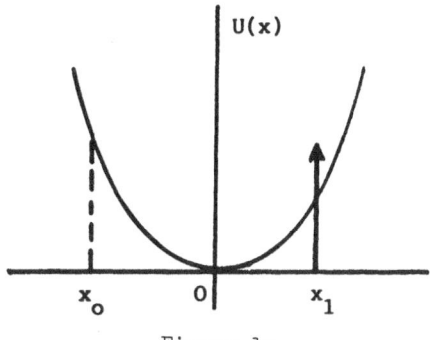

Figure 1a

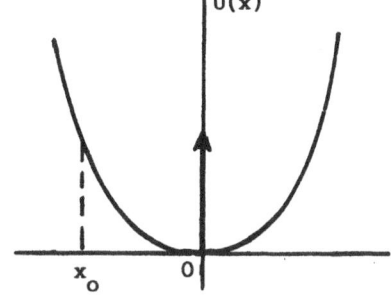

Figure 1b

Since we are interested in the description of the disappearance of the reactants, let us define the random variable T as the instant at which the reaction takes place (lifetime of the reactant). The so-called survival probability $q(t) = Pr[T > t]$ that the reactant has not undergone a reaction at time t is given by

$$q(t) = \int \overline{P}(0,x_o;t,x)dx \qquad (29)$$

For a system undergoing a reaction in the excited state, $q(t)$ is related to the population of the excited state, i.e. it represents the decay law of the fluorescence emission. It is obvious that $q(t)$ obeys the following differential equation (see below Eqs.31 and 32)

$$q(t) = -k_r q(t) - k_{nr}(t)q(t) \qquad (30)$$

where k_r is the previously defined spontaneous rate and $k_{nr}(t)$ the macroscopic rate of reaction which in the most general case is time dependent.

TIME DEPENDENT REACTION RATE

To make significant the meaning of the time dependent reaction rate $k_{nr}(t)$ it is relevant to consider the case of a bistable potential (Fig.2). Let $N(t)$ be the number of particles at time t localized in the reactant well. During the reaction process, $N(t)$ must be considered as a random function and the knowledge of the expectation value $E[N(t)]$ is of fundamental interest. Furthermore, let the random variable T -called

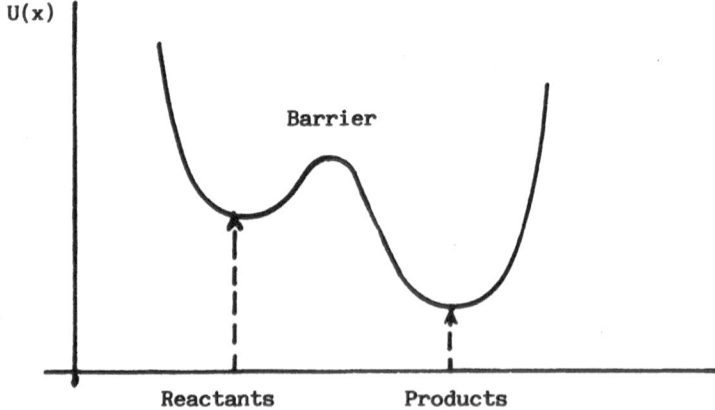

Figure 2

particle lifetime- be the instant at which the particle crosses the barrier ; its behavior is determined by the survival probability

$$Pr[T > t] = q(t) \tag{31}$$

which represents the probability of the reactant being in the well after time t.

Starting from the initial reactant well population n_o and assuming that all the particles obey the same probability, it is easy to demonstrate that (see Appendix 1)

$$E[N(t)] = n_o Pr[T > t] = n_o q(t)$$
$$\equiv n(t) \tag{32}$$

Noting that the current representing the mean number of particles crossing the barrier per unit time is given by

$$j(t) = \lim_{\Delta t \to 0} \frac{n(t) - n(t+\Delta t)}{\Delta t} \tag{33}$$

the reaction rate $k_{nr}(t)$ defined as the expected rate of disappearance is of the form of

$$k_{nr}(t) = \frac{j(t)}{n(t)} = - \frac{1}{n(t)} \frac{d}{dt} n(t) \tag{34}$$

It results the celebrated kinetic equation (in the absence of spontaneous decay)

$$\frac{d}{dt} n(t) = - k_{nr}(t)n(t) \tag{35}$$

Since we are interested in the probabilistic meaning of the reaction rate we consider Eqs.32 and 35 from whom we get

$$k_{nr}(t) = - \frac{d}{dt} Ln\ q(t) \tag{36}$$

and it is not difficult to establish that (see Appendix 2)

$$k_{nr}(t) = \lim_{\Delta t \to 0} \frac{1}{\Delta t} Pr[t < T \leqslant t+\Delta t/T > t] \tag{37}$$

The rate of escape is a conditional density describing the disappearance of a particle in the well at time t assuming its presence up to time t.

It is interesting to note that this last interpretation holds only if the same lifetime probability density $f(t) = - d \Pr[T > t]/dt$ governs the particle in the well.

The plots (Fig.3) of the survival probability $q(t)$, the lifetime density $f(t)$ and the reaction rate $k_{nr}(t)$ are shown for the well-known lognormal distribution

$$f(t) = \frac{1}{\sigma t \sqrt{2\pi}} \exp - [Lnt-m]^2/2\sigma^2$$

$$q(t) = \int_t^\infty f(u)du \qquad (38)$$

$$k_{nr}(t) = f(t)/q(t)$$

From a practical point of view, it can be recalled that the recording of the population decay of the reactants is equivalent to the determination of the survival probability.

Finally it is of importance to point out that **(a)** the necessary and sufficient condition to be in the presence of constant rate of reaction is to observe an exponential population decay, **(b)** in the situation of a

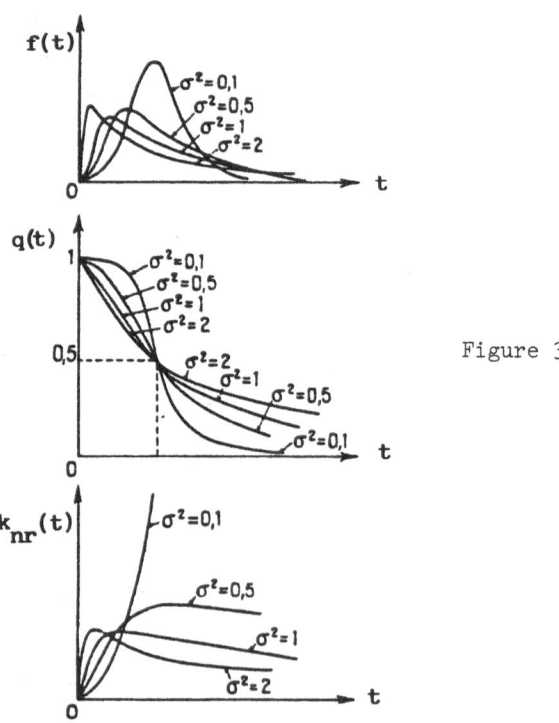

Figure 3

reaction with a constant rate the mean lifetime $E[T]$ equals k_{nr}^{-1} and the lifetime T is memoryless (see Appendix 3).

FIRST PASSAGE TIME

In the diffusion modeling related to the intramolecular motion the first-passage-time problem arises naturally, i.e. the problem of finding the probability distribution of the random time of absorption or of a given level crossing of the particle. Let X(t) be a diffusion process with the transition distribution given by Eq.5. Let $X(0) = x_o$, and let x_2 and x_1 be quantities that represent the lower and the higher levels such that $-\infty \leqslant x_2 < x_o < x_1 \leqslant +\infty$. The first passage time $T(x_o)$ is defined as the instant for the process to leave the interval (x_2,x_1) for the first time. For example, in photoisomerization processes involving large intramolecular motions X(t) represents the reaction coordinate which is related to the molecular geometry and $T(x_o)$, depending on the initial condition x_o, is the time at which the reaction occurs. Setting $x_2 = -\infty$, it is easy to prove in view of Eqs.25 and 29, and with $K_{nr} = \infty$ that $T(x_o)$ can be identified with the lifetime T, thus its probability density is f(t).

The mean first passage-time (or lifetime) can be written as

$$E[T(x_o)] = -\int_o^\infty t\, dq(t) = \int_o^\infty q(t)dt \tag{39}$$

which represents the reciprocal of the average rate of disappearance of the reactants and is interpreted as the inverse steady state rate constant. As mentioned above for a reaction process characterized by a time independent rate constant k_{nr} and in the presence of spontaneous decay, $E[T(x_o)] = (k_r + k_{nr})^{-1}$.

This quantity gives also the average time (mean lifetime) that the reactant spends in the diffusion space prior to absorption.

INITIAL CONDITIONS AND SMOLUCHOWSKI SOLUTIONS

(i) For a perfectly absorbing sink at the origin and for the initial condition $X(0) = x_o$, the solution of Eq.24 and the corresponding survival probability q(t) are given by [5-7] :

$$\bar{P}(0,x_o;t,x) = \frac{1}{\sqrt{2\pi}\,\sigma}\left\{\exp\left[-\frac{(x+y_o)^2}{2\,\sigma^2}\right] - \exp\left[-\frac{(x-y_o)^2}{2\,\sigma^2}\right]\right\}\exp(-k_r t)$$

$$(40)$$

$$q(t) = erf(y_o/\sigma\sqrt{2})\exp(-k_r t)$$

y_o and σ are defined by Eq.20. It may be emphasized that in these relations the radiative decay (k_r) has been taken into account.

A time-scale separation evidently can be made conspicuous for the reaction process

(a) $t/\tau_c^\omega \ll 1$: for short times or for high friction and/or small characteristics frequencies ω, one can write

$$q(t) \approx e^{-k_r t}\,erf\left(\frac{\tau_c^o}{t}\right)^{1/2}$$

$$(41)$$

This expression is identical to that calculated for $\omega = 0$: at short times, the nonexponential decay is the same as that of a free motion particle in the presence of a pinhole sink.

(b) If $t/\tau_c^\omega \gg 1$, which corresponds to small friction and/or large frequencies or to the long-time limit, $q(t)$ becomes

$$q(t) \simeq \frac{2}{\sqrt{\pi}}\left(\frac{2\,\tau_c^o}{\tau_c^\omega}\right)^{1/2} e^{-k_r t}\, e^{-t/\tau_c^\omega}$$

$$(42)$$

which predicts a single exponential decay.

At the beginning the system relaxes non exponentially to a quasi stationary state with the characteristic time $\tau_c^o = x_o^2/4D$; on a larger time scale it evolves exponentially with a decay time τ_c^ω (Eq.21) and is no longer dependent on the initial condition. It is instructive to remark that in the last situation the temporal and spatial variations of $P(0,x_o;t,x)$ can be separated ; indeed

$$\bar{P}(0,x_o;t,x) = \sqrt{\frac{2}{\pi}}\,\frac{|x_o|}{\sqrt{D\tau_c^\omega}}\,\frac{x}{D\tau_c^\omega}\,e^{-x^2/2D\tau_c^\omega}\, e^{-t/\tau_c^\omega}\, e^{-k_r t}$$

$$(43)$$

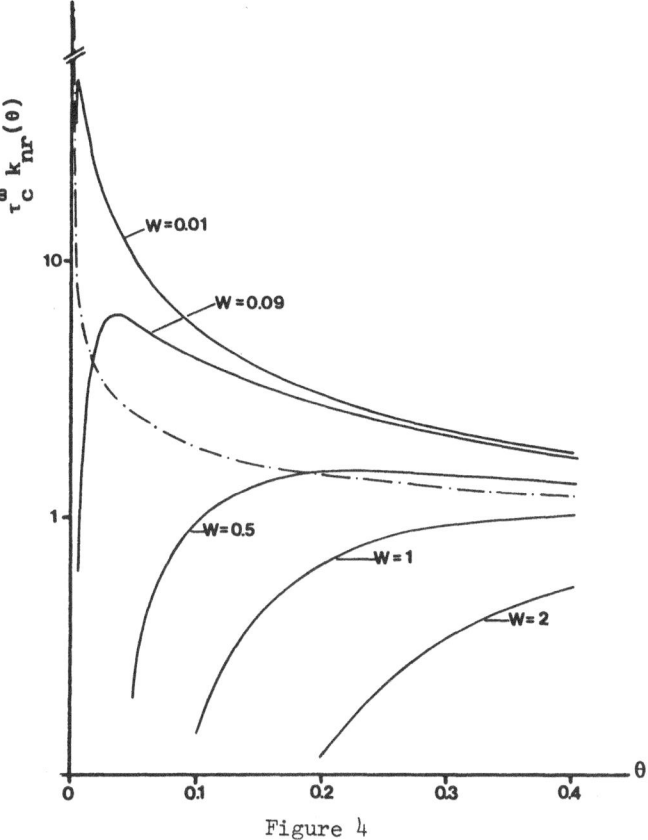

Figure 4

Physically this can be interpreted as a reaction space which is time independent (quasi stationary regime). The two scale description is also apparent on the temporal variation of the reaction rate $k_{nr}(t) = -\dfrac{d}{dt} Ln [q(t)\exp k_r t]$. The essential feature is that $k_{nr}(t)$ starts from zero at $t=0$ and equals $1/\tau^{\omega}_c$ for t tending to ∞. The variations of $_c k_{nr}(t)$ as a function of $\theta = t/\tau^{\omega}_c$ are depicted in Fig.4 for a set of values of $W = 2 \tau^0_c/\tau^{\omega}_c$ characterizing the initial potential energy $U(x_o)$. For high values of W, $k_{nr}(t)$ increases slowly to $1/\tau^{\omega}_c$, whereas a maximum is observed for $W < 1$: if W is diminished, the amplitude of the maximum is enhanced and the time at which it takes place is decreased.

The mean first-passage-time can be explicitly computed for the particular case $\omega = 0$

$$E[T(x_o)] = (1/k_r)\{1-\exp[-2(k_r \tau^0_c)^{1/2}]\} \tag{44}$$

Assuming $k_r \tau^0_c \ll 1$ (corresponding to physical situations) and setting $\tau_r = 1/k_r$

$$E[T(x_o)] = (2 \tau^0_c \tau_r)^{1/2} \tag{45}$$

The relative quantum yield Y (ratio of yields with and without reaction) is given by $Y = E[T(x_o)]/\tau_r = (4\,\tau^o{}_c/\tau_r)^{1/2}$. Recalling that $\tau^o{}_c = x_o{}^2/4D$ (with $D = k_B T/\zeta$ and $\eta \propto \zeta$), one sees that Y should vary with temperature as $D^{-1/2}$, that is, as $(\eta/T)^{1/2}$, giving an apparent activation energy of $Q_\eta/2$ (Q_η activation energy of the reciprocal of the viscosity η).

(ii) A quantity relevant to the discussion of the relaxation in the presence of a sink is the broadness of the probability density function of the initial condition. The following case where the system is made reactive $K_{nr} \neq 0$) at times $t > 0$, but is unreactive ($K_{nr} = 0$) for $t < 0$ -corresponding to physical situations- can be solved easily. Indeed, for an initial distribution $p(0,x_o)$ equal to the Boltzmann equilibrium distribution, that is,

$$\bar{p}(0,x_o) = (2\pi D\tau^\omega{}_c)^{-1/2} \exp(-x_o{}^2/2D\tau^\omega{}_c) \tag{46}$$

the survival probability q(t) has the form [6]

$$\begin{aligned} \bar{q}(t) &= \int q(t)\,\bar{p}(0,x_o)dx_o \\ &= \frac{2}{\pi}\ \text{arcsin}\ \exp(-t/\tau^\omega{}_c)\ \exp(-k_r t) \end{aligned} \tag{47}$$

This expression can be expanded in Taylor series

$$\bar{q}(t) = \frac{2}{\pi} \sum_{k=0}^{\infty} 2k!\,[2^{2k}(k!)(2k+1)]^{-1}\ e^{-(2k+1)t/\tau^\omega{}_c}\ e^{-k_r t} \tag{48}$$

and clearly shows that for times greater than $\tau^\omega{}_c$ the decay appears exponential.

For $\tau_r \gg \tau^\omega{}_c$, the relative quantum yield $Y = E[T(x_o)]/\tau_r$ equals $(\tau^\omega{}_c\ \text{Ln}\ 2)/\tau_r$ predicting that it is proportional to the viscosity and consequently giving an activation energy equal Q_η in contrast to the preceeding case. The temporal variation of $\tau^\omega{}_c k_{nr}(t)$ versus $\theta = t/\tau^\omega{}_c$ is represented in Fig.4 (dashed curve). One can remark that $k_{nr}(t) \to \infty$ for $t \to 0$, which is different from the case of a δ-shaped initial distribution for which it begins with zero value and $k_{nr}(t) \to 1/\tau^\omega{}_c$ for $t \to \infty$.

APPENDIX 1

To demonstrate Eq.32 let us define

$$\begin{aligned} Y_i(t) &= 1 &&\text{if} && T_i \geqslant t \\ &= 0 &&\text{if} && T_i < t \end{aligned}$$

It is obvious that

$$E\left[Y_i(t)\right] = 1 \times Pr\left[T_i \geqslant t\right] + 0 \times Pr\left[T_i < t\right]$$
$$= Pr\left[T_i \geqslant t)\right]$$

Noting that

$$N(t) = \sum_{i=1}^{n_0} Y_i(t)$$

it results Eq.32.

APPENDIX 2

Clearly

$$\lim_{\Delta t \to 0} \frac{1}{\Delta t} Pr\left[t < T \leqslant t+\Delta t/T > t\right] = \lim_{\Delta t \to 0} \frac{1}{\Delta t} \frac{Pr\left[t < T \leqslant t+\Delta t, T > t\right]}{Pr\left[T > t\right]}$$

$$= \lim_{\Delta t \to 0} \frac{1}{\Delta t} \frac{Pr\left[t < T \leqslant t+\Delta t\right]}{Pr\left[T > t\right]} = \frac{f(t)}{q(t)}$$

APPENDIX 3

The lifetime T (random variable) is said to be memoryless if

$$Pr\left[T > t+s/T > t\right] = Pr\left[T > s\right]$$

For $q(t) = Pr\left[T > t\right] = e^{-kt}$ this expression is verified.

REFERENCES

1. N.G. Van Kampen, Stochastic Processes in Physics and Chemistry, North-Holland, Amsterdam, New York, Oxford, 1981 ; C.W. Gardiner, Handbook of Stochastic Methods, Springer Verlag, Berlin, 1983 ; N.S. Goel and N. Richter-Dyn, Stochastic Models in Biology, Academic Press, New York, 1974.

2. R.L. Stratonowitch, Topics in the Theory of Random Noise, vols.I, II, Gordon and Breach, New York, 1963.

3. L. Arnold, Stochastic Differential Equations : Theory and Applications, Wiley, New York, 1974.

4. H.A. Kramers, Physica 7, 284 (1940).

5. S. Chandrasekhar, in Selected Papers on Noise and Stochastic Processes, N. Wax ed., Dover, New York, 1954.

6. A. Szabo, K. Schulten and Z. Schulten, J. Chem. Phys. 72, 4350 (1980) ; B. Bagchi, G.R. Fleming and D.W. Oxtoby, J. Chem. Phys. 78, 7375 (1983).

7. P.M. Morse and H. Feschbach, Methods of Theoretical Physics, Mc Graw Hill, 1953.

DISCUSSION

BORGIS - What is the importance of the "pinhole" hypothesis in the theoretical treatment ? Can you allow for a broader hole ?

MIEHE - The general behavior with a position dependent sink $s(x)$ can be modeled by a pinhole sink $K_{nr} \delta (x-x_1)$ by choosing the appropriate values of K_{nr} and x_1 (see F. Heisel and J.A. Miehé, Chem. Phys. 98 (1985) 233, 243 ; Chem. Phys. Lett. 100 (1963) 183, 134 (1987) 379). Our experimental results for DMABN in propanol at high viscosity are more similar to those predicted by the pinhole case with $K_{nr} \longrightarrow \infty$ than by a broad gaussian sink function.

MICHEAU - From the experimental point of view, you have observed non exponential decays and variations of the fluorescence lifetime with temperature. Have you taken a special care of concentration effects, reactivity of probes molecules, presence of quenching impurities in the solvent and also of the occurrence of possible intermolecular interactions giving rise to exciplexes formation ?

MIEHE - Experiments performed with different solutions have shown no influence of the solute concentration on the decay curves. Furthermore all the discussed results have been obtained with dilution high enough to avoid excimers formation.
After each run of experiments it was also controlled that the results could be reproduced, indicating that they are not altered by photodecomposition processes.
Concerning the purity of the solvents, it was checked that the solute decays are not perturbed by impurities fluorescence.

ERDI - Just a small remark on the terminology. Why did you emphasize the time-dependence of the rate. Excepting degenerate cases, rate is time-dependent.

MIEHE - In the Kramer's situations the rate of the reaction is constant and does not depend on the initial preparation. At the first times at which the initial condition plays an important role, the reaction rate is time dependent.

SCEATS - My understanding is that the torsional frequency for 1-1 binaphtyl has been measured by Wiersma in the gas phase, so that the existence of a barrier in condensed phase is not unlikely. We have not been able to detect this motion in the Raman spectrum in solution, perhaps because it is over-damped.
Also, it is my understanding that the initial transient observed by both yourself and Eisenthal and coworkers for 1-1 binaphtyl is associated with direct excitation of the relaxed conformational state by hot-band absorption to the next electronic state. Could you comment ?

MIEHE - Experimental results of Raman spectroscopy of binaphtyl in liquid solution obtained by A.R. Lacey and F.J. Craven (Chem. Phys. lett. 126 (1986) 588) led the authors to conclude that in the ground electronic state only one single conformer exists with a dihedral angle approximately equal to 90°.

COUPLING TRANSPORT AND REACTIVITY IN THE LIQUID PHASE :

EVIDENCE OF A SPACE DEPENDENT CHEMICAL RATE CONSTANT

F. Baros*, A.T. Reis E. Sousa**, and J.C. Andre*

*Grapp of UA 328 of CNRS, 1, rue Grandville
 F-54042 Nancy Cedex
**Centro Quimica Fisica Molecular, Av. Rovisco Pais
 100 Lisboa, Portugal

INTRODUCTION

Theories of reactions between two molecules A and B, where the kinetics are practically diffusion controlled have been the subject of many studies which used continuous models (Fick's law or similar molecular systems [1-4]). In a general way, if we accept the notion of "molecular chaos" (fluctuations of forces applied on a molecule are only correlated during a very short period in front of displacement time of molecules) and if we use a superposition principle, the introduction of a space dependent chemical rate contant, $k(r)$, leads, for a simple reaction such as $A + B \rightarrow C$, to :

$$\frac{\partial \phi}{\partial t} = \text{div} \left[D \text{ grad } \phi - X \frac{D}{k_B T} \phi \right] - k(r) \phi$$

with :

D : mutual diffusion coefficient of A and B
ϕ : configurational distribution function of B around A
X : potentials applied between A and B
k_B : Boltzmann constant
T : absolute temperature.

Assuming a spherical symmetry, this basic equation is the key, in a condensed fluid, to studies of coupling between transport and reactivity, at the molecular scale.

From the experimental point of view, we can study this type of reaction by choosing an electronically excited A molecule, very reactive towards B. It is then possible to follow the statistical behaviour of such molecules through their fluorescence intensity, according to the well known scheme :

$$A + h\nu \xrightarrow{\hspace{2cm}} A^*$$

$$A^* \xrightarrow{\hspace{0.5cm} 1/\tau_o \hspace{0.5cm}} A + h\nu'$$

$$A^* + B \xrightarrow{\hspace{0.5cm} k_a(t) \hspace{0.5cm}} A + B + h\nu'' \text{ (bimolecular quenching)}$$

In this paper, we use two kinds of space dependent functions $k(r)$:

- $k(r) = k$ $\sigma < r < \sigma'$ σ :   collision distance
  $k(r) = 0$ $r > \sigma'$ σ' :   reaction distance
- $k(r) = k_o \, e^{-b_o(r - \sigma)}$ according to DEXTER mechanism.

The first model allows us to compute easily two experimental parameters σ' and D. By comparison with the second model, it becomes possible to deduce the more significant parameters k_o and b.

I - SIMPLE MODEL

The assumptions are :

- a reaction range (between σ and σ') in which the reaction occurs with a rate constant k, without diffusion

- an external volume in which only diffusion occurs.

Equation governing this model are given in [5]. A great advantage of this model is that it is possible to obtain an analytical expression for $k_a^{(1)}(t)$.

II - MODEL WITH SPACE EXPONENTIAL DEPENDENCE FOR $k(r)$

[apparent rate constant of reaction $k^{(2)}(t)$]

We must now solve :

$$\frac{\partial \phi}{\partial r} = D \left(\frac{\partial^2 \phi}{\partial r^2} + \frac{2}{r} \frac{\partial \phi}{\partial r} \right) - k(r) \, \phi$$

with : $\phi(r) = 1, \quad t = 0$

$$\left(\frac{\partial \phi}{\partial r} \right)_\sigma = 0$$

Introducing $y = \rho\phi$ and with reduced coordinates $\rho = r/\sigma$, $\tau = Dt/\sigma^2$, we obtain :

$$\frac{\partial y}{\partial \tau} = \frac{d^2 y}{d\rho^2} - k(\rho) \, y$$

with $k(\rho) = \dfrac{k_o \sigma^2}{D} \, e^{b_o \sigma \rho} = K_o \, e^{-b\rho}$

To avoid problems for larger values of ρ, we have substituted the steady state solution which is easily given by solving the stationary equation $dy/d\tau = 0$ (in LAPLACE space). This problem can now be solved by using a classical Runge-Kutta-Wes algorithm.

Now, the apparent rate constant is given by :

$$k_a^{(2)}(t) = \int_\sigma^\infty 4 \pi r^2 N \, k(r) \, \phi(r,t) \, dr$$

where N is the Avogadro number and can be numerically computed.

Figure 1 shows some typical cases of time evolution of $k_a^{(2)}(\tau)$.

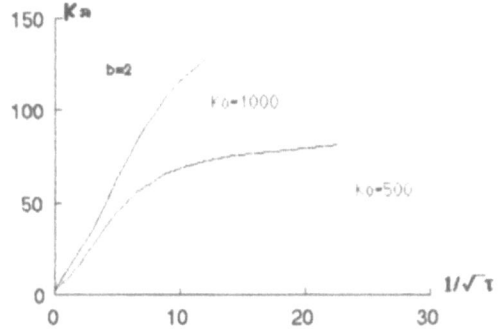

Figure 1. Time evolution of Ka(τ) for two
sets of parameters (Ko,b).

III - COMPARISON OF MODELS: CONSEQUENCES

We have fitted $k_a^{(2)}(t)$ by $k_a^{(1)}(t)$, taking the two parameters $\alpha = \sigma'/\sigma$ and $\beta = kc/4\pi N\sigma D$, for small values of the reduced time τ. Figure 2 shows an example of computed deviations between $k_a^{(1)}$ and $k_a^{(2)}$, for the best set of parameters (α, β). As a consequence, we obtained α as a function of K_o, as shown in figure 3. We can see a large increase of the values of σ' as D decreases, which is the expected result.

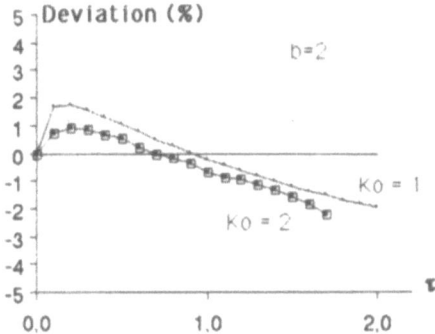

Figure 2. Deviation between $k_a^1(\tau)$ and $k_a^2(\tau)$ for
best set of parameters (α, β).

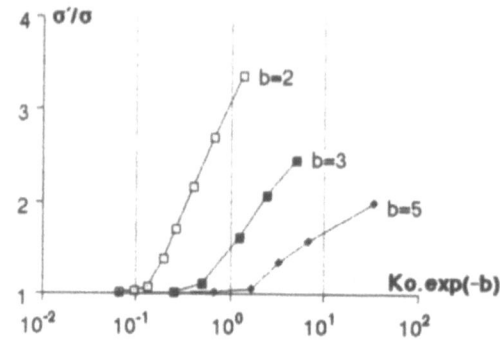

Figure 3. Evolution of the fitted parameter α
function of ln(Ko) for some values of b.

IV - EXPERIMENTAL DETERMINATION OF k(r)

We consider the system Azulen-Biacetyl. Scheme of figure 4 clearly shows that the S_2 fluorescence of Azulen (τ_0 = 1.4 ns) can be quenched by biacetyl and, by the same mechanism, the S_1 fluorescence of biacetyl (τ_0 = 8.5 ns) can be quenched by azulen. Experiments have been carried out for different solvents with a range of viscosities (0.4 to 8 cp).

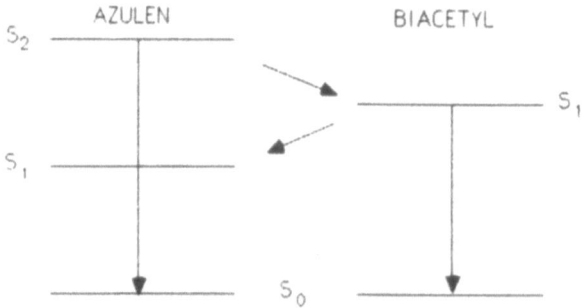

Figure 4. Schematic representation of energy levels
of azulen and biacetyl.

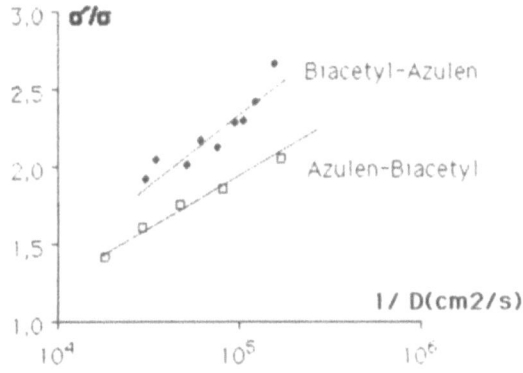

Figure 5. Experimental variations of α vs 1/D

STERN-VOLMER plots have been made for each case and we have used the simple model of [6] assuming a pure diffusional quenching and a static quenching between σ and σ'. Value of σ was estimated to 6 A. Figure 5 shows the variations of $\alpha = \sigma'/\sigma$ with Ln (1/D), in agreement with our model. Slopes of these straight lines will give the values of b. Interaction with the axis $\sigma'/\sigma = 1$ gives K_o. We find :

Azulen-Biacetyl : b $\simeq$ 5   contact value : $K_o e^{-b}$ = .88
Biacetyl-Azulen : b $\simeq$ 4 $K_o e^{-b}$ = .64

leading to values of k(r) illustrated in figure 6.

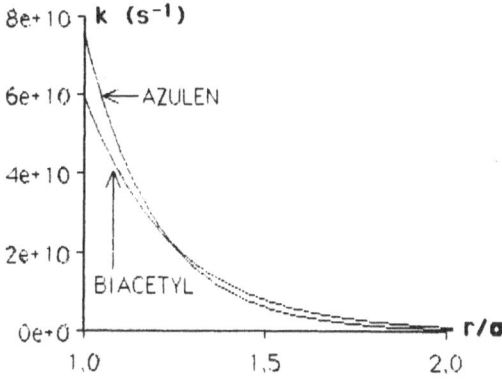

Figure 6. Experimental values of k(r)

We can see that the contact value is about 10^{11} s^{-1}, which is probably a good order of magnitude. However, it is clear, from figure 2, that the preceeding analysis can be applied for short lived molecules (such as Azulen) better than for long lived molecules. In spite of that, we obtained similar values of k_o and b for the two molecules.

CONCLUSION

Introduction of a physically significant function $k(r)$ can explain an increase of the artificial reaction distance σ'. Determination of σ' and D is easy from the experimental point of view and allows us to compute $k(r)$ with adequate set of values (k_o, b), in particular if lifetime of molecule is short and viscosity of solvent is large.

REFERENCES

[1] M.V. Smoluchowski, Z. Physik. Chem., 92 (1917) 129.
[2] F.C. Collins and G.E. Kimball, J. Colloid. Sci., 4 (1949) 425.
[3] R.M. Noyes, J. Am. Chem. Soc., 78 (1956) 5486.
[4] T.L. Nemzek and W.R. Ware, J. Chem. Phys., 62 (1975) 477.
[5] J.C. André, M. Bouchy and W.R. Ware, Chem. Phys., 37 (1979) 103.
[6] J.C. André, M. Niclause and W.R. Ware, Chem. Phys., 28 (1978) 371.

CHEMICAL INSTABILITIES IN A NON-UNIFORM ENVIRONMENT :

COUPLING BETWEEN HYDRODYNAMIC AND CHEMICAL MODES

G. Nicolis, A. Puhl, and V. Altares

Faculté des Sciences
Université Libre de Bruxelles
Campus Plaine C. P. 231
Boulevard du Triomphe
1050 Bruxelles Belgium

I. INTRODUCTION

In as much as a typical chemical reaction involves encounters between molecules, spatial distribution of the various species in a reactor should play a role in such properties as the time evolution of the concentrations or the rate of the overall process. Traditionally chemists thrive for homogeneity, in the hope that a smooth, predictable release of the reaction products in time will be achieved. To guarantee this, most of the reactors involved in laboratory experiments as well as in chemical plants in industrial scale are subjected to a vigorous stirring, whose role is to wipe out appreciable deviations from homogeneity and mix efficiently the chemicals participating in the process.

In actual fact, macroscopic inhomogeneities are always present. In the externally driven stirring process, perfect mixing is a mathematical limit rather than a reality. This is even more true in real world systems, where mixing is achieved by the system's own hydrodynamic flow. In short, contrary to what is usually assumed, a typical reaction takes place in a non-uniform environment. The question that arises naturally is, therefore, how the chemical system responds to this ubiquitous constraint.

If the reaction sequence is simple (involving, for instance, only linear steps) or if it takes place around equilibrium, deviations from homogeneity will introduce some possibly undesired quantitative changes in the functioning of the reactor, but the overall pattern of behavior will remain qualitatively unchanged. On the other hand, in the last few years, it has been firmly established that large classes of chemical systems operating far from equilibrium and involving cooperative nonlinear steps may show a variety of transition phenomena : bistability and hysteresis, explosive behavior, periodic (limit cycle) or chaotic oscillations, space patterns, propagating wavefronts etc ... These transitions are characterized by a high sensitivity of the underlying system, as a result of which small perturbations are expected to trigger large, or even qualitatively new effects.

Our purpose in the present communication is to analyse this type of sensitivity in the particular case in which the perturbation arises from the fact that the reactions take place in a non-uniform host medium. In most of our study we will be interested in non-uniformities due to incomplete stirring. More generally, however, we believe that the response to such disturbances provides also an insinght into (a), the coupling between chemical and hydrodynamic modes and (b), the role of inhomogeneous fluctuations in chemical instabilities.

We shall first review, in section 2, some experimental evidence of inhomogeneity-induced transitions. In section 3 we present a phenomenological model and apply it to two reaction systems exhibiting bistability and oscillatory behavior, respectively. Finally, in section 4, we outline a more fundamental approach to the problem of chemical reactions in a non-uniform medium, and illustrate its applicability on Semenov's model of thermal ignition.

II. EXPERIMENTAL EVIDENCE

The continuous stirred tank reactor (CSTR) has served as a major tool for experimental investigations on chemical instabilities such as bistability, oscillations or chaos. A dramatic effect of stirring on bistability in the chlorite-iodide reaction has been first discovered by Roux et al. (1983). Fig. 1 demonstrates the strong reduction of the bistability region with decreasing stirring rate. Furthermore, it turns out that only the thermodynamic branch will be considerably affected. This behavior critically depends on whether the two feedstreams are premixed (PM) or non-premixed (NPM) prior to injection. In particular, under PM-conditions the reduction of the bistability region is much less pronounced then for the NPM mode (Menzinger et al., 1985). Similar results have been obtained by Kumpinsky and Epstein (1985) for the bromate-bromide-manganous system.

Stirring-induced oscillations have been observed by Menzinger and Giraudi (1986) for the chlorite-iodide reaction. Here the stirring rate can serve as a bifurcation parameter : Fig. 2 and Fig. 3 demonstrate how oscillations can emerge at a critical stirring rate (Hopf bifurcation), grow in amplitude and period and subsequently die out as an infinite period solution with further increase of the stirring rate. Further evidence of this kind of behavior in the same chemical system has been given by Luo and Epstein (1986).

The kinetics of stirring-induced transitions is represented in Fig. 4 (Roux et al., 1984). We observe that just before the transition the variability of the concentration

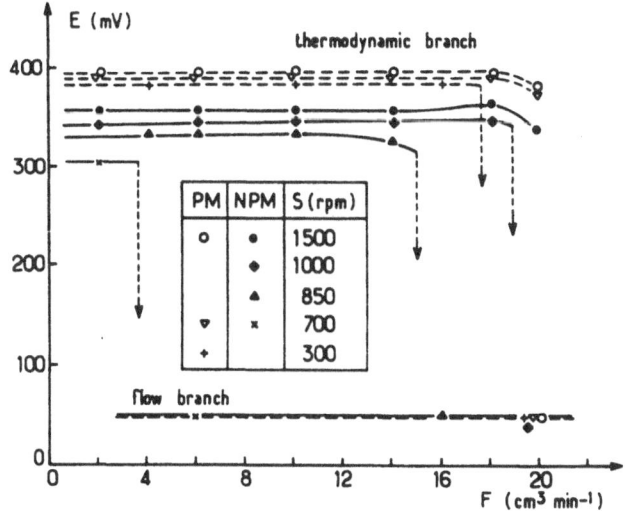

Fig. 1.   Steady-state redox potential E as a function of
total flow F for NPM conditions (full curves) and
PM conditions (broken curves) at different stirring
rates S.   Downward arrows indicate spontaneous
transitions from the thermodynamic branch to the
flow branch.   The constraint composition of the
reactor is $[NaClO_2]_0 = 1.3 \times 10^{-4}M$, $[NaOH]_0 = 1 \times 10^{-3}M$,
and $[KI]_0 = 5.0 \times 10^{-4}M$, $[H_2SO_4]_0 = 5.0 \times 10^{-3}m$, $[Na_2SO_4]_0$
$= 5 \times 10^{-2}M$ at $T_0 = 13°C$.

becomes large suggesting once again that inhomogeneities
play a significant role and act somehow as a powerful "noise
generator".

Our objective is now to develop a mathematical formalism allowing to interpret, at least qualitatively, these

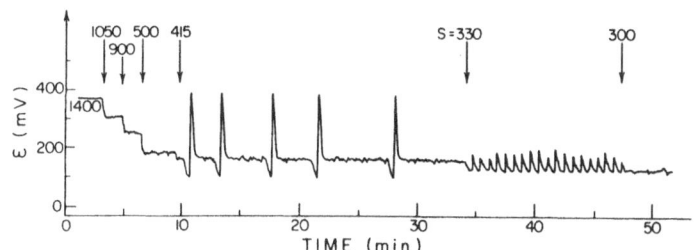

Fig. 2.   The effect of stirring on oscillations. $[KI] =$
$1.72 \times 10^{-3}M$, $[NaClO_2] = 5.7 \times 10^{-4}M$, ph 1.1 ; t_{tot}^{-1}
$= 3.1 \times 10^{-3}s^{-1}$.

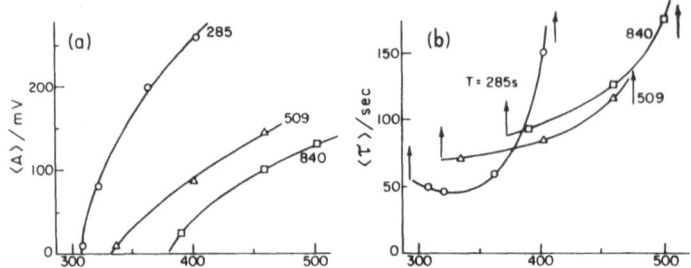

Fig. 3. Stirring rate dependence of (a) mean Amplitude $\langle A \rangle$ and (b) mean period $\langle \tau \rangle$ for three different values of residence times t_{tot} (in seconds).

observations. We proceed in two steps considering, successively, a phenomenological approach and a more fundamental theory.

III. A PHENOMENOLOGICAL APPROACH

Consider an isothermal nonlinear chemical reaction system with two variables operating under CSTR-conditions. In the limit of perfect mixing, the system is spatially homogeneous and is described by the following equations :

$$d\alpha/dt = t_{tot}^{-1}(\alpha_0 - \alpha) + R_1(\alpha, \beta)$$

$$d\beta/dt = t_{tot}^{-1}(\beta_0 - \beta) + R_2(\alpha, \beta) \qquad . \qquad (1)$$

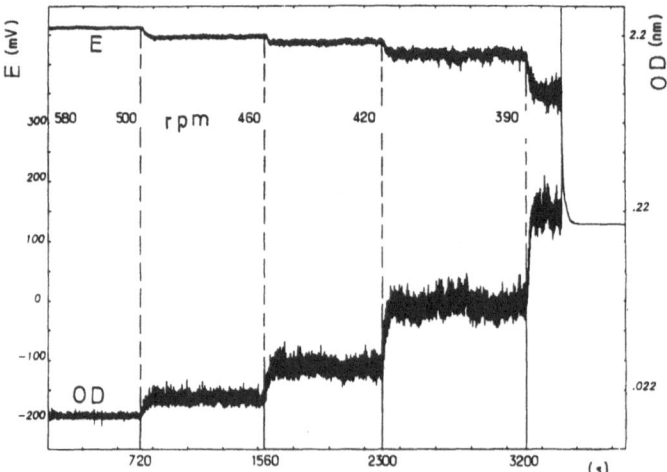

Fig. 4. Transition from the thermodynamic branch to the flow branch after 4 changes in the stirring rate.

Here α and β denote (dimensionless) concentrations, α_0 and β_0 their values in the incoming feedstreams and R_1, R_2 the chemical rate functions. t_{tot}^{-1} is the sum of the inverse residence times t_I^{-1} and t_{II}^{-1} for each input species, respectively. The restriction to two variables is adopted only for notational convenience.

Our objective is to extend the classical CSTR-equations (1) in order to incorporate the effect of inhomogeneous perturbations due to the reactor feeding - for both PM and NPM feeding modes - and the (incomplete) homogeneizing effect of turbulent mixing. For this purpose we adopt the elegant micromixing-concept of Zwietering (1984) and represent the mixing process as a dilution of incoming streams of fluid by the surrounding bulk fluid (Puhl and Nicolis, 1986). For a general review of micromixing phenomena in stirred reactors we refer to Villermaux (1986).

The dilution process of lumps of fluid injected into the bulk - which itself is assumed to remain homogeneous - will be described dynamically by a small scale time having the same order of magnitude as the characteristic mixing time

$$t_m \sim \omega^{-1} (L_s/d)^{2/3}$$

(2)

where ω is the stirring frequency, d the stirrer dimension and L_s the scalar macroscale for the correlation of the concentration fluctuations. The volume growth of entering lumps per unit time is

$$dv/d\zeta = \phi(y) = \phi_0 \exp(\zeta/t_m) \quad ,$$

(3)

and the equations for the lump concentrations, a and b, are given by

$$da/d\zeta = t_m^{-1}(\alpha - a) + R_1(a, b)$$

$$db/d\zeta = t_m^{-1}(\beta - b) + R_2(a, b)$$

(4)

where α and β represent the concentrations of the surrounding bulk material.

The aggregate volume of all lumps younger then ζ is governed by the relation

$$v(\zeta) = \int_o^\zeta \phi(\zeta')d\zeta' = V(t_m/t_{tot}) \exp\left[(\zeta/t_m)-1\right]$$

(5)

and is assumed to remain, for all time intervals considered, much smaller than the total tank volume V. This means that the lumps will not be allowed to grow and infinitum. We therefore define a "truncation age"

$$\zeta^* = \mu \cdot t_m$$

(6)

beyond which the lumps will be regarded as belonging to the bulk material α and β. This new parameter μ appearing in the model has to satisfy

$$t_m \left[\exp(\mu) - 1\right] \ll t_{tot} \quad .$$

(7)

405

The specific value of μ is, however, not critical and simply amplifies or weakens the effect of incomplete mixing without changing qualitatively the results.

The corrected equations for the bulk concentrations α and β are derived by considering the rate of change in α and β to be composed of an internal stream into the bulk due to lumps at age μ , a flux from the bulk to lumps, an outlet stream from the reactor and the chemical rates (see Puhl and Nicolis 1986). For premixed feedstreams (PM) we have :

$$d\alpha/dt = \frac{\exp(\mu)\left[a(\alpha,\beta,\mu) - \alpha\right]}{t_{tot} - t_m\left[\exp(\mu) - 1\right]} + R_1(\alpha,\beta)$$

$$d\beta/dt = \frac{\exp(\mu)\left[b(\alpha,\beta,\mu) - \beta\right]}{t_{tot} - t_m\left[\exp(\mu) - 1\right]} + R_2(\alpha,\beta) \quad . \tag{8}$$

Here $a(\alpha,\beta,\mu)$ and $b(\alpha,\beta,\mu)$ are the solutions of (4) at $\zeta^* = \mu \cdot t_m$ and for the initial conditions

$$a(\zeta = 0) = \alpha_0 , \qquad b(\zeta = 0) = \beta_0 \quad . \tag{9}$$

For two separated feedstreams (NPM) the corrected equations for the bulk material adopt the form

$$d\alpha/dt = \frac{\exp(\mu) \cdot \left[\frac{t_{tot}}{t_I} a_I(\alpha,\beta,\mu) + \frac{t_{tot}}{t_{II}} a_{II}(\alpha,\beta,\mu) - \alpha\right]}{t_{tot} - t_m\left[\exp(\mu) - 1\right]} + R_1(\alpha,\beta)$$

$$d\beta/dt = \frac{\exp(\mu) \cdot \left[\frac{t_{tot}}{t_I} b_I(\alpha,\beta,\mu) + \frac{t_{tot}}{t_{II}} b_{II}(\alpha,\beta,\mu) - \beta\right]}{t_{tot} - t_m\left[\exp(\mu) - 1\right]} + R_2(\alpha,\beta)$$

$$\tag{10}$$

where $a_I(\alpha,\beta,\mu)$ and $b_I(\alpha,\beta,\mu)$ denote the lump concentrations of feedstream I and are evaluated by the differential equations

$$da_I/d\zeta = t_m^{-1}(\alpha - a_I) + R_1(a_I, b_I)$$

$$db_I/d\zeta = t_m^{-1}(\beta - b_I) + R_2(a_I, b_I) \tag{11}$$

with the initial conditions

$$a_I(\zeta = 0) = (t_I/t_{tot})\alpha_0, \qquad b_I(\zeta = 0) = 0 \quad . \tag{12}$$

Analogously, the lump concentrations $a_{II}(\alpha,\beta,\mu)$ and $b_{II}(\alpha,\beta,\mu)$ of feedstream II are found by solving

$$da_{II}/d\zeta = t_m^{-1}(\alpha - a_{II}) + R_1(a_{II}, b_{II})$$

$$db_{II}/d\zeta = t_m^{-1}(\beta - b_{II}) + R_2(a_{II}, b_{II}) \tag{13}$$

with the initial conditions

$$a_{II}(\eta = 0) = 0 \quad , \quad b_{II}(= 0) = (t_{II}/t_{tot})\beta_0 \; .$$

$$(14)$$

Perturbative solutions for these equations are possible near the ideal stirring limit, i.e. if

$$t_m \lll t_{ch} \lll t_{tot} \qquad , \qquad (15)$$

a condition which is in agreement with (7).

Let us now outline some applications of the above model. We first consider the following isothermal autocatalytic reaction scheme :

$$2A + B \longrightarrow 3A$$

$$A \longrightarrow B \qquad . \qquad (16)$$

This system gives rise to multiple steady states for certain values of parameters. Normalising the bulk concentrations over the inlet concentration for B ($\beta_0 = 1$) and setting

$$t_{ch} = 1 \qquad (17)$$

by adequate choice of units, the overall rates become

$$R_1(\alpha, \beta) = -R_2(\alpha, \beta) = \alpha^2 \beta - \alpha/t_2 \qquad . \qquad (18)$$

Futhermore, we assume

$$t_I = t_{II} = 2t_{tot}$$

$$(19)$$

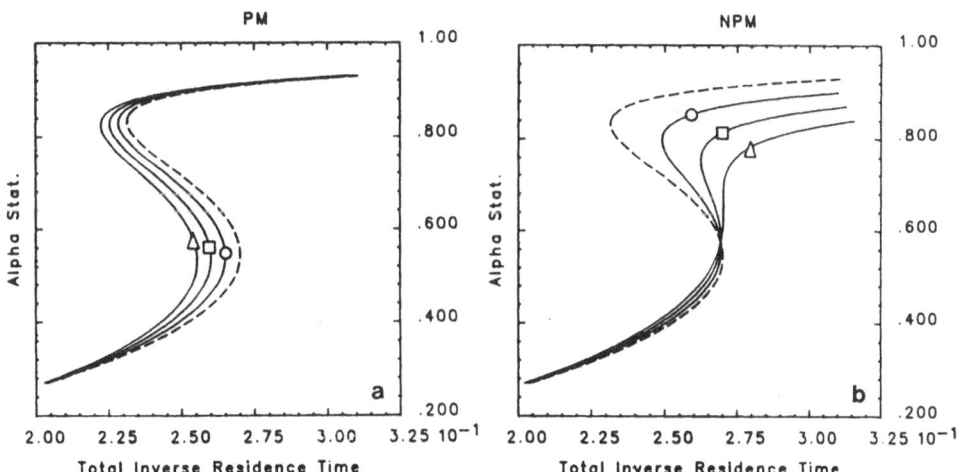

Fig. 5. Calculated steady state concentration α_S as a function of total inverse residence time t_{tot}^{-1} in (a) PM mode and (b) NPM mode. Dashed, $t_m=0$; $0, t_m=1/45$; $\square$, $t_m=2/45$; $\triangle$, $t_m= 3/45$; $\mu =3$, $\alpha_0=1$; $t_2=54/55$.

Fig. 5 depicts the steady states for α over t_{tot}^{-1} for the PM mode and for the NPM mode, respectively. The values α_0 = 1 and t_2 = 54/55 were assumed. The effect of mixing on the bistability behavior of the NPM case is remarkably pronounced (compared with the PM case) even for small deviations from the ideal stirring limit. Clearly, the transition point on the flow branch is shifted drastically to higher inverse residence times. Since the transition point on the thermodynamic branch is less affected, we observe a complete destruction of the bistability phenomenon for mixing times larger than 0.06.

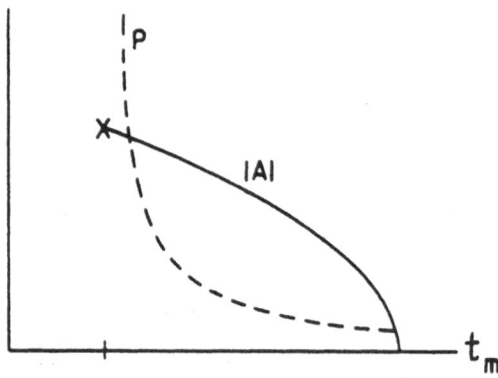

Fig. 6.   Oscillation period P and amplitude $|A|$ over mixing time t_m as a bifurcation parameter (schematically). X : onset of oscillation as an inifinite period saddle-loop bifurcation.

As a second application we study the effect of mixing on oscillations or, more precisely, on infinite period bifurcations. For this purpose the following reaction scheme will be investigated :

$$X + 2Y \longrightarrow 3Y$$
$$Y \longrightarrow \text{inert products}$$
$$C + Y \longrightarrow X + Y \qquad\qquad (20)$$

where C is held constant (Puhl and Nicolis 1987). Small parameter changes or perturbations (e.g. due to incomplete mixing) in the vicinity of "organizing centers" such as cusp points or Takens-Bogdanov points (see for instance , Guckenheimer and Holmes (1983)) can deeply affect the dynamic characteristics of a reaction system in the CSTR. It is, therefore, fruitful to explore the high sensitivity of oscillatory behavior around organizing centers via normal form analysis.

We do not go into the details of the calculations here, but rather present one of the possibilities suggested by the analysis in the form of the bifurcation diagram cf. Fig. 6. When t_m is considered a bifurcation parameter, we observe a dependence of the oscillation period P and the amplitude $|A|$ on t_m. Thus decreasing mixing efficiency (i.e., increasing t_m) first yields an infinite period homoclinic orbit (saddle loop bifurcation). The oscillations subsequently decrease in amplitude and period and finally die out as a Hopf bifurcation. This is in qualitative agreement with the measurements by Menzinger and Giraudi (1986), see Figs. 2 and 3 where oscillations occur only in a narrow window of stirring rates.

IV. TOWARD A FUNDAMENTAL APPROACH

General problem

The basic laws governing turbulent mixing in presence of chemical reactions are contained in the following transport equation :

$$\partial_t X + (\underline{u}(\underline{r},t) \cdot \nabla)X = D_0 \nabla^2 X + R(X) \tag{21}$$

which describes the transport of a scalar quantity $X(r,t)$ (such as temperature or concentration of a reacting solute) in the presence of a turbulent velocity field $\underline{u}(r,t)$ and of chemical reactions with the rate $R(X)$. D_0 denotes the molecular diffusivity. Futhermore, we assume incompressibility, i.e. div $\underline{u} = 0$.

Of practical interest is the ensemble averaged quantity $\langle X \rangle$ which is supposed to be weakly inhomogeneous for highly developed turbulence and crucially dependent on boundary conditions. Averaging (21) yields :

$$\partial_t \langle X \rangle + \text{div} \langle \underline{u}X \rangle = D_0 \nabla^2 \langle X \rangle + \langle R(X) \rangle \quad . \tag{22}$$

This equation immediately raises the closure problem : $\langle \underline{u}X \rangle$ and $\langle R(X) \rangle$ cannot be expressed straightforwardly in terms of $\langle X \rangle$, owing to the nonlinearities and the complex space-time dependence of $\underline{u}$. In the immediate proximity of the ideal mixing limit, however, the following closure is expected to provide a satisfactory zeroth order theory :

$$\langle \underline{u}X \rangle = \langle \underline{u} \rangle \langle X \rangle - D_T \nabla \langle X \rangle \tag{23a}$$

$$\langle R(X) \rangle = R(\langle X \rangle) + 1/2 \langle X^2 \rangle \cdot R_{\overline{XX}}^{--} + \ldots \tag{23b}$$

where it is understood that the mixing process is much faster than any chemical reaction step.

Example : thermal ignition

Let us consider the effect of incomplete mixing on stationary states in a closed non-adiabatic stirred tank reactor of volume V and surface area S in which the exothermic reaction A $\rightarrow$ B can take place (Nicolis and Frisch, 1985). In this case $\langle x \rangle$ represents temperature. We further neglect, following Semenov, reactant consumption. Hence

$$R(\langle x \rangle) \sim \exp(- X^*/\langle x \rangle) \tag{24}$$

where X^* denotes the activation energy divided by the gas constant.

With the closure approximations (23) and the further assumption $\langle \underline{u} \rangle = 0$ we have

$$\partial_t \langle x \rangle = (D_T + D_0') \nabla^2 \langle x \rangle + R(\langle x \rangle) + h.o.t. \tag{25}$$

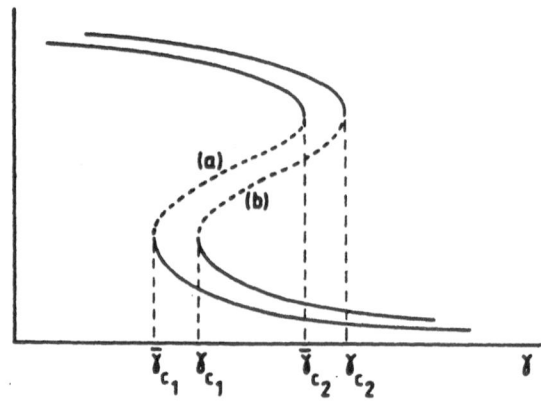

Fig. 7. Shift of ignition and extinction points. (a) completely effective stirring ; (b) incomplete stirring.

together with the (Newton cooling-) boundary conditions

$$-D_T \nabla \langle x \rangle \cdot \hat{e}_n = \gamma (\langle x \rangle - X_e) \quad \text{on S} \quad . \tag{26}$$

Eq. (25) can be solved by making the Ansatz

$$\langle x \rangle = \overline{\langle x \rangle} (t) + \xi(\underline{r},t) \tag{27}$$

where $\overline{\langle x \rangle}$ is the volume average of $\langle x \rangle$ and $\xi(\underline{r},t)$ denotes the deviations from $\langle x \rangle$. Since for large D_T the deviations $\xi(r,t)$ are expected to be small we are able to solve (25) via a perturbation technique (see Nicolis and Frisch, 1985) and we find for a spherical reactor of radius R :

$$\partial_t \langle \overline{x} \rangle = (3\gamma/R) \quad (1 - \gamma R/5D_T)(X_e - \langle \overline{x} \rangle) + R(\langle \overline{x} \rangle) \quad . \tag{28}$$

Thus (28) only amounts to a renormalisation of the heat transfer coefficient γ as a first order correction to the ideal mixing limit. In this latter limit ($D_T \rightarrow \infty$) we would recover the classical Semenov equation. From Fig. 7 we see that the critical points for ignition and extinction ($\overline{\gamma}_{c1}$ and $\overline{\gamma}_{c2}$, respectively) are both shifted to higher values of the cooling parameter γ. Clearly, inhomogeneous fluctuations destabilize (advance) thermal ignition, but stabilize (retard) thermal extention. This behavior seems to be universal as long as the deviations from ideal mixing are small.

Imperfect turbulent mixing : stochastic formulation

Our next goal will be to incorporate into the theory some more detailed information regarding the statistical properties of turbulence (Puhl, Altares and Nicolis 1987). For this purpose we model the turbulent velocity field $\underline{u}$ as a random field with zero mean ($\langle \underline{u} \rangle$ = 0). $\underline{u}$ is further assumed to be a stationary, homogeneous and isotropic Gaussian process with variance u_0^2. The following form for the correlation tensor will be postulated

$$\langle u_i(\underline{r},t)u_j(\underline{r}',t') \rangle = u_0^2 \, Q_{ij}(\underline{r} - \underline{r}') \, k(t - t') \tag{29}$$

where spherical symmetry implies (see Landau and Lifschitz, 1959)

$$Q_{ij}(\underline{R}) = (f(R) - g(R)) \, n_i n_j + g(R) \, \delta_{ij}^{Kr} \quad . \tag{30}$$

Here $\underline{R} = \underline{r} - \underline{r}'$ and n_i are the components of the unit vector $\underline{n}$ in the direction of $\underline{R}$. Futhermore, the following relation between $f(R)$ and $g(R)$ holds :

$$g(R) = f(R) + 1/2 \, R \, f'(R) \quad . \tag{31}$$

Integral scales can be defined as follows :

$$l_0^3 = \int_0^\infty R^2 \, f(R) \, dR \tag{32}$$

and

$$t_0 = \int_0^{t \rightarrow \infty} k(t - t') \, dt' = \int_0^\infty k(\tau) \, d\tau \quad . \tag{33}$$

Integration of (29) over space and time yields

$$\int_{-\infty}^\infty d\underline{R} \int_0^\infty d\tau \, \langle u_i(\underline{r}, t)u_j(\underline{r} - \underline{R}, t - \tau) \rangle = u_0^2 l_0^3 t_0 N_{ij} \tag{34}$$

where

$$N_{ij} = 3/2 \int_0^\pi d\vartheta \int_0^{2\pi} d\varphi \, \sin\vartheta \, n_i n_j - 2\pi \, \delta_{ij}^{Kr} \tag{35}$$

and

$$\underline{n} = (\sin\vartheta \, \cos\varphi \, , \, \sin\vartheta \, \sin\varphi \, , \, \cos\vartheta) \quad . \tag{36}$$

Futhermore, we assume that $\underline{u}$ -interpreted as colored noise-is rapidly varying in space and time (scales l_0 and t_0, respectively) and has rapidly decaying correlations $f(R)$ and $k(\tau)$ (l_0 and t_0 small), i. e. is strongly mixing. In the limit of an infinitely high turbulence level ($u_0^2 \to \infty$) l_0 and t_0 are assumed to approach zero and we speak of a Gaussian white noise (in space-time).

Two consequences follow from these assumptions : firstly, the convected (ensemble-) average field $\langle X \rangle$ is close to homogeneity on a scale $L \gg l_0$ and secondly, the scalar fluctuations $\delta X = X - \langle X \rangle$ are much smaller then the mean value $\langle X \rangle$. Moreover, the assumptions for $\underline{u}$ imply restriction to the domain close to the homogeneous limit (ideal mixing) :

$$t_0, \ l_0, \ \langle \delta X^2 \rangle \longrightarrow 0 \quad ; \quad L, \ u_0, \ u_0^2 t_0 \longrightarrow \infty \ . \quad (37)$$

We further assume the effective noise intensity

$$q^2 = u_0^2 l_0^3 t_0 \tag{38}$$

to be weak. From the viewpoint of a stochastic process, the homogeneous limit is therefore equivalent to the white noise limit. Hence, in the vicinity of the ideal mixing limit we are dealing with large Reynolds numbers $Re = u_0^2 t_0 / \nu$ and large Péclet numbers $Pe = u_0^2 t_0 / D_0$. Moreover, we restrict the considerations to chemical reactions which are slower than t_0^{-1}.

The mean value $\langle X \rangle$ is given to lowest order (neglecting the corrections arising from turbulent transport on the chemical rate function via the nonlinear terms in δX) by

$$\partial_t \langle X \rangle + \text{div} \langle \underline{u} \ \delta X \rangle - D_0 \nabla^2 \langle X \rangle = R(\langle X \rangle) + \ldots \quad (39)$$

which has to be solved with appropriate initial and boundary conditions for $\langle X \rangle$. In order to find the closure for $\langle \underline{u} \ \delta X \rangle$ we have to solve as well the equation for the fluctuations :

$$\partial_t \delta X + \underline{u} . \nabla \langle X \rangle + \nabla . (\underline{u} \ \delta X - \langle \underline{u} \ \delta X \rangle) - D_0 \nabla^2 \langle X \rangle$$
$$= R_X(\langle X \rangle) \ \delta X + 1/2 \ R_{XX}(\langle X \rangle)(\delta X^2 - \langle \delta X^2 \rangle) + \ldots \tag{40}$$

Eq. (40) has to be interpreted as a stochastic differential equation with multiplicative noise, and its solution is a difficult task.

Assuming small effective noise intensity q^2 we perform a stochastic weak noise expansion (Gardiner, 1983) close to the white noise limit :

$$q^2 = \varepsilon^2 q_1^2 \ , \qquad \underline{u} \to \varepsilon \underline{u} \ , \qquad \varepsilon \ll 1$$
$$\delta X = \varepsilon \ \delta X_0 + \varepsilon^2 \ \delta X_1 + \ldots \tag{41}$$

which transforms (40) into a hierarchy of linear differential equations :

$$\partial_t \, \delta x_0 - D_0 \, \nabla^2 \, \delta x_0 - R_X \, \delta x_0 = -\underline{u} \cdot \nabla \langle x \rangle \tag{42}$$

etc.

Solving (42) in the limit $D_0 \longrightarrow 0$ to lowest order and substituting into (39) yields

$$\partial_t \langle x \rangle = u_0^2 \, \nabla \cdot \int_0^t dt' \, B(t', \langle x \rangle) \, \nabla \langle x \rangle \; + \; R(\langle x \rangle) \tag{43}$$

where

$$B(t', \langle x \rangle) = k(t - t') \cdot \exp \left\{ \int_{t'}^{t} dt'' \, R_X(\langle x \rangle) \right\} \; . \tag{44}$$

The analysis of higher order terms yields the following convergence criterion for the validity of the expansion (41) :

$$u_0 t_0 \ll \lambda \tag{45}$$

where λ is the Taylor microscale of turbulence defined as

$$\lambda^{-2} = -1/2 \, f''(0) \qquad . \tag{46}$$

Notice that (43) features a weak delay effect in the time evolution of $\langle x \rangle$. This indicates an increase of the degrees of freedom of a reaction system subdued to inhomogeneous perturbations.

Seeking for stationary states ($t \longrightarrow \infty$) and assuming the turbulent noise to be an Ornstein-Uhlenbeck process in time then

$$k(t - t') = \exp \left\{ -|t - t'|/t_0 \right\} \tag{47}$$

and (43) becomes

$$\operatorname{div}(D_T \, \nabla \langle x \rangle) + R(\langle x \rangle) = 0 \tag{48}$$

where

$$D_T = \frac{u_0^2 t_0}{1 - t_0 \, R_X(\langle x \rangle)} \tag{49}$$

is now reaction dependent. For autocatalytic reactions R_X is positive. (49) predicts therefore an enhancement of the effective transport coefficient due to the kinetics.

In future work we plan to solve (48) with realistic geometries and boundary conditions, in order to model adequately the feeding and other basic characteristics of a CSTR. An extension of the theory to fast reactions compared to the mixing time would also be highly desirable.

ACKNOWLEDGMENTS

We are indebted to H. Frisch, J. Boissonade, P. De Kepper and J. Villermaux for fruitful discussions. The work of A.P. and V.A. is supported, respectively, by the Stiftung Volkswagenwerk, and the Belgian Institut pour l'encouragement de la Recherche Scientifique dans l'Industrie et l'Agriculture (IRSIA).

REFERENCES

- C. Gardiner, "Handbook of Stochastic Methods" (Springer, Berlin, 1983).
- J. Guckenheimer and P. Holmes, "Nonlinear Oscillations, Dynamical Systems, and Bifurcations of Vector Fields" (Springer, New York, 1983).
- E. Kumpinsky and I.R. Epstein, J. Chem. Phys. 82, 53 (1985).
- Y. Luo and I.R. Eptein, J. Chem. Phys. 85, 5733 (1987).
- M. Menzinger, M. Boukalouch, P. De Kepper, J. Boissonade, J.C. Roux and H. Saadaoui, J. Phys. Chem. 90, 313 (1985).
- M. Menzinger and A. Giraudi, J. Phys. Chem. 91, 4391 (1987).
- G. Nicolis and H. Frisch, Phys. Rev. A31, 439 (1985).
- A. Puhl and G. Nicolis, Chem. Engng. Sci. 41, 3111 (1986).
- A. Puhl and G. Nicolis, J. Chem. Phys. 87 (2), 1070 (1987).
- A. Puhl, V. Altares and G. Nicolis (submitted).
- J.C. Roux, P. De Kepper and J. Boissonade, Phys. Lett. 97, 168 (1983).
- J.C. Roux, H. Saadaoui, P. De Kepper and J. Boissonade, Springer Proc. Phys. 1, 70 (1984).
  J. Villermaux, Encyclopedia of Fluid Mechanics, Ch. 27, Gulf Publishing Cy. (1986).
- Th.N. Zwietering, Chem. Engng. Sci. 39, 1765 (1984).

SPONTANEOUS BREAKING Of SPATIAL HOMOGENEITY IN A BISTABLE REACTIVE

SYSTEM FAR FROM EQUILIBRIUM

D. Borgis and M. Moreau

Laboratoire de Physique Théorique des Liquides
Université Pierre et Marie Curie
75252 PARIS CEDEX 05

SUMMARY

The time evolution of an inhomogenous bistable reactive system in absence of convection is studied in the birth and death formalism. With the aid of multivariate Master Equations it is shown in simple cases that spatial homogenity can be spontaneously breaken during the passage from metastable to stable state ; a theory of nucleation is presented, which allows the evaluation of the nucleation rate.

1. INTRODUCTION

It is well known that some non-linear chemical systems driven far from equilibrium can exhibit a large variety of exotic behaviours leading to sefl-organization[1] : they belong to the class of dissipative systems in the formalism of non-linear irreversible thermodynamics. The most famous example is certainly the Belousov-Zahabotinski reaction[2], which has been extensively studied since 1958 and gives rise to periodic or chaotic oscillations, bistability, and spatial patterns such as waves, targets or spirals. The Arsenious acid-iodate reaction[3,4] is another case, recently discovered, where spatial organization has been observed.

As a consequence of the many theoretical invesgations devoted to these phenomena, it appears that the possibility of oscillations and the propagation of spatial structures can be explained by the deterministic description of chemical reactions in a diffusive medium, which only involves the classical laws of chemical kinetics (ordinarily, of the third order) and the Fick's law of diffusion : the application of the bifurcation theories to this set of coupled, non-linear differential equations leads to a number of interesting results which, at least qualitatively, account for most experimental observations.

On the other hand, the emergence of spatial inhomogeneity from an initially uniform medium cannot be understood from a purely deterministic point of view, since it results from the fluctuations of the concentrations of the chemical species. The origin of these fluctuations is still under discussion : are they due to external causes (dust, catalyst, noise ...), or can they appear spontaneously, as a result of the stochastic nature of the processes ? Up to now the experiments do not permit to decide. Walgraef et al[6] have proposed a theoretical study of internal fluctuations in the case of an oscillatory reaction, but recent experiments of Vidal et al[7] do not confirm their predictions ; instead, they seem to suggest an heterogeneous mechanism, without giving a definitive answer.

In order to obtain some clear conclusions on the spontaneous breaking of spatial homogeneity, our purpose here is to study the simplest non-linear reaction which can exhibit such phenomena, i.e. the Schlögl model, in case of bistability : we will show that the coupling of reaction and diffusion can actually produce non-uniformity by means of the intrinsec fluctuations. Two cases will be treated : the evolution of two homogeneous cells weakly coupled by diffusion, and the nucleation in an inhomogeneous system.

Before stating the model precisely, it should be pointed out that, although elementary, it is similar to the Arsenious acide-Iodate reaction[8] and thus can provide useful results.

2. AN ELEMENTARY BISTABLE REACTION : THE SCHLOGL MODEL

2.1 Schlögl reaction

The Schlögl reaction[9] may be written

$$
A + 2 X \underset{k_1}{\overset{k_2}{\rightleftharpoons}} 3X
$$

$$
B \underset{k_3}{\overset{k_4}{\rightleftharpoons}} X
$$

(1)

A and B are chemical species, the concentrations of which are maintained constant, and X is a third chemical species with variable concentration ; $k_1, \ldots k_4$ are the rate constants of the different steps. The system is kept far from equilibrium by a convenient choice of the concentrations of A and B.

2.2 Homogeneous system

(i) **Deterministic analysis.** In a homogeneous system of volume V at constant temperature, the concentration x of molecules X obeys the classical law of chemical kinetics :

$$
\frac{dx}{dt} = (k_2 \, x_A \, x^2 + k_4 \, x_B) - (k_1 \, x^3 + k_3 x \equiv F(x)
$$

(2)

x_A and x_B being the concentrations of A and B. Then the number n of molecules X satisfies the equation :

$$
\frac{dn}{dt} = w(n) - \bar{w}n)
$$

(3)

with

$$w(n) = k_2 \, n_A \, n^2/V^2 + k_4 n_B$$

$$\bar{w}(n) = k_1 \, n^3/V^2 + k_3 \, n$$

(3')

It is easily seen than eq. (3) can have 3 stationary solutions α, β and γ ($\alpha < \beta < \gamma$) where α and γ are stable, and β unstable (Fig.1).

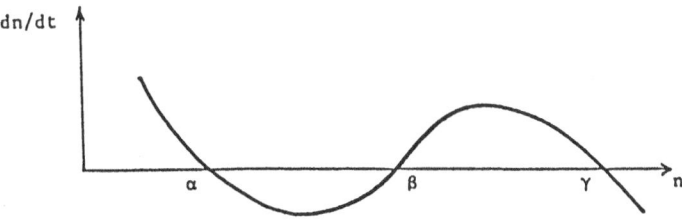

Fig.1 : reaction velocity

The concentrations corresponding to α,β,γ will be denoted

$$x_\alpha = \alpha/V \quad , \quad x_\beta = \beta/V \quad , \quad x_\gamma = \gamma/V$$

(ii) **Stochastic analysis : birth and death formalism.** According to the standard assumptions of birth and death formalism[5] (mainly : the elastic collisions are sufficient to maintain thermal equilibrium and the diffusion is sufficient to maintain uniformity) the probability $p(n,t)$ to have n molecules X at time t obeys the Master Equation :

$$\frac{d}{dt} p(n,t) = W_{n-1} \, p(n-1,t) - (W_n + \bar{W}_n) \, p(n,t) + \bar{W}_{n+1} \, p(n+1,t) \equiv R[p]$$

(4)

where W_n and $\bar{W}_n$ are, respectively, the birth and death rates from a state with n molecules :

$$W_n = k_2 \, n_A \, (n-1)/V^2 + k_4 \, n_B$$

$$\bar{W}_n = k_1 \, n(n-1) \, (n-2)/V^3 + k_3 \, n$$

(4')

The stationary solution $q_n \equiv p(n,\infty)$ of (4) satisfies the detailed balance condition $W_n . q_n = \bar{W}_{n+1} \, q_{n+1}$, which yields :

$$q_n \qquad \frac{W_0 \, W_1 \, \cdots \, W_{n-1}}{\bar{W}_1 \, \bar{W}_2 \, \cdots \, \bar{W}_n}$$

(5)

This solution presents two peaks for the deterministic stationary stable states α and γ, and a minimum for β (see Fig.2).

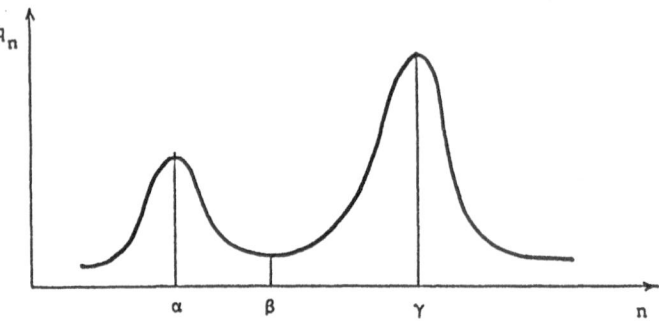

Fig.2 . stationary distribution q_n

In the case shown by Fig.2 the state γ is more probable than α in the final stage of the evolution : macroscopically, α is a metastable state, whereas γ is the stable state ; if the system in initially in the metastable state α, it will eventually pass to γ with a characteristic relaxation time τ_1, but the deterministic theory cannot explain this passage.

In fact, like many dissipative chemical systems, the Schlögl reaction exhibits two scales of time τ_1 and τ_2 ($\tau_2 \ll \tau_1$) : if the number of X molecules at time 0 is n_0, at short times ($t \sim \tau_2$) the probability distribution $p(n,t)$ evolves towards a quasi-stationary disbribution with one maximum at the nearest deterministic stable state (α in the case of Fig.2) ; however at long times ($t \gtrsim \tau_1$) $p(n,t)$ tends towards $q_n^{(n_0)}$.

The deterministic theory only accounts for the short time evolution, whereas the longer relaxation time τ_1 can be evaluated from the Master Equation (4) ; τ_1 strongly depends on the stationary distribution (q_n) in the neighbourhood of the unstable state β ; more precisely if $n_0 \sim \alpha$ and if the return fro γ is made impossible by some ad hoc device, it is found that :

$$\tau_1 \sim (\sum_{i \le \beta} q_i) \times (\sum_{n=m}^{M} 1/W_n \, q_n)$$

the second sum extending over a neighbourhood $I = [m,M]$ of β (the exact values of m and M do not matter).

On the other hand, the shorter time τ_2 is the caracteristic time of the reaction, which is controlled by the trimolecular step and can be evaluated from the kinetics laws ; thus if the system is near the metastable concentration x_α :

$$\tau_2 \sim k_1 \, (x_\alpha)^2$$

In this section we have emphasized the elementary case of a homogeneous system, because it displays some features that will appear in more general cases : the existence of several locally stable states ; the passage from the metastable state to the stable one through an unstable state, driven by the fluctuations ; the importance of the neighbourhood of the unstable state in evaluating the corresponding relaxation time. Studying these points for inhomogeneous systems will be our main task now.

2.3. Inhomogeneous system ; general formalism

For sake of simplicity we suppose that the system is unidimensional, ξ denoting the spatial coordinate.

(i) **Deterministic analysis** : for an inhomogeneous liquid system in absence of convection a diffusional term must be added to the chemical equation (2) ; assuming the validity of Fick's law we obtain the well known Reaction-Diffusion Equation for the concentration x of X :

$$\frac{\partial x}{\partial t} = F(x) + D \frac{\partial^2 x}{\partial \xi^2} \tag{6}$$

where the chemical term $F(x)$ is the 3rd order polynomial given by (2), and D is the Fick diffusion coefficient. In a time τ the X molecules diffuse over a distance of the order of $(D\tau)^{1/2}$; if τ_2 is the characteristic reaction time and if L is the size of the vessel, we conclude that the diffusion can maintain homogeneity if

$$(D\tau_2)^{1/2} \gg L \tag{7}$$

but this condition of high diffusion coefficient is hardly satisfied practically.

In the case of low and moderate diffusion coefficients the stochastic treatment will show that the diffusion can have interesting effects on the reaction ; for instance it can make easier the passage from a homogeneous metastable state to a homogeneous stable state via inhomogeneous configurations, which explains their spontaneous emergence.

(ii) **Stochastic analysis ; multivariate formalism** : the generalization of the Master Equation (4) is obtained by dividing the system into N cells of length[5] $l = L/N$ ($l \gtrsim \lambda$, mean free path of molecules X) and, assuming as usual that (the cells being numbered from 1 to N) :
- each cell remains homogeneous, with n_i molecules X in cell i ;
- only molecules of the same cell react together ;
- diffusion can only transfer a molecule X from a cell i to the neighbourhing cells $i \pm 1$, and may be assimilated to a first order "reaction" : X_i (molecule X in cell i) $X_{i \pm 1}$.

The stochastic state of the system is described by the probability $p(n_1, n_2, \ldots, n_i, \ldots n_N ; t)$ to have n_i particles X in cell i, $i = 1, 2 \ldots N$, at time t. This probality obeys the multivariate Master Equation :

$$\frac{\partial}{\partial t} p = \sum_{i=1}^{N} R_i [p] + D[p] \tag{8}$$

where
R_i is the operator representing a reaction in cell i : it acts on the variable n_i only, its expression being given by (4) ;
D is the diffusion operator :

$$D[p(n_1, \ldots n_i, n_{i+1}, \ldots n_n ; t)] = \sum_{i=1}^{N-1} \frac{D}{l^2} [(n_i+1)p(\ldots, n_i+1, n_{i+1}-1, \ldots)$$

$$+ (n_{i+1}+1)p(\ldots, n_i-1, n_{i+1}+1, \ldots) - (n_i+n_{i+1})p(\ldots, n_i, n_{i+1}, \ldots)] \tag{9}$$

the generic term in the right hand side representing the diffusion between cell i and cell i+1 (in writing it we have omitted the variables which have the same value in both sides of the equation).

419

Obviously the Multivariate Master Equation is far more complicated than the equation for homogeneous systems and generally it is not even possible to compute the stationary distribution analytically. For this reason we will use different approximations, beginning by the simplest case : two homogeneous cells coupled by diffusion.

3. TWO HOMOGENEOUS CELLS COUPLED BY DIFFUSION

This case seems rather academic ; as a matter of fact if inequality (7) is not satisfied for the whole system, it will neither be true for half the system. However it is possible that the interface between two real cells permits diffusion but with a coefficient much smaller than in the bulk, so that practical applications of our results could be expected. In any way it is nice example of spontaneous emergence of inhomogeneity in a reaction-diffusion system, and it will be useful for the theory of nucleation.

3.1. **The deterministic analysis** is based on the following coupled differential equations :

$$\frac{dn_1}{dt} = w(n_1) - \overline{w}(n_1) + D(n_2 - n_1)$$

$$(10)$$

$$\frac{dn_2}{dt} = w(n_2) - \overline{w}(n_2) + D(n_1 - n_2)$$

where n_1 and n_2 are the numbers of particles in cells 1 and 2 ; w and $\overline{w}$ are the reaction rates defined by (3') ; D is a diffusion coefficient such that the flux of particles leaving cell i per second is n_i (i = 1 or 2 ; we suppose the problem to be symmetrical) ; for instance if the crossing of the interface can be simply described as the dynamical crossing of a potential barrier of height E, we have

$$D \simeq \frac{1}{l} \left(\frac{kT}{2\pi m} \right)^{1/2} e^{-E/kT}$$

(l = length of cell ; m = mass of particles X ; T = temperature).

The study of equations (10) shows[11] that they can have up to nine stationary solutions : this case occurs if D is inferior to a critical value D_1 ; then the homogeneous solutions ($n_1 = n_2 = \alpha$ or γ) are stable nodes and $n_1 = n_2 = \beta$ is an unstable node, as it should be expected from the results on homogeneous systems ; however there are also two inhomogeneous stable nodes : $(n_1, n_2) \sim (\alpha, \gamma)$ and $(n_1, n_2) \sim (\gamma, \alpha)$ and four saddle points : $(n_1, n_2) \sim (\alpha, \beta)$, $(n_1, n_2) \sim (\gamma, \beta)$, $(n_1, n_2) \sim (\beta, \alpha)$ and $(n_1, n_2) \sim (\beta, \gamma)$.

If $D_1 \leq D \leq D_2$, D_2 being a second critical value, there are only two inhomogeneous stationary states, and if $D > D_2$ only homogeneous stationary states remain, and (β, β) becomes a saddle-point : if the system is slightly moved from (β, β), in some directions it returns to (β, β) whereas in other directions it escapes definitively.

We will mainly treat the first, most interessing case of weak diffusion, for $D < D_1$.

3.2. **The stochastic analysis** uses the Master Equation (9), with i = 1 or 2, and D replacing $\mathcal{D}/l^2$. Computer simulations yield the stationary distribution $q(n_1, n_2)$, the qualitative shape of which is given in Fig.3, and agrees with the deterministic results.

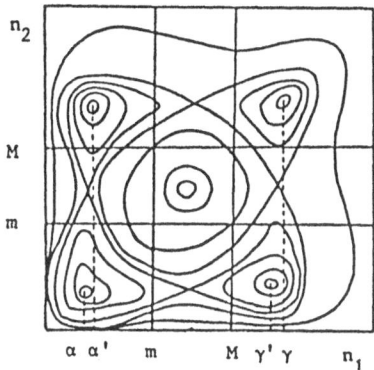

Fig. 3 . Stationary distribution $q(n_1, n_2)$

In particular :

- if $D > D_2$, the states (α, α) and (γ, γ) are maxima of q and state (β, β) is a saddle point ;

- if $D < D_1$, state (β, β) becomes a minimum of q, which suggests that the passage from (α, α) to (γ, γ) occurs through inhomogeneous states : this will be confirmed later.

As a matter of fact an approximate study of Eq.(9) can be made[10] by noticing that in each cell the passage from state α to state γ is much slower than the local relaxation towards α or γ, in spite of the diffusion between the cells, if D is small. Then we define for each cell an "unstable region" consisting of the state in the neighbourhood of β : $(\beta) = \{r : m \le r \le M\}$, as for the homogeneous case (§.2.2) ; we also define the "metastable region" $(\alpha) = \{n : n < m\}$ and the "stable region" $(\beta) = \{n : n > M\}$ (see fig.2) ; we now introduce the reduced probabilities $p(a_1, a_2)$ where (a_i) is (α) or (γ) for i = 1,2 :

$$p(a_1, a_2) = \sum_{n_1 \epsilon(a_1)} \sum_{n_2 \epsilon(a_2)} p(n_1, n_2) \qquad (11)$$

Thus $p(a_1, a_2)$ is the probability that cell 1 is in some state of region (a_1) and that cell 2 is in some state of (a_2).

With the aid of the previous remarks one can write an approximate Reduced Master Equation for $p(a_1, a_2)$. It should be pointed out that the probabilities of the intermediate states (n_1, n_2), where n_1 or n_2 belongs to the unstable region (β), have disappeared from the description by adiabatic elemination, the internal relaxation time in (ß) being very small on the scale of the long time evolution.

The coefficients of the Reduced Master Equation can be computed in terms of the parameters of the problem, and this linear differential system with a 4 x 4 matrix is easily solved analytically by the spectral method. The results strongly depend on the value of D :

(i) for large D (D > D_2), the system essentially remains homogeneous while passing from a metastable distribution concentrated in the region (α,α), to the final stationary distribution $q(n_1,n_2)$ (which is maximum at (γ,γ)), through the point (ß,ß) which is a saddle point of $q(n_1,n_2)$.

(ii) for small D (D < D_1) the system evolves from the metastable region (α,α) towards its stationary distribution, through an intermediary stage where the probability is maximum for (α,γ) or (γ,α) (see Fig.4). Macroscopically, this means that although the system passes from the homogeneous state (α,α) to the homogeneous state (γ,γ), a transitory inhomogeneous stage can be observed with a cell in (α) and the other one in (γ).

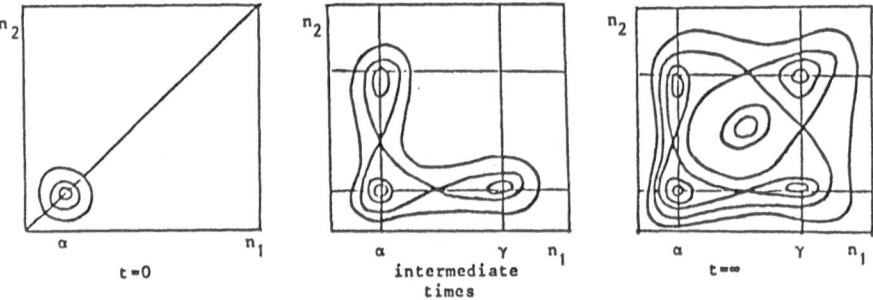

Fig.4 . evolution of $p(n_1,n_2)$ for small D

This is perhaps the simplest case where spontaneous breaking of spatial symmetry can be analysed and predicted (of course the probabilities to observe (α,γ) or (γ,α) are equal, which preserve the symmetry of the probabilistic description).

4. INHOMOGENEOUS SYSTEM ; NUCLEATION THEORY

Generally a reactive one-phase liquid in which uniformity is not maintened by artificial mixing must be described by a Reaction-Diffusion Equation, using the methods of § 2.3.

4.1. Deterministic description ; nucleation nucleus.

The Reaction-Diffusion Equation (6) is written more conveniently with the reduced coordinates :

$$\xi = \frac{x - x_\alpha}{x_\alpha} \qquad \text{(relative deviation from the metastable concentration)}$$

$$\tau = t/\tau_2 \qquad (\tau_2 : \text{characteristic reaction time}) \tag{12}$$

$$D = \mathfrak{D}\tau_2 \qquad \text{(square of the diffusion length)}$$

Then Eq.(6) becomes (ξ being the space coordinate) :

$$\frac{\partial \xi}{\partial \tau} = - b\xi + a\xi^2 - \xi^3 + D \frac{\partial^2 \xi}{\partial \varrho^2} \equiv f(\xi) + D \frac{\partial^2 \xi}{\partial \varrho^2} \tag{13}$$

where the constants a and b are reduced reaction coefficients.

It is well known[12][13][14] that an equation like (13) admits exacts time-dependent solutions:

$$\xi_\pm(\varrho,\tau) = C \left[1 + \exp(\pm\delta(\varrho - \varrho_0 \overline{+} v\tau))\right]^{-1}$$

the parameters C, v, δ being easily expressed in terms of a, b, D ; ϱ_0 is an arbitrary integration constant.

These particular "kink" and "antikink" solutions are fronts of constant profile moving at constant velocity $\pm v$ and representing the propagation of the stable state (for which $\xi = c$) against the metastable state ($\xi = 0$) (Fig.5a).

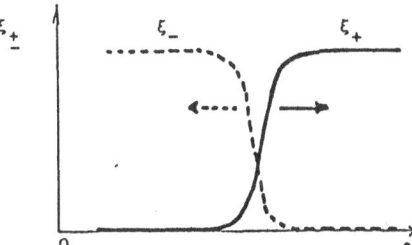

Fig.5a . kink and antikink
solutions

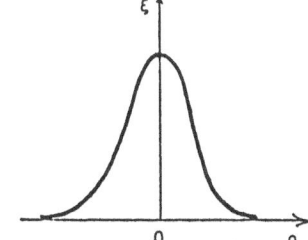

Fig.5b . nucleation nucleus

The nucleation can be described as the passage from an initial uniform metastable configuration of the medium to the uniform stable state, implying the spontaneous emergence of a local fluctuation of X concentration somewhere in the liquid. When this nucleus has just appeared, for instance at point $\varrho = 0$, the configuration $\xi(\varrho)$ of the medium has the shape shown in Fig.5b ; it can be viewed as the product of a kink and of an antikink solutions centered respectively at ϱ_0 and $-\varrho_0$:

$$\xi(\varrho) \propto \left[1 + \exp(\varrho(\varrho - \varrho_0))\right]^{-1} \times \left[1 + \exp(-\varrho(\varrho + \varrho_0))\right]^{-1} \tag{15}$$

If distance ϱ_0 is inferior to a critical value κ, the kink and antikink evolve towards mutual annihilation, and the homogeneous metastable state is eventually recovered ; on the contrary if $\varrho_0 > \kappa$ the kink and antikink separate and the system evolves towards the homogeneous stable state. Finally if $\varrho_0 = \kappa$ the profile given by (15) remains stationary : this critical configuration $\overline{\xi}(q)$ is called **nucleation nucleus**.

It is found that the parameter κ vanishes when the chemical parameters are such that the homogeneous unstable state ß tends to merge into the metastable state α ; then expression (15) becomes (if $\varrho_0 = 0$) :

$$\bar{\xi}(\varrho) = \frac{3b}{2a} \, \text{sech}^2 \, (\frac{\delta\varrho}{2}) \tag{16}$$

Expression (16) is an exact stationary solution of eq.(13) when the cubic term is discarded, which amounts to consider only the small deviations from the metastable state. With this approximation, it can be shown that if the stationary profile (16) is given a small perturbation which is not a mere translation, this perturbation will in some cases decay to 0, in other cases increase indefinitely (a translation of course leaves the profile stationary) : thus the nucleation nucleus represents a "saddle point" in the functional space of solutions, in analogy with an ordinary saddle point such as (ß,ß) in the two-dimensional space (n_1, n_2) used in the two-cells problem. However the deterministic theory does not account for the spontaneous creation of the nucleation nucleus.

4.2. Stochastic analysis

The system, according to the general formalism of § 2.3, is divided into N cells and the probability $p(n_1, \ldots n_N ; t)$ to find n_i particles X in cell i at time t, i = 1, ... N, obeys the Multivariate Master Equation (9). In order to proceed on, we have to forget the discrete nature of n_i : we describe the state of cell i by its concentration $x_i = n_i/\Omega = \epsilon n_i$, $\Omega = 1/\epsilon$ being the volume of a cell, and we consider x_i as a continuous variable. Expressing Eq.(9) with the x_i and expanding it in powers of ϵ we get an approximate continuous Multivariate Master Equation :

$$\frac{\partial p}{\partial t} ((x_i), t) = \sum_i \frac{\partial}{\partial x_i} (\frac{\partial}{\partial x_i} \, \psi p) + \frac{\epsilon}{2} \sum_j G_{ij} \frac{\partial^2}{\partial x_i \partial x_j} p + 0(\epsilon^2) \tag{17}$$

where :

(i) $- \partial\psi/\partial x_i$ is the evolution term of the deterministic equation for x_i :

$$- \frac{\partial\psi}{\partial x_i} = F(x_i) + \frac{\mathcal{D}}{l^2} (x_{i+1} + x_{i-1} - 2x_i) \tag{18}$$

(it can easily be verified that in the continuous limit $l \to 0$ expression (18) is the right hand side of eq.(6) giving $\partial x/\partial t$).

(ii) G_{ij} is a diffusion term depending on $\mathcal{D}$ and on the reaction rates W and $\bar{W}$.

Switching to the reduced variables introduced in § 4.1, eq.(12), p becomes a function of $\xi_1 \ldots \xi_N$, τ which obeys an equation similar to (17) where $\mathcal{D}$ is replaced by $D = \mathcal{D}\tau_2$; F is replaced by f of eq.(13), ψ is replaced by a function $\varphi((\xi_i))$ and G_{ij} by g_{ij}.

When the term in ϵ^2 is neglected in eq.(17) one obtains a Multivariate Fokker-Planck Equation which has been widely used to study the stochastic reaction-diffusion problem approximately. At long times, however, this approximation is only valid locally and does not yield the correct stationary solution of the Multivariate Master Equation. This stationary solution can be evaluated by another continuous approximation, developed by Lemarchand and Nicolis[16,1]; it can be written in the form :

$$P_{station.}(\{x_i\}) \equiv q(\{x_i\}) \propto \exp(-\epsilon^{-1} U(\{x_i\})) = \exp(-\epsilon^{-1} u(\{\xi_i\}))$$

(19)

The function $U(\{x_i\})$ is the **stochastic potential**. This expression supposes that when $\epsilon \to 0$ i.e. (the volume V of the system becomes infinite) the probability concentrates on the deterministic stationary profiles $x_1 = x_2 = \ldots = x_N = x_\alpha$ or x_γ (system uniformly metastable or stable) ; the stochastic potential should then have minima for these two particular distributions. The assumption (19) is of course based on the fact that the deterministic results must be recovered from the stochastic theory when $V \to \infty$.

Equivalently, the reduced stochastic potential $u(\{\xi_i\})$ must be minimum for the configuration $\xi_1 = \ldots \xi_N = 0$ (uniform metastable medium) and for the configuration $\xi_1 = \ldots \xi_N = c$ (uniform stable medium). The stochastic problem of nucleation is then to study the passage from a probability distribution initially concentrated on the first configuration (metastable medium) to the final stationary distribution, which is maximum for the uniform stable configuration : this is quite similar to the problem studied in the two-cells case (§ 3), with the difference that we have now to treat probabilities with N variables.

Thus, in analogy with the two-cells problem, we will assume that the passage from the metastable state to the stable state is made through particular critical configurations which are saddle points of the stationary probability q, or equivalently, **saddle points of the stochastic potential u.** These saddle points can hardly be computed directly, but the deterministic study of § 4.1 shows that they should correspond to the nucleation nucleus ξ given given by eq.(16), and to the translated profiles : the discretized version of ξ then gives these saddle points of u. In the neighbourhood of these points, u can be computed by the method proposed by Lemarchand and Nicolis[16,17], which eventually allows to obtain the transition rate from the metastable to the stable state.

As a matter of fact Kramers[18], Langer[19] and other authors[20,21] have derived the nucleation rate for equations of the type of (17) when (G_{ij}) is a diagonal matrix. Their method can be generalized to our problem[20,21] : calculating the probability flux through the saddle points ξ yields the nucleation rate under the form :

$$k = C \exp(-\epsilon^{-1} x_\alpha \bar{U})$$

(20)

where x_α is the metastable concentration, and $\bar{U} = u(\bar{\xi})$ is the value of the stochastic potential at the nucleation nucleus ; the prefactor C involves the eigenmodes of this potential at the metastable state and at the saddle point $\bar{\xi}$:

$$C = \frac{\lambda_0}{(2\pi)^{3/2}} \left(\frac{\epsilon}{x_\alpha}\right)^{-3/2} A B$$

(21)

with the following definitions :

Φ being the matrix of second derivatives of the deterministic potential Φ :

$$\Phi = \left(\frac{\partial^2 \Phi}{\partial \xi_r \partial \xi_{r'}} \right)$$

U being the similar matrix for the stochastic potential :

$$U = \left(\frac{\partial^2 U}{\partial \xi_r \partial \xi_{r'}} \right)$$

then :

λ_0 is only negative value of Φ at the nucleation nucleus $\bar{\xi}$;

$$\mathcal{B}^2 = \frac{\text{product of eigenvalues of } U \text{ at the metastable state}}{\text{product of positive eigenvalues of } U \text{ at } \bar{\xi}} \; ;$$

$\mathcal{A}$ is a "volume factor" due to the translational invariance of the nucleation nucleus.

Formulae (20) (21) can be considered as the main result of this work ; it may be compared with the celabrated Kramer's formula giving the transition rate of a one-dimensional system diffusing between two minima of potential, which implies the curvature of this (deterministic) potential at the minimum and at the top of the Barrier.

The present theory allows to compute k explicitly, the only parameters being the Fick coefficient $\mathbb{D}$ and the reaction rates, which are known experimentally (the formal parameter ϵ of course disappears) ; it has been extended to many dimensional systems[22,23] and thus can be of practical interest.

Numerical discussion of formula (19) shows that the observation of intrinsic homogeneous nucleation should be difficult except for reactions which imply a very fast trimolecular step and a large stationary concentration of the active species, at least for tridimensional systems with one chemical variable.

These conclusions explain why such a nucleation has not been detected up to now, and may help in designing new experiments.

LITERATURE

(1) Among the many papers on the subject, we only quote two recent articles : C. Vidal and P. Hanusse, Int. Rev. Phys. Chem., 5 (1986), 1; C. Vidal : communication at the present Conference.
(2) A.M. Zhabotinskii and A.N. Zaikin, J. Theor. Biol., 40 (1973, 45.
(3) A. Hanna, A. Saül and K. Showalter, J. Am. Chem. Soc., 104 (1982), 3838.
(4) R.P. Rastogi, I. Das, M.K. Verma and A.R. Singh, J. Non-Equil. Therm, 8, 1983), 255.
(5) See for instance
G. Nicolis and I. Prigogine, Self-Organization in Non-Equilibrium Systems, Wiley, New York (1977), and H. Haken, Synergetics : An Introduction, Springer) Verlag, Berlin (1978).

(6) D. Walgraef, G. Dewel and P. Borkmans, J. Chem. Phys., $\underline{78}$ (1983), 3043.

(7) C. Vidal, A. Pagola, J.M. Bodet, P. Hanusse and E. Bastardic, Journal de Physique, $\underline{47}$ (1986) 1999.

(8) A. Saül and K. Showalter, in : Oscillations and Travelling Waves in Chemical Systems, R. J. Fields and M. Burger ed., Wiley, New York (1985).

(9) F. Schlögl, Z. Phys. $\underline{253}$ (1972) 147.

(10) D. Borgis and M. Moreau, Physica $\underline{123A}$ (1984) 109.

(11) W. Ebeling and H. Malchow, Ann. Phys. (Leipzig) $\underline{36}$ (1979) 121.

(12) H. Metiu, K. Kitahara and J. Ross, J. Chem. Phys. $\underline{64}$ (1976) 292.

(13) E.W. Montroll, in : Statistical Mechanics, S.A. Rise, K.F. Freed and J.C. Light ed., Univ. Of Chicago Press (1972).

(14) K. Parlinski and P. Zielinski, Zeit. Phys., $\underline{B44}$ (1981), 317.

(15) M. Büttiker and R. Landauer, Phys. Rev. $\underline{A23}$ (1981) 1397.

(16) H. Lemarchand, Physica, $\underline{101A}$ (1981) 518.

(17) H. Lemarchand and G. Nicolis, J. Stat. Phys. $\underline{37}$ (1984) 609.

(18) H. Kramers, Physica, $\underline{7}$ (1940) 284.

(19) J.S. Langer, Phys. Rev. Lett., $\underline{21}$ (1968) 973.

(20) H.C. Brinkman, Physica, $\underline{22}$ (1956), 29.

(21) R. Landauer and J.A. Swanson, Phys. Rev., $\underline{121}$ (1961), 1668.

(22) D. Borgis, PhD Thesis, P. and M. Curie University, Paris (1986).

(23) D. Borgis and M. Moreau, to be published.

KAPRAL - 1) Has the full bifurcation of the deterministic dynamics of the the two-coupled cells been studied as a function of the reaction and diffusion parameters ? Since the system is two-dimensional the possibility of more complex system states exists.

2) Have you estimated the nucleation rate for the iodate-arsenous system that is modelled by the cubic kinetics of the model you studied ? Is nucleation by spontaneous fluctuations the likely mechanism for generation of nuclei in this system ?

MOREAU - 1) The bifurcations of the two-coupled cells have been studied by by various authors, in particular by Ebeling, Malchow and Frankowicz.

2) The nucleation rate of the Schlögl reaction has been explicitly expressed as a function of the concentrations, of the reaction constants and of the Fick's coefficient. The full numerical evaluation has not been attempted for the iodate-arsenious acid reaction, but it has been carried out for the Oregonator, although the values of the parameters are still in discussion. The resulting estimate shows that it should not be possible to observe internal nucleation in this case. As for the iodate-arsenious acid reaction, it follows from partial estimates that spontaneous nucleation should only be observable under special conditions, such as very fast trimolecular step.

BORGIS - The iodate-arsenious acid reaction does not give any spontaneous nucleation due to internal fluctuations, under the conditions taken be Saül and Showalter in their experiments. The phenomenon is not very far, though.

FRANKOWICZ - 1) One of the practical realizations of the Schlögl-type mechanism is the formol reaction, P. Decker, Ann. N.Y. Acad. Sci. 316, 236 (1979).

2) The complete bifurcation analysis for the system of coupled bistable cells was performed by Ebeling and Malchow, Ann. Physik 36, 121 (1979).

3) Experiments on coupled cells in a bistable region (BZ reaction) were reported by Studel and Mavek, J. Chem. Phys. 77, 2956 (1982). Inhomogeneous structures were observed and their relative stability was investigated.

4) The first-passage time between homogeneous states for a system of coupled bistable cells displays a minimum in its diffusion dependence, i.e. diffusion may facilitate the nucleation process. This effect becomes more pronounced if the number of coupled cells increases, M. Frankowitcz, Acta Phys. Pol. A, (to appear Oct. 1987).

BELLONI - My comment concerns the very starting hypothesis of your model. It seems that a three bodies reaction assumption is required to create the double stationary states. Actually, this assumption is unrealistic for an encounter of solutes and on the other hand cascade reactions will be unable to create the same situation. Is the cubic term indispensable for your conclusion ?

MOREAU - A third order kinetics is indeed necessary for the existence of two locally stable stationary states in the homogeneous case. Although I am not convinced that three-bodies interactions can never occur, it is sure that the kinetics which is used here (and in many other works) is only the overall result of several bimolecular elementary reactions (this is clear in the detailed models proposed to represent, for instance, the Belousov-Zhabotinski reaction).

BOURCEANU — Is it not enough if you say fluctuations and not internal fluctuations ? Because these are inevitable and appear only inside of the system.

MOREAU — I agree with you, but I only used the terms "internal fluctuations" to distinguish them, as usual, from the external perturbations (dust, noise,...).

AMATORE — This is a further comment on Dr. Belloni and Borgis' comments on true chemical kinetics. One can obtain apparent rate laws which are much more complex than bimolecular ones. Yet those rate laws derive from the establishment of chemical steady state for transient intermediates. Thus for analyses such as those presented, it is implicitely assumed that the time constants of the chemical steady state establishment are considerably smaller than the times of disruption observed for the overall system This should be usually true except may be for sharp transition. Thus it is possible that small local deviations from the chemical steady state will perturb significantly the prediction of the reduced model.

MOREAU — This problem is certainly interesting, but we have not studied it so far.

STOCHASTIC EVOLUTION OF INHOMOGENEOUS CHEMICAL SYSTEMS

M. Frankowicz

Faculty of Chemistry, Jagellonian University

Karasia 3, 30-060 Kraków, Poland

INTRODUCTION

Among various aspects of nonequilibrium phenomena the problem of transient behaviour is one of the most interesting ones. In particular, the relaxation from the unstable state was the subject of many studies [1]. Recent experiments [2] show that stochastic effects may be very important in explaining irreproducibility of results for some nonlinear chemical systems.

In this paper the problem of transient behaviour of stochastic models of inhomogeneous chemical systems will be considered. First, the main results concerning compartmental models will be presented. Next, a qualitative approach will be proposed. Finally, some remarks concerning "mean-field" approach and explosion in inhomogeneous system will be made.

COMPARTMENTAL MODELS

Inhomogeneous reaction-diffusion systems are often approximated by models consisting of coupled homogeneous cells (compartmental models). This simplification is very useful in numerical computations for deterministic systems as well as in stochastic simulations. Systems of coupled CSTR were also used to study experimentally chemical dissipative structures [3].

The kinetic equation for a system consisting of n cells (for simplicity we assume that the dynamics is described by one relevant variable) has the form

$$dx_i/dt = f(x_i) + \sum_i D_{ij} (x_j - x_i) \ , \quad i,j = 1,\ldots,n \qquad (1)$$

where x_i is the value of the relevant variable (i.e. concentration) in the i-th cell, D_{ij} are exchange coefficients (diffusion coefficients), $f(x)$ is a dynamical law governing the evolution of the homogeneous system. In what follows we will assume that $D_{ij}=D=$ const for adjacent cells, otherwise they are equal to zero.

In the sequel we will deal with systems with a cubic rate law (such as e.g. the Schlogl model)

$$dx/dt = -x^3 + Ax^2 - Bx + C \qquad (2)$$

Cubic systems in the homogeneous case may display one or two stable stationary states. For the systems of coupled cells, deterministic analysis shows that only fold and cusp catastrophes can appear; diffusion cannot lead to higher order catastrophes than those appearing in homogeneous systems [4]. Depending on the value of D, a system of N coupled bistable cells can have up to 3^N stationary solutions (stable nodes, unstable nodes and saddles). For large D, only homogeneous solutions are stable; for small D, stable inhomogeneous structures can appear.

The stochastic description of a compartmental system is usually performed in terms of a multivariate master equation. For a chemical system this equation has the form

$$dP(\{X_i\},t)/dt = \sum_i \left[\lambda(X_i -1)P(\ldots,X_i -1,\ldots,t) \right.$$

$$+ \mu(X_i +1)P(\ldots,X_i +1,\ldots,t) - (\lambda(X_i)+\mu(X_i))P(\{X_i\},t) \Big]$$

$$+D \sum_i \left\{ (X_i +1)\left[P(\ldots,X_i +1,X_{i+1}-1,\ldots,t)+P(\ldots,X_{i-1}-1,X_i +1,\ldots,t) \right] \right.$$

$$\left. - 2 X_i \ P(\{X_i\},t) \right\} \qquad (3)$$

$\lambda(X_i)$ and $\mu(X_i)$ are birth and death rates.

The Eq. (3) was studied numerically for the system with cubic rate law (2-,3-,4- and 8-cell models) [5,6]. In the bistable regime it was found that transitions between homogeneous states in case of small diffusional coupling occur via transient inhomogeneous states. The number of possible inhomogeneous states diminishes with the increase of diffusion . Moreover, small diffusion facilitates transitions between homogeneous states: the passage time has a minimum as function of diffusion. The increase of the number of coupled cells makes this minimum deeper; on the other hand, the passage time for large D gets longer. The numerical results for the 2-cell model were confirmed analytically by Borgis and Moreau [7]. In the sequel a qualitative argumentation confirming the above results will be proposed.

A QUALITATIVE APPROACH

In order to get some ideas about the mechanism of transitions between steady states let us consider a qualitative model, in which fluctuations are described by adding a random force term to the determnistic description. Let us also assume that the matrix of the noise correlation function is completely diagonal with noise strength equal to θ. The stationary probability distribution is then proportional to $\exp(-U(\underline{x},D)/2\theta)$ [8] where U is a kinetic potential

$$U(\underline{x},D) = \sum_i \phi(x_i) + (1/2) \sum_{i,j} D_{ij} (x_j^2 + x_i^2 - 2x_i x_j) (4)$$

and

$$-d\phi(x_i)/dt = f(x_i)$$

$\underline{x}$ is the vector of concentrations $(x_1, x_2, \ldots, x_N)$

The equation (1) can be written as

$$dx_i/dt = -\partial U/\partial x_i$$

We can then consider the stochastic evolution of a multistable dynamical system as a sequence of transitions between local minima of U through saddle points separating these minima: the transitions are driven by fluctuations.

In Fig. 1 the kinetic potential profiles for the 4-cell model with periodic boundary conditions are shown. The profiles are sketched along transition paths joining homogeneous steady states via inhomogeneous stationary states and saddle points. It is seen that the increase of D results in the following effects:
    - the potential barriers between stationary states inccrease
    - the relative stability of inhomogeneous steady states, being a function of the depth of the corresponding potential well, decreases up to the total disappearance of these states for critical values of diffusion
        - the most probable transition path (the lowest potential barriers) corresponds to the nucleation-like scenario $(0000 \to 1000 \to 1100 \to 1110 \to 1111)$

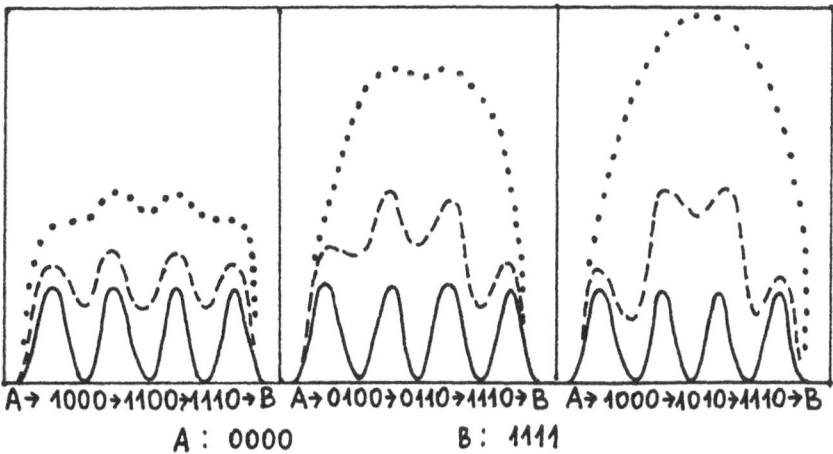

A→ 1000→1100→1110→B A→0100→0110→1110→B A→1000→1010→1110→B

A : 0000   B : 1111

Fig. 1. Profiles of the kinetic potential along paths
        joining homogeneous steady states (4-box model)
        "0" and "1" denote states of individual cells.
        The values of D: 0.0 (solid line), 0.01 (broken
        line), 0.03 (dotted line).

THE EXPLOSIVE CASE

Cubic systems near a limit point bifurcation, but outside the region of bistability, can display explosive behaviour: initial slow evolution is followed by a rapid motion
towards the final state. Stochastic effects may lead to
transient splitting of probability distribution into two-
humped form (transient bimodality) [11]. For spatially
distributed systems (e.g. the compartmental models)
transient bimodality may give rise to the creation of "hot
spots" (regions where the transition to the final state
already took place); they can be transiently stabilized by
diffusion. In [6] it was shown that for small values of
diffusion a "plateau" in the time dependence of the mean
value of concentration in the 8-box Schlogl model appears;
its position is halfway between the initial region amd the
final one. This plateau corresponds to the transient
inhomogeneous structure (4 cells near the initial state, 4
others already in the final regime).

THE MEAN FIELD APPROACH: NONLINEAR MASTER EQUATION

Prigogine, Nicolis and co-workers proposed a "nonlinear
master equation" based on a mean-field approach and on the
hypothesis of translational invariance [9]. This equation
has the form

$$dP(X,t)/dt = \lambda(X-1)P(X-1,t)+\mu(X+1)P(X+1,t)-\left[\lambda(X)+\mu(X)\right]P(X,t)$$
$$+D\left[(X+1)P(X+1,t)-XP(X,t)+<X>\left[P(X-1,t)-P(X,t)\right]\right] \quad (5)$$

The numerical investigation of this equation shows that in
the bistable regime the minimum in the diffusional
dependence of the transition time between stationary states
also occurs (as in the case of the multivariate master
equation). Fig. 2 shows the results for the Schlogl model
(the system size is 100 and 200).

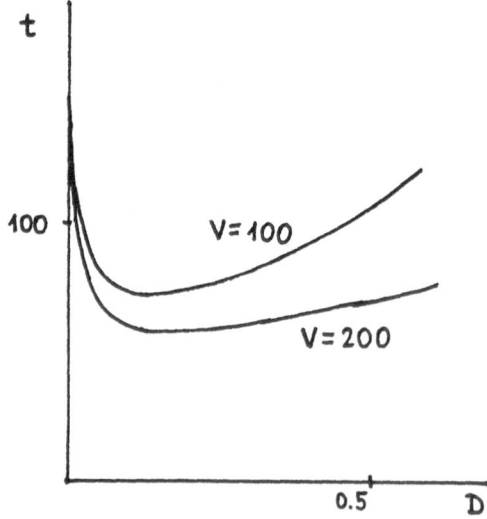

Fig. 2. The diffusion dependence of the passage time for
mean-field approach: System size V=100 and 200

In [9] it was stated that there exists a critical value of the system size for which the life time of the metastable branch is shortest (this size can be viewed as the size of a "critical nucleus"). This critical value depends on diffusion. So for given value of diffusion we have a minimum in the system size dependence of the passage time, and also for given system size we have a minimum in the diffusion dependence of the passage time. For the parameters used in the computations the critical system size should be greater than 200 (the curve for V=200 has deeper minimum than the one for v=100). Similar behaviour was found by Borgis [10].

CONCLUSIONS

Random effects, which may be caused by external noises, thermodynamic fluctuations or uncertainties in initial conditions can significantly influence the evolution of inhomogeneous nonlinear systems. For example, in the explosive regime they may lead to the appearance of transient structures (like "hot spots" which are sources of flame propagation). In the bistable regime, fluctuations drive the transitions between homogeneous states. The passage time depends on diffusion; for a certain value of D, it has a minimum. The study of stochastic models of inhomogeneous systems may be useful in understanding such effects as incomplete mixing and its influence on the effective reaction rate in chemical reactors, in studying early stages of nucleation, and in investigating the reactions in micelles [12].

ACKNOWLEDGEMENTS

The author acknowledges the financial support from the organizers. The research was partly supported by the program CPBR 03.20.

REFERENCES

1. M. Suzuki, Adv. Chem. Phys. 46:195(1981).
2. I.R. Epstein and I. Nagypal, in: "Spatial Inhomogenei- ties and Transient Behaviour in Chemical Kinetics", P. Grey and G. Nicolis, Eds.,Manchester University Press (to be published 1988).
3. I. Stuchl and M. Marek, J. Chem. Phys. 77:2956(1982).
4. W. Ebeling and H. Malchow, Ann. Phys. (Germany) 36:121 (1979).
5. M. Frankowicz, Acta Phys. Pol. A72:487(1987).
6. M. Frankowicz, Acta Phys. Pol. A73:3(1988).
7. D. Borgis and M. Moreau, J. Stat. Phys. 37:6131(1984).
8. B. Caroli, C. Caroli, B. Roulet, and J.F. Gouyet, J. Stat. Phys. 22:515(1980).
9. I. Prigogine, R. Lefever, J.S. Turner and J.W. Turner, Phys. Lett. 51A:317(1975).
10. D. Borgis, Thesis, Univ. Pierre et Marie Curie, Paris (1986).
11. M. Frankowicz and G. Nicolis, J. Stat. Phys. 33:595 (1983).
12. M. Tachiya, in: "Kinetics of Nonhomogeneous Processes", G.R. Freeman, Ed., Wiley, New York (1987).

AN EXPERIMENTAL STUDY OF THE BEHAVIOUR OF CHEMICAL SYSTEMS
FAR FROM EQUILIBRIUM

Christian Vidal

Centre de Recherche Paul Pascal
Université de Bordeaux I
Domaine Universitarie
33405 Talence Cédex, France

INTRODUCTION

Under most circumstances, chemical reactions proceed monotonically in space and time. Certain conditions must, however, be fulfilled; in particular a sufficient displacement from equilibrium. In fact, we sometimes observe non-monotonous behaviour, either with respect to time (e.g. oscillations), or space (structures), or even both (chemical waves). These phenomena were originally thought of as being exotic or even dubious, although over the past twenty years they have been the subject of a great deal of serious work. Certain are well understood nowadays, having been analysed and interpreted, whereas others retain a faint mysterious flavour. Considering the whole list of observations carried out in reactive media in the liquid phase, we note the extreme diversity of the situations encountered and identified: periodic oscillations, coexistence of stable stationary states (bistationarity) or oscillations (birhythmicity), chemical hysteresis, deterministic chaos, spatial structures and chemical waves. Although the reactions showing such behaviour known to date are still limited in number, they are of fundamental interest because of the wide range of phenomena they portray. For one thing, they irrevocably establish that there is no barrier between inert matter and the living world with regard to self-organization. Secondly, they point out that the second law does not support the conclusion according to which all evolution would be inevitably accompanied by an increase in disorganization. Finally, the exceptional sensitivity of these phenomena with regard to certain parameters opens up possibilities for bringing to light and study mixing effects, including those over a very short scale, and the possible role of fluctuations.

This article is intended to present a rapid panorama of this research field, laying its emphasis on the experimental aspects. Apart from the specialized publications, references (1), (2) and (3) contain technically far more detailed descriptions on the various questions brought up concisely in the present work.

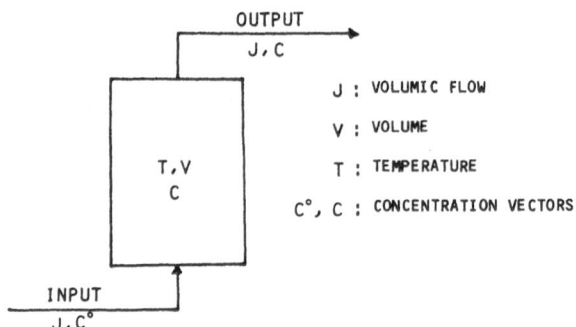

Fig. 1. Scheme of the principle of the ideal stirred reactor.

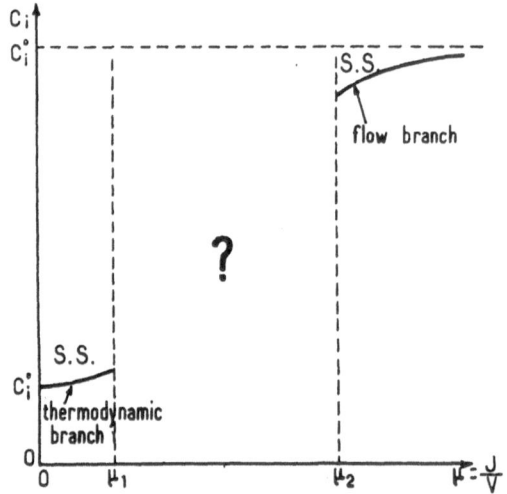

Fig. 2. Bifurcation diagram of [1] according to the value of
the parameter $\mu = J/V$.

Non-monotonous behaviour with respect to time and/or space was originally observed, with very rare exceptions, during the course of evolution leading a closed reacting system to equilibrium predicted by the laws of thermodynamics. Because of the necessarily transient character of these conditions, no detailed quantitative study can be carried out on the subject. The intervention of an experimental technique (4), the CSTR well known in chemical engineering, was a considerable step forward in the study of time-dependent behaviour, permitting the reacting system to be maintained constantly at a finite distance from equilibrium. This is also why the state of the art in this domain is relatively advanced, whilst this is not the case of reaction-diffusion phenomena for which there has not as yet been comparable progress. The attempts made to ensure the continuous renewal of the reactants require highly ingeneous processes which are, nevertheless, singularly artificial (e.g. coupling of reactors (5), setting up of a Couette flow (6)). It is not out of the question that new perspectives should open up in this field, even so (7).

Experiments in open stirred reactors have become common place these days for the study of time-dependent phenomena. By dint of the usual hypothesis of ideality, i.e. that the medium is instantaneously of homogeneous composition (^{1}), the concentrations are space-independent functions, whence a considerable simplification of the equations describing the behaviour. The time variations of the concentrations C_i within an ideal reactor (as in the scheme of fig. 1), can thus be expressed simply with the set of autonomous differential equations (8):

$$\dot{C}_i = F_i(C) + \mu\,(C_i^0 - C_i) \qquad i = 1,\ldots,L \qquad [1]$$

derived from the mass conservation law for each species. L designates the total number of chemical species and μ the reciprocal of the mean residence time τ in the reactor:

$$\tau = \frac{V}{J} = \frac{1}{\mu}$$

The functions $F_i(C)$ represent the contribution, which is in general non-linear, of the chemical reaction, whereas the linear term corresponds to the transport of matter through the reactor. Thus, it is the flux J of reactants $(C_i^0 \neq 0)$ which ensures the maintenance of the reacting medium permanently out of equilibrium.

We know that [1] can have four qualitatively different types of solution (9): stationary, periodic, quasi-periodic and chaotic. Transition from one type of solution to another gives rise to a bifurcation, which can be provoked by external variation of one of the "parameters" of [1] (that is time fixed quantities; e.g. C_i^0 or μ). The most frequently exploited experimentally is μ. In addition, it provides a method of altering, simultaneously and in the same manner, the linear transport term in all the equations forming the system [1], by the controled variation of the volumic flow J. We can readily show (8) that, in both extremities of the range of μ, the asymptotic solutions of [1], and consequently the stable regime of the reactor, are of stationary form, whatever the particular form of the functions F_i (see fig. 2). For very large μ, the concentrations remain in the neighbourhood of their inlet values, the reaction hardly having the time to take place (flow branch). For $\mu = 0$ (closed systems), the stable state is that of thermodynamic equilibrium, defined by the minimum of the Helmholtz free energy. For low, non-zero μ, the theorem of minimum entropy production guarantees the existence, uniqueness and stability of a stationary state (thermodynamic branch). As long as the amplitude of the non-linear terms of F_1 is large enough, the thermodynamic branch can become unstable above a critical value μ_1, and the flow branch below another critical value μ_2. Only in the interval between these values, when there is one (which depends above all on the chemical reaction under consideration), can the rather unusual phenomena presented in the following paragraphs occur. In order to observe them, one must choose both the reaction _and_ the conditions for setting it up appropriately, as we see.

Unquestionably slightly less conventional, another type of experiment has been developed over the last few years. It consists of submitting the reacting medium, whilst still maintained out of equilibrium, to a time dependent perturbation (10). This can be periodic or non-periodic. This is an investigation domain, which, in some sense, extends to this field the methods developed for chemical relaxation. It is sometimes refered to as "spectral kinetics". In a variable, rather than a constant environment, new instabilities can come to light, which give rise to a new phenomenology. The loss of stability of a parametric oscillator and the phenomena of frequency-locking of oscillators have indeed long been known in a different context. The few experiments published to date reveal the extraordinary diversity of behaviour which we should expect (5,11,12). The analyses and interpretation of results under these circumstances are, naturally, more delicate to carry out sucessfully. A time-dependent term must be added to the right-hand side of [1], giving a set of non-autonomous differential equations. The amplitude, frequency and phase of the perturbation each exerts its own influence; thus we are confronted with problems of intuitively evident complexity. In fact, it must be said that the bulk of the work in this field remains to be done.

The measurement techniques of the instantaneous state of a reacting medium are relatively common. They make use of receptors which have, in general, a very short response time. The most wide spread are either of the potentiometric variety, or spectroscopic (U.V. or visible, or more rarely NMR or EPR (13)). When the signal has to be given a specific numerical treatment (improvement of the signal to noise ratio, Fourier transform, or characterization of the topological properties of the attractor), computerization of the system of data acquisition becomes necessary. In the great majority of experiments, we record a single response, which takes the variations of the system in the course of time into account. This turns out to be sufficient for the purely dynamic aspects of the study, but provides very little information from a chemical point of view. Work pertaining to the simultaneous measurement of the concentrations of several intermediate species remain relatively rare, even these days (see fig. 3).

The above is entirely based on the validity of the ideal reactor approximation. This can, however, be a faulty premise, there being strongly non-linear reactions involved. Even a very vigorous stirring of the medium, creating very strong convection, could be insufficient to ensure instantaneous uniformity of composition. In the liquid phase, where hydrodynamic turbulence becomes progressively more inefficient in giving mixing over the very short distance, diffusion thus starts playing an important role. The course of the reaction is then substantially dictated by the micromixing of the medium, a state which depends on the stirring and method of injection of the reactants. Spectacular effects can ensue (e.g. modification of a domain of bistability (14,15) or of the amplitude and period of a chemical oscillator (16)), providing the proof of the elaborate

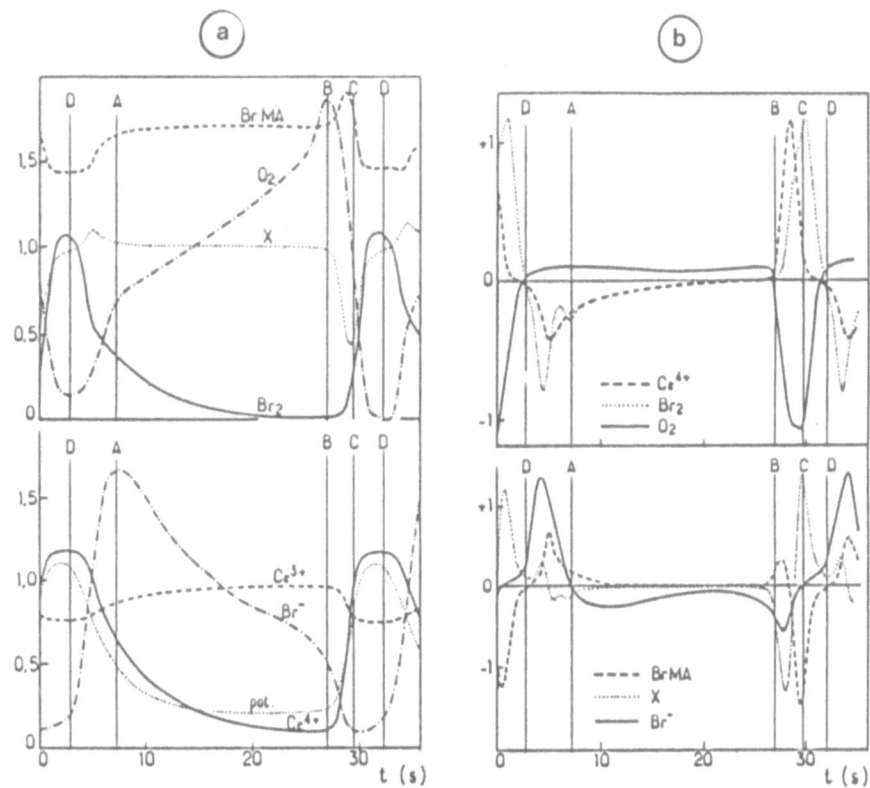

Fig. 3. Quantitative study of an oscillation of the Belousov-Zhabotinsky
reaction (62). (a) variations in the concentrations of several
species over a period (limits given by the vertical lines D).
X designates a chemically unidentified species. (b) Instantaneous
contribution of the chemical reaction to production or consumption of the various species indicated.

precautions that ought to be taken in certain cases prior to reaching the
conclusions that the medium be homogeneous[2]. By simple reversal of the
above logic, we notice that the reactions displaying such sensitivity with
respect to mixing could provide a very efficient mean to analyse this, and
to establish the performances attained in this respect by a given reactor.
This is a new example of research, with very fundamental inspiration at its
origin, which could have interesting practical implications in the field of
chemical engineering, a very unforeseeable result initially.

OSCILLATORY CHEMICAL REACTIONS

As the previous paragraph recalls, the observation of non-monotonous
behaviour with respect to time requires several conditions, in particular
the selection of a chemical reaction "ad hoc". In the majority of cases,
although not always[3], the reaction mechanism should include an intrinsic
source of instability. In a schematic manner, we can visualize this as a
feedback loop, such that the effect produced reinforces the cause provoking
it. Autocalalysis is the best known and commonest example of such a
situation, being of opposite nature to a law of moderation. It is not,

however, the only one. A classified repertoire of "chemical" instabilities in non-equilibrium stationary states was indeed proposed some years ago (18). One should thus distinguish between: direct autocatalysis (i.e. activation by the product or inhibition by the substrate), indirect autocatalysis (competition, symbiosis, positive feedback), and indirect auto-inhibition (i.e. negative feedback). As soon as a non-equilibrium stationary state becomes unstable, another situation prevails via bifurcation. The variety of situations possible a priori is quite substantial. Nevertheless, we mostly obtain the emergence either of two states (in general stationary) which are stable simultaneously, or of time-dependent oscillations (which are usually periodic).

When set up appropriately, autocatalytic chemical reactions (taking the usual meaning of the term) frequently give rise to a "bistability" (in particular bistationarity), and chemical hysteresis as its corollary. Since these reactions are relatively well documented elsewhere, it does not appear necessary to list them all there, particularly since they are quite numerous. Conversely, oscillatory chemical reactions in homogeneous chemical media are rather rarer and, apart from two exceptions, have recently been discovered. Since there is no satisfactory criterion permitting the classification of chemical oscillators at present, we can do little more than record them in a somewhat arbitrary order:

(i)   bromate/organic reducing agent (e.g.: $CH_2(COOH)_2$)/metal ion catalyst (e.g.: Ce)(this is the prototype of the series, i.e. the B.Z. reaction discovered by Belousov (19) in the 1950s, subsequently studied by Zhabotinsky, and which has been the subject of the most extensive studies).

(ii)   bromate/inorganic reducing agent (e.g.: AsO_3^{3-})/Mn^{2+}

(iii)   bromate/phenols, or aniline, or thiourea (N.B. no catalyst)

(iv)   bromate/bromide/ Ce^{3+} or Mn^{2+} (often called "the minimal oscillator")

(v)   bromate/iodide

(vi)   bromate/inorganic reducing agent (e.g.: AsO_3^{3-}) or iodide/ chlorite

(vii)   iodate/hydrogen peroxide (historically the first chemical oscillator to be discovered; Bray (20))

(viii) iodate/hydrogen peroxide/organic reducing agent (e.g.: $CH_2(COOH)_2$)/Mn^{2+} (the Briggs-Rauscher reaction (21))

(ix)   iodate/ferricyanide/sulphite

(x)   pyrosulphate/iodide/oxalic acid

(xi)   chlorite/iodate/organic reducing agent (e.g.: $CH_2(COOH)_2$) or inorganic reducing agent (e.g.: AsO_3^{3-})

(xii)   chlorite/iodide

(xiii) chlorite/iodine/inorganic reducing agent (e.g.: SO_3^{2-})

(xiv)   chlorite/iodide/inorganic oxidizing agent (e.g.: MnO_4^-)

(xv)   chlorite/thiosulphate or thiourea

(xvi)   pyrosulphate/oxalic acid/Ag^+

(xvii) thiocyanate/hydrogen peroxide/Cu^{2+}

(xviii) sulphide/oxygen/sulphite/methylene blue

(xix)   sulphide/hydrogen peroxide

(xx)   sulphide/pyrosulphate

(xxi)   benzaldehyde/oxygen/CO^{3+}/Br^-.

This list is in fact growing from year to year, and successive attempts at classification (in particular by constituent) are regularly called into question. Where the rules permit a useful ordering of this series remains, obviously, to be pinpointed. The search for new chemical oscillators can, however, be based on the method refered to as the "crossed diagram" (22), which shows how the application of a suitable feedback should allow the transition from a situation of bistability to one of relaxation oscillations. Consequently, the majority of those oscillators mentioned above results from such an approach. As for the direct transition from stationary state to periodic behaviour (i.e. limit cycle), we know that it can only operate through two types of bifurcation, with quite distinct properties: firstly, a (normal or inverse) Hopf bifurcation, and secondly a saddle-node bifurcation. In specific cases, experimental indications have been recorded, confirming that one or other such transition indeed occurs in the chemistry (23,24).

Because of the unique position that the Belousov-Zhabotinsky reaction occupies today in this whole field of research, certain additional details of it deserve a mention. It was discovered, presumably at the beginning of the 1950s, by B.P. Belousov, a Soviet biochemist working on the components of the Krebs cycle. He noticed, quite by chance, that during the oxidation of the citric acid by the bromate under acidic conditions, catalyzed by the Cerium ions, the solution alternated between being yellow and colourless. This discovery was confronted by a lot of sckepticism and Belousov did not manage to publish the result, judged at the time to be suspicious, not to say heretical. Several years later, A.M. Zhabotinsky, a young researcher beginning his career, devoted his thesis to the study of this reaction, which not only oscillates in time, but also very readily gives rise to spatial structures and chemical waves. A simple composition allowing the simultaneous observation of both at room temperature, would be the following:

$$CH_2(COOH)_2 \; : \; 0.08 \; M \qquad NaBrO_3 \; : \; 0.3 \; M$$
$$H_2SO_4 \qquad : \; 0.1 \; M \qquad ferroin \; : \; 0.004 \; M$$

The redox couple ferroin (red)/ferriin (blue) gives a far better visual contrast than the couple Ce^{3+} (colourless)/Ce^{4+} (yellow), thus lending itself better to demonstrations.

TIME-DEPENDENT BEHAVIOUR

This is a series of phenomena, which are, in effect, relatively well known in various fields, but for which it appeared for a long time that chemical reactions could not be the source. Besides the general indications given above, their complete elucidation, in terms of detailed reaction mechanisms, remains either a matter of current research or, nearly always, of conjecture. We will thus content ourselves with the list below, recalling their most noteworthy characteristics.

(i) Periodic Oscillations

The chemical oscillator is nowadays a typical oscillator. A respectable number of them are now known, and this is continually on the increase. It constitutes a dissipative oscillator (i.e. limit cycle), whose motion must be maintained by a constant supply of fresh reactants. Its amplitude and period are extremely sensitive to the conditions. An amusing application of this is the setting up of a real chemical clock (26). The observed oscillations usually have a very marked relaxational character. They frequently comprise a single arch. There is, however, also a series that results from the periodic repetition of a basic motif comprising several (occasionally more than a dozen) arches of different shapes (3).

(ii) Quasi-Periodic Oscillations

These are oscillators that bring into play several fundamental frequencies, which have no common denominator. Only biperiodic regimes have so far been identified with reasonable confidence (i.e. two basic frequencies (23)).

(iii) Chaotic Oscillations

These are characterized by an apparent incoherence of the observed (amplitude and period) variations. It has, nevertheless, been unequivocally established that such behaviour is by no means random, and is not the product of superposition of a multitude of mutually unrelated causes. On the contrary, the evolution of the chemical system is rigorously determined and involves only a limited number of degrees of freedom (i.e. deterministic chaos in the sense of the theory of dynamic dissipative systems). This provides one of the most elegant examples known of "weak turbulence", illustrating the concept of the strange attractor in a striking manner, and more precisely, the property refered to as "sensitivity to initial conditions" (27).

A rather odd phenomenon must be added to these "directly" time-dependent patterns of behaviour. It is associated with the coexistence of several dynamic states or regimes, stable or otherwise.

(iv) Excitability

This phenomenon, which is well-known in the living world, corresponds to the substantial temporary amplification of an initial perturbation when this exceeds a critical threshold value (e.g. propagation of signals along nerves). The transient behaviour that subsequently returns the system to its original stationary state is almost independent of the applied perturbation. For an initial departure below the threshold, we observe, on the other hand, an ordinary relaxation, with a monotonous decrease with respect to time (28). This process occurs when two fixed points, namely a stable node and a saddle point, are sufficiently close.

(v) Bistationarity

This is the simultaneous existence, for the same given set of conditions, of two different stable stationary states. The selection of the state actually occupied at a given time is a consequence of the system past history. By appropriate variation of a parameter (or, possibly, by a sufficiently intense perturbation), the transition from one state to the other can be triggered. Since the value of the parameter for which this occurs depends on the direction of the applied variation, a classical hysteresis phenomenon occurs, which is of the same type as that observed in mechanics or magnetism. In addition, near the boundaries of the bistationary domain, the approach to the stationary state gets slower and slower: the evolution is then subject to the so-called "critical slowing down".

(vi) Birhythmicity

This situation has many similar aspects to the previous one, except that the stable states concerned are periodic regimes in this case. In other words, for the same set of conditions, two distinct oscillations (w.r.t. amplitude, period and the form of the oscillations) are manifest. Here again, the choice of a periodic regime is a consequence of past evolution, and the change is accompanied by hysteresis (29). The number of examples of this rather peculiar situation is, admitedly, quite restricted so far.

(vii) *Multistability*

Some cases of coexistence of a stationary state and a periodic regime have of course been observed (both being stable simultaneously). Transition from the former to the latter is called "hard excitation", and is triggered by a perturbation. An example of tristability, involving two stationary states and a periodic regime, has even been discovered for the Briggs-Rauscher reaction (30), clearly showing the potentially rich source provided by chemical dynamics.

SPATIAL SELF-ORGANIZATION

In the mid 1960s, Zhabotinsky noticed that the reaction discovered by Belousov also gave rise to a spontaneous self-organization of space (31). Since then, there have been many other observations. The chemical reactions used and the conditions prevailing therein are very varied. The forms obtained no less so, as shown by the four photos presented here. Even, this is only a very restricted sample of a far richer reality.

Most experiments carried out to date concern the structures that appear temporarily during the evolution towards equilibrium of the chemical system (necessarily uniform). Efforts are, nevertheless, being made to develop experimental conditions to ensure that they persist, at least in some cases. This applies to the creation of structures in a system maintained out of equilibrium by light irradiation (32-34). Other methods are currently being explored (6,7).

The creation and existence of spatially coherent patterns on a macroscopic scale (substantially larger than the range of molecular interactions) brings up a number of fundamental problems, many remaining to be clearly explained. What is the origin of symmetry-breaking in the medium? What "driving force" maintains these systems non-equilibrium "dissipative" structures? Should one read into one or other (and which one) a prototype to the process of morphogenesis, such as Turing envisaged and suggested as a possibility (35)?

These structures bring into play, in proportions that vary a great deal depending on the case involved, three basic ingredients: the chemical reaction, diffusion (of matter and/or heat) and convection. For reasons of convenience in the presentation, we describe separately below the two archetypal situations([4]) constituted by chemico-convective structures and chemical waves (of reaction-diffusion). We should not, however, lose sight of the fact that the physical reality is a continuum, the details of which are hardly rendered by this dichotomy.

Chemico-convective Structures

They are observed in a large number of media, since they do not necessarily require the intervention of reaction with strongly non-linear dynamics: a monotonous reaction can quite readily be used. They present a wide variety of morphology, closely linked to the experimental conditions concerned. Their most remarkable characteristic is their immobility (on the

(a) Example of a photochemical chemico-convective structure (5mm layer of saturated dithizonate solution in toluene, irradiated by visible light. Courtesy of Dr. J.C. Micheau

(b) Example of mosaïc structure in the B.Z. reaction (3,5mm layer sandwiched between two plates. CNRS-CRPP photo)

(c) Example of 2D Chemical waves of the B.Z. reaction leading
    to target formation (1mm layer, sandwiched between two
    rigid plates. CNRS-CRPP photo).

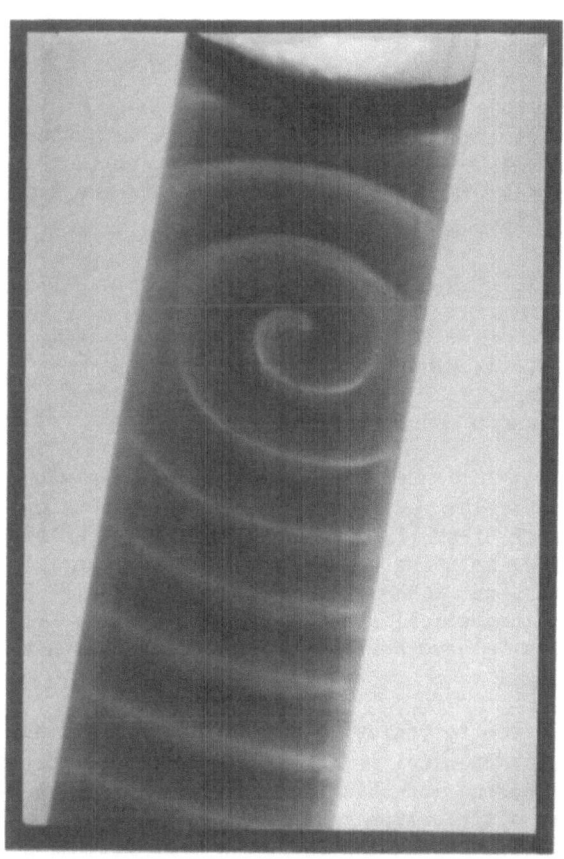

(d)
Example of a 3D chemical
wave of the B.Z. reaction
leading to helix formation
(8mm diameter test tube.
Courtesy of Pr.A.E. Burgess)

time scales and laboratory dimensions available): they nucleate, develop and dissappear in the same place in the reacting medium's container (photo (a) and (b)).

When we operate with a free surface, the set of qualitative or quantitative results available to date indicates that the structures observed are hydrodynamic in origin. Convection is triggered by Bénard and Marangoni effects (buoyancy force and surface tension force, respectively); the chemical or photochemical reaction serves to reveal the convection motion (36). Inhomogeneity of composition or temperature of the medium, surface evaporation of the solvent, endothermic reactions at the surface constitute examples of possible causes of instability.

When there is no free surface, this range is substantially reduced. Nevertheless, we still observe the formation of structures (37). Since we can no longer invoke an interfacial instability in this case, since a Bénard instability (due to a vertical gradient of temperature or composition) appears less probable, it is thus wise to envisage (38) the intervention of a double diffusion process (39). In any case, we thus enter the domain of conjecture.

Although the chemical reactions presented above are not a necessary element to the triggering of this type of structure formation, they are in no way precluded from playing a part. Several times in the past, structures of the type refered to as "mosaïcs" have been observed in an excitable or oscillatory B.Z. type medium (31,40,41). The presence of convection currents, which had been strongly suspected, has now been unequivocally established (42), although their origin has not yet been elucidated.

Chemical waves

By "Chemical waves", we mean a mode of propagation of difference(s) in concentration(s) in which a chemical reaction and the transport of matter by diffusion take part. In a more mathematical statement, we state that a chemical wave is a solution written $C(r,t)$ - which depends on space and time[5] - of the reaction-diffusion equation:

$$\frac{\partial C}{\partial t} = F(C) + D \, \Delta C \qquad [2]$$

C   : concentration vector
F(C) : rate function of the chemical reaction
D   : matrix of diffusion coefficients
ΔC : laplacian of C in the space under consideration.

The system [2] is nothing more than a local balance of matter, written with the hypothesis that Fick's second law appropriately takes diffusion transport into account. As distinct from [1], we have here a system of partial differential equations, whose solution poses a far more substantial mathematical problem (43). The dimension of physical space considered is of prime importance, as much from a theoretical standpoint as in experiment, even when we make the hypothesis, as we have here, that convection does not intervene (see photos (c) and (d)).

In order for a chemical wave to exist, the involvement of an excitable, bistable or oscillating[6] chemical reaction is a necessary (but no sufficient) requirement. The reason for this is straightforward: the instability of a spatially uniform distribution cannot, in fact, stem from a source other than the chemical reaction itself, because diffusion otherwise tends to return the medium to a homogeneous composition. A few experiments have been carried out in one-dimensional (one dimension being much longer than the other two, such as a capillary) or in three-dimensional geometry (i.e. a cell or test tube). Nevertheless, a large number of results available come from studies carried out in "bi-dimensional" geometry (one dimension much smaller than the other two: a Pétri dish). Above all, the vast majority of these studies concern the B.Z. reaction catalyzed by the ferroin/ferrin redox couple (an iron complex of 1-10-tris-phenanthroline, which is red in a reducing environment Fe^{2+} and blue when oxidized Fe^{3+}). Studies concerning other reactions (e.g. $ClO_2^-/I^-/CH_2(COOH)_2$ (44), IO_3^-/AsO_3^{3-} (45)) have received far less attention in comparison. We will therefore only refer to the most notable observations of 2-D chemical waves of the B.Z. reaction in what follows.

At the concentrations usually used, we obtain blue oxidized fronts which propagate in a red reduced medium. The opposite colour scheme only appears at lower concentrations of malonic acid. Such a front propagates radially from a "centre" in the medium, forming a ring whose radius increases continuously with time. Each centre thus emits a series of fronts in a roughly periodic manner; the geometric pattern which results is quite characteristic and is known as a "target" (see photo (c)). All the fronts move at essentially the same speed (48,49). We observe, however, a dispersion effect at the highest frequencies (50). At a given temperature, this speed depends mainly on the proportion of bromate and sulphuric acid in the medium (3,51); at room temperature, a typical order of magnitude is 100 $\mu m.s^{-1}$, that is 6 mm/min. When two oxidation fronts emitted by neighbouring centres meet, they dissappear, because the concentration gradients cancel (i.e. they are equal and opposite). In an oscillating medium, each time the bulk oscillation passes the oxidized state, we similarly observe complete annihilation of the outermost ring of each target. The rate of growth of the target is, in this case, proportional to the difference between the period of emission of the centre and the oscillation period of the medium. Experience shows that the emission frequency of the fronts varies from one centre to another (49). The velocity of propagation being nearly identical for all fronts, the target formed by the centre with the highest emission frequency gradually absorbs the others and ends up invading the entire reacting medium.

With an appropriate experimental procedure, it is possible to obtain spirals rather than rings. Hence, the rupture of an oxidation front sometimes leads to the formation of a pair of spirals, with their (spiral) arms in opposite directions. A rotating wave is thus substituted for the plane wave. Thanks to a judicious manipulation, we can even manufacture - this being the appropriate word - spirals with two, three, or four branches (52); but in is case, we are a long way from a spontaneous phenomenon.

By combining the system [2] with a plausible mechanism for the B.Z. reaction, it is possible to describe in a fairly satisfactory manner the various aspects of plane wave (53,54) or rotating wave propagation (55). Conversely, what really happens at the centre of a target or a spiral is still poorly understood, and remains a controversial subject. Certain authors consider that a "catalytic" particle (e.g. dust, surface defect) is necessary to locally modify the chemical kinetics: because of the extreme sensitivity associated with this reaction (see above), it amplifies this perturbation, hence serving as an indicator of the singularity in the boundary conditions that would otherwise go unnoticed. Others suggest that amongst all the centres there must, nevertheless, be some originating from a fluctuation-nucleation process, which are therefore endogeneous. With this hypothesis in mind, stochastic theoretical analysis leads us to expect a statistical correlation between the properties of the targets and those of the medium (56). Experimental research is at present limited to a single trial (49), which does not permit these two points of view to be distinguished, in particular because there is too great a discrepancy between the conditions accessible to the experimentalist and the approximations to which the theoretician inevitably has recourse in order to complete his calculations.

Over the past two years, microscopes have been used in observations, which have enabled the materials and procedures of automatic image processing to be marketed. Without having supplied a complete answer, these methods have provided a body of quantitative data which must now be taken into account in theoretical considerations. Hence, it has been ascertained that the heart of a spiral is a disc whose diameter is of the order of 0.35 mm, inside which the centre (i.e. the region where no variation in the absorption of light is detected) occupies a zone of 30 μm in diameter (57). Initially, the oxidation front emitted by the centre of a target has virtually zero amplitude. The latter increases linearly with the separation of the front from the centre, until a plateau is reached at about 1 mm from the centre (58). In addition, when the oxidation phase is reached, a homogeneous disc develops around the centre, until its radius is about 0.5 mm. Only then does an individual front as such become detached (59). This occurs regardless of whether a microscopically visible particle is present in the centre or not. It is possible that this process be the combined result of growth (as yet unexplained) of the front's amplitud the variation of its radius of curvature; but it is still too early to assert this. Whatever the details, the necessity of developing a global interpretation of the appearance and development of a front cannot be called into question.

CONCLUSION

This article is intentionally limited to the presentation of the facts that direct experimentation on physico-chemical systems has established. It does not consider the theoretical studies, nor the numerical simulations of the behaviour of model systems, although these has an irreplaceable role in the progress of knowledge. The abundance and variety of the results are such that nowadays a complete review constitutes the subject matter for a book rather than an article (1,8,60). Should we require an indication of the progress that has been made, it will suffice to indicate certain conclusions that are now beyond doubt, but which were far from obvious twenty years ago:

(i) the chemical reaction is the source of a vast set of phenomena from which it was considered free for a long time: oscillation, hysteresis, deterministic chaos, critical slowing down, etc.

(ii) the spontaneous structuring of a physico-chemical medium out of equilibrium is a real occurence that does not contravene the second law of thermodynamics. The corollary of a statistical interpretation of entropy leads to maximum disorder only at equilibrium;

(iii) the self-organization in time and/or space is not the prerogative of the living world. Inert matter is also susceptible to present coherent overall behaviour on macroscopic time scales or distances. There is a complete range of such phenomena, extending from disorganized states and equilibrium structures to morphogenesis, and also from stationary states and monotonous evolutions to biological rhythms.

A fundamental physical question remains in suspense: that of the role played by fluctuations in the evolution of a system. In view of the number, the variety and the complexity of theoretical studies devoted to this problem, experiment remains curiously reticent of profound results. Perhaps the extreme sensitivity of chemical reactions with "highly" non-linear dynamics could provide the means of sheding some light on this subject, which is still in need of clarification.

NOTES

(^{1})   And possibly on temperature. In what follows, we only consider cases for which the release or absorption of heat are negligible in practice. When this is not the case, we have to add the heat balance equation to the system [1]. For a detailed discussion, see (8).

(^{2})   A related notion is that of the trace impurity altering, even radically (17), the behaviour of strongly non-linear reactions discussed in this article. These reactions have the nature of being much more sensitive than others to numerous factors.

(^{3})   In particular if "physical" processes also intervene: e.g. diffusion of matter and/or heat, surface tension, convection, etc.. This article has as its principal object the presentation of what is a direct consequence of the chemical reaction itself.

(^{4})   The subject of this article assumes the intervention of a chemical reaction. Similar structures of a purely "physical" nature (Bénard cells, salt fingers, etc.) are not considered here, although they concern basically the same fundamental principles. To avoid any ambiguity, we should finally stress that none of the phenomena described in the rest do not seem to be unequivocal consequences of structural notions (morphogenetic) according to Turing's definition: the inhomogeneity (stable, durable and invariant) of the spatial distribution of the constituents, as a result of the coupling of a chemical reaction and the diffusion of matter. This is a problem still open to discussion (61), in particular the experimental aspect.

(^{5})   The time phase shift of a time-dependent solution leads to the appearance of a "pseudo-wave". We leave aside this special case (which formally corresponds to $D = 0$ in [2]), although it was originally the source of certain difficulties in interpreting certain observations (46).

(^{6})   In an oscillating medium, we should, apart from everything else, distinguish between two types of wave: trigger waves - the only ones considered in this article - and phase diffusion waves. Although their existence was predicted some time ago, the latter have only recently been produced artificially and identified (47).

REFERENCES

(1) G. NICOLIS, I. PRIGOGINE, Self-organization in nonequilibrium systems. Wiley, New York, 1977.

(2) Oscillations and traveling waves in chemical systems, R. FIELD, M. BURGER, ed., Wiley, New York, 1985.

(3) C. VIDAL, P. HANUSSE, Int. Rev. Phys. Chem., 1986, 5, 1

(4) A. PACAULT, P. DE KEPPER, P. HANUSSE, C.R. Acad. Sci. Paris, 1975, B 280, 157
A. PACAULT, P. HANUSSE, P. DE KEPPER, C. VIDAL, J. BOISSONADE, Acc. Chem. Res., 1976, 9, 438

(5) K. BAR-ELI, W. GEISELER, J. Phys. Chem., 1981, 85, 3461
I. STUCHL, M. MAREK, J. Chem. Phys., 1982, 77, 2956

(6) H. SWINNEY, J. BOISSONADE, P. DE KEPPER, J.C. ROUX, Communication privée.

(7) Z. NOSZTICZIUS, W. HORSTHEMKE, W.D. McCORMICK, H. SWINNEY, soumis à Nature

(8) C. VIDAL, H. LEMARCHAND, La réaction créatrice : dynamique des systèmes chimiques. Hermann, Paris, 1987.

(9) P. BERGE, Y. POMEAU, C. VIDAL, L'ordre dans le chaos : vers une approche déterministe de la turbulence. Hermann, Paris, 1984.
Order within chaos : towards a deterministic approach to turbulence. Wiley, New York, 1986.

(10) P. REHMUS, J. ROSS, Dans "Oscillations and traveling waves in chemical systems". Wiley, New York, 1985, 287-332

(11) K. NAKAJIMA, Y. SAWADA, J. Chem. Phys., 1980, 72, 2231

(12) M.F. CROWLEY, R.J. FIELD, J. Phys. Chem., 1986, 90, 1907
J.L. HUDSON, P. LAMBA, J.C. MANKIN, J. Phys. Chem., 1986, 90, 3430

(13) M.G. ROELOFS, J.H. JENSEN, J. Phys. Chem., 1987, 91, 3380

(14) J.C. ROUX, H. SAADAOUI, P. DE KEPPER, J. BOISSONADE, Phys. Lett., 1983, 97A, 168
M. MENZINGER, M. BOUKALOUCH, P. DE KEPPER, J. BOISSONADE, J.C. ROUX, H. SAADAOUI, J. Phys. Chem., 1986, 90, 313

(15) I. NAGYPAL, I.R. EPSTEIN, J. Phys. Chem., 1986, 90, 6285

(16) M. MENZINGER, P. JANKOWSKI, J. Phys. Chem., 1986, 90, 1217

(17) K.G. COFFMAN, W.D. McCORMICK, Z. NOSZTICZIUS, R.H. SIMOYI, H.L. SWINNEY, J. Chem. Phys., 1987, 86, 119

(18) J.J. TYSON, J. Chem. Phys., 1975, 62, 1010

(19) B.P. BELOUSOV, Sbornik referatov po Radiatsonni Meditsine 1958. Medgiz, Moscou, 1959, 145

(20) W.C. BRAY, J. Am. Chem. Soc., 1921, 43, 1262

(21) T.S. BRIGGS, W.S. RAUSCHER, J. Chem. Educ., 1973, 50, 496

(22) J. BOISSONADE, P. DE KEPPER, J. Phys. Chem., 1980, 84, 501

(23) F. ARGOUL, A. ARNEODO, P. RICHETTI, J.C. ROUX, J. Chem. Phys., 1987, 86, 3325

(24) V. GASPAR, P. GALAMBOSI, J. Phys. Chem., 1986, 90, 2222

(25) A.T. WINFREE, J. Chem. Educ., 1984, 61, 661

(26) A. PACAULT, P. DE KEPPER, P. HANUSSE, C.R. Acad. Sci. Paris, 1975, B 280, 157

(27) C. VIDAL, Le chaos dans la réaction de Belousov-Zhabotinsky, dans "Le chaos : théories et expériences" (P. Bergé, éd.). Masson, Paris, 1987.

(28) P. DE KEPPER, C.R. Acad. Sci. Paris, 1976, C 283, 25

(29) R. KHALIFEH, thèse Université de Bordeaux I, 1986.

(30) P. DE KEPPER, A. PACAULT, C.R. Acad. Sci. Paris, 1978, C 286, 5

(31) A.M. ZHABOTINSKY, A.N. ZAIKIN, J. Theor. Biol., 1973, 40, 45

(32) P. MOCKEL, Naturwis., 1977, 64, 224

(33) M. KAGAN, A. LEVI, D. AVNIR, Naturwis., 1982, 69, 548

(34) M. GIMENEZ, J.C. MICHEAU, Naturwis., 1983, 70, 90

(35) A.M. TURING, Phil. Trans. R.S. London, 1952, 237 B, 37

(36) J.C. MICHEAU, M. GIMENEZ, P. BORCKMANS, G. DEWEL, Nature, 1983, 305, 43

(37) D. AVNIR, M. KAGAN, Nature, 1984, 307, 717

(38) P. BORCKMANS, G. DEWEL, D. WALGRAEF, Y. KATAYAMA, J. Stat. Phys., sous presse

(39) J.K. PLATTEN, J.C. LEGROS, Convection in liquids, Springer-Verlag, Heidelberg, 1984.

(40) K. SHOWALTER, J. Chem. Phys., 1980, 73, 3735

(41) M. ORBAN, J. Am. Chem. Soc., 1980, 102, 4311

(42) J. RODRIGUEZ, C. VIDAL, V. PEREZ-VILLAR, in preparation

(43) P.C. FIFE, Mathematical aspects of reacting and diffusing systems, Springer-Verlag, Heidelberg, 1979.

(44) P. DE KEPPER, I.R. EPSTEIN, K. KUSTIN, J. Am. Chem. Soc., 1981, 103, 2133

(45) K. SHOWALTER, J. Phys. Chem., 1981, 85, 440
J. HARRISON, K. SHOWALTER, J. Phys. Chem., 1986, 90, 225

(46) A. PACAULT, C. VIDAL, J. Chim. Phys., 1982, 79, 691

(47) J.M. BODET, J. ROSS, C. VIDAL, J. Chem. Phys., 1987, 86, 4418

(48) D.M. WOOD, J. ROSS, J. Chem. Phys., 1985, 82, 1924

(49) C. VIDAL, A. PAGOLA, J.M. BODET, P. HANUSSE, E. BASTARDIE, J. Phys., 1986, 47, 1999

(50) A. PAGOLA, J. ROSS, C. VIDAL, J. Phys. Chem., 1988, 92, 163

(51) L. KUNHERT, H.J. KRUG, J. Phys. Chem., 1987, 91, 730

(52) K.I. AGLADZE, V.I. KRINSKY, Nature, 1982, 296, 424

(53) J.J. TYSON, P. FIFE, J. Chem. Phys., 1980, 73, 2224

(54) E.J. REUSSER, R.J. FIELD, J. Am. Chem. Soc., 1979, 101, 1063

(55) J.P. KEENER, J.J. TYSON, Physica, 1986, 21D, 307
J.D. DOCKERY, J.P. KEENER, J.J. TYSON, Physica D., in press

(56) D. WALGRAEF, G. DEWEL, P. BORCKMANS, J. Chem. Phys., 1983, 78, 3043

(57) S.C. MULLER, T. PLESSER, B. HESS, Science, 1985, 230, 661 ;
Naturwis., 1986, 73, 165 ; Physica, 1987, 24D, 71, ibid. 87

(58) A. PAGOLA, C. VIDAL, J. Phys. Chem., 1987, 91, 501
C. VIDAL, J. Stat. Phys., 1987, 48, 1017

(59) C. VIDAL, A. PAGOLA, submitted to J. Phys. Chem.

(60) A. BABLOYANTZ, Dynamics, molecules and life. Wiley, New York, 1986.

(61) J. BOISSONADE, soumis à J. Chem. Phys.
G. DEWEL, D. WALGRAEF, P. BORCKMANS, J. Chim. Phys., 1987, 84, 1335

(62) C. VIDAL, J.C. ROUX, A. ROSSI, J. Am. Chem. Soc., 1980, 102, 1241

SPECTRAL KINETICS: STUDY OF

COMPLEX REACTIONS BY EXTERNAL PERTURBATIONS*

John Ross

Chemistry Department
Stanford University
Stanford, CA 94301 U.S.A.

*This article is an extended abstract of a lecture
presented at the 42nd International Meeting of the Societe
de Francaise de Chimie, Division de Chemie Physique entitled
"Chemical Reactivity in Liquids - Fundamental Aspects." This
work was supported in part by the National Science Foundation,
the Air Force of Scientific Research and the Department of
Energy/BES Engineering Program.

The study of the response of nonlinear systems to
external periodic perturbations leads to interesting
information.[1-7] Cool-flame[8,9] oscillations occur in a number
of combustion reactions, and we discuss an experimental study
of the effect of external periodic perturbations on such
systems. The application of perturbations to a chemical
reaction can reveal important information about the stability,
kinetics, and dynamics of the reaction. This technique is
well known in the field of relaxation kinetics, in which
perturbations are applied to a chemical system at equilibrium.
In our work, periodic perturbations are first applied to the
input rates of acetaldehyde and oxygen, one at a time, in
the combustion of acetaldehyde in a CSTR. We measure periodic
responses in five entrainment bands as we vary the frequency
and amplitude of the external periodic perturbation. Outside
of entrainment bands we find quasi-periodic responses. Next-
phase maps[10,11] of the experimental results are constructed
in real time and used in the observation and interpretation
of entrainment and quasi-periodic behavior. Within the
fundamental entrainment band, we measure critical slowing
down and enhancement of the response amplitude. As the bath
temperature is increased, so that the oscillatory system
approaches a Hopf bifurcation, we observe an increase in the
amplitude enhancement. The predictions of a five-variable
thermokinetic model agree well with the experimental results.

Measurements of entrainment bands and similar dynamic
features are examples of what we call spectral kinetics, the
response of a complex kinetic system far from equilibrium
to external periodic perturbations. There is an interesting
analogy with optical spectroscopy. The structure of a

polyatomic molecule is studied by external periodic perturbation of light and in an absorption band the frequency of molecule has an integral relation to the frequency of light. From the measurement of absorption lines we get much information on the structure of polyatomic molecules. In principle an inversion procedure may be possible in which the structure is not guessed but may be deduced from optical measurements. In complex reaction kinetic systems the chemical species are the analogs of the atoms in the poly-atomic molecule. The reaction mechanism is the analog of chemical bonds and the rate coefficients are to be compared with the force constants. In the case of chemical kinetic systems we may determine the phase relations among the oscillations of the various species and perturbations, and in addition we may perturb one specie or constraint at a time. Hence there is the hope of obtaining much detailed information about chemical reaction mechanisms from the study of the response of reaction systems to external periodic perturbations.

Periodic perturbations are also applied to the input rates of acetaldehyde and oxygen simultaneously in the combustion of acetaldehyde in a CSTR. With the two perturbations at the same frequency, we measure bistability and hysteresis as a function of the phase shift between the two perturbations. The application of a perturbation in the flowrate of one reactant to the system already entrained to a perturbation of the flowrate of the second reactant can cause the system: to become quasi-periodic in both perturbations; to become entrained to both perturbations; to remain entrained, but not phase-locked, to the first perturbation; or to become quasi-periodic in the first perturbation but entrained to the second perturbation. We measure the effects of frequency-modulated and amplitude-modulated acetaldehyde flowrate perturbations; again the results compare well with predictions by a five-variable thermokinetic model.[12]

In the study of the response of nonlinear systems to external periodic perturbations there exists a dual search, that for universal relations and that for responses specific to a particular reaction mechanism. System mathematicians are, of course, intrigued by commonalities and universal relations. As an example, the similarities of alkali atoms and irons are of course remarkable. However, the chemists and biologists must also face the task of differences in the behavior of the sequence in the periodic table. Lithium carbonate controls manic depressive illness effectively, whereas the other alkali carbonates do not (nor do other alkali salts other than lithium salts). We have the same duality of interest in complex reaction mechanisms. Bifur-cations, limit cycles, critical slowing down, occur in many nonlinear systems and have common features and universal laws. To the extent that these hold we find out little about the specific reaction mechanism of a given system and we seek properties which are specific to such reaction mechanisms.

REFERENCES

1. Tomita, K., Kai, T., and Hikami, F., Prog. Theo. Phys. 57, 1159 (1977); Kai, T., and Tomita, K., Prog. Theor. Phys. 61, 54 (1979).

2. Richter, P.H. and Ross, J., J. Chem. Phys. 69, 5521 (1978); Richter, P.H., Physica 10D, 353 (1984).

3. Rehmus, P., and Ross, J., J. Chem. Phys. 78, 3747 (1983); P. Rehmus and Ross, J., in "Oscillations and Traveling Waves," edited by R.J. Field and M. Burger (Wiley, New York, 1985, p 287; Rehmus, P., Vance, W. and Ross, J., J. Chem. Phys. 80, 3373 (1984).

4. Guevara, M.R., and Glass, L., J. Math Biol. 14, 1 (1982).

5. Boiteux, A., Goldbeter, A., and Hess, B., Proc. Natl. Sci. U.S.A. 72, 3829 (1975); Markus, M., Muller, S.C. and Hess, B., Ber. Bunsenges. Phys. Chem. 89, 651 (1985); Markus, M., and Hess, B., Proc. Natl. Acad. Sci. U.S.A. 81, 4394 (1985); Marksu, M., Kuschmitz, D., and Hess, B., FEBS Lett. 172, 235 (1984).

6. Dolnik, M., Schreiber, I., and Marek, M., Phys. Lett. A100, 316 (1984); Taylor, T.W., and Geiseler, W., Ber. Buns. Phys. Chem. 89, 441 (1985); F.W. Schneider, Ann. Rev. Phys. Chem., 36, 347 (1985).

7. Bucholtz, F. and Schneider, F.W., J. Am. Chem. Soc. 105, 7450 (1983).

8. Felton, P.G., Gray, B.F., and Shank, N., "Second European Symposium on Combustion" (The Combustion Institute, Orleans, France, 1975); Gray, P., Griffiths, J.F., Hasko, S.M., and Lignola, P.-G., Combust. Flame 43, 174 (1981); Gray, P., Griffiths, J.F., Hasko, S.M., and Lignola, P.-G., Proc. R. Soc. London Ser. A374, 313 (1981); Gray, B.F. and Jones, J.C., Combust Flame 57, 3 (1984).

9. Pugh, S.A., Kim, H.-R., and Ross, J., J. Chem. Phys. 86, 776 (1987); Pugh, S.A. and Ross, J., J. Phys. Chem. 91, 2178 (1987).

10. Pugh, S.A., Schell, M. and Ross, J., J. Chem. Phys, 85, 868 (1986); Pugh. S.A., Ph.D. thesis, Stanford University, Stanford, CA 1985.

11. Guevera, M.R. Glass, L., and Shrier, A., Science 214, 1350 (1981); Glass, L., and Perez, R., Phys. Rev. Lett. 48, 1772 (1982).

12. Pugh, S.A., Dekock, B., and Ross, J., J. Chem. Phys. 85, 879 (1986).

DISCUSSION

WILSON - Do you think that the biological oscillatory systems can be explained on the basis of the non-linear phenomena seen in simpler chemical systems, or will other phenomena have to be involved ?

ROSS - Over 150 oscillatory biological systems (in vivo and many in vitro) have been reported so far. A few have been studied in detail and a chemical, non-linear analysis suffices to reproduce the observed oscillations, bistabilities, chaos, etc.

NICOLIS - 1) In the 1.1 entrainment band, analytic calculations show that both hysteresis and infinite period bifurcations are possible even in the presence of a single periodic perturbation. Is there experimental evidence of these effects in your system ?

2) Have you any experimental evidence of chaos and devil staircase in your system ?

ROSS - 1) Not yet.

2) There is some evidence of chaos (not yet solidly confirmed) but no evidence of a devil's staircase.

MICHEAU - I consider, for example, a chemical open reactor undergoing some reaction (oscillating or not, bistable, and so on...). Can it be hoped that the global yield of the reaction will be increased by pulsating the reactants ?

ROSS - Only in the case of parallel reactions, that is more than one product channel for a given set of reactants ; then external periodic perturbations can enhance the yield of one channel over the other.

ERDI - May I give a report on a specific periodically perturbed oscillatory system taking place in the nervous system. The transmitter recycling hypothesis describe the oscillatory behaviour of the concentration of acetylcholine. Coupling between the periodic uptake of choline and the ACh oscillation can lead to complex temporal patterns. There is a conjecture that the system normally operates within the main entrainment region. Even the mild impairment of neurochemical control system due to the failure of metabolism might lead to arythmicity to be associated to certain kinds of neurological disorders (see Erdi and Barno in Lect. Notes in Biomathematics, Vol. 71 (Teramoto, Yamaguti, eds., 1987).

A TECHNIQUE TO DESIGN ISOTHERMAL OSCILLATING

REACTIONS IN AQUEOUS MEDIA

Patrick De Kepper and Qi Ouyang

Centre de Recherche Paul Pascal, Université de Bordeaux I
Domaine universitaire
33405 Talence cédex (France)

INTRODUCTION

Study of liquid-phase homogeneous chemical oscillators has been an ever growing area of chemistry for the past 15 years[1]. This activity was first stimulated from the recognition that oscillating isothermal chemical reactions may be thought as simplified prototype examples for the many rhythmic behaviours found at different levels in biological systems. These reactions were also a chalenge to kineticists. More recently mathematicians and physicists have also focused their attention on these chemical systems. These reactions are then thought as rather flexible experimental system where theoreticians can test their latest conjectures on nonlinear dynamical systems with large number of degree of freedom.

There has not always been such a wide spread interest in these amazing reactions. In the late 60's the small number of oscillating reactions known were still regarded as mere laboratory curiosities, and more often as artifacts resulting from poorly controlled experimental conditions. A misinterpretation of the phenomenon, then communly spread, led to believe that such a behaviour would transgress the second principle of thermodynamics. The number of sceptics droped with the progress in thermodynamics of irreversible processes pioneered by Prigogine and his colleagues[2] showing that chemical reactions can develop autonomous oscillatory behaviour if they have appropriate nonlinear kinetics and evolve far from their thermodynamic equilibrium. The first elaborations of detailed kinetic mechanisms[3,4] accounting for the observed oscillatory dynamics further reinforced the realness of homogeneous oscillating chemical reactions. At that time all reactions of this type were the result either of serendipidious findings or that of more or less significant variants of the accidental discoveries[1].

There was a need for a practical method to produce basically different oscillating reactions. A workable procedure was proposed by one of us in 1980[5]. It makes a systematic use of continuous stirred tank reactors (CSTR) and is based on a theoretical model initially developped to analyze the relationship that may exist between bistability phenomenon and relaxation oscillations[6]. The method turned out to be very efficient and was gradually refined to account for some of the topological distorsions encountered in the complexity of real systems[7]. A tremendous expansion in the number and the variety of chemical oscillators has resulted[8].

In this paper we give a synopsis of the method as it is presently developed. It is examplified by our two latest discoveries: the oscillating oxidation reactions of sulfide and oxalic acid by persulfate.

THE BISTABILITY - OSCILLATION METHOD

There is no known general set of sufficient conditions to produce chemical oscillating reactions but from comprehensive theoretical studies a number of minimal necessary conditions can be determined. As already mentioned the reaction system must evolve far from thermodynamic equilibrium and involve feedback mechanism. Permanent deviation from equilibrium can be obtained by operating the chemical reactions in open flow reactors[9]. Feedback mechanisms can be of many different types, the most interesting are autocatalysis and substrate inhibition. In batch condition such reactions are characterized by a slow induction period followed by a more rapid evolution before reaching equilibrium. From our practical point of view the important thing with these reactions is that their rate go through a maximum as they proceed to equilibrium. It is well documented that if this kind of reaction is performed in a CSTR one may observe multistability phenomena[9,11].

Now it was shown by Boissonade and De Kepper,[6] that intrinsic bistable systems can be changed into oscillatory systems if associated to properly managed reactions which modify the steady state stability range on the initial bistable slow manifold. The goal being to have the resulting overall system undergo periodic rapid transitions at the edge of the fold alternating with slow drifts on the manifold as schematically depicted on figure 1a. The expected oscillatory time trace is shown figure 1b. Such systems

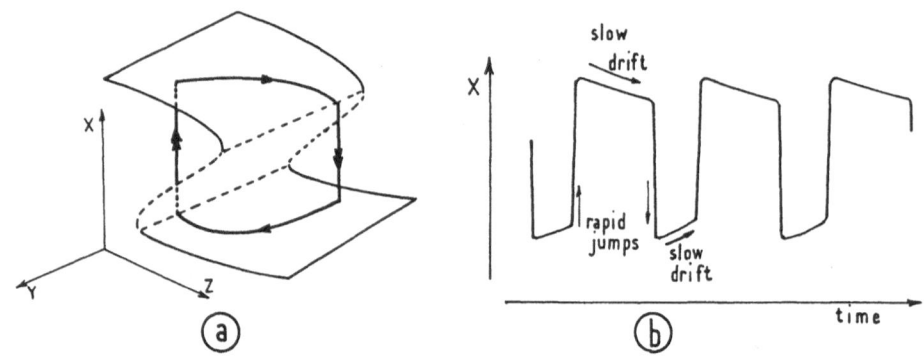

Fig. 1a. Folded manifold
1b. Schematic relaxation oscillations

generally exhibit a characteristic "cross-shaped phase diagram"[6,11] as sketched on figure 2. In this type of diagram autonomous oscillations emerge beyond a region of vanishing bistability and result from the cross-over of the stability limits L_i of the two steady states. A cross point P is only generated if the switching time, from one branch of the initial bistable to the other, is much shorter than the drift time on the branches. If this time scale ratio is not fulfilled the limit bistability point correspond to an ordinary cusp point and no oscillation is observed beyond. In real systems the phase diagram in the neighbourhood of cross point P can be extremely complex but the main topological features presented figure 2 are usually preserved[12].

Bearing the above consideration in mind the synopsis of our search method for new oscillating reaction goes as follows:

- review the literature for reactions showing marked induction period, the old chemical literature is often very instructive, or try to design your own;

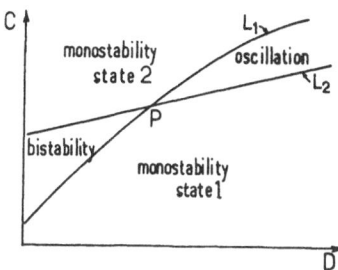

Fig. 2. Schematic cross-shaped phase diagram

- study the reactions in continuous-flow well-stirred tank reactors (CSTR)[9] as a function of the different control parameters (i.e. flow rate k_0, concentration of input reagents $|A|_0$, temperature T,...). The primary goal is to produce bistability, the ultimate is to discover autonomous oscillatory conditions by using the bistability phenomenon as a lead. Grossely three types of constraint-response curves may be obtained when the selected reactions are performed in a CSTR. The evolution of these curves as a function of the other parameters will guide our search by studying plane sections of the constraint diagram as sketchet in the chart, figure 3.

a) Continuous sigmoidal changes between two quite different chemical regimes is the topological situation most communly encountered. Starting with this situation experimental conditions are modified in order to increase the steepness of change between these two regimes and eventually find bistability conditions. In doing so one may, on rare occasions, also directly reach oscillatory conditions which may be linked or not to effective bistability conditions in other regions of the physical phase diagram. We call physical phase diagram the region of phase diagram that can be explored experimentally. The solubility of chemical species is often a severe physical limitation.

b) Bistability is observed. The regions of vanishing bistability must then be systematically explored in plane sections of the different constraints. One either reaches a cusp point or a cross point. Many of the bistable systems discovered up to now have shown to be complex enough to be able to produce oscillatory behaviour without additive reactions[1,12]. If only cusp points are observed the system must be manipulated by adding appropriate competing reactions which would modify the effective value of some species controlling the bistability according to the composition of the system. The systematic exploration of bistability limit point in search for a cross shaped diagram starts again with the modified system.

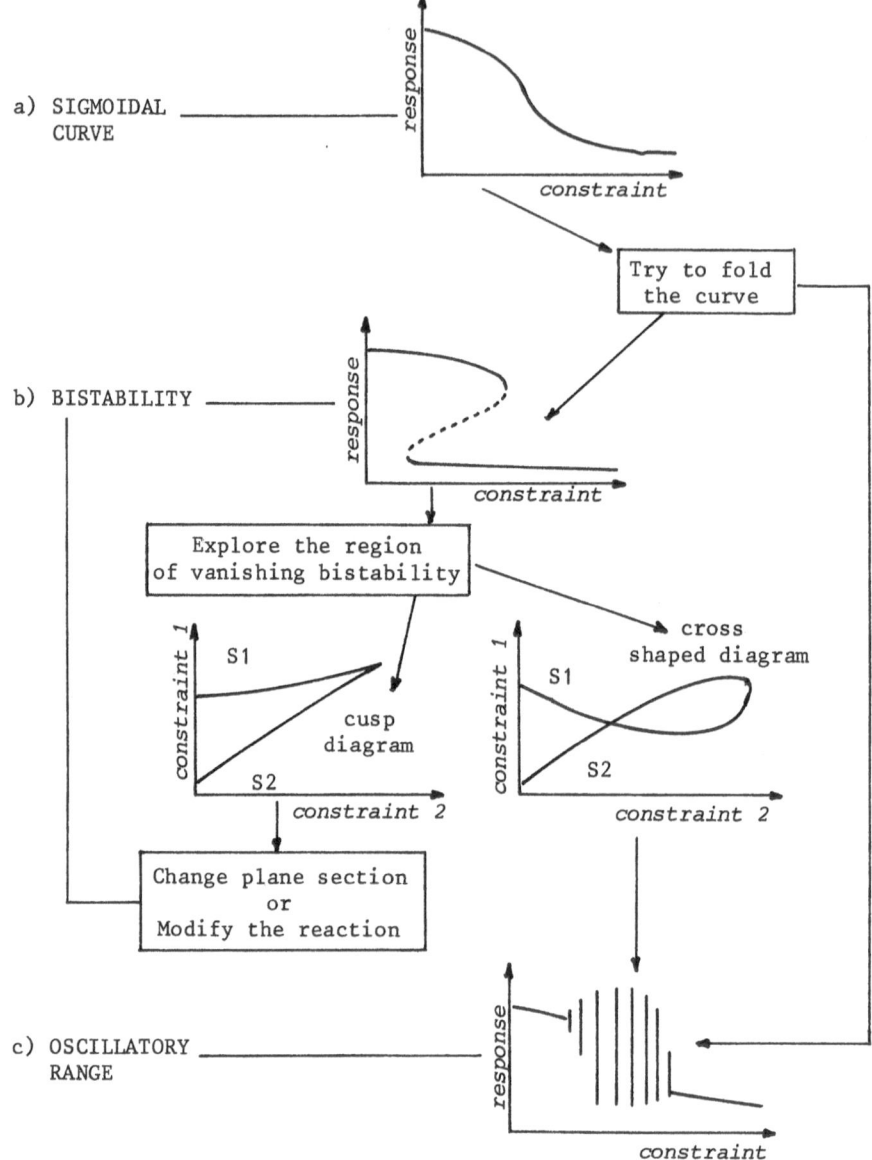

a) SIGMOIDAL CURVE

Try to fold the curve

b) BISTABILITY

Explore the region of vanishing bistability

cusp diagram

cross shaped diagram

Change plane section
or
Modify the reaction

c) OSCILLATORY RANGE

Fig. 3. Synopsis of the bistability-oscillation method

c) A range of oscillatory dynamics is immediately observed. This is a rather lucky situation since oscillatory phenomena are usually observed over relatively small domains of the constraint space. Even in this case, it is of mechanistic interest to find out if the oscillating domain can be linked to a bistability domain and to know which are the principal constraints inducing this dynamical transformation[11].

This method has the virtues of been simple and flexible enough to apply to wide range of real situations. It splits the task of finding new oscillators in more tractable subproblems, the search for bistable systems and the invention of suitable additional feedback reactions. The characteristic cross shape of the expected diagram is the most effective guide for the experimentalist in this task. It is also important to bear in mind that in real systems many distorsions of the phase diagram may occur as a result of complicated paths in the dynamical space of the system as a function of the natural control parameters[7],[12]. Multiple ranges of bistability can thus be found as in the B.Z. reaction[11] or isola phenomenon[13] as in the chlorite-iodate-arsenite system[14].

To the author's knowledge, with the exception of the benzaldehyde autoxidation oscillating reaction and the methylene blue catalyzed oxidation of sulfide by oxygen, all the new chemical oscillators discovered since 1980 are the result of the bistability-oscillation approach. This shows that, if bistability (or pleated slow manifold) is not of basic necessity for oscillatory behaviour, our method which realies on some particular type of relationship between bistability and relaxation oscillations, is presently the most efficient method to produce new chemical oscillators.

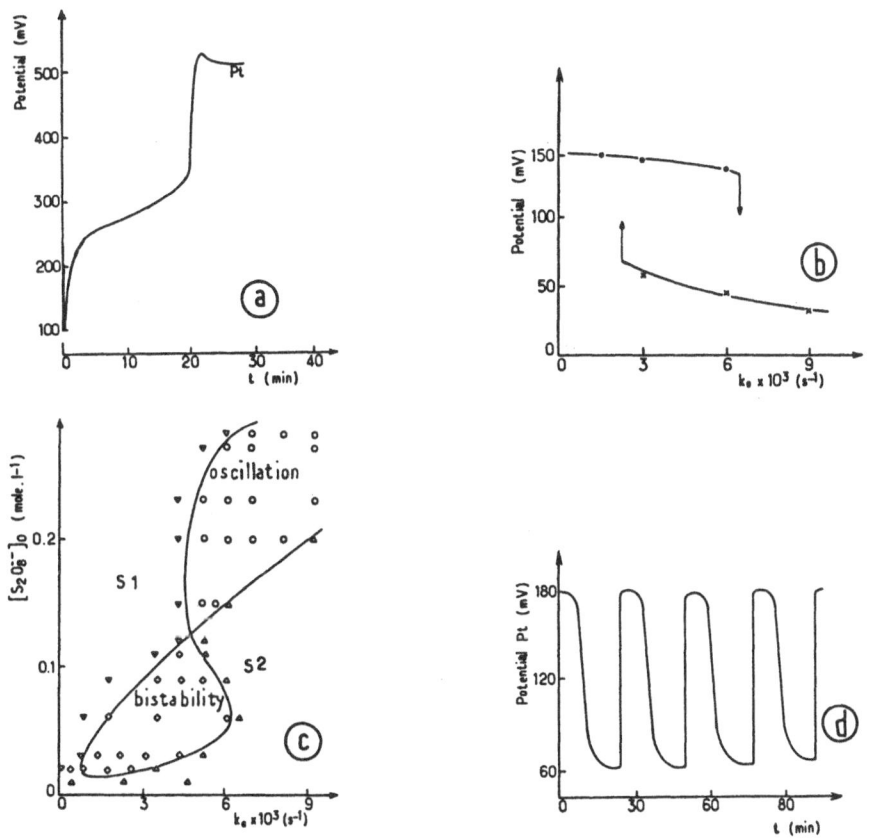

Fig. 4. Sulfide-persulfate-silver ion reaction: a) batch Pt-potenial évolution. Initial concentrations $|S^{2-}|_i = 4.2.10^{-5}$ M ; $|S_2O_8^{2-}|_i = 0.1$ M; $|Ag^+|_i = 3.6.10^{-5}$ M ; T = 25°C.
b) Pt-potential bistability versus flow rate k_0 with $|S^{2-}|_0 = 2.83.10^{-5}$ M ; $|S_2O_8^{2-}|_0 = 0.09$ M ; $|Ag^+|_0 = 2.4.10^{-5}$ M.
c) Plane section of the phase diagram with $|{}_S^{2-}|_0 = 2.83.10^{-5}$ M ; T = 30°C.
d) Typical Pt-potential oscillation trace.

EXPERIMENTAL ILLUSTRATIONS

In this paragraph we briefly present as illustrations of our method the differents steps that lead to the resent discovery of two new chemical oscillating reactions.

The sulfide-persulfate-silver ion reaction

In batch reactor we found that sulfide (S^{2-}) oxidation by persulfate ($S_2O_8^{2-}$) is strongly catalyzed by silver ions (Ag^+) and that over some range of catalyst concentration the reaction showed a marked induction period as illustrated in figure 4a. Figure 4b show the steady potential bistability observed as a function of flow rate k_0 when the same reaction is performed in a CSTR. Repeting the same type of experiments for different $|S_2O_8^{2-}|_0$ values the cross-shaped phase diagram (fig. 4c) is obtained. Figure 4d presents typical Pt-potential oscillations observed beyond the cross point. No cross shaped diagram could be reach by changing $|S^{2-}|$ or $|Ag^+|_0$ starting from the bistability condition of figure 4b[7] .

The oxalic acid-persulfate-silver ion reaction

This reaction had previously been studied in batch reactors by diffe-rent authors[16] and was known to show an induction period. The Pt-potential trace of such an experiment is shown in figure 5a. On introducing this

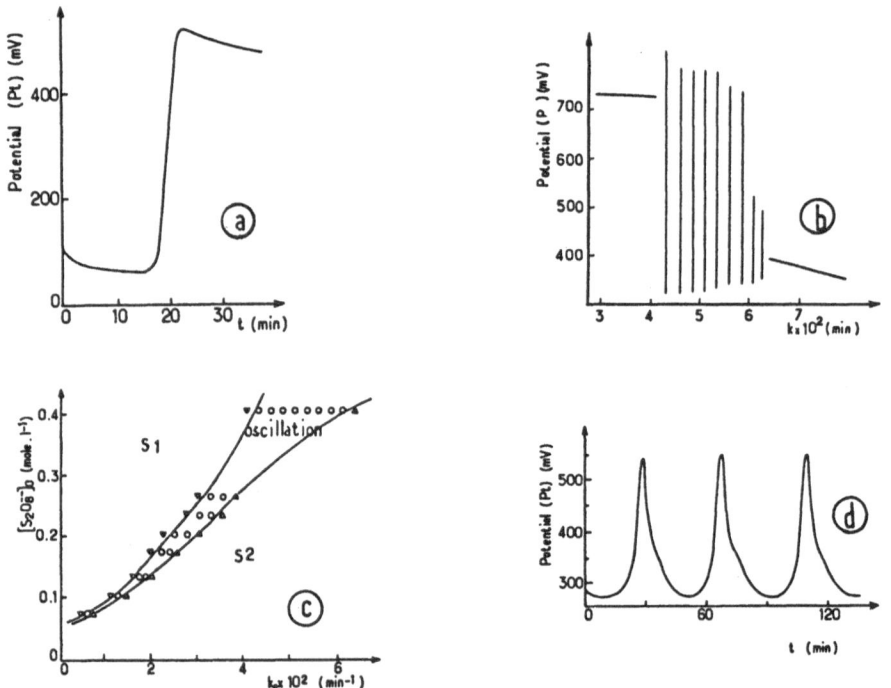

Fig. 5. Oxalic acid-persulfate-silver ion reaction. a) Batch Pt-evolution trace. Initial concentrations $|(COOH)_2|_i = 1.10^{-3}$ M, $|S_2O_8^{2-}|_i = 0.1$ M, $|Ag^+|_i = 4.5\ 10^{-5}$ M, $|H_2SO_4|_i = 0.05$ M, T = 25°C.
b) Steady state evolution and oscillation region (vertical segments) versus flow rate k_0 with $|(COOH)_2|_0 = 1.10^{-3}$ M, $|S_2O_8^{2-}|_i = 0.4$ M, $|Ag^+|_i = 1.10^{-4}$ M, $|H^+| = 0.1$.
c) Plane section of the oscillation domain with $|(COOH)_2|_0 = 1.10^{-3}$M, $|Ag^+|_0 = 1 \times 10^{-4}$ M, $|H^+|_0 = 0.1$.
d) Typical Pt-potential oscillation trace.

reaction in a CSTR oscillatory behaviour was observed over a finite range of flow rate as illustrated in figure 5b; note the steady state potential difference on the two side of the oscillatory domain. Figure 5c provides a section of the oscillation domain in the $(k_0, |S_2O_8^{2-}|_0$ plane and figure 5d presents typical oscillation traces in this region. No bistability has been found for this reaction in the region of control parameter explored[17]. Only very few reactions systems have untill now shown no bistability associated to oscillatory behaviour when performed in a CSTR.

CONCLUSION

In conclusion we would like to emphasize that our method is directly linked to the use of flow reactors and that most new chemical oscillators only oscillate in such reactors. Batch transient oscillatory reactions are now a minority since they call for higher kinetical complexity. In these latter some primary reactions must play roles analog to that of the flow in open reactor such as the supply, at appropriate rate, of active intermediate species and the scavenge of inhibitory end products over time long compared to the period of oscillations. In this respect the behaviour of the B.Z. reaction where more than 500 oscillations can be observed in batch condition is a truly exceptional situation which through of sent the early attempts to produce new homogeneous chemical oscillators in close reactors. Chemical oscillatory behaviour is probably still more ubiquitous than it appears now. There are certainly classes of oscillatory reactions not linked to bistability but there is not yet any practical systematic method to spot these reactions.

REFERENCES

1. "Oscillations and traveling waves in chemical systems", Ed. R.J. Field and M. Burger, Wiley Interscience, New York (1985)
2. P. Glansdorff and I. Prigogine, "Thermodynamic theory of structure, stability and fluctuation", Wiley Interscience, New York (1971)
    G. Nicolis and I. Prigogine, "Self-organization in nonequilibrium systems", Wiley Interscience, New York (1977)
3. R.J. Field, E. Körös and R.M. Noyes, J. Am. Chem. Soc. 94, 8649 (1972)
4. G.J. Kasperek and T.C. Bruice, Inorg. Chem. 10, 382 (1971)
5. P. De Kepper, I.R. Epstein and K. Kustin, J. Am. Chem. Soc. 103, 2133 (1981)
6. J. Boissonade and P. De Kepper, J. Phys. Chem. 84, 501 (1980)
7. Q. Ouyang, thèse Bordeaux (1987)
8. A. Pacault, Q. Ouyang and P. De Kepper, J. Stat. Phys. 48, 1008 (1987)
9. J. Villermaux, "Génie de la réaction chimique", Ed. Tech. et Doc., Lavoisier (1982)
10. K.F. Lin, Can. J. Chem. Eng. 57, 476 (1979) ; Chem. Eng. Sci. 36, 1447 (1981)
11. P. De Kepper and J. Boissonade, chapitre 7 in ref. 1.
12. P. De Kepper, Acta Chem. Com. Hung. (1987) in press
13. P. Gray and S.K. Scott, Chem. Eng. Sci. 39, 1087 (1984)
14. M. Orbán, C.E. Dateo, P. De Kepper and I.R. Epstein, J. Am. Chem. Soc. 104, 5911 (1982)
15. Q. Ouyang and P. De Kepper, J. Phys. Chem. (1987) in press
16. T.L. Allen, J. Am. Chem. Soc. 73, 3589 (1951)
    A.J. Kalb and T.L. Allen, J. Am. Chem. Soc. 86, 5107 (1964)
17. Q. Ouyang and P. De Kepper, J. Am. Chem. Soc., submitted (1987)

DISCUSSION

PRUDHOMME - Vous envisagez des réactions oscillantes en réacteur ouvert. Ce réacteur est-il toujours homogène à mélange parfait ?
Se peut-il qu'une réaction soit oscillante dans un certain type d'écoulement et non oscillante dans un autre ?

DE KEPPER - En première approximation nos réacteurs sont bien agités mais certainement pas parfaitement homogènes (cas idéal impossible à atteindre). On connait effectivement au moins un cas où le caractère oscillant de la réaction dépend entièrement de la façon d'alimenter le réacteur*.

(*) M. Boukalouch, P. De Kepper, J. Boissonade
Proceedings the International Conference : "Spatial Inhomogeneities and Transient Behaviour in Chemical Kinetics", Brussels (1987).

M. Boukalouch, J. Boissonade, P. De Kepper - J. Chim. Phys. (à paraître).

BOURCEANU - Votre diagramme s'applique-t-il uniquement pour les systèmes bistables ?

DE KEPPER - La méthode a été initialement construite sur les systèmes bistables. Ceux-ci offrent des points de départ commodes pour la procédure expérimentale. Toutefois, bien que les dynamiques oscillantes générées par notre méthode soient toujours basées sur un effet de variété lente plissée, le phénomène n'est pas toujours explicitement observé dans l'espace des paramètres de contrôle.

NICOLIS - Est-ce que les oscillations entretenues sont déjà "annoncées" dans le domaine bistable par le fait que les perturbations autour d'un des états stationnaires stables décroissent en effectuant des oscillations amorties ?

DE KEPPER - Il existe effectivement un certain nombre de "comportements annonciateurs" de l'approche d'un domaine d'oscillations entretenues. Lorsque cette approche se fait à partir du domaine bistable, ces événements annonciateurs sont souvent l'excitabilité des états stationnaires dans la région monostable et surtout l'absence de ralentissement prononcé des vitesses de transition entre états stationnaires à l'approche du "sommet" du domaine bistable. Le caractère oscillant amorti est le plus souvent caractéristique d'une approche du domaine oscillant loin du domaine bistable. Toutefois des oscillations amorties de faible amplitudes sont aussi observées dans le domaine bistable mais c'est un cas plus rare ou limité à des domaines de paramètres relativement étroits révélés ultérieurement par des études plus fines. Ceci tient probablement à notre méthode qui favorise l'observation d'oscillations de relaxations et des transitions "dures" par rapport aux oscillations quasi-sinusoïdales caractéristiques du voisinage des transitions "douces".

BARRERE - En système ouvert il y a instabilité si la réaction a lieu en deux temps (une réaction lente et une rapide) donc avec un certain délai τ), ce qui équivaut à l'introduction dans le système d'une ligne retard. Voir les instabilités de basse fréquence dans les foyers de fusée ; on suppose dans la théorie un réacteur isotherme mais la substance, qui réagit au temps (t), est injectée au temps (t-τ). L'équation : $\frac{d\varphi(t)}{dt} = A\varphi(t-\tau) - B\varphi(t)$, possède des solutions périodiques (critère de Nyquist).

DE KEPPER - Dans un réacteur chimique continu parfaitement agité, les réactifs injectés dans le réacteur réagissent immédiatement et proportionnellement à leurs concentrations instantanées dans le réacteur. Toutes les espèces intermédiaires existent déjà dans le mélange réactionnel homogène, il n'y a pas de retard explicite à la réaction. La description de tels réacteurs relève de systèmes d'équations différentielles ordinaires.

L'interprétation que vous me proposez relève plutôt du cas des réacteurs mal mélangés où ce type de retard serait introduit par le temps de mélange des réactifs entrants.

VILLERMAUX - La recherche de nouvelles réactions oscillantes concerne-t-elle seulement les systèmes autonomes (oscillations spontanées) ou peut-on aussi imaginer des systèmes incluant une boucle de rétroaction extérieure (je pense surtout à une réinjection appropriée d'informations de type électrique après traitement du signal électrochimique d'un capteur) ?

DE KEPPER - Nos études ne concernent que le cas des réactions chimiques oscillantes où l'instabilité est due au seul mécanisme cinétique. Des études de systèmes chimiques incluant des boucles de rétroaction extérieures ont été effectuées par d'autres auteurs, notamment par Zhabotinskii[*].

(*) Zhabotinskii A.M., Zaikin A.N. et Rovinskii - React. Kinet. Catal. Lett. 20 (1982).

AMATORE - Could you comment in more details on the origin of the drift of the system on the S-shaped curve. For e.g. on electrochemical bi-stable system oscillating when coupled on a capacitor will you consider it as an oscillating system or as a bistable system forced into oscillations ?

DE KEPPER - The detailed analysis of cross shaped diagrams can be found in two papers :

- J. Boissonade, P. De Kepper, J. Phys. Chem. 84, 501 (1980),
- P. De Kepper, J. Boissonade in "Oscillating and Traveling waves in Chemical Systems" Ed. R.J. Field and M. Burger Wiley (1985) Chap. 7.

I would not call the oscillating system resulting from the coupling of a capacitor to an electrochemical system bistable but an electrochemical oscillator since the oscillations result from an externally imposed delay device.

ANALYSIS OF A MARANGONI EFFECT OF A SOLUTE

S.Bekki**, E.Nakache*, M.Vignes-Adler**, and P.M.Adler**

*Laboratoire de Chimie physique, Paris VI, 13 rue P. et M.
 Curie, 75231 Paris Cedex 05
**Laboratoire d'Aérothermique du CNRS, Route des Gardes
 92190 Meudon

SUMMARY

Interfacial instabilities may occur during the transfer of a surface
active agent between two non-miscible phases. This convection results
from the coupling between physico-chemical and hydrodynamic processes.
The spontaneous stirring enhances significantly mass exchange between the
two media and thus is very efficient for interfacial transfer processes
(liquid-liquid extraction, purification of biochemical compounds..)

In order to get definite limiting conditions we chose to analyse the
motion of a nitroethane drop placed at the surface of an aqueous solution
of a cationic surfactant.

During the tiem elapsed till the partition equilibrium of the surfactant, movements of this lens, apparently chaotic, were observed. They were
visualized by ombroscopy, recorded with a C.C.D. video camera and analysed by image analysis.

Several phases of the motion, corresponding to different states of
the system - stationary, oscillating - are pointed out. The statistical
analysis of the movement and its spectral decomposition allows to precise
the nature of the phenomenon and to quantify it.

ANALYSIS OF A MARANGONI EFFECT OF A SOLUTE

S. Bekki[*+], E. Nakache[*], M.Vignes-Adler[+], and P. M. Adler[+]

[*]Laboratoire de Chimie Physique de Paris VI
75231 Paris, France

[+]Laboratoire d'Aérothermique du CNRS
92190 Meudon, France

1 INTRODUCTION

Les systèmes chimiques initialement loin de l'équilibre peuvent évoluer vers des états oscillants[1,2] si la cinétique de réaction implique des processus à caractère fortement non-linéaire. Dans les systèmes biphasiques, la non-linéarité provient de l'existence même d'interfaces et des effets convectifs associés. Ainsi, au cours du retour à l'équilibre de partage, des instabilités interfaciales peuvent apparaître entre deux phases liquides en déséquilibre de concentration d'un composé tensioactif. Leur caractère oscillant ou chaotique n'est pas sans analogie avec les oscillations ou chaos chimiques.

Ces instabilités sont le résultat d'un couplage entre des effets physicochimiques et hydrodynamiques. Le transfert du tensioactif à travers l'interface engendre des fluctuations de la densité interfaciale de tensioactif ; ces microgradients de tension interfaciale, résultant de la non-homogénéité de la couche adsorbée, entraînent les couches de fluide adjacentes à l'interface et déclenchent ainsi une microconvection interfaciale qui s'amplifie en modifiant la distribution du tensioactif près de l'interface (effet Marangoni de soluté)[3]. Cette agitation spontanée accélère considérablement les échanges de matière entre les deux milieux liquides[4,5] d'où l'intérêt de son étude dans les processus de transferts interfaciaux (extraction liquide-liquide, purification de composés biochimiques,...).

Le système eau/nitroéthane/bromure de dodécyltriméthylammonium (DTAB) que nous avons étudié a été traité dans le cas de deux solutions superposées[6]. Cependant, l'influence de la paroi et des instruments de mesures étaient alors difficilement maîtrisables. Pour s'affranchir de ces effets et avoir ainsi des conditions aux limites rigoureusement définies, nous avons choisi d'analyser expérimentalement le comportement d'une lentille de nitroéthane à la surface d'une solution aqueuse de DTAB. A l'aide des méthodes de traitement automatique des images, nous préciserons l'origine et la nature des instabilités du système. Nous comparerons les résultats à une analyse linéaire de stabilité. L'influence de phénomènes parasites tels que la dissolution d'une phase dans l'autre sera discutée.

Installation expérimentale

Le dispositif est composé d'une boîte de Petri (1) en verre de diamètre intérieur D = 14 cm contenant la solution aqueuse (2) de tensioactif. L'ensemble est placé dans une cellule étanche (3) saturée en vapeur de nitroéthane, permettant d'éliminer les déplacements dus à l'évaporation. L'éclairage axial (4) est assuré par un faisceau parallèle issu d'une lampe à vapeur de mercure. La goutte de nitroéthane de volume connu est déposée à la surface et forme une lentille (5). Ses mouvements, visualisés par ombroscopie, sont enregistrés par une caméra vidéo matricielle (6) et analysés par un système de traitement automatique d'images numériques. Les images sont stockées sous forme de matrice 256x256 pixels, sur 64 niveaux de gris.

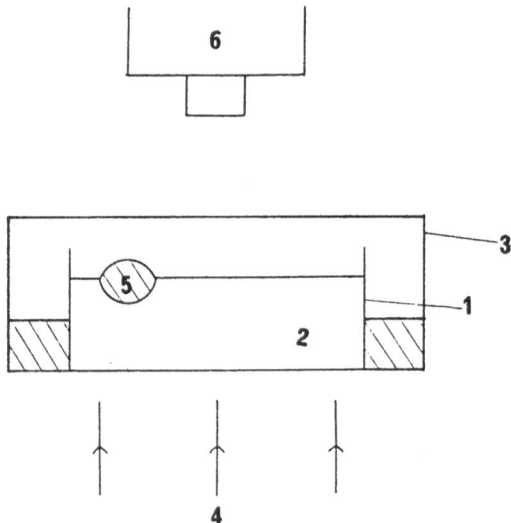

Fig. 1. Schéma du dispositif expérimental

L'erreur sur le calcul de la position de la lentille est inférieure au pixel. Les variations de forme de la lentille sont déduites des histogrammes de niveaux de gris.

Produits

Le DTAB provient des laboratoires SIGMA et a été recristallisé deux fois dans le méthanol La courbe de tension superficielle en fonction de la concentration ne présente pas de minimum près de la CCM, ce qui est considéré comme un test de pureté. L'eau est distillée. Le nitroéthane des laboratoires CARLO ERBA a été lavé deux fois à l'eau distillée. La valeur de la tension superficielle du nitroéthane ainsi nettoyé est égale à 31,9 mN/m à 24°C, ce qui est en accord avec des valeurs de la littérature[7]; celle de la tension interfaciale est de 15 mN/m à 23°C. Toute la verrerie utilisée a été nettoyée au mélange sulfochromique et rincée abondamment à

l'eau distillée. Il convient d'insister sur la rigueur de la procédure expérimentale dont dépend la reproductibilité des résultats.

Notons que la différence de densité entre les deux solvants ne dépasse pas 0,045 g/l, que le volume de la goutte est négligeable devant le volume de la solution et que toutes les expériences sont menées à température
constante (22 ± 0,5°C).

Tensions superficielle et interfaciale

Elles ont été mesurées à l'équilibre sur une interface plane avec une
méthode d'arrachement utilisant un étrier[8] au lieu d'un anneau. La précision des mesures est de 0,1 mN/m.

3 RESULTATS ET DISCUSSION

Mécanisme

A l'aide des histogrammes de niveaux de gris des images numériques,
nous avons pu mettre en évidence des changements périodiques de la forme
de la lentille et les corréler à ses déplacements (fig. 2).

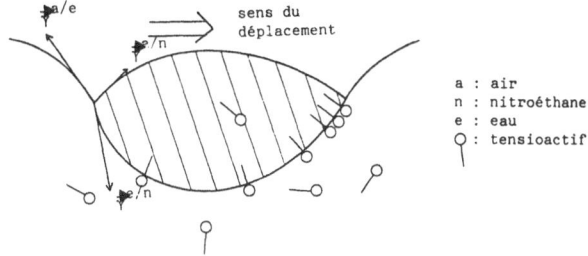

Fig. 2. Schéma de la forme de la lentille lors d'une oscillation de forte
fréquence

Nous pouvons noter que les concentrations aqueuses de DTAB étudiées se
trouvent dans une zone comprise entre 2.10^{-3}moles/l et 10^{-2}moles/l. Le
DTAB ne modifie pas, aux erreurs expérimentales près, la tension superficielle $\vec{\gamma}$ a/n du nitroéthane. La solution aqueuse de DTAB est micellaire
(la CCM du DTAB dans l'eau saturée de nitroéthane est égale à $(1,70\pm0,05)$
10^{-3}moles/l) ; sa tension superficielle $\vec{\gamma}$ a/e est pratiquement constante
dans la gamme de concentration explorée.

Le bilan des tensions sur la ligne de contact permet d'attribuer ces
modifications de forme à des variations locales de la tension interfaciale
$\vec{\gamma}$ e/n. Ce résultat confirme que l'origine du mouvement est un gradient de
tension interfaciale, résultant d'une variation de densité interfaciale
du tensioactif lors de son transfert à travers l'interface eau-nitroéthane.

Transfert de DTAB

Les solvants sont préalablement saturés mutuellement à la température
expérimentale.

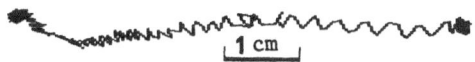

Fig. 3. Diagramme reliant les positions de la lentille toutes les 0,17sec.
(V goutte = 17,5 µℓ, C tensioactif = $7,85.10^{-3}$moles/ℓ)

La trajectoire de la lentille est reconstituée figure 3. Au début la lentille reste localisée autour de la position de dépôt, puis elle dérive en oscillant jusqu'à la paroi de la boîte de Petri. Les mouvements cessent au bout d'une soixantaine de secondes.

Analyse statistique et fréquentielle :

Le déplacement moyen est négligeable devant l'écart-type des positions.

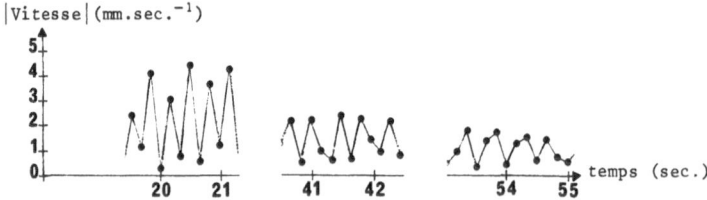

Fig. 4. Variation du module de la vitesse au cours du temps (conditions expérimentales de la figure 3)

La position et la vitesse (fig. 4) varient périodiquement avec le temps.

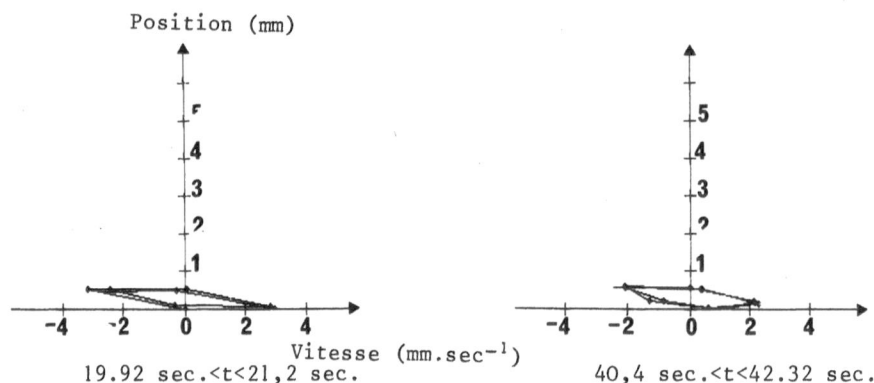

Fig. 5. Diagramme des phases à différents instants du mouvement (conditions expérimentales de la figure 3)

La projection de la trajectoire dans l'espace des phases fait apparaître des séries de cycles imbriqués très réguliers (fig. 5).

Les spectres de puissance des déplacements sur des intervalles de temps réduits ($\sim$ 5 sec.) présentent des pics bien définis.

Au cours du retour à l'équilibre de partage, la fréquence des oscillations décroît ; elles cessent lorsque les contraintes visqueuses équilibrent les gradients de tension interfaciale.

L'ensemble de ces résultats mettent en évidence la nature parfaitement périodique du mouvement de la lentille, qui se comporte comme un oscillateur amorti[9] ; son mouvement peut être caractérisé uniquement par son amplitude et sa fréquence. Des oscillations régulières amorties de tension

interfaciale ont déjà été observées à l'interface de deux solutions superposées en déséquilibre de concentration de tensioactif mais le transfert
faisait intervenir une réaction chimique interfaciale[6].

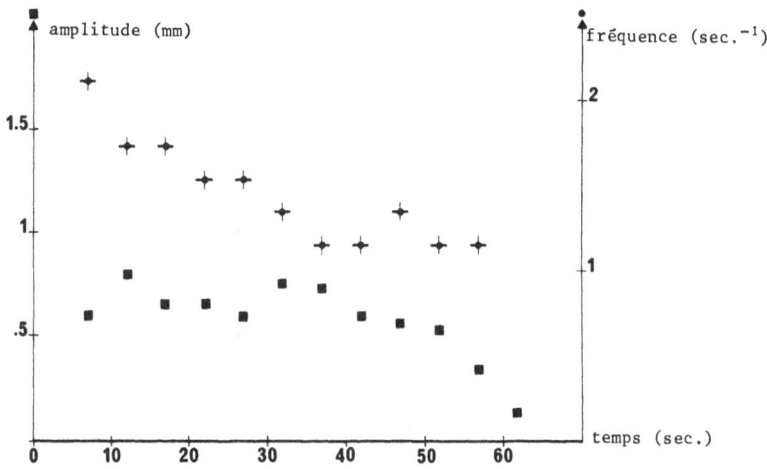

Fig. 6. Variation de l'amplitude et de la fréquence des oscillations au
cours du temps (l'amplitude a été calculée à partir de données
filtrées en fréquence et en ayant assimilé la lentille à un
oscillateur harmonique sur des intervalles de temps : $\Delta t=5,12$ sec.
conditions expérimentales de la figure 3)

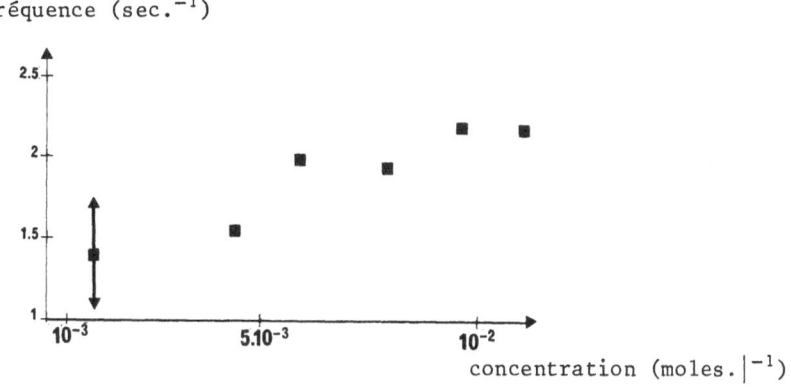

Fig. 7. Variation de la fréquence initiale du mouvement en fonction de la
concentration initiale de tensioactif (fréquence initiale calculée
sur les quarantes premières oscillations ; V goutte = 17,5 µℓ)

La fréquence initiale des oscillations est une fonction croissante de la concentration de départ de tensioactif (fig. 7). Ce résultat a déjà été obtenu dans le cas d'une goutte accrochée à un réservoir et totalement immergée dans une solution aqueuse de tensioactif pour le même système ternaire[10].

Cette diminution de la fréquence initiale des oscillations en fonction de la concentration de tensioactif est à rapprocher de la décroissance temporelle de la fréquence de la figure 6. Le déséquilibre de concentration entre les deux phases se manifeste à l'interface par un gradient normal de concentration. Si ce gradient auquel est soumise l'interface diminue, soit en abaissant la concentration initiale de tensioactif (fig. 7) soit par le retour à l'équilibre de partage (fig. 6), alors l'importance des fluctuations locales de densité interfaciale, responsables des gradients locaux de tension interfaciale, décroît, ce qui diminue la force motrice du mouvement et donc l'intensité des oscillations.

Analyse de stabilité : pour retrouver l'ordre de grandeur de la fréquence d'oscillation, nous avons utilisé les résultats d'une analyse linéaire de stabilité obtenu par Sorensen et al. [11] dans le cas d'une interface plane infinie soumise à un transfert purement diffusif de tensioactif. Pour l'appliquer au cas présent, nous avons mesuré les coefficients de diffusion du DTAB dans l'eau et le nitroéthane et tracé l'isotherme d'adsorption ; nous avons emprunté à la littérature[12,13] une valeur typique de la viscosité interfaciale ($\Gamma s = 0,1$ poise) et négligé la diffusion interfaciale. Par ailleurs, nous avons calculé le gradient de concentration à l'interface.

Pour une concentration initiale de 10^{-2}mole/ℓ et une épaisseur de couche de nitroéthane de 1 mm, les résultats sont : fréquence calculée $\sim$ 1sec.$^{-1}$ fréquence mesurée $\sim$ 2 sec.$^{-1}$. L'ordre de grandeur est convenable. L'écart peut probablement s'expliquer par les hypothèses grossières nécessaires au calcul dont les principales sont :
- interface plane infinie alors que l'interface eau-nitroéthane légèrement concave de la lentille est de dimension finie ;
- valeur non mesurée de la viscosité interfaciale.

Transfert de DTAB avec dissolution

Nous avons couplé le transfert de DTAB à la dissolution de la goutte de nitroéthane dans l'eau en saturant la solution aqueuse à une température légèrement inférieure à la température expérimentale ($\Delta t \sim 3°C$) ; la goutte de nitroéthane déposée à la surface de cette solution se dissout complètement au bout de quelques heures.

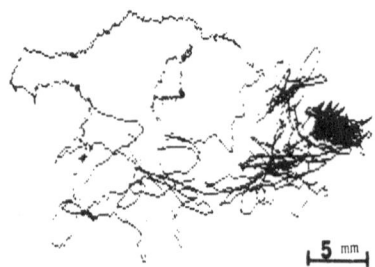

5 mm

Fig. 8. Diagramme reliant les positions de la lentille toutes les 0,2 sec.
(V goutte = 18 $\mu\ell$, C tensioactif = 8.10^{-3}moles/ℓ)

Nous avons observé des mouvements de la lentille analogues aux précédents. Cependant, la trajectoire de la lentille (fig. 8) est moins localisée que dans le cas du transfert pur de DTAB.

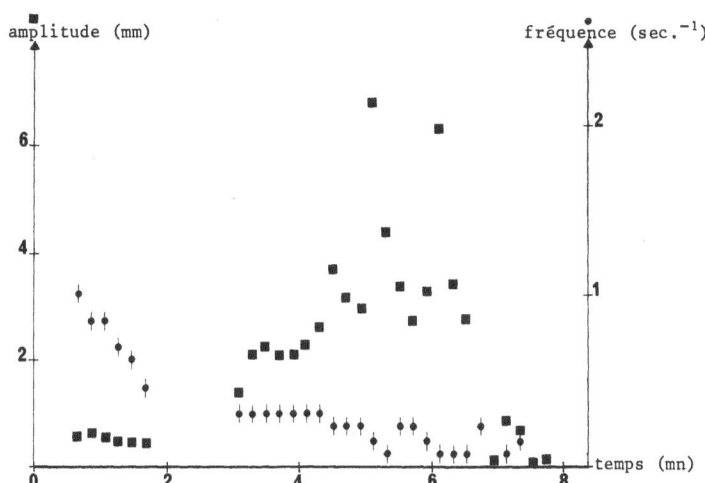

Fig. 9. Variation de l'amplitude et de la fréquence du mouvement au cours du temps (l'amplitude a été calculée à partir de données filtrées en fréquence et en ayant assimilé la lentille à un oscillateur harmonique non amorti sur des intervalles de temps : Δt=12,8 sec.; conditions expérimentales de la figure 8)

Plusieurs phases successives du mouvement correspondant aux différents états du système, sont mises en évidence (fig. 9).

Phase A : Etat instable oscillatoire (fréquence élevée, amplitude faible, ∿ 2 mn).

Les mouvements présentent les mêmes caractéristiques que ceux obtenus dans le cas du transfert pur de DTAB. La vitesse varie périodiquement avec le temps ; les diagrammes de phase présentent les séries de cycles imbriqués (fig. 10). La fréquence des oscillations décroît avec le temps mais elle est plus faible que pour le transfert pur de DTAB pour des conditions de concentration et de volume de goutte presque identiques.

Phase B : Etat stationnaire (∿ 1mn 30 sec.). La lentille ne bouge pas.

Phase C : Etat instable oscillatoire chaotique (fréquence faible, amplitude importante, ∿ 4 mn 30 sec.). La vitesse varie périodiquement avec le temps par intermittence. Des cycles simples apparaissent parfois dans les diagrammes de phase (fig. 10).

Les pics des spectres de puissance sont mal définis. Les variations d'amplitude et de fréquence des mouvements sont irrégulières. La lentille se comporte comme un oscillateur chaotique[9].

Notons que la dissolution seule peut donner lieu à des instabilités. Ainsi, pour un système à l'équilibre de partage en DTAB, le dépôt d'une

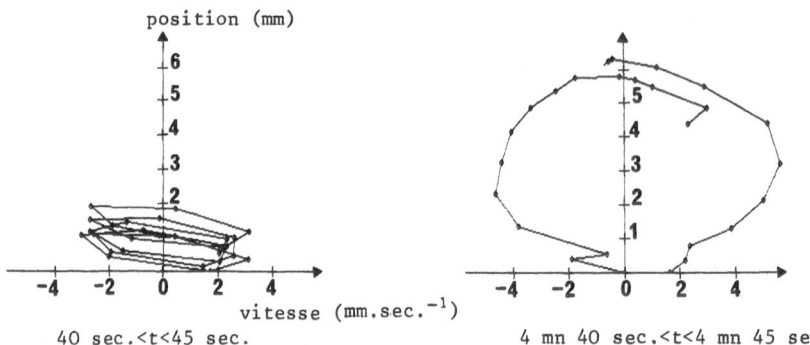

position (mm)

vitesse (mm.sec.$^{-1}$)

40 sec.<t<45 sec.   4 mn 40 sec.<t<4 mn 45 sec.

Fig. 10. Diagramme des phases à différents instants du mouvement (conditions expérimentales de la figure 8)

goutte de nitroéthane sur une solution aqueuse non saturée en nitroéthane donne naissance à des mouvements réguliers, très violents et de forte amplitude ; cette observation confirme les travaux d'Ostrovsky et al.[14] se rapportant à des gouttes de solvant organiques accrochées à un réservoir et immergées dans de l'eau non saturée. La dissolution associée au transfert de DTAB amplifie l'amplitude des déplacements et introduit du chaos dans les oscillations.

Enfin, si la surface de la solution aqueuse ou la goutte sont contaminées par des traces d'impuretés, les oscillations deviennent très irrégulières et intermittentes, comme dans la phase C.

REFERENCES

1. C. Vidal et A. Pacault, "Non-linear Phenomena in Chemical Dynamics", Springer-Verlag, Berlin (1981)
2. G. Nicolis et I. Prigogine, "Self-organization in Non-equilibrium Systems", J. Wiley, New York (1977)
3. M. Dupeyrat, J. Michel, Exp. Suppl., 269:19 (1971)
4. R.J. Whewell, M.A. Hugues, C. Hanson, J. Inorg. Nucl. Chem., 2303:37 (1975)
5. W. Nitsh, L. Navagro, in Proc. Interfacial Solvent Extraction Conf., 220:80 (1980)
6. E. Nakache, M. Dupeyrat, M. Vignes-Adler, Faraday Discuss. Chem. Soc., 13:77 (1984)
7. J.J. Jasper, J. Phys. Chem. Ref. Data, 1:4 (1972)
8. J. Guastalla, J. Chem. Phys., 822:68 (1971)
9. P. Bergé, Y. Pomeau, C. Vidal, "L'ordre dans le chaos", Hermann, Paris (1984)
10. E. Nakache, S. Raharimalala, "Interfacial convection driven by surfactant compounds at liquid interfaces. Characterization by a solutal Marangoni number", in Proc. on Physicochemical Hydrodynamics : Interfacial, Spain, July 1986. M.G. Verlade and B. Nichols, eds., Plenum Press, New York, 1987
11. T.S.Sorensen, M. Hennenberg, A. Sanfeld, J. Colloid Interface Sci., 61:62 (1976)
12. J.T. Davies, E.K. Rideal, "Interfacial Phenomena", Acad.Press, New-York (1963)
13. L. Ting, D.T. Wasan, K. Miyano, J. Colloid Interface Sci., 107:2 (1985)
14. M.V. Ostrovsky, R.M. Ostrovsky, J. Colloid Interface Sci., 392:2 (1983) .

MULLER -   1) Do the small-amplitude motions of the drop occur in a periodic or in a more irregular manner ? The large-amplitude motions at later stages of the experiment which you relate to dissolution effects appear to be more regular than the former ones.

2) How do these experiments relate to earlier reports on rather spectacular deformations with high amplitude, observed in systems with only one interface between two immiscible liquid phases. It appears that the interaction with the boundaries of the container has a substantial influence which is ruled out in these new experiments.

VIGNES-ADLER -   1) The small amplitude motions of the lens occur in a very regular manner as long as there is no contamination, change of temperature, etc. Actually, the dissolution decreases the frequency and increase the amplitude of the motions which are not much more regular than the first ones.

2) In earlier reports, the surfactant tranfer was coupled with some chemical reactions which changed drastically the desorption conditions ; moreover there was a strong interaction with the walls which amplified the motions.

However in both cases motion is due to variation of interfacial tensions related to concentration gradients.

MICHEAU - What is the final state of your system ? How many time does it work ? Have you made your experiment on a balance to see any weight loss ?

VIGNES-ADLER - In the final state, the partition equilibrium between both phases is restored and there is no motion any longer. The motions last for 1 mn up to 10 mn depending on the initial conditions (concentration, drop volume, etc.).
We have not made any measurement on a balance : however, the present motions are not due to a Bénard-Marangoni effect since the atmosphere is totally saturated at contant temperature. In absence of dissolution, the diameter of the lens remains constant : otherwise the lens disappears after a few hours.

EXPERIMENTAL STUDY OF A BISTABLE PHOTOCHEMICAL REACTION

IN AN OPEN REACTOR

X. de Senneville, D. Lavabre, J-P. Laplante*,
G. Levy and J-C. Micheau

UA 470, Univ. P. Sabatier, F-31062 Toulouse, France
* Royal Military College, K7K5LO Kingston, Ontario, Canada

INTRODUCTION

Much of the experimental work in chemical dynamics has been based on inorganic oxidation-reduction reactions in aqueous ionic solution (1). In general, these are special cases in which the oxidants are oxygenated or oxyhalogenated ions, and the reducers inorganic ions or simple organic substrates serving as electron donors (2). Such reactions have been shown to have non-linear properties (oscillations, multiple stationary states, etc.) involving numerous intermediate species, usually in complex reaction mechanisms.

We report here an investigation of an entirely new bistable system based on an organic photochemical reaction. This system thus expands the experimental scope of chemical dynamics since:
1) the reaction mechanism is simpler, and therefore easier to simulate
2) an external constraint, the incident light flux, can be readily manipulated. This stems from the purely photochemical nature of the reaction.

THE CONTINUOUS PHOTOCHEMICAL REACTOR SYSTEM

We have developed a new photochemical open reactor of the CSTR type (RéPAC for Réacteur Photochimique Agité Continu, (4)). Its main chararacteristic is that is open to both matter and energy (3,4). Reactants are supplied continuously by two linear pumps, and the reaction mixture is irradiated from a mercury vapor lamp via a fiber optic cable fitted with an interference filter. The ouput is fed directly into a microcuvette contained in a diode array spectrophotometer (HP 8541). This is connected to a computer terminal for direct representation of absorbance as a funtion of time.

PHOTOCHEMICAL REACTION

1) Advantages of a photochromic system
Photolysis of the dimer of the triphenylimidazyl radical was investigated in solution in chloroform. This reaction is photochromic with a strong change in optical density in the visible region. The reaction

progress can thus be recorded directly from the absorbance at 554 nm (AU554 nm). Its value is proportional to the concentration of the violet radical 2. This reaction, whose mechanism has been partially investigated by other workers (5), is easy to set up.

When carried out in a closed reactor for an adequate duration of irradiation, the photochromic properties of the system are modified. The violet radicals 2 disappear with the simultaneous appearance of 2,4,5-triphenylimidazole 3, which can undergo secondary reactions under UV irradiation. The nature of these processes and their influence on the primary reaction will discussed below.

2) Synthesis of dimer 1 and preparation of the photochromic solutions

Dimer 1 is obtained from 2,4,5-triphenylimidazole or lophine 3 which is commercially available (Aldrich T8;320-8) using the reaction described by White and Sonnenberg (6). It is a white powder which is stable in air (MP = 180°C). When dissolved in chloroform, however, the dimer is violet colored. It turns pale yellow when left for several hours in the dark. Before all experiments the solutions were subjected to prolonged degassing by bubbling with nitrogen saturated with chloroform vapor.

PROCEDURE AND RESULTS

1) Experimental conditions

The solution of the dimer 1 in chloroform (10^{-3} M) prepared as described above is pumped into the reactor, and irradiated (λ_{irr} = 360 nm; pass band 40 nm; photon flux I_o = $1.5.10^{-4}$ $M.s^{-1}$). A stationary state is established for all species (concentration independent of time). The violet color of the triphenylimidazyl radicals 2 (ε_{554}^2 = 10^4 $cm^{-1}.M^{-1}$) is clearly observed. In the following experiments the concentration of the dimer 1 and the light flux I_o were kept constant. The flow rate was varied from 6.60 to $7.33.10^{-3}$ ml/min which corresponds to a range of residence times from 10 to 9,000 sec in the RéPAC (vol = 1.1 ml).

2) Stationary state as a function of residence time

In these experiments the dimer 1 was pumped into the reactor at a fixed rate. The AU554 in the absence of irradiation was taken as the reference value. On irradiation, the violet radicals appear and the AU554 increases and then stabilizes. During this phase, three main types of transient regimes can be observed for different residence times (Figs. 1A, B and C).

3) Photochemical hysteresis

When the reactor was in the fluorescent stationary state, the residence time was progressively reduced by a stepwise increase in flow rate. The incident light flux was kept constant. Under these conditions, the system remained in the fluorescent stationary state for intermediate residence times (τ < 1,500 sec) which normally give rise to the violet stationary states. Further increase in flow rate led to an abrupt replacement of the fluorescent state by the violet stationary state. This was observed for residence times close to 800 sec. Figure 2 illustrates the regimes observed when the flow rate was increased from the fluorescent stationary state value.

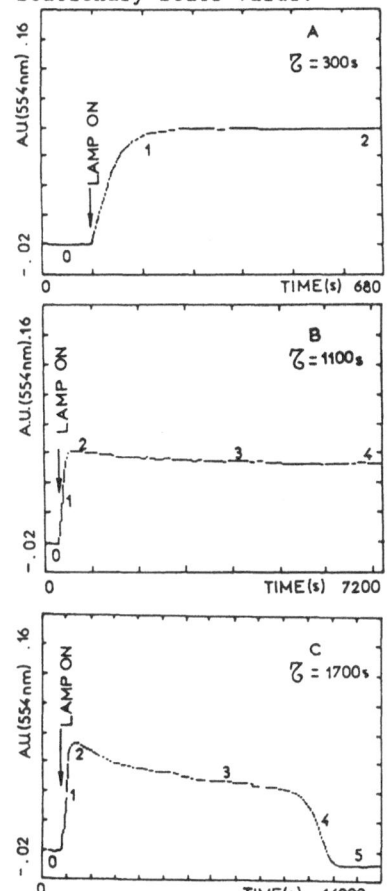

Fig.1.
A) Transient regimes between the non-irradiated state and the violet stationary state under irradiation for a residence time τ < 800 sec;
0 : non-irradiated reference;
1 : quasi-exponential regime;
2 : stationary state, high proportion of violet radicals 2;

B) Transient regimes between the non-irradiated state and the violet stationary state under irradiation for a residence time 800 < τ < 1,500 sec
0 : non-irradiated reference;
1 : fast regime;
2 : overshoot;
3 : slow exponential decay;
4 : stationary state, high proportion of violet radicals 2;

C) Transient regimes between the non-irradiated state and the fluorescent stationary state under irradiation for a residence time τ > 1,600 sec
0 : non-irradiated reference;
1 : fast regime;
2 : peak;
3 : metastable state;
4 : acceleration;
5 : fluorescent stationary state;

The results shown in figures 1B and 2D are indicative of a phenomenon of photochemical hysteresis.

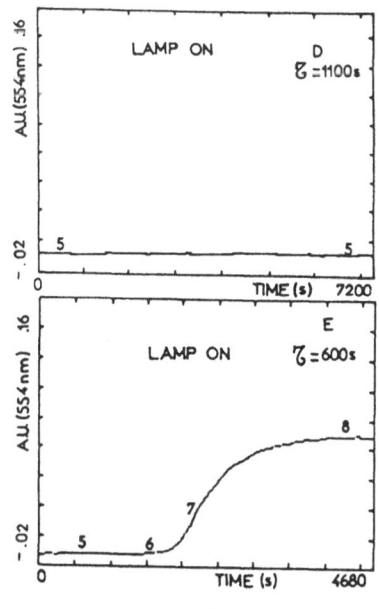

Fig. 2:
D) When the reactor is stabilized in the fluorescent stationary state, the flow rate can be increased considerably without a change in the response of the system; 5 : fluorescent stationary state

E) Beyond a certain flow rate, the fluorescent state is no longer stable, and is replaced by the violet stationary state; 5 : fluorescent state;
6 : onset of the increase in optical density; 7 : marked increase in optical density with a point of inflexion
8 : violet stationary state (stable).

483

The complete stationary state behavior of the reactor (AU554 vs τ) can
be plotted for the different values of residence time.

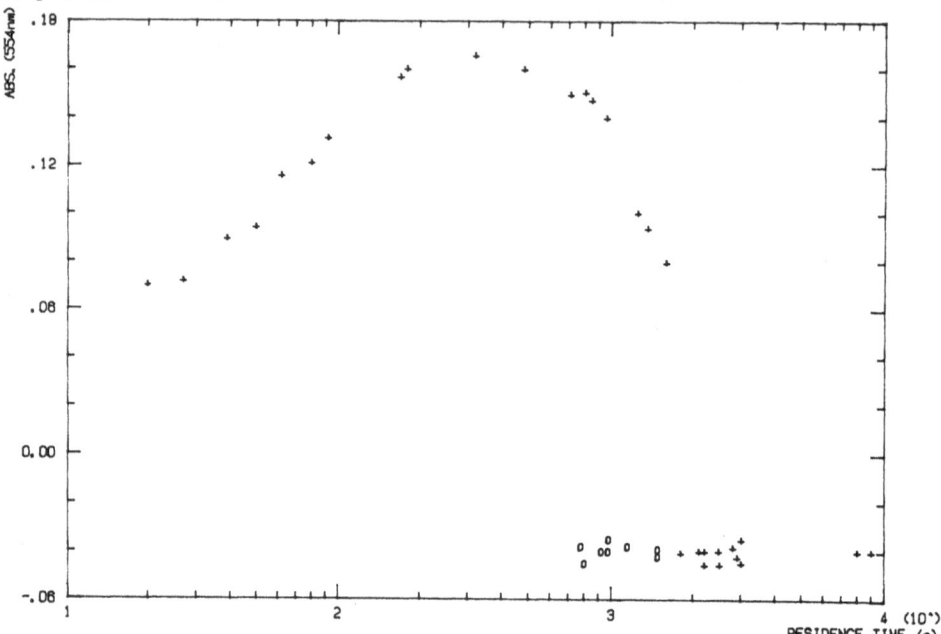

Fig.3

Plot of stationary states of the 2,4,5-triphenylimidazyl radical
(AU554) as a function of residence time
      + points for conditions of figure 1
      o points for conditions of figure 2
A region of overlap is observed for residence times between 800 and 1,600
sec corresponding to a process of chemical hysteresis.

    The following series of experiments were designed to demonstrate the
bistable nature of this reaction.

4) Experimental demonstration of bistability
a) *Effect of a change in light intensity*

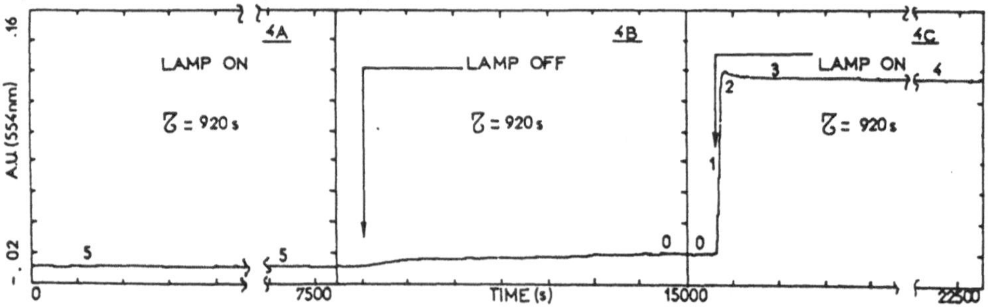

Fig.4 . Demonstration of bistability by changing light intensity
A) Reactor stabilized in the fluorescent stationary state; τ = 920 sec, Io
= $1.5.10^{-4}$ M.s^{-1}

B) Emptying of reactor in the absence of irradiation (by closing a
shutter); τ = 920 sec, Io = 0

C) Transient regime towards the violet stationary state; τ = 920 sec Io =
$1.5.10^{-4}$ M.s^{-1} .

    Although the regimes 4A and 4C were subjected to identical
constraints, different results were observed. This provides direct

484

evidence for the existence of two different stable state for identical external constraints.

b) *Effect of a pulse injection of fluorescent solution*
When the system was in the fluorescent stationary state, a solution containing the fluorescent product was collected at the output of the reactor. This solution (see above) is able to destroy the violet radicals (bleaching) at a high rate. The experiment was carried out for a residence time of 1,400 sec i.e in the hysteresis zone (800 < τ < 1,500 sec).

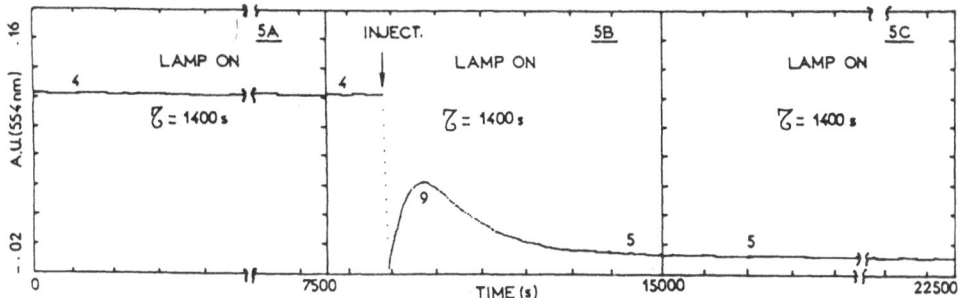

Fig.5. Demonstration of bistability after a pulse injection of fluorescent solution
A) Reactor stabilized in the violet stationary state, τ = 1,400 sec

B) Injection of 300 μl of the fluorescent solution. The reactor passes through a complex transient regime towards the fluorescent stationary state.

C) Stabilization of the reactor in the fluorescent stationary state.

Although the system in figure 5A and 5C is under identical constraints, two different stationary states can be observed, providing further evidence for bistability. The explanation of this phenomenon was derived from reaction mechanisms for photochromic and deactivation processes.

REACTION MECHANISM

1) <u>Processes involved and properties of the products</u>
Analysis by HPLC of the reaction mixture showed that the concentration of 2,4,5-triphenylimidazole <u>3</u> increased with increasing residence time. Various other highly fluorescent molecules were formed which were identified as products of the photolysis of 2,4,5-triphenylimidazole <u>3</u> itself. These products which are formed from both the dimer <u>1</u> and triphenylimidazole <u>3</u> were found to rapidly bleach the violet radicals. Measurement of the absorption at 554 nm was thus rendered impossible. During this bleaching, the radicals <u>2</u> are partially or completely transformed into 2,4,5-triphenylimidazole <u>3</u>. Phenanthrene <u>4</u> was one of the positively identified compounds among the photoproducts. The essential components of a feedback loop thus appear to be present.

2) <u>Elements of a feedback loop</u>
- The violet radicals <u>2</u> are slowly reduced to triphenylimidazole <u>3</u>, by capture of hydrogen from the solvent for example (element 1).
- Triphenylimidazole <u>3</u> is transformed into inhibitors under irradiation. During this inhibition some of the triphenylimidazole <u>3</u> is regenerated (element 2).
- When the violet radicals are bleached they uncover some of the absorption spectrum of triphenylimidazole <u>3</u> (screen effect).

Triphenylimidazole $\underline{3}$ is thus subjected to increasing irradiation as the violet radicals $\underline{2}$ are bleached (element 3).

Operation of the feedback loop can be explained in qualitative terms as follows: in the presence of a high concentration of violet radicals $\underline{2}$, the screen effect prevents irradiation of 2,4,5-triphenylimidazole $\underline{3}$, which is always present in low concentration in the starting solution of dimer $\underline{1}$. Since the effect of irradiation is reduced, only low levels of photolysis products are formed, and there is no inhibitory effect. The concentration of the radicals can thus remain elevated. In contrast, if there are only small amounts of the violet radicals $\underline{2}$, there is no screen effect and 2,4,5-triphenylimidazole $\underline{3}$ is subjected to irradiation. Photoproducts are formed leading to inhibition. The radicals $\underline{2}$ are transformed into $\underline{3}$, fuelling the generation of inhibitory photoproducts. The low level of violet radicals $\underline{2}$ is therefore maintained.
In summary, two situations can arise with respect to the concentration of violet radicals:
1 - high concentration ---> high concentration
2 - low concentration ---> low concentration
Both situations are stable in an open system. This corresponds to the bistable phenomenon (stable high or low concentration, unstable intermediate concentrations) that was observed in the RéPAC over the intermediate range of residence times.

3) Discussion and proposal of reaction scheme
This bistabilty did not appear to be due to a direct photothermic effect (8). No variations in temperature were detected with the different operating regimes (either in flux or thermodynamic parameters). This can be explained by the relatively low power output (50 mW) of our irradiation source at 360 nm. 100% of this energy is absorbed in the violet stationary state against 68% in the fluorescent stationary state. The 16 mW difference in energy between these two is insignificant given the potential for elimination of thermal ballast in our system. We concluded that a direct photothermic effect was not involved, and that the bistability could be better accounted for by the following kinetic schemes:

$$(I) \quad \underline{1} \quad \xrightarrow{\quad h\nu_1 \quad} \quad \underline{2} + \underline{2}$$

$$(II) \quad \underline{2} + \underline{2} \quad \xrightarrow{\quad k_2 \quad} \quad \underline{1}$$

$$(III) \quad \underline{2} \quad \xrightarrow{\quad k_1 \quad} \quad \underline{3}$$

$$(IV) \quad \underline{3} \quad \xrightarrow{\quad h\nu_2 \quad} \quad \underline{4} + (.X.)$$

$$(V) \quad \underline{2} + \underline{4} \text{ or } (.X.) \xrightarrow{\quad k_3 \quad} \underline{3} + \underline{4} \text{ or } (.X.)$$

This overall scheme is compatible with our findings and with literature data (5-7). However, the molecular structure of some of the photolytic products (.X.) could not be determined with certainty, nor could the the nature of the mechanism of the reaction(s) summarized in stage (V). Other elements (oxygen effect, reabsorption of fluorescence (quenching?), chemiluminescence, etc.) should also be taken into account. It should be noted that the biphotonic terms ((I) and (IV)) and inhibition by product ((III) + (IV) + (V)) have non-linear kinetic properties (9,10). The scheme proposed here can account, at least in a semi-quantitative way, for the observations after appropriate adjustment of the kinetic parameters.

CONCLUSION

We present here the first experimental demonstration of photochemical bistability in an open reactor. This bistable reaction results from the non-linear properties of a photochromic system: the dimer of the triphenylimidazyl radical 1 in chloroform. Hysteresis is observed on the plots of the stationary states of the system over a wide range of flow rates. Within this region, the system is bistable and can be made to flip from one state to the other by an external manipulation. One of the stable states is characterized by a high concentration of violet radicals 2 while in the other the violet radicals are replaced by highly fluorescent compounds. Mechanistic studies showed that this bistability was due to a positive feedback loop. This was thought to arise from the screening effect of the violet radicals 2 with respect to the irradiation of the triphenylimidazole 3 in combination with an inhibition of the violet radicals 2 by the products of photolysis of triphenylimidazole 3. These results represent the first indication of the existence of multiple stationary states for an organic photochemical reaction. Two different responses (high or low concentration of violet radicals) can be observed for the same external constraints in the reactor (identical photon flux, flow rate, and initial concentration).

REFERENCES

1. C. Vidal and P. Hanusse, Non-equilibrium behavior in isothermal liquid chemical system. Int. Rev. Phys. Chem. 5:1-55 (1986).
2. M. Orban and I.R. Epstein, Bistability in the oxidation of iron(II) by nitric acid. J. Am. Chem. Soc. 104:5918-22 (1982).
3. F. Grégoire, D. Lavabre, J.C. Micheau, M. Gimenez and J.P. Laplante, Kinetics in a continuously stirred photochemical tank reactor. J. Photochem. 28:261-71 (1985).
4. J-C. Micheau and D. Lavabre, Stirred flow reactor - A new approach to photochemical kinetics. EPA Newsletters 26:26-33 (1986).
5. K. Maeda and T. Hayashi, The mechanism of photochromism, thermochromism and piezochromism of dimers of triarylimidazyl. Bull. Chem. Soc. Jap. 43:429-38 (1970).
6. M. White and J. Sonnenberg, Oxidation of triarylimidazoles. Structures of photochromic and piezochromic dimers of triarylimidazyl radicals. J. Amer. Chem. Soc. 88:3825-9 (1966).
7. J. Hennesy and A.C. Testa, Photochemistry of imidazoles. J. Phys. Chem. 76:3362-5 (1972).
8. M. Schell and J. Ross, Effect of time delay in rate processes. J. Chem. Phys. 85:6489-6503 (1986).
6. J-C. Micheau, S. Boue and E. Vander Donckt, Theoretical analysis of biphotonic processes. J. Chem. Soc. Farad. Trans 2, 78:39-50 (1982).
7. P. de Kepper and J. Boissonnade, From bistability to sustained oscillations in homogeneous chemical systems in a flow reactor. in: "Oscillations and Travelling Waves in Chemical Systems," R.J. Field and M. Burger ed., Wiley Interscience. p.223-56 (1985).

DISCUSSION

MULLER - Do you think that this photochemically bistable system is a candidate for spatial pattern formation (without convective effects) under externally homogeneous illumination with the appropriate wavelength ? The chemical "feed-back loop" combined with diffusive transport might constitute a source of spatial symmetry-breaking, e.g. in an extended solution layer.

MICHEAU - It is now very well known that the overhelming majority of photochemical spatial patterns are in fact due to a coupling of a linear photochemical reaction with a convection process coming from an evaporative cooling phenomenon. The patterns do not appear if convection is not allowed concerning now the new photochemical bistable system : this system must be used in a CSTR under UV irradiation (360 nm) so, I don't see any special condition where spatial patterns could be experimentally observed.

WAVE PROPAGATION IN THE BELOUSOV-ZHABOTINSKII REACTION INVOLVING DIFFERENT CATALYSTS

Zs. Nagy-Ungvarai, S.C. Müller and B. Hess

Max-Planck-Institut für Ernährungsphysiologie
Rheinlanddamm 201, D-406 Dortmund 1, FRG

The most extensively studied oscillatory chemical system is the Belousov-Zhabotinskii (BZ) reaction [1] which involves the oxidation of an organic susbstrate (for example malonic acid) by bromate in the presence of a metal-ion catalyst in sulfuric acid medium.

For homogeneous stirred conditions, the mechanisms and the detailed properties of this reaction have been investigated using cerium as a catalyst [2,3], while spatial pattern formation in thin layers of the reaction has been studied mostly for the ferroin-catalyzed case [4,5]. Manganese and $Ru(bpy)_3^{2+}$ catalysts are not as frequently used as the two former ones.

The four catalysts have different redox properties, the redox potentials being 1.51 V, 1.44 V, 1.25 V and 1.10 V for manganese, cerium, $Ru(bpy)_3^{2+}$ and ferroin, respectively. Electron transfer reactions occur with the metal ions cerium and manganese by an inner sphere mechanism [6], and with the complexes ferroin and $Ru(bpy)_3^{2+}$ by an outer sphere mechanism.

It is apparent that there are differences in the behaviour of the BZ oscillators catalyzed by metal ions and metal-ion complexes, as is shown in Fig. 1 in case of temporal oscillations. Important features to be noted are : (1) The frequency of oscillations decreases with increasing redox potential. (2) The amplitude of oscillations increases with increasing redox potential. (3) There is a characteristic change in the shape of an oscillatory cycle, which means that one of the most characteristic properties of the reaction, the critical Br^--ion concentration, depends on the type of the catalyst. Also the heat evolution and the amount of bromomalonic acid produced during one oscillatory cycle were measured to be different for each of the four catalysts [7].

This different behaviour may result from differences in the oxidation of the organic substrate by the oxidized form of the catalyst and in the auto catalytic reaction between bromate and the reduced form of the catalyst.

Chemical waves in the BZ reaction travel in a reduced layer, where small amounts of Br^--ions, produced by the above mentioned organic reactions, play an important role :

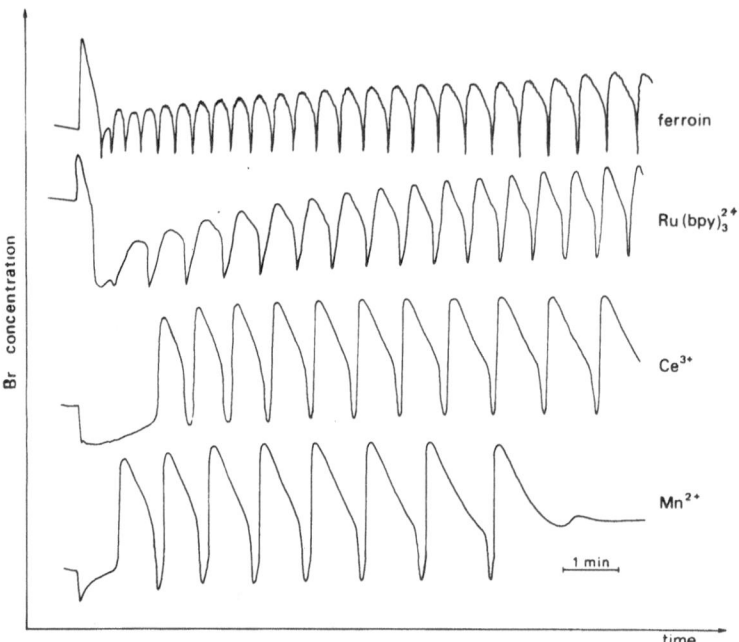

Fig. 1. Logarithmic traces of [Br$^-$]-oscillations for the four catalysts.

$$Br^- + HOBr + H^+ = Br_2 + H_2O \qquad\qquad R1$$

$$Br^- + HBrO_2 + H^+ = 2HOBr \qquad\qquad R2$$

$$Br^- + BrO_3^- + 2H^+ = HOBr + HBrO_2 \qquad\qquad R3$$

The oxidized chemical waves themselves are a consequence of the coupling between diffusion and chemical reaction. Their velocity is determined by the diffusion coefficient of the autocatalytic species $HBrO_2$ and the rate constant of its formation in the autocatalytic reaction mentioned above :

$$2HBrO_2 = BrO_3^- + HOBr + H^+ \qquad\qquad R4$$

$$HBrO_2 + BrO_3^- + H^+ = 2\ BrO_2 \cdot + H_2O \qquad\qquad R5$$

$$BrO_2 \cdot + cat^{n+} + H^+ = cat^{(n+1)+} + HBrO_2 \qquad\qquad R6$$

Consequently, differences due to the nature of the catalyst can be observed not only in the parameters of temporal oscillations, but also in the properties of travelling waves.

While ferriin- and $Ru(bpy)_3^{3+}$-waves are readily detectable in the visible range, the color changes of the Ce(IV)/Ce(III) and Mn(III)/Mn(II) redox couples are practically invisible at low concentrations, especially in thin layers of the reagent. Therefore, no spatial pattern formation in solution layers of these two systems has been directly observed yet in order to show the differences in wave propagation when different catalysts are used.

We detected various types of chemical waves in the Ce- and Mn-catalyzed BZ reaction by our UV-sensitive two-dimensional spectrophotometer [8] at 344 nm, because both Ce(IV) and Mn(III) have an absorption maximum in the UV range. Pictures a and b of Fig. 2 show a section of concentric waves in the Ce-catalyzed BZ reaction at different concentrations. Although the optical contrast in the Mn-catalyzed system is much less pronounced, photometric measurements can still be made with good precision (see spiral wave in Fig. 2c).

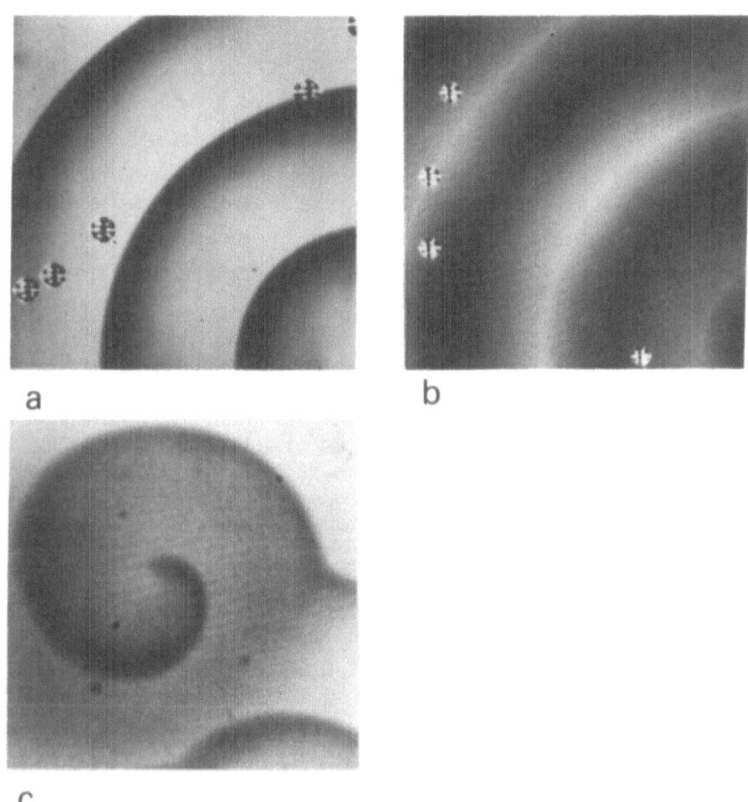

a

b

c

Fig. 2. Spatial patterns in Ce- and Mn-catalyzed BZ systems with the following initial concentrations ; a and b : 0.3 M BrO_3^-, 0.366 M (a) and 0.01 M (b) malonic acid, 0.09 M bromomalonic acid, 0.41 M H_2SO_4, 0.006 M Ce(IV) ; c : 0.48 M BrO_3^-, 0.4 M malonic acid, 0.66 M H_2SO_4, 0.03 M Mn^{2+}.

Applying this technique we were able to measure the propagation features (e.g. wavelength, amplitude, velocity) of Ce- and Mn-waves with various choices of initial concentrations. This allows a direct comparison of the measured properties as a function of the redox catalyst. The main results are as follows :

(1) The wavelength, that is the distance between the wave fronts (although it varies in a given system), is always considerably higher with cerium or manganese as compared to the ferroin or $Ru(bpy)_3^{2+}$ catalysts.

(2) The amplitude of the waves, which is a measure of the amount of the catalyst involved in one oscillatory cycle, is substantially larger in case of the high-redox-potential metal catalysts, than when using the low-redox-potential complexes.

(3) The velocity v of trigger waves in the BZ reaction can be described in first approximation by an equation of the following form :

$$v = b + a \, ([BrO_3^-] \, [H^+])^{1/2} \qquad\qquad (1)$$

which is found to be valid within a certain concentration range.

The parameters of this equation and the range of validity for three different catalysts are listed in Table 1. Within the experimental error, the parameters of the ferroin and $Ru(bpy)_3^{2+}$ equation are in excellent agreement. However, for cerium waves both intercept and slope of the linear portion of the velocity curve are smaller. Concentrations above the specified range in case of the ferroin catalyst cannot be measured. When $Ru(bpy)_3^{2+}$ is used at higher concentrations of H_2SO_4 and BrO_3^-, the system exhibits spontaneous oscillations and the velocity of waves exceeds by far the values of equation (1). In case of the cerium catalyst the velocity curve in this extended concentration range is found to have a nonlinear form in that at concentrations above the linear region the velocity of the waves approaches a saturation level. A preliminary study of the manganese-catalyzed system indicates a behaviour, which is similar to that of the cerium-catalyzed reaction.

Table 1. Parameters of equation (1) for different catalysts at 25°C.

	b	a	concentration range	Ref.
	$(mm \ min^{-1})$	$(mm \ min^{-1} \ M^{-1})$	$([BrO_3^-] \ [H^+])^{1/2}$ (M)	
ferroin	- 0.8	27.9	0.18 - 0.34	9
$Ru(bpy)_3^{2+}$	- 1.2	27.5	0.15 - 0.28	10
cerium	- 2.1	22.5	0.18 - 0.38	this work

The experimentally determined velocity curve of the low-redox-potential catalysts can be calculated from the $HBrO_2$ reaction-diffusion equation, which involves only the most important steps R5 and R2. In contrast the reaction-diffusion equation in systems with a high-redox-potential catalyst should involve the following reactions :

(1) R5 and R6 as reversible steps, since $BrO_2\cdot$ was measured in the Ce-catalyzed reaction as a stable intermediate [3] ;

(2) R2 with a real, non-adjusted $[Br^-]$ in front of the waves in these systems ; and

(3) the disproportionation reaction R4. This problem can be solved only numerically.

Acknowledgements

The technical assistance of Mrs. I. Beyer, Mrs. B. Plettenberg and Mr. U. Heidecke is gratefully acknowledged.

References

[1]   R.J. Field and M. Burger (eds.), "Oscillations and Travelling Waves
            in Chemical Systems", Wiley-Interscience, New York (1985).
[2]   R.J. Field, E. Körös and R.M. Noyes, J. Am. Chem. Soc. 94:8649 (1972).
[3]   R.J. Field, H.-D. Försterling, J. Phys. Chem. 90:5400 (1986).
[4]   A.N. Zaikin and A.M. Zhabotinskii, Nature 225:535 (1970).
[5]   A.T. Winfree, in Ref. 1, p. 441.
[6]   J.J. Jwo and R.M. Noyes, J. Am. Chem. Soc. 97:5422 (1975).
[7]   E. Körös, M. Orban and Zs. Nagy, Acta Chim. Hung. 100:449 (1979).
[8]   S.C. Müller, Th. Plesser and B. Hess, Naturwissenschaften 73:165 (1986).
[9]   R.J. Field and R.M. Noyes, J. Am. Chem. Soc. 96:2001 (1974).
[10]   L. Kuhnert and H.-J. Krug, J. Phys. Chem. 91:730 (1987).

DISCUSSION

MICHEAU - What is the relationship between the oscillation pattern and the value of the redox potential of the metal catalyst ?

NAGY-UNGVARAI - With growing redox potential the overall autocatalytic reaction step in the FKN model scheme becomes slower, because the reverse step of the rate-determining reaction can be neglected in case of the slow redo potential complexes ferroin and $Ru\,(bpy)_3^{2+}$, but becomes increasingly important for high redox potential metals, as cerium and manganese. These results in a lengthening of the oscillation period in temporal oscillations, as well as an increase of the wavelength in spatial-reaction-diffusion patterns with growing redox potential of the catalyst.

CHAOTIC PHENOMENA IN AN ENZYME REACTION UNDER ELECTRICAL

CONSTRAINTS

Jean-Marc Valleton

Laboratory"Polymères,Biopolymères,Membranes"
UA CNRS n°500 Université de Rouen
76130 Mont-Saint-Aignan France

SUMMARY

The action of electric fields on structured enzyme systems leads to a great variety of behaviors; it is possible to obtain regulations of enzyme activity or commutations; in addition it is possible to induce oscillatory behaviors in such systems with two chemical parameters; direct electric fields are able to induce periodical oscillations, whereas alternating electric fields are able to induce aperiodical oscillations (chaotic oscillations) in wide domains of the different parameters.

INTRODUCTION

Many studies have been devoted to the modelling of systems in which immobilized enzymes are involved [1,2] especially because of implications in biotechnology [3] (synthesis: bioreactors with beads, membranes, fibres; analysis: enzyme electrodes, biosensors...).

Immobilized enzyme systems, or more generally structured enzyme systems, can also be used as models for studying complex dynamics; in particular, it is possible to elaborate simple artificial enzyme systems able of generating chaotic behaviors, wheras chemical systems exhibiting analog behaviors are much more complex; in particular, the number of chemical species involved is more important[5,6].

The aim of this paper is to illustrate the fundamental aspect corresponding to the the study of complex dynamics by an example of induction of periodical or aperiodical oscillations by electric fields. Concerning the last point, previous studies showed that coupling two chemical oscillators might lead to chaotic phenomena. It was also shown that coupling an endogeneous chemical oscillator with an exogeneous oscillator (oscillating electric field, for example) might lead to the same type of results [7,8]. This is the point we want to illustrate here in the case of the reaction catalyzed by uricase.

1. THE MODEL

We have considered here a very simple model [9,10,11]; it is a "zero-dimensional" system which is constituted of a compartment in which the enzyme reaction takes place (thickness l), the enzyme being solubilized and trapped because of the presence of two membranes of thickness e (chemically inert, with no fixed charges). The different chemical species, except the enzyme, can flow through these membranes. The reactive compartment is placed between two concentration reservoirs (theoretically infinite volumes) assumed perfectly homogenized; all phenomena obtained with this type of systems will be independent of space, and will only depend on time. In this system a direction is privileged (x'x); two electrodes placed in the reservoirs allow the injection of an electrical current into the system, along direction x. Experimentally, it is necessary, for avoiding back-diffusion of chemical species produced at the electrodes, to isolate them in two supplementary compartments.

Our study has been essentially theoretical. Chemical processes have been simplified: the reaction rate is written as the product of terms assumed independent and taking into account the effects of pH and of the concentrations of substrate, cofactor, activator or inhibitor. When these species are ionic, they are called the mediators of the electrical action. For this study of complex dynamics, we have considered a system modelling the reaction catalyzed by uricase:

$$\text{Uric acid} + O_2 + H_2O \longrightarrow \text{Allantoin} + H_2O_2 + CO_2 \qquad (1)$$

This reaction is characterized by a kinetics depending on three chemical parameters: the uric acid concentration (which is the substrate and which will be denoted S), the oxygen concentration (which is the cosubstrate and which will be denoted A) and pH. To simplify the presentation, letters S and A will be used to denote the species themselves as well as their concentration. We shall only consider here the first two parameters: S and A. pH is imposed much higher than the pK_a of uric acid; in this way uric acid will always be considered as ionic. The kinetics corresponds to an inhibition by excess of substrate, and is of the first order for the cosubstrate; the reaction rate is expressed as follows:

$$v = V_m \, A \, S \, / \, (K_m + S + K_i \, S^2) \qquad (2)$$

To take into account the different interactions, the behavior of the immobilized enzyme systems, without any external force, has been modelled so far by diffusion-reaction equations in which the parameters of the reaction (maximum rate, shape of the kinetics, dependence of the kinetics on substrate and cosubstrate concentration) and diffusion parameters (diffusion-coefficients) are taken into account. Here a supplementary force is involved: the electric field, and diffusion-electrotransport-reaction equations have to be expressed. The theoretical treatment is based on the expression of the reaction kinetics, of the different fluxes (Fick's law for A and Nernst-Planck relation for S) and of their coupling by the two mass balances related to S and A. The treatment leads to a couple of interdependent and highly non linear differential equations. Except for some limit cases (steady states in particular conditions), these equations cannot be solved analytically; numerical methods are generally necessary. Verifications of the numerical calculations (variations of time increment in particular) were performed for avoiding eventuel artefacts, especially when chaotic phenomena are studied.

2. RESULTS

The results will be presented in two parts depending on the nature of the electric field imposed: direct or alternating.

2.1. Induction of periodical oscillations by direct electric fields

The idea of introducing an electric field in this type of systems for generating oscillations comes from the analysis of uric acid/uricase system: if it is theoretically possible that endogeneous oscillations occur in this system, some conditions have to be satisfied; in particular, the ratio of the values of diffusion coefficients of S and A must have a value lower than 1. However the diffusion coefficients of uric acid and of oxygen do not satisfy to this condition: $D_{oxygen}/D_{ur.ac}$ =1.2; and oscillatory behaviors cannot be obtained. The role of the electric field is to change the ratio of the transport rates of S and A; since the electric field acts only on S, this ratio can be modulated in order to place the system in a domain favourable to the existence of oscillations. This result is illustrated in figure 1 which shows the induction of stable periodical oscillations when an electric field of adequate amplitude is imposed.

For amplitudes lower than 44 V/cm, an excitation phase is observed; then the system goes back to a steady state (obviously different from a steady state which would be obtained at zero current); for amplitudes between 44 and 44.5 V/cm, the system exhibits damped oscillations; for amplitudes between 44.5 and 47.5 V/cm, the system exhibits stable oscillations; for amplitudes between 47.5 and 55 V/cm, oscillations are damped. For higher values of the amplitude the system tends directly toward a steady state. Is is interesting to note the relatively wide domain of stable oscillations.

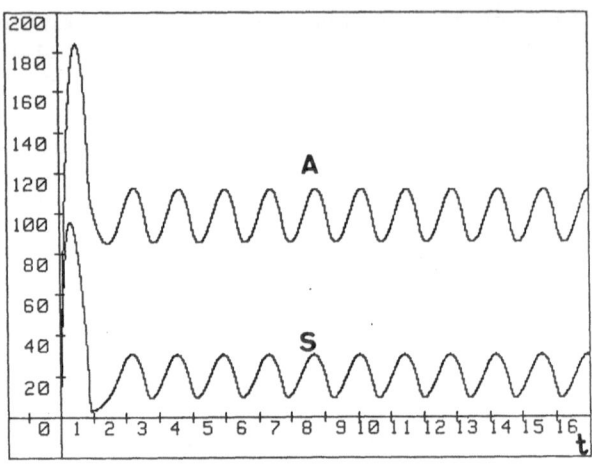

Fig.1. Induction of periodical oscillations by a direct electric field, in a structured enzyme system, whose reaction kinetics is governed by two chemical parameters.

2.2. Induction of aperiodical oscillations by alternating electric fields

When an oscillating electric field is imposed, more complex phenomena occur because of the possible interactions between the potential chemical oscillator and the exogeneous electrical oscillator.

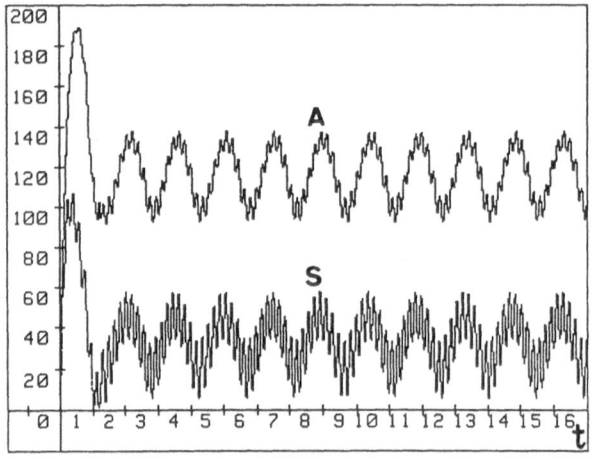

Fig.2. Induction of aperiodical oscillations by alternationg electric fields, in a structured enzyme system whose reaction kinetics is governed by two chemical parameters.

For a given value of the amplitude of the electric field, which will remain the same in this part, different behaviors are obtained depending on the value of the frequency of the imposed electric field. For low values, an oscillatory regime is obtained characterized by complex but periodical variations; for higher values of the frequency, temporary chaotic behaviors appear followed by stable oscillations; for higher values, "stable" chaotic dynamics are obtained (Fig.2).

It is difficult to see clearly these aperiodical oscillations in variations S(t) or A(t), the variations in amplitude or frequency being very small around an average value. For a better representation, it is interesting to draw the variations of S and A in a phase portrait (S,A) (Fig.3).

The attractor represented in figure 3 is constituted of trajectories which correspond to a complex oscillation almost characterized by two frequencies; however trajectories are slightly different from one revolution to an other. This result is more easily visible when a part of the attractor is enlarged (Fig.3).

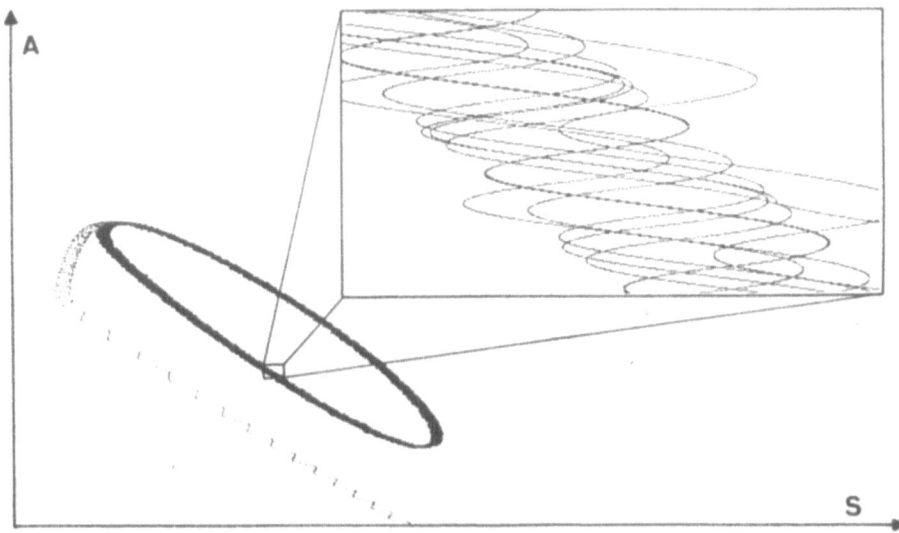

Fig.3. Representation in phase portrait (S,A) of the attractor obtained when an alternating electric field is imposed in a structured enzyme system whose reaction kinetics is governed by two chemical parameters.

DISCUSSION

The possibility of inducing complex dynamics in structured enzyme systems whose reaction kinetics is governed by two chemical parameters (substrate and co-substrate) by the action of electric fields has been illustrated here.

These first results allowed us to reveal:
- the possibility of inducing periodical oscillations by the action of direct electric fields.

- the possibility of inducing aperiodical oscillations (chaotic oscillations) by the action of alernating electric fields.

These two effects seem interesting because they can be considered in terms of monitoring processes:
- in the first case, the oscillation is obtained because an external constraint is imposed; it is an example of control of a chemical oscillator (amplitude and frequency) by the amplitude of the electric field imposed.
- in the second case, pseudo-chaotic phenomena are induced by an external electrical constraint well defined. This result might seem paradoxical, if the phenomenon obtained was really chaotic (at random); in fact, it is not the case, since the attractor of the system is only a small and well defined part of plan (S,A).

When researchers are thinking of the role of chaos in biological systems, or more generally in complex systems, a role of "research at random" is often attributed to it; for example, the movements at random of micro-organisms might be interpreted in terms of research of the local optimal medium. It is interesting to note that we have been able to show the possibility of inducing such a function of research by the action of an external electric signal easy to be controlled. This result might also have implications in the field of information processing in biomimetic systems [12,13].

REFERENCES

1. R. GOLDMAN, O. KEDEM et E. KATCHALSKI, Papain-collodion membranes. II. Analysis of the kinetic behavior of enzymes immobilized in artificial membranes. Biochemistry 7(12):4518 (1968).
2. E. SELEGNY, Some systems coupling enzymic reactions and other phenomena; energy conversions, in "Polyelectrolytes", E. SELEGNY ed., Reidel publishing company, Dordrecht, Holland (1974)
3. P. MONSAN, Application de la catalyse enzymatique à la synthèse chimique, Biosciences, 1(3):40 (1982).
4. O. DECROLY et A. GOLDBETER, Birythmicity, chaos and other patterns of temporal self-organization in a multiply regulated biochemical system, Proc. Natl. Acad. Sci. U.S.A., 79:6917 (1982).
5. J.C. ROUX et H.L. SWINNEY, Topology of chaos in a chemical reaction, in Nonlinear Phenomena in Chemical Dynamics, C. VIDAL et A. PACAULT ed., Springer-Verlag Berlin Heidelberg New-York (1981).
6. C. VIDAL, Chemical kinetics as an experimental field for studying the onset of turbulence, in Nonlinear Phenomena in Chemical Dynamics,. C. VIDAL et A. PACAULT ed., Springer-Verlag Berlin Heidelberg New-York (1981).
7. L.F. OLSEN et H. DEGN, Chaos in biological systems, Quarterly Review of Biophysics 18(2):165 (1985).
8. K. TOMITA, Chaotic Response of Nonlinear Oscillators, Physics Reports 86(3):113 (1982).
9. J.M. VALLETON, J.C. VINCENT et E. SELEGNY, Monitoring of artificial enzyme membrane systems by electric currents. I

Analytical treatment with direct current and buffered conductivity, Biophys. Chem., 15:235 (1982).

10. J.M. VALLETON, Théorie des systèmes en diffusion-électromigration-réaction. Application aux cinétiques enzymatiques, Thèse d'Etat, Université de Rouen, (1984).

11. E. SELEGNY, J.M. VALLETON et J.C. VINCENT, A.C. or D.C. imposed on enzyme systems; control of monostable systems and memory in multistable systems, Bioelectrochem. Bioenerg. 10:133 (1983).

12. J.M. VALLETON et A.J. GRODZINSKY, Conductivity mediated regulation in a compartmental enzyme system. Superactivation and conductance bistability, Biophys. Chem., 24:277 (1986)

13. J.M. VALLETON, M. THELLIER, E. SELEGNY et J.C. VINCENT, Utilisation de systèmes physico-chimiques pour le traitement de l'information. Elaboration d'un processus de mémorisation, dans Proceedings of the workshop held in Rouen, "Systèmes modèles de traitement biologique de l'information", M; THELLIER et J.M. VALLETON ed., Publications de l'Université de Rouen (1984).

THERMODYNAMIC ASPECTS OF THE EVOLUTION

OF SELF-ORGANIZED SYSTEMS

C. Beldie, G. Bourceanu and I. Grosu[*]

Physical Chemistry, Polytechnic Institute
* Physics, Polytechnic Institute, Iasi 6600,
Romania

INTRODUCTION

The school from Brussels have found that for the systems maintained far from thermodynamic equilibrium and when the kinetic laws are suitably nonlinear, the excess entropy production, $\delta_x P$, becomes negative for $t > t_0$. In this case, the nonequilibrium steady state becomes unstable and sometimes, when the internal structure permits, the system executes limit cycle oscillations [1-3].

At present, the problem is different. Assuming that a dissipative structure has been formed, one must find a thermodynamic quantity to be able to characterize this structure as uniquely as possible [3]. Moreover, having many self-organized systems introduces the question : how can we know which of the systems is more ordered ? On the other hand, is there a connection between complexity, order, and structural stability of self-organized system ? Which evolutionary principles govern the behaviour of self-organized systems ? These are some of questions which are to be solved. Mänsson [4] had investigated the entropy production in oscillating homogeneous chemical systems by analyzing the difference between the average entropy production rate, in a stable periodic oscillatory mode and in the corresponding unstable stationary state, using models with three and four intermediate species. The models exemplify both positive and negative entropy production rate differences.

Independently, at the same time, two authors of this work computed the entropy produced by a self-organized system over a period of limit cycle oscillation, S_i^τ, and the entropy produced in the steady state over the same period $S_{i,st}^\tau$, for four models with three, four, five and six intermediate species with autocatalytic properties [5].

Huberman and Hogg [6] introduce a physical measure of the complexity of a system based on its diversity.

In this paper we try to correlate the complexity, the order and the structural stability of self-organized systems with a thermodynamic quantity. We have taken for thermodynamic quantity, to characterize the triad : complexity - order - stability, the differences $\Delta_i S^\tau = S_i^\tau - S_{i,st}^\tau$, that represents a measure of the distance between organized state of system and unstable steady state in the disordered system.

NUMERICAL RESULTS

To compute S_i^T and $S_{i,st}^T$, we have taken into consideration the self-organized systems which we have suggested in a previous paper [7] with four, five and six intermediate species and, also, the self-organized system with three species suggested by Hanusse [8].

The systems of differential equations (DE) describing the temporal evolution of systems with three (DE.3), four (DE.4), five (DE.5) and six (DE.6) intermediate species are the following :

$$(DE.3)$$

$$\dot{x}_1 = Ax_1 - k_{-1}x^2 - x_1x_2 + k_{-2}x_2^2$$

$$\dot{x}_2 = x_1x_2 - k_{-2}x_2^2 - x_2x_3 + k_{-3}x_3^2$$

$$\dot{x}_3 = x_2x_3 - k_{-3}x_3^2 - x_1x_3 + k_{-4}x_1B$$

$$(DE.4)$$

$$\dot{x}_1 = Ax_1 - k_{-1}x_1^2 - x_1x_2 + k_{-2}x_2^2$$

$$\dot{x}_2 = x_1x_2 - k_{-2}x_2^2 - x_2x_3 + k_{-3}x_3^2$$

$$\dot{x}_3 = x_2x_3 - k_{-3}x_3^2 - x_3x_4 + k_{-5}x_4^2 - x_1x_3 + k_{-4}x_1B$$

$$\dot{x}_4 = x_3x_4 - k_{-5}x_4^2 - x_2x_4 + k_{-6}x_2B_1 + A_1x_4 - k_{-7}x_4^2$$

$$(DE.5)$$

$$\dot{x}_1 = Ax_1 - k_{-1}x_1^2 - x_1x_2 + k_{-2}x_2^2$$

$$\dot{x}_2 = x_1x_2 - k_{-2}x_2^2 - x_2x_3 + k_{-3}x_3^2$$

$$\dot{x}_3 = x_2x_3 - k_{-3}x_3^2 - x_3x_4 + k_{-5}x_4^2 - x_1x_3 + k_{-4}x_1B$$

$$\dot{x}_4 = x_3x_4 - k_{-5}x_4^2 - x_4x_5 + k_{-8}x_5^2 - x_2x_4 + k_{-6}x_2B_1 + A_1x_4 - k_{-7}x_4^2$$

$$\dot{x}_5 = x_4x_5 - k_{-8}x_5^2 - x_3x_5 + k_{-9}x_3B_2$$

and respectively,

$$(DE.6)$$

$$\dot{x}_1 = Ax_1 - k_{-1}x_1^2 - x_1x_2 + k_{-2}x_2^2$$

$$\dot{x}_2 = x_1x_2 - k_{-2}x_2^2 - x_2x_3 + k_{-3}x_3^2$$

$$\dot{x}_3 = x_2x_3 - k_{-3}x_3^2 - x_3x_4 + k_{-5}x_4^2 - x_1x_3 + k_{-4}x_1B$$

$$\dot{x}_4 = x_3x_4 - k_{-5}x_4^2 - x_4x_5 - x_2x_4 + k_{-6}x_2B_1 + A_1x_4 - k_{-7}x_4^2 + k_{-8}x_5^2$$

$$\dot{x}_5 = x_4x_5 - k_{-8}x_5^2 - x_5x_6 + k_{-10}x_6^2 - x_3x_5 + k_{-9}x_3B_3 + A_2x_5 - k_{-12}x_5^2$$

$$\dot{x}_6 = x_5x_6 - k_{-10}x_6^2 - x_4x_6 + k_{-11}x_4B_3$$

where A, A_1, A_2 are input reactants and B, B_1, B_2 are output products whose concentrations are kept constant and uniform inside the reaction volume, while x_i denotes concentration of intermediate species X_i (i = 1,...,6), and are allowed to vary freely.

We have assumed that rate constants forward, $k_i = 1$, and this is the reason that these don't appear in the (DE.3) - (DE.6) above and $k_{-1} = 0.1$. However, for reactions that achieves the exchange of substances with the environment, the constants of the reverse reactions have been taken : $k_{-1} = k_{-4} = k_{-6} = k_{-7} = k_{-9} = k_{-11} = k_{-12} = 10^{-15}$. We have taken into consideration these constants of reverse reaction in order to compute chemical affinity.

In a previous work [7], one of us has proved that transition from a self-organized system, less complex (with a smaller number of intermediate species and of reactions) to a more complex one, does not take place under the same concentration gradient, but only if we increase the gradient in order to maintain the new system far from the thermodynamic equilibrium. That is the reason why the concentrations of substances A, A_1 that feed the systems assume different values, depending on the complexity of the systems, but these values are close to the values corresponding to bifurcation (see Table 1). With these observations, we have computed the entropy produced along a trajectory of the limit cycle S_i^τ (of period τ) and in the corresponding unstable steady state over the same period , $S_{i.st.}^\tau$, for every system under consideration.

The entropy production per unit time and unit volume is [2,3,9]

$$d_i S/dt = P = \sum_i J_i X_i \tag{1}$$

where, J_i represents the thermodynamic flux that, in the case of forward and reverse reactions, is :

$$J_i = v_i - v_{-i} \tag{2}$$

v_i being the rate of the forward reaction i, v_{-i} the rate of the reverse reaction, and X_i represents the thermodynamic force that, in the case of chemical reactions, is :

$$X_i = \mathcal{A}_i/T \tag{3}$$

In (3) $\mathcal{A}_i$ is the chemical affinity of reaction i, defined by [10] :

$$\mathcal{A}_i = -\sum_j v_{ij}\mu_j \tag{4}$$

where v_{ij} is the stoichometric coefficient of j in reaction i, and μ_j is the chemical potential of j. Because only in real chemical systems it is possible to compute $\mathcal{A}_i$ by (4), we compute affinity with relation [1-3] :

$$\mathcal{A}_i = RT \ln(v_i/v_{-i}) \tag{5}$$

Integrating (1) over a period τ of the limit cycle oscillation, and taking into account (2), (3) and (5), we get :

$$S_i = \int_0^\tau P(t)dt = \int_0^\tau \sum_i J_i X_i dt = R \int_0^\tau \sum_i (v_i-v_{-i})\ln \frac{v_i}{v_{-i}} dt \tag{6}$$

The universal gas constant, R, was not included in the computation. By the Runge-Kutta numerical integration of the systems of differential equations : (DE.3), (DE.4), (DE.5) and (DE.6) one obtains the concentrations x_i of the X_i and then, we obtain S_i^τ for every system. We have also computed $S_{i,st}^\tau$. The results are given in Table 1. The following remarks can be made from the analysis of the results.

505

Table 1.

System	steady state	Period (τ)	S_i^τ	$S_{i,st.}^\tau$	$\Delta_i S^\tau$
(DE.3) A=1.05	(1.07,1.17,1.05)	4.42	382.71	367.82	14.89
(DE.4) A=1.00, A_1=0.75	(0.64,1.23,0.54, 0.60)	5.35	503.32	424.63	78.68
(DE.5) A=1.00, A_1=1,25	(0.64, 1.23,0.54 0.60,0.55)	5.13	646.09	519.06	127.03
(DE.6) A=1.00, A_1=1.25 A_2=1.00	(0.67,1.21,0.58 0.53,0.63,1.06)	5.07	931.25	723.44	207.81

a) For all self-organized systems under study the difference : $\Delta_i S^\tau = S_i^\tau - S_{i,st}^\tau > 0$, and this increases with the increase in the complexity of the systems, that is with the increase in the number of species $\{X_i\}$, in the number of reactions these species take part in.

b) If the order occurs far from the unstable steady state, then $\Delta_i S^\tau$, that represents a measure of the distance between organized state and the disordered unstable steady state, could constitute a thermodynamic quantity that characterizes the degree of order, namely the more ordered the system, the greater the difference $\Delta_i S^\tau$. But this difference, $\Delta_i S^\tau$, according to a) increases with the increase in complexity. Therefore :

c) The more complex a self-organized system, the more ordered.

We try to find another simple linkage between complexity and order making the assertion : a self-organized system is more ordered than another less complex self-organized system.

Indeed, let us consider two self-organized systems with K_1 and K_2 components, respectively ($K_1 > K_2$). The time dependence of the components for the two systems can be written :

$$x_i^{(1)} = x_i^{(1)}(t) \quad ; \quad i = 1,2,...,K_1 \tag{7}$$

$$x_i^{(2)} = x_i^{(2)}(t) \quad ; \quad i = 1,2,...,K_2 \tag{8}$$

The relation (7) defines a closed trajectory in the space of the K_1 variables.

If we know a component of the first system we can know the other K_1-1 components using (7).

Thus the ratio of dependent components number and total number of components can measure the relative order. Therefore we define the coefficient

$$\mathcal{N}_1 = \frac{K_1 - 1}{K_1} \tag{9}$$

as the order coefficient. If $K_1 > K_2$ then,

$$\mathcal{N}_1 > \mathcal{N}_2 = \frac{K_2 - 1}{K_2} \tag{10}$$

and we have a linkage between order and complexity.

Thus, the increase in complexity of self-organized systems determines the increasing of order.

Certainly, for a self-organized system with a certain complexity we get different closed trajectories for different concentration gradients. Consequently, we shall get different values for $\Delta_i S^T$, though the complexity of the system is the same. Indeed, we have taken different values for the concentration of component A_1 that feeds systems (DE.4)-(DE.6), and have computed $\Delta_i S^T$ (see Table 2).

Table 2.

(DE.4)		(DE.5)		(DE.6)			
A_1	$\Delta_i S^T$	A_1	$\Delta_i S^T$	A_1	$\Delta_i S^T$	A	$\Delta_i S^T$
0.50	56.37	0.75	88.81	0.75	154.27	0.75	154.05
0.60	72.45	1.00	122.09	1.00	192.58	1.00	192.58
0.75	78.67	1.25	127.03	1.25	207.81	1.25	195.53
0.85	58.22	1.50	103.55	1.50	163.27	1.50	158.58

From Table 2 we can notice that $\Delta_i S^T$ increases with the increase in the concentration of component A_1 reaching a maximum after which it decreases. In this case we can write :

$$d(\Delta_i S^T)/dA_1 = 0 \quad (\text{for } A_1 = A_1^*) \tag{11}$$

where A_1^* is the optimum concentration for which $\Delta_i S^T$ is maximum. The same results is reached if we vary the concentration of component A for the system (DE.6) (see Table 2). But, we have stated that the difference $\Delta_i S^T$ is a measure of the degree of order. Hence, it follows that a system of a certain complexity can reach a high degree of organization only for a certain concentration of the compounds that feed the self-organized systems.

Starting from (11), there follows that there is a value A_1^* (or A^*) for which $\Delta_i S^T$ is maximum.

We define the coefficient of structural stability

$$\mathcal{N}' = 1 - \left| \frac{A_1^*}{\Delta_i S^T(A_1^*)} \cdot \frac{d(\Delta_i S^T)}{dA_1} \right| \tag{12}$$

This coefficient indicates that a maximum of structural stability is obtained for maximum $\Delta_i S^T$.

As follows from numerical results for the systems which we have analysed above it seems that complexity, the order and the structural stability of self-organized systems could be characterized by the quantity $\Delta_i S^T$. However, Mänson [4] as we have already mentioned above shows that there are self-organized systems for which $\Delta_i S^T < 0$. Therefore $\Delta_i S^T > 0$ is not true in general. It remains to be seen for what classes of self-organized systems $\Delta_i S^T > 0$ and for what classes $\Delta_i S^T < 0$.

THE ENTROPY GLOBAL EQUATION FOR SELF-ORGANIZED SYSTEMS

It is now known that the evolution of dissipative structure from the simplest to the most complex ones is keeping with the principle of the creation of entropy [11].

Generally, for an open thermodynamic system and its environment with which it has an exchange of matter, one can write the balance equation for the global variation of entropy [9-12].

$$dS_g = dS + dS_m \tag{13}$$

where dS_g is the global variation of entropy ; dS is the variation of entropy of the system and dS_m is the variation of the environment.

Since the system is open, the term dS can be written as [2,3] :

$$dS = d_i S + d_e S \tag{14}$$

In equation (14) $d_i S$ represents the variation of the entropy of the system due to the irreversible processes inside the system such as the chemical reactions, heat conduction and diffusion.

The second law implies :

$$d_i S \geqslant 0 \qquad (= 0 \text{ at equilibrium}) \tag{15}$$

The second term of (14), $d_e S$ represents the flux of entropy due to exchange of matter between the system and its environment. Let us note that the term $d_e S$ in (14) can also be written as [9] :

$$d_e S = d_e S_{(s \to m)} + d_e S_{(s \to m)} \tag{16}$$

An equation similar to (14) can also be written for the variation of the entropy of the environment :

$$dS_m = d_i S_m + d_e S_m \tag{17}$$

In (17) $d_i S_m$ represents the variation of the entropy of the environment due to irreversible processes that occur, such as the reactions that lead to the recovering of a part of the amount of matter that feed the system [14].

For this case too :

$$d_i S_m \geqslant 0 \qquad (0 = \text{at equilibrium}) \tag{18}$$

The term $d_e S_m$ in (17) represents the flux of entropy exchanged by the environment only with the system under consideration. With (14) and (17), the balance equation (13) becomes :

$$dS_g = d_i S + d_e S + d_i S_m + d_e S_m \tag{19}$$

Taking into consideration (16) it is obvious that

$$d_e S = - d_e S_m \tag{20}$$

Since both within the system and its environment irreversible processes occur, and taking into account (20), equation (19) becomes :

$$dS_g = d_i S + d_i S_m > 0 \tag{21}$$

The global entropy produced per unit time is :

$$dS_g/dt = d_iS/dt + d_iS_m/dt > 0 \qquad (22)$$

and the global entropy produced over a period of limit cycle oscillation is :

$$S_g^\tau = \int_0^\tau dS_g = S_i^\tau + S_{im}^\tau > 0 \qquad (23)$$

We do not know the entropy produced by the environment (with which the system achieves exchange of substances) over the same period τ, but in keeping with (18), there follows that :

$$S_{im}^\tau \quad 0 \qquad (24)$$

Taking into consideration (23) and remark a) there follows that the global entropy S_g^τ, produced over a period of oscillation, increases with the increase in the complexity of self-organized systems. This is the very reason why the evolution of self-organized systems, from the simplest to the most complex ones is in keeping with the principle of the creation of entropy. We can say that the evolution of self-organized systems chooses the way of complexity because it corresponds to the creation of ever so great an entropy. Therefore, the apparition of dissipative structure, far from the thermodynamic equilibrium, and their evolution towards more and more complex structure appears to be a necessity. The errors in replication of the genetic material could be thought of as fluctuations [3]. These fluctuations responsible for the appearance of new species [7], represent the aleatory factor, the hazard part. As Prigogine said "Hazard and necessity cooperate instead of being opposed to each other" [11].

REFERENCES

[1]   I. Prigogine and R. Lefever, Symmetry Breaking Instabilities in Dissipative Systems. II, J. Chem. Phys. 48 : 1695 (1968).
[2]   P. Glansdorff and I. Prigogine, "Thermodynamic Theory of Structure, Stability and Fluctuation", Wiley-Interscience, New-York, N.Y. (1972).
[3]   G. Nicolis and I. Prigogine, "Self-organization in Non-equilibrium Systems", Wiley-Interscience, New York, N.Y. (1977).
[4]   B.A.G. Mänsson, "Entropy Production in Oscillating Chemical Systems", Z. Naturforsch. 40a : 877 (1985).
[5]   G. Bourceanu, I. Grosu and G. Morosanu, "Thermodynamic Aspects of the Evolution of Self-Organized Systems I", To be published.
[6]   B.A. Huberman and T. Hogg, "Complexity and Adaptation, Physica 22 D : (1986).
[7]   G. Bourceanu and G. Morosanu, "The study of the Evolution of Some Self-Organized Chemical Systems", J. Chem. Phys. 82 : 3685 (1985).
[8]   P. Hanusse, "De l'existence d'un cycle limite dans l'évolution de systèmes chimiques ouverts", C.R. Acad. Sci., Paris, Ser.C 277 : 263 (1973).
[9]   G. Bourceanu, Gh. Maftei and S. Chirita, "Thermodynamic of Evolution", ST. Cerc. Fiz. 35 : 690 (1983).
[10]   I. Prigogine and R. Defay, "Chemical Thermodynamics", Longman (1962).
[11]   I. Prigogine, "La thermodynamique de la vie", La Recherche 24 : 546 (1972).
[12]   K. Guminski, "Thermodynamika procesow nieodwracalnych", Warszawa (1962).

DISCUSSION

DE KEPPER - Have you taken into account also the diffusion ?

BOURCEANU - It is not taken into account, because was followed only the temporal order.

ANOMALOUS STOCHASTIC KINETICS

Péter Érdi[1] and János Tóth[2]

1: Central Research Institute for Physics
   Hungarian Academy of Sciences
   H-1525 Budapest, P.O.B. 49, Hungary
2: Computation and Automatation Institute
   Hungarian Academy of Sciences
   H-1502 Budapest, P.O.B. 63, Hungary

INTRODUCTION

Though traditional stochastic models of chemical reactions (e.g. McQuarrie 1967, Gardiner 1983) proved to be very efficient, the underlying (sometimes tacit) assumptions do not fulfil for arbitray chemical systems.

Both the mesoscopic level, continuous time discrete state stochastic (CDS) and the macroscopic, continuous time continuous state deterministic (CCD) models of the chemical reactions characterize the state of the system by a finite - dimensional composition vector. Spatial inhomogenity and randomness, time inhomogenity and non-Markovity, randomness in the energy distribution (and change in it at microscopic level) can not be passed into the usual mathematical framework.

The CDS model, which is a time-homogenous Markovian jump process, proved to be relevant for describing composition fluctuation phenomena both around and out of equilibrium. While the CDS model is not well-founded from microscopic point of view, the availability of new techniques for the study of fast reactions (e.g. Jonah, this volume) even at the femtosecond scale makes necessary to set up coupled microscopic - mesoscopic models. Earlier works (e.g.Gaveau and Moreau 1985, Borgis et al 1986, Moreau and Gaveau 1987) emphasizing the existence of non-Markovian collision processes tended into this direction.

In this paper the outlines of anomalous stochastic kinetic models are discussed. These models are derived by relaxing certain traditional assumptions. Furthermore, a very specific example is given to illustrate that spatial inhomogenity can lead to time inhomogenity by some lumping procedure.

0.   The traditional CDS model of a complex chemical reaction is given by

$$dP_n(t)/dt = A_{nn} P_n(t) \tag{1}$$

where $P_n(t)$ is the absolute probability distribution, and $\{A_{nn}\}$ is a time-independent constant matrix composed by the stochiometric and rate coefficients. (A more detailed mesoscopic description requires evolution equation for the transition probabilities). Assumptions of the CDS model are examined in order to neglect them if it seems to be necessary.

1.   Homogenity in space

The idea of homogenous spatial distribution of the particles is based on the concept of well-stirred reactor. However, even microscopic reactions produce local nonhomogenities, which can not be always eliminated by diffusion. There are many contraversions about the stochastic formulation of reaction - diffusion systems. Three directions in the theory of random fields seem to be able to cope with such complexity: the theory of random measures, the theory of stochastic partial differential equations relating to trajectories, and the theory of Hilbert space valued stochastic processes. The details are beyond of our scope.

Random walk treatment of reaction - diffusion systems is based on the discretization of the space. Phenomena taking place in homogenous phase are described by random walk on regular lattice. Spatial randomness, a very particular form of spatial non-homogenity might have role in heterogenous systems as well in glasses. Random walks and reactions on fractals (Zumofen et al 1985, Kopelman 1986) offer an appropriate mathematical tool for describing these kinds of situations.

2.   Homogenity in time

In deterministic framework time homogenity is associated to autonomous differential equations:

$$dx(t)/dt = f(x(t), k) \tag{2},$$

where x is the composition vector, and k is the rate constant. Traditional CCD models are specific forms of (2). Time-dependent rate constant has been introduced in the classical paper of Smoluchowski (1917) for the case of diffusion limited bimolecular reactions. Other kinds of time dependency occuring in the context of radiation chemistry have been surveyed recently by Plonka (1986).

The appropriate stochastic model for time inhomogenous Markovian systems is:

$$dP_n(t)/dt = A_{nn}(t) P_n(t) \tag{3}$$

3.   Exponential waiting time

Markovian jump processes are characterized by exponential waiting time distribution (Doob 1953, pp. 244).

The relaxation of this exponentiality condition, which precisely coincides with the Markovian assumption, leads to semi-Markov processes (Feller, 1964). What could be the chemical conditions of the occurence of non-exponential waiting times, and which distributions could have physico - chemical relevance? Though many works have been done with the method of the continuous time random walk (CTRW) mostly in connection with transport processes in amorphous media, it seems to be very hard job to associate chemical conditions to different kinds of waiting time distributions due to 'temporal disorders'.

The exponential waiting time distribution

$$w(t):=a \exp(-at), \quad a \geq 0 \tag{4}$$

was replaced by

$$w(t):=cat^{a-1} \exp(-ct^a), \quad 0 < a < 1 \tag{5}$$

and $$w(t):=ct^{-1-b}, \quad 0 < b < 1 \tag{6}$$

for particular physico-chemical systems (for review see e.g. Blumen et al 1986). In particular, the distribution is connected to long-time tail.

Formally, evolution equation for non-Markov processes can be described as integro-differential equations:

$$dP(t)/dt=\int_{t_o}^{t} A_{ts} P(s)ds \tag{7}$$

where A_{ts} is the 'kernel function'. The 'memoryless' evolution equation can be specialized from the time - convoluted master equation (see e.g. Gillespie 1977).

4. Locality in state space
    The Levy distribution and Levy process have many interesting properties, partially connected to random walks on fractals (see e.g. Lindenberg and West 1986). The right-hand-side of the evolution equation for a Levy process can not be constructed by differential operators. The schematic form of the equation is:

$$\partial P_a(x,t)/\partial t = A \int_{-\infty}^{\infty} K(x-y)P_a(y,t)dy, \quad 0 < a \leq 2 \tag{8}$$

5. Ordered energy barriers
    Usual microscopic models of chemical reaction assumes the existence of ordered energy barriers. Chemical reaction is used to be considered as diffusion in the phase space, Kramers (1940). It is not clear, whether how to switch such kinds of microscopic models to traditional CDS models. One of the main open problems in chemical kinetic is to derive CDS model from microscopic picture. It seems to be credible, that 'ordered energy barriers' can lead to traditional Markov models. Random fluctuations in the energy barrier can have ultrametric structure (e.g. Zumofen et al 1986) and may lead to much more complex, i.e. hierarchical dynamic models (Vilgis 1987). Chemical reaction can be interpreted as anomalous, i.e. 'ultradiffusion' (Huberman and Kerszberg 1985).While normal diffusion is characterized by the relation

$$R(t):=Ct^{1/2} \tag{9}$$

where $R(t)$ is the expectation of the displacement, other laws, e.g.:

$$R(t):=C\log t \tag{10}$$

define anomalous diffusion, therefore anomalous kinetic models.

SPATIAL INHOMOGENITY CAN IMPLY TIME INHOMOGENITY

The simplest extension of the homogenous kinetic models for describing spatial effects is the introduction of the two-cell model. According to this approach reactions taking place within the homogenous cells, while transport of matter between the cells are taken into account as one-dimensional diffusion. In particular, the two-cell stochastic model of the Schlögl reaction describing first-order phase transition has carefully been studied (Borgis and Moreau 1984).

Here we study a two-cells system. The reaction $A \rightleftharpoons B$ is taking place in both cell. The temperature of the two cells can be different, i.e. the rate constants not necessarily are the same. Two specific situation is studied:
(1) we have precisely one particle, and the master equation for the absolute probability distribution is calculated;
(2) the number of particles is arbitrary, equation for the temporal change of the expectation is given.

How the equations are changef after lumping the two cells, if time homogenity in the individual cells is assumed?

The lumped system is characterized for the case (1) by the lumped probability distributions

$$\hat{P}_{10}(t):=P_{1000}(t)+P_{0010}(t) \tag{11}$$

$$\hat{P}_{01}(t):=P_{0100}(t)+P_{0001}(t) \tag{12}$$

where $P....$ denotes the probability of the possible four states. (P_{1000} and P_{0010} means that the particle is in state 'A' in the first and second cell, respectively; etc.)

For the case (2) the state variables are the expectation of the number of particles:

$$a(t):=a_1(t)+a_2(t) \tag{13}$$

$$b(t):=b_1(t)+b_2(t)$$

Because of the very specific situation a common model for the two systems can be derived in the form:

$$
\begin{aligned}
x' &= -(k_1+D_1)x + k_2 y + D_2 u \\
y' &= k_1 x - (k_2+D_1)y + D_2 v \\
u' &= D_1 x - (k_1'+D_2 u)+k_2 v
\end{aligned}
\tag{14}
$$

$$v' = \qquad D_1 y \qquad +k_1'u \quad -(k_2 +D_2)v$$

Introducing the notation $x:=a_1$, $y:=b_1$, $u:=a_2$, $v:=b_2$ equation for the expectation, with the notation $x:=P_{1000}$, $y:=P_{0100}$, $u:=P_{0010}$, $v:=P_{0001}$ the master equation is obtained.

Statement 1.

If the rate constants in the two cells are the same, i.e. $k_1'=k_1$ and $k_2'=k_2$, then the time-independency of the coefficient matrix is preserved in the lumped models.

Statement 2.

If $k_1' \neq k_1$ and $k_2' \neq k_2$, then the lumped model has variable coefficient, and its form is

$$d\hat{P}_{10}(t)/dt = h(t)\hat{P}_{10}(t)$$

$$d\hat{P}_{01}(t)/dt = -h(t)\hat{P}_{01}(t)$$

(15)

DISCUSSIONS

The occurence of deviance from the usual assumptions (i.e. 'ordered' structure of time, space and energy) leads to anomalous stochastic kinetic models.
From formal point of view it is a specific open question, whether which kinds of space and energy disorders can be converted to time disorders described by semi-Markov processes? A semi-Markov process can be associated to a Markov process by identifying the expectation of the waiting time of the former with the waiting time of the latter. The stationary probability distribution of the two models is the same, but the realizations can be qualitatively different.
The concept of 'ultrametrics' has emerged as an important mathematical concept to describe complex, hierarchical dynamic structures, as spin glasses, computing devices, memory. It would be interesting to see, how to derive hierarchical chemical kinetic models by switching the microscopic and mesocopic level of the description.

ACKNOWLEDGEMENTS

Invitation and financial support coming from the organizers is highly appreciated. We thank Prof.M.Moreau, Prof. R.Schiller and Dr. M.Frankowicz for discussions.

REFERENCES

Blumen,A., Klafter,J. and Zumofen,G. 1986, In:"Optical Spectroscopy of Glasses". Physics and Chemistry of Materials with Low-Dimensional Structures, D.Reidel Publishing Company Dordrecht.
Borgis,D. and Moreau, M., 1984, On the Stochastic Theory of a Bistable Chemical Reaction. Physica 129A:109(1984)
Borgis,D., Gaveau,B. and Moreau,M., A Class of Collision Process with Memory and Application to Simple Chemical Reactions in a Solvent, J.Stat.Phys.45:319(1986)
Doob,J.L.,"Stochastic Processes" Wiley,New York (1953)

Feller,W., On Semi-Markov Processes.
    Proc.Nat.Acad.Sci. 51:653(1964)
Gardiner,C.W., "Handbook of Stochastic Mathods"
    Springer-Verlag (1983)
Gaveau,B. and Moreau,M., On a Class of Non-Markovian
    Collision Processes and their Evolution Equation,
    Lett.Math.Phys., 9:213(1985)
Gillespie,D.T., Master Equations for a Random Walks with
    Arbitrary Pausing Time Distributions, Phys.Lett.
    64A:22 (1977)
Huberman,B.A., and Kerszberg,M., Ultradiffusion: the
    relaxation of hierarchical systems, J.Phys.A.:Math.Gen.
    18:L331 (1985)
Jonah,C.D., Techniques for the Study of Fast Reactions in
    Liquids, This Volume.
Kramers,H.A., Brownian Motion in a Field of Force and the
    Diffusion Model of Chemical Reactions, Physica 7:284
    (1940)
Kopelman,R. Rate Processes on Fractals: Theory, Simulations,
    and Experiments, J.Stat.Phys. 42:185(1986)
Lindenberg,K, and West,B.J., The First, The Biggest, and
    Other Such Considerations, J.Stat.Phys.42:201 (1986)
McQuarrie,D., Stochastic Approach to Chemical Kinetic.
    Journal of Applied Probability 7:413 (1964)
Moreau,M., and Gaveau,B., Definition and General Properties
    of a Class of Non-Markovian Collision Processes,
    J.Math.Phys. 26:1430 (1987)
Plonka,A., "Time-Dependent Reactivity of Species in Condensed
    Media", Lect.Notes in Chemistry, Vol.40, Springer
    Verlag, Berlin (1986)
Schmoluchowski,M., Mathematical Theory of the Kinetics of the
    Coagulation of Colloidal Solutions, Z.Physik.Chem.
    92:129(1917)
Vilgis,T.A., On the Dynamics of Simple Hierarchical Systems,
    J.Stat.Phys. 47:133(1987)
Zumofen,G., Blumen,A., and Klafter,J., Random Walks on
    Fractals, In. Structure and Dynamics of Molecular
    Systems, Daudel,R., et al, eds., Reidel,D. Publishing
    Company, Dordrecht (1985)
Zumofen,G., Blumen, A., and Klafter, J., Reaction Kinetics
    on Ultrametric Spaces, J.Chem.Phys. 84:6679(1986).

KINETIC SYMMETRIES: SOME HINTS

János Tóth and Péter Érdi

Computer and Automation Institute
Hungarian Academy of Sciences, Budapest, Hungary
Central Research Institute for Physics
Hungarian Academy of Sciences, Budapest, Hungary

In the qualitative theory of differential equations, especially in the so called "catastrophe theory", it has turned out that differential equations of the gradient type are relatively easy to deal with /Thom, 1975/. Furthermore - and this may prove more relevant - it has been proposed sometimes that gradient systems are only worth studying in thermodynamics. /Gyarmati, 1961a, b; Edelen, 1973/. Easy treatment and physical relevance has also come up in connection with equations with other kinds of symmetries too: in connection with Hamiltonian systems and systems having similar but different specialities. In the present paper we study the relevance of these notions within the class of kinetic differential equations. The two-dimensional case is fully clarified, the multidimensional case is only sketched. Reactions are shown with each of the mentioned properties. Finally, a common framework is provided to embrace al the properties within a single definition.

GENERAL DEFINITIONS IN TWO DIMENSIONS

Let us consider the differential equation

$$\dot{x} = u \circ /x, \; y/ \qquad \dot{y} = v \circ /x, \; y/ \qquad\qquad /1/$$

with $u,v \in C^1 /R^2, R/$. This equation is said to be a <u>gradient system</u>, if there exists a scalar valued function $V \in C^1 /R^2, R/$ such that $V' = /u, \; v/$. Because the domain of $/u, \; v/$ is simply connected /1/ is known to be a gradient system if and only if $\partial_2 u = \partial_1 v$ holds. The system /1/ is said to be a <u>Shahshahani gradient system</u>, if the system with the right hand side $U/x, \; y/ := u/x, \; y//x, \quad V/x, \; y/ := v/x, \; y//y$ is a gradient system. The system /1/ is a Shahshahani gradient system, if and only if $y\partial_2 u/x, \; y/ = x \partial_1 v/x, \; y/$ holds. The system /1/ is said to be a <u>Hamiltonian system</u>, if there exists a scalar valued function $H \in C^2 /R^2, R/$ such that $u = \partial_2 H$, $v = -\partial_1 H$. This is equi-

valent to saying that $\partial_1 u = -\partial_2 v$ holds or that the system with
the right hand side /-v, u/ is a gradient system. The system
/1/ is said to be a <u>Cauchy-Riemann-Erugin</u> /or, <u>CRE</u>, for short/
system, if u+iv as a function of x+iy /i. e. the function w,
where w/z/:=u/Re z, Im z/+iv/Re z, Im z// can be considered as
an analytic function. Elementary complex analysis teaches us
that this is the case if and only if $\partial_1 u = \partial_2 v$ and $\partial_2 u = -\partial_1 v$
holds, or, equivalently if both systems with the right hand
sides /v, u/ and /u, -v/ are gradient systems. Another sufficient and necessary condition of this last property is

$$f'J = Jf' \tag{/L/}$$

where

$$f := /u, v/ \quad \text{and} \quad J := \begin{bmatrix} 0 & 1 \\ -1 & 0 \end{bmatrix} \ .$$

Now let us turn to the special case when

$$u/x, \ y/ := ax^2 + 2bxy + cy^2 + dx + ey + f \tag{/2/}$$
$$v/x, \ y/ := Ay^2 + 2Bxy + Cx^2 + Dy + Ex + F \ , \tag{/3/}$$

where a,b,...,F$\in$R are real coefficients. In accordance with
general definitions /Érdi and Tóth, in press/ this equation is
said to be a <u>kinetic one</u> /or a <u>kinetically admissible</u> one, Siré, 1986/ if c,e,f,C,E,F$\cong$0 holds. An immediate application of
the definitions shows that equation /1/ with the right hand side
/2/-/3/ is a

gradient	system	B=c, C=c, E=e
Shahshahani gradient	if and	B=b, C=c=0, E=e=0
Hamiltonian	only	A=-b, B=-a, D=-d
CRE	if	A=-C=b, B=a=-c, D=d, E=-e

holds.
    Let us consider a kinetic example for each of this notions.

Gradient	2X--→X+Y--→2X+2Y, 2Y--→2X
Shahshahani gradient	X+Y--→0--→X--→2X--→3X └--→Y--→2Y--→3Y
Hamiltonian	2X--→X, 2Y--→X+2Y, X+Y--→X
CRE	2X--→X, 2Y--→X+2Y, X+Y--→X

    If /1/ is to be <u>kinetic and gradient</u>, then B=c, E=e, F, f
$\geq$0 must hold. This, however, either excludes that /1/ be con-

servative in the Horn-Jackson sense /see e. g. Érdi and Tóth, in press/ or that /1/ is the model of second order reactions too. This is so, because conservativity would imply the existence of positive numbers r and s such that

$$/ra+sC/x^2+2/rb+sB/xy+/rc+sA/y^2+$$
$$/rd+sE/x+/re+sD/y+rf+sF = 0 \qquad\qquad /4/$$

holds at least along the trajectories of the equation /1/. This is equivalent to saying that /4/ holds for all nonnegative numbers x and y. Consequently, all the coefficients of the polynomial on the left hand side of /4/ are equal to zero, thus A=B=C=F=a=b=c=f=0, and

$$D = t^2d, \quad E = -td, \quad e=-td \quad /where\ t:=r/s/.$$

Therefore the general form of conservative gradient systems within the class /1/ of equations is

$$\dot{x} = d/x-ty/ \qquad \dot{y} = -td/x-ty/, \qquad\qquad /5/$$

thus no second order reaction can occur /cf. with Tóth, 1979/.

If /1/ is a <u>Shahshahani gradient</u> system then it is a <u>kinetic</u> system too, because the general form of these equations is

$$\dot{x} = ax^2+2bxy+dx+f \qquad \dot{y} = Ay^2+2bxy+Dy+F,$$

where f,F$\leq$0 and all the other coefficients may be of both signs. It is easy to see that the only conservative Shahshahani gradient system is

$$\dot{x} = 0 \qquad \dot{y} = 0 . \qquad\qquad /6/$$

If /1/ is to be <u>kinetic and Hamiltonian</u>, then c,e,f,C,E, F$\leq$0, A=-b, B=-a, D=-d must hold. This, however, excludes that /1/ be conservative too, except the trivial - zero right hand side - case. This is so, because conservativity would imply the existence of positive numbers r and s such that /4/ holds for all nonnegative numbers x and y. Consequently, all the coefficients of the polynomial on the left hand side of /4/ are equal to zero. Again, a short investigation of the coefficients shows, that a=...=F=0; thus the only conservative kinetic Hamiltonian system is /6/.

It may be worth spelling out the general form of CRE type kinetic equations at first:

$$\dot{x} = -cx^2-2Cxy+cy^2+dx+f \qquad \dot{y} = -Cy^2-2cxy+Cx^2+dy+F .$$

Here d is arbitrary, all the other coefficients are nonnegative, and the equation for z:=x+iy is

$$\dot{z} = gz^2+dz+h,$$

where g:=c+iC, h:=f+iF. As /7/ is a separable equation, its solution can be given in explicit form.

If /1/ is to be <u>kinetic and of the CRE type</u> too, then A=-C=b, B=a=-c, E=-e, F, f$\leq$0 and D=d must hold. These, however, only admit the trivial case /6/, as again a short investigation of the conditions shows.

REMARKS ON THE MULTIDIMENSIONAL CASE

The general definitions may be applied with slight modifications./or: appropriate generalization of the definitions/ in the multidimensional case. Hamiltonian and CRE systems can only be defined for an even number of variables. Some other notions can also be introduced, the most important among these is that div f=0, a property equivalent to being Hamiltonian in the two-dimensional case. This property and the existence of /global, time independent/ first integrals are closely connected with the existence of periodic solutions /Tóth, 1987/. Several <u>conjectures</u> will be formulated here.

1. Conservative gradient systems among those kinetic equations which have a second order right hand side can only be of the following form:

$$\dot{x} = Ax \ ,$$

where A is symmetric with nonnegative offdiagonal elements, and nonpositive diagonal elements such that there exists a positive vector r for which Ar = 0.

2. Conservative Hamiltonian systems among the set of polynomial kinetic equations /of any order/ can only have the trivial

$$\dot{x} = 0$$

form.

3. Conservative CRE systems among those kinetic equations which have a second order right hand side can only be of the form

$$\dot{x} = Ax,$$

where A is of a special form.

A COMMON FRAMEWORK

Here we present a common definition of the properties encountered up to now. This will be done using an idea of the basic paper by Horn and Jackson /1972/. They used a similar characterization to describe different type of equilibria of kinetic differential equations.

Let Φ be an operator acting on the right han side f of a differential equation and let us suppose that the following relation holds:

$$\Phi/f/' = \Phi/f/'^{T} \ . \tag{/8/}$$

Let us suppose that $f:=/u, v/$. The equation is seen to to show the above properties if Φ is appropriately chosen, as in the table below:

Symmetry type	$\Phi/u, v/:=$
Gradient	$/u, v/$
Shahshahani gradient	$/u/pr_1, v/pr_2/$
Hamiltonian	$/-v, u/$
CRE	$\begin{bmatrix} v & u \\ u & -v \end{bmatrix}$

As Φ is linear in all the cases it can be given a linear algebraic representation similar to $/L/$.

One of the next steps should be to formulate a variational principle based upon $/8/$ that is able to select between different reactions according to the appropriate choice of O.

REFERENCES

Akin, E., 1979, "The Geometry of Population Genetics", Springer-Verlag, Berlin-Heidelberg-New York /Lecture Notes in Biomathematics 31/.
Bonchev, D., Kamenski, D., and Kamenska, V., 1976, Symmetry and information content of chemical structures, Bull. Math. Biol., 38:119.
Edelen, D. G. B., 1973, Asymptotic stability, Onsager fluxes and reaction kinetics, Int. J. Engng. Sci., 11:819.
Érdi, P., and Tóth, J., in press, "Mathematical Models of Chemical Reactions", Manchester University Press, Manchester.
Gyarmati, I., 1961a, On the phenomenological basis of irreversible thermodynamics, I. Onsager's theory, Per. Polytechn., 5:219.
Gyarmati, I., 1961b, On the phenomenological basis of irreversible thermodynamics, II. On a possible nonlinear theory, Per. Polytechn., 5:321.
Hárs, V., and Tóth, J., 1981, On the inverse problem of reaction kinetics, in: "Qualitative Theory of Differential Equations" /Coll. Soc. János Bolyai 30/ M. Farkas ed., North-Holland Publ. Co., Amsterdam-London-New York.
Horn, F., and Jackson, R., 1972, General mass action kinetics, Arch. Ratl. Mech. Anal., 47:81.
Kerner, E. H., 1971, Statistical-mechanical theories in biology, Adv. Chem. Phys.,19:325.
Ronkin, L. I., 1977, "Elements of the Theory of Multivariable Analytic Functions", Naukova Dumka, Kiev /In Russian/.
Siré, E.-O., 1986, On topological-dynamical equivalent representation of reaction networks: The omega-equation and a canonical class of mass action kinetics, Ber. Bunsenges. Phys. Chem., 90:1087.

Thom, R., 1975, "Structural Stability and Morphogenesis", W. A. Benjamin, Inc., Reading, MA.

Tóth, J., 1987, Bendixson-type theorems with applications, <u>Z. Angew. Math. u. Mech.</u>, 67:31.

Tuljapurkar, S. D., and Semura, J. S., 1979, Liapunov functions: geometry and stability, <u>J. Math. Biol.</u>, 8:25.

MODELLING IN CHEMICAL DYNAMICS

Patrick Hanusse

Centre de Recherche Paul Pascal - CNRS
Université de Bordeaux I,
F - 33405 Talence Cedex, France

INTRODUCTION

It would be interesting to trace the use of the expression "Chemical Dynamics" over the last twenty years. Today, it is used essentially by people working in the field of exotic behaviours of chemical reactions. Does it really imply something different from "Chemical Kinetics"?. In both cases, chemical reaction is involved, and in both cases, one is interested in the evolution of concentrations. It is only a question of perspective and focus of attention that differentiate the reality that each vocable evokes.

In Chemical Kinetics one considers reactions evolving close to equilibrium. Attention is focused on detailed reaction mechanisms by which given initial species are transformed into final products, and equilibrate with them. The dynamics is rather simple, not in term in reaction mechanism, but in term of trajectory complexity. Most of the time, one essentially observes relaxation towards equilibrium. It is precisely the outcome of Thermodynamics that equilibrium is not only stable, but also that, because of the principle of detailed balance, which is part of the equilibrium concept, it is stable in a way that precludes any really complex behaviour. Relaxation towards equilibrium has to be monotonous[1].

Far from equilibrium, on the contrary, this particularity is no longer present. There, one enters the field of Chemical Dynamics. Attention is focused more on the phenomenological dynamical behaviour, rather than on detailed chemical machinery. Non-linear effects manifest their full consequences through a whelm of complex dynamical behaviours, attractors, and coexisting regimes. As such, Chemical Dynamics is part of the general field of Dynamical Systems, and somewhat, chemistry remains in the shadow of behaviour, although it is of considerable importance that chemistry is able to produce such behaviours. Nevertheless, trajectories, attractors, their topology and stability, and bifurcations of solutions, are the elements of the central concern in Chemical Dynamics. It worth pointing out that chemistry has proved to be one of the best fields of experiment for "dynamicians"[2].

As a direct consequence of the particular role of Dynamics, as such, in the study of non-equilibrium behaviour of chemical systems, two classes of models are to be considered, depending on which aspect one is insisting on. Formal models, of mathematical or chemical-like nature, are designed to exhibit specific dynamical behaviours, without too much concern about chemical significance. Their aim is to provide examples of evolution equations of chemical reacting systems, as described by mass action kinetics, that are able to produce those exotic behaviours, such as bistability or multistability, between various types of attractors, like steady states, oscillations or deterministic chaos. A typical historical model of that kind is the "Brusselator"[1]:

$$A \rightarrow X$$
$$2X + Y \rightarrow 3X$$
$$B + X \rightarrow Y + D$$
$$X \rightarrow E$$

where X and Y are intermediate species, and A,B and D are constraint species, which are supposed to be kept at constant concentration. Those, along with the various rate constants, constitute the set of control parameters. For a given set of their values, one follows the evolution of species X and Y. This is a two dimension model, meaning that the trajectory of the system is two dimensional.

This model was used to study the bifurcation of a stable steady state to a limit cycle oscillation, as well as Türing instability leading to spatial structures.

Other models of that kind may result from the reduction of a more realistic and complex reaction scheme. This is the case for the "Oregonator"[3,4]:

$$A + Y \rightarrow X + P$$
$$X + Y \rightarrow 2P$$
$$A + X \rightarrow 2X + 2Z$$
$$2X \rightarrow A + P$$
$$Z \rightarrow hY$$

There are two reasons to so. A practical reason is due the impossibility to analyze the behaviour of complex reaction schemes. A more theoretical reason is that Dynamical Systems Theory shows that most of the behaviours that one is likely to observe, involve only low dimensional interactions. In other words, from the mathematical point of view, most dynamical behaviours can be described in three or four dimensions. This, at the same time, establishes the validity of the reduction project, but not that of a particular reduction procedure.

Formal models, with two to four variables, the only ones that might be tractable, may have a stange chemical look. Yet, a mathematician would probably not even bother to keep any chemical structure to his models, when looking, for instance, for a particular type of bifurcation. In the most reduced form they are known to be described by "normal forms"[5,6], which no longer bear any chemical appearance. Nevertheless, although chemistry, through mass action kinetics, provides all the ingredients necessary to exhibit all those exotic behaviours, it may also impose a number of constraints, that make the reaction or reaction-diffusion equations a subset of dyamical equations in general, with special properties or limitations[7]. From the chemist point of view, it is of course of some importance that dynamics speaks is own language, or, at least, that the description of dynamical behaviours can be cast in a chemical-like language.

Precisely, the models that a chemist would propose, to describe a particular reaction, on the basis of chemical knowledge, what we call realistic models or reaction schemes, are far more complex than formal models. The chemical relevance of reacting species and reaction processes, is his first concern, eventhough, from the beginning, he is aware of the lack of detailed knowledge about what is really going on in the reacting system. An example of a realistic chemical reaction scheme is, for instance, the following[8]:

```
CSTR K0 Br-, BrO3-, ArOH, H2O, H+

BrO3-  + Br- + 2H+ = HBrO2 + HOBr        ; 2.1, 1.D+4  (R4,R5)
HBrO2  + Br- + H+  = 2HOBr               ; 2.D9        (R6)
HOBr   + Br- + H+  = Br2 + H2O           ; 8.D9, 110.  (R7,R8)
BrO3-  + HBrO2 + H+ = 2BrO2. + H2O       ; 1.D4, 2.D7  (R9,R10)
2HBrO2             = BrO3- + HOBr + H+   ; 4.D7        (R11)
BrO2.  + ArOH      = ArO. + HBrO2        ; K1          (R12)
BrO2.  + BrArOH    = BrArO. + HBrO2      ; K2          (R13)
Br2    + ArOH      = BrArOH + Br- + H+   ; K3          (R14)
HOBr   + ArOH      = BrArOH + H2O        ; K4          (R15)
BrArO. + ArO.      = DH + Br-            ; K5          (R16)
BrArOH + ArO.      = BrArO. + ArOH       ; K6,K7       (R17,R18)
```

It is given in TSR notation[9], which is rather self explanatory. The CSTR statement declares that the reaction is taking place in a open flow reactor, with flow rate K0, and input species Br-, BrO3-, ArOH, H2O and H+. This model contains 13 intermediate species, 33 processes, including the input and output flow processes, and 21 control parameters.

GOALS AND ISSUES

The evolution equations of such reactions schemes, has a general form called "Reaction-Diffusion Equations":

$$\frac{\partial X}{\partial t} = F(X) + D\frac{\partial^2 X}{\partial r^2}$$

where X is the N-dimensional concentration vector of intermediate species (N=13 in the example given above), F is the reaction term, derived from mass action kinetics, D the diffusion, usually diagonal, matrix, and r is the space coordinate.

When the reaction in run a in well stirred reactor, for instance a CSTR, the diffusion term does not contribute, as we assume that, in this case, the concentration gradients vanish. This is referred to as the uniform or homogeneous situation. Incidently, one is now aware of the limitations of such assumptions due to mixing effects [10], when some important reaction processes are faster than mixing time. Keeping the diffusion term, leads to non-homogeneous behaviour, characterized by front propagation, chemical waves, target patterns and other propagation phenomena [11]. Finally, heterogeneous effects may be considered, in which

boundary or surfaces or other localized influences play some role. This is
usually achieved by considering a spatial dependence of the reaction term.

To study such equations, in particular in the homogeneous case, for
which mathematical tools are far more developed than is the non-homogeneous
case, one will first look for the steady states of the system, X_s, defined
by $F(X_s)=0$. There is unfortunately no way to perform an exhaustive search
of all existing steady states, which is due to nonlinear character of
reaction equations. Once a steady state is known, which is hardly
expectable in an analytic way with a realistic reaction scheme such as the
one presented above, one has to study its stability. Local stability, that
is the way the system reacts against small disturbances from the steady
state, can be analyzed on linearized equations. It amounts to computing the
eigenvalues of the Jacobian matrix at steady state. But local stability
does bring enough knowledge, particularly about global behaviour. Global
stability describes the the way the various attractors interact, and the
existence of their basin of attraction. That, not only concerns
zero-dimension attractors, like steady states, but also higher dimensional
attractors like limit cycles, torii, stange attractors[5,6].

This very brief overview of the way one should proceed when studying a
reaction scheme, reveals only part the formidable complexity of this
endeavour. One is usually faced with the following alternative: Observation
or Understanding. Observation means simulation of the reaction scheme
behaviour, which is easy to do, even for complex models, but not very much
informative. One is then more or less in the same position as an
experimentalist. They are too many control parameters and too many possible
regimes, to be able to form a consistent picture of the dynamical structure
of the reaction scheme. Understanding is precisely related to our a priori
ability to analyze the structure of the reaction scheme, and to determine
the source or origin of instabilities, oscillations or even chaotic
behaviour [12] . That requires a reduction of the complete model to its
relevant "dynamical core", a formal chemical-like model that still carries
some chemical significance, next, to an even more reduced mathematical
form, like the "normal forms" of bifurcation theory[5,6] . At this point,
chemical meaning is lost, or only present through the intricate
contribution of the various reaction steps to the parameters of the normal
form.

Achieving this reduction is not a trivial step. It may require a high
level mathematical expertise, and no systematic and practical methodology
to go thus far, is available yet.

PRESENT NEEDS AND FUTURE TOOLS

The tools that we need to help the modelling of complex reaction
systems have to fill the gap between chemical and mathematical modelling.
They also should allow the chemist to gain practical and effecient access
to the required mathematical knowledge. Finally, those tools should be able
to take into account the specific features and properties of chemical
reaction equations, and, at the same time, to do this using a chemical
language to describe the expected behaviour and dynamical structure of the
model, for instance, in terms of chemical network, reaction processes,
autocatalysis, activation or inhibition. We are far from that, which
indicates that we are still lacking theoretical methods to handle those
problems. Moreover, even well established mathematical theories are still
not usually implemented in effecient practical procedures.

From a practical point of view, as we already mentioned, the simulation of the behaviour of complex systems is not too difficult, and various easy-to-use systems are available[13,9]. Too few operational methods are available to go beyond mere simulation. Let us mention the analysis of chemical network stability by B.Clarke[14]. More recently, using an approach coming from pattern recognition, we have designed a reduction method based on the analysis of the morphology of the trajectory of the system in intermediates concentration space or in the space of reaction rates, that allows a rational and systematic reduction of model complexity[15]. The starting point of this method is a given trajectory obtained by simulation, which exhibits one of the dynamical behaviours of interest, for instance a relaxation oscillation. The main idea behind this method is that the complexity of the minimal model that is able to describe such a behaviour should be in relation and in proportion to the complexity of behaviour itself. If one is able to define a quantitative decription of the information content of the morphology of a trajectory, then, one is able to assign to contribution of each reaction step to each morphological feature. In this way, simplifications, realtionships or correlations between reacting species and/or reaction processes, can be detected, and quantatitively justified. When applied to the realistic reaction scheme given above, this approach is able to propose the following formal model as quantitatively undistinguishable from the complete one, in the circumstances that have been considered, not in general of course:

$$X = Y; \qquad (P1)$$
$$Y + X = ; \qquad (P2)$$
$$Y = 2 \ Z; \qquad (P3,P4)$$
$$2 \ Y = ; \qquad (P5)$$
$$Z = W + Y ; \qquad (P6)$$
$$W = 1/2 \ X; \qquad (P7)$$

where $X=Br-$, $Y=HBrO2$, $Z=BrO2$. and $W=ArO$. . There are 4 variables and 7 processes. This reduced model is not only simpler, it allows the description of the various regimes that characterize the morphology of the trajectory, and that, in return, establishes a firm ground for the understanding of the origin of oscillations in this model system.

The method sketched above is only one the tools that are required to perform efficient modelling in chemical dynamics. As future developements, we think that there is a need of integration of available, or still to be designed, methods that contribute to define a firm basis for a real modelling methodology. With the TSR/SED system[9], which is presently being extended, we propose such an integration, through a Model Definition Language, or, more generally a Dynamical System Definition Language, which, by providing an easy way to specify the structure and the dynamics of a chemical system, under various conditions, enables the automatic generation of the algorithms that are needed for the simlation and the analysis of complex reaction schemes. The idea, which is already used in TSR (Reaction Scheme Translator), is to use a symbolic manipulation of the model specification, to generate computer programs. Among the algorithms that can be needed, without trying to be complete, let us mention the system of differential equations and the Jacobian associated to the reaction scheme, this is already done by TSR, algorithms for steady states and singularities search, stability and bifurcation analysis, unfolding of high codimension singularities, specific model-dependent functions, for instance, the complex impedance in electrode reactions, and finally, the algorithms for complexity reduction, such as the trajectory morphology analysis, referred to above.

527

Various people are involved in such developments, and we think that the definition of model representation standards would allow easier connection and integration of the various methods. One of the difficulty of the theoretical developments, whether they come from pure mathematics or chemical dynamics, is that they are not easyly manageable by chemists, unless some interfacing and integration is provided to implement them.

Modelling is chemical dynamics is a frontier of complexity, the complexity of non-linear and non-equilibrium, somewhat non-standard, behaviours. It is also a cross-disciplinary field, as soon as one pretends to apply it, and derive pratically all its consequences. It involves chemical knowledge, mathematical theories, such as Dynamical System theory, Bifurcation and Singularity theory, Catastrophe theory, but also advanced Computer Science tools and methods, such as description languages and compilation, symbolic manipulation, and even, undoubtfully, Artificial Intelligence, through software engineering, but also as way to specify and manipulate complex knowledge and interpretation[16]. Modelling in chemical dynamics is a field of research by itself, and requires the development and the use of advanced methods in the various fields just mentioned, unless one considers that modelling is only the theoretical token to be paid as an appendix to any good experimental study.

References

1. I. Prigogine and G. Nicolis, "Self-organization in Non-Equilibrium Systems", (1977) Wiley, N.Y.
2. C. Vidal and P. Hanusse, Int. Rev Phys. Chem. (1986) 5,1-55.
3. R. J. Field, E. Körös and E. M. Noyes J. Am. Chem. Soc. (1972) 94, 8649.
4. R. J. Field and R. M. Noyes, J. Chem. Phys. (1974) 60, 1877.
5. J. Guckenheimer and P. Holmes, "Nonlinear Oscillations, Dynamical Systems, and Bifurcations of Vector Fields", Applied Mathematical Science 42, (1983) Springer-Verlag.
6. M. Golubitsky, D.G. Schaeffer, "Singularities and Groups in Bifurcation Theory", vol I., Applied Mathematical Science 51, (1985) Springer-Verlag.
7. P. Hanusse, C.R. Acad. Sci. (1973) 277C, 263.
8. P. Richetti, Thesis, University of Bordeaux I, (1972);
   P. Richetti and A. Arneodo, Phys. Letters, (1985) 109A, 359.
9. P. Hanusse and P. Richetti, J. Chim. Phys. (1985) 82, 637.
10. J. Villermaux, in the present issue.
11. P. Hanusse, in "Spatial Inhomogeneities and transient behaviours in Chemical Kinetics", Proceedings, Brussels, (1987).
12. F. Argoul, A. Arneodo, P. Richetti and J.C. Roux, in the present issue.
13. B. J. Gottwald, "The KISS system", Proceedings of the Aachen meeting of the Deutsche Bunsengesellshaft fùr Physikalische Chemie, 1979.
14. B. L. Clarke, Adv. Chem. Phys. (1980) 53, 1 ;
    B. D. Aguda and B. Clarke, J. Chem. Phys. (1987) 6, 3461.
15. P. Hanusse, J. Chim. Phys. (1988) to appear.
16. L. Györgyi, T. Deutsch and E. Körös, Int. J. Chem. Kin. (1987) 19, 435.

VILLERMAUX - Complex systems involving many species and simultaneous reactions can be characterized by the matrix of stoechiometric coefficients v_{ij}.
Can the structure of this matrix be used to discuss the dynamic properties of the reaction network ?

HANUSSE - Indeed the stoichiometric matrix can be used and is used, in particular in the approach designed by B. Clarke (Adv. Chem. Phys. 43, 1980 et B. Clare in "Non-equilibrium Dynamics in chemical systems", C. Vidal et A. Pacault, Ed. Springer, 1984) to study the stability of chemical networks. But in reality one cannot say that the structure of this matrix is used as such ; rather it is used as one possible representation of the chemical network.

MICHEAU - Your experimental results are generally provided by CSTR experiments using a probe, for example electrochemical or spectrophotometric. Do you think that the nature of the probe i.e. the shape of your recorded time signal could influence the structure of the dynamical core of the reduced model which you will determine using your pattern recognition data treatment ?

HANUSSE - In fact the morphology analysis is performed on the complete trajectory given by the simulation of the full model. One interesting result is that the results do not seem to depend on the space in which you work (concentrations, reaction rates, combinations of them). You obtain the same essential dynamical features. This stability of the solution is to be compared to that of the reconstruction procedure of three dimensional attractors from a unique time series, in studies of chaotic behaviour.
In conclusion, that would indicate that the information that can be extracted by analyzing the morphology of experimental data is not crucially dependent on the kind of probe that is used, at least concerning the essential dynamical features.

BARRERE - Connaissez-vous le travail de Coullet (Nice) pour classer les types d'instabilité ?
Il sépare la partie linéaire de la partie non linéaire :

$$\frac{d\underline{x}}{dt} = \underline{\underline{J}} \ \underline{X} \ + \ g \ (\underline{X})$$

apparaît alors une matrice de Jordan $\underline{\underline{J}}$ dont l'expression permet de classer les types d'instabilités.

HANUSSE - En effet, la méthode qu'il utilise, comme beaucoup d'autres d'ailleurs, est essentiellement destinée à analyser la stabilité d'un état stationnaire d'un système différentiel ou d'une application du plan. Dans ce cas, c'est bien la partie linéaire qui intervient (matrice Jacobienne ou de Jordan selon le cas). Mais ceci ne constitue qu'une partie de l'étude. Il faut déjà avoir pu calculer les états stationnaires, problème qui n'a pas de solution pratique générale, puis être capable d'analyser ces matrices lorsque leur taille devient grande. Enfin, ceci ne constitue qu'une étude de stabilité locale. Il faut encore étudier l'interaction d'états stationnaires et tous les effets globaux qui s'y rattachent.

MULLER - Can this concept of characterizing a "dynamical core" of a complex chemical system be extended to other fields of nonlinear dynamics with attractors of complicated topology (hydrodynamics, meteorology, etc.) in order to facilitate communication and the unveiling of common underlying principles ?

HANUSSE - In principal this method could be applied to a large class of systems, provided one is able to exhibit a trajectory in the complete phase space, or more specifically that of a particular attractor.
The method can be viewed as a particular implementation of the projection techniques based on time scales separation. In fact any reduction technique of dynamical systems has, one way or the other, to project out fast variables. The interesting point here is that, starting from a complex chemical model one obtains also a chemical model, rather than a purely mathematical one.

ISSUES AND CHALLENGES IN CHEMICAL ENGINEERING OF LIQUID PHASE REACTIONS

Jacques Villermaux

Laboratoire des Sciences du Génie Chimique CNRS
Ecole Nationale Supérieure des Industries Chimiques
Institut National Polytechnique de Lorraine
1, rue Grandville - 54042 Nancy, France

The aim of this paper is to review some problems encountered in the analysis, design, optimization and scale-up of liquid phase reactions from the
laboratory to the industrial production scale. Many important processes are
based on such reactions : organic synthesis, production of specialty
chemicals, polymerizations, bioprocesses - with the additional complication
that in most cases, several phases are actually involved in the process.

1. "La Voie Royale"

Fig. 1 shows the ideal scheme for the design and scale up of a commercial reactor from laboratory experiments. In the industrial reactor, chemical reactions are most of the time strongly coupled with physical processes
such as mixing, heat and mass transfer, hydrodynamics of the reacting
mixture. The purpose of the different steps in Fig. 1 is to try to uncouple
these processes in order to study them separately in appropriate devices.
Such equipments are specifically designed to measure the parameters required
for writing a model of the industrial reactor. Of course, following all these
steps requires much effort, but this is the only way to come up with safe
and efficient models for scale-up ("La Voie Royale"[17]). This has actually
been done for large scale petrochemical processes involving sophisticated
techniques (e.g. fluidized bed reactors). In practice, chemical engineers
often by-pass some of the steps in order to save time and money. However,
the scheme in Fig. 1 provides useful guidelines for discussing the main problems a chemical engineer has to face when he wants to design and/or optimize
a chemical reactor. The present paper is limited to the case of liquid phase
reactions.

2. Kinetics of homogeneous reactions in the liquid phase

In order to derive a good model for the reactor, kinetic data free
from mixing, heat transfer and diffusion effects must be available. In
contrast with thermodynamic data, which can now be found in data banks, or
calculated from existing theories, kinetic data are very scarce in the literature. When they exist, they are mainly concerned with mechanistic studies
performed in dilute media. Phenomenologic laws, expressing the rate of
reaction as a function of observable quantities are lacking in most cases
such as

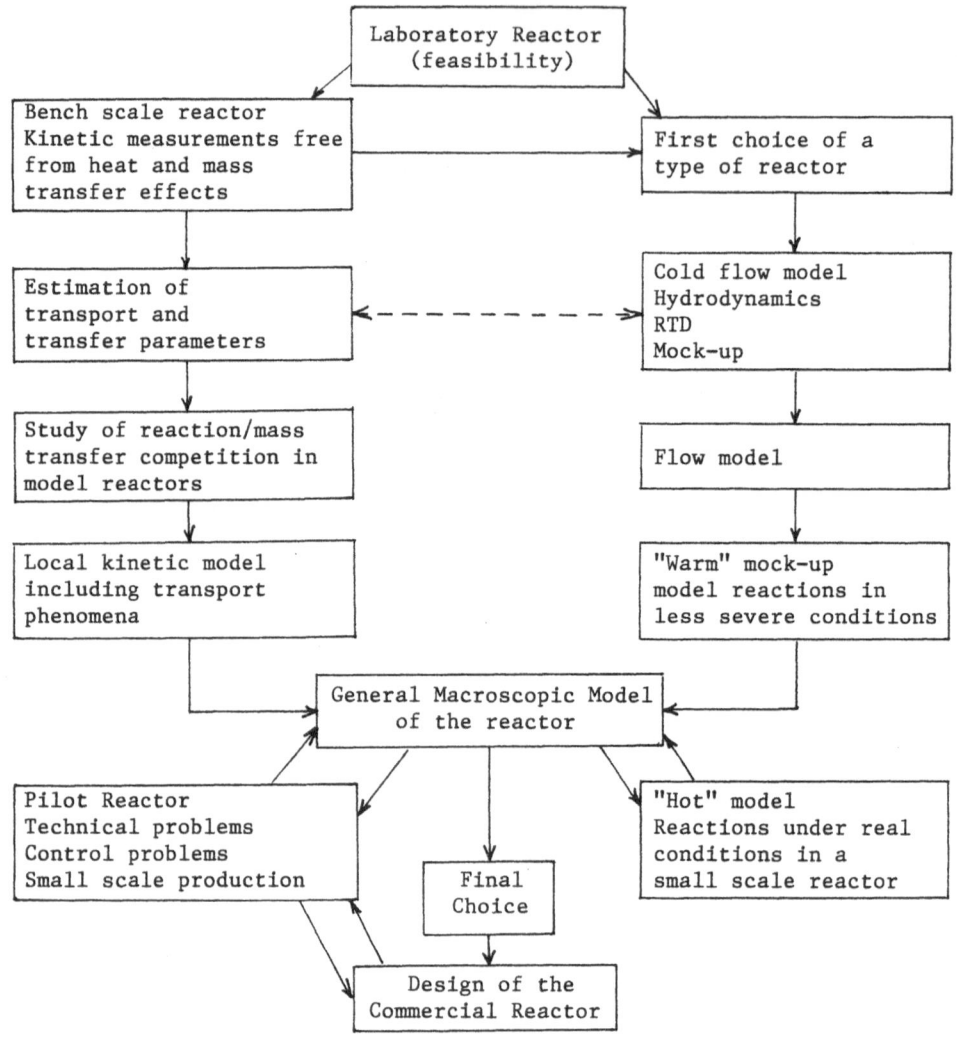

Fig. 1. "La Voie Royale" (The Royal Way) for the
design of a commercial reactor from
laboratory experiments

- complex organic reactions
- polymerization and polycondensation reactions (decomposition of
chemical initiators, propagation, chain transfer, termination, etc...)
- precipitation reactions (rate of nucleation, crystal growth, agglomeration as a function of local supersaturation)
- enzyme catalysis
- microbial growth and inhibition by various substrates.

For want of existing data, chemical engineers have to determine the
kinetics they need by use of appropriate bench scale reactors specially
designed to avoid spurious effects. In spite of this, physical chemists
should be encouraged to take some interest in the determination of kinetic
laws under realistic industrial conditions, especially in concentrated media.

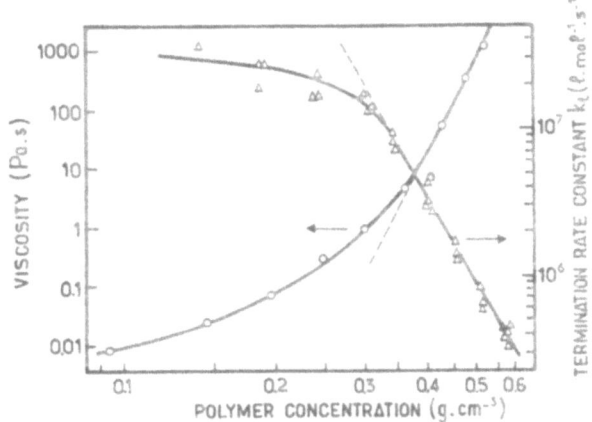

Fig. 2. Dependence of viscosity (25°C) and termination rate
constant (60°) on polymer concentration. Methyl
Polymethacrylate (M_W = 105 000). After reference[1].

Among the many problems deserving further research, two important
issues can be mentioned.

A first problem is the influence of the change of the medium composition on kinetic parameters as reaction proceeds. Actually, the so-called
rate "constants" may be considered as independent of the properties of the
reacting medium only in dilute solutions. This is no longer true in concentrated media.

Fig. 2 shows for instance the variation of the kinetic "constant" k_t
for chain termination as a function of the concentration C of polymer, observed by Tirrel and Tulig,[1] in the case of methylmethacrylate polymerization.
When the concentration is high, k_t varies as C^{-6}. This is due to the fact
that the recombination of macroradicals is controlled by diffusion, which is
slowed down by the viscosity increase. All kinetic parameters undergo similar
changes. Such effects are very important in industrial reactors where the
viscosity currently goes up by a factor of 10^6 during the reaction. In these
conditions, one may wonder if the classical formalism for rates of elementary
mechanisms is still adequate !

A second problem is that of "distributed systems" involving a large
number of molecular species which undergo thousands of reactions. Such a
situation is encountered in oil refining, in broth fermentation and again
in polymerization. It is almost impossible to follow all elementary species
individually : this would require huge computer models and would generate
superabundant and useless information. Instead, a first method consists in
"lumping" the complex mixture into key components engaged in "pseudo-
reactions" which behave on the average as the real system. Efficient methods
have been designed for lumping oil cuts, especially in the case of first
order reactions but general methods still have to be devised for more complex
kinetics. A second method consists in forgetting the underlying molecular
reality, and in describing the mixture in terms of characteristic properties
considered as "pseudo-components". This has been done successfully for the
high pressure free-radical polymerization of ethylene[2,3,4]. Instead of considering individual macromolecules, a model was derived, based on pseudo-
components such as : the first moments of the molecular weight distribution,
long chain branching points, short chain branching points, vinyl and
vinylidene double bonds, the overall concentration of free radicals and
macromolecules, etc... Inspired by molecular mechanisms, relationships

between these quantities were established and the corresponding "rate constants" were determined from bench scale data. This allowed a reasonable representation of industrial results.

Following the same idea, promising results were obtained to describe crystallisation and coagulation processes.

Should such a method be generalized, or should a new kind of approach be sought ? This is an open problem.

3. Macro and Micromixing

Macromixing is the process whereby parts of a fluid having different histories come into contact and mix-up on a macroscopic scale. It is thus a consequence of the macroscopic hydrodynamic pattern. The development of powerful computer codes make it possible now to determine the average velocity pattern in any kind of equipment, at least in single phase newtonian fluids. This probably won't eliminate methods based on the characterization of complex flow by internal age distributions and residence time distributions (RTD), which can be determined with tracers. Such methods have proven their efficiency for building up flow models well adapted to scale up and to calculation of chemical conversion,[17].

However, the macromixing pattern is sufficient to predict chemical conversion for reactions of known kinetics only in the case of first order processes. For more complex kinetics, information on mixing on the molecular scale - i.e. micromixing - is required. Micromixing is also a controlling factor of the initial contacting of reactants. The entering fluid may be viewed as formed of segregated lumps or "aggregates" (macrofluid). Upon mixing, segregation is reduced by exchange between the aggregates and reduction of their size. The process ends up with a fluid mixed on the molecular scale (microfluid).

Let us take the example of stirred vessels.

Mixing of newtonian fluids in conventional stirred tanks is now reasonably well understood, although most models rely on the assumption of homogeneous isotropic turbulence. This is only a rough approximation. For instance, the fluctuating components u'_r and u'_θ of the radial and tangential velocity were recently measured by Laser-Doppler Anemometry in a 6.3 dm^3 reactor,[5]. Values of the Reynolds Tension

$$\frac{\overline{u'_r u'_\theta}}{(\overline{u'_r{}^2})^{1/2} (\overline{u'_\theta{}^2})^{1/2}}$$

as high as 50 % were found in the discharge flow of the turbine, whereas this quantity should be nil if the turbulence were isotropic. However, for want of a more refined description, the isotropy assumption is still the basis of most existing models. Without going into details, which can be found elsewhere[6,7,8,9], the successive stages for mixing may be described as follows.

Unmixed fluids are first distributed through each other, producing an uniformization of average composition without significant decrease of local concentration variations. This typically pertains to macromixing. In a stirred tank, this process is mainly controlled by recirculation. The average recirculation time $\overline{t_c}$ is proportional to the reciprocal of the stirring speed N. Macromixing is achieved after $t_M = 4 \overline{t_c}$, a value which may be much higher in non-newtonian fluids. The size of regions of uniform composition

534

is then reduced and simultaneously, contact areas between regions of different composition increase. In laminar mixing, this occurs by streching and
folding of aggregates which reduce the striation thickness between segregated
laminae. The characteristic time for this process is $(\nu/\varepsilon)^{1/2}$ where ν is the
kinematic viscosity and ε the power dissipation per unit mass. In turbulent
mixing, the dissipation of segregation occurs by exchange between eddies
which gradually decrease in size. The characteristic time is
$\tau_S \sim 2(L_S^2/\varepsilon)^{1/3}$, where L_S is the size of larger eddies (macroscale). The
process may also be viewed as an erosion of big aggregates (time
$t_e = l_0(\nu^3/\varepsilon)^{1/4} Sc^{-1/2}/\mathcal{D}$, l_0 = size of initial aggregates). The last stage
is achieved by molecular diffusion, which takes place within smallest eddies,
with a time constant of the order of $\tau_D \sim \lambda_K^2/\mathcal{D}$, where $\lambda_K = (\nu^3/\varepsilon)^{1/4}$ is
the Kolmogorov microscale.

Micromixing plays a significant role on chemical conversion when one
of the controlling rates of reaction is of the same order of magnitude as
the rate of mixing. The main areas in which mixing effects are encountered
are

- Fast and complex synthesis reactions. For instance, in the case of
the bromination of resorcin, owing to complex pH effects, it was found that
the amount of 2-4 dibromoresorcin in the di-isomer depended on the location
of the feed pipe and might vary from 30 % to 60 % when the stirring speed
was varied from 0 to 360 RPM,[10].

- Polymerization and polycondensation reactions. Micromixing may
control the efficiency of initiator utilization,[11] and molecular weight distributions[7,12].

- Precipitation reactions. Owing to the strong dependence of nucleation,
growth and agglomeration rates on local values of supersaturation, micromixing may seriously influence the Crystal Size Distribution,[13].

- Aerobic fermentations, when oxygen uptake by microorganisms is
very fast. The availability of dissolved oxygen may then be controlled by
micromixing, especially in large size fermentors.

When one single micromixing stage is involved (micromixing time t_m),
t_R being the controlling reaction time, the relative amount of segregated
fluid (X_S) is roughly given by the "rule of thumb"

$$\frac{1 - X_S}{X_S} \sim \frac{t_R}{t_m} \tag{3-1}$$

Actually, the interpretation of the whole process may involve several
simultaneous mechanisms. For instance, the following model for micromixing-
controlled reactions in a stirred tank was proposed,[14] : The cloud of incoming fluid is convected along a recirculation path ($\overline{t_c}$) and its size l
increases upon penetration of fluid from the bulk by turbulent diffusion
($t_{macro} \sim l/u'$). The cloud contents is mixed up by periodic creation of
vortices ($t_{inc} \sim 10(\nu/\varepsilon)^{1/2}$). Micromixing is then achieved by interaction
between the small aggregates thus formed ($\tau_S \sim 2(L_S^2/\varepsilon)^{1/3}$). This model was
successfully used for interpreting the results of tests with consecutive
competing reactions where, owing to spatial distribution of u' and ε, the
segregation index was found to vary both with the location of the point of
injection of reactants and on the stirring speed (Fig. 3).

It is encouraging to see that simple chemical engineering models using
lumped parameters yield surprisingly good results.

However, progress in this kind of description requires further investigation of the local state of turbulence, especially with respect to concentration fluctuations.

4. Complex fluids

Mixing effects as well as other processes involved in liquid phase reactions can be reasonably well predicted in standard equipment for fluids having simple rheological properties. Unfortunately, industry has often to deal with "complex fluids". This term denotes very viscous or strongly non newtonian fluids, dispersions, suspensions, emulsions, slurries, foams, etc.. having an exotic rheological behaviour. They are encountered for instance in bioprocesses, in the food industry, in the manufacture of polymers. In homogeneous complex fluids, the rheological properties (e.g. the apparent viscosity) depend both on temperature and composition so that for exothermic reactions, it may be impossible to uncouple chemical reaction, heat evolution

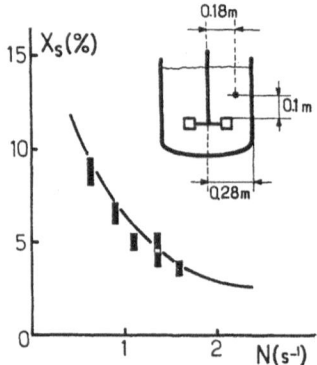

Fig. 3.   Segregation index X_S as a function of stirring speed. X_S was measured by a chemical method based on the precipitation of barium sulphate from an EDTA-complex. The reaction is of the consecutive-competing type $A + B \rightarrow R$, $R + B \rightarrow S$ and X_S can be deduced from the yield in S. The solid line represents model values. Details in reference,[14].

and hydrodynamics. This is the case for instance in reactors for the bulk polymerization of styrene : the heat of reaction induces temperature gradients. This affects the local viscosity which in turn controls the flow pattern and thus heat transfer coefficients at the wall. Heat transfer controls the temperature field and thence the rate of reaction : a complete closed loop which is very difficult to deal with when accurate parameters for the individual processes are not known.

In multiphase complex fluids, a new problem arises : should the fluid be considered as an equivalent pseudo-homogeneous medium with "effective" properties averaged over the phases ? Or should the fine texture be considered, with different composition and temperature in each phase in spite of their strong interaction ?

A tentative answer to this question may be sought from characteristic parameters, namely the reaction time t_R ; an intraphase mass transfer time $t_D = 1^2/\mathcal{D}$ or $1\delta/\mathcal{D}$ where $1 = 1/a$ is the reciprocal of the specific interfacial area of the dispersion, $\mathcal{D}$ the diffusivity and δ the equivalent diffusional film thickness ; the heat transfer time $t_h = 1^2/\alpha$ or $1\delta/\alpha$ where α is the heat diffusivity ; the adiabatic temperature rise $J = (-\Delta H)C_o/\rho c_p$. From these parameters the following criteria may be tentatively proposed,[15]. Local concentration and temperature gradients are negligible and the pseudo-homogeneous assumption is valid if

$$t_D/t_R < 5 \times 10^{-2} \quad , \quad J\, t_h/t_R < 1K \qquad (4-1)$$

It is interesting to notice that t_D/t_R contains as special cases the well known criteria of Thiele and Hatta (see below).

Complex fluids offer a new area for research. Novel approaches should perhaps be imagined, different from the two terms of the alternative mentioned here.

5. Thermal stability

The feedback effect of heat evolution on the rate of exothermic reactions may cause thermal runaway. This is a major issue in the operation of industrial reactors, as the loss of control of a chemical reactor constitutes a serious hazard : everybody has in mind the SEVESO accident or those which occured recently in the Swiss industry. Thermal instability is due to the irreducible coupling between heat accumulation and the quasi-exponential increase of reaction rate as a function of temperature accounted for by Arrhenius equation. This problem can be studied by the methods of non linear dynamics. Here again, characteristic times make it possible to establish simple criteria which give at least an order of magnitude for dangerous and safe ranges of operation.

Let us consider for instance an exothermic reaction $A + B \rightarrow$ products carried out in a semi-batch reactor cooled from the wall (T_o). The tank initially containing A, B is added at a constant rate during the time t_I. The following characteristic times are defined t_R = reaction time as usual. For instance, for a nth order reaction $t_R = 1/k\, C_{Ao}{}^{n-1}$; $t_H = (\rho\, c_p\, V)/UA$ = heat transfer time where V is the reactor volume, A the heat transfer area, U the overall heat transfer coefficient and $\rho\, c_p$ the heat capacity per unit volume ; $t_A - t_{Ro}\, (RT_o{}^2/JE)$ is the adiabatic runaway time, i.e. the time required for spontaneous runaway of an isolated volume of reacting mixture at temperature T_o ; $J = ((-\Delta H)C_{Ao})/\rho\, c_p$ is the adiabatic temperature rise previously defined, E the activation energy, $t_{Ro} = t_R$ at T_o and R is the gas constant.

Now, a simple analysis of the batch operation (addition of B in a very short time $t_I \simeq 0$) proves that thermal stability is achieved provided that

$$t_A \gg t_H \quad \text{i.e.} \quad U \gg \frac{V}{A} \cdot \frac{k\, C_{Ao}{}^n(-\Delta H)E}{R\, T_o{}^2} \qquad (5-1)$$

This condition has to be interpreted in a very conservative way. Of course, if the introduction time t_I is increased (semi-batch operation), a better stability is achieved and the reactor may be run with security even if (5-1) is not respected. Fig. 4 shows examples of such a simulated behaviour. In addition, it has been assumed that B was not instantaneously mixed (mixing time t_M). The simulation clearly reveals that partial segregation may also influence stability.

These problems can be easily studied by numerical simulation provided that good values for heat transfer and activation parameters are available.

6. Reactions in multiphase media

Gas-liquid, liquid-liquid, liquid-solid or gas-liquid-solid reactions constitute the major basis of industrial processes. This is a vast subject that can be only briefly discussed here, in spite of its practical importance,[16].

The main problem is the representation of the competition between interphase mass transfer, diffusion within phases and chemical reaction. Let us take the example of gas liquid reactions. Most engineering calculations are based on the film model of Whitman,[17] which assumes that interphase

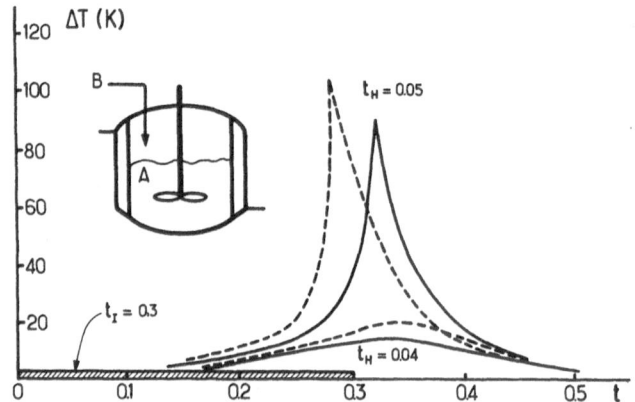

Fig. 4.   Thermal runaway of a semi-batch reactor for the exothermic reaction A + B → products. All characteristic times are normalized by the reaction time $t_R = 1/k\ C_{Ao}$. Introduction time $t_I = 0.3$. Notice the sensitivity to the heat transfer cooling time t_H. Solid lines : perfect mixing. Dashed lines : imperfect mixing (macromixing time $t_M = 0.1$, micromixing time $t_m = 0.01$). J = 200 K ; E/R = 10 000 K ; T_o = 300 K.

mass transfer occurs across a rigid film of thickness δ . On the liquid side, the mass transfer coefficient is thus $k_L = \mathcal{D}/\delta$. This is known to be a wrong picture of reality. Instead, Hiby and Danckwerts have suggested that mass transfer takes place via a transient process whereby fluid aggregates from the bulk are randomly exposed to transfer at the interface during a mean contact time t_e , yielding $k_L \approx (\mathcal{D}/t_e)^{1/2}$. In the presence of a simultaneous chemical reaction, the calculations are much more tedious with this model than with the simple film theory. For a simple reaction A (gas) + B (liq) → products, both models yield almost the same results. This explains that chemical engineers continue to prefer the old film model, owing to its simplicity. The case of more complex gas-liquid reactions (e.g. consecutive competing reactions A(g) + B(l) → R(l), A(g) + R(l) → S(l), etc..) is presently under investigation and a first

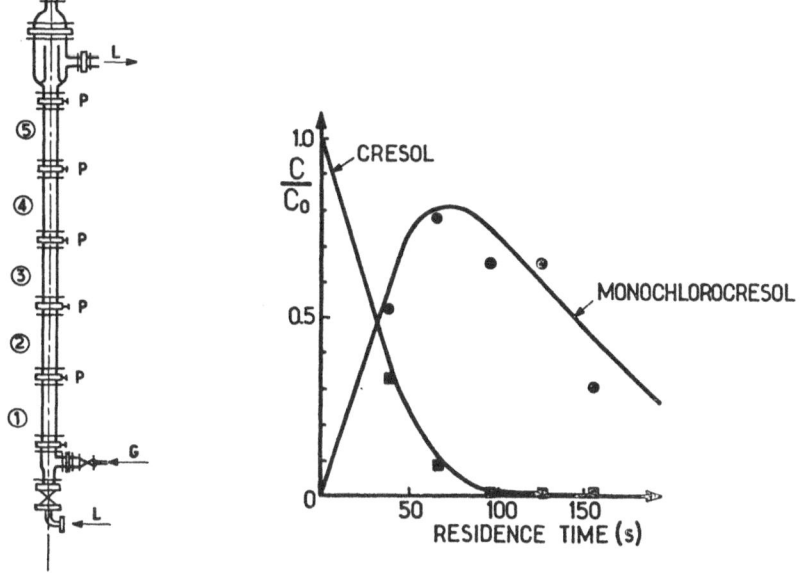

Fig. 5.   Selectivity of a complex gas-liquid reaction :
          the chlorination of cresol in a bubble column.
          Experimental data and model calculations (after
          Th. Darde, Thesis INPL Nancy, 1985)

series of models allows a satisfactory representation of experimental data
(Fig. 5). The future will tell if the wrong but useful film model must be
abandoned or if it still constitutes a good basis for practical calculations.

In the frame of this model, the mass transfer time for A across the
film is $t_\delta = \delta^2/\mathcal{D}_A$ and, assuming a second order reaction, the reaction time
is $t_R = 1/(k\ C_B)$. The competition between transfer and reaction is controlled
by the Hatta number such that

$$Ha^2 = \frac{t_\delta}{t_R} = \frac{k\ C_B\ \delta^2}{\mathcal{D}_A} = \frac{k\ C_B\ \mathcal{D}_A}{k_L^2} \qquad (6-1)$$

When Ha is very small, the reaction is slow with respect to transfer
and takes place in the bulk liquid. When Ha is very high, the reaction is
quasi-instantaneous and takes place at the interface.

Let "a" be the specific interfacial area $(a = A/V)$ and C_{Ai} the interfacial concentration of dissolved A, assumed to be at equilibrium with the
gas, the rate of reaction per unit gas-liquid dispersion volume is written

$$r = k_L a\ C_{Ai}\ E_A(Ha) \qquad (6-2)$$

E_A is an acceleration factor, function of Ha, which expresses the driving effect of the chemical reaction with respect to a simple transfer across
the totally polarised film $(r = k_L a\ C_{Ai})$. This is substantially the theory
developped by Van Krevelen and Hoftyzer for simple gas-liquid reactions,
which is widely used in practice.

Physicochemical data are needed for its application to reactor design :
   - interfacial areas (a)
   - mass transfer coefficients (k_L)

- solubilities (C_{Ai}), which are not straightforward to determine in the case of very reactive species.
- diffusivities of dissolved species ($\mathcal{D}_A$)
- rates of intrinsic reactions (k)

Most of these data must be obtained from experience in appropriate devices, either by physical or chemical methods making use of test reactions. Of course, theories yielding a priori predictions would be welcome.

Again, most published data are concerned with aqueous systems, of academic interest. As many industrial systems involve organic media and/or complex fluids, research in this area is urgently requested.

7. Conclusion

Three points can be emphasized at the end of this brief review

- Whereas the contribution of various processes occuring in liquid phase reactions can be isolated in academic conditions (dilute and/or aqueous systems), they appear to be coupled, sometimes irreducibly in real life systems, and this coupling between chemical reaction, mixing, hydrodynamics, heat and mass transfer, constitutes a research subject per se.

- All physicochemical processes mentioned above can be characterized by typical time constants. A comparison between these times constitutes a powerful method to determine the controlling processes and to generate dimensionless numbers governing the reaction regimes.

- Complex fluids constitute a new and promising area where fundamental research is needed, with large possibilities of practical application. More generally, research should be encouraged on more realistic systems under conditions prevailing in industrial reactors, which are unfortunately far from those dealt with in most academic studies available in the literature.

References

1. M. Tirrell and T.J. Tulig, in "Polymer Reaction Engineering. Influence of reaction engineering on polymer properties, p. 247 Edited by K.H. Reichert and W. Geiseler, Carl Hauser Verlag, München 1983.
2. J. Villermaux, L. Blavier and M. Pons, in "Polymer Reaction Engineering", Ibid., p. 2.
3. J. Villermaux and L. Blavier, Chem. Eng. Sci. 39 (1984), 87.
4. M. Pons, L. Blavier, J.L. Greffe and J. Villermaux, Proceed. World Congress III of Chemical Engineering Tokyo, 1986, Paper 9a-254, Vol IV p. 108.
5. F. Bouget, Diplôme d'Etudes Approfondies, Nancy, 1987.
6. J. Villermaux, "Mixing in Chemical Reactors", ACS Symposium Series 226 (1983), 135.
7. J. Villermaux, "Micromixing phenomena in stirred reactors", Encyclopedia of Fluid Mechanics Ch 27, Gulf Publishing Cy (1986)
8. J.R. Bourne, "Micromixing revisited", ISCRE 8, I. Chem. E. Symposium Series, 87 (1984), 797.
9. J. Baldyga and J.R. Bourne, Chem. Eng. Comm. 28 (1984), 231.
10. J.R. Bourne, P. Rys and K. Suter, Chem. Eng. Sci. 32 (1977), 711.
11. J. Villermaux, M. Pons and L. Blavier, ISCRE 8, I. Chem. Eng. Symp. Series 87 (1984), 553.

12. M.R.N. Costa and J. Villermaux, in "Polymer Reaction Engineering", p. 205 ; Edited by K.H. Reichert and W. Geiseler, Hüthig and Wepf, Basel, 1986.
13. J. Villermaux and R. David, J. Chim. Phys., in preparation.
14. R. David and J. Villermaux, Chem. Eng. Comm. (1987), in press.
15. J. Villermaux and N. Midoux, Revue Gén. de Thermique n° 279, mars 1985, 207.
16. J.C. Charpentier, "Mass transfer rates in gas-liquid absorbers and reactors", Adv. Chem. Eng., 11 (1981), 1.
17. J. Villermaux, "Génie de la Réaction Chimique, Conception et fonctionnement des réacteurs", Technique et Documentation, 2e éd., Paris 1985.

DISCUSSION

MICHEAU - As microemulsions are known as segregated reaction mixtures, could effects of the stirring rate be expected in such media ?

VILLERMAUX - Yes, but I am not aware of such experiments, at least in studies reported in the Chemical Engineering literature. This would probably be an interesting project.

PACAULT - On sait actuellement utiliser les microémulsions pour faire des poudres très fines souvent très colorées.

VILLERMAUX - Il serait intéressant d'étudier l'effet du micromélange de ces poudres, bien que leur taille soit en dessous du domaine de la turbulence (et donc probablement réglée par la diffusion moléculaire).

NAKACHE - Ma question se rapporte aux transferts entre deux phases liquides, en particulier dans les conditions où apparaît un effet Marangoni. Pouvez-vous nous dire s'il existe en génie chimique des méthodes ou des paramètres caractéristiques qui permettent d'identifier ce type d'échange.

VILLERMAUX - Il a été observé que la stabilité de certaines dispersions liquide-liquide dépendait du sens dans lequel se faisait le transfert (lits denses en particulier). Ce phénomène a été expliqué par l'effet Marangoni mais à ma connaissance, il n'y a pas eu d'interprétation quantitative.

BARRERE - Est-ce que le moussage est encore utilisé ?

VILLERMAUX - Oui, le moussage-essorage est une opération unitaire encore utilisée, notamment pour certains problèmes d'épuration.

BORGHI - La méthode classique de représentation de la mécanique des fluides dans les réacteurs du génie chimique consiste à utiliser des assemblages de réacteurs simples (CSTR, plug flow reactors...). Par ailleurs vous avez montré le calcul détaillé des champs de vitesses dans un réacteur. Est-ce que ces calculs détaillés sont de plus en plus utilisés, ou non, dans l'industrie pour la conception des réacteurs ?

VILLERMAUX - Oui, certains industriels songent sérieusement à les utiliser en remplacement des essais en maquette froide. Mais ces calculs ne sont actuellement faisables que pour des fluides homogènes et les codes de calcul disponibles commercialement coûtent très cher. Pour les systèmes polyphasiques, les méthodes classiques du génie chimique seront certainement utilisées pendant longtemps encore pour l'extrapolation.

Marcel Barrère

Office National d'Etudes et de Recherches Aérospatiales
(ONERA)

29 avenue de la Division Leclerc - 92320 Chatillon   (France)

INTRODUCTION

To begin with we would like to indicate the role of classical thermodynamics in the study of large systems. We use a modern approach aimed at obtaining competitive products, taking economic, material and energetic constraints into account.

The engineer, with the current aim of the search for efficiency, is confronted by choices which are facilitated by the use of thermodynamics and in particular the second law. This choice is then carried out in a rational fashion from solution to the problem of locating the optimum for a given set of constraints and a definite objective.

To underline the importance of this approach, we have limited our discourse to a few cases that apear in the field of chemical engineering where mass, energy, and momentum exchanges unite with chemical transformations.

To keep the discussion as short as possible, we have limited it to two sections:
1) the search for the minimum internal entropy production,
2) the use of exergy in chemical engineering to obtain the best performances for the minimum cost.

RECAP OF THERMODYNAMICS [1]

The use of the second law of thermodynamics in chemical engineering to solve certain problems of large systems is based on the knowledge of certain functions of which the two most important are: entropy and its exergic production corresponding to the maximum use of energy. The entropy balance takes into account both the internal and external contributions of the volume $\mathscr{V}$ under consideration. The external portion is characterized by an entropy flux J_s and the internal part by a production $\dot{\sigma}$ per unit volume.

We thus write this condition as:

$$\rho \frac{ds}{dt} = - \underline{\nabla} \, \underline{J}_s + \dot{\sigma}$$

The GIBBS function linking the internal energy u, the density ρ, and the mass fraction Y_j taking the form:

$$\rho \frac{du}{dt} = \rho T \frac{d\sigma}{dt} + \frac{p}{\rho}\frac{d\rho}{dt} + \sum_{1}^{N} \rho \mu_j \frac{dY_j}{dt}$$

μ_j corresponds to the chemical potential of the species j. The energy balance equations, those of mass balance of injected species in the Gibbs equation lead to the determination of the entropy balance:

$$\rho \frac{d\sigma}{dt} = -\nabla \cdot \left(q - \sum_{1}^{N} \mu_j \, \underline{J}_{Dj}\right)/T - \frac{1}{T^2}\left(q \cdot \nabla T\right) - \frac{1}{T}\left(\underline{\underline{\tau}} : \nabla \underline{v}\right)$$

$$-\frac{1}{T}\sum_{1}^{N}\underline{J}_{Dj}\left(-T\nabla\frac{\mu_j}{T} - f_j\right) - \frac{1}{T}\sum_{1}^{N}\mu_j\dot{W}_j$$

In this expression the flux $J_S = (q - \Sigma_1^N \mu j \, \underline{J}_{Dj})/T$ where q is the heat flux by conduction and $\underline{J}_{Dj}$ is the diffusion flux $\underline{J}_{Dj} = \rho\,V_j$, V_j, being the rate of diffusion, the production $\dot{\sigma}$ can be written:

$$\dot{\sigma} = \frac{1}{T^2}\left(q \cdot \nabla T\right) - \frac{1}{T}\left(\underline{\underline{\tau}} : \nabla \underline{v}\right) - \frac{1}{T}\sum_{1}^{N}\underline{J}_{Dj}\left(T\nabla\frac{\mu_j}{T} - f_j\right) -$$

$$-\frac{1}{T}\sum_{1}^{N}\mu_j\dot{W}_j \quad , $$

The different terms in the energy production are:

(i) the heat flux by conduction corresponding to the entropy flux: $1/T\,\nabla\,q$. This composed of two parts: an exchange with the outside $\nabla\,q/T$ and $-1/T^2(q\,\nabla\,T)$ an entropy production associated with energy transfer.

(ii) viscosity effects corresponding to an irreversible production of entropy $-1/T(\underline{\underline{\tau}}:\nabla\,\underline{v})$ in which the viscosity tensor intervenes

$$\underline{\underline{\tau}} = -\mu\left(\nabla\underline{v} + \nabla^T\underline{v}\right) - \eta\left(\nabla\cdot\underline{v}\right)\underline{\underline{1}}$$

with the two viscosity coefficients μ and $\eta = k - 2/3\,\mu$, the tensor product to be doubly contracted ($\underline{\underline{\tau}}:\nabla\,\underline{v}$) corresponding to the dissipation function $\mu\Phi = -\underline{\underline{\tau}}:\nabla\,\underline{v}$

(iii) the entropy production by diffusion of each species j:

corresponds to a diffused flux $\rho_j\,V_j$ referring to the gradient of the potential,

and the specific external forces acting on each species j, f_j. If the force derives from a potential $f_j = -\nabla\,\Psi_j$, the entropy production for each species j becomes:

$$-\frac{\rho_j V_j}{T}\left[\nabla\left(\mu_j + \Psi_j\right) - \frac{\mu_j}{T}\nabla T\right]$$

(iv) the last term corresponds to the chemical entropy production so that finally the entropy production becomes:

$$\dot{\sigma} = \frac{1}{T}\left[\frac{q}{T}\nabla T - \sum_{1}^{N}\rho_j V_j \frac{\mu_j}{T}\nabla T + \sum_{1}^{N}\rho_j Y_j \nabla\left(\mu_j + \Psi_j\right) + \right.$$

$$\left. + \left(\underline{\underline{\tau}} : \nabla\underline{v}\right) + \sum_{1}^{N}\mu_j\dot{W}_j\right]$$

The first four terms correspond to the three transport phenomena: energy, mass, momentum and the fourth to the chemical reaction.

The evaluation of the term $\dot{\sigma}$, i.e., the production of entropy for a system in the course of evolution, permits to various losses due to transport phenomena and to chemical evolution away from equilibrium.

The second function taken into account in our analysis is the exergy:

$$E = h - T_o s - \sum_1^N \mu_{j,o} \ Y_j / \mathcal{M}_j$$

where h and s are the specified enthalpy and entropy $(\mu_{j,o})$ the chemical potential under standard conditions of environment of the species j, Y_j the mass fraction and $\mathcal{M}_j$ the molar mass of this species.

The use of exergy in an evolving system allows certain properties of the system under consideration to be evaluated for a given environment of energy, mass and momentum. The only condition is that the system be small compared with its surroundings, so that any exchanges with the surroundings will leave it practically unchanged. The exergy variation thus represents the maximum usable energy.

This is a fundamental parameter in order to obtain a high yield system with optimal functionning and adjustment as well as a thermal economy such that the product is globally the most cost effective post effective possible.

Firstly, we shall examine the importance of evaluating the entropy balance in some concrete examples in order to demonstrate the advantages of this procedure. Then, we will designate the parameter p evaluated per unit time $\dot{p}$, per unity length p', per unit surface p'' and per unit volume p''', in Spalding's terminology.

Optimisation of Transfer in a Limiting Layer [2]

Let us examine the rate of entropy production inside a system in the case of a one dimensional limiting layer, the lamellar flow not being confined, as figure 1 illustrates.

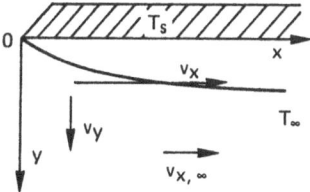

Figure 1. Rate of entropy production in a limiting layer.

The rate of entropy production per unit volume for an incompressible fluid evaluated at a point on the surface of the plate takes the form:

$$\dot{\sigma}''' = \frac{k}{T^2} \nabla T \cdot \underline{\nabla} T + (-\underline{\underline{\tau}}' : \underline{\nabla} v)/T$$

$$= \frac{k}{T^2}\left[\left(\frac{\partial T}{\partial x}\right)^2 + \left(\frac{\partial T}{\partial y}\right)^2 \right] + \frac{\mu}{T}\left\{ 2\left[\left(\frac{\partial v_x}{\partial x}\right)^2 + \left(\frac{\partial v_y}{\partial y}\right)^2\right] + \right.$$

$$\left. + \left(\frac{\partial v_x}{\partial y} + \frac{\partial v_y}{\partial x}\right)^2\right\} \ .$$

In the case of unit Prandtl number, and for laminar flow, σ''', assuming that the thickness of the limiting layer δ is small compared to the abscissa x, can be expressed as: $\delta < < x$ (T_S is the temperature at a point on the plate's surface).

$$\dot{\sigma}''' \simeq \frac{k}{T_s^2}\left(\frac{\partial T}{\partial y}\right)^2 + \frac{\mu}{T_s}\left(\frac{\partial v_x}{\partial y}\right)^2, \quad \frac{\partial v_x}{\partial y} = v_{x,\infty}\left[\frac{v_{x,\infty}}{v\,x}\,f''(0)\right]^{1/2},$$

In this case: v being the kinematic viscosity $v = \mu/\rho$ and f Blasius' function, in addition:

$$\frac{\partial T}{\partial y} = \left[\frac{v_{x,\infty}}{v\,x}\right]^{1/2}(T_s - T_\infty)\,f''(0)$$

with $f''(0) = 0.332$.

Introducing the Reynolds number $R_{e,x} = v_{x,\infty}\,x/v$, σ''' becomes:

$$\dot{\sigma}''' = \left[\frac{k}{T_s^2}(T_s - T_\infty)^2 + \frac{\mu}{T_s}v_{x,\infty}^2\right]\frac{v_{x,\infty}^2}{v^2}\cdot\frac{0,332}{R_{e,x}}.$$

If the Prandtl number is different from unity:

$$\dot{\sigma}''' = \left[\frac{k}{T_s^2}(T_s - T_\infty)^2\,P_r^{2/3} + \frac{\mu}{T_s}v_{x,\infty}^2\right]\frac{v_{x,\infty}^2}{v^2}\cdot\frac{0,110}{R_{e,x}}.$$

The local rate of production of entropy decreases when the Reynolds number increases along the plate, but there is no minimum even if we integrate σ''' over the whole volume of the limiting layer.

Evaluation of the Entropy Production Rate for Confined Flow [2]

In this paragraph, we consider the element of an exchanger, either a tube, or a flow between plates, or even a more complex configuration incorporating a hydraulic radius (i.e., the ratio of the cross-section of the passage to the wet perimeter).

To simplify the problem, we only consider a flow tube with a transit of constant $\dot{m}$ exchanging the constant quantity $\dot{q}'$ of heat per unit length.

The search for minimum losses is based on the rate or irreversible entropy production per unit length of tube:

$$\dot{\sigma}' = \frac{\dot{q}'\nabla T}{T^2} + \frac{\dot{E}'_v}{T}$$

$\dot{E}'_v$ being the rate of dissipation of mechanical energy per unit length of tube, i.e., frictional losses.

In general, $\dot{E}_v$ is determined by a volume $\mathcal{V}$, by:

$$\dot{E}_v = \int_v (-\underline{\underline{\tau}}:\nabla\,\underline{v})\,d\mathcal{V}$$

i.e., from the dissipation function.

In the tube configuration we further obtain: $\dot{E}_v = \frac{1}{2}\langle v\rangle^2\frac{L}{R_h}\,f\,\dot{m}$

$\langle v\rangle$ being the average speed in the tube and f the friction factor as a function of the Reynolds number $R_{ep} = \frac{\langle v\rangle D}{v} = \frac{4\dot{m}}{\mu\,\pi D}$, D being the diameter of the tube and R_h the hydraulic radius $R_h = D/4$, L is the length of the tube.

Hence:

$$\dot{E}_v = \frac{1}{2}\cdot\frac{16\,\dot{m}^2}{\pi^2 D^4\rho^2}\cdot\frac{L}{D}\cdot f\cdot\dot{m},$$

this finally gives since: $\dot{q}' = \pi D L h \, \Delta T/L$, h being the convection coefficient,

$$\dot{\sigma}' = \frac{\dot{q}'^2}{T^2 \pi h D} + \frac{32 \, \dot{m}^3}{\pi^2 T P^2} \cdot \frac{1}{D^5} f\left(R_{e,D} \right) ,$$

introducing the Nusselt number, $Nu = h \, D/k$

$$\dot{\sigma}' = \frac{\dot{q}'^2}{\pi T^2 k \, N_u} + \frac{32 \, \dot{m}^3}{\pi^2 T P^2} \cdot \frac{1}{D^5} f\left(R_{e,D} \right) ,$$

Hence expressing the rate of irreversible entropy production in terms of a tube length.

The discussion of extrema is not easy, although $\dot{m}$ and $\dot{q}'$ can be assumed constant and D variable.

This problem of extrema then reduces to the variation of a parameter of the Reynolds number

$$R_{e,D} = \frac{4 \, \dot{m}}{\mu \, \pi} \cdot \frac{1}{D}$$

The rate of production is then equal to:

$$\dot{\sigma}' = a / R_{e,D}^{\alpha} + b \, R_{e,D}^{\beta}$$

The optimum is hence obtained on deriving $d\dot{\sigma}'/d R_{eD} = 0$

$$-\alpha a \, R_{e,D}^{-\alpha - 1} + \beta \, R_{e,D}^{\beta - 1} = 0$$

where

$$\left(R_{e,D} \right)_{opt} = \left(a\,\alpha / b\,\beta \right)^{1/(\alpha+\beta)} ,$$

corresponding to the optimum value of R_{eD} or defining the optimum tube diameter.

Note that for confined flow, the appearance of a critical Reynolds number or diameter: for $R_{eD} < R_{eD \, opt}$, the contribution to the rate of entropy production of heat transfer is large in comparison with the irreversibilities due to speed $\sigma'_q > \sigma'_v$. The situation is reversed for $R_{eD} > R_{eD \, opt}$. In this case, viscosity losses predominate $\dot{\sigma}'_v > \dot{\sigma}'_q$. This indicates that the rate or irreversible entropy production in the frictional losses (defining the drop in pressure) increase more rapidly with speed than the losses due to heat transfer by convection.

The pressure decrease in tube per unit length is equal to:
$\Delta p' = (32 \, m^2/\rho \, \pi^2 \, D^5) \, f$
thus there is a direct relation to the rate of irreversible entropy production due to friction.

Entropy Production an Exchanger; Loss Optimisation

Let us consider a classical counter-current type exchanger represented in Figure 2, constituted by two coaxial tubes.

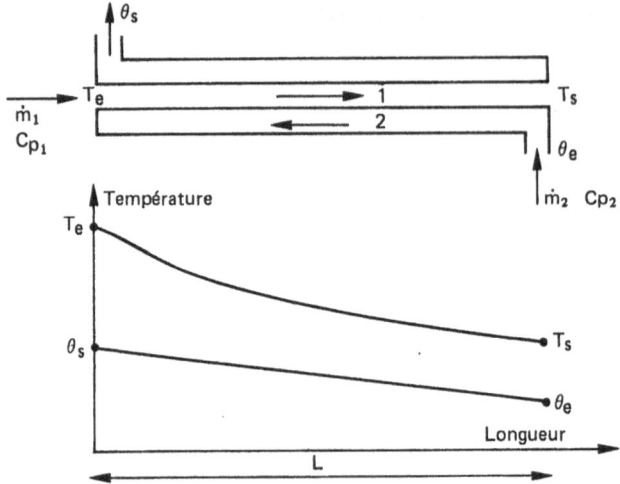

Figure 2. Counter current exchanger.

We recall some results concerning this type of exchanger.
The heat exchanged along the latter is equal to: in circuit(1)

$$q_1 = m_1\, C_{p1}(T_e - T_S) = C_1(T_e - T_S),$$
and in circuit (2)

$$q_2 = m_2\, C_{p2}(\theta_S - \theta_e) = C_2(\theta_S - \theta_e),$$
with: $q_1 = q_2$.

We assume that C_1 is a minimum $C_1 = C_{min}$, hence C_2 maximum
$C_2 = C_{max}$ such that:

$$C_{min}(T_e - T_S) + C_{max}(\theta_e - \theta_S) = 0.$$

The exchange occurs through the wall of the cylinder of inner radius r_1 and
outer radius r_2, the thickness of the cylinder is $r_2 - r_1 = e$ and the coefficient of
thermal conductivity is k. The coefficient of convection exchange is h_1 on side 1
and h_2 on side 2. The global coefficient in the circuit (1) is thus

$$u_1 = 1 \Big/ \left(\frac{1}{h_1} + \frac{r_1}{k} \ln (r_2/r_1) + \frac{r_1}{r_2} \cdot \frac{1}{h_2} \right),$$

and in circuit (2)

$$u_2 = 1 \Big/ \left(\frac{1}{h_2} + \frac{r_2}{k} \ln (r_2/r_1) + \frac{r_2}{r_1} \cdot \frac{1}{h_1} \right),$$

with:

$$A_1\, u_1 = A_2\, u_2$$

A_1 being the exchange surface on side 1, and A_2 that of side 2. In the
configuration chosen: $A_1 = 2\pi\, r_1\, L$ and $A_2 = 2\pi\, r_2\, L$.
We have chosen $C_1 < C_2$ so the maximum flux is:

$$q_{max} = C_1(T_e - \theta_e) = C_{min}(T_e - \theta_e) \qquad \text{and}$$

$$q = \varepsilon \, C_{min}(T_e - \theta_e)$$

introducing the efficiency:

$$\varepsilon = \left(T_e - T_s\right) / \left(T_e - \theta_e\right) \, .$$

We also define the notion of the number of transfer thermal unit:

$$N_{uT} = U A / C_{min}$$

such that the efficiency can be expressed as a function of N_{uT} and the ratio:

$$\Psi = \frac{C_{min}}{C_{max}} = \frac{\theta_s - \theta_e}{T_e - T_s} \, .$$

In the counter-current exchanger, the efficiency ε is given by the expression:

$$\varepsilon = \left\{ 1 - exp\left[- N_{uT}\left(1 - \Psi\right)\right]\right\} / \left\{1 - \Psi \, exp\left[- N_{uT}\left(1 - \Psi\right)\right]\right\}$$

Hence:

$$\varepsilon = F(N_{uT}, \Psi)$$

The quantity of heat exchanged can also be expressed as:

$$\dot{q} = A \, U \, \overline{\Delta T}$$

where:

$$\overline{\Delta T} = \frac{\Delta T_s - \Delta T_e}{\ln \frac{\Delta T_s}{\Delta T_e}} \quad , \text{with} \begin{cases} \Delta T_e = T_e - \theta_s \\ \Delta T_s = T_s - \theta_e \end{cases}$$

These relations allow an exchanger of this type to be defined, its geometry calculated and its performances predetermined. The two problems to be solved being:

(a)   to determine the geometry of the exchanger from a knowledge of its working parameters.

(b)   to determine the set of parameters dictating the performance of the exchanger from a knowledge of its geometry and certain input parameters.

If we are now required to optimise the exchanger, i.e., to ensure that it is as economic as possible in energy and consumption of starting materials, with the help of the rate of drop in pressure Δp, $\dot{\sigma}'_v = \dot{\sigma}'_{\Delta p}$, the other $\dot{\sigma}'_q$ but indicated by $\dot{\sigma}'_{\Delta T}$ corresponding to the temperature change ΔT due to the exchange. We transform to the reduced rate of production $\dot{\sigma}^*$ true over the whole length L of the exchanger and normalised to C_{max}:

$$\dot{\sigma}^* = \dot{\sigma}' \tau / C_{max}$$

a) Evaluation of $\dot{\sigma}^*_{\Delta p}$:

We have seen that

$$\dot{\sigma}_r = \dot{\sigma}_{\Delta p} = \frac{32 \, \dot{m}^3}{\Pi^2 \, \tau \, \rho^2} \cdot \frac{1}{D^5} \cdot f(R_{e,D})$$

using the equation of state $p = \rho n R T$ to explicitly factorise the pressure drop

$$\dot{\sigma}_{\Delta p} = \frac{\Delta p}{p} \cdot \frac{n R}{C_p} \cdot \dot{m} \, C_p$$

corresponding to:

$$\Delta p = \frac{32 \, \dot{m}^2}{\Pi^2 \, D^4 \rho} \cdot \frac{L}{D} \, f(R_{e,D}) = \frac{8 \, \dot{m}^2}{\Pi^2 \, D^4 \rho} \cdot \frac{L}{R_h} \cdot f(R_{e,D})$$

incorporating the hydraulic radius $R_h = D/4$

Such that:

$$\dot{\sigma}^*_{\Delta p} = \frac{\Delta p}{p} \cdot \frac{n R}{C_p} \cdot \frac{\dot{m} \, C_p}{C_{max}}$$

which gives for circuit (1)

$$\dot{\sigma}^*_{\Delta p,1} = \left(\frac{\Delta p}{p}\right)_1 \cdot \left(\frac{n R}{C_p}\right)_1 \cdot \psi \quad,$$

and circuit (2)

$$\dot{\sigma}^*_{\Delta p,2} = \left(\frac{\Delta p}{p}\right)_2 \cdot \left(\frac{n R}{C_p}\right)_2$$

thus in all:

$$\dot{\sigma}^*_{\Delta p} = \dot{\sigma}^*_{\Delta p,1} + \dot{\sigma}^*_{\Delta p,2} \quad.$$

This calculation should be compared with the direct calculation of the rate of entropy production:

$$\dot{S}_{\Delta p} = C_{min} \left[\left(\frac{n R}{C_p}\right)_1 \ln\left(\frac{p_{1,e}}{p_{1,s}}\right)\right] + C_{max} \left[\left(\frac{n R}{C_p}\right)_2 \ln\left(\frac{p_{2,e}}{p_{2,s}}\right)\right]$$

on dividing by C_{max}

$$\dot{S}^*_{\Delta p} = -\psi \left\{\left(\frac{n R}{C_p}\right)_1 \ln\left[1 - \left(\frac{\Delta p}{p}\right)_1\right]\right\} - \left\{\left(\frac{n R}{C_p}\right)_2 \ln\left[1 - \left(\frac{\Delta p}{p}\right)_2\right]\right\}$$

after series expansion

$$\dot{S}^*_{\Delta p} = \dot{\sigma}^*_{\Delta p} \simeq \psi\left(\frac{n R}{C_p}\right)_1 \left(\frac{\Delta p}{p}\right)_1 + \left(\frac{n R}{C_p}\right)_2 \left(\frac{\Delta p}{p}\right)_2$$

which yields the same result as the other method.

In the calculation of $\dot{\sigma}^*_{\Delta p}$, the sets of variables chosen must correspond to dimensionaless parameters of the exchanger:

(i) e.g. L/R_h is the ratio of the exchange surface to that of the passage cross-section of the fluids

$$\frac{A_p}{A_c} = \frac{\pi D L}{\frac{\pi D^2}{4}} = \frac{L}{R_h}$$

(ii) the dimensionless parameter Ψ can be defined by comparing the momentum $\rho\,v^2$ to the static pressure p in the study of flow in flux circuits. This is a characteristic of the flow and corresponds to $\gamma\,M^2$ for a gas, the square of the Mach number M multiplied by the ratio γ of specific heats:

$$\gamma = \frac{\rho v^2}{p} = \gamma\, v^2/a^2 = \frac{16\,\dot m^2}{\pi^2 D^4 \rho p}.$$

The parameters L/D_h, $\phi = \gamma\,M^2$, Re_D are three dimensionless numbers characteristic of a given exchanger project

$$\dot\sigma^*_{\Delta p} = a\,\phi\,\frac{L}{D_h}\,f(Re_{,D})$$

$a =$ constant dependent on the circuit chosen in the exchanger

$$a_1 = 1/2 \cdot \Psi\left(\frac{\dot m R}{C_p}\right)_1 \quad , \quad a_2 = 1/2 \cdot \left(\frac{\dot m R}{C_p}\right)_2 ,$$

for a gas

$$a_1 = \frac{\Psi}{2}\frac{(\gamma-1)}{\gamma} \quad , \quad a_2 = \frac{1}{2}\frac{(\gamma-1)}{\gamma},\ \gamma = \frac{C_p}{C_p}.$$

b) Evaluation of $\dot\sigma^*_{\Delta p}$

For this calculation, we could use the results of the previous paragraph but it is preferable to start from the global rates of entropy production for the whole exchanger, having already demonstrated the equivalence for $\dot\sigma^*_{\Delta p}$.

The rate of entropy production corresponding to the temperature variation is equal to:

$$\dot S_{\Delta T} = C_{min}\,\ell n\,\frac{T_s}{T_e} + C_{max}\,\ell n\,\frac{\theta_s}{\theta_e} ,$$

with

$$T_S = T_e - \varepsilon(T_e - \theta_e),$$

$$\theta_S = \theta_e + \Psi\,\varepsilon(T_e - \theta_e)$$

such that:

$$\dot S_{\Delta T} = C_{min}\,\ell n\left[1 - \varepsilon\left(1 - \frac{\theta_e}{T_e}\right)\right] + C_{max}\,\ell n\left[1 + \Psi\varepsilon\left(\frac{T_e}{\theta_e}-1\right)\right]$$

dividing by C_{max}

$$\dot S^*_{\Delta T} = \Psi\,\ell n\left[1 - \varepsilon\left(1 - \frac{\theta_e}{T_e}\right)\right] + \ell n\left[1 + \Psi\,\varepsilon\left(\frac{T_e}{\theta_e}-1\right)\right]$$

where

$$\dot S^*_{\Delta T} = \dot S^*_{\Delta T,1} + \dot S^*_{\Delta T,2} .$$

The calculation of $S^*_{\Delta T}$ is possible if the input conditions T_e and θ_e are known. The efficiency ε can be evaluated as a function of N_{uT} and Ψ:

$$\varepsilon = f(N_{uT}, \Psi)$$

The value of N_{uT} for the set of circuits (1) and (2) being:

$$N_{uT}^{-1} = N_{uT,1}^{-1} + N_{uT,2}^{-1}$$

The number of transport units can be written in terms of the exchange parameters, e.g.:

$$N_{uT} = A_e u / C_{min} = (A_e / A_p)(A_p u / \dot{m} C_p)$$

involving:

$$\frac{L}{R_h} = \frac{A_e}{A_p} \quad ,$$

(A_e being the exchange surface and A_p the passage cross-section of the fluid) and the Stanton number:

$$S_T = u A_p / \dot{m} C_p$$

Hence:

$$N_{uT} = (L/R_h) \cdot S_T$$

The result is that the efficiency depends on the dimensionaless parameters:

$$\varepsilon = f\left(\frac{L}{R_h} , S_T , \Psi \right) .$$

In the case of the counter-current exchanger for an efficiency close to unity, when we consider the irreversible portion of $S^*_{\Delta T}$, we obtain by linearising:

$$\dot{\sigma}^*_{\Delta T} \simeq (T_e - \theta_e)(\theta_e^{-1} - T_e^{-1}) \left[\frac{1}{\left(\frac{L}{R_h}\right)_1 S_{T,1}} + \frac{\Psi}{\left(\frac{1}{R_h}\right)_2 S_{T,2}} \right]$$

We thus see that for each circuit a possible simplified relationship appears. Indeed, for circuit (1), we get:

$$\dot{\sigma}^*_1 = \dot{\sigma}^*_{\Delta p, 1} + \dot{\sigma}^*_{\Delta T, 1} = \frac{\alpha}{\left(\frac{L}{R_h}\right)_1} + \beta\left(\frac{L}{R_h}\right)_2$$

and the optimum corresponds to:

$$-\alpha / \frac{L}{R_h} + \beta = 0$$

for (1) or for (2). Hence:

$$\left(L/R_h\right)_{opt} = (\alpha/\beta)^{1/2}$$

and

$$\dot{\sigma}^{*}_{opt} = 2\,(\alpha\beta)^{1/2}$$

with

$$\alpha_1 = \frac{\psi}{2}\left(\frac{m\ell}{C_p}\right)_1 \phi_1\, f_1\left(R_{e,\mathcal{D}}\right) \quad,\quad \beta_1 = g\left(T_e, \theta_e\right)/S_{T,1}$$

$$\alpha_2 = \frac{1}{2}\left(\frac{m R}{C_p}\right)_2 \phi_2\, f_2\left(R_{e,\mathcal{D}}\right)\,,\quad \beta_2 = g\left(T_e, \theta_e\right)/S_{T,2}$$

Which leads to the following results:

$$\left(\frac{L}{R_h}\right)_{opt,1} = 2\left[\frac{\psi}{2}\left(\frac{m R}{C_p}\right)_1 \phi_1\, f_1\,\frac{S_{T,1}}{g}\right]^{1/2}$$

$$\left(\frac{L}{R_h}\right)_{opt,2} = 2\left[\frac{1}{2}\left(\frac{m R}{C_p}\right)_2 \phi_2\, f_2\,\frac{S_{T,2}}{\psi\, g}\right]^{1/2}$$

and the optimal rates of entropy production

$$\left(\dot{\sigma}^{*}_{opt}\right)_1 = 2\left[\frac{\psi}{2}\cdot\left(\frac{m R}{C_p}\right)_1 \phi_1\, f_1\,\frac{g_1}{S_{T,1}}\right]^{1/2}$$

$$\left(\dot{\sigma}^{*}_{opt}\right)_2 = 2\left[\frac{1}{2}\left(\frac{m R}{C_p}\right)_2 \phi_2\, f_2\,\frac{g\,\psi}{S_{T,2}}\right]^{1/2}$$

These relationships define the minimal losses in terms of the three dimensionless parameters.

The optimum is displaced in the $[\sigma^{*}, L/R_h]$ by the variation of ϕ; for increasing ϕ, $(L/R_h)_{opt}$ decreases (figure 3). The Reynolds number does not play a decisive role, in the event, because of Reynold's analogy $(f/\partial = S_T)$ defining a constant ratio for the two functions that depend on this number, f and S_T.

Using this technique, the constraints imposed by the geometry of the exchanger can be discussed.

The exchange surface, A_e, for example can be determined in terms of ϕ and L/R_h, $A_e[(p\;\rho)^{1/2}/\dot{m}] = (L/R_h)\times\phi^{1/2}$ and economical considerations can impose certain limits on A_e, which can be evaluated reliably in this way.

These results are only those for the exchanger alone subject to certain constraints. The exchanger placed in certain systems can lead to different optimum values.

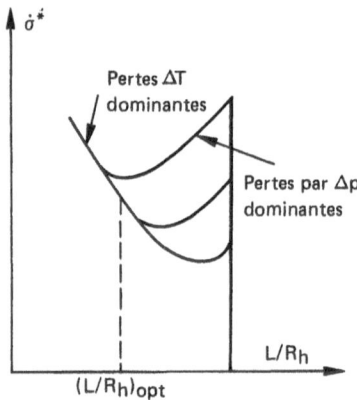

Figure 3. Evolution of the rate of entropy production versus L/Rh for various values of φ.

<u>Entropy Production in a Chemical Reactor without a Temperature Gradient</u>

In this case, the expression for irreversible entropy production simplifies since $\nabla T = 0$, the resulting for $\dot{\sigma}$ is:

$$\dot{\sigma} = \frac{1}{T} \left[\sum_1^N \left(\rho D \, \underline{\nabla} \, Y_j \cdot \underline{\nabla} \mu_j - \mu_j \overset{\circ}{\dot{W}_j} \right) - \underline{\underline{\tau}} : \underline{\nabla} \, \underline{v} \right]$$

with the chemical potential equal to e.g.

$$\mu_j = h_j - T s_j + RT \, \ell n \left(p_j / p_o \right)$$

for a perfect gas.

h_j includes the heat of formation for species j. In the equation for $\dot{\sigma}$, three terms appear: the transport of species with coefficient of diffusion D, with the gradients of the mass fractions ∇Y_j and the gradient of the chemical potentials $\nabla \mu_j$, the generation of entropy $-\sum_1^N \mu_j \dot{w}_j / T$ with the chemical production $\dot{W}_j$, the mechanical effect corresponding to outflow losses or those due to mechanical mixing (stirrers).

The term due to chemical reaction is:

$$-\sum_1^N \frac{\mu_j}{T} \, \dot{w}_j$$

and is evaluated in the following manner, for a reaction of general formula:

$$\nu_a \, A + \nu_b \, B \rightarrow \nu_p \, P + \nu_q \, Q$$

at the degree of reaction ξ

$$\frac{dm_a}{-\nu_a} = \frac{dm_b}{-\nu_b} = \frac{dm_p}{\nu_p} = \frac{dm_q}{\nu_q} = \frac{d\xi}{1}$$

we obtain

$$-\sum_1^N \frac{\mu_j \, \dot{W}_T}{T} = \frac{A}{T} \, \dot{\xi}$$

with the reaction affinity:

$$A = \nu_a \mu_a + \nu_b \mu_b - \nu_p \mu_p - \nu_q \mu_q$$

and ξ the production common to all species:

$$\dot{\xi} = \frac{\dot{W}_a}{-\nu_a m_a} = \frac{\dot{W}_b}{-\nu_b m_b} = \frac{\dot{W}_p}{\nu_p m_p} = \frac{\dot{W}_q}{\nu_q m_q}$$

This term is only involved away from equilibrium because at equilibrium $A \equiv 0$.

The contribution of diffusion losses of the species can be evaluated in the same way as the heat loss by conduction that we elaborated above. The loss by momentum or irreversible entropy production depends on the aplication: losses in pipes, in mechanical mixers, in parallel flux mixers... we have only considered the loss of momentum in an outflow determining the production $\dot{\sigma}_{\Delta p}$. The mixture depends both on the external forces f_j applied to each species j, and if f_j derives from a potential there is the following term $f_j = -\nabla \Psi_j$, it thus possible to define a genealised chemical potential

$$\mu^*_j = \mu_j + \Psi_j,$$

which can simplify the calculation under certain circumstances. The evaluation of losses in mixers which are in fact mass exchangers involve dimensionless numbers characterising aggregates and a balance equation for the mixed aggregates: micro or macro mixing, dispersion and collision phenomena. This complex structure of the mixture must be known in order to calculate the irreversible entropy production. Different models have been advanced such as that of Villermaux Devillon [4, 5], which involve a characteristic time defining the mass exchange between aggregates and allowing the balance equation for the species to be expressed in a simple form. In the field of mixing, the second law is of considerable use because it permits the total losses to be compared for complex systems and allows a classification of the different methods of mixing to be established and their efficiency evaluated.

EXERGY AND CHEMICAL ENGINEERING

The evaluation of this parameter is directly linked to the second law and to entropy production. It is of importance in at least two applications: the evaluation of performance and its optimisation, on the one hand, cost determination and its optimisation on the other, since for any process the exergy is related to the economics of the systems.

a) In a steady-flow processes in an open control volume, we can write the mass balance equation as:

$$\sum_j \dot{m}_j = 0$$

and that of energy (first law)

$$\sum_j \dot{m}_j h_j + \sum_k{}' \dot{q}_k + \sum_\ell{}' \dot{G}_\ell = 0$$

of exergy (second law)

$$\sum_j \dot{m}_j E_j + \sum_k{}' \left(1 - \frac{T_0}{T_k} \right) \dot{q}_k + \sum_\ell \dot{G}_\ell = E_I$$

where $\dot{m}_j$ is the mass flux entering the volume $\mathcal{V}$, h_j is the specific enthalpy and E_j the specific exergy:

$$E = h + T_0 \Delta - \sum_i \mu_{i,0} Y_i / m_i ,$$

$\dot{q}_k$ is the heat transfer along path k.

$\mathcal{E}$. is the power exchanged by the system of nature ℓ and E_I the exergy of the system corresponding to irreversible transformations hence $E_I > 0$

$$E = H - T_0 S - \sum \mu_{i,0} n_i$$

n_i being the number of moles of species i, suffix 0 corresponding to the surroundings whose characteristics remain unchanged during the evolution of the system since it is very large in comparison: we make the classic assumption that everything entering the system is positive (mass, heat, work, etc); with this sign convention the yield is given by:

$$\eta = - \dot{E}s / \dot{E}_c = E_s / E_e$$

E_S being the exergy leaving the system and E_e being the exergy leaving the system and E_e that entering.

Since:

$$\dot{E}_I = \dot{E}_e + \dot{E}_s$$

$$\eta = 1 - \frac{\dot{E}_r}{\dot{E}_e} = 1 / \left[1 - (\dot{E}_I / \dot{E}_s) \right] .$$

To illustrate the importance of this approach in chemical engineering, we shall take as a simple example an isothermal heat source which forms an element of a chemical system exchanging heat with another sub-system which heats a liquid of flux $\dot{m}_c$ (figure 4) [3]. The three balance equations that we have just stated are written:

- mass:
$$\dot{m}_f + \dot{m}_G + \dot{m}_p = 0$$

$$\dot{m}_{ce} + \dot{m}_{cS} = 0$$

- first law:
$$\dot{m}_f h_f + \dot{m}_a h_a + \dot{m}_P h_P + \dot{q}(T_p) = 0$$

$\dot{q}(T_p) = \dot{m}_f q(T_p)$ is the heat provided by combustion at the temperature of the heat source T_p which obeys a linear law

$$q(T_p) = q(298) \left[(T_{ad} - T_p) / (T_{ad} - 298) \right]$$

where T_{ad} is the temperature of adiabatic combustion; having eliminated the enthalpy term we obtain, for the source:

$$\dot{E}_p = T_0 \dot{m}_f \left\{ \left(1 + \frac{a}{f} \right) \Delta_p - \Delta_f - \left(\frac{a}{f} \right) \Delta_a \right\} +$$
$$+ A U (T_p - T_{c,s}) \frac{T_0}{T_p}$$

where $a/f = \dot{m}_a/\dot{m}_f$, A is the exchange surface , U the conductance of the exchanger, T_P the temperature of the oven, $T_{c,S}$ the exit temperature of the secondary fluid from the exchanger.

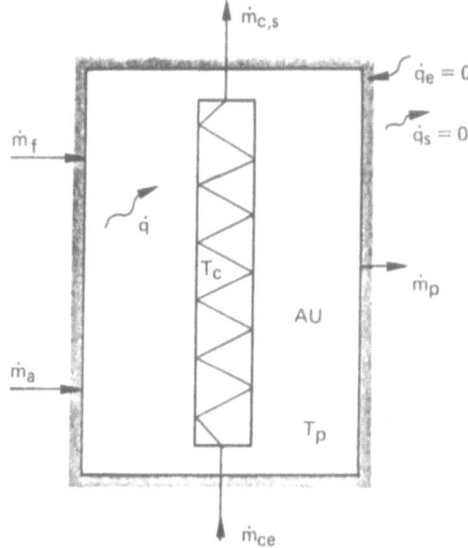

Figure 4. Isothermal oven.

For the secondary flux c, we obtain on eliminating enthalpy between entrance and exit, the exergy to be:

$$\dot{E}_c = \dot{m}_{c,e}\left\{T_0\left(\Delta_{c,e} - \Delta_{c,s}\right) + \left(\sum_{c,s}\mu_{i,o}\frac{Y_{c_i}}{n_{c_i}} - \sum_{c,e}\mu_{i,o}\frac{Y_{v_i}}{n_{c_i}}\right)\right\} - AU\left(T_p - T_{c,s}\right)\frac{T_0}{T_s}$$

The irreversible exergy E_I must be minimised if we require the maximum yield of the process:

$$\dot{E}_I = \hat{E}_p + \dot{E}_e + AU\,T_0\left(T_p - T_{c,s}\right)\left(\frac{1}{T_{c,s}} - \frac{1}{T_p}\right).$$

b)   The second important point concerning the use of exergy is the search for the process costing least. The objective function to be minimised is distinct from the irreversible entropy production, this function is the global cost Γ

$$\dot{\Gamma} = \sum_1^r \dot{\Gamma}_r + \sum_1^\sigma \dot{Z}_r$$

for a component r,which should be minimised, $\dot{\Gamma}_r$ is a constant cost of the project which is a function of $E_{r,j}$ and $\dot{Z}$ is the operating cost with the damping or evolution of the prices of certain portions which could be altered, $\dot{Z}$ is also a function of exergy $E_{r,j}$; in the other words part of the total cost. $\dot{\Gamma}_r$ is the cost independent of time whilst $\dot{Z}_r$ is the time-dependent cost the objective function can be written:

$$\dot{\Gamma} = \sum_1^r \dot{\Gamma}_r\left(\mathcal{P}_r , E_{j,r}\right) + \sum_1^\sigma \dot{Z}_r\left(E_{j,r}\right)$$

The constraints are usually functions of the type

$g(p_r, E_{jr}) = 0$ or the inequalities

$g(p_r, E_{jr}) <>0.$

At present, there exists numerous programmed methods for the location of extrema of $\dot{\Gamma}$ taking constraints of the type $g(p_r, E_{jr})$ into account. The simplest is the Lagrange method forming the function:

$$L = \dot{\Gamma} + \lambda g$$

where λ is the Lagrange multiplier corresponding to each constraint. The optimal solution is

$$\frac{\partial L}{\partial p_r} = \frac{\partial L}{\partial E_{jr}} = \frac{\partial L}{\partial \lambda} = 0$$

The Lagrange method can readily be extended to several constraints.

As we emphasized earlier there are numerous programmes for locating the extremum such as that of the conjugate gradient. In the technique or chemical reactor operation this serach for extrema is very frequently used, e.g. for optimal control of a reactor based on one or several parameters.

CONCLUSION

This selection of examples illustrate the importance of the second law of thermodynamics for solving the problems relating to the choice faced more and more in the procedural engineer's role, since he is faced with progressively more draconian restrictions.

The thermodynamics of irreversible processes should be considered a part of the scientific culture necessary on entry to the harsh world of competitiveness.

In this field the work of fundamentalists is currently well advanced and substantial such that this subject has become classical. Empiricism should make way for a more rational conception of the systems involving exchange in chemical processes.

We have deliberately left asside an entire coherent topic of fundamental research in irreversible thermodynamics based on the non-linear stability of these chemical systems. This is a field which is in progress and incorporates the major discoveries of recent times.

We would like to further emphasize the classical nature of our aproach which was already in vogue at the beginning of the century, subsequently abandonned but which is again gathering pace thanks to the progress made in three directions:

1) Modelisation on a more and more scientific basis.
2) The development of powerful computers which allow serious studies to be completed for reasonable calculation times.
3) The development of precise laser measurement techniques to validate the models.

REFERENCES

1. M. Barrère, R. Prud'homme
    Aérothermochimie des écoulements homogènes. Mémoire des Sciences
    Physiques. Fascicule 68, Gauthier-Villars.

2. A. Bejean
    Second law analysis in heat transfer. Energy Pergamon Press. Volume 5 n°
    8, August, September 1980, pp. 721-733.

3. T.A. Brzustowski
    Toward a second law taxonomy of combustion process. Energy Pergamon
    Press. Volume 5 n° 8, August, September 1980, pp. 743-757.

4. J. Villermaux
    Génie de la réaction chimique conception et fonctionnement des réacteurs.
    Technique et Documentation, 1985.

5. R. David
    Turbulent reactive flows of liquids in isothermal stirred tanks. Workshop
    "Combustion turbulente". Rouen - juillet 1987.

DISCUSSION

VILLERMAUX - Trois commentaires :

1) L'analyse entropique des procédés (proposée par R. Gibert dès 1955) est intéressante mais l'optimum technicoéconomique pratique est souvent loin de celui qui correspond au minimum de production d'entropie en raison de l'intervention d'autres critères (coût, qualité, etc.) dans la fonction objectif.

2) En revanche, cette analyse est utile pour identifier les endroits où l'entropie est principalement créée et révéler ainsi où doivent se porter les efforts de recherche pour réduire les causes de pertes par irréversibilité.

3) En liaison avec l'exposé de M. Barrère, il faut signaler un théorème récemment démontré par D. Tondeur et coll. à Nancy selon lequel la structure optimale d'un système complexe est celle où est réalisée l'équipartition spatiale de la création d'entropie. Ce principe prend ainsi une valeur opératoire très intéressante.

BARRERE - 1) Je suis d'accord que l'optimum correspondant au coût est différent de celui relatif à la performance, mais il est bon de connaître l'écart entre les deux optima et d'étudier sur chacun d'eux l'influence des principaux paramètres sur lesquels il est possible d'agir.

2) Je suis intéressé par la manière dont les systèmes complexes sont abordés et en particulier ceux rencontrés en génie chimique ; le calcul de la production d'entropie est un moyen de comparer les différents éléments et les différents phénomènes, c'est donc un moyen d'étude comparative des pertes sur chacun des éléments du système et de chacun des processus (transfert de masse, de quantité de mouvement, de chaleur, réactions chimiques, etc.). Je suis d'accord avec votre remarque car ce calcul permet d'identifier les pertes et d'essayer de les réduire.

3) Je suis d'accord avec la méthodologie de D. Tondeur concernant la structure optimale d'un système complexe qui est obtenue par l'équipartition spatiale de la création d'entropie.

PRUD'HOMME - Pour certains phénomènes (ondes de détonation par exemple) le minimum de production d'entropie correspond à l'état le plus stable.
Quelle relation possible y a-t-il entre production d'entropie optimale et stabilité ?

BARRERE - Le problème de la production d'entropie optimale pour étudier la stabilité n'est pas abordé dans mon exposé qui ne considère que le régime permanent du fonctionnement des systèmes complexes ; c'est un problème important, c'est en particulier tout le domaine abordé par Prigogine. La solution stable correspondant à une détonation par exemple est donnée par la production minimale d'entropie, la recherche de la stabilité des rouleaux de Von Karman, Ménard correspond également à une production optimale d'entropie ; mais comme je l'indique dans ma conclusion : "j'ai laissé volontairement de côté tout un ensemble cohérent de recherches fondamentales de la thermodynamique irréversible basée sur la stabilité non linéaire des systèmes chimiques" et des systèmes en général.

METHODS FOR THE MACROSCOPIC

SIMULATION OF TURBULENT REACTIVE MEDIA

R. Borghi

UA CNRS 230
Faculté des Sciences de Rouen
76130 Mont-Saint-Aignan

SUMMARY

In many practical devices and many natural phenomena, chemical reactions and convection are playing together ; in addition, the velocity of the flow is most often such that turbulence takes place. In such cases, the study and prediction of reaction rates need some type of "modelling" of the macroscopic randomness due to the turbulence ; indeed the fluctuations induced by the turbulence have generally a first order influence on the non linearity of the chemical rates.

A synthetic review of the existing methods of approach for such a modelling is presented ; three main categories have been distinguished, where the probability density function (p.d.f.) of the reactive species concentrations plays a crucial role :

i) the "computed p.d.f." methods, where a modelled balance equation is proposed and numerically solved for this p.d.f. ;

ii) the "reacting laminae models", or "flamelets models", where the internal structure of reacting zone is assumed coherent enough in spite of turbulent motions ;

iii) the langrangian, or mechanistic models, where the different assumptions can be more physically introduced into langrangian equations for the "life" of fluid particles or parcels of fluid.

A discussion of the weakest points of the models, and a presentation of analytical researches now in progress to improve these points are finally made.

I - INTRODUCTION

In many practical devices, as well in many natural phenomena, chemical reactions, turbulent diffusion, and convection are playing together. Theoretical or numerical methods for the prediction of the macroscopic properties of these turbulent reactive media are then of large interest.

It is now a common view to consider a turbulent flow as an ensemble of random flow fields ; each of these fields does satisfy the classical macroscopic balance equations with a particular set of initial and boundary conditions, but each is different due to differences in initial and boundary conditions ; and each flow can be very different from the other ones, because the sensitivity of the equations to very small changes of boundary conditions is quite large, within the "turbulent regime". That point of view is classical for non reacting turbulent flows, and the "turbulent regime" occurs very often when a characteristic Reynolds number (a convection time divided by a viscous time) is large enough. Other types of instabilities leading to "turbulence", which are characterised with other numbers, could occur , especially for reacting flows where additionnal highly non linear terms appear within the equations ; but the actual mechanism leading to turbulence is believed to be less important when the turbulence is well developped, far from the limits of the "turbulent regime".

The computation of every realizations of this ensemble of flow fields, even if the usual macroscopic balance equations are valid, is impossible in practice, even simply with a brute numerical method : the time scales and length scales, that we know to exist within the turbulent regime, are so small with respect to the time or length scales in which we are interested, that we would need an incredible amount of computer memories and an incredible amount of computer time. In addition the computation of just one or few realizations is without interest : we would not be able to perform experiments with the same initial and boundary conditions. Indeed only statistical quantities are of practical meaning in order to describe the randomness : we need first mean values, then variances, correlations, and, at the best, probability density functions.

During the past one or two decades, great efforts have been devoted to establish macroscopic balance equations for these statistical quantities : mean, variances... etc, in such a way that, knowing only "mean initial and boundary conditions", it would be possible to obtain directly these mean quantities, after analytical or numerical integration, at every point of the flow, and for any type of flows. These new statistical macroscopic balance equations cannot be obtained from the primitive ones without additional assumptions, called modelling assumptions, or closure assumptions ; that is due to the non linearity of the primitive equations. In the case of reacting flows, the reaction terms being highly non linear, the problem is quite increased.

The purpose of this paper is to present a synthetic picture of the work done up to now concerning this "modelling". We will first explain the basic principles, and we will show that, for a multireactive mixture, the joint probability function of the species concentrations is a key quantity. Secondly, we intend to briefly review the three groups of models that are presently under investigation. Finally, we will emphasize the weakest point of all these models, the one which has to attract all our efforts in the future.

The methods and works that we shall review here are not particular to liquid flows ; they are relevant for the general domain of one phase turbulent reacting fluid flows. In fact, the field of applications of our own work has been the turbulent flows of gases and particularly turbulent combustion.

Owen to such "modelled" macroscopic equations, the numerical simulation of any type of turbulent reacting flow is possible, provided that we use convenient numerical algorithms (that we now know almost sufficently). Classical chemical reactors, like stirred reactors or plug flows, in transient or steady operations, can be simulated, but also more detailled studies, giving local quantities within more complex flows can be performed. That open the door to a large range of interesting applications.

II - THE BASIC PRINCIPLES OF THE MODELLING PROCEDURE

II.1 - The basic macroscopic balance equation with reactive flows are the well known Navier Stokes equations, together the continuity equation and the diffusion-reaction equations for each of the involved reactive species ; in the case of exothermic or endothermic reactions, an energy balance equation has to be added, but we will not emphasize this point here.

We shall write here only the diffusion reaction equation, using as variables describing the composition of the fluid, the mass fractions Y_i,

$i = 1... \ n - 1$; it writes
$$\frac{\partial}{\partial t} Y_i + \frac{\partial}{\partial x_\alpha} u_\alpha Y_i = \frac{\partial}{\partial x_\alpha} J_\alpha^i + \dot{w}_i \qquad (1)$$

with u_α the mass velocity, J_α^i the diffusion flux (the velocity of the species i is just $-J_\alpha^i / Y_i$) and $\dot{w}_i$ the rate of reactions (mass of i per unit mass and per unit time) ; in the case of a dilute medium, J_α^i can be

written simply :
$$J_\alpha^i = d_i \frac{\partial Y_i}{\partial x_\alpha} \qquad (2)$$

d_i is a mass diffusion coefficient of the species i within the diluant. The density of the fluid is supposed to be a constant here, but it is not a problem to use other state equations, even with quite strong effect of compressibility.

The occuring chemical mechanism, and its chemical kinetics, allows the knowledge of the rate $\dot{w}_i$, as a non linear function of all Y_j. We shall consider here as examples only three simple situations.

a) The case of a single bimolecular reaction $A + \nu B \to (1 + \nu) C$, giving simply $\varrho \dot{w}_i = - \nu_i \varrho^2 Y_A Y_B$, for $i = A$ or B ($\nu_A = 1$ and $\nu_B = \nu$ being "massic coefficients") ; we shall write
$$\dot{w}_i = K_i Y_A Y_B \qquad (3)$$

b) The case of a single bimolecular reaction, but with a highly non linear rate :
$$\dot{w}_i = - K_i Y_B f(Y_A), \text{ with } f(0) \text{ and } f(Y_{A_0}) = 0.$$
This rate does not correspond to a given mechanism, but is chosen here to emphasize the effects of highly non linear rates.

c) The case of two successive reactions :
$$A + \nu_1 B \to (1 + \nu_1)C,$$
$$C + \nu_2 B \to (1 + \nu_2)D,$$

with :
$$\dot{w}_A = - K_1 Y_A Y_B \, , \quad \dot{w}_B = - \nu_1 K_1 Y_A Y_B - \nu_2 K_2 Y_C K_B$$
$$\dot{w}_C = + K_1 (1 + \nu_1) \, Y_A Y_B - K_2 Y_A Y_C$$

II.2 - The balance equation for the statistical mean value $\overline{Y}_i$ can directly found from (1) ; following Reynolds, first we define Y'_i and u'_α, called the fluctuations, with :
$$Y_i = \overline{Y}_i + Y'_i \, , \quad u_\alpha = \overline{u}_\alpha + u'_\alpha \tag{4}$$
and then we replace (3) in (1) and average ; it comes :

$$\frac{\partial}{\partial t} \overline{Y}_i + \frac{\partial}{\partial x_\alpha} \overline{u}_\alpha \, \overline{Y}_i = \frac{\partial}{\partial x_\alpha} \left(\overline{J^i_\alpha} - \overline{u'_\alpha Y'_i} \right) + \dot{\overline{w}}_i \tag{5}$$

This balance equation cannot be used directly, because it involves new unknown quantities : a) the diffusion fluxe $\overline{u'_\alpha Y'_i}$, which is usually much larger than the mean molecular one $\overline{J^i_\alpha}$ (in turbulent regime...) ; it involves correlations between fluctuations and not only mean values ;
b) the mean reaction rate ; when we know the molecular reaction rate, for instance following one of the formulae given just previously, we cannot express the mean reaction rate only in function of the mean mass fractions : $\dot{\overline{w}}_i \neq \dot{\overline{w}}_i \left(\overline{Y}_j \right)$.

For instance, for a single bimolecular reaction :

$$\dot{\overline{w}}_i = - K_i \left(\overline{Y}_A \, \overline{Y}_B + \overline{Y'_A Y'_B} \right)$$

we need to know the correlation $\overline{Y'_A Y'_B}$ and not only $\overline{Y}_A$ and $\overline{Y}_B$ in order to compute $\dot{\overline{w}}_i$.

The equation (5) is then unclosed ; it involves new unknowns. It is possible to derive from the primitive balance equation, (1), and the Navier-Stokes equations, new balance equations for $\overline{u'_\alpha Y'_i}$, or $\overline{Y'_i Y'_j}$; but these equations would involve new terms, like $\overline{u'_\alpha Y'^2_i}$, or $\overline{Y'^2_i Y'_j}$, and also $\overline{Y'_i p'}$ and others ; in fact more and more unknown correlations would be generated, as new balance equation are derived. Then the only possibility for using these equations is to close at some level the hierarchy of equations by some asumptions ; the set of closed equations that will be obtained will constitute the "macroscopic model".

II.3 - A model for the turbulent diffusion fluxes, which is very commonly used, is the following :

By analogy with the Fick Law (2), one assumes as closure assumption :

$$\overline{u'_\alpha Y'_i} = - d_t \frac{\partial \overline{Y}_i}{\partial x_\alpha} \tag{6}$$

But d_t is not a property of the fluid ; the experiments show, in particular, that it is independant of the nature of the species ; d_t has to be a property of the turbulent flow and then, at the minimum, does depend on a length scale and a time scale of the turbulence. The simplest way to characterize the turbulence is to consider only two quantities, l_t its integral scale and k its kinetic energy (the time scale is nothing but $l_t/k^{1/2}$...), and by dimensional analysis it follows :

$$d_t = C_d \, k^{1/2} \, l_t, \text{ where } C_d \text{ is a contant} \tag{7}$$

The last step now is to provide the knowledge of $k_{1/2}$ and l_t ; that is the role of the so called "turbulence models", and there are several ones.

One of the simplest and mostly used is the Prandtl mixing length model, where $k^{1/2}$ is put proportional to $l_t \left| \dfrac{\partial \overline{u}}{\partial y} \right|$ ($\dfrac{\partial \overline{u}}{\partial y}$ is the largest mean velocity gradient), and l_t is prescribed with algebraic formula following the geometrical shape of the flow. A more general model (and more flexible with respect to any flow configuration) is the so called "k-ϵ" model, where k is given by a macroscopic balance equation similar to (4) ; l_t is related to ϵ, the dissipation rate of k, through $\epsilon = k^{3/2}/l_t$, and ϵ itself follows another balance equation. The different terms included in the balance equations for k and ϵ have a physical meaning ; their form are given by dimensional analysis after a small number of physical choices...

The Prandtl mixing length model, as well as the k-ϵ model, coupled with (8) and (5), have given quite good results predicting turbulent diffusion fluxes is a number of cases ; however quite clear discrepancies of (5) have been emphasized in other important cases, and other models can now be proposed. For more details concerning turbulence models, a good basic book is the one of Tennekes and Lumley [1].

II.4 - A model for the mean reaction rates is needed, similarly as the turbulent diffusion fluxes. This model can be quite complicated, because the molecular reaction rate itself can include a large number of species, depending on the mechanism ; it has to include the effects of fluctuations of species mass fractions, and of correlations of these fluctuations, but without direct influence of velocity fluctuations.

The effect of these fluctuations is not expected to be low ; one can verify easily that, if the fluctuations of the two reactive species A and B are in phase opposition, $\overline{Y'_A Y'_B}$ tends to be negative, and may reach $- \overline{Y}_A \overline{Y}_B$ in case of perfect segregation of A and B ; in this limiting case

$$\overline{\dot{w}}_A = - K \overline{Y_A Y_B} = 0, \text{ while } - K \overline{Y}_A \overline{Y}_B \neq 0...$$

In the case of a single bimolecular reaction, the problem of $\overline{\dot{w}_i}$ is just the problem of computing the correlation $\overline{Y'_A Y'_B}$; consequently the first studies (Toor, O'Brien) have been performed trying to approximate the moments $\overline{Y'^m_i Y'^n_j}$. A good review concerning these studies can be found in [2], by J.C. Hill. But with a more non linear rate, or with a more complex mechanism, $\overline{\dot{w}_i}$ is not so simple ; its right definition is :

$$\overline{\dot{w}_i} = \int_{Y_1} \int_{Y_n} \dot{w}_i (Y_j, j = 1 \ldots n) \ P(Y_j, j = 1 \ldots n) \ dY_1 dY_2 \ldots dY_n \tag{8}$$

where P is the joint probability density function (p.d.f.) of all the mass fractions involved in $\dot{w}_i$, and the Y_i are random variables associated to each $\overline{Y_i}$. P provide us the more general way to compute $\overline{\dot{w}_i}$; of course, the knowledge of P is equivalent to the one of any moments of the type $\overline{Y'^n_j}$ or $\overline{Y'^n_i Y'^m_j}$, or $\overline{Y'^n_i Y'^m_j Y'^p_k} \ldots$, and we could prefer (as in the past) to use these moments instead of P. But the moments do not exist always, and, in addition, they have to verify a lot of inequalities :

$$\overline{Y'^2_j} \geqslant 0, \quad \left| \overline{Y'_i Y'_j} \right| < \sqrt{\overline{Y'^2_j}} . \sqrt{\overline{Y'^2_i}},$$ and similar others (see [2]). Additional inequalities are to be satisfied due to the fact that P can be defined only on a restricted domain of the phase space... All that renders more attractive finding directly close forms of P instead of that of the moments.

In order to emphasize the existence of a domain supporting P, let us consider a reactor, in which the reaction $A + \nu B \rightarrow C$ has to be produced, feeded through two inlets E_1 and E_2 (see fig. 1a). At every point of the flow we can plot a phase plane (Y_A, Y_B) ; at the inlet E_1, the composition of the flow is assumed non fluctuating, and represented by the point E_1, at the inlet E_2, also non fluctuating, the composition is represented by E_2 ; with the simple reaction rate $\dot{w}_i = - K_1 Y_A Y_B$, E_1 and E_2 have to lie on the axis in order that the reaction rate is zero at the inlets, but with a more general law (like in § II.1.b), it can be conceived that $\dot{w}_i$ is zero even with a non zero value of Y_B, if $Y_A = Y_{A_0}$, and then we obtain E_2 outside the $Y_A = 0$ axis.

It can be demonstrated that, with impervious boundaries for A as well as for B, and with equal molecular diffusivities of A and B, the p.d.f. P of the fluctuations of Y_A and Y_B is non zero only within the polygon plotted fig. 1b : the limiting lines of this polygon are : the axis, the pure mixing line joining E_1 to E_2, and the pure reaction line (whose slope is just ν).

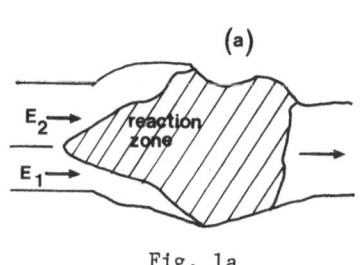

(a)

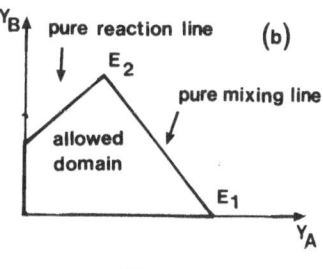

(b)

Fig. 1a   Fig. 1b

More details on this property can be found in [3]. In case where E_1 and E_2 are on the reaction line, the domain is even only a segment, as shown fig. 1c and a deterministic relation between Y_A and Y_B holds.

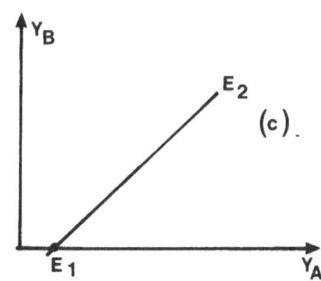

Fig. 1c

One can discuss the shape of this domain, following the chemical mechanisms involved. That as been attempted in [4]. Of particular interest is the domain permitted f $P(Y_A, Y_B, Y_C)$, in the case of a two step mechanism (like in § II.1.C).

The volume offered to non zero P is schematized fig. 2a : it is obtained plotting pure reaction lines obtained from any point of the segment $E_1 E_2$, and taking the surface that is their enveloppe.

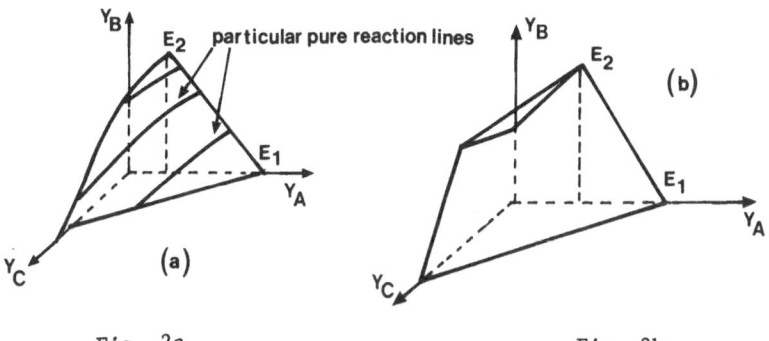

Fig. 2a   Fig. 2b

This volume can be approximated (in excess) assuming that only the first reaction is occuring ; one obtains a polyedron as in fig. 2b.

III - A GLOBAL PICTURE OF THE EXISTING MODELS

III.1 - The problem with very fast reactions (with a very small reaction time with respect to any convection or diffusion times), has been adressed first. This case has been discussed and almost solved by Toor [5], as early as 1963, followed by Lin and O'Brien [6].

The solution is quite clear when we look at the joint p.d.f. of Y_A and Y_B, the two reactants of a single step reaction. When the reaction is very fast, the coefficients K_i of § II.1a are infinite in such a way that the mean reaction rate $\dot{w}_i = K_i \overline{Y_A Y_B}$ is finite only if $\overline{Y_A Y_B} = 0$, i.e. either $Y_A = 0$ or $Y_B = 0$ (and not $\overline{Y_A} = 0$ or $\overline{Y_B} = 0$!). It follows then that $P(Y_A, Y_B)$ within the phase plane (Y_A, Y_B) is supported only by the axis $Y_A = 0$ and $Y_B = 0$; the p.d.f. is pushed against one boundary of its permitted domain.

The result is then that the p.d.f. is one-dimensional instead of two-dimensional ; in addition if we define $\phi = \nu Y_A - Y_B$, ϕ satisfy a balance equation without reaction term, and, because Y_A and Y_B cannot be non zero at the same time, ϕ represents either νY_A, if positive, or $- Y_B$, if negative ; one can write :

$$\begin{cases} \overline{Y_A} = \displaystyle\int_0^\infty \frac{\phi}{\nu} P(\phi) \, d\phi \\[4mm] \overline{Y_B} = - \displaystyle\int_{-\infty}^0 \phi \, P(\phi) \, d\phi \end{cases} \qquad (9)$$

and similarly for any moment $\overline{Y_A^n}$ or $\overline{Y_B^m}$. The knowledge of $P(\phi)$ allows us to solve the problem completely.

If we assume that the molecular diffusion of A and B are identical, the balance equation for ϕ writes : $\dfrac{\partial}{\partial t} \phi + \dfrac{\partial}{\partial x_\alpha} u_\alpha \phi = \dfrac{\partial}{\partial x_\alpha}\left(d \dfrac{\partial \phi}{\partial x_\alpha} \right)$ (10)

The problem of reactive turbulent flow is then reduced to the computation of the p.d.f. of the fluctuations of a single inert scalar diffusing into the turbulent flow. That problem, however, was not adressed by specialists of turbulence, at the time of Toor's studies ; it has been studied since that, and now we have at our disposal some modelled balance equations for $P(\phi)$ itself ; as P is usefull even when the chemistry is not very fast, we will consider this problem with this more general case.

The generalization of (9) to the case of a complex mechanism is quite obvious when the diffusion coefficients of each species involved can be assumed almost equal ; in this case one finds that each species Y_i is deterministically related with a single inert ϕ, conveniently defined (see Bilger [7]), and we can write : $Y_i = \displaystyle\int_{-\infty}^{+\infty} Y_i(\phi) \, P(\phi) \, d\phi$ (11)

For the case where the diffusion coefficients are not equal, it is necessary to consider several inert scalars.

III.2 - Methods using modelled balance equations for P. They appears as a generalization of the approach of Toor when the reactions are not very fast. We have seen in § II.3 that the knowledge of P allowed the exact computation of $\dot{\overline{w}}_1$. This is the reason why, when searching for a balance equation for P, the reaction term appears exactly, without any need of closure assumption ; this property has been pointed out for the first time by C. Dopazo and E.E. O'Brien [8].

Since this pionniering work, a quantity of researchers have used this approach ; they have proposed first modelled equations for the p.d.f. of one inert or reactive species, then for the joint p.d.f. of several Y_i, and later for the joint p.d.f. of Y_i and the velocity components u_α.

The balance equation for $P(\phi)$ (ϕ being an inert scalar), can be found from the primitive equations, i.e. the Navier Stokes equations and the diffusion equation (10). The derivation is not as easy as the one of the moments equations for $\overline{\phi}$ or $\overline{\phi'^2}$; there are two ways to obtain it, as explained in [9] ; the equation obtained is unclosed, due to two types of things : first, the correlations with the fluctuations of velocity (again the problem of turbulent diffusion), and second the correlations with the fluctuations of ϕ but at another close point : this effect is called the micromixing.

The model that has been first proposed by Dopazo and O'Brien, gives :

$$\frac{\partial}{\partial t} P(\phi) + \frac{\partial}{\partial x_\alpha} \overline{u}_\alpha P(\phi) = \frac{\partial}{\partial x_\alpha}\left(D_P \frac{\partial P(\phi)}{\partial x_\alpha}\right) - \frac{\partial}{\partial \phi}\left(\frac{6d}{\lambda^2}(\overline{\phi} - \phi)P(\phi)\right) \quad (12)$$

The turbulent diffusion is modelled through a diffusion coefficient D_p , which has to be related to the turbulence, similarly to d_t in § II.3. The micromixing term, the last one, involves λ^2, the Taylor microscale of the fluctuations ϕ, which is also to be prescribed ; in turbulence of large Re_τ, $\dfrac{6d}{\lambda^2}$ is a time scale that can be related to the time scale τ_t of the turbulence itself, eventually with a dependance on the Schmidt number (see [10]).

Another model for the micromixing term can be derived from a mechanistic model, the coalescence-dispersion model of Curl [11] ; it gives :

$$\frac{\partial}{\partial t} (P(\phi)) + \frac{\partial}{\partial x_\alpha}\left(\overline{u}_\alpha P(\phi)\right) = \frac{\partial}{\partial x_\alpha}\left(D_P \frac{\partial P(\phi)}{\partial x_\alpha}\right) - \frac{C_D}{\tau_t}\int_{-\phi}^{\phi} 2P(\phi')$$

$$\left(\int_{\phi}^{\phi^+} P(\phi'') \delta\left(\frac{\phi' + \phi''}{2} - \phi\right) d\phi'' - P(\phi)\right) d\phi' \quad (13)$$

C_D , here, is a constant relating the micromixing time and the turbulence time. This new model has been improved also by W. Kollman [12].

The generalization of such models to $P(Y_j, j = 1...n)$, where Y_j are several reactive species, is straigh forward ; for instance with (12) we

$$\text{get}: \frac{\partial}{\partial t} P(Y_j) + \frac{\partial}{\partial x_\alpha}\left(u_\alpha P\right) = \frac{\partial}{\partial x_\alpha}\left(D_P \frac{\partial P}{\partial x_\alpha}\right) - \sum_{j=1,n} \frac{\partial}{\partial Y_j}\left(\frac{6d}{\lambda_j^2}\left(\bar{Y}_j - Y_j\right)P + \dot{w}_j P\right) \quad (14)$$

The closure problem for the turbulent diffusion can be avoided if we look at the pdf as Y_j and u_α ; but there are other closure problems, related to the dependance in the velocity. S. Pope [13] has been working along this line from several years and has been able to propose a model for this new pdf ; in fact, this new pdf contains also a part of a turbulence model (the k equation, in particular) ; S. Pope verified the agreement of his model with the existing turbulence models.

The use for numerical integration of balance equations for $P(\phi)$, such as (12), is slightly less easy than eq. (5) : there is one dimension more ; the use of multidimensional equation, like eq. (14), however, is quite difficult ; with classical numerical methods, based on a grid, only three dimensions are feasible ; with Monte Carlo simulations Jones and Kollmann [14] can handle, at maximum, 4 species in a two dimensional, steady flow.

The results of computations, even with the simplest model (12), are quite realistic. The ref. [15] gives calculations of a single bimolecular reaction within a well stirred reactor. The ref. [16] shows the shape of $P(\phi)$ computed with the Curl model and the values of mean mass fraction that are obtained, with the assumption of very fast reactions, within a diffusion jet flame ; the agreement with experiments appears quite satisfactory.

Additional improvements of the method are currently under investigation by using conditional pdf, where the peaks are separated from the continuous part of the p.d.f. (see [17]).

III.3 - The "Reacting laminae" models, are attempting to include more physics into modelling assumptions. The idea has been proposed a long time ago by Mao and Torr, in 1970 [18] and has been used under several forms, as can be seen in the review of J. Villermaux [19]. It has became very popular, more recently, and is often called "Flamelets approach" in the field of combustion.

The method is based on a particular picture of the turbulent reacting flows ; it is assumed that it consist in a collection of reaction diffusion layers, which are continuously displaced and stretched within the turbulent medium. For the sake of simplicity, the layers are often assumed to be one dimensional, that is almost planar, with a thickness very small with respect to their main curvature radii, and similar to study laminar diffusion reaction layers. It is consequently possible to study separately these layers, or "reacting laminae", or "flamelets", and to compute their internal structure with all chemical details, like in a laminar flow. After that, it is necessary to build the turbulent field by a convenient collection of such a "lamina".

With this common feature, many types of models have been devised ; we will describe here only two of them, because they are directly usable within the mean balance equations. The first one has been proposed by Broadwell and Marble [20], and is based on a key-quantity, the mean flame (or reaction) surface by unit of volume, Σ. Indeed, the mean reaction rate

is given by :
$$\dot{W}_i = V_{Di}\Sigma \quad (15)$$

where V_{Di} is a velocity : it is the volumetric flow rate per unit of surface of the reactant i disappearing through the "reacting laminae". V_{Di} is a quantity that can be computed independantly of the turbulence, just computing the "lamina". At the contrary, Σ is a turbulent quantity ; Marble and Broadwell have postulated a balance equation for it, following the works of Batchelor about surface within turbulent flows.

The second type of approach, followed by Liew and Bray [21] and Peters [22], consists in extracting from laminar calculations $Y_i(\phi)$, where ϕ is again an inert scalar, and then the turbulence can be taken into account simply through $P(\phi)$. Indeed we can write for any species

$$\overline{Y}_i = \int_{-\infty}^{+\infty} Y_i(\phi) P(\phi) d\phi \tag{16}$$

The formula (16) has been used directly in [21], but Peters proposed to take into account an additional parameter characterizing the turbulent medium : the stretch rate γ. In this case a library of laminar stretched reaction layer have to be built, giving $Y_i(\phi,\gamma)$, and any mean value Y_i is

computed through : $$\overline{Y}_i = \int \int P(\phi,\gamma) \ Y_i(\phi,\gamma) \ d\phi d\gamma \tag{16'}$$

In the first case, only $P(\phi)$ is needed, and we already know the mean to compute it (see § III.2) ; but now, with (16'), we need $P(\phi,\gamma)$, and our present turbulence knowledge is not able to give us this joint pdf. Peters proposed then to assume a statistical independancy between ϕ and γ, to assume that $P(\gamma)$ is lognormal, and to compute $P(\phi)$ as previously. An interesting example of application of this method has been recently published for a diffusion flame, in [23].

The "reacting laminae" models, whatever can be its exact form, allows to use the complex mechanism, that is actually playing ; they are based on physically clear assumptions. However, in their present form, they neglect important physical phenomena : the picture of the turbulent reacting flow on which they are based is not complete ; the curvature of laminae is clearly neglected, as well as their transient evolution between the several states of strain that are imposed by the turbulence ; in addition interactions of adjacent laminae, are likely to occur, in particular within coherent vortices present in many mixing layers (see for instance [24]). In our opinion, these approaches are then restricted to the cases where these phenomena are negligible, that is essentially for fast enough reactions and not to large turbulence Reynolds number ; they appear then quite complementary with the previous pdf models. It is possible, however, that their validity can be extended including the more complex phenomena we have mentionned ; a step toward that may be [25].

III.4 - The Lagrangian models constitutes the third category of models. The common features of such models is to try to represent the evolution of fluid particles, or fluid elements, along their life and trajectory within the reacting medium. Such an approach can, in principle, avoid the use of Eulerian balance equations, like eq. (5), and does actually, in some simple cases ; but in more complex cases it can be used jointly with Eulerian equations, in order to build good approximations of the joint pdf $P(Y_j, j = 1...n)$.

The simplest example of Lagrangian models is the IEM model, supported by Villermaux et al [26], and others in the field of chemical engineering [27]. It consist to replace the primitive equation (1) by :

$$\frac{d}{dt} Y_i \left(= \frac{\partial Y_i}{\partial t} + \frac{\partial}{\partial x_\alpha} u_\alpha Y_i \right) = \frac{\overline{Y_i} - Y_i}{\tau_{ex}} + \dot{w}_i \tag{17}$$

The first term on the RHS is a model for the molecular diffusion fluxes within the turbulent flow, and represents them as Exchanges by Interaction with the Mean value (then justifying the name...) ; τ_{ex}, here, is an exchange time assumed to be simply proportional to the turbulence time τ_t.

Indeed, when from (17) we try to derive a balance equation for $\overline{Y_i}$, we obtain the right one, (5), in which J^i_α is neglected ; similarly it can be derived equations for $\overline{Y'^2_i}$ and $\overline{Y'_i Y'_j}$, that are the same as modelled equations proposed many years ago by people using the moments approach (see [2]). Finally, it is possible from (17) to find the pdf equation (14), with the model of Dopazo and O'Brien ; we have just to postulate the existence of D_P (which is not assumed in (17)) and to identify $\frac{1}{\tau_{ex}}$ and $\frac{6d}{\lambda^2_\alpha}$.

This model allows us, by numerical time integration of (17), to take into account a quite complex chemistry. The first use of this model has been within simple reactors, in which the residence time distribution f(t) was known ; then, together with (17), the equation defining $\overline{Y_i}$ may be used :

$$\overline{Y_i} = \int_0^\infty Y_i(t) \, f(t) \, dt \tag{18}$$

where $Y_i(t)$ is the solution of (17) (depending, of course of $\overline{Y_i}$ itself and of initial conditions, to be given...). An extension of the use of this model for complex flow has been proposed in [28] ; instead of (18), Y_i is then given by the Eulerian balance equation (5), and the curves $Y_i(t)$, with conveniently chosen initial conditions, are used to build an approximation of the multidimensional p.d.f. $P(Y_j, j = 1...n)$ at each point of the flow.

Other Lagrangian models have been proposed, in which the lagrangian equation is intended to represent fluid elements bigger than a fluid particle, with an internal structure ; for example the "ESCIMO" model of D.B. Spalding [29], or the "three environment model" [30], are of this type. For each of these model, the physical picture necessarily neglects some physical phenomena... with this respect these models appear similar to the "reacting laminae" models.

Perhaps the most detailed lagrandian model is the one of Durbin and coworkers [31], first devised in order to simulate the turbulent diffusion only, and later extended to take into account a two species bimolecular reaction [32]. It is based on a stochastic simulation of the random walk of fluid particles, but is able also to provide the probability density function of the position (and then of the composition), within the entrance section of the reactor, of two fluid particles which would be at the same later time at a given point within the reactor. Owen to this new crucial quantity, the calculation of the second moment $\overline{Y'_i Y'_j}$ is possible, and then the modelling of the mean reaction rate of a single bimolecular reaction is performed. This model is quite interesting, but unfortunately the knowledge of the full p.d.f. would require more informations (about the position of three, four... particles) than provided by the present model.

IV - THE WEAK POINTS OF THE MODELLING

IV.1 - What is the agreement between the previously described models and the experiments ? With a qualitative point of view, all the models clearly give the right tendancies. Concerning the mean reaction rates, the expected result of the influence of fluctuations is to lead to a much smaller sensitivity with respect to chemical entries, K_i for instance ; that effect is obtained, and, in the case of very fast chemistry, it is clear that the mean reaction rate is turbulence-controlled. Several experimental results concerning the mean reaction rate, or more exactly the consumption of the reactants or apparition of products, have been compared with the differents models, either in plug flow reactors or well mixed reactors (see in [2] or [19] , or [3]). The general impression is that the agreement is good, but there is always one or two empirical parameters (and particularly C_D or the one related to τ_{ex}) whose adjustment guarantee the agreement ; it is not quite clear that these parameters have a large range of validity. On the other hand, the experiment results concern always mean concentations ; very rare, if not absent, are results concerning fluctuations and p.d.f. of fluctuations ; so, the experimental data do not constitute a test of the shape of this p.d.f. as directly induced by the closure assumptions. Our conclusion at this time is that the comparisons with experiments are quite encouraging, but that we need more comparisons with well chosen set of experiments (for instance in which a chemical time can be varied step by step with respect to the turbulence time τ_t), and with more detailed measurements about the p.d.f.'s.

IV.2 - Anyway, a simple inspection of the formulae and equations of the proposed models shows that the result will be crucially influenced by one key modelling assumption, and one modelling constant. It is the one concerning the time scale the dissipation rate of the fluctuations of Y_i :

for the p.d.f. approaches, it is $\dfrac{6d}{\lambda_i^2}$, or $\dfrac{C_D}{\tau_t}$, for the IEM model, it is τ_{ex}.

For the model of Marble and Broadwell, the crucial quantity is Σ, not a time scale, but a reciprocal length scale, which could be translated in to a time scale : $(k^{1/2}\Sigma)^{-1}$. A weak point of all these models appears as follows : why this time scale (say τ_{ex}) could be just proportional to the integral turbulence time, without any influence of chemistry ?

Any experimental validation of that assumption is difficult, but recently direct numerical simulation of turbulent reactive flows appeared possible, for very moderate Reynolds number, however. Direct numerical simulations use the primitive equations themselves, the continuity and Navier Stokes equations, jointly with one or two diffusion-reaction equations like (1) ; a very big computer is required for the integration (Cray 1 or Cray 2), the maximum possible value of turbulence Reynolds number is limited to 20 or 50 (while well developed turbulence usually involves Reynolds numbers up to 500...) ; finally only very simple geometries can be taken into account, mainly homogeneous turbulence. Given a particular set of randomly chosen initial conditions, the direct simulation follows a particular realization of turbulent flows along its life ; volume averages are calculated during the calculation, that are identified to the ensemble averages, due to the spatial homogeneity of the field.

Leonard and Hill [34] have recently performed such a simulation, in the case of a single bimolecular reaction, starting from an initial state with A and B perfectly segregated. The fig. 3 shows the time evolution of

$\langle \varepsilon_A \rangle = d_A \overline{\dfrac{\partial Y'_A}{\partial x_\alpha}} \, \overline{\dfrac{\partial Y'_A}{\partial x_\alpha}}$ and λ_A (the taylor microscale), for different Damköhler number, that is different time scale ratio ($D_a = \tau_t /$chemical time scale $= (K_A Y_{A_0})^{-1}$).

Picart et al [35] have studied a similar case, but with a monomolecular very non linear reaction : $\dot{w}_A = - K_A Y_A (1 - Y_A)^5$.

The fig. 4 shows the evolution with time of several quantities. First the decrease of $\overline{Y_A}$, compared with the one of $\overline{Y_{lam}}$, shows the effect of the fluctuations : $\overline{Y_{lam}}$ is just the solution of

$\dfrac{dY_{lam}}{dt} = - K_A \overline{Y_{lam}} \left(1 - \overline{Y_{lam}}\right)^5$, while $\overline{Y_A}$ is the solution of

$\dfrac{d\overline{Y_A}}{dt} = - \overline{K_A Y_A (1 - Y_A)^5} = \dot{\overline{w}}_A$. The curves labelled λ_τ and λ_c are the Taylor microscales of the velocity field and the Y_A field, respectively ; they are directly related to the dissipation time scales : $\tau_t \propto \dfrac{\lambda_\tau^2}{\nu}$ and $\tau_Y \propto \dfrac{\lambda_c^2}{d_A}$.

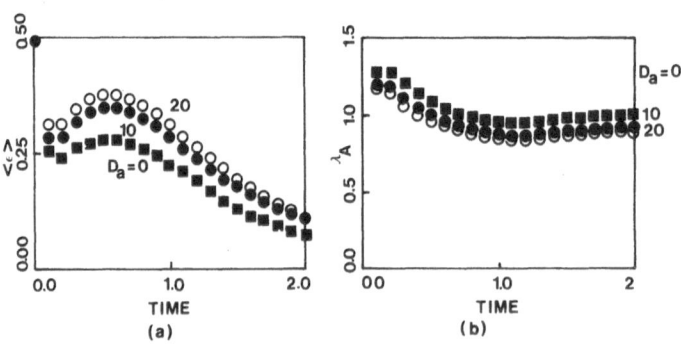

Fig. 3

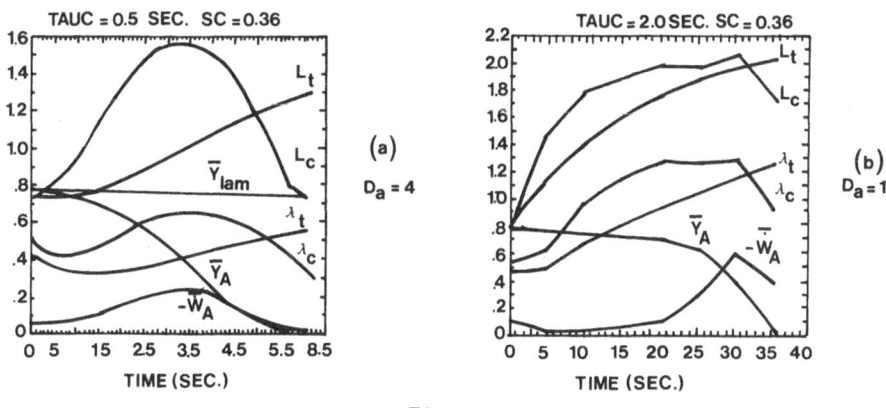

Fig. 4

Also plotted are the integral length scales of the velocity field, L_t, and of the scalar field, L_c. The figures a and b are relative to different values of D_a.

The results of these simulations concerning the question under discussion are clear :
- in case of a single bimolecular reaction, the assumption $\tau_Y \propto \tau_t$ independantly of D_a is quite acceptable : less than 5 % of error is found for different values of D_a ;
- however, in the case of a strongly non linear reaction, this assumption is not acceptable : λ_t and λ_c are not at all parallel. Here a strong influence of the reaction on λ_c is displayed, due to the modifications of the spectrum of the scalar by the reaction, which can be also emphasized with these simulations [35].

Direct numerical simulations cannot be used for practical reactors ; we have then to improve the usual models in order to take into account this new phenomenon. If we follow the classical approach we have described here, we would be led to add in our model one more equation for τ_Y, or

$$\overline{\varepsilon}_Y = d \, \overline{\frac{\partial Y'}{\partial x_\alpha} \frac{\partial Y'}{\partial x_\alpha}} \; ;$$ a first step in this direction has been made in [36].

IV.3 - Another weak point of the previous models is the modelling of the turbulent diffusion of reactive species. Many models are using again the gradient transport assumption explained in § II.3, with the same diffusion coefficient as for a non reactive species. The pdf model of Pope (with a joint pdf including velocity fluctuations) and the Lagrangian model of Durbin et al are the only ones able to avoid this type of assumption. The question arising here is the following : why the diffusion coefficient, if existing, could not depend on the chemistry ?

A first answer to this question can be obtained considering the balance equation for $u'_\alpha Y'_i$, the diffusion flux itself ; it can be obtained from the Navier Stokes equations and the diffusion reaction equation ; one finds :

$$\frac{\partial}{\partial t} \overline{u'_\alpha Y'_i} + \frac{\partial}{\partial x}\left(u \overline{u'_\alpha Y'_i}\right) + \frac{\partial}{\partial x_\beta}\left(\overline{u'_\alpha u'_\beta Y'_i + p' Y'_i \delta_{\alpha\beta}}\right) =$$

$$- \overline{u'_\alpha u'_\beta} \frac{\partial Y_i}{\partial x_\beta} - \overline{u'_\beta Y'_i} \frac{\partial u_\alpha}{\partial x_\beta} + \overline{\frac{p'}{\rho} \frac{\partial Y'_i}{\partial x_\alpha}} + \overline{u'_\alpha \dot{w}_i} \qquad (19)$$

The last term on the RHS is a reaction term : the reaction has an explicit effect on $\overline{u'_\alpha Y'_i}$.

When the LHS terms are neglected (homogeneous and steady approximation), and for simple flows, the eq. (18) can be reduced to a gradient formulation, like eq. (5) ; if that is done in our reacting case, we will recover a formula like (5), but with an explicit effect of reaction within the diffusion coefficient.

Such a study has been made for the case of a single species non linear reaction by Borghi and Dutoya [36] ; the crucial point is to find a closure assumption for $\overline{u'_\alpha \dot{w}_i}$, and an assumption similar to the one of

Dopazo and O'Brien leads to :
$$\overline{u'_\alpha \dot{w}_i} = \frac{\overline{Y'_i \dot{w}_i}}{\overline{Y'^2_i}} \cdot \overline{u'_\alpha Y'_i}$$

giving finally
$$d_{ti} = C_d \, k^{1/2} l_t \left/ \left(1 - C'_d \, \tau_t \frac{\overline{\dot{w}_i Y'_i}}{\overline{Y'^2_i}}\right)\right. \qquad (20)$$

instead of (7).

The formula (20) shows two opposite effects of the reaction : usually when $\overline{\dot{w}_i Y'_i}$ is large, $\overline{Y'^2_i}$ is also large, and then the later compensate the former.

A good way to test closure assumptions for that problem, as well as for the previous aspect of dissipation time, could be again the direct simulation ; but here, the concentration field must resile the spatial homogenity. That appears to be possible with the present experience in direct simulation ; a first attempt has been done in [37].

V - CONCLUDING REMARKS

Modelled macroscopic equations for the numerical simulations of many types of different turbulent reacting flows are now available. The present state of knowledge concerning the modelling assumptions is quite encouraging ; the base of the models appears well founded, even if more work is needed in order to improve the modelling assumptions in more complicated cases. There are several types of models, but they are more complementary than conflicting. There has been over the two past decades a quite sensible progress, and, owen to the new computers, a number of experimental or practical reactor could be simulated now with these new methods.

More detailed comparisons between experiments and predictions are now needed. No doubt that they will contribute to both the improvement of the models and their extended use by chemical engineers.

REFERENCES

1. H. Tennekes, J.L. Lumley, A first course in turbulence, The MIT Press, Cambridge, Mass, 3e Ed (1974).
2. J.C. Hill, Homogeneous turbulent mixing with chemical reaction, Annual review of fluid Mechanics, vol. 8, pp. 135-161 (1976).
3. R. Borghi, Réactions chimiques en milieu turbulent, Publication ONERA n° 1979-2 (1979).
4. R. Borghi, Models of Turbulent combustion for Numerical predictions, in : Prediction methods for turbulent flows, Ed. W. Kollmann, Avon Karman Institute Book, Hemisphere Pub. Corp. (1980).
5. H.L. Toor, A.I. Ch. E. Journal, vol. 8, n° 1, p. 79 (12962).
6. C.H. Lin, E.E. O'Brien, J.F. Mech., vol. 64, part 1, pp. 195-206 (1974).
7. R.W. Bilger, Turbulent diffusion flames, in : Turbulent Reacting Flows, Ed. P.A. Libby, F.A. Williams, Topics in Applied Physics, vol. 44, Springer-Verlag, Heidelberg (1980).
8. C. Dopazo, E.E. O'Brien, A probabilistic approach to the autoignition of reactive turbulent mixtures, Acta Astronautica, vol. 1, p. 1239 (1974).
9. E.E. O'Brien, The probability density function approach to reacting turbulent flows, in : Turbulent reacting Flows, Ed. P.A. Libby, F.A. Williams, Topics in Applied Physics, vol. 44, Springer-Verlag, Heidelberg (1980).
10. E.E. O'Brien, Recent contributions to the statistical theory of chemical reactants in turbulent flows, PCH Journal, vol. 7, n° 1, pp. 1-15 (1986).
11. R.I. Curl, Dispersed phase mixing theory and effects in simple reactors, A.I. Ch. E.J., $\underline{9}$ (2), pp. 175-181 (1963).
12. W. Kollman, J. Janicka, The probability density function of a passive scalar in turbulent shear flows, Phys. Fluids, vol. 25, n° 10, pp. 1755-1769 (1982).
13. S.B. Pope, Transport equation for the joint probability density function of velocity and scalars in turbulent flow, Phys. Fluids, vol. 24, n° 4, pp. 588-596 (1981).
14. W.P. Jones, W. Kollmann, Multiscalar P.d.f. transport equations for turbulent diffusion flames, 3th Turbulent shear flows conference, Cornell Univ., Aug. (1985).
15. C. Bonniot, R. Borghi, Joint probability density function in turbulent combustion, Acta Astronautica, vol. 6, pp. 309-327 (1979).
16. J. Janicka, W. Kollmann, A Two-variables formalism for the treatment of chemical reactions in turbulent H_2-Air diffusion flame, 17th Symp. (Int.) on Combustion, pp. 421-430, The Comb. Inst., Pittsburgh (1979).
17. M.Z. Wu et E.E. O'Brien, Prediction of single point temperature statistics in a Half Heated grid flow, Comb. Sc. Tech., vol. 29, pp. 53-66 (1982).

18. K.W. Mao, H.L. Toor, A diffusion Model for reactions with turbulent mixing, A.I. Ch. E. Journal, vol. 16, pp. 49-52 (1970)

19. J. Villermaux, Micromixing phenomena in stirred reactors, Encyclopedia of Fluid Mechanics, Gulf Pub. Co., West Orange NJ, 707 (1986).

20. F.E. Marble, J.E. Broadwell, The Coherent flame model for turbulent chimical reactions, Project squid tech. Rep., TRW-9-PU, Pindue University, West Lafayette, Indiana (1977).

21. S.K. Liew, K.N.C. Bray, J.B. Moss, A flamelet model of turbulent non premixed combustion, Comb. Sci. Tech., vol. 27, pp. 69-73 (1981).

22. N. Peters, Laminar flamelets concepts in turbulent combustion, 21th Symp. (Int.) on Combustion, Munich, Aug. (1986).

23. B. Rogg, F. Behrendt, J. Warnatz, Turbulent non premixed combustion in partially premixed diffusion flamelets with detailed chemistry, 21st Symp. (Int.) on Combustion, München, Aug. (1986).

24. R.E. Breidenthal, Structure in turbulent mixing layers and waves using a chemical reaction, J. Fluid Mech., vol. 109, n° 1 (1982).

25. J.E. Broadwell, R.E. Breidenthal, A simple model of mixing and chemical reaction in a turbulent Shear layer, J. Fluid Mech., vol. 125, pp. 397-410 (1982).

26. R. David, J. Villermaux, Chem. Eng. Sci., vol. 30, p. 1309 (1975).

27. J.M. Tarbell, R.V. Mehta, Mechanistic models of mixing and chemical reaction with a turbulence analogy, Physicochemical Hydrodynamics, vol. 7, pp. 17-32 (1986).

28. R. Borghi, E. Pourbaix, Lagrangian models for turbulent combustion, 4th Symp. on Turbulent Shear flows, Karlsruhe, Sept. (1983).

29. D.B. Spalding, A general theory of turbulent combustion, J. Energy, vol. 2, n° 1, pp. 16-32 (1977).

30. R.V. Mehta, J.M. Tarbell, A four environment model of mixing and chemical reaction, Part II : comparison with experiments, A.I. Ch. E. Journal, vol. 65, p. 125 (1983).

31. P.A. Durbin, A stochastic model of two particle dispersion and concentration fluctuations in homogeneous turbulence, J.F.M., vol. 100, Part 2, pp. 279-302 (1980).

32. B.L. Sawford, J.C.R. Hunt, Effects of turbulence structure, molecular diffusion and source size on scalar fluctuations in homogeneous turbulence, J. F. Mech., vol. 165, pp. 373-400 (1986).

33. R.S. Brodkey, in : Fluid Mechanics of Mixing, Ed. E. Millram, V.W. Goldschmidt, pp. 1-13, ASME, New-York.

34. A.D. Leonard, J.C. Hill, Direct Numerical simulation and simple closure theory for a chemical reaction in homogeneous turbulence, Workshop France USA on "Turbulent reacting flows", Ronen, July (1987).

35. A. Picart, R. Borghi, J.P. Chollet, Numerical simulation of turbulent reactive flows : new results and discussion, 6th Symp. on Turbulent Shear flows, Toulouse, Sept. (1987).

36. R. Borghi, D. Dutoya, On the scales of the fluctuations in turbulent combustion, 17th Symp. (Int.) on Combustion, The Comb. Inst., Pittsburgh, pp. 235-244 (1978).

37. W.H. Jou, J.J. Riley, On direct numerical simulations of turbulent reacting flows, AIAA paper 87-1324, AIAA 19th Fluid dynamics, Plasma dynamics and laser conferences, Honolulu, June (1987).

VILLERMAUX - Two comments :

1) In the IEM Model, the micromixing time (or exchange time) is a phenomenological parameter. It may actually represent various processes including molecular diffusion (and not only exchange in the Taylor microscale range).

2) In practice, it is often necessary to invoke several simultaneous mechanisms of micromixing (several steps) in order to account for experimental observations.

3) For simple reactions (e.g. A+B⟶C) different mixing models may yield equivalent results. However the results may strongly depend on the structure of models in the case of complex reactions involving autocatalytic steps. These are thus excellent candidate test reactions for discriminating between these models.

BORGHI - I agree fully with your comments and especially with the third one : complex reaction schemes with very non linear behaviour are now raising the most appealing challenges.

Concerning the IEM model, my purpose was to emphasize that, with the exchange time related for the Taylor microscale, it leads to the same eulerian equations as the ones used very classically in the field of turbulence modelling. Then, at least with this interpretation, the IEM model has not to be considered as poor with respect to these models of turbulence.

MIXING, DIFFUSION AND CHEMICAL REACTION OF LIQUIDS

IN A VORTEX FIELD

Frank E. Marble[*]

California Institute of Technology

ABSTRACT

The process of mixing two liquids to the molecular level is extremely complex, even under conditions of laminar motion, and takes on additional complexity when the liquids react chemically. As a consequence the relatively simple example treated here has the advantage of providing a transparent, but incisive, description of the processes that are essential in more complex situations. We consider a two-dimensional motion in which, initially, two different liquids are situated respectively in the upper and lower half-planes. At the start of motion a vortex is imposed at the origin. The interface between the two liquids, across which molecular diffusion is taking place, is distorted and stretched in a manner resembling a conventional stirring process in liquids.

It is shown that the mixing process in an ideal vortex is described by the similarity variable $r/(\Gamma^{2/3} D^{1/6} t)^{1/2}$. The departure from this result due to viscous stresses is described by a new dimensionless group $\mu_a \equiv \Gamma^{1/3} D^{1/6}/\nu$, the ratio of diffusive to viscous spreading rates. The mixing is strongly reduced when $\mu_a < 5$; for $\mu_a > 20$ the mixing is identical with that in an ideal vortex.

For very slow chemical reactions, a well-mixed core develops, the radius r_* of which grows as $(\Gamma^{2/3} D^{1/3} t)^{1/2}$. The product of reaction is to a large part contained within this core and the total product formed scales as $\Gamma^{2/3} D^{1/4} t_c / \beta_*$ where β_* is a value determined by the mixing, dependent principally upon the value of μ_a and very weakly upon the Reynolds number of the vortex.

For very fast chemical reactions, the behavior at the interface resembles a diffusion flame; the production of reaction products replaces molecular mixing because, in the case of very fast chemistry, the two reactants cannot coexist. Then the totally reacted core of radius r_* grows as $(\Gamma^{2/3} D^{1/4} t)^{1/2}$. Of interest in this case also is the augmentation of reactant consumption rate induced by the vortex. The augmentation of product generation rate is independent of time and very nearly proportional to $\Gamma^{3/4} D^{1/3}$.

---

[*] Richard L. and Dorothy M. Hayman Professor,
Mechanical Engineering and Jet Propulsion

1. INTRODUCTION

By mixing two liquids we shall imply mixing at the molecular level
and therefore molecular diffusion is the essential process. This is in
contrast with the mixing by large scale motions of the liquids, perhaps
turbulent motions, in which macroscopic regions of the space may be
occupied by small masses of both liquids, but no diffusion occurs across
the interface. In mixing processes of technological interest, even for
such primative ones as stirring with an object, both the large scale
motions and diffusive motions are involved. The large scale motions are
recognized to be essential to hasten mixing and they do so by extending
the interface between the two liquids and provide greater area through
which diffusive mixing may take place.

The importance of mixing at a molecular level was emphasized by the
study of mixing processes in which chemical reactions played an essential
role. Investigations into combustion problems have been the most
influential. In this particular example, because the chemistry is
frequently very fast in comparison with other characteristic time scales,
separation of the process into interfacial extension by large scale
motions and interfacial reaction involving molecular transport was
particularly evident. Perhaps the first clear statement of these ideas in
an analytical framework is that of Marble & Broadwell (1977), although the
general intuitive decomposition of the problem in this manner is much
older.

It was to demonstrate the details of the elements of the model
outlined in the previous reference that the studies of flames in the field
of a vortex was initiated. The results which are given in the latter part
of this paper were first obtained by Marble (1985) and were reformulated
and extended by Karagozian & Marble (1986) to account for flames in
vortices that undergo a stretching motion along their axis. This type of
study has been extended to the case of the vortex pair by Karagozian &
Manda (1986) and Karagozian et al (1986). Subsequent detailed numerical
computations, Laverdant & Candel (1987) and Rehm et al (1987) have
extended the range of the results and have justified the analytical
approximations introduced by Marble (1987).

Although the model, as well as the analytical techniques, were
conceived particularly for the problem of the diffusion flame, it was
clear that they would encompass both the case of slow chemistry in a
vortex field and the diffusive mixing problem in general. The following
work uses the vortex distortion model to study in detail the mixing of two
liquids, their rapid chemical reaction and their slow chemical reaction,
exhibiting those fluid mechanical features that characterize laminar
mixing processes.

2. VORTEX KINEMATICS

Consider the interface, lying along the horizontal axis, between a
liquid in the upper half plane of initial concentration
$K_1 = 1$ and a different liquid in the lower half plane of initial
concentration $K_2 = 1$. At the time $t = 0$ a viscous point vortex is
established at the origin; the velocity components associated with this
vortex are, Fig. 1,

$$V_r = 0 \qquad\qquad\qquad 2.1$$

$$V_\theta = \frac{\Gamma}{2\pi r}\left(1 - e^{-r^2/4\gamma t}\right) \qquad\qquad 2.2$$

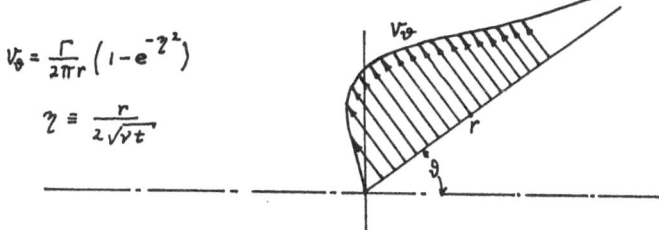

$$V_\theta = \frac{\Gamma}{2\pi r}\left(1-e^{-z^2}\right)$$

$$z \equiv \frac{r}{2\sqrt{\nu t}}$$

1. Velocity Distribution in a Viscous Vortex

where Γ is the circulation of the vortex and ν is the kinematic viscosity which we take to be the same for both pure species as well as for their mixture. As a result of this velocity field, the original interface between the two liquids is "wound up" into spirals, Fig. 2, which become more closely spaced as the time increases and as the circulation is increased.

Because the radial velocity component vanishes, fluid elements move along circular paths about the origin; an element of liquid interface initially coincident with the horizontal axis moves between two concentric circles, Fig.3. The element makes an angle ψ with respect to the local tangent to the circle and

$$-r\frac{\partial \vartheta}{\partial r} = ctn\,\psi \equiv \Lambda \qquad\qquad 2.3$$

By integrating V_ϑ/r , from Eq. 2.2, we obtain the remarkably simple result

$$\Lambda = 2V_\vartheta\, t/r \qquad\qquad 2.4$$

The length ds of the element of interface is

$$ds = \left\{1 + \left(r\frac{\partial \vartheta}{\partial r}\right)^2\right\}^{1/2} dr = (1+\Lambda^2)^{1/2}\,dr \qquad\qquad 2.5$$

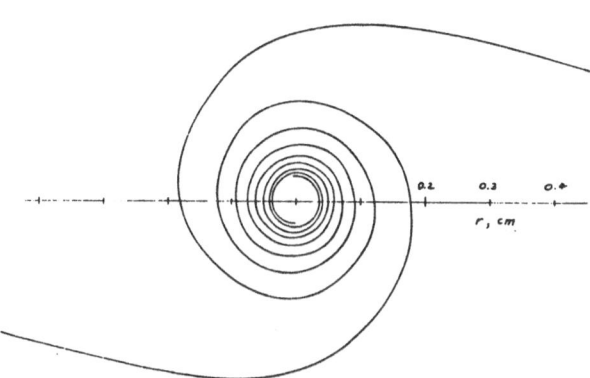

2. Distorted Contour of Initial Interface Between Two Liquids

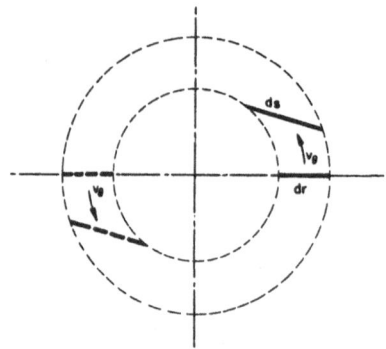

3. Strain and Rotation of
   Element Initially on
   Horizontal Axis

4. Two Interfacial Contours
   Developed Within Annulus

The liquid motion reduces ψ with time, the interface is continuously being stretched along its length, and at some time later makes a complete loop about the origin, Fig. 4. When these spirals are closely spaced, the angle is nearly constant and Eq. 2.5 may be integrated about a complete circle giving

$$2\pi r = (1+\Lambda^2)^{1/2} (r_2-r_1) \qquad\qquad 2.6$$

where (r_2-r_1) is the radial extent of the circular annulus in which the angle of the deformed interface is ψ. In actuality this strip contains two interfacial elements, Fig. 4, one originating on the right horizontal axis, the other on the left. One may interpret Eq. 2.6 as giving the width of an annulus containing a single stretched line element circling the origin,

$$\Delta r = \pi r / (1+\Lambda^2)^{1/2} \qquad\qquad 2.7$$

It is convenient to construct a local coordinate system, Fig.5, and to express the local flow field in these coordinates. The local x and y velocity components are

$$u = \left\{ (\tfrac{1}{2} \sin 2\psi)x + (\cos 2\psi)y \right\} \frac{\partial \Lambda}{\partial t} \qquad\qquad 2.8$$

$$v = -(\tfrac{1}{2} \sin 2\psi)y \frac{\partial \Lambda}{\partial t} \qquad\qquad 2.9$$

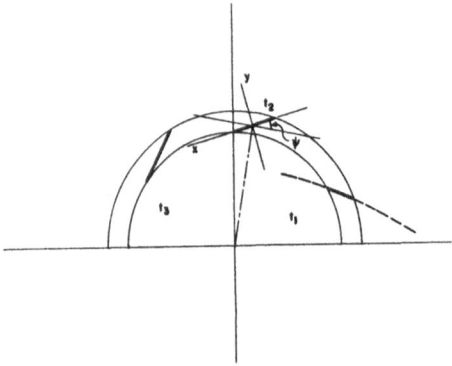

5. Coordinate System for Local
   Analysis

6. Representation of Local
   Structure in a Tightly Wound
   Vortex

The interface is always aligned with the moving x - axis and the straining rate along the interface is

$$\varepsilon \equiv \partial u / \partial x = (\tfrac{1}{2} \sin 2\psi)\frac{\partial \Lambda}{\partial t} = \Lambda / (1 + \Lambda^2) \frac{\partial \Lambda}{\partial t} \qquad 2.10$$

3. LOCAL DIFFUSIVE MIXING IN A VORTEX

The progress of diffusive mixing between two liquids distorted by a vortex motion may be analyzed conveniently using the framework of the "wound-up" interface described in the previous section. An example of this interface, shown in Fig. 5, suggests that the local situation in the spiral resembles a set of plane strips of alternating composition, Fig. 6. More specifically, to analyze the diffusive mixing at a radius r and time t in the vortex, the width of the strip is defined locally, Eq. 2.7, and we construct our array from strips of this width. Additionally, the local straining rate $\varepsilon(t)$ is assumed uniform over the entire array.

At any point within this array the velocity field is then

$u = \varepsilon x$, $v = -\varepsilon y$, the appropriate diffusion equation is

$$\frac{\partial K_1}{\partial t} - \varepsilon y \frac{\partial K_1}{\partial y} = D \frac{\partial^2 K_1}{\partial y^2} \qquad 3.1$$

and under the transformation

$$\bar{y} = y \exp\left\{ \int_0^t \varepsilon \, dt \right\} \qquad ; \qquad \mathcal{T} = \int_0^t \left[\exp\left\{ 2 \int_0^{t_2} \varepsilon \, dt_1 \right\} \right] dt_2 \qquad 3.2$$

becomes

$$\frac{\partial K_1}{\partial \mathcal{T}} = D \frac{\partial^2 K_1}{\partial \bar{y}^2} \qquad 3.3$$

The initial conditions on the diffusion problem are: for n even, $n\ell < \bar{y} < (n+1)\ell$, $K_1 = 1$; for n odd, $n\ell < \bar{y} < (n+1)\ell$, $K_1 = 0$. Under the transformation 3.2, the initial positions of the interfaces remain fixed in time.

For a short time after the start of mixing, the various strips have neglegible interaction and, although the solution may be written down exactly it may be treated in an approximate manner to give a result more convenient for extension to later times.

Integrate Eq. 3.3 over the diffusion layer, of thickness $\delta(\mathcal{T})$, Fig. 7,

$$\int_0^{\delta(\mathcal{T})} \frac{\partial K_1}{\partial \mathcal{T}} \, d\bar{y} = D \int_0^{d(\mathcal{T})} \frac{\partial^2 K_1}{\partial \bar{y}^2} \, d\bar{y}$$

$$3.4$$

under the conditions

$$K_1 (0, \mathcal{T}) = 1/2 \quad ; \quad K_1 (d, \mathcal{T}) = 1 \; ; \; \frac{\partial K_1}{\partial \mathcal{T}} (d, \mathcal{T}) = 0 \qquad 3.5$$

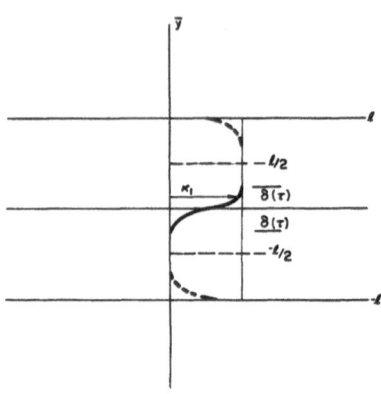

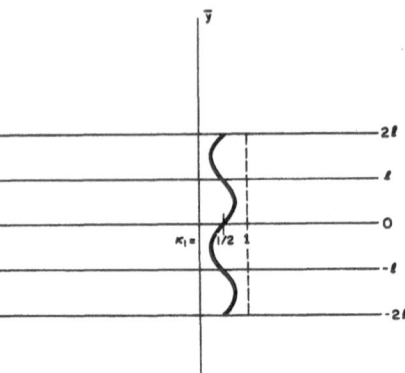

7. Development of Diffusion
   Layer Between Strips;
   Early Stage

8. Development of Diffusion
   Pattern; Later Interacting
   Stage

The result is

$$\frac{d}{d\Im} \int_0^\delta K_1 \, d\bar{y} \; - \; \frac{d\delta}{d\Im} \; = \; - \, D \frac{\partial K_1}{\partial \bar{y}}(0,\Im)$$

3.6

A convenient representation for $K_1(\bar{y},\Im)$ which satisfies the conditions is

$$K_1(\bar{y},\Im) \; = \; \frac{1}{2} \left(1 + \sin\frac{\pi}{2}\frac{\bar{y}}{\delta} \right)$$

3.7

Substitution into the integral relation Eq. 3.7 yields an ordinary differential equation for the diffusion zone thickness which may be solved, under the initial condition that $\delta(0)$ vanishes, to give

$$\delta \; = \; \left\{ D\Im/(\pi-2) \right\}^{1/2}$$

3.8

This solution is valid up to the time $\Im_*$ when the thickness has grown to half the strip width $\ell/2$, that is,

$$\sqrt{D\Im_*} \; = \; \sqrt{(\pi-2)} \; \ell/2$$

3.9

For values of $\Im$ greater than $\Im_*$ the diffusion processes in neighboring strips interact, Fig. 8, and the appropriate solution of Eq. 3.3 may be written

$$K_1 \; = \; \frac{1}{2} \left\{ 1 + \sum_1^\infty a_n \sin\frac{n\pi\bar{y}}{\ell} \, e^{-\left(\frac{n\pi}{\ell}\right)^2 D\Im} \right\}$$

3.10

The coefficients a_n are determined by the initial conditions.

If we start this solution at the time $\mathcal{J} = \mathcal{J}_*$, the initial distribution of composition is, from Eq. 3.8,

$$K_1(\bar{y}, \mathcal{J}_*) = \frac{1}{2}\left(1 + \sin\frac{\pi\bar{y}}{\ell}\right) \qquad 3.11$$

and the Fourier series consists of a single term. The complete local solution for the diffusion problem is

$$K_1 = \frac{1}{2}\left(1 + \sin\frac{\pi\bar{y}}{\ell}\sqrt{\frac{\mathcal{J}_*}{\mathcal{J}}}\right); \qquad 0 < \mathcal{J} \leq \mathcal{J}_*$$

$$K_1 = \frac{1}{2}\left(1 + \sin\frac{\pi\bar{y}}{\ell}\, e^{-\left(\frac{\pi}{\ell}\right)^2 D(\mathcal{J}-\mathcal{J}_*)}\right); \qquad \mathcal{J} \geq \mathcal{J}_*$$

$$3.12$$

4. SOLUTION FOR THE COMPLETE VORTEX

In order to interpret the local solutions we have just obtained in terms of the liquid distribution in the complete vortex it is required to transform our results back to physical variables. The local length scale may be recovered by inverting Eq. 3.2 and expressing the local strain rate in terms of the vortex kinematics using Eq. 2.10. This gives

$$\bar{y} = y(1 + \Lambda^2)^{1/2} \qquad 4.1$$

Similarly the time variable may be written

$$\mathcal{J} = \int_0^t \left\{\exp\left[\ln(1 + \Lambda^2)\right]\right\}dt = \int_0^t (1 + \Lambda^2)\,dt \qquad 4.2$$

The annulus width, denoted ℓ in the local analysis, is given by Eq. 2.7 and the scale transformation as

$$\Delta r(1 + \Lambda^2)^{1/2} = \left\{\frac{\pi r}{(1 + \Lambda^2)^{1/2}}\right\}(1 + \Lambda^2)^{1/2} = \pi r \qquad 4.3$$

Now the condition that the diffusion distance $\delta(\mathcal{J})$ has grown to the edge of the strip, expressed by Eq. 3.9, may be written

$$D\int_0^{t_*} (1 + \Lambda^2)\,dt = (\pi - 2)\left(\frac{\pi r}{2}\right)^2 \qquad 4.4$$

Recall that Λ may be expressed in terms of r and t and therefore Eq. 4.4 gives the time t_* when the diffusion layer has just spread to fill the annulus containing the interface at a radius r. Alternately the relation gives, at any time, the radius r_* within which neighboring strips interact and outside of which the diffusion zones are still independent.

This relationship may be clarified by considering an inviscid vortex, for which $V_\theta = \Gamma/2\pi r$ and thus $\Lambda = \Gamma t/\pi r^2$. For reasonably large values of Γ Eq. 4.4 gives the approximate result

$$\frac{\Gamma^{2/3} D^{1/3} t}{\pi r_*^2} = \left\{ \frac{3\pi (\pi-2)}{4} \right\}^{1/2} = 1.640 \qquad 4.5$$

and the radius r_* of the mixed core is

$$r_* = 0.440 \left(\Gamma^{2/3} D^{1/3} t \right)^{1/2} \qquad 4.6$$

Here the group $\Gamma^{2/3} D^{1/3}$ may be considered a composite transport property, generally much larger than the coefficient of molecular diffusion. The approximation Eq. 4.6 is good when the mixed core radius grows much more rapidly than the viscous core of the vortex. This latter rate follows directly from the tangential velocity, Eq.2.2, which shows that the viscous core grows as $(\gamma t)^{1/2}$. The ratio which compares these growth rates is then

$$\mu_a = \Gamma^{2/3} D^{1/3}/\gamma \qquad 4.7$$

and suggests that for large values of μ_a the mixed core radius grows much more rapidly than that of the viscous core and and hence follows the law given by Eq. 4.6.

To investigate the general solution of Eq. 4.4 it will be advantageous to introduce as a variable

$$\beta \equiv \Gamma^{2/3} D^{1/3} t / \pi r^2 \qquad 4.8$$

and to rewrite Eq. 4.4 in the form, introducing explicitly the critical ratio given in Eq. 4.7.

$$\int_0^{\beta/\mu_a} \left\{ \left(\frac{1}{2\pi R} \right)^2 + \zeta^2 \left(1 - e^{-\frac{1}{4R\zeta}} \right)^2 \right\} d\zeta = I\left(\frac{\beta}{\mu_a}, R \right) = \frac{\pi(\pi-2)}{4\mu_a^3} \qquad 4.9$$

Two asymptotic limits for this relation may be examined quite easily for large values of the Reynolds number. When the value of μ_a is large, the result is approximately

$$\beta \cong \left\{ \frac{3\pi (\pi-2)}{4} \right\}^{1/3} \qquad 4.10$$

which is identical with the result of Eq. 4.5 . On the other hand, for small values of μ_a, the result is approximately

$$\beta \cong 4\pi^3 (\pi-2) / \mu_a^2 \qquad 4.11$$

From these two results the rate of growth of the mixed core is

$$r_* = 0.440 \left(\Gamma^{2/3} D^{1/3} t \right)^{1/2} \qquad 4.12$$

for μ_a large, and

$$r_* = 0.047 \mu_a \left(\Gamma^{2/3} D^{1/3} t \right)^{1/2} \qquad 4.13$$

for μ_a small.

The numerical solution of Eq. 4.9 is presented in Fig.9 giving $r_*/\left(\Gamma^{2/3} D^{1/3} t \right)^{1/2}$ versus μ_a for fixed values of the vortex Reynolds number. Two features may be noted. First, the two asymptotic results, Eq. 4.12, 4.13, are completely adequate above $\mu_a = 20$ and below $\mu_a = 4$ respectively. Second, the result is Reynolds number dependent only at the very large values of μ_a. More generally it may be reasoned from Eq. 4.9 that Reynolds number dependence will be evident only for $\mu_a > 2\pi R$.

The physical origin of the result $\beta = const$ for high μ_a has been discussed, Eq. 4.5 et seq , and is a consequence of the fact that the diffusive action lies at radii where the vortex is essentially ideal. The behavior of the mixed core radius in the low μ_a range originates in the fact that the viscous core grows more rapidly than the diffusive mixing region. As a consequence, for the radius r_* at which the diffusion zones merge, the term $r/(4\nu t)^{1/2}$ in the velocity distribution, Eq.2.2, is small and the tangential velocity is approximately

$$V_\vartheta \cong \frac{\Gamma}{2\pi r} \left(\frac{r^2}{4\nu t} + \cdots \right) = \left(\frac{\Gamma}{8\pi\nu t} \right) r + \cdots \qquad 4.14$$

The surfaces across which diffusion is taking place therefore are moving with a solid body rotation and hence are not being deformed further. Their shape under these conditions may be deduced from

$$- r \frac{\partial \vartheta}{\partial r} \equiv \Lambda = R/2 \qquad 4.15$$

which integrates into a set of logarithmic spirals the winding of which becomes tighter as the Reynolds number increases.

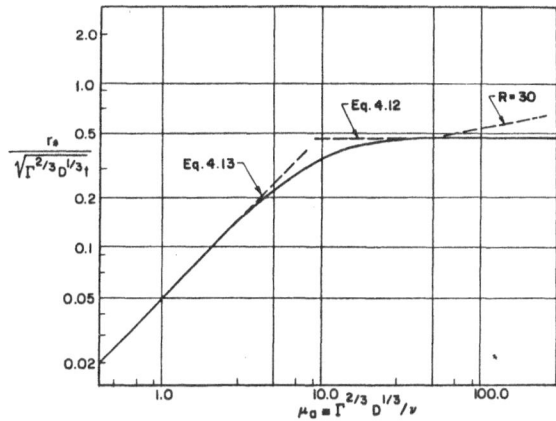

9. Dimensionless Radius of Mixed Core as a Function of $\mu_a \equiv \Gamma^{2/3} D^{1/3}/\nu$ for Constant R

5. DEGREE OF MIXING IN A VORTEX

Two liquids may be considered "mixed" when both κ_1 and κ_2 coexist in a region of molecular dimensions, that is, they are molecularly mixed. This fact may be expressed also by the statement that $\kappa_1 \kappa_2 = \kappa_1 (1-\kappa_1)$ is positive and this may be used as the basis of a measure of the degree of mixing.

Suppose that within a macroscopic volume V, $\kappa_1 = 1$ in a volume αV, and $\kappa_2 = 1$ within the remainder of the volume, $(1-\alpha)V$. Then if this macroscopic region achieved a uniform mixture on the molecular level, the species concentrations would be $\kappa_1 = \alpha$ and $\kappa_2 = (1-\alpha)$. In particular, the integral

$$\frac{1}{V} \int \kappa_1 \kappa_2 \, dV \qquad\qquad 5.1$$

extended over the macroscopic region vanishes when the liquids are unmixed and increases to the value $\alpha(1-\alpha)$ when local diffusive equilibrium has been achieved. We shall normalize this maximum value to unity and define the integral

$$\frac{1}{\alpha(1-\alpha)V} \int \kappa_1 \kappa_2 \, dV \equiv M_x \qquad\qquad 5.2$$

as the degree of mixing. Although this definition will be adequate for our purpose it has the difficulty that the result may, under some circumstances, be sensitive to the choice of initial macroscopic volume.

In applying this measure of mixing to present problem, consider a region localized about r, t, centered at an interface between two strips, one initially containing κ, the other κ_2. It will be necessary to treat separately the cases when $r > r_*$ and $r < r_*$. When $r \gtrsim r_*$, the concentrations are given in terms of the local transformed coordinates by Eq. 3.8,

$$\kappa_1(\bar{y},\tau) = 1 - \kappa_2(\bar{y},\tau) = \frac{1}{2}\left(1 + \sin\frac{\pi}{2}\frac{\bar{y}}{\delta(\tau)}\right) \qquad 5.3$$

Then the integral Eq. 5.2 may be written, using the fact that $\alpha = \frac{1}{2}$,

$$\frac{4}{l} \int_{-\delta}^{\delta} \kappa_1 (1-\kappa_1) \, d\bar{y} = \frac{2\delta}{\pi l} \int_{-\pi/2}^{\pi/2} (1 - \sin^2\varphi) \, d\varphi = \frac{\delta}{l} \qquad\qquad 5.4$$

Note that at the radius $r = r_*$, the diffusion layer thickness had grown to the value $l/2$ and hence the degree of mixing at that location is $M_x = 1/2$.

In general, the strip thickness l is the quantity which has been calculated in Eq. 4.3 and the diffusion zone thickness follows from Eq. 3.9. The degree of mixing is then

$$M_x = \frac{1}{\pi r} \left(\frac{D\gamma}{(\pi - 2)} \right)^{1/2} \qquad 5.5$$

valid only for $r \geq r_*$ which implies that $M_x \leq 1/2$. The
transformed time scale γ , written explicitly in Eq. 4.2, may be
expressed using the variables employed in Eq. 4.9, allowing the
degree of mixing to be written

$$M_x^2 = \frac{\mu_a^3}{\pi (\pi - 2)} I \left(\frac{\beta}{\mu_a}, R \right) \qquad 5.6$$

This allows expression of the degree of mixing as a function of
$r/(r^{2/3} D^{1/3} t)^{1/2}$ and μ_a , the variables employed in defining the core
radius r_* in Fig. 9. Now we have shown that the degree of mixing at
the core radius is $M_x = 0.5$ and hence we may think of the curve in
Fig. 9 as the contour of constant mixing index, $M_x = 0.5$. Utilizing
the expression given in Eq. 5.6 we may construct a diagram giving
the contours of constant mixing index, $M_x \leq 0.5$, in terms of these
same dimensionless quantities. This has been done and the result is
shown in Fig. 10. Note that for any fixed value of μ_a , the
dimensionless group $r/(r^{2/3} D^{1/3} t)^{1/2}$ determines the degree of mixing. Thus
there exists a similarity between distributions of mixed liquid at
various times; the locations at which $M_x = 0.5$ correspond to the mixed
core radius and the present results correspond only to points
outside this radius.

For points within the core, that is for radii $r < r_*$, we must
employ the second of the expressions given of Eq. 3.10. Treating the
mixing integral in a manner similar to that employed previously, it
follows that

$$M_\gamma = 1 - \frac{1}{2} exp \left\{ \frac{-2D (\gamma - \gamma_*)}{r^2} \right\} \qquad 5.7$$

Moreover the expression for the transformed time, Eq. 4.2, may be
written in terms of the variables β/μ_a and R ,

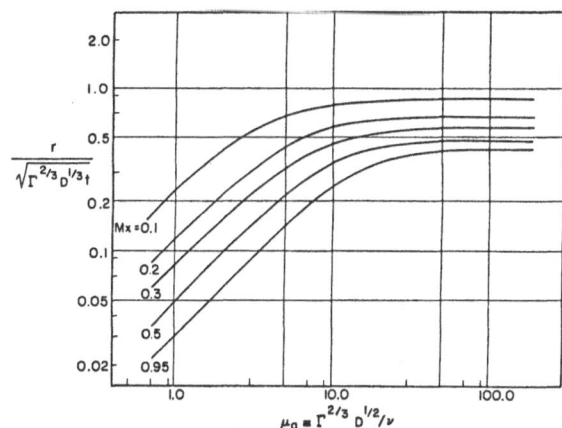

10. Contours of Constant Mixing Index Exterior
and Interior to the Mixed Core

$$\frac{D \mathcal{J}}{r^2} = \pi \mu_a^3 \, I \left(\beta / \mu_a , R \right) \qquad\qquad 5.8$$

and the particular value when the "mixed core" has grown to the radius r_* , Eq. 3.10, is

$$\frac{D \mathcal{J}_*}{r^2} = \frac{\pi^2 (\pi - 2)}{4} \qquad\qquad 5.9$$

Equation 5.7 is now more conveniently expressed as a relation between β and μ_a for a fixed degree of mixing, that is

$$\mu_a^3 \, I \left(\frac{\beta}{\mu_a} , R \right) = \frac{\pi (\pi - 2)}{4} \left\{ 1 - \frac{2}{\pi} \ln \left[2 (1 - M_*) \right] \right\} \qquad 5.10$$

It is then a straightforward calculation to find the values of $r / (r^{2/3} D^{1/3} t)^{1/2}$ and μ_a corresponding to a given degree of mixing, $1.0 > M_* > 0.5$. These contours of constant M_* in this range have been added to Fig. 10. It is particularly to be noted that the degree of mixing increases quite rapidly at radii below r_* and has increased to $M_* = 0.95$ at radii of $r_*/2$. Hence, the term "mixed core" for the region within r_* is indeed appropriate.

6. SLOW CHEMICAL REACTION IN A VORTEX FIELD

Because the primary interest here is to assess the augmentation of chemical reaction by the vortex field, we shall limit the discussion to a simple bimolecular reaction.

$$\frac{d \kappa_1}{dt} = - (1/t_c) \, \kappa_1 \kappa_2 \qquad\qquad 6.1$$

Referring to the analysis of mixing in the vortex, Section 5, the composition will be assumed initially uniformly mixed for $r < r_*$ and unmixed for $r > r_*$. At a local radius r , the reaction begins at the time when the mixed core radius passes over it. It will be assumed further that the stoichiometry of the reaction is unity and, as a consequence, $\kappa_1 = \kappa_2 = 1/2$ at the start of reaction and $\kappa_1 = \kappa_2 < 1/2$ throughout the time of reaction.

Thus at any radius $r < r_*$, Eq. 6.1 integrates to

$$- \frac{1}{\kappa_1} + \frac{1}{1/2} = - \left\{ \frac{t}{t_c} - \frac{\pi r^2}{r^{2/3} D^{1/3} t_c} \, \beta \left(\mu_a , R \right) \right\}$$

and the distribution of reactant within the core is

$$\kappa_1 = \kappa_2 = \frac{1}{2} \left\{ 1 + \frac{1}{2} \left(\frac{t}{t_c} - \frac{\pi r^2}{r^{2/3} D^{1/3} t_c} \, \beta_* \right) \right\}^{-1} \qquad 6.2$$

for all $\dfrac{t}{t_c} > \dfrac{\pi r^2}{r^{2/3} D^{1/3} t_c}$.

592

The product which has been formed within an annulus $2\pi r\,dr$ is just $(1-2\kappa_1)2\pi r\,dr$ and the total product which has been formed within the mixed core is

$$\int_0^{r_*} (1 - 2\kappa_1)2\pi r\,dr$$

6.3

This integral is evaluated for a fixed time t using the result of Eq. 6.2 and noting that r_* and the fixed time are related according to $\Gamma^{2/3}D^{1/3}t/\pi r_*^2 = \beta_*$. The result of the integration is then

$$\pi r_*^2 \sim 2\,\frac{\Gamma^{2/3}D^{1/3}t_c}{\beta_*}\,\ln\left(1 + \frac{1}{2}\,\frac{t}{t_c}\right)$$

which may be written more conveniently as

$$\text{Total Product} = \frac{\Gamma^{2/3}D^{1/3}t_c}{\beta_*}\left\{\frac{t}{t_c} - 2\ln\left(1 + \frac{1}{2}\frac{t}{t_c}\right)\right\}$$

6.4

This remarkably simple result shows that for a given physical situation for which β_* and t_c are fixed, the total product generated within a time t increases with stirring as $\Gamma^{2/3}$ where Γ is the vortex strength generated in the stirring process.

7. FAST CHEMICAL REACTION IN A VORTEX FIELD

By "fast chemical reaction rates" it is implied that the two reactants cannot coexist on the time scale of interest in the problem. Physically, this requires that the chemical time be short in comparison to the time associated with diffusive mixing. This feature introduces two significant changes on the analysis: i) There is negligible effect of remote strips of liquid upon the local one under consideration, and ii) the process is complete in a given strip when the reactants are consumed, which occurs in a finite time.

For the local analysis, therefore, it is sufficient to consider a single in an infinite plane domain with κ_1 initially in the upper half-plane and κ_2 in the lower half-plane. As time progresses, the two reactants diffuse toward their interface, at which κ_1 and κ_2 both vanish because of fast chemistry, and react in their stoichiometric ratio to produce the product. The product, in turn, diffuses back into the upper and lower half-planes. If there were no strain rate, the reactant distribution would be

$$\kappa_1,\ \kappa_2 = \pm\,\mathrm{erf}\left\{y/(4Dt)^{1/2}\right\}$$

7.1

where a stoichiometric ratio of unity has been assumed. The rate at which the liquid κ_1 is consumed at the reaction interface is

$$D\,\frac{\partial\kappa_1}{\partial y}(0,t) = (D/\pi t)^{1/2}$$

7.2

On the other hand, when the local interfacial region is being strained, as discussed in Section 3, the diffusion equation takes the form Eq. 3.1 and the reduction to the form Eq. 3.3 is appropriate. Then the solution for K_1 is

$$K_1 = erf\left\{ \frac{\gamma}{(4D\gamma)^{1/2}} \right\} \qquad 7.3$$

and the consumption rate of reactant 1 per unit area of interface is

$$\left(\frac{D}{\pi \gamma} \right)^{1/2} exp \int_0^t \varepsilon(t)\, dt \qquad 7.4$$

Recalling Eq. 2.10 giving the strain rate $\varepsilon(t)$ in the vortex, and the transformed time variable, Eq. 4.2, the rate at which the reactant K_1 is being consumed per unit area of interface is

$$\dot{V}(r,t) = \left(\frac{D}{\pi} \right)^{1/2} \left\{ \int_0^t (1+\Lambda^2)\, dt \right\}^{-1/2} (1+\Lambda^2)^{1/2} \qquad 7.5$$

Now an element of interface originally of length dr has been stretched into a length ds, Eq. 2.5, an hence the reactant consumption rate by this element is

$$\dot{V}(r,t)\, ds = \left(\frac{D}{\pi} \right)^{1/2} \left\{ \int_0^t (1+\Lambda^2)\, dt \right\}^{-\frac{1}{2}} (1+\Lambda^2)\, dr \qquad 7.6$$

In the unreacted state an annulus $2\pi r\, dr$ contains half reactant 1 with concentration $K_1 = 1$, and half by reactant 2 with concentration $K_2 = 1$. Moreover, as was pointed out in connection with Fig. 4, that there are two interfaces within this annulus. Thus the reactant consumption rate, Eq. 7.6, integrated over time, will at some time t_* consume all of the reactant 1 within the annulus. Expressed analytically,

$$2 \int_0^{t_*} \left(\dot{V}(r,t)\, ds \right) dt = \pi r\, dr \qquad 7.7$$

This relation may be written in detail using Eq. 7.6. Rewriting the resulting expression in the manner of Eq. 4.7-4.9, the result may be cast in the form

$$\int_0^{\beta_*/\mu_a} \left(I(s) \right)^{-\frac{1}{2}} \left\{ \left(\frac{1}{2\pi R} \right)^2 + s^2 \left(1 - e^{-\frac{1}{4\pi s}} \right)^2 \right\} ds = \frac{\pi}{2} \mu_a^{-3/2} \qquad 7.8$$

This expression, Eq 7.8, is convenient for computating the relationship between β_* and μ_a where β_* is the value of β at the radius r_* of the completely reacted core,

$$\beta_* = \frac{r^{2/3} D^{1/3} t}{\pi r_*^2} \qquad 7.9$$

Because β_* is a number $\beta_*(\mu_a, R)$, Eq. 7.10 gives the reacted core radius at any time t after the start of motion.

The result is independent of Reynolds number except for very large values of μ_a. The asymptotic behavior for large μ_a is

$$r_*/(\Gamma^{2/3} D^{1/3} t)^{\frac{1}{2}} = \left\{ \frac{\pi}{2} \left(\frac{3\pi^2}{2} \right)^{1/3} \right\}^{-\frac{1}{2}} = 0.509 \qquad 7.10$$

The corresponding result for low values of μ_a is

$$r_*/(\Gamma^{2/3} D^{1/3} t)^{\frac{1}{2}} = \mu_a (\pi)^{-5/2} = 0.057 \mu_a \qquad 7.11$$

Comparison of the results given by Eqn. 7.11, 7.12 with Eqn. 4.12, 4.13 as well as with the computations plotted in Fig. 9, show the close similarity between the growth of the core region for fast chemical reaction and that for the first stage of diffusive mixing. The facts that both processes are diffusion controlled and that neighboring annuli are non-interacting are the features which lead to this similarity.

NOTATION

D — Coefficient of molecular diffusion

dr — Interfacial element length before distortion

ds — Interfacial element length after distortion

I — Defined in Eq. 4.9

l — Initial width of liquid layers

M_x — Mixing index, defined in Eq. 5.2

r, ϑ — Polar coordinates for vortex flow

r_*, ϑ_* — Defined in Eq. 4.4, Eq. 3.10

t — Time

t_c — Chemical time

v_r, v_ϑ — Radial and tangential velocity components

$\dot{V}$ — Reactant consumption rate per unit area of interface

x, y — Local cartesian coordinates of interfacial element

$\bar{y}, \mathcal{J}$ — Transformed length and time scales, Eq. 3.2

β — Defined in Eq. 4.8

Γ — Circulation of vortex

δ — Local thickness of diffusion zone

ε Element strain rate, Eq. 2.10

κ_1, κ_2 Concentration of Liquid 1 and 2 respectively

Λ Defined by Eq. 2.4

μ_a Ratio of diffusive spreading to viscous spreading, Eq. 4.7

γ Kinematic viscosity

ψ Angle of interfacial element, Eq. 4.7

REFERENCES

Laverdant, A.M. & Candel, S.M. (1987), "A Numerical Analysis of a Diffusion Flame-Vortex Interaction" (In Publication)

Karagozian, A.R. & Marble, F.E. (1986), "Study of a Diffusion Flame in a Stretched Vortex." Combustion Science and Technology V. 45, pp 65-84.

Karagozian, A.R. & Manda, B.V.S. (1986), "Flame Structure and Fuel Consumption in the Field of a Vortex Pair." Combustion Science and Technology, V. 49, pp 185-200.

Karagozian, A.R., Suganuma, Y, & Strom, B.D. (1987), "Experimental Studies in Vortex Pair Motion Coincident with a Liquid Reaction" Submitted to Physics of Fluids.

Marble, F.E. & Broadwell, J.E. (1977) "The Coherent Flame Model for Turbulent Chemical Reactions." Project Squid, Technical Report TRW-9-PU

Marble, F.E. (1985) "Growth of a Diffusion Flame in the Field of a Vortex." Recent Advances in the Aerospace Sciences Ed. C. Casci. pp 395-413.

Rehm, R.G., Baum, H.R. & Lozier, D.W. (1987) To Appear in Studies in Applied Mathematics.

REACTIVE FLOW IN THE VICINITY OF SURFACES

Roger Prud'homme

Laboratoire d'Aérothermique
du Centre National de la Recherche Scientifique
92190 Meudon, France

INTRODUCTION

When we observe certain phenomena on a sufficiently large scale they seem to occur within a physical surface. On either side of such a surface, certain properties of the medium under consideration vary continuously and others show discontinuities on crossing the surface. If we stretch the scale over which observations are made to a sufficient extent perpendicular to the surface, this surface takes on the appearance of a three dimensional medium within which the properties vary continuously. This is at least true on a microscopic scale. Indeed, in certain cases, it is not possible ro reach continuity by increasing the scale of observation and the surface is discontinuous up to the molecular scale.

The "interfacial process" is currently acceptable as far as capillary and adsorption-desorption surfaces[1,2,3], shock waves, combustion and thin layers are concerned[5,6].

For shock-waves or slip-surfaces, it is usually adequate to impose the conservation laws for the relative normal flux of mass, momentum and total energy across the surface. The obtained relationships suffice for setting up a suitable model for the phenomenon and allow us to calculate the downstream conditions on the wave knowing those on the upstream section, assumed stationary. The same procedure applied to combustion waves leads to Rankine-Hugoniot theory. This theory only leads to a complete solution for Chapman-Jouguet detonations for which we assume minimum entropy production. Conversely, in the case of deflagration waves, we cannot solve, i.e. determine the combustion rate, unless we study the interior of the wave, seat of dissipation phenomena and chemical reactions. The calculations become rapidly more complex when the classical hypotheses breakdown. (Straight stationary wave, Lewis number equal to one, single chemical reaction).[7]

With capillary surfaces, the normal relative flux of momentum is no longer conserved and the surface tension appears as a tangential flux.

In a more general manner, we would be tempted to assume that the physical surface presents characteristic surface quantities, as for a volume, in addition to internal flux and sources.

This hypothesis, adopted by numerous authors, poses the problem of the constitutive relations apt to exist between these surface states parameters, on the one hand and between the fluxes and sources and state parameters and their gradients on the other. These constitutive relations are apt to bring jumps in volume properties into play as well.

In certain situations, as with capillary surfaces, we assume that the Gibbs relation is obeyed by the surface state variables. The form of the complementary laws can then be deduced from the principles or irreversible thermodynamics by writing linear expressions relating the fluxes and generalized forces intervening in the expression for the entropy yield.

This method is nevertheless limited. In particular, it does not seem to be applicable to deflagration waves. Hence, it is necessary, if we require an interfacial description of the phenomenon, to deduce it and the corresponding constitutive relations from a detailed study of the interfacial medium considered to be three dimensional. It is the integration over the normal coordinate to the interface that supplies us with the balance equations and the required constitutive relations. We will see that this method leads to acceptable results by means of certain hypotheses. In particular, the form of the balance equations obtained is quite characteristic and sufficiently general to be considered as the basis for treating a wide variety of cases. We remain, however, less satisfied as far as the constitutive relations are concerned, since their determination implies the study of a three dimensional medium each time, over the thickness of the interface. There would be no gain in time or simplicity if this work had to be repeated in each case.

The procedure indeed appears analogous to that leading to transport coefficients and to specific reaction rates per unit volume. In this case it is a microscopic scale study, or experiment that permits the constitutive relation to be determined. The calculations are complex but the results obtained lead to the expression of the diffusion coefficient, to give but one example. The simplification in comparison with the microscopic equations is considerable.

With certain interfaces, this again proves to be the case. In addition elementary dimensional analysis can lead directly to satisfactory results. We should, finally consider experimental determination. This procedure is routine for certain phenomena.

In what follows, we shall fist define interfacial movement and field deformation, then we shall establish the general form of the interfacial balance by a suitable integration of the volume balance across the interface, considered, at first, as a continuous three dimensional medium.

From this general form, we shall establish the constitutive relations for the interface in a certain number of cases.

MOVEMENT AND DEFORMATION FIELD IN THE INTERFACIAL ZONE

On the macroscopic scale of the continuous medium, the interface appears as a discontinuity, i.e. the surface equation is unknown.

On a smaller scale, of order of magnitude equal to the thickness of the interface, it is a volume. We consider a basis surface to exist, S_0, which is mobile and deformable, along which the properties of the medium vary negligeably at a given instant.

From this surface, and for every instant, we can construct a system of orthogonal curvilinear coordinates :

$$
(1) \quad \begin{cases} x = x\ (x_1, x_2, x_3, t) \\ y = y\ (x_1, x_2, x_3, t) \\ z = z\ (x_1, x_2, x_3, t) \end{cases}
$$

At time t, the surface S_0 is determined by :

$$(2) \quad x_3 = 0$$

At time goes on, the coordinate system is deformed with the surface $S_{0(t)}$ continuously. We choose the system in order that each surface $S_{0(t)}$ corresponds to x_3 = Cte, i.e. parallel to $S_{0(t)}$ and such that the distance between them is time independent (Fig. 1).

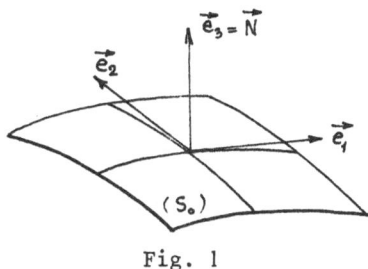

Fig. 1

The surface $S_{0(t)}$ can also be defined implicitly by an equation of the form :

$$(3) \quad \mathcal{F}\ (x, y, z, t) = 0$$

The function $\mathcal{F}$ hence allows us to construct a vector field :

$$(4) \quad \vec{N}(x, y, z, t) = \vec{\nabla}\mathcal{F} / |\vec{\nabla}\mathcal{F}|$$

and two scalar fields which are dependent on the orientation of the normal:

$$(5) \quad w = -(\partial\mathcal{F}/\partial t) / |\vec{\nabla}\mathcal{F}|$$

$$(6) \quad \vec{\nabla}.\vec{N}$$

When equation (3) is satisfied, the field (4) is that of the unitary normals to the interface with a given orientation, the field (5) is that of the normal speeds and the field (6) is that of curvatures.

If V is a vecteur satisfying :

$$(7) \quad V_\perp = \vec{V}.\vec{N} = w$$

and such that :

$$(8) \quad \vec{V}_{/\!/} = \vec{V} - V_\perp \vec{N}$$

characterizes the material movement tangentially to the surface, hence :

$$(9) \quad \vec{\nabla}_{/\!/} . \vec{V}$$

represents the stretch of $S_{o(t)}$[8].

In each case, we separate the tangential and normal parts of the tensors or operators under consideration. Thus :

(10)
$$\vec{\nabla} = \vec{\nabla}_{/\!/} + \vec{\nabla}_{\perp}$$

with :

(11)
$$\vec{\nabla}_{/\!/} = (\vec{1} - \vec{N} \otimes \vec{N}) . \vec{\nabla}$$
$$\vec{\nabla}_{\perp} = (\vec{N} \otimes \vec{N}) . \vec{\nabla}$$

All these quantities can be expressed in the orthogonal curvilinear coordinates defined above[9]. We show, in particular that :

(12)
$$\vec{\nabla} . \vec{N} = \vec{\nabla}_{/\!/} . \vec{N}$$
$$\vec{\nabla}_{/\!/} . \vec{V} = \vec{\nabla}_{/\!/} . \vec{V}_{/\!/} + w \ \vec{\nabla} . \vec{N}$$
$$\nabla^2 a = \vec{\nabla} . (\vec{\nabla}_{/\!/} \, a) + \vec{\nabla} . \vec{N} \frac{\partial a}{\partial N} + \frac{\partial^2 a}{\partial N^2}$$
$$\vec{\nabla} . \vec{v} = \vec{\nabla}_{/\!/} . \vec{v} + v_{\perp} \ \vec{\nabla} . \vec{N} + \frac{\partial v_{\perp}}{\partial N}$$

where a is a scalar, $\vec{v}$ a vector and $\partial/\partial N$ is the normal derivative :

(13)
$$\frac{\partial}{\partial N} = \vec{N} . \vec{\nabla} = \nabla_{\perp}$$

In the case of a fluid, the above definitions suffice when we consider $\vec{v}$ to be its particle velocity and a its pressure or temperature etc. A choice must, nevertheless be made for the parallel component of $\vec{v}$, the perpendicular component being defined by (7). For flames, the speed $\vec{v}_{/\!/}$ is often conserved according to a first order approximation, such that :

$$\vec{V} = w \vec{N} + \vec{v}_{/\!/}$$

For the boundary layers, we choose $\vec{v}$ to be the local speed of the solid surface.

For deformable solids, the velocity field depends on the chosen and allowable hypotheses. We can, in some cases assume that of a velocity distributor (Fig. 2).

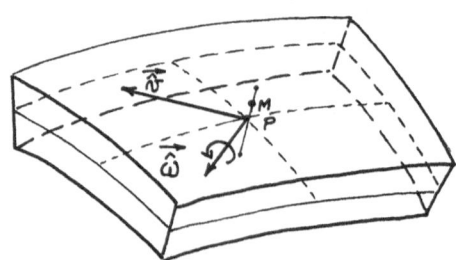

Fig. 2

We then obtain :

$$(14) \qquad \vec{v_M} = \vec{\hat{v}} + \vec{\hat{\omega}} \wedge \vec{PM}$$

where $\vec{\hat{v}} = \vec{V}$ and $\vec{\hat{\omega}}$ are functions of the position of the point P on the surface S_o and time only. The field of rotation vectors is thus added to that of velocities.

The degrees of deformation thus constitute a distributor $\{\hat{\mathfrak{D}}\}$:

$$(15) \qquad \{\hat{\mathfrak{D}}\} = \left\{ \begin{array}{c} \mathcal{S}(\hat{\omega}_{j,i}) \\ \mathcal{S}(\hat{v}_{j,i} + \hat{\ell}_i N_j) \end{array} \right\} \qquad (\mathcal{S}: \text{symetrical part})$$

where $\vec{\hat{\ell}} = \vec{\hat{\omega}} \wedge \vec{N}$.

GENERAL FORM OF THE INTERFACIAL BALANCE OF A FLUID MEDIUM

The equation for volumetric balance of an arbitrary extensive property F can be written :

$$(16) \qquad d_v \rho_F / dt + \rho_F \vec{\nabla}_{/\!/} \cdot \vec{V} + \vec{\nabla}_{/\!/} \cdot \vec{J}_{VF} + \partial J_{VF\perp} / \partial N = \dot{W}'_F$$

The terms involved have the following significance :

$$(17) \qquad \rho_F = \rho(f - f_o)$$

where f is the property F per unit mass and f_o a reference field for f (e.g. ρ_o would be the field of densities for cool gases in a flame),

$$(18) \qquad \left\{ \begin{array}{l} d_v / dt = \partial / \partial t + \vec{V} \cdot \vec{\nabla} \\ \vec{J}_{VF} = \vec{J}_F - \vec{J}_{Fo} + \rho_F \vec{u} \\ \dot{W}'_F = \dot{W}_F - \dot{W}_{Fo} - (\rho - \rho_o) \partial f_o / \partial t - (\rho \vec{V} - \rho_o \vec{V}_o) \cdot \vec{\nabla} f_o \end{array} \right.$$

where $\vec{J}_F$ is the flux of F following the fluid motion, $\vec{J}_{Fo}$ is a field of reference flux, $\vec{u}$ is equal to $\vec{V} - \vec{V}$; $\dot{W}_F$ is the rate of production of F, $\dot{W}_{Fo}$ its field of reference values.

Integration of equation (16) along the normal to the interface is possible by means of convergence conditions resulting from a judicious choice of f_o and if the velocity and differential operators d_v / dt and $\vec{\nabla}_{/\!/}$ are conservative along N.

We thus obtain the following general equation of interfacial balance :

$$(19) \qquad d_v \hat{\rho}_F / dt + \hat{\rho}_F \vec{\nabla}_{/\!/} \cdot \vec{V} + \vec{\nabla}_{/\!/} \cdot \vec{\hat{J}}_{VF} + [\![J_{VF\perp}]\!] = \hat{W}_F$$

with :

$$(20) \qquad \left\{ \begin{array}{l} \hat{\rho}_F = \displaystyle\int_{N-}^{N+} \rho_F \, dN \\ \vec{\hat{J}}_{VF} = \displaystyle\int_{N-}^{N+} \vec{J}_{VF} \, dN \\ \hat{W}_F = \displaystyle\int_{N-}^{N+} \dot{W}'_F \, dN \end{array} \right.$$

N- and N+ are the limits of the interfacial zone. For a flame, for example, we put, for complete combustion :

(21) $\qquad$ $dN = \varepsilon\, dn$

where ε is the ratio of the characteristic thickness of the flame to the hydrodynamic length. We then have :

(22) $\qquad$ $\begin{cases} n_- = - \infty \\ n_+ = 0 \end{cases}$

The choice :

(23) $\qquad$ $f_0\,(x,y,z,t) = f_-(x,y,z,t)$

thus assures the convergence of the integrals (Fig. 3).

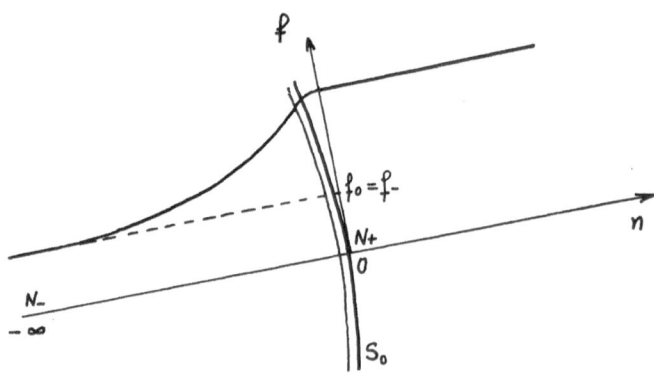

Fig. 3

CONSTITUTIVE RELATIONS

Gibbs interfaces

Whether we consider surfaces assimilable on a small scale to continua or interfaces whose characteristic thickness is of the same order of magnitude as molecular dimensions, equation (19) is satisfied. The velocity $\vec{V}$ is then the physical velocity of the surface. We then assume that the Gibbs relation is satisfied by the interfacial variables (S : entropy ; E : internal energy and j : chemical species) :

(24) $\qquad$ $d\hat{\rho}_S = (1/\hat{T})\, d\hat{\rho}_E - \sum_{j} (\hat{g}_j/\hat{T})\, d\hat{\rho}_j$

We thus introduce the surface temperature $\hat{T}$ and the chemical potentials $\hat{g}_j$ for each species as well as a surface tension σ such that :

(25) $\qquad$ $\hat{\rho}_S = (1/\hat{T})\hat{\rho}_E - \sigma/\hat{T} - \sum_{j} (\hat{g}_j/\hat{T})\hat{\rho}_j$

The rates of chemical production $\widehat{W}_j$ are obtained from the laws of chemical kinetics. The adsorption desorption and chemisorption laws lead us to similar consequences.

The different flux expressions are obtained by irreversible thermodynamics from the rate of entropy production[9,10] :

$$\widehat{W}_S = (1/\widehat{T}) \overset{\Rightarrow}{\Sigma} : \vec{\nabla}_{/\!/} \otimes \vec{v} + \vec{q} \cdot \vec{\nabla}_{/\!/}(1/\widehat{T}) - \sum_d \vec{\widehat{J}}_d \cdot [\vec{\nabla}_{/\!/}(\widehat{g}_d/\widehat{T}) - \vec{\widehat{f}}_d/\widehat{T}]$$

(26)
$$- \sum_d (\widehat{g}_d/\widehat{T}) \widehat{W}_d + [\![(q_\perp + \rho_H u_\perp)[(1/T) - (1/\widehat{T})] - \sum_d (J_{d\perp} + \rho_d u_\perp)$$

$$[(g_d/T) - (\widehat{g}_d/\widehat{T})] - (1/\widehat{T})(\vec{v} - \vec{\widehat{v}}) \cdot \overset{\Rightarrow}{\Pi} \cdot \vec{N}]\!]$$

Thin premixed flames with high activation energies

Assuming complete adiabatic combustion, uniform concentrations and temperatures of the cool gases and a constant, uniform pressure, equation (19) becomes (Fig. 4) :

(27)
$$d_v \widehat{\rho}_F/dt + \widehat{\rho}_F \vec{\nabla}_{/\!/} \cdot \vec{v} + [\![J_{VF}]\!] = \widehat{W}_F$$

with :

$$\vec{V} = \vec{v}_{/\!/} + w\vec{N}, \text{ and } \vec{\nabla}_{/\!/} \cdot \vec{V}$$

assumed to be of order one on a hydrodynamic scale.

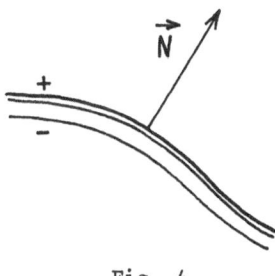

Fig. 4

Dimensional analysis for the dissipative variables (energy and concentration) gives us[11] :

(28)
$$\widehat{\rho}_F = L_F \rho_0 [\![f]\!] D_F/\rho u$$

where L_F is a phenomenological coefficient, ρu the combustion rate of the flame (constant in a zero order analysis), D_F the volume transport coefficient of the property F.

If the density ρ is assumed constant (a mixture diluted in a neutral gas, of low energy) we simply find that L_F is equal to 1 for the concentrations and energy.

If the density is variable within the flame, a calculation by the method of matched asymptotic expansions leads to :

$$(29) \quad \begin{cases} L_T = [T_0/(1-T_0)] \, \text{Log}\,(1/T_0) \\ L_Y = [L_e T_0/(1-T_0)] \left\{ \text{Log}\,(1/T_0) - (L_e-1)[T_0/(1-T_0)]^{L_e-1} \int_0^{(1-T_0)/T_0} [\text{Log}\,(1+\xi)/\xi^{2-L_e}] d\xi \right\} \end{cases}$$

where T_0 is the ratio of the cool gas temperature to that of the burning gases of the adiabatic laminar flame and where Le is the Lewis number.

Boundary layers above a flat plate (incompressible fluid)

In the frame of reference of the plate (time τ), the velocity $\vec{V}$ is chosen to be zero, equation (19) becomes :

$$(30) \quad \partial \hat{\rho}_F/\partial \tau + \vec{\nabla}_{/\!/} \cdot \vec{J}_F + [\![J_{Fl}]\!] = \hat{W}_F$$

Dimensional analysis leads to satisfactory constitutive relations.

For a stationary boundary layer :

$$(31) \quad \hat{J}_u = - K_u \; \rho U_\infty \; x \, [\![u]\!] \; \Psi \; (\mu/\rho \; U_\infty \; x)$$

where u is the component of velocity parallel to the plate (in the x-direction) equal to U_∞ at infinity.

For a Blasius profile, we obtain :

$$(32) \quad \begin{cases} \Psi \, (\mu/\rho \, U_\infty x) = (\mu/\rho \, U_\infty x)^{1/2} \\ k_u = 0,665 \end{cases}$$

In the presence of diffusion (mass fraction Y of reactants) and for a fast chemical reaction at the wall :

$$(33) \quad \hat{J}_Y = - K_Y \; \rho U_\infty \; x \, [\![Y]\!] \; \Psi \; [(\rho U_\infty x/\mu), \, (\mu/\rho \mathcal{D})]$$

The function Ψ depends on the regime under consideration (laminar or turbulent flow) and on the order of magnitude of the Schmidt number, it is of the form :

$$(34) \quad \Psi = R_{ex}^a \; S_c^b$$

If the Schmidt number is very small compared to one the exact calculation gives :

$$(35) \quad K_Y = \sqrt{4/\pi}$$

SHELLS AND PLATES

In the special case of a deformation field defined by (14), the method of virtual powers leads to the following equations for the interfacial momenta (effort and moment) :

$$(36) \quad \begin{cases} \hat{\rho} \, d\vec{v}/dt - \vec{\nabla}_{/\!/} \cdot (\vec{\sigma} + \vec{N} \otimes \vec{g}) - \vec{P} = 0 \\ - \vec{\nabla}_{/\!/} \cdot (\vec{M} + \vec{N} \otimes \vec{m}) + \vec{N} \wedge \vec{g} - \vec{\varphi} = 0 \end{cases}$$

604

At the boundaries of the schell (Fig. 5) :

(37)
$$
\begin{cases}
(\vec{\sigma} + \vec{N} \otimes \vec{g}) . \vec{\eta} = \vec{T} \\
\vec{M} . \vec{\eta} = \vec{\phi}
\end{cases}
$$

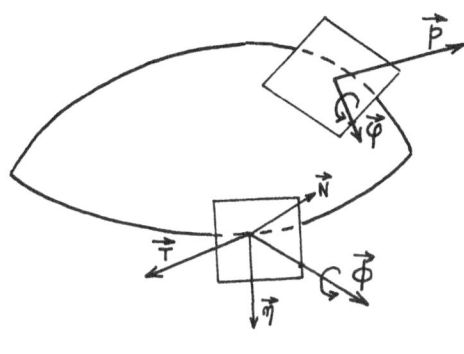

Fig. 5

In these equations, we have :

$\vec{p}$: surface force density
$\vec{\varphi}$: surface couple density
$\vec{T}$: line force density
$\vec{\phi}$: line couple density
$\vec{\sigma}$ and $\vec{g}$: surface (tensors) constraints
$\vec{M}$ and $\vec{m}$: directional moments.

In the case of zero moments, we regain the flexible surface equations.

CONCLUSIONS

In spite of the wide variety of interfacial media encountered, it is possible in many cases to obtain surface balance equations, which are representative and unrestrictive in comparison to actual phenomena. The constitutive laws can be established by dimensional analysis. They relate to surface variables, to flux and rates of production. The phenomenological coefficients can be determined experimentally, or by detailed analysis over the interfacial thickness. The solution implies a coupling between these equations and those for volumes in contact. In the examples presented here, the results are in perfect agreement with those of simple classical theory.

BIBLIOGRAPHY

1. R. Defay, I. Prigogine, "Tension superficielle et adsorption", Ed. Desoer, Liége.
2. V.G. Levich, "Physico-Chemical Hydrodynamics", Prentice-Hall, 1962.
3. L. Landau, E. Lifschitz, "Mécanique des fluides", Ed. Mir, Moscou, 1971.
4. A.L. Jaumotte, "Chocs et ondes de choc", Masson, 1971.
5. P. Germain, "Mécanique", Ellipses, Ecole Polytechnique, 1986.
6. W.T. Koiter, "The theory of thin elastic shells", North-Holland Pub., Amsterdam, 1960.

7. P. Clavin, G. Joulin, "Premixed flames in large and high intensity turbulent flow", J. de Phys., Lettres, 44, 1, January 1983.
8. A.M. Klimov, Zhur. Prikl. Mekh. Techn. Fiz. 49, 3, 1963.
9. R. Prud'homme, "Phénomènes d'interfaces", Rapport 87-1 du laboratoire d'Aérothermique du CNRS, january 1987.
10. G. Bertrand, R. Prud'homme, "Possibilities of surface reactions coupling with transport phenomena", Int. J. Quant. Chem., 12, suppl. 2, 1977, 159-168.
11. R. Prud'homme, F. Baillot, "Equations de bilan d'interface pour les flammes de prémélange", C.R. Acad. Sc., Paris, t. 304, série II, n° 17, 1987.

DISCUSSION

BRATOS - The constitutive laws you mention contain a number of parameters. Are statistical (molecular) expressions available at least for some of them ?

PRUD'HOMME - Les méthodes statistiques présenteront un intérêt pour déterminer les lois de comportement d'interface dans deux cas :

1) L'épaisseur interfaciale est de l'ordre de quelques dimensions moléculaires. Cela inclut la cinétique chimique de surface.

2) L'épaisseur de l'interface est macroscopique, mais une turbulence à petite échelle y est présente.

Dans les cas où l'interface est macroscopique en épaisseur, il est en effet inutile de partir du niveau moléculaire, car on retrouvera les équations volumiques classiques (Navier-Stokes, etc.).

BARRERE -   1) Votre démarche simplifie-t-elle les méthodes numériques lorsque le milieu possède des surfaces libres séparant par exemple un milieu compressible d'un milieu incompressible ?

2) Lorsque des réactions catalytiques de surface ont lieu, cela avez-vous dit, simplifie le problème ; je pense, en particulier, au problème de la rentrée d'Hermès pour lequel la tenue thermique des parois dépend des réactions catalytiques de surface.

PRUD'HOMME -   1) Si l'on sait résoudre les problèmes numériques avec surface de discontinuité d'équation inconnue (par exemple problème de Stéfan), il y a simplification. Sinon, cette démarche ne simplifie pas les choses.

2) Je pense que la démarche présentée ici serait certainement utile pour traiter des problèmes avec réaction catalytique de surface tels que ceux posés pour la navette Hermès. La question sera certainement de déterminer correctement les vitesses spécifiques de réaction aux parois.

PREDICTION AND SIMULATION OF THE BEHAVIOUR OF DISCRETE

PARTICLES TRANSPORTED BY TURBULENT FLOWS: A REVIEW PAPER

G. Gouesbet, A. Berlemont, P. Desjonquères

Laboratoire d'Energétique des Systèmes et Procédés
INSA de Rouen, UA. CNRS. n° 230,
BP 08, 76 130, Mont-Saint-Aignan

INTRODUCTION

The realm of multiphase flows and problems is very vast
and indeed potentially infinite. In the Universe fluids are
nearly everywhere, and when they occur they almost invariably
contain particles. Inside our bodies we can take the example
of blood transporting a vital procession of red and white
cells. Around us, we can find various particles in the air we
breathe, bubbles in the champagne or the soda we drink, or
natural and artificial (polluting!) particles in the lakes we
swim in. These examples correspond to the case when discrete
particles are transported by flows. They could be filed as
suspensions of particles in fluids, bubbly flows and drop
flows. But considering only these cases would be forgetting
about slug and annular flows, and the various combinations
and transitions between all these regimes. Numerous models
have been designed to understand and study all these
situations, from empirical correlations and homogeneous
models to more sophisticated approaches : separated flows,
drift-flux models, differential or integral analysis... It is
impossible to do justice to such a large topic in a limited
review paper. Consequently we shall limit our discussion to
the case of discrete particles transported by turbulent flows
for which our personal knowledge is a bit more than scholar.

The dispersion of discrete particles in turbulent flows
has become a major domain of research, of increasing
interest, with challenging fundamental aspects and various
applications in geophysical systems, or in laboratory and
industrial applications : coal pulverized or droplet flames,
Diesel engines, centrifugation, some rocket exhausts with
applications to guidance control and radar detection,
particulate pollution transport...Surprisingly enough for
such an important field with numerous applications, modern
research remains nowadays limited to a few laboratories,
probably because of the involved difficulties. However, the
international effort in this domain seems to increase rather

fast and we might well be located at the beginning of an exponential curve.

Before the expected development of this curve, this conference is certainly a good opportunity to present the results obtained in Rouen in the last ten years, and also to stress some prospective ideas. This work is the outcome of four thesis dissertations (A.Berlemont, 1981, A.Picart, 1984, P.Desjonquères, 1987 and A.Berlemont, 1987). Most of the material has been published elsewhere and we shall refer to these published works to discuss the topic, avoiding too much details which would overcome us and overflow the available room. Although unfair to other workers, the reference list will be limited to Rouen publications, from which the interested reader can reach other works, up to the tips of issued ramifications and tentacles, located in various laboratories and expanding over several decades.

TURBULENCE PREDICTIONS

Whatever the approach (Eulerian or Lagrangian) for the prediction of particle dispersion, the turbulence field in which the particles disperse must be known. To test the dispersion model itself, this knowledge can be acquired through accurate experiments. However, for practical purposes, another option is to predict the turbulence properties in a first module of the computational codes. This provides the user with a complete computational package, but it is clear that any inaccuracy of the turbulence model and predictions will then be echoed to the dispersion predictions.

The chosen model is certainly a very good compromise between a relative simplicity and a good accuracy, at least for non-exotic situations. It is a $K-\epsilon$ model, supplemented by algebraic stress relations due to Rodi for the prediction of the Reynolds tensor. The governing equations are the continuity equation, the momentum equation, the transport equations for the turbulence energy and for its dissipation, a closure equation to evaluate the turbulent viscosity, plus the algebraic stress relations mentioned above. These equations are given in Berlemont (1981), Berlemont et al (1986), Picart et al (1986) for the $K-\epsilon$ model, with details for the algebraic relations in Berlemont and Gouesbet (1981), Gouesbet and Berlemont (1981) and Berlemont (1981). The resolution algorithms for the algebraic relations are discussed in Picart (1984).

Validation of these models and of our computer code for turbulence predictions has been carried out on the following test-cases : round free jet in the self-preserving domain and also from the discharge section, using an expanding grid (Berlemont and Gouesbet, 1981, Gouesbet and Berlemont, 1981), pipe flows for the fully established regime and also for the developing domain (Berlemont, 1981, Gouesbet and Berlemont, 1981), channel flows and mixing layers (Picart, 1984).

Another required ingredient is the knowledge of the Lagrangian correlation function of fluid particle velocities that we shall designate by R_{fL}. The present state of art in turbulence modelling does not allow a prediction of this function. In a first approach, a fluid Lagrangian spectrum E_{fL} was built from the turbulence predictions by a rather intuitive method, from which R_{fL} could be obtained by Fourier transforming (Berlemont, 1981, Gouesbet and Berlemont, 1981, Berlemont and Gouesbet, 1982). In a second approach, we relied on a semi-empirical expression given by Frenkiel, in terms of (i) the Lagrangian time macroscale (ii) a loop parameter linked to the occurrence and to the significance of negative loops in the correlation function. When the loop parameter is set equal to O, we recover the classical exponential decrease for R_{fL}. From various arguments, theoretical and experimental, we however recommend the value 1, as discussed in Berlemont, 1987. The existence of negative loops in R_{fL} leads to the possibility of a dispersion (discrete particles) more efficient than a diffusion (fluid particles), even for dense particles in homogeneous turbulence (Gouesbet et al, 1982).

Other turbulent ingredients are the knowledge of Eulerian and Lagrangian macroscales which can be evaluated by relying on turbulence theory and modelling. However, the state of art in these evaluations is not very satisfactory because poorly known constants are involved in them. Consequently efficient tests of the particulate models must preferably rely on scale experimental determinations.

THE EQUATION OF MOTION OF THE PARTICLES

In the Eulerian approach, the equation of motion of an individual particle does not appear explicitly but is hidden in the formulation to express the dispersion tensor. We used the Tchen equation, although it is actually controversial. The use of another equation would require a modification of the details of the Eulerian approach. Such a modification is certainly not possible for any equation of motion, especially if it is not linear.

In the Lagrangian approach, we used an equation due to Riley (from which we can deduce the Tchen equation as a special case) modified to account for non-zero particulate Reynolds numbers, according to expressions given by Odar and Hamilton. In this approach, the equation of motion appears explicitly and serves to build trajectories. Any equation of motion can be used without any modification of the formulation.

THE EULERIAN APPROACH

The Eulerian approach relies on solving a transport equation for the particle concentration or number-density. The particles are not considered individually but as a continuous field akin to heat or chemical species, but furthermore characterized by inertia phenomena. The transport equation contains a convection term in which the mean velocities are known from the turbulence model, and also a dispersion term which is at the core of the problem. This dispersion term involves a dispersion tensor or, in a 1D-formulation, a dispersion coefficient that we have to determine.

To determine the dispersion coefficient, it is not reliable to use experimental correlations. It is preferable to rely on a standard theory of dispersion which, although of limited range of validity, provides however a firm basis for discussion. This dispersion theory is expressed as a synthesis between the Batchelor theory for fluid particle diffusion and the Tchen theory for particle dispersion. The Batchelor formulation can be immediately adapted to the dispersion of discrete particles if some extra-assumptions concerning the particle behaviour are stated. One of these assumptions is the so-called stationary assumption for homogeneous turbulence which states that the particle statistics at the dispersion initial time must be the same as the one which would be exhibited, for arbitrary initial conditions, at an infinite dispersion time (asymptotic behaviour). The essence of this assumption is that, in the Eulerian approach, the particle response must only depend on turbulence local properties so that the concept of dispersion coefficient may hold. The theory is explained in Gouesbet and Berlemont, 1979, Berlemont and Gouesbet, 1982, Berlemont et al, 1983, and in a final detailed form in Gouesbet et al, 1984. The special case when the so-called Basset term in the equation of motion is neglected, and the examination of conditions when that term can actually be neglected, is given in Picart et al, 1982, Gouesbet et al, 1984, Picart et al, 1986, Desjonquères et al, 1986. Other assumptions are described in Picart, 1984, Desjonquères et al, 1986, Desjonquères, 1987.

The above theory requires that a discrete particle remains in the same fluid particle during its motion (no over-shooting). This is certainly a good approximation of reality for non-buoyant particles, but is certainly not true in numerous cases, for instance when the particle drifts due to external forces, or when it is injected in the carrier fluid at a velocity not equal to the fluid one. In these cases, a fluid particle and a discrete particle having the same location at the initial time will have different trajectories, leading to what is called Crossing-Trajectory-Effects (CTE). The CTE produce an extra-loss of correlations along the particle path, consequently decreasing the dispersion coefficients. In the Eulerian approach, the CTE are taken into account by modifying the dispersion coefficients, using a semi-empirical expression. This correcting process is described in Picart, 1984, and Picart et al, 1986.

The results obtained using the Eulerian approach concern :

-The evolution of dispersion coefficients for a given turbulent field, versus particle diameters, dispersion time, and the value of the Lagrangian macroscale (Berlemont, 1981, Berlemont and Gouesbet, 1981).
-The evolution of a cloud of particles in a fully established turbulent pipe flow, with emphasis on the coupling between dispersion and convection (Berlemont, 1981, Berlemont and Gouesbet, 1982).

-The influence of the Basset term on dispersion coefficients and phenomena (Picart et al, 1982, Picart, 1984).

-The comparison between dispersion and diffusion, showing that dense discrete particles may disperse faster than fluid particles, even in the framework of the Tchen theory (Gouesbet et al, 1982, Gouesbet et al, 1983, Picart, 1984).

-Influence of the CTE with comparisons against experimental results in grid turbulences (Picart, 1984, Picart et al, 1986).

-Prediction of the dispersion in a pipe flow, and comparisons with experimental results (Picart, 1984, Picart et al, 1986).

Computations have been carried out using the codes DISCO (DISpersion COmputing), in its first version DISCO-1 then in its second version DISCO-2. Comparisons with expriments are satisfactory and show evidence for the validity of the Eulerian approach, at least when underlying assumptions are satisfied.

THE LAGRANGIAN APPROACH FOR FLUID PARTICLE DIFFUSION

The Eulerian approach is limited to restricted situations due to numerous underlying assumptions. For instance, only the Tchen equation of motion is implanted in the formulation. Furthermore, the CTE cannot be accounted for in an elegant way. Although we expect that some assumptions could be relaxed, it would lead to a difficult formal work to generalize the approach. Also, complex phenomena such as coalescence or break-ups, particle vaporization and combustion, would certainly be very difficult to study. Consequently, we also developped a Lagrangian approach which is more flexible.

This section is devoted to the Lagrangian approach for the diffusion of fluid particles. Fluid particle trajectories are simulated by the computer program, and required statistics are computed by averaging on these trajectories.

To simulate one fluid particle trajectory, we need know the velocity of the fluid at the considered location and time when a small trajectory increment is computed. This velocity is equal to the mean fluid velocity computed from the turbulence model plus the velocity fluctuation determined from a random numerical process complying with the following assumption and requirements :

-The velocity fluctuation probability density function is normal. That assumption can easily be relaxed if indications on the actual pdf are known.
-The one-point velocity correlations known from the turbulence model (or experiments) must be satisfied.
-The Lagrangian correlation function along the path of the fluid particle must also be satisfied. In the present case, the Frenkiel family with the loop parameter equal to 1 is recommended. For an exponential decrease, the loop

parameter must be set to 0. Such an exponential correlation can also be generated using a Poisson process as done by Ormancey and Martinon (Ecole des Mines, Paris), and by other individuals such as Gosman, Shuen, Durst and Milojevic (eddy life time).

The generating random process we used is based on a rather subtle mathematical technique that we cannot describe here. Basically, we start from a symmetric, positive definite, correlation matrix A from which we deduce an accessory matrix B using the Cholesky method. The required vector U whose the components are the correlated velocity fluctuations is then equal to the matrix B multiplied by a vector whose components are uncorrelated, centered, normal variables of variances unity. The procedure first designed for an 1D-formulation has been extended to 2D-problems. Mean turbulence inhomogeneities can be accounted for in the process. Details can be found in Desjonquères, 1987, Berlemont, 1987, Gouesbet et al, 1987, Berlemont et al, 1987, Desjonquères et al, 1987.

Computations are carried out using the code PALAS (PArticle LAgrangian Simulation). Comparisons have been made with experimental results in grid turbulences, theoretical results according to the Taylor theory of diffusion, analytical results for the diffusion from a point-source in homogeneous, isotropic, turbulence, and numerical predictions from the Eulerian approach (Desjonquères, 1987, Berlemont, 1987, Gouesbet et al, 1987, Berlemont et al, 1987, Desjonquères et al, 1987). The high degree of agreement between PALAS simulations and the other results leads to a very good validation of our Lagrangian approach.

THE LAGRANGIAN APPROACH FOR DISCRETE PARTICLE DISPERSION

For dispersion, discrete particle trajectories are built using the fundamental law of dynamics, relying on an equation of motion (in the present case, on the modified Riley equation, see section III). However, the problem of determining the fluid particle velocity fluctuation at the discrete particle location, necessary to compute the instantaneous force acting on the particle, is more complex than in section V due to the CTE. To the purpose, a fluid particle and a discrete particle are simultaneously launched at the same location at the initial time. The fluid particle trajectory is built according to the process described in the previous section. To know the fluid particle velocity fluctuation at the discrete particle location, at a given time step, we start from the velocity fluctuation at the fluid particle location on the fluid particle trajectory, and generate the fluid particle velocity fluctuation at the discrete particle location using a random process satisfying spatial Eulerian correlations between the two trajectories simultaneously followed. Furthermore, at each time step, the fluid particle is surrounded by a correlation domain. When the discrete particle leaves this correlation domain attached to the fluid particle, the fluid particle trajectory is given up and a new fluid particle trajectory is launched from the discrete particle location at this time step. This process permits a direct simulation of the CTE.

The above process is used for small mass loading ratios when the influence of the particles on the turbulence can be neglected. When the mass loading ratio increases, the simulation must be completed to account for the influence of the particles on the turbulence (two-way coupling). This is carried out by modifying some governing equations in the turbulence model, more specifically by adding source terms in the momentum equation, and in the equations for the turbulence energy and for its dissipation. These source terms account for momentum and energy exchanges between the two phases in presence. From a numerical point of view, the computer program PALAS then carried out iterations between the turbulence computations and the Lagrangian simulations, up to the convergence of the process. This convergence is obtained after a small number of iterations, usually three or four.

Lagrangian simulations have been compared with Eulerian predictions, and with experimental results in the following test-cases :

    -grid turbulence, the external force acting on the particles being gravity.
    -grid turbulence, the external force acting on the particles being gravity plus an electric force. Using an extra electric force allows for CTE-control.
    -discrete particle point-source on the axis of a fully established turbulent pipe flow.
    -particle dispersion in a turbulent round free jet, with high mass loading ratios to test the two-way coupling model (reciprocal influence between turbulence and particles).

Results are given in Desjonquères, 1987, Berlemont, 1987, Gouesbet et al, 1987, Berlemont et al, 1987, Desjonquères et al, 1987. The high agreement between Lagrangian simulations, Eulerian predictions, and experimental results, provides a good validation of the Lagrangian approach we used.

CONCLUSION

The present paper provides the reader with a summary of the effort carried out in Rouen to predict the behaviour of discrete particles transported in turbulent flows. Two approaches (Eulerian and Lagrangian) are described. Comparisons between Eulerian predictions, Lagrangian simulations (including the influence of particles on turbulence) and analytical, or experimental results, in a large number of situations, show that these methods are very well suited for solving the addressed problem.

The two approaches however are not equivalent. They own respective advantages and disadvantages that we must now point out in this conclusion.

The main advantage of the Eulerian approach is that it is not time-consuming in terms of CPU, and consequently not costly. As a matter of fact, the code DISCO-2, including turbulence predictions, can run on a personal computer in a reasonable time. This efficiency is certainly the consequence of a previous formal effort which permits to carry out statistical averages on trajectories, in a implicit way, through pure physics and mathematics. However, this formal effort has been successful only by stating restrictive assumptions producing the disadvantage of the Eulerian approach, namely that, rigorously speaking, it is limited to restricted situations. In the Lagrangian approach, the fact that the trajectories are explicitly simulated permits to extend the range of applications. The main advantage of the Lagrangian approach is consequently the very large domain of situations it can apprehend. But it requires a large number of trajectories to build (typically 10 000) before averaging to extract the statistical quantities of interest. Consequently, mainframe computers must be used and the procedure becomes time-consuming and costly. However, this disadvantage must be attenuated by recalling that progress in hardware is going on very fast. More and more complex situations will be studied in the future and, at the present step, we are apparently very far from encountering computer limitations. The effort to apprehend and simulate such more complex situations is now on the way.

REFERENCES

A. Berlemont, Modélisation et prédiction du comportement de particules discrètes dans un écoulement turbulent. Thèse de 3ème cycle, Rouen, 1981.

A. Berlemont. Modélisation eulérienne et lagrangienne de la dispersion particulaire en écoulement turbulent. Thèse d'Etat, Rouen, 1987.

A. Berlemont, P. Desjonquères. A Lagrangian approach for the prediction of particle dispersion in turbulent flows. Third workshop on two phase flow predictions. Belgrade. June 1986.

A. Berlemont, A. Picart, G. Gouesbet. Prediction of turbulence fields and particle dispersion using the code DISCO-2. Computational techniques for fluid flow, chapter 10, 281-313, Pineridge Press, Swansea, 1986.

A. Berlemont, G. Gouesbet. Prediction of the turbulent round free jet, including fluctuating velocity correlations, by means of a simplified second order closure scheme. Lett. Heat and Mass Transfer, 8, 207-217, 1981.

A. Berlemont, G. Gouesbet. Prediction of the behaviour of a cloud of discrete particles released in a fully developed turbulent pipe flow, using a nondiscrete dispersive approach, Second Int. Conf. on Numerical Methods in Laminar and Turbulent Flow, Venice, Italy, pp 327-338, Pineridge, Swansea, 1981.

A. Berlemont, G. Gouesbet. The dispersion of a cloud of particles released in a fully developed pipe flow, using the code DISCO-1, Lett. Heat and Mass Transfer, 9, 407-419, 1982.

A. Berlemont, A. Picart, G. Gouesbet. The code DISCO-2 for predicting the behaviour of discrete particles in turbulent flows and its comparisons against the code DISCO-1, and experiments. Third Int. Conference on Numerical Methods in Laminar and Turbulent Flow. Seattle, USA, pp 963-973, Pineridge, Swansea, 1983.

A. Berlemont, G. Gouesbet, P. Desjonquères. Lagrangian simulation of particle dispersion, United States-France joint workshop on turbulent reactive flows. July 6-10, Rouen, 1987.

P. Desjonquères. Modélisation lagrangienne du comportement de particules discrètes en écoulement turbulent. Thèse de 3ᵉᵐᵉ cycle, Rouen, 1987.

P. Desjonquères, G. Gouesbet, A. Berlemont, A. Picart. Dispersion of discrete particles by continuous turbulent motions : new results and discussions. Phys. Fluids, 29, 2147-2151, 1986.

P. Desjonquères, A. Berlemont, G. Gouesbet. Lagrangian simulation of a two phase turbulent round jet. Sixth Symposium on turbulent shear flows. Toulouse, France, Sept 7-9, 1987.

G. Gouesbet, A. Berlemont. Une approche dispersive pour la modélisation du comportement de particules dans un champ turbulent, C. R. Acad. Sci., Paris, Série A, t 288, 961-964, 1979.

G. Gouesbet, A. Berlemont. Prediction of turbulent fields, including fluctuating velocity correlations and approximate spectra, by means of a simplified second order closure scheme : the round free jet and developped pipe flow. Second Int. Conf. on Numerical Methods in Laminar and Turbulent flow, Venice, Italy, pp 205-216, Pineridge, Swansea, 1981.

G. Gouesbet, A. Berlemont, A. Picart. On the Tchen's theory of discrete particle dispersion : can dense discrete particles disperse faster than fluid particles? Lett. Heat and Mass transfer, 9, 407-419, 1982.

G. Gouesbet, P. Desjonquères, A. Berlemont. Eulerian and Lagrangian approaches to turbulent dispersion of particles. Séminaire International 'Transient Phenomena in Multiphase Flow', ICHMT, Dubrovnik, 1987.

G. Gouesbet, A. Picart, A. Berlemont, Dispersion of discrete particles by turbulent continuous motions using a Frenkiel's family of Lagrangian correlation functions, in the non-discrete dispersive approach. Third Int. Conf. on Numerical Methods in Laminar and Turbulent Flow, Seattle, USA, pp 996-1005, Pineridge, Swansea, 1983.

G. Gouesbet, A. Berlemont, A. Picart. Dispersion of discrete particles by turbulent continuous motions. Extensive discussion of the Tchen's theory using a two parameter family of Lagrangian correlation functions. Phys. Fluids, 27, 827-837, 1984.

A. Picart. Le code DISCO-2 pour la prédiction du comportement de particules dans un écoulement turbulent. Thèse de 3ᵉᵐᵉ cycle, Rouen, 1984.

A. Picart, A. Berlemont, G. Gouesbet. De l'influence du terme de Basset sur la dispersion de particules discrètes dans le cadre de la théorie de Tchen, C. R. Acad. Sci., Paris, Série II, t 295, 305-308, 1982.

A. Picart, A. Berlemont, G. Gouesbet. Modelling and predicting turbulence fields and the dispersion of discrete particles transported by turbulent flows, Int. J. Multiphase Flow, 12, 2, 237-261, 1986.

DISCUSSION

PRUD'HOMME - 1) Avez-vous appliqué ces codes numériques à des cas réactifs ou avec changement de phase ? Si oui cela confirme-t-il les avantages de la méthode Lagrangienne ?

2) Pour prolonger la question de Barrère : avez-vous envisagé la présence de pression interparticulaire ?

GOUESBET - 1) We have not yet applied these codes to non-isothermal conditions, with phase changes and reactions, although these applications are currently developed. However, it is quite possible to state that the Eulerian approach would not be fully suitable under such conditions, because the formulation behind it would collapse. It might be useful for trends if the accuracy is not the priority, probably not for precise figures. Conversely, the Lagrangian approach is well fitted to handle complex situations.

2) We are effectively going to introduce forces on particles due to the presence of the others. However, we are just at the beginning of this story and, although we have some ideas concerning the procedures to do it, it is not possible to say something more precise at the present time.

BARRERE - Jusqu'à quelle concentration de particules marche votre théorie ?

GOUESBET - We are not able to answer this question at the present time. The answer will probably come in part from future computations with increasing concentrations, and comparisons with experiments. Surely, when the concentration is becoming very high, we have to use other basic equations involving the particle volume fraction, or the quality, as used for instance in the separated flow model for two-phase flows. As a rule of thumb, I would guess that the critical volume fraction might be 10%. Future works will expectedly permit to know whether this figure is optimistic or pessimistic.

CONCLUDING REMARKS

A. Pacault

Centre de Recherches Paul Pascal
Domaine Universitaire
33405 Talence, France

Juste après le sommet de la francophonie au Canada, le compromis est dans le double langage puisque nous ne sommes pas tous bilingues.

So, it is in English that I begin my concluding remarks.

Twenty years ago, in 1967, Stig Claesson opened the Nobel Symposium at the Royal Academy of sciences of Sweden at which I was invited. He was asked why is the study of fast reactions so important ?

Twenty years later, it is still important.

It is important because the understanding of the elementary reaction is still a challenge and the primary processes are in all the fields of science the basis of knowledge.

According to Manfred Eigen "the term primary process might have a quite different meaning depending on whether a physicist, chemist or biologist is using it. What a chemist usually calls a primary process might be a whole series of physical events. On the other hand the biologist's use of the word primary may even include complex chemical reaction sequences".

"A chemical primary process will always include a transition between two atomic or molecular states whih are separated by a finite potential barrier. Because of the immense variety of chemical compounds that can be transformed into one another the range above 10^{-12} s is continuously covered by chemical time constants ; each being characteristic of a certain reaction".

This duration was given in the Eigen's paper in 1967. We will see later the evolution of the instant.

But, let me go back because the research of the instant has a long story philosophical as well as scientific.

Je rappelle ici pour mémoire ou je conseille la lecture de "la Nouvelle Siloe" de Roupnel si bien analysée par Bachelard dans "L'intuition de l'instant".

From a scientific point of view we recall that already in 1858, the phosphoroscope of Edmond Becquerel could be used for times as short as

10^{-4} s. He discovered by this way that the decay of phosphorescence is exponential. By another method he measured 10^{-8} s for the lifetime of fluorescence from solutions.

In 1863 Esselbach found the duration of light emission from Uranium glass equal to $0,5 \cdot 10^{-3}$ s. But, we do not forget that the chemical reactions begin by the mixing of the reactants and until the 1950's it remained the main obstacle to measuring a big time rate constant.

Then the relaxation methods allowed us to progress towards the kinetics of chemical reaction.

These methods clearly appear in a discussion of the Faraday Society in 1954 about the study of fast reactions. Eigen, at this time, was looking for a word for this kind of reaction and friends suggested to him : "damned rapid reaction" when their duration was 10^{-6} to 10^{-9} s.

In 1967, a reaction time of 10^{-6} s was usual. Claesson thought that we were rapidly approaching a situation where time resolution of the order of the nanosecond would be easily attainable. He said that, in fact, 10^{-10} was measurable in physics but in chemistry 10^{-7} s was the limit.

What is, after this meeting the real situation ?

We observe a big development of the time measurement techniques because of modern electronics, the use of the laser and the discovery of new spectroscopic methods.

The use of mathematics and principally computers, as powerful tools, is also responsible for the progress in our understanding of chemical reaction in the liquid state during there last years.

A few examples :

One can investigate the reactivity of a precursor of the hydrated electron with biological molecules. So, one can measure a lifetime in the femtosecond (10^{-15} s.) range.

One can see that some weak electron donor-acceptor complexes have a lifetime of the order of 10^{-12} s. During these very short times - 10^{-9} s or less - one can observe the diffusion of molecules into a solvent, ion pairs formation and dissociation, the ion solvation in certain dilute solvent which takes place by solvent molecule interaction.

One measures the time dependent reaction rates and shows the viscosity dependence of non radiative lifetimes.

A good competition and complementarity exist now between the spectroscopic and electrochemical methods which are able to create and monitor the chemical evolution of transient intermediates.

Therefore, the development of the pulsed techniques, the measurement of very short times, allowed us to verify the validity of the models proposed to explain the primary processes and obliged us, as we will see now, to continue the further development of the theories.

Quantum mechanics and statistical mechanisms did not change. Starting from the basic equations - Schrödinger, Langevin, Fokker-Planck - one has developed very good methods of solution as the simulation methods which take into account most of the observed features and, perhaps, some imaginary ones.

All the models are more complicated than before. One accounts for the breakdown of the Born-Oppenheimer approximation. Another shows the possibility of proton transfer by tunnel effect. We see that the role of the solvent cavity size and electron coupling was no longer negligible. It was shown that the water molecule ionisation and hydrogen bonding polarisation play an important role in the ionic equilibrium of water.

A synthesis of these many results, principally in the field of electron transfer is absolutely necessary but I am not able to make it in a so short time.

However, the important features which seem to emerge from the theoretical papers are :

- the possibility of computing accurate values of the main parameters which are needed for the traditional theories of the chemical rate constants,

- the importance of the time evolution of the solvent around the reacting species using friction coefficient,

- the emergence of quantum Monte-Carlo techniques which allow the study of electrons as reacting species.

The word equilibrium has not been used very often during the first days of the meeting.

However, an evolution critically depends on the distance from the equilibrium state.

So, a transition appears in this meeting and consequently in my talk and in my language.

Bien après l'acte primaire, si rapide, comme l'ont montré des expériences difficiles et des calculs compliqués, la réaction chimique continue dans l'espace et dans le temps.

Si un ensemble pertinent de contraintes - variables contrôlées par l'observateur - maintient l'évolution loin de l'équilibre, des phénomènes qui parurent étranges il y a vingt ans environ, peuvent se produire.

Ceux, relatifs à la structuration du temps commencent à être connus, même des étudiants. Citons les oscillations chimiques, les multistabilités, l'excitabilité et plus récemment, le chaos chimique. C'est en effet en 1974, à la réunion de la Société de Chimie physique à Dijon que j'avais montré une horloge chimique utilisant la réaction de Bray. Mais depuis, aidé par la méthode dite du diagramme croisé, on a pu trouver des centaines de réactions présentant ces phénomènes.

L'iode et le brome y jouent cependant un rôle prédominant mais on a vu que la chimie du soufre est sur les rangs. La plupart de ces réactions sont des réactions d'oxydo-réduction mais des réactions photochimiques viennent maintenant s'ajouter à la classification qui regroupe ces réactions par familles.

L'étude de la structuration de l'espace était très en retard sur celle de la structuration du temps. Ce colloque et celui de Bruxelles, la semaine dernière, montrent à l'évidence les énormes progrès faits en quelques années.

On connaît la fameuse expérience de Zhabotinski faite en 1967 mais quinze ans passèrent sans pouvoir analyser qualitativement cette structuration colorée du milieu liquide. Le dévelopement des techniques d'observation - microscopie et analyse informatique d'images - sont à l'origine des progrès récents.

En mettant à part les pseudo ondes de gradient de phase et de gradient de fréquence, on sait maintenant distinguer les ondes de phases et les ondes déclenchées. On peut mesurer le gradient de concentration dans le front d'onde, suivre la forme du front, voir l'annihilation de deux fronts. On sait produire des spirales et on n'observe rien dans leur coeur dont le diamètre est de l'ordre de 20 à 30 µ.

Les premières expériences décrivant ces cinétiques inhabituelles furent faites en supposant le milieu liquide homogène dans des réacteurs continus et agités supposés idéaux. Les ingénieurs du génie chimique savaient bien qu'ils ne l'étaient pas. La rencontre de chercheurs venant d'horizons différents a été fructueuse et ce colloque officialise cette rencontre.

Après ce que nous avons entendu, nul ne peut ignorer l'influence sur la réaction chimique des macro et des micro-mélanges, de la turbulence et du bruit. Des modèles théoriques issus de concepts différents vont se fertiliser les uns les autres.

L'importance industrielle de ces phénomènes en particulier au voisinage des surfaces, catalytiques ou non requiert une intensification des recherches et une confrontation des résultats de la dynamique moléculaire et du génie chimique.

Rappelons enfin que c'est en 1863 que Clausius inventa l'entropie. On notera qu'au cours de ce colloque, quelques rares auteurs seulement en ont parlé. Le mot après avoir fait florès semble devenir rétrograde.

On me dit parfois que l'entropie n'est pas un paramètre pertinent. Il nous manque évidemment un entropimètre performant mais il n'est pas sûr que la dynamique gagne à oublier la thermodynamique.

C'est en effet une autre vue du monde que nous aurions aujourd'hui sans les concepts fondamentaux qui fondèrent la thermodynamique des processus irréversibles.

Puis-je me permettre, en terminant, d'exprimer quelques regrets. Le projet des organisateurs, que je dois remercier tout particulièrement pour vous, était clair : réunir des spécialistes regardant la réactivité chimique sous des aspects divers.

Or, ceux qui n'envisagent que l'équilibre ont souvent relaxé avant que n'arrivent ceux qui étaient en non-équilibre. Nous avons observé des flux périodiques, des macro-mélanges et à dire vrai peu de turbulence. Puisse un prochain colloque être un attracteur suffisamment étrange pour que tous les participants y décrivent toutes les trajectoires possibles.

PARTICIPANTS

AMAR J.G./ Thermophysics Division/ National Bureau of Standards/ Physics-A 105/ GAITHERSBURG, MD,(USA)

AMATORE C./ ENS Chimie/ 24, rue Lhomond/F-75005 PARIS

ARVIS M./CEN/SACLAY/IRDI/DESICP/DPC/F-91191 GIF SUR YVETTE

BAILLOT F. Mrs/ Lab. Aerothermique/CNRS/ 4 ter route des Gardes/F-92190 MEUDON

BAROS F./ENSIC/INPL/ 1, rue Grandville/F-54042 NANCY

BARRERE M./ ONERA/ av. de la Division Leclerc/F-92320 CHATILLON

BARTHEL J./ Inst. fur Physikalische und Theoretische Chemie/ der Universitat Regensburg/ Universitatsstrasse 31/ D. 8400 REGENSBURG (FRG)

BEAUFILS J.P./ INP/ILL 156X/F-38042 GRENOBLE

BEKKI S./ Chimie physique/ 11, rue P. et M. Curie/F-75231 PARIS Cedex 05

BELLONI J. Mrs/ Physico chimie des rayonnements/ Univ. Paris-Sud/Bât. 350/F-91405 ORSAY

BELLONI L./ CEN SACLAY/IRDI/DESICP/DPC/F-91191 GIF SUR YVETTE

BELOEIL J.C./ ICNS/ CNRS/F-91190 GIF SUR YVETTE

BENZINEB K./ Univ. René Descartes/ Chimie physique/ 45, rue des Sts Pères-/F-75006 PARIS

BERTRAN J./ Universitat autonoma de Barcelona/ 08193-BELLATERRA/ Barcelona/(Espagne)

BESNARD M./ Spectroscopie moléculaire et cristalline/ Univ. Bordeaux I/ 351, cours de la Libération/F-33405 TALENCE

BLUM L./ Physics Depart/ P.O. Box AT/ Univ. of Puerto Rico/ RIO PIEDRAS 00931/(USA)

BORGHI R./ Fac. des Sciences et des Techniques/ P.B. 67/ 76130 MONT ST AIGNAN

BORGIS D./Physique théorique des liquides/ Tour 16/ 4, place Jussieu/F-75005 PARIS

BOTTER R./DPC/CEA/CEN SACLAY/F-91191 GIF SUR YVETTE

BOURCEANU G./ Physical Chemistry/ Polytechnic Institute/IASI 6600 (Roumanie)

BRATOS S./ Physique théorique des liquides/ Tour 16/4, place Jussieu/F-75252 PARIS Cedex 05

CANDEL S./ Ecole Centrale/ Grande voie des vignes/F-92290 CHATENAY MALABRY

CHIRAT R./ Centre d'études de Vaujours/ B.P. 7/F-77181 COURTRY

CICCOTI G./ Dipart. di Fisica/ Univ. "La Sapienza" Piazzale Aldo Moro, 2/ 00185 ROMA (Italie)

CONDAMINES N. Miss/ CEA/ DGR/SEP/SCPR/ B.P. 6/F-92265 FONTENAY AUX ROSES

DE KEPPER P./ CRPP/ Domaine universitaire/ Univ. Bordeaux I/F-33405 TALENCE

DELAIRE J.A./ Physico-Chimie des Rayonnements/ Univ. Paris Sud/F-91405 ORSAY

DIANTOUBA B.A./ E.H.I.C.S./ Lab. Chimie minérale/1, rue Blaise Pascal-/F-67008 STRASBOURG Cedex

DIGUET R./ Univ. Nancy I/ Chimie théorique UA 510/ B.P. 239/F-54506 VANDOEUVRE LES NANCY

DRIFFORD M./ CEN SACLAY/IRDI/DESICP/DPC/F-91191 GIF SUR YVETTE Cedex

DUPEYRAT M. Mrs/ Chimie physique/ 11, rue P. et M. Curie/F-75231 PARIS Cedex 05

DURAND S. Mrs/ I.S.F./ Résidence Albert ler/ 47 bis rue Albert ler/F-41000 BLOIS

ECKERT C. Miss/CRN/ 23, rue du Loess/F-67037 STRASBOURG

ERDI P./ Central Research/ Inst. Physics, Hungarian acad. of Sciences/ P.O. Box 49/ H-1525 BUDAPEST

FERRADINI C. Miss/ Chimie physique/ Univ. René Descartes/ 45, rue des Saints Pères/F-75006 PARIS

FRANKOWICZ M./ Fac. of Chemistry/ Jagellonian Univ./ Wydzial Chemii UJ/ ul. Karasia 3/ 30-060 CRACOW(Pologne)

FRIES P.H./ CEN GRENOBLE/ DRF-Chimie/ 85X/F-38041 GRENOBLE Cedex

GARDES-ALBERT M. Mrs/ Lab. Chimie physique/ Univ. René Descartes/ 45, rue des Sts Pères/F-75006 PARIS

GAUDUEL Y./ Optique apliquée/ Ecole Polytechnique/ ENSTA/F-91128 PALAISEAU

GAUMANN T./ Inst. Chimie Physique/ EPFL-Ecublens/ CH-1015 LAUSANNE (Suisse)

GERSCHEL A./ CPMA/ Univ. Paris Sud/F-91405 ORSAY

GOUESBET G./ LESP/ INSA/ CNRS 230/ B.P.08/F-76130 MONT ST AIGNAN

HAERTL E./ Fasanenweg 2/ D-6070 LANGEN (FRG)

HANUSSE P./ CRPP/ Univ. Bordeaux/F-33405 TALENCE

HAYOUN M./ CEA/CEN SACLAY/ Technologie/F-91191 GIF SUR YVETTE

HEISEL F. Miss/CRN/PMOA/23, rue du Loess/F-67200 STRASBOURG

HICKEL B./ CEN SACLAY/IRDI/DESICP/DPC/F-91191 GIF SUR YVETTE

HYNES J.T./ Depart. of Chemistry/ Univ. of Colorado/ BOULDER, CO
80309-0215 (USA)

JONAH C.D./ Chemistry Division/ Argonne National Lab./ 9700 S Cass Ave./
ARGONNE IL 60439 (USA)

JORGENSEN W.L./ Depart. of Chemistry/ Purdue Univ./ WEST LAFAYETTE,
IN 47907 (USA)

KAPRAL R./ Univ. of Toronto/ Depart. of Chemistry/ TORONTO, Ontario/
(Canada M5S 1A1)

KLEIN M.L./Department of Chemistry/Univ. of Pennsylvania/ PHILADELPHIA
PA 19104-6323(USA)

KUENTZMANN P./ ONERA/ Fort de Palaiseau/F-91120 PALAISEAU

LALLEMAND J.Y./ Chimie, Ecole Polytechnique/F-91128 PALAISEAU

LASCOMBE J./ Univ. Bordeaux I/F-33405 TALENCE

LAUNAY J.P./ L.C.M.T./Univ. P. et M. Curie/Bât. F./4, place Jussieu/F-75252
PARIS Cedex 05

LESQUIBE F. Miss/ 15, rue A. Salel/F-92260 FONTENAY AUX ROSES

LLUCH J.M./ Depart. de Quimica/ Univ. Barcelona/ 08193 BELLATERRA
Barcelona (Espagne)

LUCAS M./ CEA-IPSN/DERS-SESRU/CEA Cadarache/B.P. 1/F-13115 ST PAUL
LEZ DURANCE

MACOVEI V./ Chem. Physics/ Facultatea de Technologie/ Str. Splai Bahlui-
-Stinga, 71/6600 IASI (Roumanie)

MALLIAVIN T. Miss/ 10, rue Saint Louis en L'Ile/ 75004 PARIS

MARBLE F.E./ Karman Lab. of Fluid Mechanics/ Jet Propulsion/ Caltech/-
PASADENA, CA 91125 (USA)

MARESCHAL M./ Chimie physique II/ U.L.B./ C.P. 231, Campus Plaine, Bd
du Triomphe, B 1050 BRUXELLES(Belgique)

MARX R. Mrs/ L.P.C.R./ Univ. Paris Sud/ Bât. 350/ F-91405 ORSAY

MIALOCQ J.C./ CEN SACLAY/ IRDI/DESICP/DPC/F-91191 GIF SUR YVETTE

MICHEAU J.C./Univ. Paul Sabatier/I.M.R.C.P./118, route de Narbonne/F-31062 TOULOUSE Cedex

MIEHE J.A./ CRN et ULP/ 23, rue du Loess/F-67037 STRASBOURG

MOREAU M./ Physique théorique des liquides/ Tour 16/ 4, place Jussieu /F-75252 PARIS Cedex 05

MOREAU T./ Electrochimie interfaciale/ CNRS/ 1, place A. Briand/F-92195 MEUDON

MOSTAFAVI M./ Physico Chimie des rayonnements/ Univ. Paris Sud/ Bât. 350/F-91405 ORSAY

MULLER S./ M.P.I.f. Ernahrungsphysiologie/ Rheinlanddamm 201/ D-4600 DORTMUND 1 (FRG)

MUSIKAS C./ CEA/DGR/SEP/SCPR/B.P. 6/F-92265 FONTENAY AUX ROSES

NAGY-UNGVARAI Z. Mrs/M.P.I.f. Ernahrungsphysiologie/ Rheinlanddamm 201/ D-4600 DORTMUND 1 (FRG)

NAKACHE E. Mrs/ Chimie physique/ 11, rue P. et M. Curie/F-75231 PARIS Cedex 05

NEWTON M.D./ Brookhaven National Lab/ UPTON, NY 11973 (USA)

NICOLIS G./ Chimie physique II/ U.L.B./ C.P. 231/ Bd du Triomphe/ 1050 BRUXELLES (Belgique)

NORDIO P.L./ Dipart. Chimica Fisica/ Univ. di Padova/ Via Loredan 2/ 35131 PADOVA (Italie)

PACAULT A./ C.R.P.P./ Univ. de Bordeaux/F-33405 TALENCE

PILENI M.P. Mrs/ CEN SACLAY/ IRDI/DPC/DESICP/F-91191 GIF SUR YVETTE

PRUD'HOMME R./ Lab. d'Aerothermique/ CNRS/ 4, ter route des Gardes- /F-92190 MEUDON

RICHETTI P./ C.R.P.P./ Univ. de Bordeaux/F-33405 TALENCE

RIVAIL J.L./ Lab. de Chimie théorique/ Fac des Sciences/ B.P. 239/F-54506 VANDOEUVRE LES NANCY

ROSS J/ Chemistry Depart./ Stanford Univ./STANFORD, CA 94305 (USA)

SALAMITO B/ CEN Grenoble/DRF Chimie/85 X/F-38041 GRENOBLE

SAMPOLI M/ Dipartimento di Chimica Fisica/Universita di Venezia/Dorso Duro 2137/-I-30123 VENEZIA (Italie)

SAUMAGNE P./ Univ. Bretagne occidentale/ Spectrochimie moléculaire/ Thermodynamique chimique/ 6 av. le Gorgeu/F-29287 BREST

SCEATS M.G./ School of Chemistry/ Univ. of Sydney/ N.S.W.2006 SYDNEY (Australia)

SIMONIN J.P./ Electrochimie/UPMC/Bât. F./ 4, place Jussieu/F-75232 PARIS Cedex 5

TACHIYA M./ National Chemical Lab./ Yatabe, IBARAKI 305 (Japan)

TAUBE D./ RORER Group, INC/Process R. and D./ 1, scarsdale Rd/ TUCKAHOE N.Y. 10707 (USA)

TOTH J./ Computer and Automation Inst./ Hungarian Academy of Sciences/ P.o.B. 63/ H.1502 BUDAPEST(Hongrie)

TROYANOWSKY C./ SFC-DIVISION DE CHIMIE PHYSIQUE/ 10, rue Vauquelin/ F-75005 PARIS

TURQ P./ Electrochimie/UPMC/Bât F./ 4, place Jussieu/ F-75252 PARIS Cedex 05

VALLETON J.M./ Polymères, Biopol. Membranes/ Univ. de Rouen/ F-76130 MONT ST AIGNAN

VANHOVE D./ Catalyse organique/ UA CNRS 231/UCB/ESCIL/43, Bd du 11 Novembre 1918/ F-69622 VILLEURBANNE Cedex

VIDAL C./ CRPP/Univ. de Bordeaux/ F-33405 TALENCE

VILLERMAUX J./ LSGC/CNRS/ENSIC/INPL/ 1, rue Grandville/ F-54042 NANCY

VIOT P./ Physique théorique des liquides/ Tour 16/ 4, pl. Jussieu/ F-75005 PARIS

WILSON K.R./ Depart. of Chemistry/ B.014/ Univ. of California/ San Diego/ LA JOLLA CA 92093 (USA)

BURSHTEIN A.I./ Institute of Chemical Kinetics and Combustion of the USSR Academy of Sciences/ NOVOSIBIRSK 630 090 (USSR)

CSTR, 405

Debye equation, 113, 327
Density profile
  at interface, 269, 279
Deterministic chaos
  and external noise, 311
Dielectric
  constant, 32,108,188,263
  permittivity, 58
  relaxation, 31, 55
Diffusion
  multivariate, 220
  operator, 214
  reaction control by, 80, 113
  rotational, 218
Diffusive mixing
  in a vortex, 585
Dynamical equation
  microscopic, 300
Dynamics
  intermediate, 293
  stochastic, 307

Effective potentials, 348
Electric field, 3
  effect, 373
  external, 371, 372
  high frequency, 55
Electrochemistry
  cyclic voltammetry, 75
  indirect methods, 83
  organic, 73
  organometallic, 73
  redox catalysis, 83
  ultramicroelectrodes, 78
Electron
  aqueous solvation of, 15
  orbital models, 165, 187
  reactivity of, 21
  transfer reactions, 113, 157,
      175, 197
  trapping, 25
Electron transfer
  fast, 321
  intervalence, 320
  intramolecular, 315
  spontaneous, 321
  thermal, 315
Electronic curve crossing, 232
Electronic spectra, 137, 232
Electronic structure
  calculation, 157
Encounter dynamics, 347
Energy
  relaxation, 232
  controlled rate constant, 330
  diffusion, 354
  splitting, 208
Ensemble
  isothermic-isobaric, 254

Entropy production
  and exchangers, 547
  in chemical reactor, 554
  rate of, 546
Enzyme reactions
  electric constraint, 495
Enzymes work, 237
ESR, 10
Eulerian approach, 609
Evolution equation
  stochastic, 301
EXAFS, 14
Experimental study
  far from equilibrium, 437
Explosive case, 434

Fast reaction
  femtosecond, 1, 15
  in a vortex, 593
  picosecond, 1, 73
  techniques for, 1, 73
Feedback loop, 485
First passage time, 389
Fischer-Tropsch, 147
Flash photolysis, 5
Flow techniques, 4
Fluctuation dissipation, 302
Fokker Planck, 381
  equation, 299
  multivariate, 219
Free energy surface
  conformational, 255
  in solution, 253
Friction
  coefficient, 220, 274
  constant, 227
  dielectric, 222
  model
    frequency dependent, 334
  short time, 227
  time dependent, 299
  viscous, 230

High friction limit, 383, 384
Hopf bifurcations, 307
Hopping reorientation, 368
Hydrodynamic model, 333
Hydrodynamics, 217
Hydrogen bond inner shell, 161
Hydrogen bonding in water, 241
Hydrophobic effet
  trans-gauche, 256
Hyteresis photochemical, 482

Infrared, 33
Inhomogeneity in time, 514
Instabilities, Chemical, 401
Instability
  hydrodynamical, 265
  interfacial, 470
Instantaneous flux, 304